Organic Conductors

Organic Conductors

Editor

Toshio Naito

MDPI • Basel • Beijing • Wuhan • Barcelona • Belgrade • Manchester • Tokyo • Cluj • Tianjin

Editor
Toshio Naito
Ehime University
Japan

Editorial Office
MDPI
St. Alban-Anlage 66
4052 Basel, Switzerland

This is a reprint of articles from the Special Issue published online in the open access journal *Crystals* (ISSN 2073-4352) (available at: https://www.mdpi.com/journal/crystals/special_issues/Organic_Conductors?authAll=true).

For citation purposes, cite each article independently as indicated on the article page online and as indicated below:

LastName, A.A.; LastName, B.B.; LastName, C.C. Article Title. *Journal Name* **Year**, *Volume Number*, Page Range.

ISBN 978-3-0365-4559-2 (Hbk)
ISBN 978-3-0365-4560-8 (PDF)

Contents

About the Editor

Toshio Naito

Toshio Naito graduated from The University of Tokyo, Japan, in 1988. Then he studied organic conductors at the same university as a master's course student in 1988-1990. His PhD research was mainly conducted at Toho University, Japan, as a faculty staff member (an assistant in 1990-1995, then a lecturer in 1995. In 1995, he received a PhD degree (Doctor of Science) from The University of Tokyo, Japan. Afterwards, he moved to Hokkaido University (as an Associate Professor in 1996-2011), Japan; Ehime University, Japan (a Professor in 2011-present). Since 2011, he has been leading several groups at Ehime University belonging to different faculties/departments/research centers at the same time. The group investigates the structural, electrical, magnetic, and optical properties of different types of organic conductors, particularly their dynamics after photoexcitation.

crystals

Editorial

Organic Conductors

Toshio Naito [1,2,3]

[1] Graduate School of Science and Engineering, Ehime University, Matsuyama 790-8577, Japan; tnaito@ehime-u.ac.jp
[2] Research Unit for Development of Organic Superconductors, Ehime University, Matsuyama 790-8577, Japan
[3] Geodynamics Research Center (GRC), Ehime University, Matsuyama 790-8577, Japan

Since the pioneering work concerning organic semiconductors in the middle of the 1900s, organic conductor research has experienced a series of milestones, from metallic to superconducting charge-transfer complexes (1980s–1990s). The history of these materials is reviewed in this Special Issue by considering a body of references [1].

The publication "Organic Conductors" covers various solids containing both organic and inorganic species with a variety of shapes and dimensions. The physical properties of these materials include electrical, magnetic, structural, optical, dielectric, and mechanical. The most prominent feature of organic conductors and related materials is the wide variety of degrees of freedom, which enables peculiar electronic states, physical properties, and phase transitions that are otherwise unobserved. As a result, even an insulating organic crystalline material can be a center of interest in this field if it provides an important piece of information concerning the mechanism of superconductivity, for example, or sheds light on other subjects of broad interest from the abovementioned point of view.

In 2022, researchers in this ever-expanding and developing field are looking for further interesting and exciting targets that are yet to be explored. On such a memorial occasion, this Special Issue, entitled "Organic Conductors", has collated 25 papers (plus editorials) from 11 countries. Readers will enjoy the latest developments in new materials, ideas, and methodologies, which will propel the field in a new direction and age.

The Special Issue consists of the following papers and reviews:

1. The synthesis of new molecules and organic conductors:
    - Kubo et al. reported a series of nickel–dithiolene complexes fused with bulky cycloalkane substituents to study their steric-based effects on molecular arrangements [2].
    - Kadoya et al. reported the structural and physical properties of a new organic Mott insulator with a θ-type molecular arrangement [3].
    - Mroweh et al. reported new chiral conducting salts based on ethylenedithio-tetrathiafulvalene (EDT-TTF) derivatives, presenting their crystal structures, extended Hückel band structures, and electrical properties [4].
    - Sakaguchi et al. reported a new single-component molecular conductor, $[Au(etdt)_2]\cdot THF$ (etdt = ethylenedithiotetrathiafulvalenedithiolate), that is, the synthesis, structure, and physical properties of a neutral gold-dithiolene-complex [5].
    - Akutsu et al. reported three types of new organic conductors containing $HOC_2H_4SO_3^-$ anions and discussed their electrical properties in terms of the Madelung energies [6].
    - Sato et al. reported new stable neutral radical species, $M^{III}(Pc)Cl_2\cdot THF$ (M = Co or Fe, Pc = phthalocyanine, THF = tetrahydrofuran), as three-dimensional, single-component molecular conductors [7].
    - Koyama et al. reported the synthesis and structural, spectroscopic, and electrical properties of a new molecular conductor with a weak hydrogen-bonding network [8].

Citation: Naito, T. Organic Conductors. *Crystals* **2022**, *12*, 523. https://doi.org/10.3390/cryst12040523

Received: 14 March 2022
Accepted: 8 April 2022
Published: 9 April 2022

2. Structural, optical, magnetic, electrical, and other related properties:
    - Rohwer et al. reported the effects of deuteration on the transport properties of quasi-one-dimensional Fabre salts, $(TMTTF)_2X$ (TMTTF = tetramethyltetrathiafulvalene, X = Br, PF_6, and ClO_4), and discovered various conduction properties such as Mott insulators, variable range hopping, and activated band transport with a temperature-dependent bandgap [9].
    - Kitou et al. reported an experimental method for use in estimating valence electron densities, that is, frontier orbitals, in the solid state based on X-ray diffraction data [10].
    - Yoshino et al. reported the crystal structures and electrical resistivities of a series of TMTSF (TMTSF = tetramethyltetraselenafulvalene) salts with unusual stoichiometries and determined the valence state of TMTSF in each type of salt using quantum chemistry calculations [11].
3. Physical property measurements with new techniques and/or under extreme conditions:
    - Yamamoto et al. reported a new system for measuring the pyroelectricity of small ferroelectric single crystals [12].
    - Hino et al. reported the current- and voltage-dependence of the heat capacity of a single crystal of a charge glass compound, θ-$(BEDT\text{-}TTF)_2CsZn(SCN)_4$ (BEDT-TTF = bis(ethylenedithio)-tetrathiafulvalene) [13].
4. Spectroscopic studies concerning molecular functional crystals:
    - Hiraki et al. reported ^{77}Se-NMR studies on λ-type BETS superconductors, λ-$(BETS)_2Fe_{1-x}Ga_xCl_4$ (BETS = bis(ethyleneditho)tetraselenafulvalene), to examine the π-spin polarization affected by the localized 3*d* spins on Fe atoms [14].
5. Theoretical studies on organic conductors:
    - Tsumuraya et al. reported first-principle density functional theory calculations of the charge-ordered phase of α-$(BEDT\text{-}TTF)_2I_3$, which is closely related to Dirac electron systems [15].
    - Roy et al. reported accurate zero-temperature density matrix renormalization group calculations for κ-$(BEDT\text{-}TTF)_2X$, the most studied family of organic superconductors, and concluded that magnetic fluctuations in the effective half-filled band approach do not drive superconductivity in these and related materials [16].
    - Ménard et al. reported a one-dimensional alternating extended Hubbard model at quarter-filling based on a renormalization group method to examine structural instabilities in Fabre and Bechgaard salts and related organic conductors [17].
    - Suzumura et al. calculated electrical transport in nodal line semimetals of single-component molecular conductors to examine the effects of acoustic phonon scattering on electrical conductivity [18].
    - Naito et al. reported a method of calculating intermolecular interactions in disordered molecular charge-transfer complexes of STF (STF = bis(ethylenedithio)-diselenadithiafulvalene) by proposing a new interpretation or usage of wavefunctions [19].
    - Kesharpu et al. calculated the evolution of the shape and volume fraction of superconducting domains in relation to temperature and anion disorder in a highly anisotropic organic superconductor, $(TMTSF)_2ClO_4$ [20].
6. Molecular π-d, Dirac, and strongly correlated electron systems:
    - Cui et al. reported the high-pressure crystal structure and magnetoresistance of a single-component molecular conductor $[Pt(dddt)_2]$ (dddt = 5,6-dihydro-1,4-dithiin-2,3-dithiolate), a theoretically expected three-dimensional Dirac electron system, under high pressure [21].
    - Yasuzuka reviewed the interplay between the vortex dynamics and superconducting gap structures in layered organic superconductors containing strongly correlated electron systems. Particular attention was paid to the symmetry of

the superconducting gaps, which is a key feature in understanding the pairing mechanism [22].

7. Organic magnets, dielectrics, semiconductors, superconductors, and photoconductors:
   - Manabe et al. reported reversible structural and magnetic transformations in inorganic ladder compounds; these transformations are dependent on the gas-phase chemical species absorbed in single crystals [23].
8. Molecular electronics, optoelectronics, spintronics, devices, and related topics:
   - Kawasugi et al. reported doping effects using field-effect transistors consisting of electric double layers of charge-ordered insulators, α-$(BEDT\text{-}TTF)_2I_3$, and α-$(BETS)_2I_3$ [24].
   - Gou et al. reported experimental and modelling studies of a series of high-performance inorganic semiconductor devices based on doped GaAs and related materials [25].

The Guest Editor would like to express his sincere thanks to all of the authors for their contributions to this Special Issue, as well as to all of the reviewers for the time and effort expended to provide valuable feedback to the authors. Special thanks also go to the *Crystals* editorial office members, Ms. Debbie Yang in particular, for their continuous collaboration, timely communication, and efficient support, along with their friendly and professional attitude; without their contributions, the publication of this Special Issue would not have been possible. The Guest Editor hopes that this Special Issue will encourage more and more scientists to join this field to further expand the horizon and to discover new possibilities of molecular conductors.

Funding: The author is grateful for the financial support from a Grant-in-Aid for Challenging Exploratory Research (18K19061) of JSPS, the Iketani Science and Technology Foundation (ISTF; 0331005-A), the Research Grant Program of the Futaba Foundation, the Casio Science Promotion Foundation, and an Ehime University Grant for Project for the Promotion of Industry/University Cooperation.

Conflicts of Interest: The author declares no conflict of interest.

References

1. Naito, T. Modern History of Organic Conductors: An Overview. *Crystals* **2021**, *11*, 838. [CrossRef]
2. Kubo, K.; Sadahiro, M.; Arata, S.; Hoshino, N.; Kadoya, T.; Akutagawa, T.; Kato, R.; Yamada, J.-I. Donor-Type Nickel–Dithiolene Complexes Fused with Bulky Cycloalkane Substituents and Their Application in Molecular Conductors. *Crystals* **2021**, *11*, 1154. [CrossRef]
3. Kadoya, T.; Sugiura, S.; Higashino, T.; Tahara, K.; Kubo, K.; Sasaki, T.; Takimiya, K.; Yamada, J.-I. Dihedral-Angle Dependence of Intermolecular Transfer Integrals in BEDT-BDT-Based Radical-Cation Salts with θ-Type Molecular Arrangements. *Crystals* **2021**, *11*, 868. [CrossRef]
4. Mroweh, N.; Auban-Senzier, P.; Vanthuyne, N.; Lopes, E.B.; Almeida, M.; Canadell, E.; Avarvari, N. Chiral Conducting Me-EDT-TTF and Et-EDT-TTF-Based Radical Cation Salts with the Perchlorate Anion. *Crystals* **2020**, *10*, 1069. [CrossRef]
5. Sakaguchi, K.; Zhou, B.; Idobata, Y.; Kamebuchi, H.; Kobayashi, A. Syntheses, Structures, and Physical Properties of Neutral Gold Dithiolate Complex, $[Au(etdt)_2]$·THF. *Crystals* **2020**, *10*, 1001. [CrossRef]
6. Akutsu, H.; Koyama, Y.; Turner, S.S.; Furuta, K.; Nakazawa, Y. Structures and Properties of New Organic Conductors: BEDT-TTF, BEST and BETS Salts of the $HOC_2H_4SO_3^-$ Anion. *Crystals* **2020**, *10*, 775. [CrossRef]
7. Sato, R.; Matsuda, M. Formation of Three-Dimensional Electronic Networks Using Axially Ligated Metal Phthalocyanines as Stable Neutral Radicals. *Crystals* **2020**, *10*, 747. [CrossRef]
8. Koyama, S.; Kawai, M.; Takaishi, S.; Yamashita, M.; Hoshino, N.; Akutagawa, T.; Kanno, M.; Iguchi, H. Synthesis, Structure and Physical Properties of $(trans\text{-}TTF\text{-}py_2)_{1.5}(PF_6)$·EtOH: A Molecular Conductor with Weak CH···N Hydrogen Bondings. *Crystals* **2020**, *10*, 1081. [CrossRef]
9. Rohwer, A.; Dressel, M.; Nakamura, T. Deuteration Effects on the Transport Properties of $(TMTTF)_2X$ Salts. *Crystals* **2020**, *10*, 1085. [CrossRef]
10. Kitou, S.; Hosogi, Y.; Kitaura, R.; Naito, T.; Nakamura, T.; Sawa, H. Direct Observation of Molecular Orbitals Using Synchrotron X-ray Diffraction. *Crystals* **2020**, *10*, 998. [CrossRef]
11. Yoshino, H.; Iwasaki, Y.; Tanaka, R.; Tsujimoto, Y.; Matsuoka, C. Crystal Structures and Electrical Resistivity of Three Exotic TMTSF Salts with I 3−: Determination of Valence by DFT and MP2 Calculations. *Crystals* **2020**, *10*, 1119. [CrossRef]

12. Yamamoto, K.; Kawasaki, A.; Chinen, T.; Ryugo, K. Temperature-Modulated Pyroelectricity Measurements of a Thin Ferroelectric Crystal with In-Plane Polarization and the Thermal Analysis Based on One-Dimensional Layer Models. *Crystals* **2021**, *11*, 880. [CrossRef]
13. Hino, K.; Nomoto, T.; Yamashita, S.; Nakazawa, Y. Single Crystal Heat Capacity Measurement of Charge Glass Compound θ-(BEDT-TTF)$_2$CsZn(SCN)$_4$ Performed under Current and Voltage Application. *Crystals* **2020**, *10*, 1060. [CrossRef]
14. Hiraki, K.; Takahashi, T.; Akiba, H.; Nishio, Y.; Zhou, B. Microscopic Observation of π Spin Polarization by d Localized Spin in λ Type BETS Based Organic Superconductors. *Crystals* **2020**, *10*, 1055. [CrossRef]
15. Tsumuraya, T.; Seo, H.; Miyazaki, T. First-Principles Study on the Stability and Electronic Structure of the Charge-Ordered Phase in α-(BEDT-TTF)$_2$I$_3$. *Crystals* **2021**, *11*, 1109. [CrossRef]
16. Roy, D.; Clay, R.T.; Mazumdar, S. Absence of Superconductivity in the Hubbard Dimer Model for κ-(BEDT-TTF)$_2$X. *Crystals* **2021**, *11*, 580. [CrossRef]
17. Ménard, M.; Bourbonnais, C. One-Dimensional Alternating Extended Hubbard Model at Quarter-Filling and Its Applications to Structural Instabilities of Organic Conductors. *Crystals* **2020**, *10*, 942. [CrossRef]
18. Suzumura, Y.; Kato, R.; Ogata, M. Electric Transport of Nodal Line Semimetals in Single-Component Molecular Conductors. *Crystals* **2020**, *10*, 862. [CrossRef]
19. Naito, T.; Suzumura, Y. Theoretical Model for Novel Electronic State in a Dirac Electron System Close to Merging: An Imaginary Element between Sulphur and Selenium. *Crystals* **2022**, *12*, 346. [CrossRef]
20. Kesharpu, K.K.; Kochev, V.D.; Grigoriev, P.D. Evolution of Shape and Volume Fraction of Superconducting Domains with Temperature and Anion Disorder in (TMTSF)$_2$ClO$_4$. *Crystals* **2021**, *11*, 72. [CrossRef]
21. Cui, H.; Yeung, H.H.-M.; Kawasugi, Y.; Minamidate, T.; Saunders, L.K.; Kato, R. High-Pressure Crystal Structure and Unusual Magnetoresistance of a Single-Component Molecular Conductor [Pd(dddt)$_2$] (dddt = 5,6-dihydro-1,4-dithiin-2,3-dithiolate). *Crystals* **2021**, *11*, 534. [CrossRef]
22. Yasuzuka, S. Interplay between Vortex Dynamics and Superconducting Gap Structure in Layered Organic Superconductors. *Crystals* **2021**, *11*, 600. [CrossRef]
23. Manabe, J.; Nishida, K.; Zhang, X.; Nakano, Y.; Fujibayashi, M.; Cosquer, G.; Inoue, K.; Shimono, S.; Ishibashi, H.; Kubota, Y.; et al. Gas-Dependent Reversible Structural and Magnetic Transformation between Two Ladder Compounds. *Crystals* **2020**, *10*, 841. [CrossRef]
24. Kawasugi, Y.; Masuda, H.; Pu, J.; Takenobu, T.; Yamamoto, H.M.; Kato, R.; Tajima, N. Electric Double Layer Doping of Charge-Ordered Insulators α-(BEDT-TTF)$_2$I$_3$ and α-(BETS)$_2$I$_3$. *Crystals* **2021**, *11*, 791. [CrossRef]
25. Gou, Y.; Wang, J.; Cheng, Y.; Guo, Y.; Xiao, X.; Liu, H.; Tan, S.; Zhou, L.; Yang, H.; Deng, G.; et al. Experimental and Modeling Study on the High-Performance p^{++}-GaAs/n^{++}-GaAs Tunnel Junctions with Silicon and Tellurium Co-Doped InGaAs Quantum Well Inserted. *Crystals* **2020**, *10*, 1092. [CrossRef]

crystals

MDPI

Article

Theoretical Model for a Novel Electronic State in a Dirac Electron System Close to Merging: An Imaginary Element between Sulphur and Selenium

Toshio Naito [1,2,3,*] and Yoshikazu Suzumura [4]

[1] Graduate School of Science and Engineering, Ehime University, Matsuyama 790-8577, Japan
[2] Research Unit for Development of Organic Superconductors, Ehime University, Matsuyama 790-8577, Japan
[3] Geodynamics Research Center (GRC), Ehime University, Matsuyama 790-8577, Japan
[4] Department of Physics, Nagoya University, Nagoya 464-8602, Japan; suzumura@s.phys.nagoya-u.ac.jp
* Correspondence: tnaito@ehime-u.ac.jp

Abstract: Topological materials with Dirac electron systems have been extensively studied. Organic crystalline materials form a unique group of such compounds with well-defined crystal structures. While most organic compounds require high pressures to exhibit Dirac-cone-type band structures, the title compound, α-STF_2I_3, has garnered increasing interest due to its Dirac-cone-type band structure under ambient pressure. Various experiments have been conducted under ambient pressure; their results can be compared with those of theoretical calculations to obtain insights into Dirac electron systems. However, structural disorder peculiar to the STF molecules in the solid-state has prevented any type of theoretical calculation of the states. In this study, we report a new method for calculating intermolecular interactions in disordered systems based on the extended Hückel approximation. This method enables band calculations, suggesting that this material is a rare example of a system close to merging. The obtained band structure indicates that the characteristic disorder in the STF solids distributed electrons equally on the sulphur and selenium atoms as if they belong to an imaginary element between sulphur and selenium and are arranged without disorder.

Keywords: crystalline organic charge-transfer complexes; disordered systems; overlap integrals; extended Hückel approximation; Dirac electrons; zero-gap semiconductors; merging of Dirac cones

Citation: Naito, T.; Suzumura, Y. Theoretical Model for a Novel Electronic State in a Dirac Electron System Close to Merging: An Imaginary Element between Sulphur and Selenium. *Crystals* **2022**, *12*, 346. https://doi.org/10.3390/cryst12030346

Academic Editor: Martin Dressel

Received: 11 February 2022
Accepted: 28 February 2022
Published: 2 March 2022

1. Introduction

Recently, Dirac electron systems (DESs), which are characterised by massless relativistic electrons with speeds of 1/100–1/1000th of the velocity of light, have been widely studied. The majority of the studies on DESs are theoretical because of the limited availability of DES materials [1–9]. Concerning the general molecular systems, the electronic structures including those under high pressure are calculated by DFT methods [10–12]. Although DESs were initially found in graphene [13,14] and some inorganic compounds [15], those found in organic compounds [16] possess clear advantages over their inorganic counterparts. For example, most organic DESs are found in bulk systems with well-defined crystal structures and chemical stoichiometries, whereas inorganic DESs often lack either of these two features. However, organic DESs have some disadvantages; for example, most become DES only under high pressure [17–20], unlike inorganic DESs. Performing detailed, accurate experiments and measurements (including crystal structure analyses) at high pressures is more complicated than those at ambient pressure. In this regard, the organic charge-transfer (CT) complex, α-STF_2I_3 (Figure 1; STF = bis(ethylenedithio)-diselenadithiafulvalene), is unique as it contains Dirac electrons at ambient pressure [21–26]; that is, it possesses the advantages of both organic and inorganic DESs.

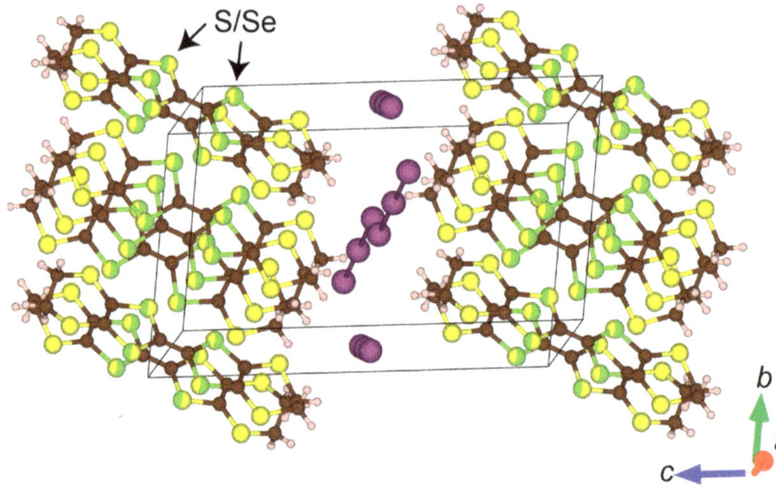

Figure 1. Crystal structure of α-STF_2I_3. The brown, yellow, green, blank, and violet spheres denote C, S, Se, H, and I atoms, respectively. The double-coloured (yellow and green) atoms are disordered atoms between S and Se (S: Se = 50%:50%).

All the electrical, magnetic, and optical properties of the DESs are governed by the intermolecular interactions between the STF molecules in α-STF_2I_3. In principle, intermolecular interactions, band structures, and various physical properties of crystalline materials can be calculated when all the atomic positions, namely atomic parameters, are known, assuming perfect three-dimensional periodic arrangements of the atoms in the framework of well-established band theories. However, it is challenging to calculate the intermolecular interactions, *that is*, the overlap and transfer integrals between the STF molecules in α-STF_2I_3, because all the STF molecules in this CT complex are randomly disordered between the two patterns shown in Figure 2. The calculations of the electronic structures and related quantities in the solid states require well-defined atomic parameters for all the atoms involved. Although DES materials with the disorder are seldom reported and have garnered considerable attention [27–29], the lack of the abovementioned calculation results has seriously hindered further studies on this CT complex and a clear understanding of the general DESs.

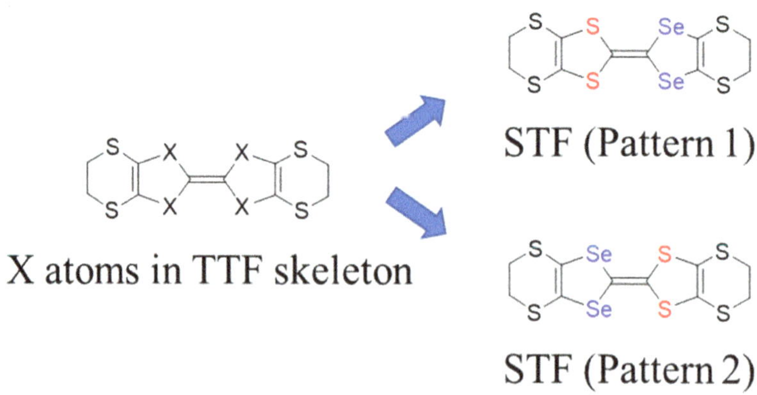

Figure 2. Two possible packing patterns of the STF molecule at a given site in α-STF_2I_3 (X = S or Se).

We have recently proposed a tight-binding band structure of α-STF_2I_3 based on the overlapping integrals S_{STF}, assuming statistically averaged structures between all the possible molecular arrangements at each site in the disordered crystal (Figure 3; see Appendix A for details) in addition to the assumption of t_{STF}/eV = $-10S_{STF}$, where t_{STF} represents the corresponding transfer integrals [21]. The band structure obtained in our previous study was characterised by less tilted and more isotropic Dirac cones compared to that in the present study.

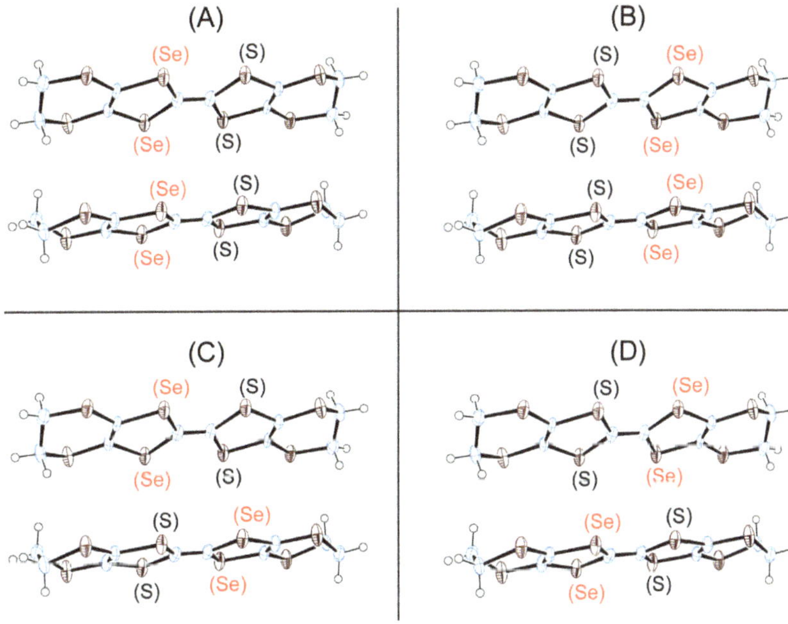

Figure 3. Four possible packing patterns (**A–D**) of the interacting pair of STF molecules at a given site in α-STF_2I_3. All four patterns should occur in the crystal with equal probability to reproduce the observed structure. Adapted from [21] and modified with permission.

The calculated band structure failed to reproduce the observed anisotropic temperature dependence of resistivity, despite being qualitatively consistent with the nearly temperature-independent electrical resistivity. In this paper, we propose a new method based on molecular orbitals (MOs) to estimate the overlap and transfer integrals in such disordered systems more clearly. More importantly, we found that solid-state STF molecules form unique electron configurations, which are qualitatively different from those in other disordered crystalline materials. The differences between the STF solids and other disordered systems include observed qualitative differences in electrical, magnetic, spectroscopic, and structural properties. The STF solids almost always behave as if there is no disorder in their crystal structures. In contrast, the slightest disorder frequently alters qualitative behaviours compared with those of related or corresponding materials without the disorder in common crystalline materials. To explain this fact, we propose a new idea of wavefunctions analogous to molecular orbitals and based on the original meanings of wavefunctions.

2. Calculation Methods

Single-crystal X-ray structural analyses demonstrated that all four chalcogen atoms (X: Figure 2) in the solid-state STF molecules possessed equal electron densities [21–26]. This trend is not only the case with conducting STF CT complexes, but also the case with insulating STF CT complexes and even with the neutral STF molecular crystals, irrespective of the crystal structures and the molecular symmetries required for the positions in the unit cells [26]. This trend is peculiar to the STF molecule in the solid-state, which cannot be explained by the known quantum stabilisation effects, such as delocalisation or intermolecular interactions. Such electron configurations require that all the atomic positions and the electron densities on each X atom are fixed in each molecule, regardless of the packing pattern. Thus, only the degrees of freedom remain in the assignment of S and Se atoms at fixed X positions with equal electron densities. In other words, the electron densities were averaged for all four X atoms as if the X atoms belonged to imaginary element between sulphur and selenium. This is probably because the differences in the atomic energy levels are sufficiently small between S and Se atoms, favouring such averaged electron configuration. Ultimately, this configuration would lead to the reduction of Coulombic repulsion between electrons, in addition to stabilisation due to the increased kinetic energy of electrons. As wavefunctions for electrons are associated with electron densities, we assumed that the wavefunction of an STF molecule (Equation (1)) in the solid state (ϕ_{STF} (MO of STF)) should be described using the two wavefunctions corresponding to the average of the two patterns in Figure 2.

$$\phi_{\mathrm{STF}} = \frac{1}{\sqrt{2}}\phi_1 + \frac{1}{\sqrt{2}}\phi_2 \tag{1}$$

where suffixes 1 and 2 denote the different patterns shown in Figure 2. To equally distribute the electrons on the X atoms in the STF molecules, both coefficients must be equal, that is, $1/\sqrt{2}$ if Equation (1) is normalised and if the cross-terms are ignored. The cross-terms between ϕ_1 and ϕ_2 are ignored because the two orientations (ϕ_1 and ϕ_2) never coincide at a given site for a single STF molecule. In other words, such an averaged electron configuration does not involve the exchange or oscillating movement between the S and Se nuclei in the STF molecules. Equation (1) describes the electronic configuration of the X atoms in Figure 2, corresponding to the imaginary element between S and Se. Notably, it appears that Equation (1) describes the interaction between the two states or the fluctuation originating from the quantum interference between the two states. However, this is not the case. As we will see by substituting atomic orbitals in Equation (1), a "conjugate" equation, *that is*, a state corresponding to Equation (2) does not describe the electronic configuration of the solid STF molecules (Equation (6)).

$$\phi_{\mathrm{STF}}' = \frac{1}{\sqrt{2}}\phi_1 - \frac{1}{\sqrt{2}}\phi_2 \tag{2}$$

All the wavefunctions below are assumed to be normalised. We use the same coefficients even after altering the expression, except when it is misleading.

The molecular orbital ϕ_{i} (i = 1, 2) was substituted into Equation (1) through a linear combination of atomic orbitals, φ_{j},

$$\phi_{\mathrm{i}} = \sum_{\mathrm{j}} c_{\mathrm{j}}\varphi_{\mathrm{j}} \tag{3}$$

where c_{j} and φ_{j} denote the coefficient and atomic orbital of atom j, respectively. For example, φ_{H} designates the 1s orbital of a hydrogen atom. For j = C (carbon), φ_{j} should describe one of the C 1*s*, C 2*s*, C 2p_x, C 2p_y, and C 2p_z orbitals. Note that all the atomic positions, except for the X atoms, $\vec{r_{\mathrm{j}}}$, and electron densities on all the atoms in an STF molecule are common between ϕ_1 and ϕ_2.

The molecular orbitals ϕ_{i} (i = 1, 2) obtained in this way are as follows:

$$\phi_1 = \sum_{\mathrm{j}} a_{\mathrm{Y}}\varphi_{\mathrm{Y}}\left(\vec{r_{\mathrm{J}}}\right) + \sum_{\mathrm{k}} b_{\mathrm{S}}\varphi_{\mathrm{S}}\left(\vec{r_{\mathrm{k}}}\right) + \sum_{\mathrm{l}} c_{\mathrm{Se}}\varphi_{\mathrm{Se}}\left(\vec{r_{\mathrm{l}}}\right) \tag{4}$$

$$\phi_2 = \sum_{\mathrm{j}} a_{\mathrm{Y}}\varphi_{\mathrm{Y}}\left(\vec{r_{\mathrm{J}}}\right) + \sum_{\mathrm{k}} b_{\mathrm{Se}}\varphi_{\mathrm{Se}}\left(\vec{r_{\mathrm{k}}}\right) + \sum_{\mathrm{l}} c_{\mathrm{S}}\varphi_{\mathrm{S}}\left(\vec{r_{\mathrm{l}}}\right) \tag{5}$$

where suffixes j, k, and l indicate the atomic positions in the STF molecule. Suffix Y indicates atoms other than X atoms in the STF molecule. Notably, the first terms in Equations (4) and (5) are identical. Meanwhile, the coordinates $\vec{r_k}$ and $\vec{r_l}$ for S and Se atoms indicate the X atoms on the left and right sides of each STF molecule, respectively. Accordingly, the two equations of ϕ_1 and ϕ_2 are identical, except for the coordinates $\vec{r_k}$ and $\vec{r_l}$ for S and Se atoms (the position vectors in real space for X atoms), respectively. By substituting Equations (4) and (5) in Equation (2), we obtain

$$\begin{aligned}\phi_{STF}' &= \frac{1}{\sqrt{2}}\phi_1 - \frac{1}{\sqrt{2}}\phi_2 \\ &= \frac{1}{\sqrt{2}}\left\{\sum_k b_S\varphi_S\left(\vec{r_k}\right) + \sum_l c_{Se}\varphi_{Se}\left(\vec{r_l}\right)\right\} \\ &-\frac{1}{\sqrt{2}}\left\{\sum_k b_{Se}\varphi_{Se}\left(\vec{r_k}\right) + \sum_l c_S\varphi_S\left(\vec{r_l}\right)\right\}\end{aligned} \quad (6)$$

which evidently does not describe the electronic configuration required for the solid STF molecules.

By substituting Equations (4) and (5) in Equation (1), we obtain

$$\begin{aligned}\phi_{STF} &= \frac{1}{\sqrt{2}}\left\{\sum_j a_Y\varphi_Y\left(\vec{r_j}\right) + \sum_k b_S\varphi_S\left(\vec{r_k}\right) + \sum_l c_{Se}\varphi_{Se}\left(\vec{r_l}\right)\right\} \\ &+\frac{1}{\sqrt{2}}\left\{\sum_j a_Y\varphi_Y\left(\vec{r_j}\right) + \sum_k b_{Se}\varphi_{Se}\left(\vec{r_k}\right) + \sum_l c_S\varphi_S\left(\vec{r_l}\right)\right\} \\ &= \frac{1}{\sqrt{2}}\left\{\sum_j a_Y\varphi_Y\left(\vec{r_j}\right) + \sum_k b_S\varphi_S\left(\vec{r_k}\right) + \sum_l c_S\varphi_S\left(\vec{r_l}\right)\right\} \\ &+\frac{1}{\sqrt{2}}\left\{\sum_j a_Y\varphi_Y\left(\vec{r_j}\right) + \sum_k b_{Se}\varphi_{Se}\left(\vec{r_k}\right) + \sum_l c_{Se}\varphi_{Se}\left(\vec{r_l}\right)\right\}\end{aligned} \quad (7)$$

Therefore,

$$\phi_{STF} = \frac{1}{\sqrt{2}}\phi_{(X=S)} + \frac{1}{\sqrt{2}}\phi_{(X=Se)} \quad (8)$$

where $\phi_{(X=S)}$ and $\phi_{(X=Se)}$ represent the following MOs, respectively.

$$\phi_{(X=S)} = \sum_j a_Y\varphi_Y\left(\vec{r_j}\right) + \sum_k b_S\varphi_S\left(\vec{r_k}\right) + \sum_l c_S\varphi_S\left(\vec{r_l}\right) \quad (9)$$

$$\phi_{(X=Se)} = \sum_j a_Y\varphi_Y\left(\vec{r_j}\right) + \sum_k b_{Se}\varphi_{Se}\left(\vec{r_k}\right) + \sum_l c_{Se}\varphi_{Se}\left(\vec{r_l}\right) \quad (10)$$

In short, $\phi_{(X=S)}$ and $\phi_{(X=Se)}$ are molecular orbitals, assuming that all the X atoms in Figure 2 should be either S or Se atoms with identical atomic positions, respectively. As such, the wavefunctions (MOs) of disordered solid-state STF molecules can be described by the equal contributions of the MOs of $\phi_{(X=S)}$ and $\phi_{(X=Se)}$, as shown in Equation (8).

Next, we discuss the electron densities of the STF sites in the solid state using Equation (8). The contribution of MO, ϕ_{STF}, to the electron densities of a given STF site in a solid state is described as follows:

$$\begin{aligned}2\int\left|\phi_{STF}\right|^2 dV &= 2\int\left(\frac{1}{\sqrt{2}}\phi_{(X=S)} + \frac{1}{\sqrt{2}}\phi_{(X=Se)}\right)^*\left(\frac{1}{\sqrt{2}}\phi_{(X=S)} + \frac{1}{\sqrt{2}}\phi_{(X=Se)}\right)dV \\ &= \left(\int\left|\phi_{(X=S)}\right|^2 dV + \int\left|\phi_{(X=Se)}\right|^2 dV\right) \\ &+\left(\int\phi^*_{(X=Se)}\phi_{(X=S)}dV + \int\phi^*_{(X=S)}\phi_{(X=Se)}dV\right)\end{aligned} \quad (11)$$

where the integral is calculated over the entire space. The coefficient "2" is required by spin multiplicity. Thus, by noting that ϕ_{STF}, $\phi_{(X=S)}$, and $\phi_{(X=Se)}$ are normalised,

$$2 = 2 + \left(\int \phi^*_{(X=Se)} \phi_{(X=S)} dV + \int \phi^*_{(X=S)} \phi_{(X=Se)} dV \right)$$

Then,

$$\int \phi^*_{(X=Se)} \phi_{(X=S)} dV + \int \phi^*_{(X=S)} \phi_{(X=Se)} dV = 0 \tag{12}$$

In other words,

$$\int |\phi_{STF}|^2 dV = \frac{1}{2} \left(\int \left|\phi_{(X=S)}\right|^2 dV + \int \left|\phi_{(X=Se)}\right|^2 dV \right) \tag{13}$$

Equation (13) indicates that the electron densities of disordered STF molecules can be described by the average of the electron densities of the $\phi_{(X=S)}$ and $\phi_{(X=Se)}$ eigenstates, consistent with the observations.

The discussion thus far indicates that intermolecular interactions between disordered STF molecules in the solid state can be described by replacing the wavefunction of STF, ϕ_{STF}, with an average of the wavefunctions of $\phi_{(X=S)}$ and $\phi_{(X=Se)}$ (Equation (8)). In disordered STF CT complexes, one can similarly calculate the overlap (S) and transfer (t) integrals between STF molecules based on Equation (8). The overlap (S_{STF}) and transfer (t_{STF}) integrals between two STF molecules located at $\vec{R_1}$ and $\vec{R_2}$, respectively, are:

$$\begin{aligned} S_{STF} &= \int \phi_{STF} * (\vec{R_1}) \phi_{STF}(\vec{R_2}) dV \\ &= \int \left(\tfrac{1}{\sqrt{2}} \phi_{(X=S)}(\vec{R_1}) + \tfrac{1}{\sqrt{2}} \phi_{(X=Se)}(\vec{R_1}) \right)^* \left(\tfrac{1}{\sqrt{2}} \phi_{(X=S)}(\vec{R_2}) \right. \\ &\left. + \tfrac{1}{\sqrt{2}} \phi_{(X=Se)}(\vec{R_2}) \right) dV \; (\because \text{Equation (8)}) \\ &\approx \tfrac{1}{2} \left\{ \int \phi_{(X=S)} * (\vec{R_1}) \phi_{(X=S)}(\vec{R_2}) dV + \int \phi_{(X=Se)} * (\vec{R_1}) \phi_{(X=Se)}(\vec{R_2}) dV \right\} \\ &= \tfrac{1}{2} \left\{ S_{(X=S)} + S_{(X=Se)} \right\}, \end{aligned} \tag{14}$$

where

$$S_{(X=S)} = \int \phi_{(X=S)} * (\vec{R_1}) \phi_{(X=S)}(\vec{R_2}) dV \tag{15}$$

$$S_{(X=Se)} = \int \phi_{(X=Se)} * (\vec{R_1}) \phi_{(X=Se)}(\vec{R_2}) dV \tag{16}$$

If the cross terms are ignored (Equation (17)):

$$\int \phi^*_{(X=Se)}(\vec{R_1}) \phi_{(X=S)}(\vec{R_2}) dV + \int \phi^*_{(X=S)}(\vec{R_1}) \phi_{(X=Se)}(\vec{R_2}) dV = 0 \tag{17}$$

The validity for Equation (17) is discussed after deriving the transfer integrals. Similarly,

$$\begin{aligned} t_{STF} &= \int \phi_{STF} * \left(\vec{R_1}\right) \hat{H} \phi_{STF} \left(\vec{R_2}\right) dV \\ &= \tfrac{1}{2} \left\{ E_{(X=S)} S_{(X=S)} + E_{(X=Se)} S_{(X=Se)} \right\} \\ &+ \tfrac{1}{2} \left\{ E_{(X=Se)} \int \phi_{(X=S)} * \left(\vec{R_1}\right) \phi_{(X=Se)} \left(\vec{R_2}\right) dV \right. \\ &\left. + E_{(X=S)} \int \phi_{(X=Se)} * \left(\vec{R_1}\right) \phi_{(X=S)} \left(\vec{R_2}\right) dV \right\} \qquad (\because \text{Equation(14)}) \end{aligned} \tag{18}$$

where $E_{(X=S)}$ and $E_{(X=Se)}$ are the energies of the orbitals of interest, for example, the HOMO. Below, we limit our discussion to the HOMO of an STF molecule, which dominates the physical properties of α-STF_2I_3.

$$\hat{H}\phi_{(X=S)} = E_{(X=S)}\phi_{(X=S)},\ \hat{H}\phi_{(X=Se)} = E_{(X=Se)}\phi_{(X=Se)}, \quad (19)$$

Therefore, Equation (18) is further altered to

$$t_{STF} = \frac{1}{2}\left\{t_{(X=S)} + t_{(X=Se)}\right\}, \quad (20)$$

where

$$t_{(X=S)} = E_{(X=S)}S_{(X=S)},\ t_{(X=Se)} = E_{(X=Se)}S_{(X=Se)} \quad (21)$$

and we approximated

$$E_{(X=Se)}\int \phi_{(X=S)} * \left(\vec{R_1}\right)\phi_{(X=Se)}\left(\vec{R_2}\right)dV + E_{(X=S)}\int \phi_{(X=Se)} * \left(\vec{R_1}\right)\phi_{(X=S)}\left(\vec{R_2}\right)dV \approx 0 \quad (22)$$

$$E_{(X=Se)} \approx E_{(X=S)} = -10\ (\text{eV}) \quad (23)$$

3. Results and Discussion

The obtained S_{STF} overlap integrals are listed in Table 1. Figure 4 shows the interacting pairs of the STF molecules corresponding to the S_{STF} values in Table 1.

Table 1. Estimated overlap integrals ($S_{STF} \times 10^3$/dimensionless) in α-STF_2I_3. HOMO-HOMO overlaps between adjacent STF radical cations.

Pairs [(a)]	X = Se	X = S	X = Se or S (50%)
g_1	+9.0	−19.7	−5.35
g_2	−8.50	−17.9	−13.2
g_3	+2.2	−11.7	−4.75
h_1	+31.8	−25.9	+2.95
h_2	−33.7	−25.3	−29.5
h_3	−15.0	−13.3	−14.15
h_4	+6.8	−8.6	−0.9

[(a)] For the interacting STF pairs in the unit cell, see Figure 1. The extended Hückel parameters utilised in the calculations are reported in [21]. For reference, the values of $S_{(X=Se)}$ and $S_{(X=S)}$ are listed in Table 1.

The validities of Equations (17) and (22) were not proven in this general discussion. However, the derived parameters, S_{STF} (Table 1) and t_{STF}, semi-quantitatively reproduced the observed electrical properties. In contrast, those based on our former estimation [21] failed to reproduce them even qualitatively. Namely, the anisotropy in conductivity in the *ab*-plane was observed as $\sigma_b > \sigma_a$ [22], which probably agrees with the contour plot of $E_1 - E_2$ in this study (Figure 5b), exhibiting an oval elongated approximately in the k_a-direction. Note that the ratio between the anisotropic conductivities in the *ab*-plane is proportional to the ratio between the square of the velocities in the corresponding directions [30]. In our previous method, we did not consider the details of the molecular orbitals shown in Equations (4)–(7). We simply averaged all the terms appearing in Equations (14) and (18) between the four interacting patterns (Figure 3). Consequently, the obtained overlap S_{STF} and transfer t_{STF} integrals in our previous study gave the contour plot of $E_1 - E_2$, exhibiting an oval elongated approximately in the k_b-direction [21], which qualitatively contradict the observed anisotropy in conductivity in the *ab*-plane. Once the overlap S_{STF} and transfer t_{STF} integrals are obtained in this study, we can calculate a tight-binding band structure using the model in our previous study [21]. The band structure obtained (Figure 5) is unique and essential because it indicates that α-STF_2I_3 has a pair of Dirac cones close to merging. The term "merging" implies that the Dirac electrons are close to transforming into normal electrons. Such cases have been extensively investigated theoretically [7,31–34]. However, to the best of our knowledge, actual examples have never been reported. The Dirac cones could not exhibit such a large anisotropy in resistivity, as observed, except when they are close to merging. Accordingly, the anisotropy observed in the resistivity of α-STF_2I_3 was exclusively explained by this band structure.

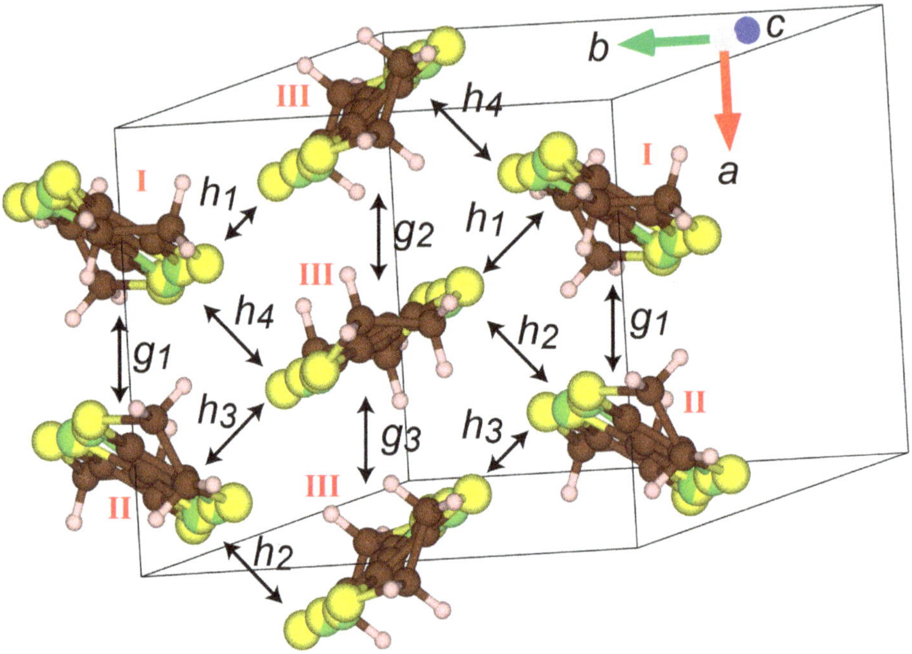

Figure 4. Intermolecular interactions between STF molecules in α-STF_2I_3. The unit cell contains three independent STF molecules, two halves (I and II) and one whole (III) STF molecule. All the I and II STF molecules possess an inversion centre on the central C=C bond, which means that atomic positions and electron densities are the same between Patterns 1 and 2 in Figure 2 for molecules I and II. Although molecule III does not possess an inversion centre, the atomic positions and electron densities for X in molecule III are also identical between Patterns 1 and 2.

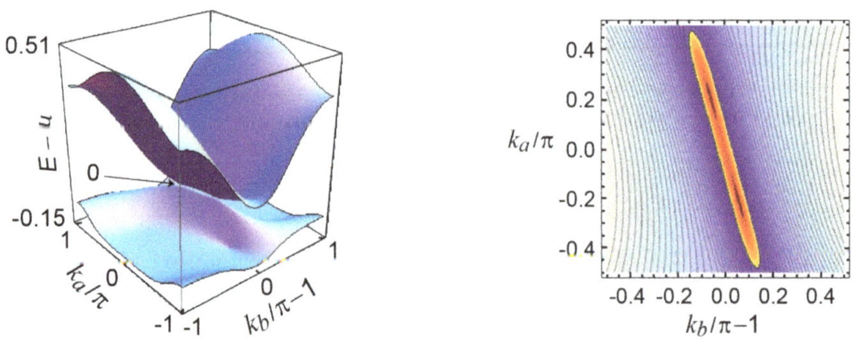

(a) (b)

Figure 5. Structure of the top two bands in three-quarter filled bands based on the tight-binding model, where transfer energies t_{STF} were assumed as $t_{STF}/eV = -10S_{STF,}$ and the values of S_{STF} were newly estimated using Equation (14) (Table 1) in this study. The tight-binding model, which is required in the band calculation using the values of t_{STF}, is identical to that in our previous study [21]. The obtained band structure is different from our previous study [21]. (**a**) Conduction and valence bands given by E_1 (upper band; $0 \leq E_1 \leq 0.51$ eV) and E_2 (lower band; $-0.15 \leq E_2 \leq 0$ eV) as the function of wavevector $\boldsymbol{k} = (k_a, k_b)$. The lattice constant of the square lattice is taken as unity. The energy (eV) is measured from the chemical potential μ. The two bands contact with each other at the Dirac points $(k_a/\pi, k_b/\pi - 1) = \pm\,(0.21, -0.06)$ with the energy $E_D = 0.173$ eV, which coincides with μ. (**b**) Contour plot of $E_1 - E_2$ with the energy range of $0 \leq E_1 - E_2 \leq 0.51$ (eV). The Dirac points exist in the two darkest points in the orange region, which indicates $0 \leq E_1 - E_2 \leq 0.05$ eV. The two Dirac points are close to merging at one of the TRIMs (time-reversal invariant momentum) X = (0, π) since the contour is elongated toward the X point and $E_1 - E_2$ is much smaller than that at other TRIMs.

4. Conclusions

We proposed a calculation method for intermolecular interactions in disordered systems based on the extended Hückel approximation. Despite the generally negative impression of the effects of disorder on cooperative properties, the physical properties and band structures of α-STF_2I_3 indicated that the fine tuning of intermolecular interactions using disorder in organic CT complexes is an effective new method. This method enables electron densities between S and Se atoms as if there was no disorder in the solid state, which is otherwise impossible. The reconciliation of the established band theories with the disordered systems has remained uninvestigated for a long time. The calculation method proposed in this study was used to investigate the band structures of such systems in detail. It has not only revealed that α-STF_2I_3 is a rare example of a system close to merging, but also that there is a novel electronic state requiring a wavefunction of the type of Equation (1) and the relevant assumptions.

Author Contributions: T.N. developed the approximated calculation method of overlap and transfer integrals (Equations (1)–(23)) with discussion with Y.S. Y.S. performed the band calculation (Figure 5) using the overlap integrals estimated by Equation (14). All authors have read and agreed to the published version of the manuscript.

Funding: This work was partially funded by a Grant-in-Aid for Challenging Exploratory Research (18K19061) of JSPS, the Iketani Science and Technology Foundation (ISTF; 0331005-A), Research Grant Program of the Futaba Foundation, Casio Science Promotion Foundation, and an Ehime University Grant for Project for the Promotion of Industry/University Cooperation.

Institutional Review Board Statement: Not applicable.

Informed Consent Statement: Not applicable.

Data Availability Statement: Not applicable.

Conflicts of Interest: The authors declare no conflict of interest.

Appendix A

Details of Our Previous Method of Calculation of Overlap and Transfer Integrals in α-STF_2I_3

According to the four possible patterns of relative orientation of the STF molecules in an interacting pair shown in Figure 3, we consider the average of four overlap and transfer integrals, respectively. Using the wavefunctions, ϕ_1 and ϕ_2, for Patterns 1 and 2, respectively, in Figure 2, the overlap integrals for Patterns A–D in Figure 3 are described as follows.

$$S_A = \int \phi_2\left(\vec{R_1}\right)^* \phi_2\left(\vec{R_2}\right) dV \tag{A1}$$

$$S_B = \int \phi_1\left(\vec{R_1}\right)^* \phi_1\left(\vec{R_2}\right) dV \tag{A2}$$

$$S_C = \int \phi_2\left(\vec{R_1}\right)^* \phi_1\left(\vec{R_2}\right) dV \tag{A3}$$

$$S_D = \int \phi_1\left(\vec{R_1}\right)^* \phi_2\left(\vec{R_2}\right) dV \quad \text{(A4)}$$

where we assume the two STF molecules are located at $\vec{R_1}$ and $\vec{R_2}$, respectively. In this paper, we use Equations (4) and (5) in the main text with the present notation.

$$\phi_1\left(\vec{R_0}\right) = \sum_j a_Y \varphi_Y\left(\vec{R_0} + \vec{r_j}\right) + \sum_k b_S \varphi_S\left(\vec{R_0} + \vec{r_k}\right) + \sum_l c_{Se} \varphi_{Se}\left(\vec{R_0} + \vec{r_l}\right) \quad \text{(A5)}$$

$$\phi_2\left(\vec{R_0}\right) = \sum_j a_Y \varphi_Y\left(\vec{R_0} + \vec{r_j}\right) + \sum_k b_{Se} \varphi_{Se}\left(\vec{R_0} + \vec{r_k}\right) + \sum_l c_S \varphi_S\left(\vec{R_0} + \vec{r_l}\right) \quad \text{(A6)}$$

where $\vec{R_0}$ indicates the location of the STF molecule serving as the origin for atomic position vectors in the same molecule. Based on the origin $\vec{R_0}$, $\vec{r_k}$ and $\vec{r_l}$ respectively indicate the position vectors of atoms in the STF molecule of the right-hand and left-hand side chalcogen atoms X in the STF molecule. Similarly, the position vectors of the remaining atoms in the STF molecule are indicated by $\vec{r_j}$. Note that replacement of the X atoms between $\vec{r_k}$ and $\vec{r_l}$ do not change the position vectors $\vec{r_j}$, because there is an only degree of freedom in assignment of X atoms between S or Se for an interacting pair of STF molecules, and additionally because all the atomic positions are fixed regardless of the assignment of S/Se.

Substituting Equation (A1) by Equation (A6),

$$\begin{aligned}
S_A &= \int \phi_2\left(\vec{R_1}\right)^* \phi_2\left(\vec{R_2}\right) dV \\
&= \int \left\{ \sum_j a_Y \varphi_Y\left(\vec{R_1} + \vec{r_j}\right) + \sum_k b_{Se} \varphi_{Se}\left(\vec{R_1} + \vec{r_k}\right) \right. \\
&\left. + \sum_l c_S \varphi_S\left(\vec{R_1} + \vec{r_l}\right) \right\}^* \left\{ \sum_j a_Y \varphi_Y\left(\vec{R_2} + \vec{r_j}\right) + \sum_k b_{Se} \varphi_{Se}\left(\vec{R_2} + \vec{r_k}\right) \right. \\
&\left. + \sum_l c_S \varphi_S\left(\vec{R_2} + \vec{r_l}\right) \right\} dV \\
&= \int \left\{ \sum_j a_Y \varphi_Y\left(\vec{R_1} + \vec{r_j}\right) \right\}^* \left\{ \sum_j a_Y \varphi_Y\left(\vec{R_2} + \vec{r_j}\right) \right\} dV \\
&+ \int \left\{ \sum_k b_{Se} \varphi_{Se}\left(\vec{R_1} + \vec{r_k}\right) \right\}^* \left\{ \sum_k b_{Se} \varphi_{Se}\left(\vec{R_2} + \vec{r_k}\right) \right\} dV \\
&+ \int \left\{ \sum_l c_S \varphi_S\left(\vec{R_1} + \vec{r_l}\right) \right\}^* \left\{ \sum_l c_S \varphi_S\left(\vec{R_2} + \vec{r_l}\right) \right\} dV \\
&+ \int \left\{ \sum_j a_Y \varphi_Y\left(\vec{R_1} + \vec{r_j}\right) \right\}^* \left\{ \sum_k b_{Se} \varphi_{Se}\left(\vec{R_2} + \vec{r_k}\right) \right\} dV \\
&+ \int \left\{ \sum_j a_Y \varphi_Y\left(\vec{R_1} + \vec{r_j}\right) \right\}^* \left\{ \sum_l c_S \varphi_S\left(\vec{R_2} + \vec{r_l}\right) \right\} dV \\
&+ \int \left\{ \sum_k b_{Se} \varphi_{Se}\left(\vec{R_1} + \vec{r_k}\right) \right\}^* \left\{ \sum_j a_Y \varphi_Y\left(\vec{R_2} + \vec{r_j}\right) \right\} dV \\
&+ \int \left\{ \sum_k b_{Se} \varphi_{Se}\left(\vec{R_1} + \vec{r_k}\right) \right\}^* \left\{ \sum_l c_S \varphi_S\left(\vec{R_2} + \vec{r_l}\right) \right\} dV \\
&+ \int \left\{ \sum_l c_S \varphi_S\left(\vec{R_1} + \vec{r_l}\right) \right\}^* \left\{ \sum_j a_Y \varphi_Y\left(\vec{R_2} + \vec{r_j}\right) \right\} dV \\
&+ \int \left\{ \sum_l c_S \varphi_S\left(\vec{R_1} + \vec{r_l}\right) \right\}^* \left\{ \sum_k b_{Se} \varphi_{Se}\left(\vec{R_2} + \vec{r_k}\right) \right\} dV
\end{aligned} \quad \text{(A7)}$$

Similar substitutions for S_B–S_D give an average S_{avg} of all the contributions from possible patterns A–D in Figure 3.

$$\begin{aligned} S_{avg} &= \tfrac{1}{4}\{S_A + S_B + S_C + S_D\} \\ &= \tfrac{1}{4}\Big\{\int \phi_2\left(\vec{R_1}\right)^* \phi_2\left(\vec{R_2}\right)\mathrm{d}V + \int \phi_1\left(\vec{R_1}\right)^* \phi_1\left(\vec{R_2}\right)\mathrm{d}V \\ &+ \int \phi_2\left(\vec{R_1}\right)^* \phi_1\left(\vec{R_2}\right)\mathrm{d}V + \int \phi_1\left(\vec{R_1}\right)^* \phi_2\left(\vec{R_2}\right)\mathrm{d}V\Big\} \end{aligned} \tag{A8}$$

Before substituting Equations (A5) and (A6) for ϕ_1 and ϕ_2 in Equation (A8), i.e., Equations (A1)–(A4), we note

$$\begin{aligned} \int \Bigg[& \left\{\sum_j a_Y\varphi_Y\left(\vec{R_1}+\vec{r_J}\right)\right\}^* \left\{\sum_j a_Y\varphi_Y\left(\vec{R_2}+\vec{r_J}\right)\right\} \\ &+ \left\{\sum_k b_S\varphi_S\left(\vec{R_1}+\vec{r_k}\right)\right\}^* \left\{\sum_k b_S\varphi_S\left(\vec{R_2}+\vec{r_k}\right)\right\} \\ &+ \left\{\sum_l c_S\varphi_S\left(\vec{R_1}+\vec{r_l}\right)\right\}^* \left\{\sum_l c_S\varphi_S\left(\vec{R_2}+\vec{r_l}\right)\right\} \\ &+ \left\{\sum_j a_Y\varphi_Y\left(\vec{R_1}+\vec{r_J}\right)\right\}^* \left\{\sum_k b_S\varphi_S\left(\vec{R_2}+\vec{r_k}\right)\right\} \\ &+ \left\{\sum_k b_S\varphi_S\left(\vec{R_1}+\vec{r_k}\right)\right\}^* \left\{\sum_j a_Y\varphi_Y\left(\vec{R_2}+\vec{r_J}\right)\right\} \\ &+ \left\{\sum_k b_S\varphi_S\left(\vec{R_1}+\vec{r_k}\right)\right\}^* \left\{\sum_l c_S\varphi_S\left(\vec{R_2}+\vec{r_l}\right)\right\} \\ &+ \left\{\sum_l c_S\varphi_S\left(\vec{R_1}+\vec{r_l}\right)\right\}^* \left\{\sum_k b_S\varphi_S\left(\vec{R_2}+\vec{r_k}\right)\right\} \\ &+ \left\{\sum_j a_Y\varphi_Y\left(\vec{R_1}+\vec{r_J}\right)\right\}^* \left\{\sum_l c_S\varphi_S\left(\vec{R_2}+\vec{r_l}\right)\right\} \\ &+ \left\{\sum_l c_S\varphi_S\left(\vec{R_1}+\vec{r_l}\right)\right\}^* \left\{\sum_j a_Y\varphi_Y\left(\vec{R_2}+\vec{r_J}\right)\right\}\Bigg] dV = S_{(X=S)} \end{aligned} \tag{A9}$$

and

$$\begin{aligned} \int \Bigg[& \left\{\sum_j a_Y\varphi_Y\left(\vec{R_1}+\vec{r_J}\right)\right\}^* \left\{\sum_j a_Y\varphi_Y\left(\vec{R_2}+\vec{r_J}\right)\right\} \\ &+ \left\{\sum_k b_{Se}\varphi_{Se}\left(\vec{R_1}+\vec{r_k}\right)\right\}^* \left\{\sum_k b_{Se}\varphi_{Se}\left(\vec{R_2}+\vec{r_k}\right)\right\} \\ &+ \left\{\sum_l c_{Se}\varphi_{Se}\left(\vec{R_1}+\vec{r_l}\right)\right\}^* \left\{\sum_l c_{Se}\varphi_{Se}\left(\vec{R_2}+\vec{r_l}\right)\right\} \\ &+ \left\{\sum_j a_Y\varphi_Y\left(\vec{R_1}+\vec{r_J}\right)\right\}^* \left\{\sum_k b_{Se}\varphi_{Se}\left(\vec{R_2}+\vec{r_k}\right)\right\} \\ &+ \left\{\sum_k b_{Se}\varphi_{Se}\left(\vec{R_1}+\vec{r_k}\right)\right\}^* \left\{\sum_j a_Y\varphi_Y\left(\vec{R_2}+\vec{r_J}\right)\right\} \\ &+ \left\{\sum_k b_{Se}\varphi_{Se}\left(\vec{R_1}+\vec{r_k}\right)\right\}^* \left\{\sum_l c_{Se}\varphi_{Se}\left(\vec{R_2}+\vec{r_l}\right)\right\} \\ &+ \left\{\sum_l c_{Se}\varphi_{Se}\left(\vec{R_1}+\vec{r_l}\right)\right\}^* \left\{\sum_k b_{Se}\varphi_{Se}\left(\vec{R_2}+\vec{r_k}\right)\right\} \\ &+ \left\{\sum_j a_Y\varphi_Y\left(\vec{R_1}+\vec{r_J}\right)\right\}^* \left\{\sum_l c_{Se}\varphi_{Se}\left(\vec{R_2}+\vec{r_l}\right)\right\} \\ &+ \left\{\sum_l c_{Se}\varphi_{Se}\left(\vec{R_1}+\vec{r_l}\right)\right\}^* \left\{\sum_j a_Y\varphi_Y\left(\vec{R_2}+\vec{r_J}\right)\right\}\Bigg] dV = S_{(X=Se)} \end{aligned} \tag{A10}$$

By substituting Equations (A9) and (A10) for the corresponding terms in Equation (A8),

$$
\begin{aligned}
S_{\mathrm{avg}} &= \tfrac{1}{4}\{S_{\mathrm{A}} + S_{\mathrm{B}} + S_{\mathrm{C}} + S_{\mathrm{D}}\} \\
&= \tfrac{1}{4}\int \Bigg[4\left\{\sum_{\mathrm{j}} a_{\mathrm{Y}}\varphi_{\mathrm{Y}}\left(\vec{R_1}+\vec{r_{\mathrm{J}}}\right)\right\}^{*}\left\{\sum_{\mathrm{j}} a_{\mathrm{Y}}\varphi_{\mathrm{Y}}\left(\vec{R_2}+\vec{r_{\mathrm{J}}}\right)\right\} \\
&+\left\{\sum_{\mathrm{k}} b_{\mathrm{S}}\varphi_{\mathrm{S}}\left(\vec{R_1}+\vec{r_{\mathrm{k}}}\right)\right\}^{*}\left\{\sum_{\mathrm{k}} b_{\mathrm{S}}\varphi_{\mathrm{S}}\left(\vec{R_2}+\vec{r_{\mathrm{k}}}\right)\right\} \\
&+\left\{\sum_{\mathrm{l}} c_{\mathrm{S}}\varphi_{\mathrm{S}}\left(\vec{R_1}+\vec{r_{\mathrm{l}}}\right)\right\}^{*}\left\{\sum_{\mathrm{l}} c_{\mathrm{S}}\varphi_{\mathrm{S}}\left(\vec{R_2}+\vec{r_{\mathrm{l}}}\right)\right\} \\
&+2\left\{\sum_{\mathrm{j}} a_{\mathrm{Y}}\varphi_{\mathrm{Y}}\left(\vec{R_1}+\vec{r_{\mathrm{J}}}\right)\right\}^{*}\left\{\sum_{\mathrm{k}} b_{\mathrm{S}}\varphi_{\mathrm{S}}\left(\vec{R_2}+\vec{r_{\mathrm{k}}}\right)\right\} \\
&+2\left\{\sum_{\mathrm{k}} b_{\mathrm{S}}\varphi_{\mathrm{S}}\left(\vec{R_1}+\vec{r_{\mathrm{k}}}\right)\right\}^{*}\left\{\sum_{\mathrm{j}} a_{\mathrm{Y}}\varphi_{\mathrm{Y}}\left(\vec{R_2}+\vec{r_{\mathrm{J}}}\right)\right\} \\
&+\left\{\sum_{\mathrm{k}} b_{\mathrm{S}}\varphi_{\mathrm{S}}\left(\vec{R_1}+\vec{r_{\mathrm{k}}}\right)\right\}^{*}\left\{\sum_{\mathrm{l}} c_{\mathrm{S}}\varphi_{\mathrm{S}}\left(\vec{R_2}+\vec{r_{\mathrm{l}}}\right)\right\} \\
&+\left\{\sum_{\mathrm{l}} c_{\mathrm{S}}\varphi_{\mathrm{S}}\left(\vec{R_1}+\vec{r_{\mathrm{l}}}\right)\right\}^{*}\left\{\sum_{\mathrm{k}} b_{\mathrm{S}}\varphi_{\mathrm{S}}\left(\vec{R_2}+\vec{r_{\mathrm{k}}}\right)\right\} \\
&+2\left\{\sum_{\mathrm{j}} a_{\mathrm{Y}}\varphi_{\mathrm{Y}}\left(\vec{R_1}+\vec{r_{\mathrm{J}}}\right)\right\}^{*}\left\{\sum_{\mathrm{l}} c_{\mathrm{S}}\varphi_{\mathrm{S}}\left(\vec{R_2}+\vec{r_{\mathrm{l}}}\right)\right\} \\
&+2\left\{\sum_{\mathrm{l}} c_{\mathrm{S}}\varphi_{\mathrm{S}}\left(\vec{R_1}+\vec{r_{\mathrm{l}}}\right)\right\}^{*}\left\{\sum_{\mathrm{j}} a_{\mathrm{Y}}\varphi_{\mathrm{Y}}\left(\vec{R_2}+\vec{r_{\mathrm{J}}}\right)\right\} \\
&+\left\{\sum_{\mathrm{k}} b_{\mathrm{Se}}\varphi_{\mathrm{Se}}\left(\vec{R_1}+\vec{r_{\mathrm{k}}}\right)\right\}^{*}\left\{\sum_{\mathrm{k}} b_{\mathrm{Se}}\varphi_{\mathrm{Se}}\left(\vec{R_2}+\vec{r_{\mathrm{k}}}\right)\right\} \\
&+\left\{\sum_{\mathrm{l}} c_{\mathrm{Se}}\varphi_{\mathrm{Se}}\left(\vec{R_1}+\vec{r_{\mathrm{l}}}\right)\right\}^{*}\left\{\sum_{\mathrm{l}} c_{\mathrm{Se}}\varphi_{\mathrm{Se}}\left(\vec{R_2}+\vec{r_{\mathrm{l}}}\right)\right\} \\
&+2\left\{\sum_{\mathrm{j}} a_{\mathrm{Y}}\varphi_{\mathrm{Y}}\left(\vec{R_1}+\vec{r_{\mathrm{J}}}\right)\right\}^{*}\left\{\sum_{\mathrm{k}} b_{\mathrm{Se}}\varphi_{\mathrm{Se}}\left(\vec{R_2}+\vec{r_{\mathrm{k}}}\right)\right\} \\
&+2\left\{\sum_{\mathrm{k}} b_{\mathrm{Se}}\varphi_{\mathrm{Se}}\left(\vec{R_1}+\vec{r_{\mathrm{k}}}\right)\right\}^{*}\left\{\sum_{\mathrm{j}} a_{\mathrm{Y}}\varphi_{\mathrm{Y}}\left(\vec{R_2}+\vec{r_{\mathrm{J}}}\right)\right\} \\
&+\left\{\sum_{\mathrm{k}} b_{\mathrm{Se}}\varphi_{\mathrm{Se}}\left(\vec{R_1}+\vec{r_{\mathrm{k}}}\right)\right\}^{*}\left\{\sum_{\mathrm{l}} c_{\mathrm{Se}}\varphi_{\mathrm{Se}}\left(\vec{R_2}+\vec{r_{\mathrm{l}}}\right)\right\} \\
&+\left\{\sum_{\mathrm{l}} c_{\mathrm{Se}}\varphi_{\mathrm{Se}}\left(\vec{R_1}+\vec{r_{\mathrm{l}}}\right)\right\}^{*}\left\{\sum_{\mathrm{k}} b_{\mathrm{Se}}\varphi_{\mathrm{Se}}\left(\vec{R_2}+\vec{r_{\mathrm{k}}}\right)\right\} \\
&+2\left\{\sum_{\mathrm{j}} a_{\mathrm{Y}}\varphi_{\mathrm{Y}}\left(\vec{R_1}+\vec{r_{\mathrm{J}}}\right)\right\}^{*}\left\{\sum_{\mathrm{l}} c_{\mathrm{Se}}\varphi_{\mathrm{Se}}\left(\vec{R_2}+\vec{r_{\mathrm{l}}}\right)\right\} \\
&+2\left\{\sum_{\mathrm{l}} c_{\mathrm{Se}}\varphi_{\mathrm{Se}}\left(\vec{R_1}+\vec{r_{\mathrm{l}}}\right)\right\}^{*}\left\{\sum_{\mathrm{j}} a_{\mathrm{Y}}\varphi_{\mathrm{Y}}\left(\vec{R_2}+\vec{r_{\mathrm{J}}}\right)\right\}\Bigg] dV \\
&+\tfrac{1}{4}\int(\text{eight cross terms between } \varphi_{\mathrm{S}} \text{ and } \varphi_{\mathrm{Se}})dV \\
&= \tfrac{1}{4}\left\{S_{(\mathrm{X=S})} + S_{(\mathrm{X=Se})}\right\} + \tfrac{1}{4}\int 2\left\{\sum_{\mathrm{j}} a_{\mathrm{Y}}\varphi_{\mathrm{Y}}\left(\vec{R_1}+\vec{r_{\mathrm{J}}}\right)\right\}^{*}\left\{\sum_{\mathrm{j}} a_{\mathrm{Y}}\varphi_{\mathrm{Y}}\left(\vec{R_2}+\vec{r_{\mathrm{J}}}\right)\right\}dV \\
&+\tfrac{1}{4}\int(\text{four cross terms between } \varphi_{\mathrm{Y}} \text{ and } \varphi_{\mathrm{S}})dV \\
&+\tfrac{1}{4}\int(\text{four cross terms between } \varphi_{\mathrm{Y}} \text{ and } \varphi_{\mathrm{Se}})dV \\
&+\tfrac{1}{4}\int(\text{eight cross terms between } \varphi_{\mathrm{S}} \text{ and } \varphi_{\mathrm{Se}})dV
\end{aligned}
\tag{A11}
$$

Comparing Equation (A11) with Equation (14) in the main text, the difference between our previous and present calculations lies in the coefficient of the average of $S_{(\mathrm{X=S})}$ and $S_{(\mathrm{X=Se})}$ in addition to the contribution from the remaining terms in Equation (A11).

The differences between our previous and present values of t_{STF} are shown to lie in the coefficient of the average of $t_{(\mathrm{X=S})}$ and $t_{(\mathrm{X=Se})}$ in addition to additional terms by similar discussion. Based on the assumptions and calculations thus far, the coefficient of the average of $t_{(X = S)}$ and $t_{(X = Se)}$ is altered from 1/4 in Equation (A11) (in our previous work) to 1/2 in Equation (14) (in this work).

References

1. Shwetha, G.; Chandra, S.; Chandra Shekar, N.V.; Kalavathi, S. Existence of spin-polarized Dirac cone in Sc_2CrB_6: A DFT study. *Phys. B Cond. Mat.* **2022**, *624*, 413369. [CrossRef]
2. Yi, W.; Jiang, X.; Wang, Z.; Yang, T.; Yang, B.; Liu, X. ABX_6 monolayers: A new Dirac material family containing high Fermi velocities and topological properties. *Appl. Surf. Sci.* **2021**, *570*, 151237. [CrossRef]
3. Li, M.; Wang, Q.; Wang, G.; Yuan, Z.; Song, W.; Lou, R.; Liu, Z.; Huang, Y.; Liu, Z.; Lei, H.; et al. Dirac cone, flat band and saddle point in kagome magnet YMn_6Sn_6. *Nat. Commun.* **2021**, *12*, 3129. [CrossRef]
4. Kitayama, K.; Tanaka, Y.; Ogata, M.; Mochizuki, M. Floquet theory of photoinduced topological phase transitions in the organic salt α-(BEDT-TTF)$_2$I$_3$ irradiated with elliptically polarized light. *J. Phys. Soc. Jpn.* **2021**, *90*, 104705. [CrossRef]
5. Li, P.; Guo, Z.-X. The Dirac half-semimetal and quantum anomalous Hall effect in two-dimensional Janus $Mn_2X_3Y_3$ (X, Y = F, Cl, Br, I). *Phys. Chem. Chem. Phys.* **2021**, *23*, 19673–19679. [CrossRef] [PubMed]
6. Bittner, N.; Golež, D.; Casula, M.; Werner, P. Photoinduced Dirac-cone flattening in $BaNiS_2$. *Phys. Rev. B* **2021**, *104*, 115138. [CrossRef]
7. Herrera, S.A.; Naumis, G.G. Optoelectronic fingerprints of interference between different charge carriers and band flattening in graphene superlattices. *Phys. Rev. B* **2021**, *104*, 115424. [CrossRef]
8. Tanaka, Y.; Mochizuki, M. Real-time dynamics of the photoinduced topological state in the organic conductor α-(BEDT-TTF)$_2$I$_3$ under continuous-wave and pulse excitations. *Phys. Rev. B* **2021**, *104*, 085123. [CrossRef]
9. Jiang, W.; Ni, X.; Liu, F. Exotic topological bands and quantum states in metal-organic and covalent-organic frameworks. *Acc. Chem. Res.* **2021**, *54*, 416–426. [CrossRef] [PubMed]
10. Kino, H.; Miyazaki, T. First-Principles Study of Electronic Structure in α-(BEDT-TTF)$_2$I$_3$ at Ambient Pressure and with Uniaxial Strain. *J. Phys. Soc. Jpn.* **2006**, *75*, 034704. [CrossRef]
11. Tsumuraya, T.; Suzumura, Y. First-principles study of the effective Hamiltonian for Dirac fermions with spin-orbit coupling in two-dimensional molecular conductor α-(BETS)$_2$I$_3$. *Eur. Phys. J. B* **2021**, *94*, 17. [CrossRef]
12. Briganti, M.; Totti, F. Magnetic anisotropy on demand exploiting high-pressure as remote control: An ab initio proof of concept. *Dalton Trans.* **2021**, *50*, 10621–10628. [CrossRef] [PubMed]
13. Novoselov, K.S.; Jiang, D.; Schedin, F.; Booth, T.J.; Khotkevich, V.V.; Morozov, S.V.; Geim, A.K. Two-dimensional atomic crystals. *Proc. Natl. Acad. Sci. USA* **2005**, *102*, 10451–10453. [CrossRef] [PubMed]
14. Novoselov, K.S.; Geim, A.K.; Morozov, S.V.; Jiang, D.; Katsnelson, M.I.; Grigorieva, I.V.; Dubonos, S.V.; Firsov, A.A. Two-dimensional gas of massless Dirac fermions in graphene. *Nature* **2005**, *438*, 197–200. [CrossRef] [PubMed]
15. Wehling, T.O.; Black-Schaffer, A.M.; Balatsky, A.V. Dirac materials. *Adv. Phys.* **2014**, *63*, 1–76. [CrossRef]
16. Naito, T. Modern History of Organic Conductors: An Overview. *Crystals* **2021**, *11*, 838. [CrossRef]
17. Kato, R.; Suzumura, Y. Novel Dirac electron in single-component molecular conductor [Pd(dddt)$_2$] (dddt = 5,6-dihydro-1,4-dithiin-2,3-dithiolate). *J. Phys. Soc. Jpn.* **2017**, *86*, 064705. [CrossRef]
18. Kato, R.; Cui, H.-B.; Tsumuraya, T.; Miyazaki, T.; Suzumura, Y. Emergence of the Dirac electron system in a single-component molecular conductor under high pressure. *J. Am. Chem. Soc.* **2017**, *139*, 1770–1773. [CrossRef] [PubMed]
19. Tajima, N.; Sugawara, S.; Tamura, M.; Kato, R.; Nishio, Y.; Kajita, K. Transport properties of massless Dirac Fermions in an organic conductor α-(BEDT-TTF)$_2$I$_3$ under pressure. *EPL* **2007**, *80*, 47002. [CrossRef]
20. Kobayashi, A.; Suzumura, Y.; Fukuyama, H. Hall effect and orbital diamagnetism in zerogap state of molecular conductor α-(BEDT-TTF)$_2$I$_3$. *J. Phys. Soc. Jpn.* **2008**, *77*, 064718. [CrossRef]
21. Naito, T.; Doi, R.; Suzumura, Y. Exotic Dirac Cones on the Band Structure of α-STF$_2$I$_3$ at Ambient Temperature and Pressure. *J. Phys. Soc. Jpn.* **2020**, *89*, 023701. [CrossRef]
22. Naito, T.; Doi, R. Band Structure and Physical Properties of α-STF$_2$I$_3$: Dirac Electrons in Disordered Conduction Sheets. *Crystals* **2020**, *10*, 270. [CrossRef]
23. Kobara, R.; Igarashi, S.; Kawasugi, Y.; Doi, R.; Naito, T.; Tamura, M.; Kato, R.; Nishio, Y.; Kajita, K.; Tajima, N. Universal Behavior of Magnetoresistance in Organic Dirac Electron Systems. *J. Phys. Soc. Jpn.* **2020**, *89*, 113703. [CrossRef]
24. Hiraki, K.-I.; Harada, S.; Arai, K.; Takano, Y.; Takahashi, T.; Tajima, N.; Kato, R.; Naito, T. Local Spin Susceptibility of α-D$_2$I$_3$ (D = bis(ethylendithio)tetraselenafulvalene (BETS) and bis(ethylendithio)dithiadiselenafulvalene (BEDT-STF)) Studied by ^{77}Se NMR. *J. Phys. Soc. Jpn.* **2011**, *80*, 014715. [CrossRef]
25. Inokuchi, M.; Tajima, H.; Kobayashi, A.; Ohta, T.; Kuroda, H.; Kato, R.; Naito, T.; Kobayashi, H. Electrical and Optical Properties of α-(BETS)$_2$I$_3$ and α-(BEDT-STF)$_2$I$_3$. *Bull. Chem Soc. Jpn.* **1995**, *68*, 547–553. [CrossRef]
26. Naito, T.; Kobayashi, H.; Kobayashi, A. The Electrical Behavior of Charge-Transfer Salts Based on an Unsymmetrical Donor Bis(ethylenedithio)diselenadithiafulvalene (STF): Disorder Effect on the Transport Properties. *Bull. Chem. Soc. Jpn.* **1997**, *70*, 107–114. [CrossRef]
27. Zhao, P.-L.; Wang, A.-M. Interplay between tilt, disorder, and Coulomb interaction in type-I Dirac fermions. *Phys. Rev. B* **2019**, *100*, 125138. [CrossRef]

28. Papaj, M.; Isobe, H.; Fu, L. Nodal arc of disordered Dirac fermions and non-Hermitian band theory. *Phys. Rev. B* **2019**, *99*, 201107. [CrossRef]
29. Martínez-Velarte, M.C.; Kretz, B.; Moro-Lagares, M.; Aguirre, M.H.; Riedemann, T.M.; Lograsso, T.A.; Morellón, L.; Ibarra, M.R.; Garcia-Lekue, A.; Serrate, D. Chemical disorder in topological insulators: A route to magnetism tolerant topological surface states. *Nano Lett.* **2017**, *17*, 4047–4054. [CrossRef]
30. Suzumura, Y.; Proskurin, I.; Ogata, M. Effect of tilting on the in-plane conductivity of Dirac electrons in organic conductor. *J. Phys. Soc. Jpn.* **2014**, *83*, 023701. [CrossRef]
31. Montambaux, G.; Piéchon, F.; Fuchs, J.-N.; Goerbig, M.O. Merging of Dirac points in a two-dimensional crystal. *Phys. Rev. B* **2009**, *80*, 153412. [CrossRef]
32. Pellegrino, F.M.D.; Angilella, G.G.N.; Pucci, R. Strain effect on the optical conductivity of graphene. *Phys. Rev. B* **2010**, *81*, 035411. [CrossRef]
33. Kobayashi, A.; Suzumura, Y.; Piéchon, F.; Montambaux, G. Emergence of Dirac electron pair in the charge-ordered state of the organic conductor α-(BEDT-TTF)$_2$I$_3$. *Phys. Rev. B* **2011**, *84*, 075450. [CrossRef]
34. Samal, S.S.; Nandy, S.; Saha, K. Nonlinear transport without spin-orbit coupling or warping in two-dimensional Dirac semimetals. *Phys. Rev. B* **2021**, *103*, L201202. [CrossRef]

crystals

MDPI

Article

Donor-Type Nickel–Dithiolene Complexes Fused with Bulky Cycloalkane Substituents and Their Application in Molecular Conductors

Kazuya Kubo [1,*], Mamoru Sadahiro [1], Sonomi Arata [1], Norihisa Hoshino [2], Tomofumi Kadoya [1], Tomoyuki Akutagawa [2], Reizo Kato [3] and Jun-ichi Yamada [1]

1 Graduate School of Science, University of Hyogo, 3-2-1, Kouto, Kamigori-cho, Akou-gun, Kobe 678-1297, Japan; g1127sm@gmail.com (M.S.); rinrin520sono48@gmail.com (S.A.); kadoya.t@sci.u-hyogo.ac.jp (T.K.); yamada@sci.u-hyogo.ac.jp (J.-i.Y.)
2 Institute of Multidisciplinary Research for Advanced Materials, Tohoku University, 2-1-1 Katahira, Aoba-ku, Sendai 980-8577, Japan; norihisa.hoshino.b5@tohoku.ac.jp (N.H.); akutagawa@tohoku.ac.jp (T.A.)
3 RIKEN, 2-1 Hirosawa, Wako-shi, Saitama 351-0023, Japan; reizo@riken.jp
* Correspondence: kubo@sci.u-hyogo.ac.jp; Tel.: +81-791-58-0163

Abstract: The effects of substituents on the arrangement of metal–dithiolene complexes based on π-conjugated systems, which are extensively used to synthesize various functional materials, have not been studied adequately. New donor-type nickel–dithiolene complexes fused with bulky cycloalkane substituents [Ni(C_n-dddt)$_2$] (C_5-dddt = 4a,5,6,6a-pentahydro-1,4-benzodithiin-2,3-dithiolate; C_6-dddt = 4a,5,6,7,8,8a-hexahydro-1,4-benzodithiin-2,3-dithiolate; C_7-dddt = 4a,5,6,7,8,9,9a-heptahydro-1,4-benzodithiin-2,3-dithiolate; and C_8-dddt = 4a,5,6,7,8,9,10,10a-octahydro-1,4-benzodithiin-2,3-dithiolate) were synthesized in this study. All the complexes were crystallized in cis-[Ni(cis-C_n-dddt)$_2$] conformations with cis-oriented (*R*,*S*) conformations around the cycloalkylene groups in the neutral state. Unique molecular arrangements with a three-dimensional network, a one-dimensional column, and a helical molecular arrangement were formed in the crystals owing to the flexible cycloalkane moieties. New 2:1 cation radical crystals of [Ni(C_5-dddt)$_2$]$_2$(X) (X = ClO_4^- or PF_6^-), obtained by electrochemical crystallization, exhibited semiconducting behaviors (ρ_{rt} = 0.8 Ω cm, E_a = 0.09 eV for the ClO_4^- crystal; 4.0 Ω cm, 0.13 eV for the PF_6^- crystal) under ambient pressure due to spin-singlet states between the dimers of the donor, which were in accordance with the conducting behaviors under hydrostatic pressure (ρ_{rt} = 0.2 Ω cm, E_a = 0.07 eV for the ClO_4^- crystal; 1.0 Ω cm, 0.12 eV for the PF_6^- crystal at 2.0 GPa).

Keywords: nickel–dithiolene complex; cycloalkane substituent; molecular conductor

Citation: Kubo, K.; Sadahiro, M.; Arata, S.; Hoshino, N.; Kadoya, T.; Akutagawa, T.; Kato, R.; Yamada, J.-i. Donor-Type Nickel–Dithiolene Complexes Fused with Bulky Cycloalkane Substituents and Their Application in Molecular Conductors. *Crystals* **2021**, *11*, 1154. https://doi.org/10.3390/cryst11101154

Academic Editor: Alberto Girlando

Received: 26 August 2021
Accepted: 14 September 2021
Published: 23 September 2021

Publisher's Note: MDPI stays neutral with regard to jurisdictional claims in published maps and institutional affiliations.

1. Introduction

Organic π-conjugated molecules, including tetrathiafulvalene (TTF) skeletons, are among the most common building blocks for the realization of various functional materials [1]. The TTF molecule has an electron donor ability, which can help realize various charge transfers and produce radical cation crystals. The TTF-TCNQ (TCNQ: 7,7,8,8-tetracyanoquinodimethane) is a prototype of charge-transfer compounds, where the highest occupied molecular orbital (HOMO) and lowest unoccupied molecular orbital (LUMO) bands of the open-shell donors and acceptors, respectively, contribute to the conduction [2,3]. It is the first organic conductor to exhibit a metallic conductivity within a wide temperature range with a minimum of 59 K, where a sharp metal-to-insulator transition is observed [4]. Moreover, TTF has attracted significant research attention with the discovery of superconducting salts [5–8] based on TTF derivatives such as atetramethyltetraselenafulvalene (TMTSF) [9], bis(ethylenedithio)tetrathiafulvalene (BEDT-TTF) [10,11], and dimethyl(ethylenedithio)diselenadithiafulvalene (DMET) [12,13]. The conducting properties of molecular-based materials are significantly influenced by their molecular arrangements in crystal form. The cation radicals of BEDT-TTF and bis(ethylenedithio)tetraselenafulvalene (BETS) can crystallize in various donor packing motifs, including α-, β-, θ-, κ-, and λ-type arrangements by combination with inorganic, organic, and organometallic anions [14–18]. Alternative strategies for tuning the molecular arrangements in organic conductors were reported based on organic synthesis techniques. A potential method involves the direct introduction of various substituents into the TTF or BEDT-TTF skeletons [19–42]. Kimura et al. reported that meso-2-(5,6-dihydro-1,3-dithiolo[4,5-b][1,4]dithiin-2-ylidene)-5,6-dihydro-5,6-dimethyl-1,3-dithiolo[4,5-b][1,4]dithiin (meso-DMBEDT-TTF) forms a superconducting crystal β-(meso-DMBEDT-TTF)$_2$PF$_6$ with a transition temperature at 4.3 K under a hydrostatic pressure of 4.0 kbar due to a checkerboard-type charge order state on the donor molecules [43,44]. Another alternative method involves the introduction of bulky cycloalkane moieties into the donor molecules to determine the stacking of the donor molecules; thus, the ratio of the on-site Coulomb repulsion energy U to the bandwidth W can be controlled by changing W for the organic Mott insulators based on the enhanced steric hindrance in the radical cation crystals of CnDT-EDO-TTF (n = 5, 6, 7, 8; *cis*-1,2-cycloalkylene-1,2-dithio)ethylenedioxytetrathiafulvalene] and its derivatives (Scheme 1) [45,46].

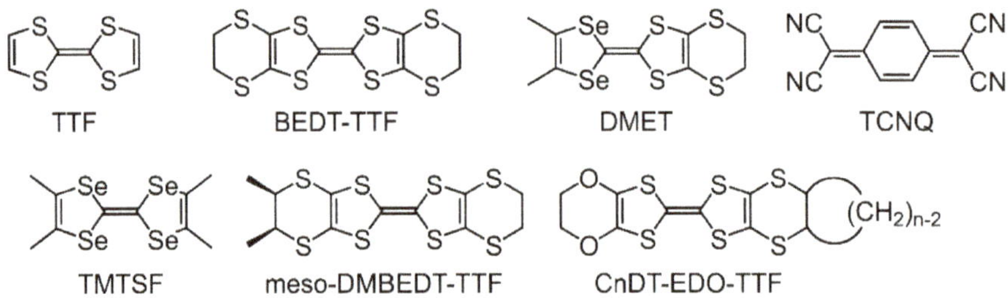

Scheme 1. Organic components of molecular conductors.

Metal bis-dithiolene complexes can exhibit various oxidation states [47] and have smaller HOMO–LUMO gaps than those of the organic donors [48,49], which is advantageous for the components of molecular conductors. Electron-acceptor type [M(dmit)$_2$] (M = Pd, Pt; dmit = 1,3-dithiole-2-thion-4,5-dithiolete) complexes with small gaps between the HOMO and LUMO form dimer structures. These structures induce crossing HOMO–LUMO band inversions through metal–metal interactions in the solid state, which provide various conducting, magnetic, and other physical properties by combining with closed-shell cations [47,50–52]. The electron-donor-type metal bis-dithiolene complexes [M(dddt)$_2$] (M = Ni, Pd, Pt, and Au; dddt = 5,6-dihydro-1,4-dithiine-2,3-ditholate) are candidate materials for the development of molecular conductors due to their orbital symmetries, which are similar to those of BEDT-TTF [53–56]. A cation radical crystal, [Ni(dddt)$_2$]$_3$(AuBr$_2$)$_2$, is the first metal based on the donor-type metal–dithiolene complexes exhibiting metallic behavior down to at least 1.3 K [57]. Recently, Kato et al. reported a single component molecular Dirac electron system based on Pd(dddt)$_2$ that exhibits temperature-independent resistivity under high hydrostatic pressures [58–63]. The conducting and magnetic properties exhibited by the molecular crystals based on the dithiolene complexes are significantly dependent on the molecular arrangements of the metal–dithiolene complexes and organic molecular conductors. In the case of molecular conductors based on metal–dithiolene complexes, the molecular arrangements of the complexes are dependent on the combination of counter cations and anions in the crystals [47,48]. However, there are few reports on the substituent effects on the molecular arrangement, which influences the physical properties, despite the advantages of the components of molecular conductors [64,65]. Thus, we designed novel donor-type metal bis-dithiolene complexes [Ni(C$_n$-dddt)$_2$] formed by a [Ni(dddt)$_2$] skeleton fused with bulky cycloalkane rings and investigated the substituent effects on the arrangement of the complexes and their physical properties (Scheme 2).

Donor organic molecules fused with bulky substituents such as cycloalkane and dioxane moieties can form unique molecular arrangements in the crystals [66], which exhibit superconducting and metallic behaviors [45,46,67–69].

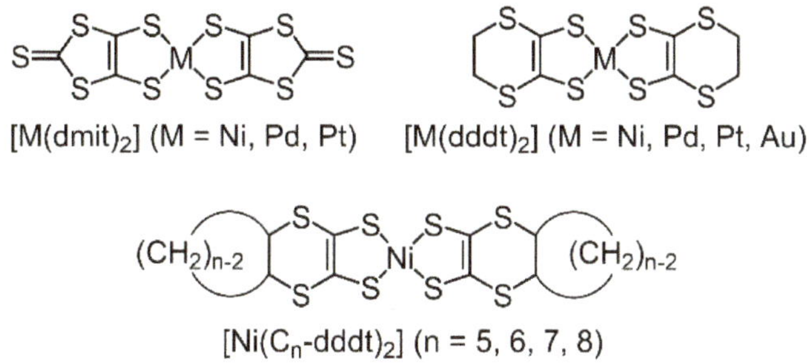

Scheme 2. Metal–dithiolene complexes as components of molecular conductors.

In this paper, we report the donor abilities of novel donor complexes [Ni(C_n-dddt)$_2$] with two cycloalkane rings (C_5-dddt = 4a,5,6,6a-pentahydrro-1,4-benzodithiin-2,3-dithiolate; C_6-dddt = 4a,5,6,7,8,8a-hexahydro-1,4-benzodithiin-2,3-dithiolate; C_7-dddt = 4a,5,6,7,8,9,9a-heptahydro-1,4-benzodithiin-2,3-dithiolate, and C_8-dddt = 4a,5,6,7,8,9,10,10a-octahydro-1,4-benzodithiin-2,3-dithiolate) based on electrochemical measurements and electronic absorption spectra. The packing motifs of the neutral [Ni(C_n-dddt)$_2$] (n = 5, 6, 7, 8) were discussed based on X-ray crystallographic analysis results. Furthermore, electronic features of new cation radical crystals [Ni(C_5-dddt)$_2$]$_2$[X] (X = ClO_4^- and PF_6^-) were revealed using electrical resistivity, magnetic measurements, and band calculations based on the crystal structures.

2. Materials and Methods

2.1. Materials

All the solvents were of analytical grade and were used without further purification. All the reactions were performed in a nitrogen atmosphere. Oligo(1,3-dithiole-2,4,5-trithione) ((C_3S_5)$_x$) [70] and (Bu_4N)[Ni(dddt)$_2$] [56] were prepared using a previously reported method.

2.2. Synthesis

2.2.1. Synthesis of 4,5-Cis(cyclopentalenedithio)-1,3-dithiole-2-thione (L1), 4,5-Cis(cyclohexylenedithio)-1,3-dithiole-2-thione (L2), 4,5-Cis(cycloheptalenedithio)-1,3-dithiole-2-thione (L3), and 4,5-Cis(cyclooctalenedithio)-1,3-dithiole-2-thione (L4)

In a nitrogen atmosphere, cyclopentene (25 mmol) was dissolved in 1,2-dichloroethane (25 mL), and oligo(1,3-dithiole-2,4,5-trithione) (10 mmol) was added. The resulting suspension was heated to 135 °C for 14 h with stirring, and the reaction mixture was filtered to remove insoluble precipitates. The evaporation of the filtration in vacuo, followed by purification of the residue by column chromatography (SiO_2, carbon disulfide (CS_2)-hexane 1:3), yielded 67.6% 4,5-cis(cyclopentalenedithio)-1,3-dithiole-2-thione (**L1**) as yellow crystals after recrystallization from hot ethanol (EtOH). Moreover, 4,5-cis(cyclohexylenedithio)-1,3-dithiole-2-thione (**L2**), 4,5-cis(cycloheptalenedithio)-1,3-dithiole-2-thione (**L3**), and 4,5-cis(cyclooctalenedithio)-1,3-dithiole-2-thione (**L4**) were obtained using a similar procedure with corresponding cycloalkenes as starting materials, yielding 32.7% **L2**, 77.6% **L3**, and 69.2% **L4**.

2.2.2. Synthesis of 4,5-Cis(cyclopentalenedithio)-1,3-dithiole-2-one (**L1′**) and 4,5-Cis(cyclohexylenedithio)-1,3-dithiole-2-one (**L2′**), 4,5-Cis(cycloheptalenedithio)-1,3-dithiole-2-one (**L3′**), and 4,5-Cis(cyclooctalenedithio)-1,3-dithiole-2-one (**L4′**)

Mercuric acetate (1.5 mmol) was added to a solution of **L1′** (1 mmol) in tetrahydrofuran-acetic acid (2:1, 45 mL), and the resulting white suspension was stirred for 30 min at 20 °C. The resulting white precipitate was removed by filtration using Celite and washed thoroughly using chloroform. The combined organic phases were washed using water (2 × 100 mL), saturated sodium hydrogen carbonate (2 × 100 mL), and water (1 × 100 mL), and then they were dried using magnesium sulfate. The removal of the solvent in vacuo yielded 98% 4,5-cis(cyclopentalenedithio)-1,3-dithiole-2-one (**L1′**) as large pale yellow needle crystals. Moreover, 4,5-cis(cyclohexylenedithio)-1,3-dithiole-2-one (**L2′**), 4,5-cis(cycloheptalenedithio)-1,3-dithiole-2-one (**L3′**), and 4,5-cis(cyclooctalenedithio)-1,3-dithiole-2-one (**L4′**) were obtained using a similar procedure; thus yielding 97% **L2′**, 98% **L3′**, and 97% **L4′**.

2.2.3. Synthesis of Monoanionic Complexes $(Bu_4N)[Ni(C_n\text{-dddt})_2]$ (n = 5, **1**; n = 6, **2**; n = 7, **3**; and n = 8, **4**)

Sodium (3.0 mmol) and **Lm′** (m = 1–4) (1.2 mmol) were dissolved in EtOH (15 mL). The resulting clear orange solution was dropped into an EtOH solution (15 mL) of $NiCl_2{\cdot}6H_2O$ (0.6 mmol) for 30 min with stirring. The resulting solution had a dark purple color. It was stirred in the air for a further 30 min, and the color changed to dark green as the reaction proceeded. After filtration to remove insoluble precipitates, one equivalent tetra-butylammonium bromide was added to the solution, and a dark green solid precipitated immediately. The green solids of monoanionic complexes **1–4** were collected by filtration and recrystallized using acetone. The yields for **1**, **2**, **3**, and **4** were 79.6%, 63.0%, 46.4%, and 38.8%, respectively.

2.2.4. Synthesis of Neutral Complexes $[Ni(C_n\text{-dddt})_2]$ (n = 5, **5**; n = 6, **6**; n = 7, **7**; and n = 8, **8**)

An acetonitrile (CH_3CN) (50 mL) solution of iodine (1.3 mmol) was added to a CH_3CN solution (100 mL) of monoanionic complexes **1–4** (0.23 mmol). Black precipitates of **5–8** were obtained immediately. The precipitates of the neutral complexes were collected by filtration, washed using CH_3CN and methanol, and dried under reduced pressure. Black block-shaped single crystals of the neutral complexes **5–8**, which were suitable for X-ray crystallographic analyses, were obtained by recrystallization from their carbon disulfide solutions. The yields for **5**, **6**, **7**, and **8** were 69.2%, 31.5%, 62.9%, and 36.8%, respectively. The elemental analysis results for these complexes are summarized in Table 1.

Table 1. Results of the elemental analyses of 1–8.

Complex	Observed (Calculation Result) (%)			Complex	Observed (Calculation Result) (%)			
Formula	C	H	N	Formula	C	H	Ni	S
1 $C_{30}H_{52}NNiS_8$	48.32 (48.56)	6.91 (7.06)	1.89 (1.89)	**5** $C_{14}H_{16}NiS_8$	33.59 (33.66)	3.14 (3.23)	11.58 (11.75)	51.33 (51.36)
2 $C_{32}H_{56}NNiS_8$	49.65 (49.91)	7.26 (7.33)	1.85 (1.82)	**6** $C_{16}H_{20}NiS_8$	36.21 (36.43)	3.70 (3.82)	11.08 (11.13)	48.60 (48.63)
3 $C_{34}H_{60}NNiS_8$	50.88 (51.17)	7.50 (7.58)	1.67 (1.76)	**7** $C_{18}H_{24}NiS_8$	38.87 (38.91)	4.26 (4.35)	10.44 (10.56)	46.40 (46.17)
4 $C_{36}H_{64}NNiS_8$	51.90 (52.34)	7.71 (7.81)	1.70 (1.70)	**8** $C_{20}H_{28}NiS_8$	41.00 (41.16)	4.81 (4.84)	9.81 (10.06)	43.86 (43.95)

2.2.5. Synthesis of Radical Cation Crystals $[Ni(C_5\text{-dddt})_2]_2[ClO_4]$ (**9**) and $[Ni(C_5\text{-dddt})_2]_2[PF_6]$ (**10**)

Dichloromethane (CH_2Cl_2) solutions (10 mL) of **5** (0.03 mmol) containing $(Bu_4N)(X)$ (X = ClO_4 or PF_6) (0.09 mmol) as electrolytes were electrolyzed under constant-current conditions (3 μA) for three days in an H-shaped cell equipped with platinum wire electrodes in an argon atmosphere at 20 °C. Black crystals of **9** and **10** were formed on the anode surface. The chemical formulas of **9** and **10** were confirmed by X-ray crystal structural analyses.

2.3. Physical Measurements

2.3.1. Elemental Analyses

The analyses of the carbon, hydrogen, and nitrogen elements were performed using a PERKIN-ELMER 240 C elemental analyzer for **1**, **3**, **7**; and a Yanako MT-6 CHN CORDER for **2**, **4**, **5**, **6**, and **8**. The elemental analyses of the nickel and sulfur atoms of the neutral complexes **5–8** were performed using a SHIMADZU ICPS-1000IV by the Chemical Analysis Team of AD&SD in RIKEN.

2.3.2. Electrochemical Measurements

Cyclic voltammograms of monoanionic complexes **1–4** were obtained using acetonitrile solutions (1×10^{-3} mol L^{-1}) with $(Bu_4N)(ClO_4)$ (0.1 M) as a supporting electrolyte and recorded using a BAS ALS/[H] CH Instruments Model 610 electrochemical analyzer combined with an Ag/Ag^+ reference electrode, platinum counter electrode, and glassy carbon working electrode at room temperature. The scan rate of the measurements was 100 mV/s.

2.3.3. Electronic Absorption Spectra

The electronic absorption spectra (300–1600 nm) of **1–8** in CH_2Cl_2, CH_3CN, and benzene solutions (5×10^{-5} mol L^{-1}) at room temperature were measured by the Chemical Analysis Team of AD&SD in RIKEN using a Shimadzu UV-3100 system.

2.3.4. Electrical Resistivity

The temperature dependence of the electrical resistivity was measured using the standard four-probe method at ambient pressure for **9** and **10**. Gold wires (15 μm in diameter) were attached to the crystal with a carbon paste. Resistivity measurements under the hydrostatic pressure of the salt were performed within the range from 0.3–1.8 GPa using a clamp-type piston-cylinder high-pressure cell [71,72].

2.3.5. Magnetic Susceptibility

The magnetic susceptibilities of **9** (15 mg) and **10** (20 mg) were measured in a magnetic field of 1 T using a superconducting quantum interference device (SQUID) magnetometer (Quantum Design, MPMS-XL7) within the temperature range of 2–300 K.

2.3.6. Band Calculation

The intermolecular overlap integrals (*S*) between the frontier orbitals were calculated based on the extended Hückel molecular orbital (MO) method. The semi-empirical parameters for the Slater-type atomic orbitals were obtained from the literature [73–75].

2.3.7. Crystal Structure Determinations of **L2–L4**, Neutral Complexes **5–8**, and Radical Cation Crystals **9** and **10**

Crystallographic data for the single crystals of **L2–L4**, **5–8**, and **9–10** were obtained using a Rigaku MicroMax-007 diffractometer with a multilayer mirror monochromated Mo-Kα (λ = 0.71073 Å) and charge-coupled device detector at 93 K. Structural refinements were carried out using the full-matrix least-squares method on F^2 [76]. The structures were determined and refined using SHELXL-2018 in the Olex2 software package [77,78]. The parameters were refined using anisotropic temperature factors, with the exception of the hydrogen atoms, which were parametrically refined using the riding model with a fixed C-H bond distance of 0.95 Å. The crystallographic data for **L2–L4**, **5–8**, and **9–10** are summarized in Tables 2–4, respectively.

Table 2. Crystallographic data for neutral complexes **L2–L4**.

	L2	**L3**	**L4**
Chemical formula	$C_{7.2}H_8S_4$	$C_8H_{9.60}S_4$	$C_{7.34}H_{9.34}S_{3.34}$
Crystal size/mm^3	0.40 × 0.10 × 0.04	0.30 × 0.06 × 0.03	0.40 × 0.03 × 0.02
Formula weight	222.78	234.00	204.45
Crystal system	Monoclinic	Monoclinic	Monoclinic
Space group	$P2_1/c$	$P2_1/c$	$P2/c$
a/Å	12.7079(3)	5.3224(3)	11.2835(7)
b/Å	9.8356(2)	21.7855(10)	6.8878(3)
c/Å	18.7846(4)	10.5687(5)	17.1727(13)
α/°	90	90	90
β/°	105.640(2)	95.793(5)	106.533(7)
γ/°	90	90	90
V/Å^3	2260.95(9)	1219.19(11)	1279.46(14)
Z	10	5	6
T/K	93	93	93
μ(Mo Kα)/mm^{-1}	0.980	0.913	0.874
D_{calc}/g cm^{-3}	1.636	1.594	1.592
F(000)	1152	608	640
λ (Mo-Kα)/Å	0.71073	0.71073	0.71073
Measured 2θ range/°	4.75–60.98	3.938–60.722	3.766–60.41
No. of reflections collected	22,206	10,849	11,351
Independent reflections	16,862	7823	6705
Observed reflections with $I > 2.00\ \sigma(I)$	5456	2882	2973
R_{int}	0.0191	0.0291	0.0255
$R(F^2)$ ($I > 2.00\sigma(I)$) [a]	0.0211	0.0293	0.0267
$wR(F^2)$ (all data) [b]	0.0537	0.0740	0.0654
Goodness of fit (GOF)	1.064	1.056	1.041
CCDC number	2,103,657	2,103,658	2,103,659

[a] $R(F^2) = \Sigma\ (F_o^2 - F_c^2)/\Sigma\ F_o^2$; [b] $wR(F^2) = [\Sigma\ w(F_o^2 - F_c^2)^2/\Sigma\ w(F_o^2)^2]^{1/2}$ and $w^{-1} = [\Sigma^2\ (F^2) + (0.0278P)^2 + 0.7843P]$ for **L2**; $w^{-1} = [\Sigma^2\ (F^2) + (0.0450P)^2 + 0.3773P]$ for **L3**; and $w^{-1} = [\Sigma^2\ (F^2) + (0.0379\ P)^2 + 0.4043P]$ for **L4** (where $P = (F_o^2 + 2F_c^2)/3$).

Table 3. Crystallographic data for neutral complexes **5–8**.

	5	**6**	**7**	**8**
Chemical formula	$C_{14}H_{16}NiS_8$	$C_{16}H_{20}NiS_8$	$C_{7.2}H_{9.6}Ni_{0.4}S_{3.2}$	$C_{6.67}H_{9.32}Ni_{0.33}S_{2.66}$
Crystal size/mm^3	0.14 × 0.11 × 0.06	0.16 × 0.09 × 0.06	0.07 × 0.04 × 0.01	0.18 × 0.07 × 0.07
Formula weight	499.46	527.51	222.22	194.41
Crystal system	Monoclinic	Monoclinic	Monoclinic	Triclinic
Space group	$P2_1/n$	$P2_1$	$P2_1/n$	P-1
a/Å	6.2830(2)	8.0303(3)	12.5305(6)	6.8973(10)
b/Å	9.8825(3)	8.8084(4)	6.7500(3)	8.6190(13)
c/Å	15.2978(4)	14.8128(5)	13.6312(6)	11.5730(15)
α/°	90	90	90	69.507(13)
β/°	100.685(3)	105.033(4)	108.057(5)	85.237(11)
γ/°	90	90	90	71.298(13)
V/Å^3	933.40(5)	1011.91(7)	1096.15(9)	610.09(16)
Z	2	2	5	3
T/K	93	93	93	93
μ(Mo Kα)/mm^{-1}	1.928	1.783	1.651	1.486
D_{calc}/g cm^{-3}	1.777	1.731	1.683	1.587

Table 3. *Cont.*

	5	6	7	8
$F(000)$	512	544	576	304
λ (Mo-Kα)/Å	0.71073	0.71073	0.71073	0.71073
Measured 2θ range/°	4.938–60.77	5.224–60.742	5.376–61.08	3.75–60.606
No. of reflections collected	10,661	9330	9850	9364
Independent reflections	7606	6979	6434	3699
Observed reflections with $I > 2.00\sigma(I)$	2215	4857	2436	2568
R_{int}	0.0182	0.0187	0.0226	0.0267
$R(F^2)$ ($I > 2.00\sigma(I)$) [a]	0.0237	0.0499	0.0252	0.0827
$wR(F^2)$ (all data) [b]	0.0595	0.1305	0.0544	0.1977
GOF	1.056	1.215	1.051	1.185
Flack parameter		0.328(7)		
CCDC number	2,103,660	2,103,661	2,103,662	2,103,663

[a] $R(F^2) = \Sigma (F_o^2 - F_c^2)/\Sigma F_o^2$, [b] $wR(F^2) = [\Sigma w(F_o^2 - F_c^2)^2/\Sigma w(F_o^2)^2]^{1/2}$ and $w^{-1} = [\Sigma^2 (F^2) + (0.0341P)^2 + 0.4506P]$ for **5**; $w^{-1} = [\Sigma^2 (F^2) + (0.0110P)^2 + 9.3099P]$ for **6**; $w^{-1} = [\Sigma^2 (F^2) + (0.0275P)^2 + 0.2769P]$ for **7**; and $w^{-1} = [\Sigma^2 (F^2) + (0.0195P)^2 + 10.6333P]$ for **8** (where $P = (F_o^2 + 2F_c^2)/3$).

Table 4. Crystallographic data for the radical cation crystals **9** and **10**.

	9	10
Chemical formula	$C_7H_8Cl_{0.25}Ni_{0.50}OS_4$	$C_7H_8F_{1.5}Ni_{0.5}P_{0.25}S_4$
Crystal size/mm^3	0.12 × 0.12 × 0.08	0.14 × 0.07 × 0.01
Formula weight	274.59	285.97
Crystal system	Monoclinic	Monoclinic
Space group	$P2/c$	$P2/c$
a/Å	12.6888(3)	12.6865(4)
b/Å	6.25150(10)	6.3020(2)
c/Å	24.1229(6)	24.3145(8)
α/°	90	90
β/°	93.462(2)	92.900(3)
γ/°	90	90
V/Å^3	1910.03(7)	1941.46(11)
Z	8	8
T/K	93	93
μ(Mo Kα)/mm^{-1}	1.969	1.928
D_{calc}/g cm^{-3}	1.910	1.957
$F(000)$	1122	1162
λ (Mo-Kα)/Å	0.71073	0.71073
Measured 2θ range/°	4.782–60.788	4.518–60.73
No. of reflections collected	16,993	17,389
Independent reflections	12,771	9840
Observed reflections with $I > 2.00\sigma(I)$	4756	4142
R_{int}	0.0135	0.0338
$R(F^2)$ ($I > 2.00\sigma(I)$) [a]	0.0302	0.0335
$wR(F^2)$ (all data) [b]	0.0659	0.0737
GOF	1.068	1.024
CCDC number	2,103,664	2,103,665

[a] $R(F^2) = \Sigma (F_o^2 - F_c^2)/\Sigma F_o^2$; [b] $wR(F^2) = [\Sigma w(F_o^2 - F_c^2)^2/\Sigma w(F_o^2)^2]^{1/2}$ and $w^{-1} = [\Sigma^2 (F^2) + (0.0186P)^2 + 4.1733P]$ for **9**; and $w^{-1} = [\Sigma^2(F^2) + (0.0403P)^2 + 1.0162P]$ for **10** (where $P = (F_o^2 + 2F_c^2)/3$).

3. Results and Discussion

3.1. Synthesis of Complexes 1–8

3.1.1. Syntheses and Molecular Structures of Precursors L1–L4

One of the most effective methods for synthesizing various TTF derivatives and their precursors involves [2 + 4] cycloadditions of $(C_3S_5)_x$ (as 4π) and a molecule that involves one of its external double bonds (as 2π) [79]. This reaction is stereospecific, similar to the [2 + 4] cycloaddition of tetrathiooxalate with alkanes, thus providing a cis-oriented compound [80]. Precursors **L1–L4** were synthesized using a modified version of a method outlined in a previous study, by the [2 + 4] cycloadditions of $(C_3S_5)_x$ and cycloheptene, cyclohexene, cycloheptene, and cyclooctene [81]. The cycloalkylene residues were cis-oriented with respect to the newly formed 1,4-dithiene rings. Single crystals of **L1–L4** were obtained by recrystallization from their dichloromethane solutions, which were suitable for X-ray crystallographic analyses, with the exception of **L1**. Figure 1 presents the molecular structures of two crystallographically independent molecules A and B in the crystal of **L2**. The fused cyclohexane moieties in molecules A and B formed cis-oriented (*R*,*S*) conformations at the two carbons C(4)-C(5) and C(13)-C(14), respectively. Stable chair-type configurations of the cyclohexylene groups were observed for molecules A and B. A similar cis-oriented conformation was observed in the crystals of 4,5-(l,4-dioxanediyl-2,3-dithio)-l,3-dithiole-2-thione and 4,5-(cis-cyclohexylenedithio)-1,3-dithiole-2-one [82,83]. The bond distances of C=C and C=S were 1.349 Å and 1.648 Å for molecule A, and 1.352 Å and 1.648 Å for molecule B. The angles between the planes formed by S(1)-C(2)-C(3)-S(2) and S(3)-C(2)-C(3)-S(4) in molecule A and S(6)-C(11)-C(12)-S(7) and S(8)-C(11)-C(12)-S(9) in molecule B were 3.86° and 3.08°, respectively. The bond distances and angles indicate that the C_3S_5 moieties of molecules A and B formed similar planar structures. The torsion angles around the fused cyclohexyl group in molecule A were −71.20°, −6.76°, and 56.18° for C(2)-S(1)-C(4)-C(5), C(3)-S(2)-C(5)-C(4), and S(1)-C(4)-C(5)-S(2), respectively. The corresponding torsion angles in molecule B were −62.56°, −26.37°, and 63.91° for C(11)–S(6)–C(13)-C(14), C(12)-S(7)-C(14)-C(13), and S(6)-C(13)-C(14)-S(7), respectively. The comparison of the torsion angles between molecules A and B revealed that the fused cyclohexylene group in molecule A deviated more from the C_3S_5 plane at the S(1) and S(2) positions than that of molecule B.

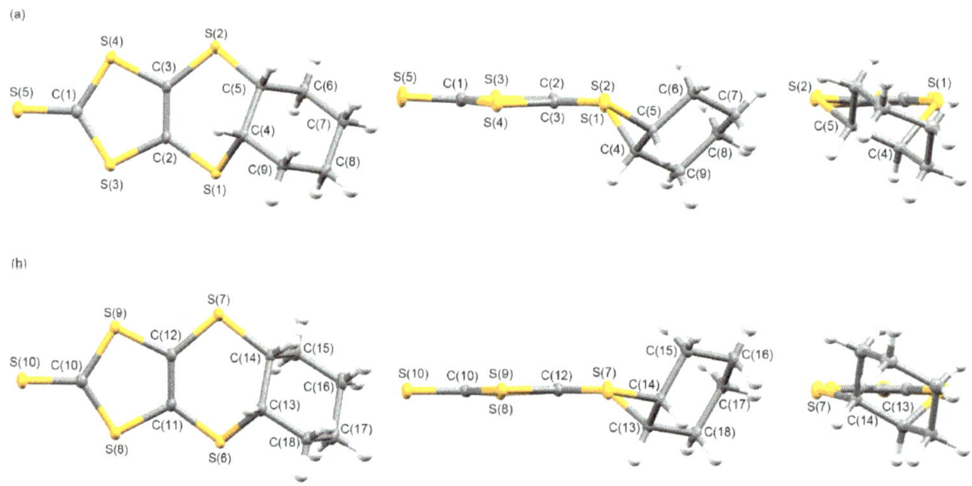

Figure 1. Molecular structures of two crystallographically independent molecules, (**a**) **A** and (**b**) **B**, in the crystal of **L2** along the vertical, parallel, and end-on directions on the C_3S_5 molecular plane with atom labels.

Figure 2 presents the molecular structures of **L3** and **L4**. The fused cycloheptane and cyclooctane moieties formed cis-oriented (*R*,*S*) conformations at the two carbons C(4) and C(5), respectively, in addition to molecule **L2**. Twist-chair- and boat-chair-type configurations of the cycloheptylene and cyclooctylene groups were formed in **L3** and **L4**, which were observed in the crystals of CnDT-EDO-TTF ($n = 7, 8$) [45]. The C_3S_5 moieties of **L3** and **L4** were planar structures, as observed in the crystal of **L2**; where the bond lengths C(1)=S(5) and C(2)=C(3) and the angles between the planes formed by S(1)-C(2)-C(3)-S(2) and S(3)-C(2)-C(3)-S(4) were 1.656 Å, 1.355 Å, and 2.89° for **L3**; and 1.647 Å, 1.344 Å, and 2.75° for **L4**, respectively. The torsion angles around the fused cycloalkylene groups of C(2)-S(1)-C(4)-C(5), C(3)-S(2)-C(5)-C(4), and S(1)-C(4)-C(5)-S(2) were 57.68°, 42.60°, and −72.47° for **L3**; and 58.00°, 41.11°, and −72.18° for **L4**, respectively.

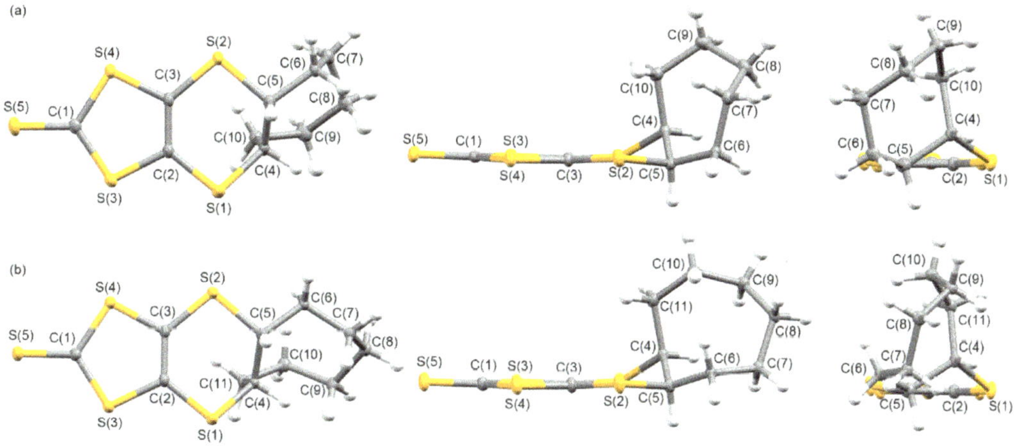

Figure 2. Molecular structures of (**a**) **L3** and (**b**) **L4** along the vertical, parallel, and end-on directions on the C_3S_5 molecular plane with atom labels.

3.1.2. Molecular and Crystal Structures of Neutral Complexes **5–8**

We successfully obtained single crystals of the neutral complexes that were suitable for X-ray crystallographic analysis by recrystallization from CS_2 solutions of complexes **5–8**. The synthetic procedure for the neutral complexes is shown in Scheme 3. The precursors of **L1–L4** formed cis-oriented (*R*,*S*) conformations around the cycloalkylene groups due to the [2 + 4] cycloadditions. Using the synthetic procedures shown in Scheme 3, Monoanionic complexes **1–4** were obtained as mixtures, including diastereomeric isomers (Bu_4N)\{trans-[Ni(cis-Cn-dddt)$_2$]\} and (Bu_4N)\{cis-[Ni(cis-Cn-dddt)$_2$]\}, due to the conformations between the two cycloalkylene groups in the molecules. The oxidation of Monoanionic complexes **1–4** resulted in the formation of mixtures of diastereomeric isomers for neutral complexes **5–8** (Scheme 4). From the recrystallization processes, single crystals of neutral complexes **5–8** were formed by the trans-isomers.

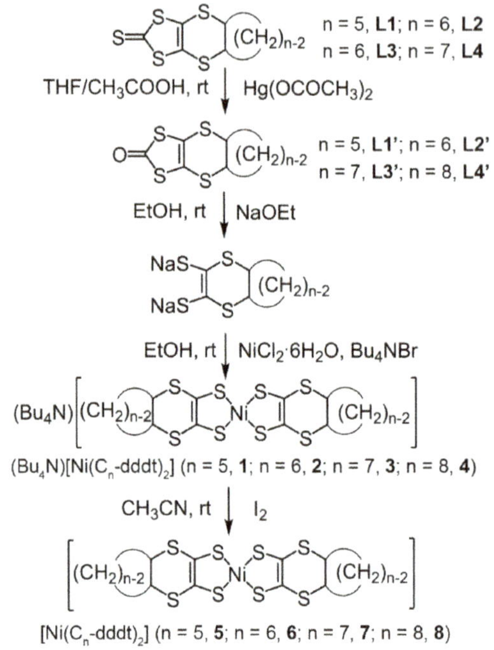

Scheme 3. Synthetic procedure of neutral [Ni(Cn-dddt)$_2$] complexes.

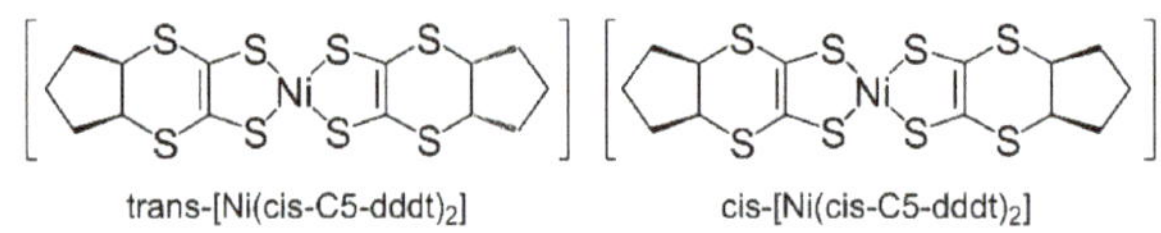

Scheme 4. Diastereomeric isomers of the trans- and cis-[Ni(cis-C5-dddt)$_2$] complexes.

The selected bond lengths and angles of the neutral complexes in crystals **5**–**8** are summarized in Table 5. In molecules of **5**–**8**, square planar co-ordinations were constructed around the Ni atoms, similar to previously reported square planar Ni-dithiolene complexes such as [Ni(dddt)$_2$] in a neutral state [84–91].

Table 5. Selected bond lengths and angles of neutral complexes **5**–**8**.

	5	6	7	8
Ni(1)-S(1)	2.1291(4)	2.134(2)	2.1350(4)	2.1341(18)
Ni(1)-S(2)	2.1340(4)	2.134(2)	2.1343(4)	2.1404(17)
Ni(1)-S(5)		2.137(2)		
Ni(1)-S(6)		2.138(2)		
S(1)-C(1)	1.7147(15)	1.705(8)	1.7143(15)	1.719(7)
S(2)-C(2)	1.7153(15)	1.716(9)	1.7084(16)	1.708(7)
S(3)-C(1)	1.7406(15)	1.736(9)	1.7393(16)	1.748(7)
S(4)-C(2)	1.7406(15)	1.733(9)	1.7418(15)	1.750(7)
S(5)-C(9)		1.715(9)		
S(6)-C(10)		1.699(9)		
S(7)-C(9)		1.739(8)		
S(8)-C(10)		1.755(9)		

Table 5. *Cont.*

	5	6	7	8
S(3)-C(3)	1.8239(15)	1.819(9)	1.8486(15)	1.829(7)
S(4)-C(4)	1.8126(15)	1.823(9)	1.8291(16)	1.820(7)
C(1)-C(2)	1.386(2)	1.413(12)	1.389(2)	1.380(10)
C(3)-C(4)	1.524(2)	1.549(12)	1.525(2)	1.508(10)
S(7)-C(11)		1.818(9)		
S(8)-C(12)		1.832(9)		
C(9)-C(10)		1.400(12)		
C(11)-C(12)		1.523(12)		
S(1)-Ni(1)-S(2)	91.652(14)	91.79(10)	91.959(15)	91.68(7)
S(5)-Ni(1)-S(6)		91.50(9)		
S(1)-Ni(1)-S(2′) [a)]	88.347(14)		88.040(15)	88.32(7)
S(1)-Ni(1)-S(5)		88.79(9)		
S(1)-Ni(1)-S(6)		87.97(9)		
C(1)-C(2)-S(2)	119.03(11)	118.3(7)	119.86(12)	119.5(5)
C(1)-C(2)-S(4)	127.43(12)	127.8(7)	122.23(12)	126.2(5)
C(2)-C(1)-S(1)	119.29(11)	119.4(7)	118.98(12)	119.5(5)
C(2)-C(1)-S(3)	126.84(12)	125.6(7)	126.44(12)	126.6(5)
C(9)-C(10)-S(6)		120.3(7)		
C(9)-C(10)-S(8)		125.6(7)		
C(10)-C(9)-S(5)		117.8(6)		
C(10)-C(9)-S(7)		128.3(7)		

[a)] The S(2′) atoms were positioned by the following symmetry operations of S(2) atoms: (2-x, 1-y, 1-z) for **5**, (1-x, 1-y, -z) for **7**, and (-x, 1-y, 1-z) for **8**.

Figure 3 presents the molecular structure of **5**. One half of the trans-[Ni(cis-C5-dddt)$_2$] molecule was crystallographically independent with an inversion center on the nickel atom in the crystals of **5**. The torsion angles around the fused cyclopentylene group of the molecule were 45.93°, 42.06°, and −62.81° for C(1)-S(3)-C(3)-C(4), C(2)-S(4)-C(4)-C(3), and S(3)-C(3)-C(4)-S(4), respectively. The approximate geometry of the [Ni(dddt)$_2$] moiety in crystal **5** was highly similar to those of an anionic [Ni(dddt)$_2$]$^-$ molecule [92]. The cyclopentylene ring of **5** adopted a folded conformation from the [Ni(dddt)$_2$] moiety with the trans-orientation, which is different from the conformation observed in (Ph$_4$P){trans-[Ni(cis-C5-dddt)$_2$]} (an envelope conformation) [93]

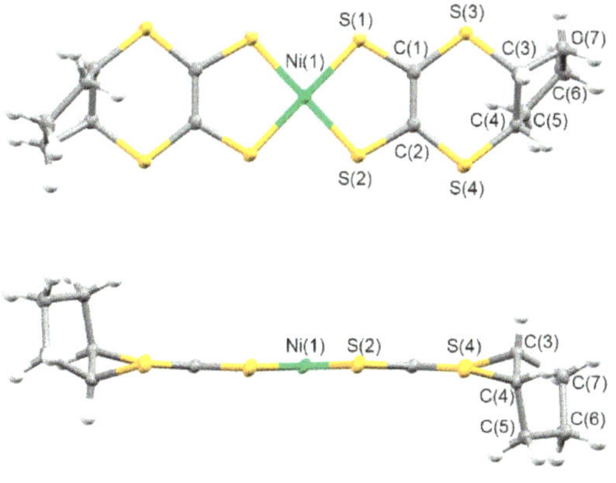

Figure 3. Molecular structure of **5** along the vertical and parallel directions on the molecular plane, with atom labels.

The packing diagrams of **5** are shown in Figure 4. Due to the folded geometry, the columnar structures of the complex were restricted. The shortest Ni–Ni distance was 6.283 Å, which was the length of the *a axis*. The intermolecular S···S contacts shorter than the sum of the van der Waals radius (3.70 Å) were observed between the S(1)–S(1) atoms (3.574 Å) and S(1)–S(4′) atoms (3.604 Å), which formed a three-dimensional network structure in crystal **5**.

(a)

(b)

(c)

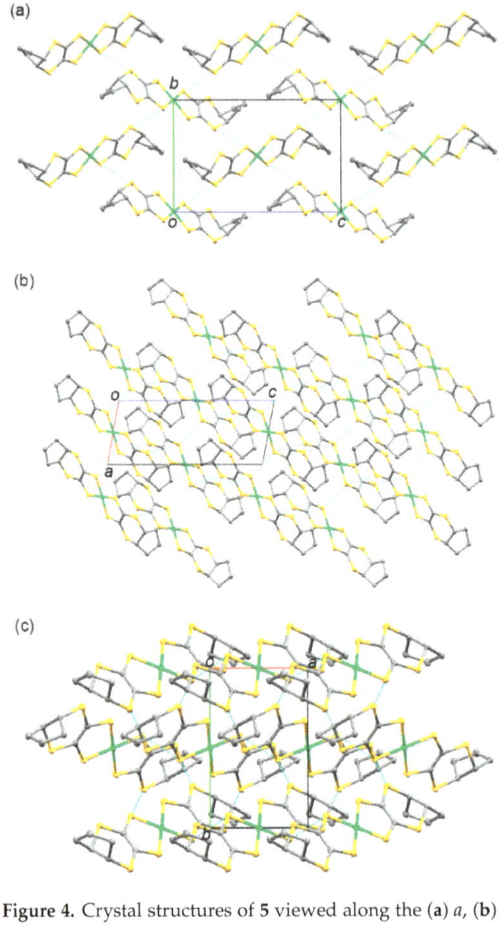

Figure 4. Crystal structures of **5** viewed along the (**a**) *a*, (**b**) *b*, and (**c**) *c* axes. Hydrogen atoms are omitted for clarity. The blue dashed lines indicate the intermolecular S···S contacts shorter than the sum of the van der Waals radius (3.70 Å).

The molecular structure of compound **6** is shown in Figure 5. One trans-[Ni(cis-C6-dddt)$_2$] molecule was present in the asymmetric unit of the crystal. The torsion angles around the fused cyclohexylene group of the molecule were 43.79°, −68.30°, and −62.81° for C(1)-S(3)-C(3)-C(4), C(2)-S(4)-C(4)-C(3), and S(3)-C(3)-C(4)-S(4), respectively. The corresponding torsion angles in the opposite site of the molecule were −45.87°, −47.04°, and 66.65° for C(9)-S(7)-C(11)-C(12), C(10)-S(8)-C(12)-C(11), and S(7)-C(11)-C(12)-S(8), respectively. The stable chair-type configurations of the cyclohexylene groups were observed. However, the folded conformations of the cyclohexylene groups from the planar [Ni(dddt)$_2$] core were different from those of **L4**, which exhibited larger flexures around the terminal S-C bonds (Figure 1).

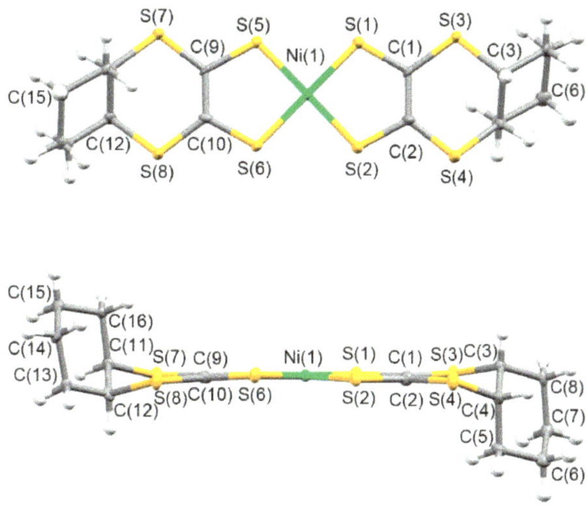

Figure 5. Molecular structure of **6** along the vertical and parallel directions on the molecular plane with atom labels.

A similar molecular geometry of trans-[Ni(cis-C6-dddt)$_2$] was reported by Bai et al. [94]. However, the packing motif of crystal **6** was significantly different from that of the reported crystal constructed by the centrosymmetric molecular arrangement of trans-[Ni(cis-C6-dddt)$_2$] with an inversion center on the nickel atom. In the case of crystal **6**, a non-centrosymmetric molecular arrangement with a helix axis along the *b* axis formed by the trans-[Ni(cis-C6-dddt)$_2$] molecules was observed (Figure 6). The cyclohexylene rings prevented the formation of stacks of the complexes and intermolecular interactions through the S···S contacts. The shortest intermolecular S···S distance was 3.749 Å between S(1) and S(8). The shortest Ni–Ni distance was 8.030 Å, which was the length of the *a axis*.

One half of the trans-[Ni(cis-C7-dddt)$_2$] molecule was present in the asymmetric unit in Crystal **7** (Figure 7). The twist chair-type configurations of the cycloheptylene groups in **L3** were maintained in the crystals after the coordination reaction of **L3′** and the nickel source (Scheme 3). The twist chair-type configurations of the cycloheptylene groups around the terminal S–C bonds in the molecule exhibited larger flexures (Figure 2), with torsion angles around the fused cycloheptylene group of 10.13°, 68.04°, and −55.70° for C(1)-S(3)-C(3)-C(4), C(2)-S(4)-C(4)-C(3), and S(3)-C(3)-C(4)-S(4), respectively. Figure 8 presents the crystal structure of **7**. The cycloheptylene moieties were in contact along the *c* axis. There was no intermolecular S···S contact, given that the [Ni(dddt)$_2$] cores were completely separated by the steric hindrance of the cycloheptane moieties.

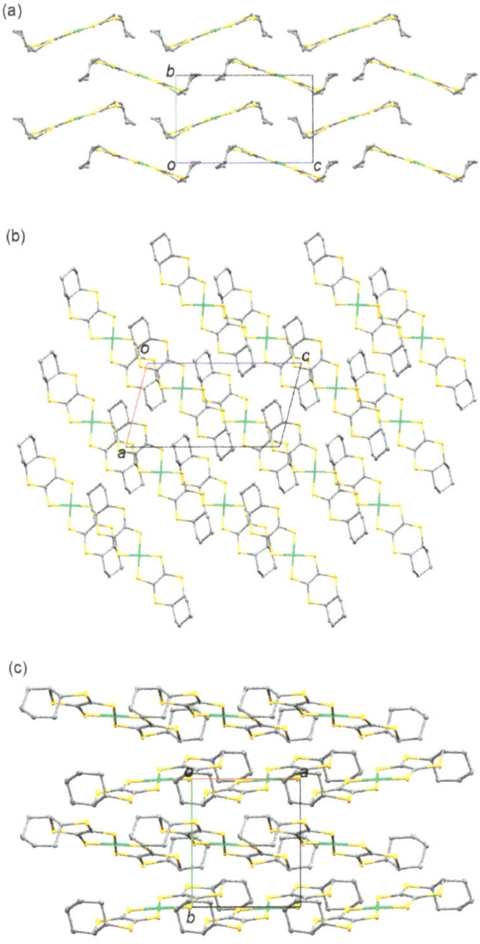

Figure 6. Crystal structure of **6** along the (**a**) *a*, (**b**) *b*, and (**c**) *c* axes. Hydrogen atoms are omitted for clarity.

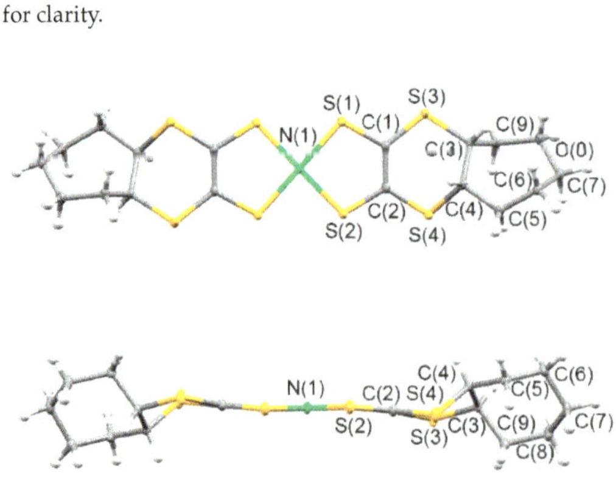

Figure 7. Molecular structure of **7** along the vertical and parallel directions on the molecular plane with atom labels.

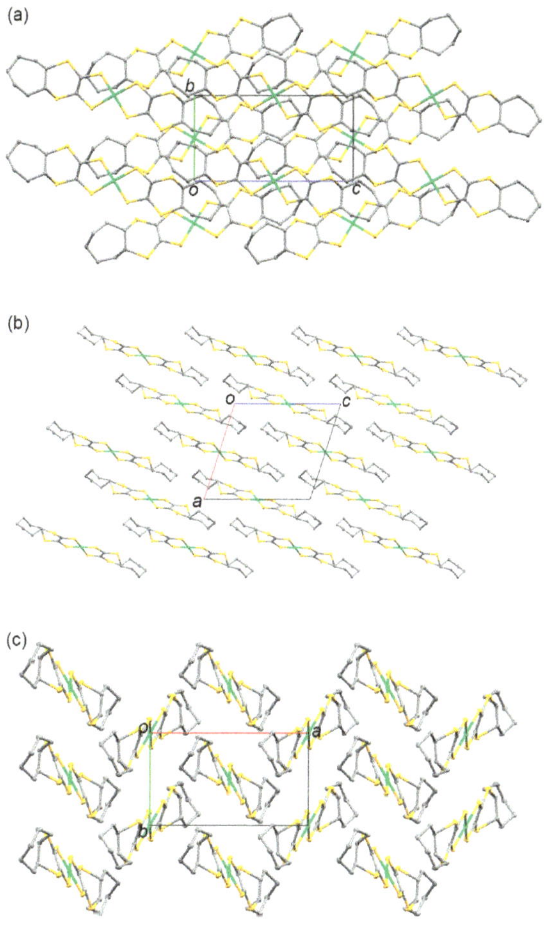

Figure 8. Crystal structure of **7** along the (**a**) *a*, (**b**) *b*, and (**c**) *c* axes. Hydrogen atoms are omitted for clarity.

Figures 9 and 10 present the molecular and crystal structures of **8**. The centrosymmetric crystal was formed by one-half of the trans-[Ni(cis-C8-dddt)$_2$] molecule in the asymmetric unit. The boat-chair-type configurations of the cyclooctylene groups formed in **L4** remained in molecule **6** with torsion angles of 54.74°, 46.88°, and −70.84° for C(1)-S(1)-C(3)-C(4), C(2)-S(4)-C(4)-C(3), and S(3)-C(3)-C(4)-S(4), respectively. One-dimensional columns along the *a* axis were constructed by ring-over-ring overlaps between the molecules sliding to the molecular longitudinal axis. There were no intercolumnar interactions through the S–S contacts. The shortest distance was 3.775 Å between S(2) and S(3′) in the column. The columns interacted through the side-by-side intermolecular S···S contacts between S(3) and S(3′) with a distance of 3.444 Å.

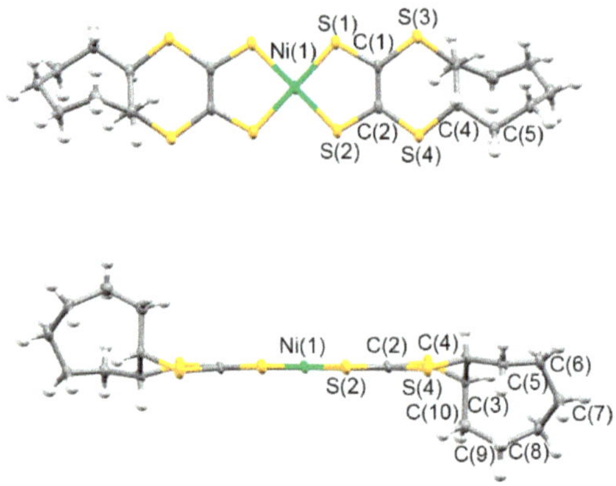

Figure 9. Molecular structure of **8** along the vertical and parallel directions on the molecular plane with atom labels.

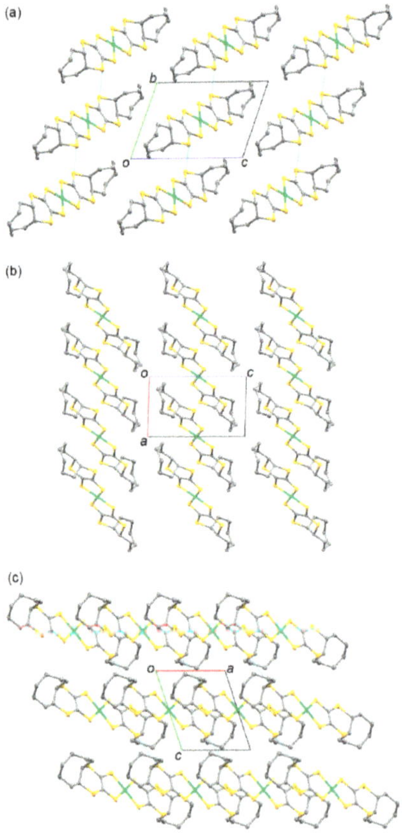

Figure 10. Crystal structure of **8** along the (**a**) *a*, (**b**) *b*, and (**c**) *c* axes. Hydrogen atoms are omitted for clarity. The blue dashed lines indicate the intermolecular S···S contacts shorter than the sum of the van der Waals radius (3.70 Å).

3.2. Electrochemical Properties of **1–4**

Cyclic voltammograms of monoanionic complexes **1–4** and $(Bu_4N)[Ni(dddt)_2]$ in MeCN were obtained (Figure 1). The voltammograms of $(Bu_4N)[Ni(dddt)_2]$ exhibited two quasi-reversible couples and one irreversible redox couple in the recorded region (Figure 1). The quasi-reversible redox peaks were assigned to $[Ni(dddt)_2]^0/[Ni(dddt)_2]^+$ ($E_{1/2}{}^1 = -1.24$ V) and $[Ni(dddt)_2]^-/[Ni(dddt)_2]^0$ ($E_{1/2}{}^2 = -1.02$ V) (Figure 11a). An irreversible oxidation peak was observed at 0.28 V, which indicates a reduction process between the dianionic $[Ni(dddt)_2]^-$ and monoanionic $[Ni(dddt)_2]^{2-}$ states. The electrochemical features of **1–4** were similar to those of $(Bu_4N)[Ni(dddt)_2]$. The redox potentials $E_{1/2}{}^1$ ($[Ni(C_n\text{-dddt})_2]^0/[Ni(C_n\text{-dddt})_2]^+$) and $E_{1/2}{}^2$ ($[Ni(C_n\text{-dddt})_2]^-Ni(C_n\text{-dddt})_2]^0$) of **1–4** are summarized in Table 6. Potentials E_1 and E_2 were correlated to the donor abilities of the complexes. The ΔE (= $E_{1/2}{}^2 - E_{1/2}{}^1$) values could be considered a measure of the on-site Coulomb energies [47,64]. Based on the $E_{1/2}{}^1$, $E_{1/2}{}^2$, and ΔE values, the complexes with fused cycloalkanes maintained the donor ability and on-site Coulomb energy of $(Bu_4N)[Ni(dddt)_2]$.

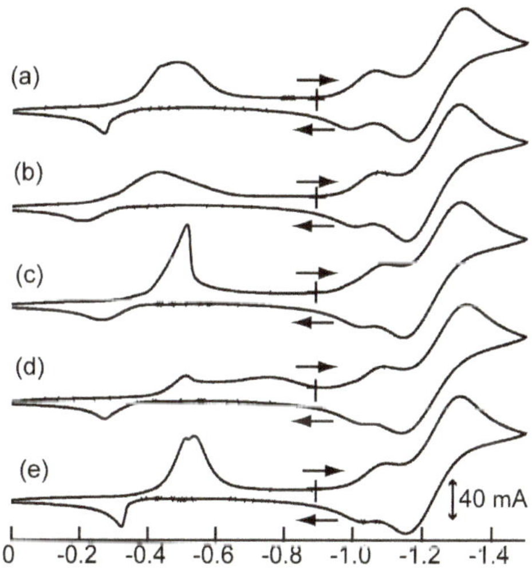

Figure 11. Cyclic voltammograms of (**a**) $(Bu_4N)[Ni(dddt)_2]$, (**b**) **1**, (**c**) **2**, (**d**) **3**, and (**e**) **4**. Conditions: MeCN (1×10^{-3} M), supporting electrolyte. $(Bu_4N)(ClO_4)$ (0.1 M), scan rate: 100 mV/s, reference electrode: Ag/Ag^+, counter electrode: Pt wire, working electrode: glassy carbon at room temperature.

Table 6. Redox potentials $E_{1/2}{}^1$/V (0, −1), $E_{1/2}{}^2$/V (−1, −2), and ΔE (= $E_{1/2}{}^2 - E_{1/2}{}^1$)/V of $(Bu_4N)[Ni(dddt)_2]$ and Complexes **1–4**.

	$E_{1/2}{}^1$/V	$E_{1/2}{}^2$/V	ΔE/V
$(Bu4N)[Ni(dddt)_2]$	−1.24	−1.02	0.22
1	−1.23	−1.03	0.20
2	−1.23	−1.05	0.18
3	−1.23	−1.05	0.18
4	−1.23	−1.06	0.17

Conditions: MeCN (1×10^{-3} M), supporting electrolyte: $(Bu_4N)(ClO_4)$ (0.1 M), scan rate: 100 mV/s, reference electrode: Ag/Ag^+, counter electrode: Pt wire, working electrode: glassy carbon at room temperature.

3.3. Electronic Absorption Spectra of **1–8**

The electronic absorption spectra of monoanionic complexes **1–4** and neutral complexes **5–8** in CH_2Cl_2 were measured. Monoanionic complexes **1–4** exhibited a strong and broad absorption band at approximately 1036–1201 nm in the near-infrared region, which can be attributed to a π–π * transition between their HOMOs and LUMOs (Table 7) [95,96]. With respect to neutral complexes **5–8**, the HOMO–LUMO absorption bands were shifted to the higher energy region (982–1040 nm), which corresponds to the energy gap between HOMO-1 and the HOMOs of the monoanionic complexes due to the weak shoulder absorptions observed at approximately 1000 nm. For example, monoanionic complex **4** exhibited two absorptions at 1201 nm and 1040 nm, which can be attributed to the HOMO–LUMO and HOMO-1 to HOMO gaps, respectively (Figure 12). The HOMO–LUMO gaps of the neutral complexes estimated by the spectra were within the range 1.19–1.26 eV. The spectroscopic behaviors of the complexes were similar to those of other bis(ethylene-1,2-dithiolato) complexes and $M(S_2C_2R_2)_2$ [85–89,93,94,97,98].

Table 7. Wavelength at absorption maxima (λ_{max}) and absorption coefficients (ε) for the π–π* transitions of **1–8** in CH_2Cl_2.

	λ_{max}/nm	Energy gap/eV	ε/dm^3 mol^{-1}
1	1036	1.19	12,000
2	1198	1.03	10,000
3	1144	1.08	13,000
4	1201	1.03	13,000
5	982	1.26	32,000
6	1035	1.20	32,000
7	1031	1.20	39,000
8	1040	1.19	49,000

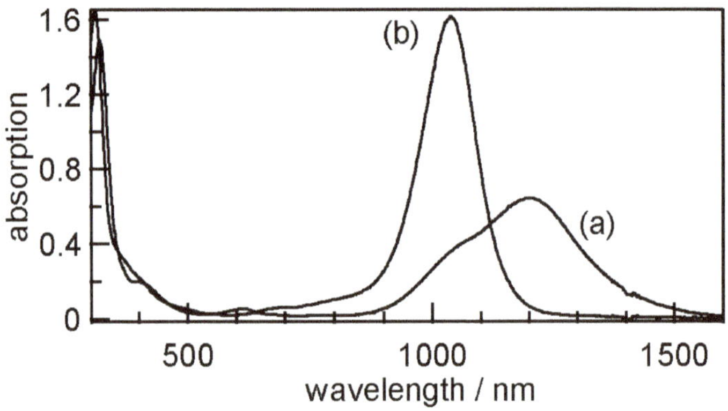

Figure 12. Electronic absorption spectra of (a) **4** (4.82×10^{-5} M) and (b) **8** (3.40×10^{-5} M) in CH_2Cl_2 solutions.

3.4. *Structures and Physical Properties of Radical Cation Crystals* **9** *and* **10**

3.4.1. Molecular and Crystal Structures of **9** and **10**

The electrochemical oxidation of CH_2Cl_2 solutions, including neutral complex **5** and the corresponding supporting electrolytes $(Bu_4N)X$ ($X = ClO_4^-$ and PF_6^-), resulted in the crystallization of 2:1 [Ni(cis-C5-dddt)$_2$]$_2$(X) salts. The donor/acceptor ratios of the 2:1 crystals were different from those of the previously reported radical cation crystals based on the [Ni(dddt)$_2$] complex without bulky substituents, constructing the 3:2 crystals of [Ni(dddt)$_2$]$_3$(A)$_2$ (A = $AuBr_2^-$, BF_4^-, ClO_4^-, HSO_4^-, and $HSeO_4^-$ [55,56,99–103]. The crystals of **9** and **10** included one crystallographically independent donor and half of the ClO_4^- or PF_6^- anions, respectively. Figure 13 and (Supplementary Materials) Figure S1 present the molecular structures of the donors in **9** and **10**. In the case of the radical cation crystals, the cis-[Ni(cis-C5-dddt)$_2$] conformation was formed preferentially, although the corresponding neutral complex exhibited the trans-[Ni(cis-C5-dddt)$_2$] conformation

(Scheme 4). The donors exhibited a square planar coordination around the Ni atoms. Moreover, the bond distances of Ni-S = 2.1438(5)–2.1481(5) Å, S–C = 1.703(2)–1.7077(19) Å, and C=C = 1.406(3)–1.408(3) Å for **9**; and Ni-S = 2.1435(6)–2.1498(6) Å, S-C = 1.702(2)–1.705(2) Å, and C=C = 1.404(3)–1.414(3) Å for **10** in the NiS_4C_4 framework were slightly larger than those of the neutral trans-[Ni(cis-C5-dddt)$_2$]. Figure 14 and Figure S2 present the crystal structures of **9** and **10**. In the crystals, similar molecular arrangements of the donors and anions were observed with a two-fold axis on one of the Cl-O and P-F bonds of the anions along the *b* axis, respectively. In **9**, the perchlorate anions were disordered at the two sites by rotation around the two-fold axis. The cis-[Ni(cis-C5-dddt)$_2$] conformation would induce dimerized structures of the donors in **9** and **10**, which differ from the molecular arrangement in the cases of the 3:2 crystals of [Ni(dddt)$_2$]$_3$(A)$_2$ significantly. The dimerized structures of the donors were formed by the S···S contacts (3.665 Å and 3.664 Å for **9** and **10**, respectively), and the dimers were arranged into columnar structures through the S···S contacts (3.682 Å and 3.556 Å for **9** and **10**, respectively) along the *a-c* axis directions in **9** and **10**. Intermolecular side-by-side interactions through the S···S contacts (3.507–3.637 Å and 3.500–3.629 Å for **9** and **10**, respectively) between the columns were observed along the *b* axis, thus providing two-dimensional donor sheets parallel to the *bc* planes. Although the donor molecule is fused with the bulky cyclopentane moieties, the two-dimensional donor sheet arrangements through the S···S contacts preferentially are constructed in the radical cation crystals, which are also observed in the organic radical cation crystals based on the BEDT-TTF and BETS molecules exhibiting metallic and superconducting behaviors [14–18].

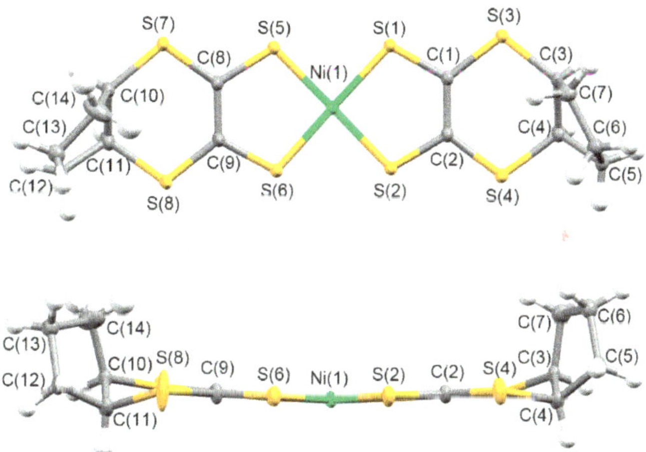

Figure 13. Molecular structure of **9** along the vertical and parallel directions on the molecular plane with atom labels.

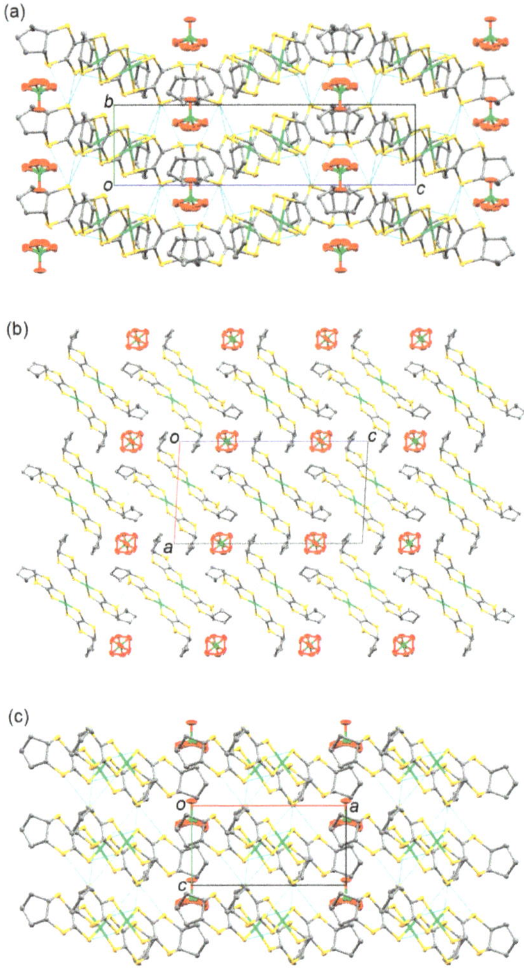

Figure 14. Crystal structure of **9** along (**a**) the *a*, (**b**) *b*, and (**c**) *c* axes. Hydrogen atoms are omitted for clarity. The dashed lines indicate sulfur–sulfur distances shorter than the sum of the van der Waals radius (<3.7 Å).

3.4.2. Molecular Orbitals and Energy Band Calculations of **9** and **10**

To reveal the electronic features of the salts, molecular orbitals (MO) and energy band structures were determined using the extended Hückel and tight-binding methods. Figure 15 presents the determined HOMOs and LUMOs of the donor molecules of **9** and **10** in their neutral states. The highly symmetric HOMOs and LUMOs with the nickel *d* orbitals of the donors were in good agreement with the molecular orbitals of the [M(dddt)$_2$] complex calculated based on the first-principles density-functional theory [56–62].

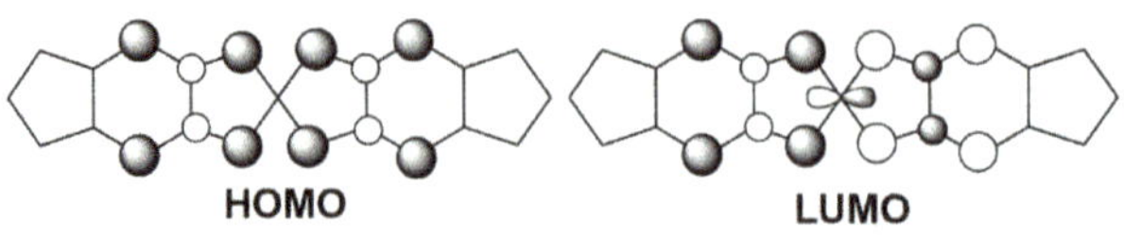

Figure 15. The HOMO and LUMO of [Ni(C$_5$-dddt)$_2$].

The calculated overlap integrals (S) of the crystals between the HOMOs are summarized in Table 8 with a schematic diagram of their donor arrangements. The symbols a1–b4 indicate directions of intermolecular interactions between the donors. Figure S3 presents the corresponding molecular arrangement of the donor molecules viewed along the end-on direction in the crystal. The significant dimerization of the crystals was observed, given that the face-to-face interactions (a2) were significantly stronger than the other interactions (a1) in the columns. Dominant side-by-side interactions (b1 and b4) connected the dimers, thus forming two-dimensional sheets in the crystals. Figure 16 presents the calculated band structure of **9**, which is similar to that of crystal **10**. Both the upper bands crossed the Fermi level, thus forming the small two-dimensional Fermi surfaces of the crystals.

Table 8. Overlap integrals (S) between the HOMOs and a schematic diagram of the donor arrangements for Crystals **9** and **10**.

S ($\times 10^3$)	9	10	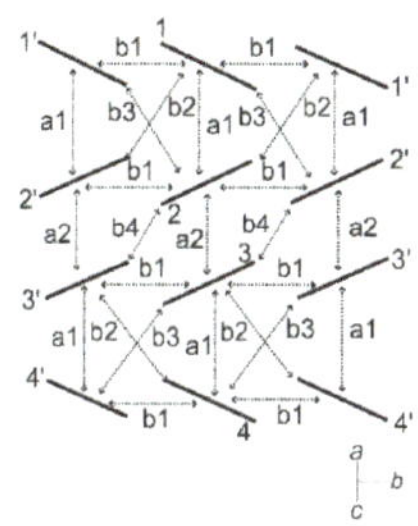
a1	−2.7	−1.9	
a2	−10	−10	
b1	−6.9	−3.8	
b2	−0.14	−0.7	
b3	−0.14	−0.7	
b4	1.2	−0.3	

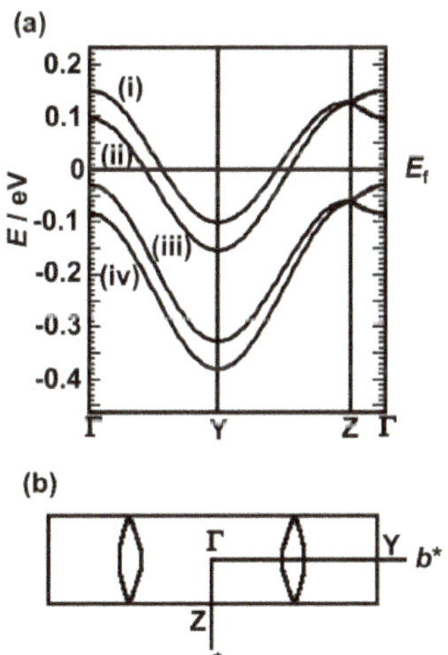

Figure 16. (**a**) Energy band structure and (**b**) Fermi surface of **9**. The characters b^* and c^* denote reciprocal lattice axes.

3.4.3. Electrical Conductivities and Magnetic Properties of **9** and **10**

The temperature dependence of the resistivities was measured for single crystals of **9** and **10** K using a standard four-probe direct current (DC) method (Figure 17). The salts exhibited semiconducting behaviors (**9**: ρ_{rt} = 0.8 Ω cm and E_a = 0.09 eV; **10**: ρ_{rt} = 4.0 Ω cm and E_a = 0.13 eV). Although the energy band calculations for **9** and **10** suggested a metallic nature, crystals **9** and **10** exhibited semiconducting behaviors under ambient pressure. The results of the resistivity measurements and the energy band calculations indicate that the crystals forms a Mott insulating state or a spin-singlet state under ambient pressure [104]. In order to reveal the electronic structures of the crystals, temperature dependence of the resistivities under hydrostatic pressures and magnetic susceptibilities of **9** and **10** were measured. The resistivity measurements under various hydrostatic pressures (0.3–1.8 GPa) exhibited slightly smaller activation energies than those under ambient pressure (**9**: ρ_{rt} = 0.2 Ω cm and E_a = 0.07 eV; **10**: ρ_{rt} = 1.0 Ω cm and E_a = 0.12 eV at 1.8 GPa). The temperature dependencies of the magnetic susceptibilities of **9** and **10** were in accordance with the Curie–Wiess law. However, significantly small χ values at 300 K indicate diamagnetic features ($\chi = 2.9 \times 10^{-5}$ cm^3 mol^{-1} for **9** and $\chi = 2.3 \times 10^{-5}$ cm^3 mol^{-1} for **10**). These results suggest that the crystals formed spin-singlet states between the dimers through the S···S contacts to the b1 direction (Table 8), which is in accordance with their conducting behaviors.

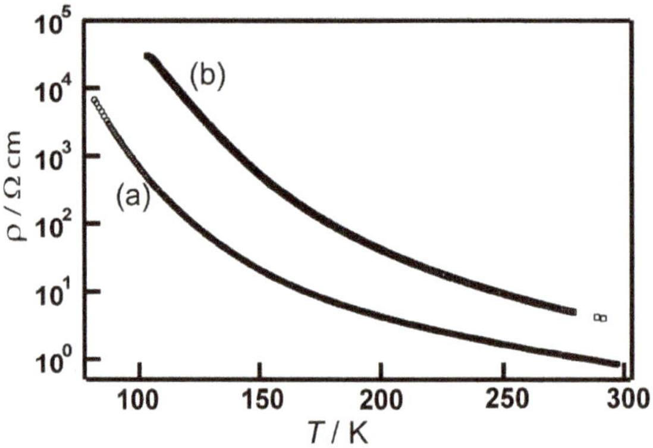

Figure 17. Temperature dependence of electrical resistivities of (a) **9** and (b) **10**.

4. Conclusions

New donor-type nickel–dithiolene complexes based on the [Ni(dddt)$_2$] skeleton fused with bulky cycloalkane substituents were synthesized as new components of molecular conductors. All the neutral crystals were crystallized in cis-[Ni(cis-Cn-dddt)$_2$] conformations with cis-oriented (*R*,*S*) conformations around the cycloalkylene groups due to the [2 + 4] cycloadditions. The flexible cycloalkane moieties in the molecules can induce unique molecular arrangements in their neutral crystals. The three-dimensional network and one-dimensional columnar structures were constructed through intermolecular S···S contacts in crystals **5** and **8**. No intermolecular S···S contacts were observed in **6** and **7**. However, [Ni(C$_6$-dddt)$_2$] formed a helical molecular arrangement in the crystal. New cation radical crystals [Ni(C$_3$-dddt)$_2$]$_2$(ClO_4^- or PF_6^-) with dimerized structures were successfully obtained by electrochemical crystallization, which formed different electronic structures of the [Ni(dddt)$_2$]$_3$(X)$_2$ crystals [55,57,99–103]. The radical cation crystals were insulators due to the spin singlet states between the dimers through the strong intermolecular S···S contacts. However, the donors with cycloalkane substituents can adjust the proportions of the inter- and intra-dimer interactions through the S···S contacts in their radical cation crystals by modification of the cycloalkane moieties, and allow for the realization of various cation radical crystals with unique donor arrangements by introducing linear, tetrahedral, and octahedral anions. Moreover, neutral crystals are potential candidates for new single-component molecular conductors [58–63].

Supplementary Materials: The following are available online at https://www.mdpi.com/article/10.3390/cryst11101154/s1, Figure S1: Molecular structure of **10** along vertical and parallel directions on the molecular plane with atom labels. Figure S2: Crystal structure of **10** along the (a) *a*, (b) *b*, and (c) *c* axes. Hydrogen atoms are omitted for clarity. The dashed lines indicate sulfur–sulfur distances shorter than the sum of the van der Waals radius (<3.70 Å), CIFs: CCDC numbers 2103657–2103665 for **5–8**, **9**, and **10**.

Author Contributions: Conceptualization, project administration, writing—original draft preparation, writing—review and editing, and funding acquisition, K.K.; investigation and data curation, M.S., S.A., N.H., and T.K.; supervision, T.A., R.K., and J.-i.Y. All authors have read and agreed to the published version of the manuscript.

Funding: This research was funded by a Grant-in-Aid for Scientific Research (B) [grant number JP18H01956]; and the Research Program of the Dynamic Alliance for Open Innovation Bridging Human, Environment and Materials of the Network Joint Research Center for Materials and Devices [grant number 20214021].

Institutional Review Board Statement: Not applicable.

Informed Consent Statement: Not applicable.

Acknowledgments: We thank Takeo Fukunaga from RIKEN and Yoshitaka Kawasugi from Toho University for technical support related to the X-ray crystallographic analyses and the resistivity measurements under hydrostatic pressures.

Conflicts of Interest: The authors declare no conflict of interest.

References

1. Narita, M.; Pittman, C.U., Jr. Preparation of tetrathiafulvalenes (TTF) and their selenium analogs -Tetraselenafulvalenes (TSeF). *Synthesis* **1976**, 489–514. [CrossRef]
2. Kistenmacher, T.J.; Phillips, T.E.; Cowan, D.O. The crystal structure of the 1:1 radical cation-radical anion salt of 2,2′-bis-l,3-dithiole (TTF) and 7,7,8,8-tetracyanoquinodimethane (TCNQ). *Acta Cryst.* **1974**, *B30*, 763–768. [CrossRef]
3. Coleman, L.B.; Cohen, M.J.; Sandman, D.J.; Yamagishi, F.G.; Garito, A.F.; Heeger, A.J. Superconducting fluctuations and the peierls instability in an organic solid. *Solid State Commun.* **1973**, *12*, 1125–1132. [CrossRef]
4. Ferraris, J.; Cowan, D.O.; Walatka, V.; Perlstein, J.H. Electron transfer in a new highly conducting donor-acceptor complex. *J. Am. Chem. Soc.* **1973**, *95*, 948–949. [CrossRef]
5. Jerome, D.; Mazaud, A.; Ribault, M.; Bechgaard, K. Superconductivity in a synthetic organic conductor $(TMTSF)_2PF_6$. *J. Phys. Lett.* **1980**, *41*, L95. [CrossRef]
6. Williams, J.M.; Ferraro, J.R.; Thorn, R.J.; Carlson, K.D.; Geiser, U.; Wang, H.H.; Kini, A.M.; Whangbo, M.-H. *Organic Superconductors (Including Fullerenes)*; Prentice Hall: Englewood Cliffs, NJ, USA, 1992. [CrossRef]
7. Ishiguro, T.; Yamaji, K.; Saito, G. *Organic Superconductors*, 2nd ed.; Springer Ser. Solid-State Sci.; Springer: Berlin, Germany, 1998; Volume 18.
8. Fabre, J.M. From the organic molecule to superconductors. *J. Phys. IV* **2000**, *10*, Pr3-19–Pr3-33. [CrossRef]
9. Bechgaard, K.; Jacobsen, C.S.; Mortensen, V.; Thorup, N. The properties of five highly conducting salts: $(TMTSF)_2X$, X = PF_6^-, AsF_6^-, SbF_6^-, BF_4^- and NO_3^-, derived from tetramethyltetraselenafulvalene (TMTSF). *J. Solid State Commun.* **1980**, *33*, 1119–1125. [CrossRef]
10. Urayama, H.; Yamochi, H.; Saito, G.; Nozawa, K.; Sugano, T.; Kinoshita, M.; Sato, S.; Oshima, K.; Kawamoto, A.; Tanaka, J. A New ambient pressure organic superconductor based on BEDT-TTF with Tc higher than 10K (T_c = 10.4 K). *Chem. Lett.* **1988**, 55–58. [CrossRef]
11. Williams, J.M.; Kini, A.M.; Wang, H.H.; Carlson, K.D.; Geiser, U.; Montgomery, L.K.; Pyrka, G.J.; Watkins, D.M.; Kommers, J.K.; Boryschuk, S.J.; et al. Semiconductor-semiconductor transition (42 K) to the highest-Tc organic superconductor, $\varkappa$-$(ET)_2Cu[N(CN)_2]Cl$ (T_c = 12.5 K). *Inorg. Chem.* **1990**, *29*, 3272–3274. [CrossRef]
12. Papavassiliou, G.C.; Mousdis, G.A.; Zambounis, J.S.; Terzis, A.; Hountas, A.; Hilti, B.; Mayer, C.W.; Pfeiffer, J. Low temperature measurements of the electrical conductivities of some charge transfer salts with the asymmetric donors MDT-TTF, EDT-TTF and EDT-DSDTF. $(MDT\text{-}TTF)_2AuI_2$, a new superconductor (T_c = 3.5 K at ambient pressure). *Synth. Met.* **1988**, *27*, B379–B383. [CrossRef]
13. Kikuchi, K.; Murata, K.; Honda, Y.; Namiki, T.; Saito, K.; Ishiguro, T.; Kobayashi, K.; Ikemoto, I. On ambient-pressure superconductivity in organic conductors: Electrical properties of $(DMET)_2I_3$, $(DMET)_2I_2Br$ and $(DMET)_2IBr_2$. *J. Phys. Soc. Jpn.* **1987**, *56*, 3436–3439. [CrossRef]
14. Shibaeva, R.P.; Yagubskii, E.B. Molecular conductors and superconductors based on trihalides of BEDT-TTF and some of Its analogues. *Chem. Rev.* **2004**, *104*, 5347–5378. [CrossRef]
15. Geiser, U.; Schlueter, J.A. Conducting organic radical cation salts with organic and organometallic anions. *Chem. Rev.* **2004**, *104*, 5203–5242. [CrossRef]
16. Kobayashi, H.; Cui, H.-B.; Kobayashi, A. Organic metals and superconductors based on BETS (BETS = Bis(ethylenedithio) tetraselenafulvalene). *Chem. Rev.* **2004**, *104*, 5265–5288. [CrossRef] [PubMed]
17. Uji, S.; Shinagawa, H.; Terashima, T.; Yakabe, T.; Terai, Y.; Tokumoto, M.; Kobayashi, A.; Tanaka, H.; Kobayashi, H. Magnetic-Field-induced superconductivity in a two-dimensional organic conductor. *Nature* **2001**, *410*, 908–910. [CrossRef]

18. Kobayashi, H.; Tomita, H.; Naito, T.; Kobayashi, A.; Sakai, F.; Watanabe, T.; Cassoux, P. New BETS Conductors with Magnetic Anions (BETS) bis(ethylenedithio)tetraselenafulvalene). *J. Am. Chem. Soc.* **1996**, *118*, 368–377. [CrossRef]
19. Aoyagi, I.; Katsuhara, M.; Mori, T. Syntheses and structures of highly soluble bis(ethylenedithio)tetrathiafulvalene molecules with alkyl chains. *Sci. Tech. Adv. Mater.* **2004**, *5*, 443–447. [CrossRef]
20. Matsumiya, S.; Izuoka, A.; Sugawara, T.; Taruishi, T.; Kawada, Y. Effect of methyl substitution on conformation and molecular arrangement of BEDT-TTF derivatives in the crystalline environment. *Bull. Chem. Soc. Jpn.* **1993**, *66*, 513–522. [CrossRef]
21. Riobé, F.; Avarvari, N. Electroactive oxazoline ligands. *Coord. Chem. Rev.* **2010**, *254*, 1523–1533. [CrossRef]
22. Awheda, I.; Krivickas, J.; Yang, S.; Martin, L.; Guziak, M.A.; Brooks, A.C.; Pelletier, F.; Le Kerneau, M.; Day, P.; Horton, P.N.; et al. Synthesis of new chiral organosulfur donors with hydrogen bonding functionality and their first charge transfer salts. *Tetrahedron* **2013**, *69*, 8738–8750. [CrossRef]
23. Biet, T.; Fihey, A.; Cauchy, T.; Vanthuyne, N.; Roussel, C.; Crassous, J.; Avarvar, N. Ethylenedithio-tetrathiafulvalene-helicenes: Electroactive helical precursors with switchable chiroptical properties. *Chem. A Eur. J.* **2013**, *19*, 13160–13167. [CrossRef]
24. Yang, S.; Pop, F.; Melan, C.; Brooks, A.C.; Martin, L.; Horton, P.; Auban-Senzier, P.; Rikken, G.L.J.A.; Avarvari, N.; Wallis, J.D. Charge transfer complexes and radical cation salts of chiral methylated organosulfur donors. *CrystEngComm* **2014**, *16*, 3906–3916. [CrossRef]
25. Chas, M.; Lemarié, M.; Guleab, M.; Avarvari, N. Chemo- and enantioselective sulfoxidation of bis(ethylenedithio)-tetrathiafulvalene (BEDT-TTF) into chiral BEDT-TTF-sulfoxide. *Chem. Commun.* **2008**, 220–222. [CrossRef]
26. Chas, M.; Riobé, F.; Sancho, R.; Minguíllon, C.; Avarvari, N. Selective monosulfoxidation of tetrathiafulvalenes into chiral TTF-sulfoxides. *Chirality* **2009**, *21*, 818–825. [CrossRef]
27. Karrer, A.; Wallis, J.D.; Dunitz, J.D.; Hilti, B.; Mayer, C.W.; Biirkle, M.; Pfeiffer, J. Structures and electrical properties of some new organic conductors derived from the donor molecule TMET (*S,S,S,S*- Bis(dimethylethy1enedithio)tetrathiafulvalene). *Helv. Chim. Acta* **1987**, *70*, 942–953. [CrossRef]
28. Krivickas, S.J.; Ichikawa, A.; Takahashi, K.; Tajima, H.; Wallis, J.D.; Mori, H. One-dimensional antiferromagnetic behavior of a chiral molecular crystal, α'-(*S,S*-DMBEDT-TTF)$_2$PF$_6$. *Synth. Met.* **2011**, *161*, 1563–1565. [CrossRef]
29. Krivickas, S.J.; Hashimoto, C.; Yoshida, J.; Ueda, A.; Takahashi, K.; Wallis, J.D.; Mori, H. Synthesis of racemic and chiral BEDT-TTF derivatives possessing hydroxy groups and their achiral and chiral charge transfer complexes. *Beilstein J. Org. Chem.* **2015**, *11*, 1561–1569. [CrossRef]
30. Madalan, A.M.; Réthoré, C.; Fourmigué, M.; Canadell, E.; Lopes, E.B.; Almeida, M.; Auban-Senzier, P.; Avarvari, N. Order versus disorder in chiral tetrathiafulvalene-oxazoline radical-cation salts: Structural and theoretical investigations and physical properties. *Chem. Eur. J.* **2010**, *16*, 528–537. [CrossRef]
31. Matsumiya, S.; Izuoka, A.; Sugawara, T.; Taruishi, T.; Kawada, Y.; Tokumoto, M. Crystal ctructure and conductivity of chiral radical ion salts (Me$_2$ET)$_2$X. *Bull. Chem. Soc. Jpn.* **1993**, *66*, 1949–1954. [CrossRef]
32. Réthoré, C.; Avarvari, N.; Canadell, E.; Auban-Senzier, P.; Fourmigué, M. Chiral molecular metals: Syntheses, structures, and properties of the AsF$_6$-salts of racemic ($\pm$)-, (R)-, and (S)-tetrathiafulvalene-oxazoline derivatives. *J. Am. Chem. Soc.* **2005**, *127*, 5748–5749. [CrossRef] [PubMed]
33. Pop, F.; Allain, M.; Auban-Senzier, P.; Martínez-Lillo, J.; Lloret, F.; Julve, M.; Canadell, E.; Avarvari, N. Enantiopure conducting salts of dimethylbis(ethylenedithio)tetrathiafulvalene (DM-BEDT-TTF) with the hexachlororhenate(IV) anion. *Eur. J. Inorg. Chem.* **2014**, *24*, 3855–3862. [CrossRef]
34. Avarvari, N.; Wallis, J.D. Strategies towards chiral molecular conductors. *J. Mater. Chem.* **2009**, *19*, 4061–4076. [CrossRef]
35. Zambounis, J.S.; Muyer, C.W.; Huuenstein, K.; Hilti, B.; HoJherr, W.; Pfegfer, J.; Biirkle, M.; Rihs, G. Crystal structure and electrical properties of κ-((*S,S*)-DMSEDT-TTF)$_2$ClO$_4$. *Adv. Mater.* **1992**, *4*, 33–35. [CrossRef]
36. Yang, S.; Brooks, A.C.; Martin, L.; Day, P.; Li, H.; Horton, P.; Malec, L.; Wallis, J.D. Novel enantiopure bis(pyrrolo)tetrathiafulvalene donors exhibiting chiral crystal packing arrangements. *CrystEngComm* **2009**, *11*, 993–996. [CrossRef]
37. Wallis, J.D.; Karrer, A.; Dunitz, J.D. Chiral metals? A chiral substrate for organic conductors and superconductors. *Helv. Chim. Acta* **1986**, *69*, 69–70. [CrossRef]
38. Short, J.L.; Blundell, T.J.; Krivickas, S.J.; Yang, S.; Wallis, J.D.; Akutsu, H.; Nakazawa, Y.; Martin, L. Chiral molecular conductor with an insulator–metal transition close to room temperature. *Chem. Commun.* **2020**, *56*, 9497–9500. [CrossRef]
39. Saad, A.; Jeannin, O.; Fourmigué, M. Chiral, flexible binaphthol-substituted tetrathiafulvalenes. *Tetrahedron* **2011**, *67*, 3820–3829. [CrossRef]
40. Pop, F.; Auban-Senzier, P.; Canadell, E.; Avarvari, N. Anion size control of the packing in the metallic versus semiconducting chiral radical cation salts (DM-EDT-TTF)$_2$XF$_6$ (X = P, As, Sb). *Chem. Commun.* **2016**, *52*, 12438–12441. [CrossRef]
41. Pop, F.; Auban-Senzier, P.; Frąckowiak, A.; Ptaszyński, K.; Olejniczak, I.; Wallis, J.D.; Canadell, E.; Avarvari, N. Chirality driven metallic versus semiconducting behavior in a complete series of radical cation salts based on dimethyl-ethylenedithio-tetrathiafulvalene (DM-EDT-TTF). *J. Am. Chem. Soc.* **2013**, *135*, 17176–17186. [CrossRef] [PubMed]
42. Galán-Mascarós, J.R.; Coronado, E.; Goddard, P.A.; Singleton, J.; Coldea, A.I.; Wallis, J.D.; Coles, S.J.; Alberola, A. A chiral ferromagnetic molecular metal. *J. Am. Chem. Soc.* **2010**, *132*, 9271–9273. [CrossRef] [PubMed]
43. Kimura, S.; Maejima, T.; Suzuki, H.; Chiba, R.; Mori, H.; Kawamoto, T.; Mori, T.; Moriyama, H.; Nishio, Y.; Kajita, K. A new organic superconductor β-(meso-DMBEDT-TTF)$_2$PF$_6$. *Chem. Commun.* **2004**, 2454–2455. [CrossRef]

44. Kimura, S.; Suzuki, H.; Maejima, T.; Mori, H.; Yamaura, J.; Kakiuchi, T.; Sawa, H.; Moriyama, H. Checkerboard-type charge-ordered state of a pressure-induced superconductor, β-(meso-DMBEDT-TTF)$_2$PF$_6$. *J. Am. Chem. Soc.* **2006**, *128*, 1456–1457. [CrossRef] [PubMed]
45. Suzuki, H.; Yamashita, K.; Suto, M.; Maejima, T.; Kimura, S.; Mori, H.; Nishio, Y.; Kajita, K.; Moriyama, H. Control of electronic state in the 2D hydrogen-bonded system: β″-(CnDT-EDO-TTF)2(PF6)x (*n* = 5, 6, 7, 8). *Synth. Met.* **2004**, *144*, 89–95. [CrossRef]
46. Kimura, S.; Suzuki, H.; Maejima, T.; Suto, M.; Yamashita, K.; Ichikawa, S.; Mori, H.; Moriyama, H.; Mochida, T.; Nishio, Y.; et al. Crystal structures and electrical conductivities controlled by CH/n interactions. *J. Phys. IV* **2004**, *114*, 521–522. [CrossRef]
47. Kato, R. Conducting metal dithiolene complexes: Structural and electronic properties. *Chem. Rev.* **2004**, *104*, 5319–5346. [CrossRef]
48. Kobayashi, A.; Fujiwara, E.; Kobayashi, H. Single-component molecular metals with extended-TTF Dithiolate ligands. *Chem. Rev.* **2004**, *104*, 5243–5264. [CrossRef] [PubMed]
49. Tanaka, H.; Okano, Y.; Kobayashi, H.; Suzuki, W.; Kobayashi, A. A three-dimensional synthetic metallic crystal composed of single-component molecules. *Science* **2001**, *291*, 285–287. [CrossRef] [PubMed]
50. Ni, J.M.; Pan, B.L.; Song, B.Q.; Huang, Y.Y.; Zeng, J.Y.; Yu, Y.J.; Cheng, E.J.; Wang, L.S.; Dai, D.Z.; Kato, R.; et al. Absence of magnetic thermal conductivity in the quantum spin liquid candidate EtMe$_3$Sb[Pd(dmit)$_2$]$_2$. *Phys. Rev. Lett.* **2019**, *123*, 247204. [CrossRef]
51. Bourgeois-Hope, P.; Laliberté, F.; Lefrançois, E.; Grissonnanche, G.; de Cotret, S.R.; Gordon, R.; Kitou, S.; Sawa, H.; Cui, H.; Kato, R.; et al. Thermal conductivity of the quantum spin liquid candidate EtMe$_3$Sb[Pd(dmit)$_2$]$_2$: No evidence of mobile gapless excitations. *Phys. Rev. X* **2019**, *9*, 041051. [CrossRef]
52. Itou, T.; Watanabe, E.; Maegawa, S.; Tajima, A.; Tajima, N.; Kubo, K.; Kato, R.; Kanoda, K. Slow dynamics of electrons at a metal–Mott insulator boundary in an organic system with disorder. *Sci. Adv.* **2017**, *3*, e1601594. [CrossRef] [PubMed]
53. Miyagawa, K.; Kanoda, K. NMR studies on two-dimensional molecular conductors and superconductors: Mott transition in κ-(BEDT-TTF)$_2$X. *Chem. Rev.* **2004**, *104*, 5635–5653. [CrossRef]
54. Kato, R.; Cui, H.B.; Tsumuraya, T.; Miyazaki, T.; Suzumura, Y. Emergence of the Dirac electron system in a single-component molecular conductor under high pressure. *J. Am. Chem. Soc.* **2017**, *139*, 1770–1773. [CrossRef] [PubMed]
55. Wang, H.H.; Fox, S.B.; Yagubskii, E.B.; Kushch, L.A.; Kotov, A.I.; Whangbo, M.-H. Direct observation of the electron-donating property of the 5,6-dihydro-1,4-dithiin-2,3-dithiolate (dddt) ligands in square planar M(dddt)$_2$ complexes (M = Ni, Pd, Pt, and Au). *J. Am. Chem. Soc.* **1997**, *119*, 7601–7602. [CrossRef]
56. Cui, H.; Tsumuraya, T.; Yeung, H.H.-M.; Coates, C.S.; Warren, M.R.; Kato, R. High Pressure Crystal Structure and Electrical Properties of a Single Component Molecular Crystal [Ni(dddt)$_2$] (dddt = 5,6-dihydro-1,4-dithiin-2,3-dithiolate). *Molecules* **2019**, *24*, 1843. [CrossRef] [PubMed]
57. Kushch, L.A.; Gritsenko, V.V.; Buravov, L.I.; Khomenko, A.G.; Shilov, G.V.; Dyachenko, O.A.; Merzhanov, V.A.; Yagubskii, E.B.; Rousseau, R.; Canadell, E. The M(dddt)$_2$ family of conducting complexes: [Ni(dddt)$_2$]$_3$(AuBr$_2$)$_2$, the first quasi-two-dimensional metal stable down to at least 1.3 K. *J. Mater. Chem.* **1995**, *5*, 1633–1638. [CrossRef]
58. Tsumuraya, T.; Kato, R.; Suzumura, Y. Effective Hamiltonian of topological nodal line semimetal in single-component molecular conductor [Pd(dddt)$_2$] from first-principles. *J. Phys. Soc. Jpn.* **2018**, *87*, 113701-1–113701-5. [CrossRef]
59. Kato, R.; Suzumura, Y. Novel Dirac electron in single-component molecular conductor [Pd(dddt)$_2$] (dddt = 5,6-dihydro-1,4-dithiin-2,3-dithiolate). *J. Phys. Soc. Jpn.* **2017**, *86*, 064705. [CrossRef]
60. Suzumura, Y.; Cui, H.B.; Kato, R. Conductivity and resistivity of Dirac electrons in single-component molecular conductor [Pd(dddt)$_2$]. *J. Phys. Soc. Jpn.* **2018**, *87*, 084702. [CrossRef]
61. Liu, Z.; Wang, H.; Wang, Z.F.; Yang, J.; Liu, F. Pressure-induced organic topological nodal-line semimetal in the three-dimensional molecular crystal Pd(dddt)$_2$. *Phys. Rev. B* **2018**, *97*, 155138. [CrossRef]
62. Suzumura, Y.; Tsumuraya, T.; Kato, R.; Matsuura, H.; Ogata, M. Role of Velocity Field and Principal Axis of Tilted Dirac Cones in Effective Hamiltonian of Non-Coplanar Nodal Loop. *J. Phys. Soc. Jpn.* **2019**, *88*, 124704. [CrossRef]
63. Suzumura, Y.; Kato, R.; Ogata, M. Electric transport of nodal line semimetals in single-component molecular conductors. *Crystals* **2020**, *10*, 862. [CrossRef]
64. Kato, R.; Kashimura, Y.; Sawa, H.; Okano, Y. Synthesis, structure, and electrochemical properties of new "unsymmetrical" metal dithiolate complexes. *Chem. Lett.* **1997**, *26*, 921–922. [CrossRef]
65. Kashimura, Y.; Okano, Y.; Yamaura, J.; Kato, R. Synthesis, structures, and physical properties of molecular conductors based on unsymmetrical metal-dithiolene complexes. *Synth. Met.* **1999**, *103*, 2123–2124. [CrossRef]
66. Kini, A.M.; Geiser, U.; Wang, H.-H.; Lykke, K.R.; Williams, J.M.; Campana, C.F. Alternative synthesis of (z),(z)-4,5;4′,5′-bis(1,4-dioxanediyl-2,3-dithio)tetrathiafulvalene (*cis,cis*-BDDT-TTF), and preparation, epr properties and structure of the radical cation salt (*cis,cis*-BDDT-TTF)$_2$I$_3$. *J. Mater. Chem.* **1995**, *5*, 1647–1652. [CrossRef]
67. Kotov, A.I.; Buravov, L.I.; Konovalikhin, S.V.; Dyachenko, O.A.; Yagubskii, E.B.; Malfant, I.; Courcet, T.; Cassoux, P.; Akimoto, J.; Honda, K.; et al. BEDT-TTF Derivatives with one and two dioxane rings and their conductive salts. *Synth. Met.* **1999**, *102*, 1630–1631. [CrossRef]
68. Nishikawa, H.; Morimoto, T.; Kodama, T.; Ikemoto, I.; Kikuchi, K.; Yamada, J.; Yoshino, H.; Murata, K. New organic superconductors consisting of an unprecedented π-electron donor. *J. Am. Chem. Soc.* **2002**, *124*, 730–731. [CrossRef] [PubMed]
69. Nishikawa, H.; Sato, Y.; Kikuchi, K.; Kodama, T.; Ikemoto, I.; Yamada, J.; Oshio, H.; Kondo, R.; Kagoshima, S. Charge ordering and pressure-induced superconductivity in β″-(DODHT)$_2$PF$_6$. *Phys. Rev. B* **2005**, *72*, 052510-1–052510-4. [CrossRef]

70. Svenstrup, N.; Becher, J. The organic chemistry of 1,3-Dithiole-2-thione-4,5-dithiolate (DMIT). *Synthesis* **1995**, 215–235. [CrossRef]
71. Walker, I.R. Nonmagnetic piston–cylinder pressure cell for use at 35 kbar and above. *Rev. Sci. Instrum.* **1999**, *70*, 3402–3412. [CrossRef]
72. Ishii, Y.; Tamura, M.; Kato, R.; Hedo, M.; Uwatoko, Y.; Môri, N. Effect of hydrostatic pressure on molecular conductors, Pd(dmit)$_2$ salts. *Synth. Met.* **2005**, *152*, 389–392. [CrossRef]
73. Albright, T.A.; Hoffmann, P.; Hoffmann, R. Conformational preferences and rotational barriers in polyene-ML3 transition metal complexes. *J. Am. Chem. Soc.* **1977**, *99*, 7546–7557. [CrossRef]
74. Chen, M.M.L.; Hoffmann, R. Sulfuranes. Theoretical aspects of bonding, substituent site preferences, and geometrical distortions. *J. Am. Chem. Soc.* **1976**, *98*, 1647–1653. [CrossRef]
75. Komiya, S.; Albright, T.A.; Hoffmann, R. Reductive elimination and isomerization of organogold complexes. Theoretical studies of trialkylgold species as reactive intermediates. *J. Am. Chem. Soc.* **1976**, *98*, 7255–7265. [CrossRef]
76. Rigaku Oxford Diffraction. *CrysAlisPro Software system, version 1.171.40.67a*; Rigaku Corporation: Wroclaw, Poland, 2012.
77. Dolomanov, O.V.; Bourhis, L.J.; Gildea, R.J.; Howard, J.A.K.; Puschmann, H. OLEX2: A complete structure solution, refinement and analysis program. *J. Appl. Cryst.* **2009**, *42*, 339–341. [CrossRef]
78. Sheldrick, G.M. SHELXT—Integrated space-group and crystal-structure determination. *Acta Cryst.* **2015**, *C71*, 3–8. [CrossRef]
79. Khodorkovsky, V.Y.; Becker, J.Y.; Bernstein, J. [2 + 4] Cycloaddition to tetrathiafulvalene (TTF): A new route to multisulfur TTF derivatives. *Synthesis* **1992**, 1071–1072. [CrossRef]
80. Gillman, J.; Hartke, K. Thion- und Dithioester, 40. [2 + 4]-Cycloadditionen von Tetrathiooxalsäuredimethylester an cyclisch konjugierte, mehrfach ungesättigte Verbindungen. *Chem. Ber.* **1986**, *119*, 2859–2867. [CrossRef]
81. Medne, R.S.; Katsens, Y.Y.; Kraupsha, I.L.; Neiland, O.Y. Bis(1,2-cyclopentylenedithio)- and bis(1,2-cyclohexylenedithio) tetrathiafulvalenes. *Chem. Heterocycl. Compd.* **1991**, *27*, 1053–1055. [CrossRef]
82. Kotov, A.I.; Faulmann, C.; Cassoux, P.; Yagubskii, E.B. New π-donor precursor molecule for the preparation of organic metals, 4,5;4′,5′-bis(l,4-dioxanediyl-2,3-dithio)-tetrathiafulvalene: Synthesis and properties. *J. Org. Chem.* **1994**, *59*, 2626–2629. [CrossRef]
83. Bai, J.-F.; Zuo, J.-L.; Shi, F.-N.; Jin, R.; Shen, Z.; You, X.-Z. Synthesis and crystal structure of 4,5-(cis-cyclohexylenedithio)-1,3-dithiole-2-one. *J. Chem. Crystallogr.* **1999**, *29*, 719–723. [CrossRef]
84. Kim, H.; Kobayashi, A.; Sasaki, Y.; Kato, R.; Kobayashi, H.; Nakamura, T.; Nogami, T.; Shirota, Y. Crystal and molecular structure of tetrabutylammonium bis(1,4-dithiin-2,3-dithiolato)nickelate(III), (Bun4N)[Ni(ddt)$_2$]. *Bull. Chem. Soc. Jpn.* **1988**, *61*, 2559–2562. [CrossRef]
85. Bigoli, F.; Deplano, P.; Devillanova, F.A.; Lippolis, V.; Lukes, P.J.; Mercuri, M.L.; Pellinghelli, M.A.; Trogu, E.F. New neutral nickel dithiolene complexes derived from 1,3-dialkylimidazolidine-2,4,5-trithione, showing remarkable near-IR absorption. *J. Chem. Soc. Chem. Commun.* **1995**, 371–372. [CrossRef]
86. Bigoli, F.; Deplano, P.; Mercuri, M.L.; Pellinghelli, M.A.; Pintus, G.; Trogu, E.F.; Zonnedda, G.; Wang, H.H.; Williams, J.M. Novel oxidation and reduction products of the neutral nickel-dithiolene Ni(Pr2itimdt)2 (Pr2itimdt is the monoanion of 1,3-diisopropylimidazolidine-2,4,5-trithione). *Inorg. Chim. Acta* **1998**, *273*, 175–183. [CrossRef]
87. Bigoli, F.; Deplano, P.; Devillanova, F.A.; Ferraro, J.R.; Lippolis, V.; Lukes, P.J.; Mercuri, M.L.; Pellinghelli, M.A.; Trogu, E.F.; Williams, J.M. Syntheses, X-ray crystal structures, and spectroscopic properties of new nickel-dithiolenes and related compounds. *Inorg. Chem.* **1997**, *36*, 1218–1226. [CrossRef] [PubMed]
88. Charlton, A.; Hill, C.A.S.; Underhill, A.E.; Malik, K.M.A.; Hursthouse, M.B.; Karaulov, A.I.; Møller, J. Synthesis, physical properties and x-ray crystal structures of a series of nickel complexes based on n-alkylthio-substituted ethylene-1,2-dithiolene ligands. *J. Mater. Chem.* **1994**, *4*, 1861–1866. [CrossRef]
89. Zuo, J.-L.; Yao, T.-M.; You, F.; You, X.-Z.; Fun, H.-K.; Yip, B.-C. Syntheses, characterization and non-linear optical properties of nickel complexes of multi-sulfur 1,2-dithioiene with strong near-IR absorption. *J. Mater. Chem.* **1996**, *6*, 1633–1637. [CrossRef]
90. Zuo, J.-L.; You, F.; You, X.-Z.; Fun, H.-K. Syntheses and properties of the neutral nickel complexes of 1,2-dithiolates MEDT and PHDT. The crystal structure of [Ni(MEDT)$_2$]. *Polyhedron* **1997**, *16*, 1465–1469. [CrossRef]
91. Proctor, K.A.; Boyle, P.D.; Bereman, R.D. The synthesis and characterization of new neutral metal complexes containing the tetrathioethylene unit. *J. Coord. Chem.* **1996**, *39*, 13–58. [CrossRef]
92. Schultz, A.J.; Wang, H.H.; Soderholm, L.C.; Sifter, T.L.; Williams, J.M.; Bechgaard, K.; Whangbo, M.H. Crystal structures of bis(5,6-dihydro-1,4-dithiin-2,3-dithiolato)gold and tetrabutylammonium bis(5,6-dihydro-1,4-dithiin-2,3-dithiolato)nickelate(1-) and the ligandlike character of the isoelectronic radicals [Au(DDDT)$_2$]^0 and [Ni(DDDT)$_2$]-. *Inorg. Chem.* **1987**, *26*, 3757–3761. [CrossRef]
93. Ji, Y.; Zuo, J.-L.; Chen, L.; Tian, Y.-Q.; Song, Y.; Li, Y.-Z.; You, X.-Z. Synthesis, characterization and optical limiting effect of nickel complexes of multi-sulfur 1,2-dithiolene. *J. Phys. Chem. Solids* **2005**, *66*, 207–212. [CrossRef]
94. Bai, J.-F.; Zuo, J.-L.; Tan, W.-L.; Ji, W.; Shen, Z.; Fun, H.-K.; Chinnakali, K.; Razak, I.A.; You, X.-Z.; Che, C.-M. Synthesis, structure and optical limiting effect of two new nickel complexes containing strongly bound geometrically fixed multi-sulfur 1,2-dithiolene ligands showing remarkable near-IR absorption. *J. Mater. Chem.* **1999**, *9*, 2419–2423. [CrossRef]
95. Mueller-Westerhoff, U.T.; Vance, B.; Yoon, D.I. The synthesis of dithiolene dyes with strong near-IR absorption. *Tetrahedron* **1991**, *47*, 909–932. [CrossRef]
96. Herman, Z.S.; Kirchner, R.F.; Loew, G.H.; Mueller-Westerhoff, U.T.; Nazzal, A.; Zerner, M.C. Electronic spectra and structure of bis(ethylene-1,2-dithiolato)nickel and bis-(propene-3-thione-1-thiolato)nickel. *Inorg. Chem.* **1982**, *21*, 46–56. [CrossRef]

97. Tan, W.L.; Ji, W.; Zuo, J.L.; Bai, J.F.; You, X.Z.; Lim, J.H.; Yang, S.; Hagan, D.J.; Van Stryland, E.W. Optical-limiting properties of neutral nickel-dithiolenes. *Appl. Phys. B* **2000**, *70*, 809–812. [CrossRef]
98. Lee, H.-J.; Noh, D.-Y. Electrochemistry and strong near-IR absorption of the $[Ni(dphdt)_2]^n$ complexes (n = −1, 0; dphdt = 5,6-diphenyl-1,4-dithiin-2,3-dithiolate); X-ray crystal structure of $(Ph_4P)[Ni(dphdt)_2](CH_2Cl_2)$. *Polyhedron* **2000**, *19*, 425–429. [CrossRef]
99. Nagapetyan, S.S.; Shklover, V.E.; Struchkov, Y.T.; Kotov, A.I.; Yagubskii, E.B.; Ukhin, L.Y. Synthesis andthe X-ray study of $[Ni(DDDT)_2]_3(ClO_4)_2$. Trimeric cationic dithiolate complex—A precursor of low-dimensional metals of a new type. *Dokl. Akad. Nauk SSSR* **1990**, *310*, 94–98.
100. Yagubskii, E.B. Effect of metal (M) and counter ion nature on crystal structure and conducting properties of $M(dddt)_2$ cation complexes. *Synth. Met.* **2003**, *133–134*, 385–387. [CrossRef]
101. Kotov, A.I.; Buravov, L.I.; Yagubskii, E.B.; Khasanov, S.S.; Zorina, L.V.; Shibaeva, R.P.; Canadell, E. Comparative study of BEDT-TTF and $Ni(dddt)_2$ electroconducting salts with the HXO_4 (X = Se, S) anions. *Synth. Met.* **2001**, *124*, 357–362. [CrossRef]
102. Yagubskii, E.B.; Kotov, A.I.; Khomenko, A.G.; Buravov, L.I.; Schegolev, A.I.; Shibaeva, R.P. $[Ni(dddt)_2]_3(HSO_4)_2$, the first metal among the $M(dddt)_2$ salts. *Synth. Met.* **1992**, *46*, 255–259. [CrossRef]
103. Yagubskii, E.B.; Kotov, A.I.; Laukhina, E.E.; Ignatiev, A.A.; Buravov, L.I.; Khomenko, A.G.; Shklover, V.E.; Nagapetyan, S.S.; Struchkov, Y.T. New class of metal BIS-dithiolene electroconducting solids: Cation complexes of metals with dddt. *Synth. Met.* **1991**, *42*, 2515–2522. [CrossRef]
104. Kubo, K.; Nakao, A.; Ishii, Y.; Yamamoto, T.; Tamura, M.; Kato, R.; Yakushi, K.; Matsubayashi, G. Electrical Properties and Electronic States of Molecular Conductors Based on Unsymmetrical Organometallic-Dithiolene Gold(III) Complexes. *Inorg. Chem.* **2008**, *47*, 5495–5502. [CrossRef] [PubMed]

crystals

MDPI

Article

First-Principles Study on the Stability and Electronic Structure of the Charge-Ordered Phase in α-(BEDT-TTF)$_2$I$_3$

Takao Tsumuraya [1,*], Hitoshi Seo [2,3] and Tsuyoshi Miyazaki [4]

1 Priority Organization for Innovation and Excellence, Kumamoto University, 2-39-1 Kurokami, Kumamoto 860-8555, Japan
2 Condensed Matter Theory Laboratory, RIKEN, 2-1 Hirosawa, Wako 351-0198, Japan; seo@riken.jp
3 Center of Emergent Matter Science, RIKEN, 2-1 Hirosawa, Wako 351-0198, Japan
4 International Center for Materials Nanoarchitectonics, National Institute for Materials Science (WPI-MANA) 1-1 Namiki, Tsukuba 305-0044, Japan; miyazaki.tsuyoshi@nims.go.jp
* Correspondence: tsumu@kumamoto-u.ac.jp

Abstract: We theoretically study the structural and electronic properties of a molecular conductor, α-(BEDT-TTF)$_2$I$_3$, using first-principles density-functional theory calculations, especially in its low-temperature charge-ordered state at ambient pressure. We apply a hybrid functional approach and compare the results with a conventional exchange-correlation functional within the generalized gradient approximation. By performing structural optimization, we found a stable charge-ordered solution for the former, in contrast to the latter approach where the magnitude of the charge imbalance becomes considerably small compared to that when the experimental structure is adopted. The electronic band structure near the Fermi level, with and without structural optimization, as well as the molecule-dependent local density of states of the charge-ordered state are discussed.

Keywords: molecular conductor; first-principles calculation; density-functional theory; charge ordering; hybrid functional; electronic structure

Citation: Tsumuraya, T.; Seo, H.; Miyazaki, T. First-Principles Study on the Stability and Electronic Structure of the Charge-Ordered Phase in α-(BEDT-TTF)$_2$I$_3$. *Crystals* **2021**, *11*, 1109. https://doi.org/10.3390/cryst11091109

Academic Editor: Toshio Naito

Received: 18 August 2021
Accepted: 10 September 2021
Published: 12 September 2021

1. Introduction

Molecular conductors show a rich variety of electronic properties originating from the interplay between the diversity in their crystal structures and the strongly correlated electrons in the valence bands [1,2]. Among them, α-(ET)$_2$I$_3$, where ET is the abbreviation for BEDT-TTF = bis(ethylenedithio)tetrathia-fulvalene, is a peculiar example where a (semi)metallic state, in which a massless Dirac-type electronic dispersion is discussed to be realized, competes with a nonmagnetic charge-ordered (CO) insulating state [3,4]. At ambient pressure, the compound undergoes a phase transition by varying the temperature, between the high-temperature metallic and low-temperature insulating state at $T_{CO} = 135$ K [5,6]; the magnetic susceptibility shows a sharp drop below T_{CO}, indicating a nonmagnetic ground state [7]. Model calculations predicted a charge ordering owing to Coulomb repulsion between the valence electrons as the origin of this phase transition, with the so-called horizontal CO pattern resulting in a spin-singlet formation between electron spins localized on two molecules in the unit cell [8–12]; such a picture has now been confirmed by different experiments [13–16]. An interesting consequence of the CO is that the pattern breaks the inversion symmetry at high temperatures, and then, the concept called electronic ferroelectricity was developed [17–22].

Although such calculations, using the extended Hubbard model, which is a typical effective model to study strongly correlated electron systems where the basis set is the HOMO of ET molecules, are widely performed, first-principles calculations based on the density-functional theory (DFT) face a difficulty in treating the CO phase. Calculations adopting the crystal structure of the low-temperature CO phase [23,24] succeed in reproducing the insulating gap, as well as the charge disproportionation (CD) among the ET molecules; however, once structural optimization is performed, the degree of CD becomes considerably small or even vanishes within the numerical accuracy. This is a common problem in molecular conductors showing charge ordering. Recently, we applied a hybrid functional approach using the exchange-correlation functional by Heyd, Scuseria, and Ernzerhof (HSE06) [25–27] to a hydrogen-bonded molecular conductor system κ-D_3(Cat-EDT-TTF/ST)$_2$, where Cat-EDT-TTF is catechol with ethylenedithio-tetrathiafulvalene and Cat-EDT-ST is its selenium-substituted analog [28]. There, the structural stability of the CO insulating phase, which is coupled to the displacements of the deuteriums forming hydrogen bonds, can be reproduced, compared to the widely used DFT method based on the generalized gradient approximation (GGA). This is owed to the more localized nature of the wave functions in the hybrid-functional method; therefore, it is expected that this approach will provide more reliable results than GGA in this class of materials with a pronounced electron-correlation effect.

Here, we apply the HSE06 functional to α-(ET)$_2$I$_3$ and compare the results with the conventional exchange-correlation functional by Perdew, Burke, and Ernzerhof (PBE) within GGA, commonly used at present. We investigate the structural stability of the CO phase and, indeed, find a larger degree of CD with the use of HSE06 than GGA-PBE, judging from the central C=C bond of the ET molecules, whose length differences among nonequivalent molecules are closer to, but still smaller than the experimental situation. We further show the calculated electronic structures for the two different functionals and the local density of states (LDOS) probing the charge densities on each ET molecule. The LDOS is obtained as the summation of the density of state (DOS) projected on C and S atoms in each molecule in order to understand the degree of CD between the four molecules in the unit cell.

2. Crystal Structure

We refer to the experimental structure at 30 K and 150 K [29], below and above T_{CO}, respectively, for which structural optimization for the hydrogen positions was performed. We show the crystal structure at 30 K in Figure 1. Figure 1a shows the view along the a-axis where ET layers alternate with layers of iodine ions, I_3^-. In Figure 1b, a layer of ET molecules in the ac-plane is shown, where the ET molecules form a herringbone pattern consisting of two chains, one with A and A$'$ molecules stacked along the a-axis and the other in which B and C molecules are stacked.

Above T_{CO}, the space group is $P\bar{1}$; while the center of mass of the B and C molecules lie on the inversion centers, A and A$'$ are connected with the inversion operations, e.g., with respect to the inversion center in the middle of A and A$'$ at (0.5, 0, 0), and these two are crystallographically equivalent. On the other hand, below T_{CO}, the system loses the spatial inversion symmetry, and the molecules A and A$'$ become crystallographically independent, which is considered a consequence of the charge ordering, which is schematically shown in Figure 1b. The structure has a noncentrosymmetric space group of $P1$ [16,29,30].

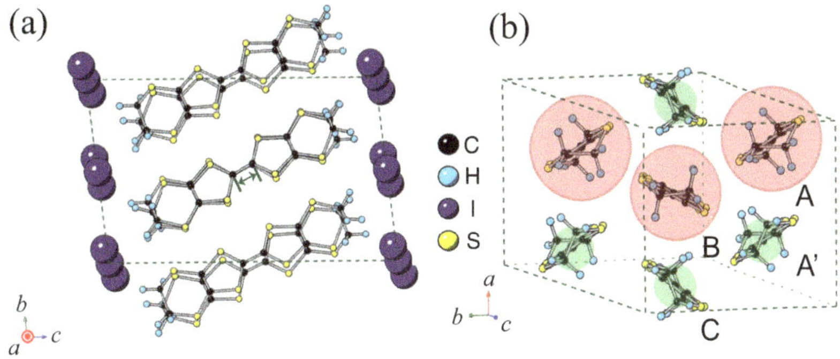

Figure 1. Crystal structure of α-(BEDT-TTF$_2$)I$_3$ at 30 K [29]. The unit cell is shown by the dashed lines. (**a**) A view along the *a*-axis; (**b**) ET molecules in the *ab*-plane, where four independent molecules, A, A′, B, and C, are shown. The shaded circles schematically show the charge order, whose diameters are drawn to be proportional to the estimated charge values of the molecules in [16]: A = +0.82, A′ = +0.29, B = +0.73, and C = +0.26.

3. Calculation Methods

In the present DFT calculations, the Kohn–Sham equations were solved using the pseudopotential method based on the projected augmented plane wave (PAW) method with plane wave basis sets implemented in the Vienna Ab initio Simulation Package [31,32]. In the HSE06 calculations, we first obtained a converged charge density from the self-consistent calculations within GGA, and then, using the GGA charge density as the initial state, the self-consistent hybrid functional calculations were performed. The common **k**-point sampling, including that for the structural optimization, was set as $4 \times 4 \times 2$. The cutoff energy for plane waves was 29.4 Ry for the HSE06 calculations. The range-separation parameter in the HSE06 calculations was 0.2 Å^{-1}, and 25% of the exact exchange was mixed to the GGA exchange for the short-range interactions. We set the **k**-point mesh for calculating DOS as $8 \times 8 \times 2$ in both the HSE06 and GGA-PBE calculations.

4. Structural Properties

First, we discuss the structural properties by optimizing the internal coordinates using the GGA-PBE and HSE06 functionals. Table 1 lists the optimized bond parameters of the central C=C bond distance of the ET molecules, which reflects the charge density on them [16], together with the experimental values. For reference, we also list the valence of the molecules estimated by an empirical method used in the literature [16,29,33].

For the centrosymmetric ($P\bar{1}$) case above T_{CO}, one can see two points compared to the experiments: the average value in GGA-PBE gives a closer value than HSE06. We note that the experimental values are for a high temperature where thermal vibration should be significant, whereas the DFT calculations are basically for the ground state; we should be cautious about a direct comparison. More importantly, HSE06 captures the feature in the experimental values of the C=C bond lengths that shows A $\simeq$ B > C, reflecting the hole density, i.e., the amount of charge on each molecule, in clear contrast to the GGA-PBE case showing A > B $\simeq$ C. We note that the charge densities using the room temperature structure were evaluated [23,34] by Mulliken charge analyses, which agreed with the experimental estimations.

Now, let us discuss the results for the low-temperature CO phase. In the experiments, owing to the charge ordering, CD takes place among the two equivalent A molecules and results in two distinct A and A′ with a difference in the C=C bond lengths of about 0.02–0.03 Å. As mentioned in the Introduction, the present results with GGA-PBE indeed show almost the same values for the A and A′ molecules. In clear contrast, when we use the HSE06 hybrid functional, the difference in the C=C bond lengths between the A and A′

molecules is about 0.005 Å. Moreover, similar to the high-temperature phase, HSE06 gives a shorter C=C bond length for the C molecule than that for the B molecule; in GGA-PBE, they become almost the same. These HSE06 results are consistent with the CO pattern suggested in previous works, theoretically and experimentally, where the charge rich (poor) sites are A and B (A′ and C) with longer (shorter) C=C bond lengths [16,29,35]. We should note that discrepancies still exist from the experimental values, as seen in Table 1: longer C=C bond lengths of C molecules in general seen in the valence as well, a small change in that of B molecule across the phase transition, etc. However, we consider that the HSE06 functional provides the structural stability of the CO state in α-$(ET)_2I_3$, i.e., a reasonable difference in the A and A′ molecules upon structural optimization.

Table 1. Experimental and theoretically-optimized structural parameters for the central C=C double bonds (d) and the valence estimated by an empirical method (Q) [33] in ET molecules, above (top rows) and below (bottom rows) T_{CO}.

Method	A d (Å)	A′ d (Å)	B d (Å)	C d (Å)	A Q	A′ Q	B Q	C Q
GGA-PBE	1.376	(= A)	1.373	1.373	0.75	(= A)	0.75	0.71
HSE06	1.366	(= A)	1.364	1.360	0.80	(= A)	0.83	0.71
Exp. (150 K) [29]	1.375	(= A)	1.376	1.368	0.71	(= A)	0.74	0.48
GGA-PBE	1.375	1.374	1.373	1.374	0.79	0.74	0.76	0.68
HSE06	1.368	1.363	1.364	1.361	0.94	0.79	0.84	0.73
Exp. (20 K) [16]	1.384	1.361	1.385	1.360	1.00	0.43	0.86	0.32
Exp. (30 K) [29]	1.382	1.363	1.381	1.361	0.90	0.45	0.89	0.39

5. Electronic Structure

Next, we discuss the electronic states, for the experimental structures above and below T_{CO} and the optimized structures, comparing the two functionals. Figure 2a,b shows the band structure and the total DOS calculated with HSE06 and GGA-PBE, respectively, using the experimental structure above T_{CO}, at 150 K. The main features are consistent with the previously reported results, including the extended Hückel approach [36]: the four bands close to the Fermi level are basically attributed to the HOMO of the four constituent ET molecules in the unit cell [23,34,37,38]. Although it is not the focus of this study, we note that there is no consensus in the detailed band structure of this material in the high-temperature phase even within GGA-PBE. Some show a semimetallic state [23,34,38] as in the extended Hückel results, while others suggest a massless Dirac electron-type dispersion with the Fermi level at the Dirac point [29,37]; our results are consistent with the latter.

Comparing the band structures obtained by HSE06 and GGA-PBE, we found that the HSE06 band structure had a wider bandwidth, of about 20%, in spite of the fact that the atomic positions used in the calculations were the same. This trend was in agreement with our results for a proton-coupled system in κ-D_3(Cat-EDT-TTF/ST)$_2$ [28]. As for the case of the optimized atomic coordinates, the results for the HSE06 and GGA-PBE functionals are plotted in Figure 2c,d, respectively. For both functionals, structural optimization does not change the band structures much; the essential features in the band dispersions and the DOS were maintained. The tendency that the HSE06 results give a wider bandwidth than the GGA-PBE was also the same as in the experimental structure.

We then show the band structure and DOS in the low-temperature phase below T_{CO} using the experimental structure, with the HSE06 and PBE-GGA functionals, in Figure 3a,b, respectively. Compared to the result above, the size of the band gap increase with both functionals. The size of the band gap below T_{CO} with HSE06, about 0.06 eV, is larger than the value by GGA-PBE, about 0.04 eV. Similar to the case above T_{CO}, the bandwidth of the HSE06 band structure, evaluated from the DOS, is wider than that of GGA-PBE. Despite these quantitative differences in the band gaps and the bandwidths, GGA-PBE and HSE06 provide similar band structures. Figure 3c,d shows the HSE06 and GGA-PBE band structures for the optimized structure. By structural optimization, the band gap is reduced from the results using the experimental structure: in the HSE06 case, 0.06 eV→0.05 eV, and in GGA-PBE, 0.04 eV→0.02 eV. The smallness of the band gap in GGA-PBE is consistent with the structural properties discussed in the previous section, i.e., the small difference in the central C=C bond lengths among the ET molecules. We note the experimental estimates [20]: from optical measurements, the gap is estimated as 0.075 eV without noticeable in-plane directional dependence, while the DC electric conductivity gave 0.04 eV and 0.08 eV depending on the field direction.

Figure 4 shows the LDOS below T_{CO}, which explicitly indicates the CD among the molecules in the unit cell. As mentioned above, we obtained the LDOS as a summation of the projected DOS on the orbitals of C and S atoms in each ET molecule. Figure 4a,b shows the results for HSE06 without and with structural optimization, respectively, and Figure 4c,d shows the same for GGA-PBE. The CD is seen by the larger LDOS in the top band for A and B molecules than A′ and C molecules and vice versa for the second band from the top. Since the Fermi level resides between the two bands, these features directly indicate the larger (smaller) electron density in the A′ and C (A and B) molecules: the CO state is formed by the horizontal stripes along the hole-rich A and B sites. We show the estimations of the valence of each molecule in Table 2, based on the integration of the LDOS. Note that these values are different from Q in Table 1. The GGA-PBE values for the experimental structure were in good agreement with a previous evaluation using a different DFT approach [23]. Comparing the HSE06 LDOS without and with optimization, the CD between A and A′ molecules remains after structural optimization. On the other hand, in the case of GGA-PBE, the CD between A and A′ is smaller after the optimization, where the LDOS, especially in the unoccupied state, is almost the same after the structural optimization (Figure 4d).

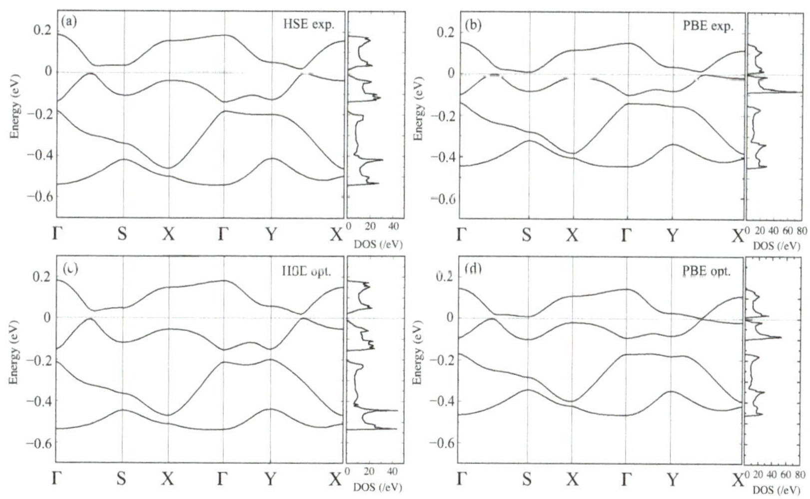

Figure 2. The band structure and the total DOS for the experimental structure at 150 K above the CO phase transition calculated with the (**a**) HSE06 and (**b**) GGA-PBE functionals and those for the geometrically optimized structure with the (**c**) HSE06 and (**d**) GGA-PBE functionals.

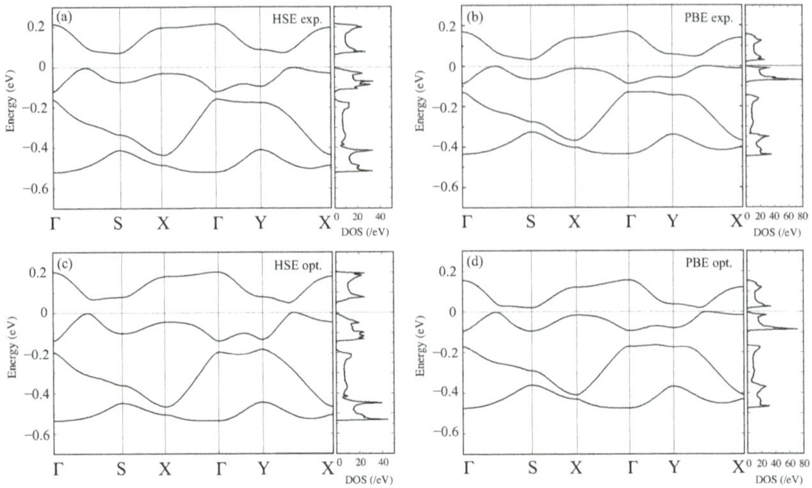

Figure 3. The band structure and the total DOS for the experimental structure at 30 K, in the CO phase calculated with the (**a**) HSE06 and (**b**) GGA-PBE functionals and those for the geometrically optimized structure with the (**c**) HSE06 and (**d**) GGA-PBE functionals.

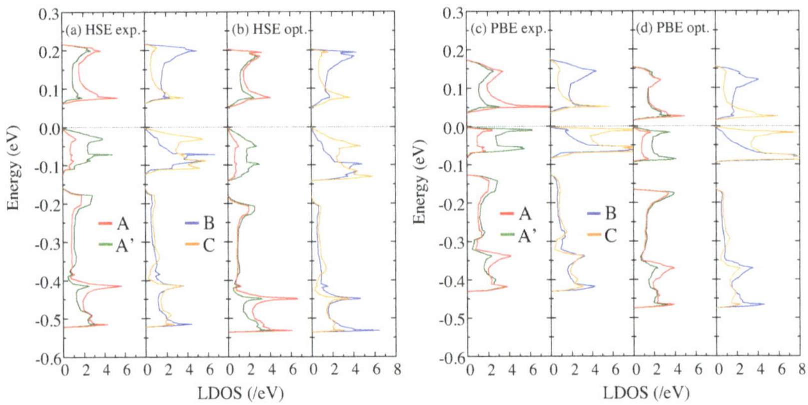

Figure 4. The LDOS in the low-temperature CO phase for the four molecules in the unit cell A, A′, B, and C. (**a**) The LDOS calculated with HSE06 for the experimental structure and (**b**) for the geometrically optimized structure. (**c**) The LDOS calculated with GGA-PBE for the experimental structure and (**d**) for the geometrically optimized structure. The plotted LDOS contains four parts: the red, green, blue, and orange curves indicate that of A, A′, B, and C, respectively. The dotted lines at 0 eV show the top of the valence bands.

Table 2. The valence of the four molecules in the unit cell, i.e., the hole density obtained by integrating the LDOS shown in Figure 4.

Method	Structure	A	A′	B	C
GGA-PBE	Exp.	0.72	0.42	0.64	0.32
	Opt.	0.58	0.51	0.58	0.42
HSE06	Exp.	0.62	0.40	0.68	0.32
	Opt.	0.64	0.50	0.58	0.38

6. Conclusions

We studied the structural and electronic properties of the molecular conductor α-$(ET)_2I_3$ using GGA-PBE and a range-separated hybrid functional of HSE06 and discussed the electronic structure above and below T_{CO}. Below T_{CO}, HSE06 captures the experimental feature of different molecules (A and A′) showing different C=C bond lengths, reflecting the charge density on each molecule, compared to the GGA-PBE case, where the charge ordering below T_{CO} is obscured when structural optimization is performed. The electronic structure calculations are consistent with such results, showing clear charge disproportionation between A and A′ in the case of HSE06 after structural optimization.

Author Contributions: T.T. performed all the calculations and data analyses. All the authors (T.T., H.S. and T.M.) equally contributed to the planning, discussion, and writing of the manuscript. All authors have read and agreed to the published version of the manuscript.

Funding: This research was funded by a Grant-in-Aid for Scientific Research (JP16K17756, JP19K04988, JP19K21860, JP19K03723, JP20H00121, JP20H04463) from the Japan Society for the Promotion of Science (JSPS) and JST, CREST Grant Number JPMJCR2094, Japan. This work was performed under the GIMRT Program of the Institute for Materials Research (IMR), Tohoku University (Proposal No. 18K0090). T.T. was supported in part by the Leading Initiative for Excellent Young Researchers (LEADER), a program of the Ministry of Education, Culture, Sports, Science and Technology, Japan (MEXT).

Data Availability Statement: Not applicable.

Acknowledgments: The study's computations were mainly conducted using the computer facilities of ITO at Kyushu University, MASAMUNE at IMR, Tohoku University, and ISSP, University of Tokyo, Japan.

Conflicts of Interest: The authors declare no conflict of interest.

Sample Availability: The experimental structures used in this study are available from the CCDC database, where the deposition numbers are 2008980 and 2008981.

References

1. Lebed, A.G. *The Physics of Organic Superconductors and Conductors*; Springer: Berlin/Heidelberg, Germany, 2008; Volume 110.
2. Seo, H.; Hotta, C.; Fukuyama, H. Toward systematic understanding of diversity of electronic properties in low-dimensional molecular solids. *Chem. Rev.* **2004**, *104*, 5005. [CrossRef]
3. Kajita, K.; Nishio, Y.; Tajima, N.; Suzumura, Y.; Kobayashi, A. Molecular Dirac Fermion Systems—Theoretical and Experimental Approaches. *J. Phys. Soc. Jpn.* **2014**, *83*, 072002. [CrossRef]
4. Liu, D.; Ishikawa, K.; Takehara, R.; Miyagawa, K.; Tamura, M.; Kanoda, K. Insulating Nature of Strongly Correlated Massless Dirac Fermions in an Organic Crystal. *Phys. Rev. Lett.* **2016**, *116*, 226401. [CrossRef] [PubMed]
5. Bender, K.; Hennig, I.; Schweitzer, D.; Dietz, K.; Endres, H.; Keller, H.J. Synthesis, structure and physical properties of a two-dimensional organic metal, di [bis (ethylenedithiolo) tetrathiofulvalene] triiodide,$(BEDT\text{-}TTF)_2^+I_3^-$. *Mol. Cryst. Liq. Cryst.* **1984**, *108*, 359–371. [CrossRef]
6. Bender, K.; Dietz, K.; Endres, H.; Helberg, H.W.; Hennig, I.; Keller, H.J.; Schäfer, H.W.; Schweitzer, D. $(BEDT\text{-}TTF)^+{}_2J^-{}_3$: A Two-Dimensional Organic Metal. *Mol. Cryst. Liq. Cryst.* **1984**, *107*, 45–53. [CrossRef]
7. Rothaemel, B.; Forró, L.; Cooper, J.R.; Schilling, J.S.; Weger, M.; Bele, P.; Brunner, H.; Schweitzer, D.; Keller, H.J. Magnetic susceptibility of α and β phases of di[bis(ethylenediothiolo)tetrathiafulvalene] tri-iodide $[(BEDT-TTF)_2I_3]$ under pressure. *Phys. Rev. B* **1986**, *34*, 704–712. [CrossRef]
8. Kino, H.; Fukuyama, H. On the Phase Transition of α-(ET)2I3. *J. Phys. Soc. Jpn.* **1995**, *64*, 1877–1880. [CrossRef]
9. Kino, H.; Fukuyama, H. Phase Diagram of Two-Dimensional Organic Conductors: (BEDT-TTF) 2X. *J. Phys. Soc. Jpn.* **1996**, *65*, 2158–2169. [CrossRef]
10. Seo, H. Charge Ordering in Organic ET Compounds. *J. Phys. Soc. Jpn.* **2000**, *69*, 805–820. [CrossRef]
11. Seo, H.; Merino, J.; Yoshioka, H.; Ogata, M. Theoretical Aspects of Charge Ordering in Molecular Conductors. *J. Phys. Soc. Jpn.* **2006**, *75*, 051009. [CrossRef]
12. Tanaka, Y.; Ogata, M. Correlation Effects on Charge Order and Zero-Gap State in the Organic Conductor α-$(BEDT\text{-}TTF)_2I_3$. *J. Phys. Soc. Jpn.* **2016**, *85*, 104706. [CrossRef]
13. Takano, Y.; Hiraki, K.; Yamamoto, H.; Nakamura, T.; Takahashi, T. Charge disproportionation in the organic conductor, α-$(BEDT\text{-}TTF)_2I_3$. *J. Phys. Chem. Solid.* **2001**, *62*, 393–395. [CrossRef]
14. Wojciechowski, R.; Yamamoto, K.; Yakushi, K.; Inokuchi, M.; Kawamoto, A. High-pressure Raman study of the charge ordering in α-$(BEDT\text{-}TTF)_2I_3$. *Phys. Rev. B* **2003**, *67*, 224105. [CrossRef]

15. Takahashi, T.; Nogami, Y.; Yakushi, K. Charge Ordering in Organic Conductors. *J. Phys. Soc. Jpn.* **2006**, *75*, 051008. [CrossRef]
16. Kakiuchi, T.; Wakabayashi, Y.; Sawa, H.; Takahashi, T.; Nakamura, T. Charge Ordering in α-(BEDT-TTF)$_2$I$_3$ by Synchrotron X-ray Diffraction. *J. Phys. Soc. Jpn.* **2007**, *76*, 113702. [CrossRef]
17. Yamamoto, K.; Iwai, S.; Boyko, S.; Kashiwazaki, A.; Hiramatsu, F.; Okabe, C.; Nishi, N.; Yakushi, K. Strong Optical Nonlinearity and its Ultrafast Response Associated with Electron Ferroelectricity in an Organic Conductor. *J. Phys. Soc. Jpn.* **2008**, *77*, 074709. [CrossRef]
18. Yamamoto, K.; Kowalska, A.A.; Yakushi, K. Direct observation of ferroelectric domains created by Wigner crystallization of electrons in α-[bis(ethylenedithio)tetrathiafulvalene]$_2$I$_3$. *Appl. Phys. Lett.* **2010**, *96*, 122901. [CrossRef]
19. Ivek, T.; Korin-Hamzić, B.; Milat, O.; Tomić, S.; Clauss, C.; Drichko, N.; Schweitzer, D.; Dressel, M. Collective Excitations in the Charge-Ordered Phase of $\alpha-(\mathrm{BEDT-TTF})_2I_3$. *Phys. Rev. Lett.* **2010**, *104*, 206406. [CrossRef] [PubMed]
20. Ivek, T.; Korin-Hamzić, B.; Milat, O.; Tomić, S.; Clauss, C.; Drichko, N.; Schweitzer, D.; Dressel, M. Electrodynamic response of the charge ordering phase: Dielectric and optical studies of α-(BEDT-TTF)$_2$I$_3$. *Phys. Rev. B* **2011**, *83*, 165128. [CrossRef]
21. Ishihara, S. Electronic ferroelectricity in molecular organic crystals. *J. Phys. Cond. Matt.* **2014**, *26*, 493201. [CrossRef] [PubMed]
22. Lunkenheimer, P.; Hartmann, B.; Lang, M.; Müller, J.; Schweitzer, D.; Krohns, S.; Loidl, A. Ferroelectric properties of charge-ordered $\alpha-(\mathrm{BEDT-TTF})_2I_3$. *Phys. Rev. B* **2015**, *91*, 245132. [CrossRef]
23. Alemany, P.; Pouget, J.P.; Canadell, E. Essential role of anions in the charge ordering transition of α-(BEDT-TTF)$_2$I$_3$. *Phys. Rev. B* **2012**, *85*, 195118. [CrossRef]
24. Yamakawa, H.; Miyamoto, T.; Morimoto, T.; Yada, H.; Kinoshita, Y.; Sotome, M.; Kida, N.; Yamamoto, K.; Iwano, K.; Matsumoto, Y.; et al. Novel electronic ferroelectricity in an organic charge-order insulator investigated with terahertz-pump optical-probe spectroscopy. *Sci. Rep.* **2016**, *6*, 20571. [CrossRef] [PubMed]
25. Heyd, J.; Scuseria, G.E.; Ernzerhof, M. Hybrid functionals based on a screened Coulomb potential. *J. Chem. Phys.* **2003**, *118*, 8207–8215. [CrossRef]
26. Heyd, J.; Scuseria, G.E.; Ernzerhof, M. Hybrid functionals based on a screened Coulomb potential. *J. Chem. Phys.* **2003**, *118*, 8207–8215; Erratum in **2006**, *124*, 219906. [CrossRef]
27. Heyd, J.; Scuseria, G.E. Efficient hybrid density functional calculations in solids: Assessment of the Heyd–Scuseria–Ernzerhof screened Coulomb hybrid functional. *J. Chem. Phys.* **2004**, *121*, 1187–1192. [CrossRef]
28. Tsumuraya, T.; Seo, H.; Miyazaki, T. First-principles study of the charge ordered phase in $\kappa-\mathrm{D_3(Cat\text{-}EDT\text{-}TTF/ST)_2}$: Stability of π-electron deuterium coupled ordering in hydrogen-bonded molecular conductors. *Phys. Rev. B* **2020**, *101*, 045114. [CrossRef]
29. Kitou, S.; Tsumuraya, T.; Sawahata, H.; Ishii, F.; Hiraki, K.I.; Nakamura, T.; Katayama, N.; Sawa, H. Ambient-pressure Dirac electron system in the quasi-two-dimensional molecular conductor $\alpha-(\mathrm{BETS})_2I_3$. *Phys. Rev. B* **2021**, *103*, 035135. [CrossRef]
30. Meneghetti, M.; Bozio, R.; Pecile, C. Electron-molecular vibration coupling in 2-D organic conductors: High and low temperature phases of α-(BEDT-TTF)$_2$I$_3$. *J. Phys. Fr.* **1986**, *47*, 1377–1387. [CrossRef]
31. Kresse, G.; Furthmüller, J. Efficient iterative schemes for ab initio total-energy calculations using a plane-wave basis set. *Phys. Rev. B* **1996**, *54*, 11169–11186. [CrossRef]
32. Kresse, G.; Joubert, D. From ultrasoft pseudopotentials to the projector augmented-wave method. *Phys. Rev. B* **1999**, *59*, 1758–1775. [CrossRef]
33. Guionneau, P.; Kepert, C.; Bravic, G.; Chasseau, D.; Truter, M.; Kurmoo, M.; Day, P. Determining the charge distribution in BEDT-TTF salts. *Syn. Met.* **1997**, *86*, 1973–1974. [CrossRef]
34. Ishibashi, S.; Tamura, T.; Kohyama, M.; Terakura, K. Ab initio electronic-structure calculations for α-(BEDT-TTF)$_2$I$_3$. *J. Phys. Soc. Jpn.* **2005**, *75*, 015005. [CrossRef]
35. Yue, Y.; Yamamoto, K.; Uruichi, M.; Nakano, C.; Yakushi, K.; Yamada, S.; Hiejima, T.; Kawamoto, A. Nonuniform site-charge distribution and fluctuations of charge order in the metallic state of $\alpha-(\mathrm{BEDT\text{-}TTF})_2I_3$. *Phys. Rev. B* **2010**, *82*, 075134. [CrossRef]
36. Mori, T.; Kobayashi, A.; Sasaki, Y.; Kobayashi, H.; Saito, G.; Inokuchi, H. Band Structure of Two Types of (BEDT-TTF)$_2$I$_3$. *Chem. Lett.* **1984**, *13*, 957–960. [CrossRef]
37. Kino, H.; Miyazaki, T. First-Principles Study of Electronic Structure in α-(BEDT-TTF)$_2$I$_3$ at Ambient Pressure and with Uniaxial Strain. *J. Phys. Soc. Jpn.* **2006**, *75*, 034704. [CrossRef]
38. Beyer, R.; Dengl, A.; Peterseim, T.; Wackerow, S.; Ivek, T.; Pronin, A.V.; Schweitzer, D.; Dressel, M. Pressure-dependent optical investigations of $\alpha-(\mathrm{BEDT\text{-}TTF})_2I_3$: Tuning charge order and narrow gap towards a Dirac semimetal. *Phys. Rev. B* **2016**, *93*, 195116. [CrossRef]

crystals

MDPI

Article

Temperature-Modulated Pyroelectricity Measurements of a Thin Ferroelectric Crystal with In-Plane Polarization and the Thermal Analysis Based on One-Dimensional Layer Models

Kaoru Yamamoto [1,*], Ayumi Kawasaki [1], Takumi Chinen [1] and Kayo Ryugo [1,2]

[1] Department of Applied Physics, Okayama University of Science, Okayama 700-0005, Japan; s19pm02ka@ous.jp (A.K.); s19pm03ct@ous.jp (T.C.); k-ryuugo@furukawadenchi.co.jp (K.R.)
[2] Research & Development Institute, The Furukawa Battery Co. Ltd., Iwaki, Fukushima 972-8501, Japan
* Correspondence: yamamoto@dap.ous.ac.jp; Tel.: +81-564-256-9470

Abstract: A temperature-modulated pyroelectricity measurement system for a small single crystal is developed and applied to standard sample measurements performed on a thin single crystal of lithium niobate. The modulation measurement is based on the AC technique, in which the temperature of the sample is periodically oscillated, and the synchronized pyroelectric signal is extracted using a lock-in amplifier. Temperature modulation is applied by irradiating periodic light on the sample placed in the heat exchange gas. To apply this technique to the transparent reference sample, a commercially available black resin is coated on the sample's surface to absorb the light energy and transmits it to the specimen. The experimental results are analyzed using a two-layer heat transfer model to verify the effect of the light-absorbing layer as well as the radiative temperature modulation system.

Keywords: pyroelectricity; temperature modulation; molecular ferroelectrics; radiative temperature control; thermal diffusion model; lithium niobate

Citation: Yamamoto, K.; Kawasaki, A.; Chinen, T.; Ryugo, K. Temperature-Modulated Pyroelectricity Measurements of a Thin Ferroelectric Crystal with In-Plane Polarization and the Thermal Analysis Based on One-Dimensional Layer Models. *Crystals* **2021**, *11*, 880. https://doi.org/10.3390/cryst11080880

Academic Editor: Toshio Naito

Received: 6 June 2021
Accepted: 26 July 2021
Published: 28 July 2021

Publisher's Note: MDPI stays neutral with regard to jurisdictional claims in published maps and institutional affiliations.

1. Introduction

Ferroelectrics and pyroelectrics are advanced materials with a variety of functional physical properties. While the dielectric matters have already established an industrial position in thermal sensors and transducers applications, the scope of material surveys goes beyond the traditional realm to include substances that have not been treated in this field, such as magnetic compounds [1–3], electric conductors [4,5], and unconventional synthesized solids [6,7]. To understand the potential applications, the electrical polarization of these new ferroelectrics properties needs to be investigated. However, most of these crystals, particularly molecular crystals, are so small and fragile that they cannot be formed into a shape suitable for applying strong electric fields. Therefore, the direct observation technique such as the polarization reversal is uneasy to carry out. Moreover, a class of charge-transfer molecular complexes, which has recently attracted attention as a new type of ferroelectric, is highly conductive in nature; thus, it cannot be loaded with sufficient fields for polarization conversions [4,8,9].

For these materials, pyroelectricity measurement can be an effective alternative. Pyroelectricity stems from the temperature dependence of the polarization, and its observation does not require strong electric fields, except for the poling treatment of the sample. The classical heat-to-electricity conversion phenomenon has long been utilized as an important function of ferroelectrics [10]. Lately, it has attracted new attention from the perspective of energy harvesting technology [11–13]. Therefore, measuring pyroelectricity would offer multiple benefits to material researchers who are developing new ferroelectrics.

Nevertheless, the signals of pyroelectricity generating under realistic conditions are very weak; thus, the measurements are usually performed for a sufficiently large sample, such as a polled film or a piece of a crystal wafer. For molecular crystals, such a large specimen can hardly be obtained. Moreover, their single crystals tend to have highly anisotropic shapes, such as thin plates, and if the polarization is oriented in-plane direction of the thin crystal, only tiny electrodes can be attached. To deal with small molecular crystals, we developed a high sensitivity measuring system based on a temperature modulation technique [14] in the present study. In the AC measurements performed here, a periodic light was irradiated to modulate the temperature of the sample in a helium-gas chamber. The radiative temperature control, coupled with the use of a removable sample holder, provides stable and reproducible heat transfer, even for as-grown single crystals with irregularities in the crystal shape.

To evaluate the performance of this system, we carried out standard sample measurements for a typical ferroelectric material, lithium niobate ($LiNbO_3$). In the standard sample measurements, we employed a small piece of a thin wafer of $LiNbO_3$ in which the electric polarization resides along the crystal plane, assuming that the system would be applied to such tiny crystals with in-plane polarization. Since $LiNbO_3$ is transparent, we covered the specimen with synthetic black paint to transfer the light energy into heat at the sample's surface. A major difficulty with modulation schemes lies in the complexity of quantifying the experimental results. In the present study, we simulated the experimental data using a two-layer model that took into account the presence of an absorbing layer and achieved a reasonable agreement with the observed results.

2. Materials and Methods

Figure 1 shows the schematic drawing of the temperature-modulated pyroelectricity measurement system developed in the present study. The system comprises a diode laser, a heat-exchanging gas chamber of an optical cryostat (Oxford Instruments plc, OptistatCF), microscope components, and electric current detection instruments. The irradiation light was generated using a diode laser (Thorlabs, L520P090, $\lambda = 520$ nm). A crystal wafer of X-cut $LiNbO_3$ (10×10 mm^2, $t = 100$ µm) was purchased from MTI Co. As a light-absorbing layer, commercial black lacquer (Asahipen, Aspen Creative Color, Gross Black) was diluted with diethyl ether and coated on the surface of the crystal using a spin coater. The thickness of the light-absorbing layer was confirmed to be ca. 11 µm with a homemade surface step measurement apparatus using a high precision micrometer (Mitutoyo, MHF2-13). To mark the crystal orientation, another surface was lightly scratched with diamond sandpaper and smashed into small pieces ca. 0.5×0.5 mm^2 wide to adjust its size for the diameter of the laser beam. The crystal edges perpendicular to the Z-axis were coated with silver paste, and gold wires ($d = 30$ µm) were attached. The specimen was suspended on the IC socket with the two lead wires, and the socket was plugged into a socket receptor fixed on the sample rod, then loaded into the gas chamber in the optical cryostat. The sample was heated by the laser irradiation: sinusoidal waves modulated using a diode controller (Thorlabs, ITC4005, $f = 0.01 \sim 30$ Hz) or rectangular waves by the optical chopper ($f = 20 \sim 1000$ Hz). Since square waves contain $1.273 \sin(\omega t)$ in its Fourier series, the generated pyroelectricity is overestimated by the factor of the sine function compared to the signal induced with the sine waves of the same amplitude. The thermally induced current generated by the temperature modulation was led out via a triaxial cable from the cryostat and converted into voltage with a current preamplifier (Stanford Research Systems, SR570). This preamplifier has a built-in line filter with a selectable frequency range. Before amplifying the raw signal, various noises far away from the temperature modulation frequency were removed to preserve the dynamic reserve of the lock-in amplifier. The voltage-converted signal was detected using a two-phase digital lock-in amplifier (Signal Recovery, Model 7265).

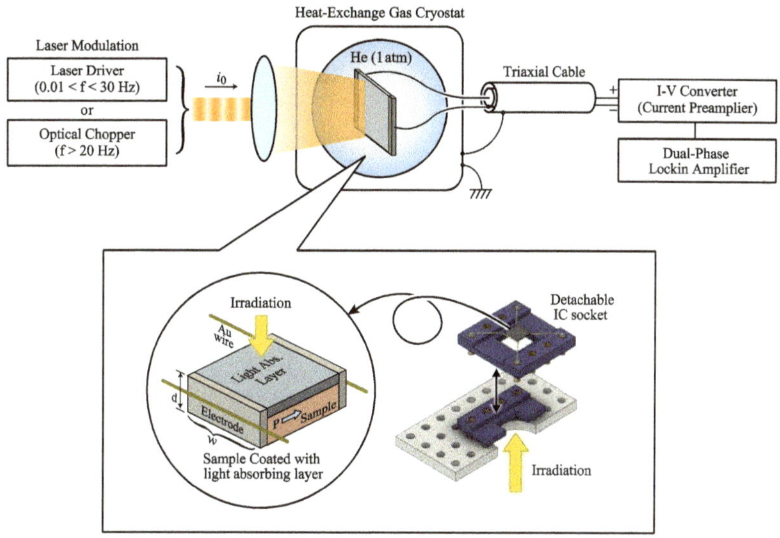

Figure 1. The setup of the temperature-modulated measurement of pyroelectricity. The standard sample with in-plane electric polarization is coated with a light-absorbing layer on the front surface. A pair of conductive-paste electrodes are attached to the crystal edges to collect the pyroelectric signal. The sample suspended upside down in the IC socket by the gold leads is irradiated through the hole in the sample holder and the base plate.

3. Results and Discussions

3.1. The AC Measurement of Pyroelectricity

Pyroelectricity is a variation in electric polarization P or that in the surface charge density of a polar material due to temperature variation and is extracted from electrodes attached on the surfaces of the sample. For a specimen of a capacitor form with an electrode area of S, the pyroelectricity I_{pyr} is defined as follows:

$$I_{\text{pyr}} = S\frac{dP}{dt} = S\frac{dP}{dT}\frac{dT}{dt} = \alpha S\frac{dT}{dt}, \tag{1}$$

where α denotes the pyroelectric coefficient. Since this formula includes the time derivative of temperature, pyroelectric measurement requires a temperature change, which is conventionally performed by sweeping the temperature of the sample at a constant rate. In the DC measurement, it is necessary to increase the speed to achieve an enhanced signal. However, the rate of temperature change is experimentally limited because the non-uniformity of the temperature distribution due to rapid temperature change causes mechanical stress, leading to sample damage.

As an alternative option to measure pyroelectricity, there is an AC technique in which the sample temperature is periodically modulated instead of the DC sweep [15]. Suppose that the temperature of the sample is modulated with an angular frequency of ω and an amplitude $T(\omega)$ around an equilibrium temperature T_0, i.e.,

$$T(t) = T(\omega)\sin(\omega t) + T_0. \tag{2}$$

From Equation (1), the pyroelectric current induced by the temperature modulation is given as a cosine wave

$$I_{\text{pyr}}(t) = \alpha S\omega T(\omega)\cos(\omega t), \tag{3}$$

and the Fourier coefficient $\alpha S\omega T(\omega)$ is detected by a lock-in amplifier.

The phase-sensitive detection allows the signal synchronized with the temperature modulation to be separated from various background noises. In addition, whereas the sweep rate of temperature determines the acquisition time in the DC method, the modulation method allows us to freely control the acquisition time at each data point, allowing fine temperature control. Furthermore, the AC method is advantageous in distinguishing true pyroelectricity from non-essential signals such as the trap-charge current; whereas the extrinsic signals are emitted from trap sites only at the early heating cycles of the modulation, intrinsic signal permanently generates along with temperature changes to dominate the averaged signal intensity [16].

3.2. *Radiative Temperature Modulation Using Periodic Light*

Molecular crystals have larger thermal expansion coefficients compared with inorganic metals. Therefore, when temperature modulation is performed by attaching a metal block to a sample and heating it, crystal distortions will occur. These distortions will be particularly severe for variable temperature measurements. To minimize such mechanical stress in the present measurement, the temperature of the sample was modulated by periodic light irradiation in a heat-exchange gas using an optical cryostat. The sample was suspended in the gas chamber with thin gold threads attached to the sample as the lead wires. The radiative heating of a freely suspended sample in the heat-exchange gas eliminates thermally induced strain that can occur in the case of heater heating, and, thus, the measurement system can be immediately applied to temperature variation measurements. In this experiment, the consistency of the thermal environment surrounding the sample is highly important. To facilitate sample exchange while maintaining the temperature environment, an IC socket was used as a removable sample holder that can be inserted and removed from the receptor installed in the gas chamber.

When measuring transparent materials or semiconductor samples that exhibit photoconductivity, it is necessary to cover the sample with a light-absorbing layer. The use of such an absorber has been widely studied to improve the performance of pyroelectric infrared sensors [17,18]. The most commonly used material would be gold black [17]. However, its deposition requires high-temperature heating of gold in gas and can damage the heat-sensitive molecular crystals. Therefore, in this study, we applied a synthetic resin paint to the transparent standard sample, as described in the Materials and Experiments section.

3.3. *Pyroelectricity Measurements on $LiNbO_3$*

To verify the performance of the radiative temperature modulation system, we conducted standard sample experiments by employing $LiNbO_3$. The widely used ferroelectric compound belongs to the trigonal crystal system, in which the spontaneous polarization points along the Z-axis. The measurements were performed for a thin plate crystal in which the polar axis points along the plane at room temperature. The modulated light was irradiated onto the absorber coating on the sample surface, and the pyroelectricity generated along the polarization axis was extracted from the electrodes attached to the edge of the crystal.

Figure 2 shows the frequency dependence of the thus observed pyroelectricity. According to Equation (3), the signal intensity measured using the AC method is expected to be proportional to the frequency of temperature modulation. However, the results obtained are very complex, and the frequency dependence can be divided into the following three regions: a low-frequency region where the data increases with frequency, an intermediate region where the data plateaus, and a high-frequency region where the data decreases.

Since pyroelectricity is a temperature variation of the electric polarization, the deviation from the linear frequency dependence indicates a discrepancy in the temperature of the sample expected by Equation (3). To correctly estimate the temperature of the sample, we performed the following thermal diffusion analysis.

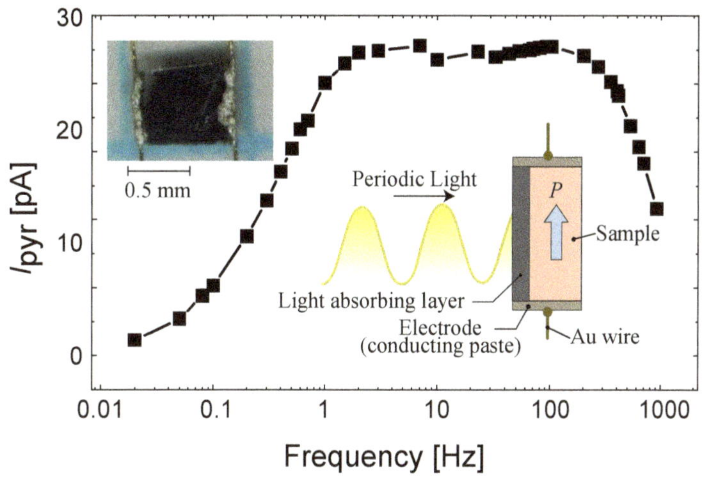

Figure 2. Frequency dependence of the pyroelectricity of $LiNbO_3$ measured using the temperature modulation method. The upper left inset is the photograph of the front surface of the sample coated with the light-absorbing layer. The inset in the center sketches the experimental layout.

3.4. Dynamical Temperature Distribution Calculated by One-Dimensional Layer Models

Suppose the surface of the sample is wide enough with respect to the thickness. In that case, the heat uniformly supplied by the irradiation light will propagate toward the back surface as a one-dimensional flow. The resultant temperature distribution in the sample should be expressed as a function of a depth from the front surface x, i.e., $T(x)$. Using the derivative of $T(x)$, heat flow $j(x)$ is written via Fourier's law as follows:

$$j(x) = -K\frac{\partial}{\partial x}T(x), \tag{4}$$

where K denotes the thermal conductivity of the sample. If the applied heat is time-modulated, the heat flux flowing into a unit volume is given by the following continuity equation:

$$c\rho\frac{\partial}{\partial t}T(x,t) = -\frac{\partial}{\partial x}j(x,t), \tag{5}$$

where c and ρ are the specific heat and density of the substance, respectively. Equations (4) and (5) give the following thermal diffusion equation:

$$\frac{\partial}{\partial t}T(x,t) = \kappa\frac{\partial^2}{\partial x^2}T(x,t), \tag{6}$$

where κ denotes the thermal diffusion constant ($\kappa = K/c\rho$). Pyroelectricity I_{pyr} generating in a thin layer with a thickness of δx at a depth of x is written as follows:

$$\delta I_{\mathrm{pyr}}(x,t) = \alpha\frac{\partial T(x,t)}{\partial t}L\delta x, \tag{7}$$

where L is the length of the electrodes. Although α is temperature dependent, it is assumed to be uniform within the sample because of the small temperature distribution in a thin sample.

In the present study, the sample is heated with the periodic light. Thereby, its temperature should be modulated in accordance with the oscillation of the light, i.e.,

$T(x,t) = \Delta T(x,\omega)\sin(\omega t) + T_0(x,\omega)$. The total pyroelectricity collected by the electrodes is thus given by the following:

$$I_{\text{pyr}}(\omega) = \alpha L \int_0^d \frac{\partial T(x,t)}{\partial t} dx = \alpha L \omega \int_0^d \Delta T(x,\omega) dx = \alpha \omega S \Delta T_{\text{m}}(\omega), \tag{8}$$

where $\Delta T_{\text{m}}(\omega)$ is the temperature modulation averaged over the crystal thickness, as follows:

$$\Delta T_{\text{m}}(\omega) = \frac{1}{d} \int_0^d T(x,\omega) dx. \tag{9}$$

Equation (8) denotes that if $\Delta T_{\text{m}}(\omega)$ has been known, α can be calculated from the experimentally observed pyroelectric current, $I_{\text{pyr}}(\omega)$. Since it is practically impossible to measure the dynamically varying temperature distribution, we estimated the distribution by the following thermodynamic analyses. The analytical expression of pyroelectricity has been systematically discussed by Holeman [14]. Following the formalism, we first treat a bare thin-plate crystal sample and then a system with a light-absorbing layer coated on the sample surface.

3.4.1. Single-Layer Model

Consider the pyroelectric current in a thin plate sample irradiated with periodic light in a heat exchange gas, as shown in Figure 3. The temperature distribution is spatially distributed according to the flow of heat from the surface to the back, thus $T(x,\omega,t)$ can be written as follows:

$$T(x,\omega,t) = \Delta T(x,\omega)\exp(i\omega t) + T_{\text{shift}}(x) + T_{\text{gas}}, \tag{10}$$

where the first term indicates the damping wave of temperature with the amplitude of ΔT. $T_{\text{shift}}(x)$ and T_{gas} represent the steady-state temperature shift and the temperature of the heat-exchange gas, respectively. Although the heat is only given from the front surface, the reflected wave from the backside is superposed to the forward wave to form a steady heat flow in equilibrium. Thus, ΔT is written as follows:

$$\Delta T(x,\omega) = C(\omega)e^{(i\frac{2\pi}{\lambda}-\gamma)x} + D(\omega)e^{-(i\frac{2\pi}{\lambda}-\gamma)x}, \tag{11}$$

where C and D are the complex coefficients representing the intensity and phase shift from the irradiation light for the forward and the reflected waves, respectively. The parameters λ and γ in the exponents denote the wavelength and the damping coefficient of the heatwave, respectively, and determine the envelope of the damped waveform. The coefficient of x in the exponential functions can be rewritten as follows:

$$k(\omega) = i\frac{2\pi}{\lambda} - \gamma = (1+i)\sqrt{\frac{\omega}{2\kappa}}. \tag{12}$$

This indicates the presence of the following relationship between λ and γ:

$$\frac{\lambda}{2\pi} = \gamma^{-1} = \sqrt{\frac{2\kappa}{\omega}}. \tag{13}$$

The coefficients C and D in Equation (11) are determined from the boundary conditions. In the steady state during measurement, the heat of irradiation is applied to the first boundary, i.e., the front surface of the sample. Some of it is lost by heat transfer to the gas, while the remaining heat passes through the sample and is finally transferred to the gas at the second boundary (the backside). Rigorously, the contribution of thermal radiation, which is approximately given as $4\sigma T^3$ [19] (σ is the Stephan Boltzmann coefficient), needs to be taken into account. However, its effect at room temperature is estimated as 6.1 W/m^{2}K, which is much smaller than the impact of the heat-exchange gas calculated in the following discussion. Thus, its effects are ignored in the following discussion. If we denote the heat flow at x as $j(x)$, the amplitude and angular frequency of the irradiation as i_0 and ω_0, the temperature of the heat exchange gas as T_{gas}, the heat-transfer coefficient between the sample and the gas as h, and the thickness of the crystal as d, the continuity of the heat flow on the front and back surfaces can be expressed as follows:

$$\begin{cases} i_0(1+\sin(\omega t)) - h(T(0,t) - T_{gas}) = -K\frac{\partial}{\partial x}T(0,t) \\ -K\frac{\partial}{\partial x}T(d,t) = h(T(d,t) - T_{gas}). \end{cases} \quad (14)$$

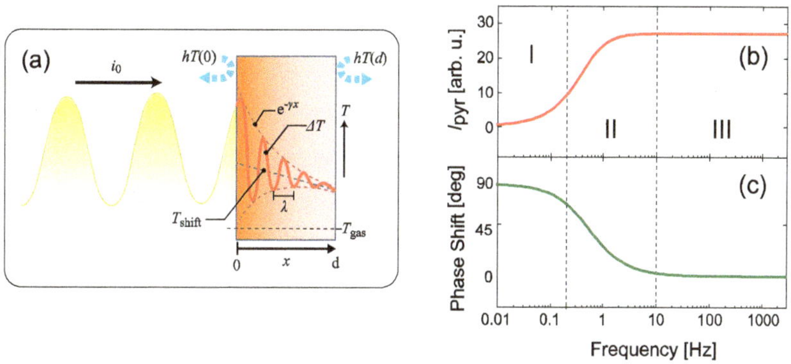

Figure 3. (**a**) Schematic drawing of temperature distribution in the single-layer model heated with a modulated light in heat exchange gas. The distribution is assumed to be a single damped wave with a time-independent temperature shift by omitting the presence of a reverse wave. (**b**,**c**) The magnitude of pyroelectricity and the phase shift from the oscillation of the irradiated light calculated using the single-layer model. The physical parameters of $LiNbO_3$ used in the calculation are listed in Table 1.

Table 1. Physical quantities for $LiNbO_3$ and heat transfer coefficient between He gas and the sample used in the simulations.

$LiNbO_3$						Heat Transfer
c (J/gK)	ρ (g/cm^3)	κ (mm^2/s)	α (μC/m^{2}K)	d (μm)	w (mm)	h (W/m^{2}K)
0.601 [1]	4.659 [1]	1.264 [1]	83 [2]	100	0.57	500 [3]

[1] [20]; [2] [21]; [3] present work.

By substituting the general solution Equation (11) into these boundary conditions, we obtain the coefficients C and D and T_{shift}.

$$T_{shift}(x) = -\frac{i_0}{2c\kappa\rho + dh}x + \frac{c\kappa\rho + dh}{2c\kappa\rho + dh}\frac{i_0}{h} \quad (15)$$

Since the time-independent term T_{shift} does not affect the pyroelectricity, the effect is ignored in most of the literature. However, when experiments are performed close to a transition point, this term significantly affects the physical properties. In the present experiments, for example, T_{shift} is estimated as 2.0 K at the surface $(x = 0)$ by substituting the experimental and calculated values for i_0 and h in Equation (15), respectively. If a higher temperature resolution than this value is required, the experimental conditions need to be adjusted.

The pyroelectricity is obtained from the time-dependent term in Equation (10). By substituting C and D into Equation (11), the solution of Equation (9) is given as follows:

$$\Delta T_{\mathrm{m}} = \frac{i_0(Kk\sinh(kd) + h(\cosh(kd) - 1))}{dk((K^2k^2 + h^2)\sinh(kd) + 2Khk(\cosh(kd) - 1) + 2Khk)}. \tag{16}$$

The pyroelectricity can be obtained by substituting ΔT_{m} into Equation (8). Figure 3b,c show the frequency plot of the intensity of the pyroelectricity $I_{\mathrm{pyr}}(\omega)$ and the phase difference from the irradiated light wave, respectively. Since ΔT_{m} varies with the modulation frequency $f = \omega/2\pi$ via the argument of the hyperbolic functions, i.e., $k(\omega)d$, the characteristics of ΔT_m depend largely on the magnitude of the argument. Assuming that the threshold frequency f_{th} is as follows:

$$f_{\mathrm{th}} = \frac{1}{\pi}\frac{\kappa}{d^2}, \tag{17}$$

the value of $I_{\mathrm{pyr}}(\omega)$ is approximated as follows [11,14]:

$$(f < f_{\mathrm{th}})\ I_{\mathrm{pyr}}(\omega) \sim \alpha S\left(\frac{i_0}{2h}\right)\omega, \tag{18}$$

$$(f > f_{\mathrm{th}})\ I_{\mathrm{pyr}}(\omega) \sim \frac{\alpha S}{c\rho d} i_0. \tag{19}$$

We divided the frequency domain of the experimental data into three regions according to the conditions given by Equations (17) and (18) and tried four different fitting methods to examine the agreement of the experimental results with the physical parameters of the standard material. In the following, the physical quantities other than the fitting parameters for each process were assumed to be the experimentally determined or reported values shown in Table 1.

1. Region I:

Equation (18) implies that the signal strength increases in proportion to the frequency in the low-frequency region. Figure 4a shows the signal intensity in the low-frequency region. The two parameters S and i_0 are given by the experimental conditions. As we will see later, α is determined by the analysis of Region III; thus, we can determine the value of h from the gradient for the low-frequency data. As is shown in Figure 4a, we found that the calculated curve for $h = 500\ \mathrm{W/m^2K}$ gives the best result among the several trials. A heat transfer coefficient is a parameter governed by the convection of the heat exchange gas. Therefore, this value obtained here can be applied as a device-specific parameter if samples with similar shape and surface conditions are used.

2. Region II:

Next, we examine the region between Region I and III, where the modulation frequency f approaches f_{th}. In this region, the hyperbolic functions cannot be simplified by the limit approximations as performed in Equations (18) and (19). Since f_{th} is defined by Equation (17), the frequency dependence of the signal in this region is suggested to be deeply related to the values of κ and d. If d is experimentally determined, fitting with κ as a parameter is possible, as shown in Figure 4b. On the other hand, if κ has been obtained by other studies, it is possible to fit the experimental value with d as a parameter, as shown in Figure 4c. In the case of the present study, κ of the standard sample is known, and the curve calculated by adopting the literature value agrees well with the experiment data, as shown in Figure 4b. Similarly, the simulation results for a sample thickness of 100 μm also agree well with the experimental values, as shown in Figure 4c.

3. Region III:

Finally, we discuss the frequency region where $f > f_{\mathrm{th}}$. Equation (19) clearly shows no frequency dependence of the signal strength in this frequency region. Therefore, the signal indicates a simple plateau, and its value is proportional to α; hence α can be estimated from the plateau level if the specific heat and density of the sample are known. The value of α in $LiNbO_3$ has been investigated in several studies [22–27], and the values at room temperature vary widely from 18 [23] to 95 [27] μC/m^{2}K. Figure 4d shows the comparison of the calculation with the experimental data and indicates that the curve calculated for α = 80 μC/m^{2}K shows good agreement with the experiment. Thus, we adopted the literature value 83 μC/m^{2}K reported in [27] for other calculations.

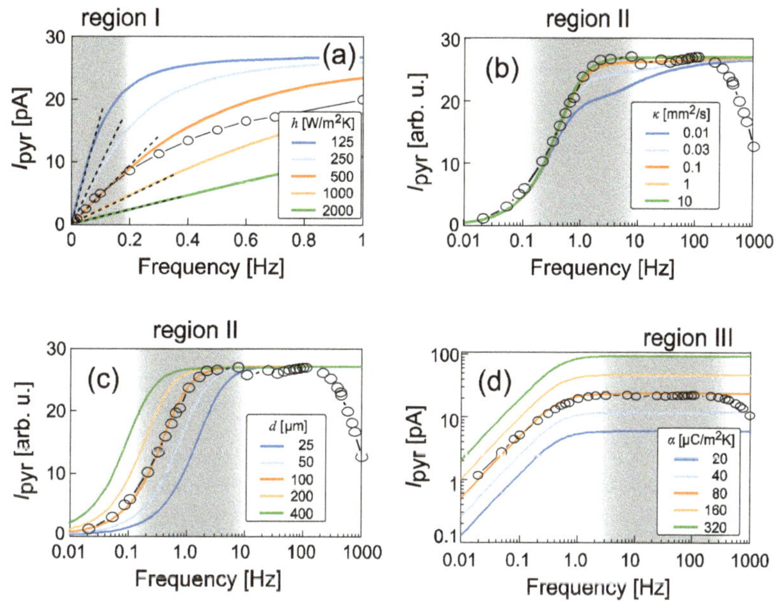

Figure 4. Comparisons between the observed pyroelectricity of the $LiNbO_3$ sample and the results of the single-layer model calculation plotted as a function of the temperature-modulation frequency. Four panels (**a–d**) show the dependence on h, κ, d, and α, respectively. Light intensity irradiated to the sample i_0 was 1.6 mW/mm^2, estimated from the difference of light entering and exiting the cryostat. Physical quantities for $LiNbO_3$ other than the parameter selected as the variable in each plot are listed in Table 1.

3.4.2. Two-Layer Model

From the above discussion, it seems that the single-layer model can reproduce the observed pyroelectricity. However, as shown in Figure 4d, the decay profile appearing above 300 Hz in the experiment is not reproduced in the calculation, which raises questions about the validity of the simulation. A candidate for this problem is the presence of a light-absorbing layer, which has not been considered in previous models. Therefore, we consider a two-layer model that includes the presence of the light-absorbing layer, as shown in Figure 5a. Such multilayer models have been investigated in detail by Peterson et al. for vertically poled pyroelectric polymer films [19]. The application has also been reported for a system in which the sample is bonded to a thermal bath [11].

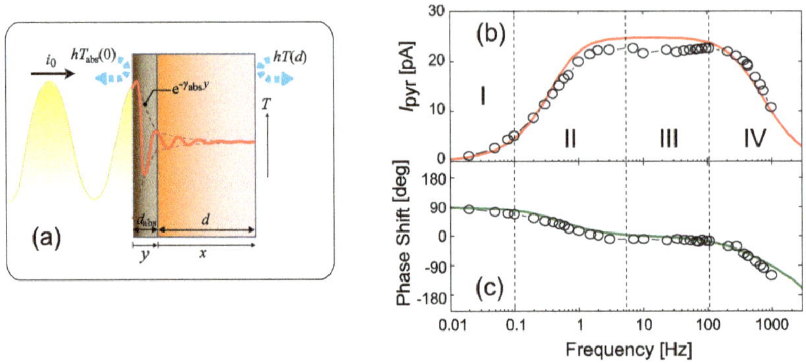

Figure 5. (**a**) Schematic diagram of the temperature distribution of a two-layer model heated by modulated light in a heat exchange gas. The distribution is represented as a two-step decaying wave, in which the presence of the inverse wave and time-independent terms are omitted. (**b**,**c**) Comparison between the experimental data and the simulated pyroelectricity and the phase shift calculated using the two-layer model.

The two-layer model consists of an infinite absorbent layer of thickness d_{abs} on the front surface of the previous one-layer model. Fourier's law of Equation (4) and the heat conduction equation of Equation (6) has yielded the complex wavenumber k that defines the heat flow in the pyroelectric material, as shown in Equation (12). By replacing the physical parameters in these formulas with the values of the light-absorbing layer, the complex wavenumber for the light-absorbing layer k_{abs} is obtained. By imposing the boundary condition of light irradiation on the surface of the light-absorbing layer and matching the temperature and heat flow at the interface of the two layers, the coupled differential equations for the two-layer model are obtained as follows:

$$\begin{cases} i_0(1+\sin(\omega t)) - h\left(T_{abs}(0,t) - T_{gas}\right) = -K_{abs}\frac{\partial}{\partial x}T_{abs}(0,t) \\ -K_{abs}\frac{\partial}{\partial x}T_{abs}(d_{abs},t) = -K\frac{\partial}{\partial x}T(0,t) \\ T_{abs}(d_{abs},t) = T(0,t) \\ -K\frac{\partial}{\partial x}T(d,t) = h\left(T(d,t) - T_{gas}\right) \end{cases} \tag{20}$$

In the above equations, the temperature distribution of the coating layer is denoted as $T_{abs}(y)$, where y indicates the depth from the surface of the layer, and the other physical parameters of the absorbing layer are represented as c_{abs}, ρ_{abs}, K_{abs}, and d_{abs}. The solutions of ΔT and ΔT_{abs} satisfying the boundary conditions are obtained using the same procedure described in the discussion of the single-layer model. By substituting the obtained ΔT into Equation (8), we obtained I_{pyr} of the two-layer model and performed the following calculations.

We should note that the frequency dependence of I_{pyr} of the one-layer model was governed by the magnitude of dk, the argument of the hyperbolic function. In addition to these hyperbolic functions, additional hyperbolic functions with $d_{abs}k_{abs}$ as an argument appear in I_{pyr} for the two-layer model. Thereby, we can infer that there should be a new threshold $f_{th,abs}$ that classifies the frequency dependence of I_{pyr} for the two-layer model, i.e.,

$$f_{th,abs} = \frac{1}{\pi}\frac{\kappa_{abs}}{{d_{abs}}^2}. \tag{21}$$

In fact, Figure 5b, the simulation result of I_{pyr} discussed below, shows a decaying region for $f > f_{th,abs}$ in addition to the three regions described in the one-layer model.

Since this threshold is a function of κ_{abs} and d_{abs}, the width of Region IV changes when either parameter is changed. As shown in Figure 6a,b, we calculated the κ_{abs} and d_{abs} dependence based on the two-layer model. These two plots show that the optical absorption layer not only caused Region IV, but also affected the signal intensity in Regions I–III. However, in the present case, the effects were limited because the layer was sufficiently thin; thereby, the parameters obtained from the one-layer model could be directly applied to the fitting of the two-layer model. Aside from these parameters adopted in the single-layer calculations, we assumed the following physical parameters for the absorbing layer: 1.4 J/gK, which is a typical value for polymers, for c_{abs}, and 0.76 g/cm^3 for ρ_{abs}, which is the disclosed density value of the commercial paint. If one of κ_{abs} and d_{abs} is known, the other can be obtained by fitting the experimental values. In this case, d_{abs} had been determined to be 11 μm using the surface step measurement, and we determined κ_{abs} to be 0.16 mm^2/s using the fitting in Figure 6a. This κ_{abs} value is consistent with typical data for the thermal diffusivity of polymer resins [28]. Although it may vary slightly depending on the surface condition of the sample and the coating method, it can be used as a reference value when the light-absorbing layer is prepared under similar conditions.

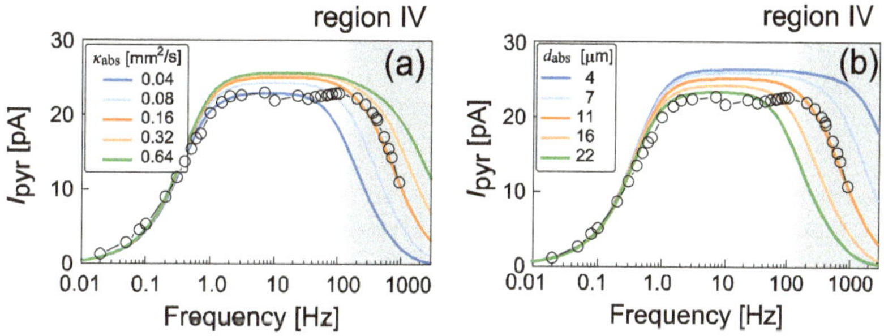

Figure 6. Comparisons between the observed pyroelectricity of $LiNbO_3$ and the results of the two-layer model calculation plotted as a function of the temperature-modulation frequency. Panels (**a**) and (**b**) show the dependence on κ_{abs} and d_{abs}, respectively. In the calculations, c_{abs} and ρ_{abs} are supposed as 1.4 J/gK and 0.76 g/cm^3, respectively. In (b), κ_{abs} is assumed as 0.16 mm^2/s.

As seen in Figure 4d, the pyroelectric coefficient α of the sample is estimated from the height of the plateau in Region III. However, this plateau structure will be squeezed out as Region IV widens and approaches Region II. Moreover, since the amplitude of the temperature wave decreases according to the decay function $\exp(-\gamma_{abs}d_{abs})$, as shown in Figure 5a, the pyroelectricity signal is also weakened by this decay. Therefore, to estimate α correctly, $f_{th,abs}$ needs to be sufficiently larger than f_{th} to protect the plateau structure; because κ_{abs} is a material-specific parameter that cannot be adjusted, the light-absorbing layer must be deposited as thinly as possible.

The parameters h, κ, and κ_{abs} determined using the above procedure, together with the literature values of $LiNbO_3$, were substituted into the equation of I_{pyr}. The resulting pyroelectricity and the phase shift are shown in Figure 5b,c, respectively. The calculation results show excellent agreement in both signal intensity and phase, confirming the reliability of the experimental data collected by the system and the validity of the two-layer model calculations.

Since the main source of uncertainty in quantitative analysis in the present modulation measurements depends on the estimation error of the sample temperature, the validity of the heat conduction model is of crucial importance. The successful reproduction of the experimental data strongly suggests the correctness of the one-dimensional thermal model. When we use naturally grown single crystals, the shape of the sample cannot be freely specified. If the size of the light-irradiated area of the sample is not sufficiently large compared to its thickness, heat diffusion in the transverse direction parallel to the light-irradiated plane cannot be neglected, and pure one-dimensionality is lost. In such cases, the heat transfer from the gold wire attached to the crystal edges cannot be neglected; thus, thinner gold wires or resistive wires should be used. When a one-dimensional model is applied to a system where lateral thermal diffusion cannot be neglected, the amplitude of the thermal wave is overestimated compared to its true value. Since thermal diffusion is caused by thermal gradients, this overestimation will be more pronounced in the high-frequency region where deviations from a uniform temperature distribution in the sample are more pronounced. If the simulation results show such behavior, the sample should be replaced with a more suitable geometry.

Finally, we would like to mention the sensitivity limit of this measurement system. If the purpose of this temperature modulation measurement is to find only the pyroelectric coefficient of a sample, there is no need to measure a wide range of modulation frequency dependence as we performed in the present study. Instead, one needs only to collect data for a single point at a frequency in the plateau region. To calculate the electric polarization from the pyroelectric coefficient, one should perform that experiment over a wide temperature range, including the transition point. In modulation measurements in the acoustic frequency domain such as the present study, the dominant noise is the 1/f noise generated by electronic components such as operational amplifiers. Therefore, experiments should be performed at as high a modulation frequency as possible in the plateau region to suppress the noise. However, in the case of using an optical absorption layer, as in this study, the plateau region is limited by Region IV, which imposes an upper limit on the frequency that can be selected. Even without the optical absorption layer, the amplifiers that can operate at high frequencies have a high input impedance; therefore, they cannot be used for semiconductor samples with low impedance. The researchers need to ensure these conditions for their sample and then conduct the experiment at the appropriate frequency. In this study, the modulation frequency was fixed at 103 Hz, which is close to the upper limit of Region III, and the dependence of the signal intensity on the light intensity was investigated. As a result, we confirmed that the signal could be measured, even down to 3% of the experimental condition. This indicates that the experimental method can be applied to samples with widths in the order of μm for samples with a pyroelectric coefficient comparable to that of $LiNbO_3$. As for the thickness, as discussed in Figure 4c, the signal strength does not depend on the thickness of the sample, and, therefore, one can use as thin a sample as is experimentally possible.

4. Conclusions

In this study, a temperature-modulated measurement system was developed to measure the pyroelectricity of a small sample with high sensitivity. The experimental results of the standard sample were analyzed using a one-dimensional heat transfer model. In the developed system, the temperature of the sample can be modulated by periodic light irradiation in the heat exchange gas. Since the measurement is performed almost in a non-contact manner, there is no mechanical stress on the sample during the experiment, which is helpful for fragile single crystals such as molecular crystals. If the temperature of the heat-exchange gas is controlled using a refrigerator or heater, this system can be directly applied to the variable temperature measurements. Unlike the DC method, which requires sweeping the temperature at a constant rate, this modulation method allows the rate of temperature change to be controlled freely. This makes it possible to investigate critical phenomena such as phase transitions with high-temperature resolution. False signals caused by the discharge of trapped charges, which is a problem in pyroelectric measurement, can be eliminated by periodic temperature changes. Therefore, the measurement is effective for basic research that requires an accurate understanding of physical properties. Although $LiNbO_3$ used as a standard sample is a typical ferroelectric material, there is a significant variation in the values of α reported thus far. Therefore, the results of this study are also important as information for examining the validity of these reported values.

Author Contributions: Project administration and writing, K.Y.; experiments and investigation, A.K.; development of measuring system, K.R. and T.C.; illustrations, T.C. All authors have read and agreed to the published version of the manuscript.

Funding: This research was supported by JSPS KAKENHI Grant Numbers JP 23550170 and JP18H01144, and by Wesco Scientific Promotion Foundation.

Acknowledgments: The authors thank Science Dream Laboratory of OUS for their cooperation.

Conflicts of Interest: The authors declare no conflict of interest.

References

1. Manfred, F.; Lottermoser, T.; Meier, D.; Trassin, M. The evolution of multiferroics. *Nat. Rev. Mater.* **2016**, *1*, 16046.
2. Kimura, T.; Goto, T.; Shintani, H.; Ishizaka, K.; Arima, T.; Tokura, Y. Magnetic control of ferroelectric polarization. *Nature* **2003**, *426*, 55–58. [CrossRef]
3. Sato, O. Dynamic molecular crystals with switchable physical properties. *Nat. Chem.* **2016**, *8*, 644–656. [CrossRef] [PubMed]
4. Ishihara, S. Novel States of Matter Induced by Frustration Electronic Ferroelectricity and Frustration. *J. Phys. Soc. Jpn.* **2010**, *79*, 11010. [CrossRef]
5. Kagawa, F.; Horiuchi, S.; Tokunaga, M.; Fujioka, J.; Tokura, Y. Ferroelectricity in a one-dimensional organic quantum magnet. *Nat. Phys.* **2010**, *6*, 169–172. [CrossRef]
6. Zhang, W.; Xiong, R.G. Ferroelectric Metal–Organic Frameworks. *Chem. Rev.* **2012**, *112*, 1163–1195.
7. Tayi, A.S.; Kaeser, A.; Matsumoto, M.; Aida, T.; Stupp, S.I. Supramolecular ferroelectrics. *Nat. Chem.* **2015**, *7*, 281–294. [CrossRef]
8. Yamamoto, K.; Iwai, S.; Boyko, S.; Kashiwazaki, A.; Hiramatsu, F.; Okabe, C.; Nishi, N.; Yakushi, K. Strong Optical Nonlinearity and its Ultrafast Response Associated with Electron Ferroelectricity in an Organic Conductor. *J. Phys. Soc. Jpn.* **2008**, *77*, 74709. [CrossRef]
9. Yamamoto, K.; Kowalska, A.A.; Nakano, C.; Yakushi, K. Inhomogeneous ferroelectric polarization in α'-(BEDT-TTF)$_2$IBr$_2$ revealed by second-harmonic generation microscopy. *Phys. B Cond. Mat.* **2010**, *405*, S363–S364. [CrossRef]
10. Hamilton, A.C.; Day, G.W.; Phelan, R.J. *An Electrically Calibrated Pyroelectric Radiometer System*; Technical Note no. 678; National Bureau of Standards: Boulder, CO, USA, 1976; p. 50.
11. Lubomirsky, I. Stafsudd, O. Practical guide for pyroelectric measurements. *Rev. Sci. Instrum.* **2012**, *83*, 51101. [CrossRef]
12. Jachalke, S.; Mehner, E.; Stöcker, H.; Hanzig, J.; Sonntag, M.; Weigel, T.; Leisegang, T.; Meyer, D.C. How to measure the pyroelectric coefficient? *Appl. Phys. Rev.* **2017**, *4*, 21303. [CrossRef]
13. Huang, H.; Scott, J.F. *Ferroelectric Materials for Energy Applications*; Wiley: Hoboken, NJ, USA, 2018.
14. Holeman, B.R. Sinusoidally modulated heat flow and the pyroelectric effect. *Infrared Phys.* **1972**, *12*, 125–135. [CrossRef]
15. Chynoweth, A.G. Dynamic Method for Measuring the Pyroelectric Effect with Special Reference to Barium Titanate. *J. Appl. Phys.* **1956**, *27*, 78–84. [CrossRef]
16. Garn, L.E.; Sharp, E.J. Use of low-frequency sinusoidal temperature waves to separate pyroelectric currents from nonpyroelectric currents. Part I. Theory. *J. Appl. Phys.* **1982**, *53*, 8974–8979. [CrossRef]
17. Blevin, W.R.; Brown, W.J. Black Coatings for Absolute Radiometers. *Metrologia* **1966**, *2*, 139–143. [CrossRef]
18. Lehman, J.; Sanders, A.; Hanssen, L.; Wilthan, B.; Zeng, J.; Jensen, C. Very Black Infrared Detector from Vertically Aligned Carbon Nanotubes and Electric-Field Poling of Lithium Tantalate. *Nano Lett.* **2010**, *10*, 3261–3266. [CrossRef] [PubMed]
19. Peterson, R.L.; Day, G.W.; Gruzensky, P.M.; Phelan, R.J., Jr. Analysis of response of pyroelectric optical detectors. *J. Appl. Phys.* **1974**, *45*, 3296. [CrossRef]
20. Yao, S.; Wang, J.; Liu, H.; Hu, X.; Zhang, H.; Cheng, X.; Ling, Z. Growth, optical and thermal properties of near-stoichiometric $LiNbO_3$ single crystal. *J. Alloys Compd.* **2008**, *455*, 501–505. [CrossRef]
21. Lang, S.B.; Das-Gupta, D.K. *Handbook of Advanced Electronic and Photonic Materials and Devices*; Nalwa, H.S., Ed.; Academic Press. San Diego, CA, USA, 2001; Volume 4, p. 1.
22. Savage, A. Pyroelectricity and Spontaneous Polarization in $LiNbO_3$. *J. Appl. Phys.* **1966**, *37*, 3071–3072. [CrossRef]
23. Glass, A.M.; Lines, M.E. Low-temperature behavior of spontaneous polarization in $LiNbO_3$ and $LiTaO_3$. *Phys. Rev. B* **1976**, *13*, 180–191. [CrossRef]
24. Barbosa, P.C.; de Paiva, J.A.C.; Filho, J.M.; Hernandes, A.C.; Andreeta, J.P.; Sombra, A.S.B. Dielectric Relaxation Process and Pyroelectric Currents in $LiNbO_3$: Fe Single Crystals. *Phys. Stat. Sol.* **1991**, *125*, 723–729. [CrossRef]
25. Bartholomäus, T.; Buse, K.; Deuper, C.; Krätzig, E. Pyroelectric coefficients of $LiNbO_3$ crystals of different compositions. *Phys. Stat. Sol.* **1994**, *142*, K55–K57. [CrossRef]
26. Newnham, R.E. *Properties of Materials: Anisotropy, Symmetry, Structure*; Oxford University Press: New York, NY, USA, 2005.
27. Popescu, S.T.; Petris, A.; Vlad, V.I. Interferometric measurement of the pyroelectric coefficient in lithium niobate. *J. Appl. Phys.* **2013**, *113*, 43101. [CrossRef]
28. Blumm, J.; Lindemann, A. Characterization of the thermophysical properties of molten polymers and liquids using the flash technique. *High Temp. High Press* **2003**, *35*, 627. [CrossRef]

crystals

MDPI

Article

Dihedral-Angle Dependence of Intermolecular Transfer Integrals in BEDT-BDT-Based Radical-Cation Salts with θ-Type Molecular Arrangements

Tomofumi Kadoya [1,*], Shiori Sugiura [2], Toshiki Higashino [3], Keishiro Tahara [1], Kazuya Kubo [1], Takahiko Sasaki [2], Kazuo Takimiya [4,5] and Jun-ichi Yamada [1]

1 Department of Material Science, Graduate School of Science, University of Hyogo, 3-2-1 Kouto, Kamigori-cho, Ako-gun, Hyogo 678-1297, Japan; taharak@sci.u-hyogo.ac.jp (K.T.); kubo@sci.u-hyogo.ac.jp (K.K.); yamada@sci.u-hyogo.ac.jp (J.-i.Y.)
2 Institute for Materials Research, Tohoku University, 2-1-1 Katahira, Aoba-ku, Sendai, Miyagi 980-8577, Japan; s.sugiura@imr.tohoku.ac.jp (S.S.); takahiko@imr.tohoku.ac.jp (T.S.)
3 National Institute of Advanced Industrial Science and Technology (AIST), Tsukuba, Ibaraki 305-8565, Japan; t-higashino@aist.go.jp
4 RIKEN CEMS, 2-1 Hirosawa, Wako, Saitama 351-0198, Japan; takimiya@riken.jp
5 Graduate School of Science, Tohoku University, 6-3 Aramaki Aza-Aoba, Aoba-ku, Sendai, Miyagi 980-8578, Japan
* Correspondence: kadoya.t@sci.u-hyogo.ac.jp; Tel.: +81-(791)-58-0164

Abstract: We report the structural and physical properties of a new organic Mott insulator (BEDT-BDT)AsF_6 (BEDT-BDT: benzo[1,2-g:4,5-g′]bis(thieno[2,3-b][1,4dithiin). This AsF_6 salt has the same structure as the PF_6 salt. Although the anions are disordered, the donor molecules form a θ-type arrangement. The temperature dependence of the resistivity exhibits semiconducting behavior. The static magnetic susceptibility follows Curie–Weiss law over a wide temperature range; however, below 25 K, the magnetic susceptibility is in agreement with a one-dimensional chain model with the exchange coupling J = 7.4 K. No structural phase transition was observed down to 93 K. At 270 K, the Fermi surface calculated by the tight-binding approximation is a two-dimensional cylinder; however, it is significantly distorted at 93 K. This is because the dihedral angles between the BEDT-BDT molecules become larger due to lattice shrinkage at low temperatures, which results in a smaller transfer integral (t_1) along the stack direction. This slight change in the dihedral angle gives rise to a significant change in the electronic structure of the AsF_6 salt. Radical-cation salts using BEDT-BDT, in which the highest occupied molecular orbital does not have a dominant sign throughout the molecule, are sensitive to slight differences in the overlap between the molecules, and their electronic structures are more variable than those of conventional θ-type conductors.

Keywords: organic conductors; organic semiconductors; molecular orbitals

Citation: Kadoya, T.; Sugiura, S.; Higashino, T.; Tahara, K.; Kubo, K.; Sasaki, T.; Takimiya, K.; Yamada, J.-i. Dihedral-Angle Dependence of Intermolecular Transfer Integrals in BEDT-BDT-Based Radical-Cation Salts with θ-Type Molecular Arrangements. *Crystals* **2021**, *11*, 868. https://doi.org/10.3390/cryst11080868

Academic Editor: Toshio Naito

Received: 30 June 2021
Accepted: 20 July 2021
Published: 27 July 2021

1. Introduction

Molecular conductors have a rich phase diagram, including metallic, charge-density-wave, charge-ordering, spin-Peierls, spin-density-wave, and superconducting states [1]. In these organic conductors, the transfer integrals have a significant influence on the electronic characteristics. This occurs because the shapes of the molecular orbitals and the way they overlap determine the energy bandwidth and dimensionality of the electronic structure [2,3]. Generally, molecular conductors that have a one-dimensional (1D) electronic structure are prone to Peierls transition, which distorts their periodic structure [4]. Therefore, it is necessary to form a two-dimensional (2D) electronic structure to obtain a particular electronic state, such as superconductivity or a spin-liquid phase. The most successful example of this molecular strategy is bis(ethylenedithio)tetrathiafulvalene (BEDT-TTF), which is shown in Scheme 1a [5,6]. Due to the sulfur atoms in the ethylenedithio group, the molecular orbitals are expected to overlap not only in the stack direction but also between the stacks. The strain at the ethylene site at the end of the molecule prevents monotonic stacking. Thus, radical-cation (RC) salts using BEDT-TTF form >200 types of molecular conductors and realize a wide variety of electronic states [2]. Based on the success of BEDT-TTF, donor molecules with tetrathiafulvalene (TTF) and tetrathiapentalene (TTP) skeletons are frequently used in organic conductors, especially in RC salts [6]. The highest occupied molecular orbitals (HOMOs) in these molecules are largely localized on the sulfur atoms in the 1,3-dithiol-2-ylidene (DY) unit. Therefore, TTF and TTP derivatives have very similar HOMO shapes [5,7].

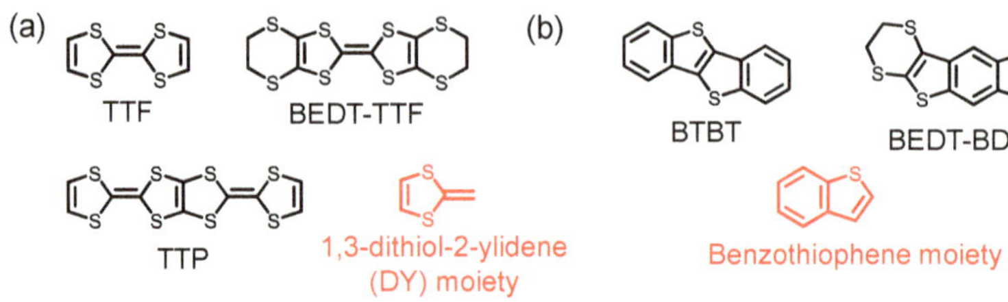

Scheme 1. Structural formula of (**a**) DY- and (**b**) benzothiophene-based electron donors.

The molecular orbitals of organic materials are unique to each molecule; therefore, theoretically, there are nearly infinite options. However, in the RC salts studied to date, only a limited number of donor molecules—including the DY moiety—have been used frequently, and few research examples focus on the shape and symmetry of the molecular orbital. In other words, very few examples use new molecular skeletons, which are essential for the future development of organic conductors. To address this issue, we have been developing RC salts using non-TTF molecules. We have developed molecular conductors using benzothioenobenzothiophene (BTBT), which is the basic skeleton of an excellent p-type semiconductor, for example, $(BTBT)_2XF_6$ (X = P, As, Sb, and Ta) [8–11]. They exhibit a characteristically high conductivity and have a high power factor as thermoelectric elements. This result has introduced the possibility of thermoelectric devices that use molecular conductors. BTBT is formed from the benzothiophene skeleton (Scheme 1b), in which the HOMO is largely localized on the sulfur atoms. This property is common to DY units such as TTF. Therefore, we have focused on the benzothiophene skeleton to develop new RC salts. Recently, we developed a 2D Mott insulator θ-(BEDT-BDT)PF_6, where BEDT-BDT is benzo[1,2-g:4,5-g′]bis(thieno[2,3-b][1,4]dithiin), and reported its structural and physical properties [12]. Because this material remains paramagnetic even at 2 K, the ground state may be a quantum spin liquid. BEDT-BDT is a molecule in which an ethylenedithio group is introduced at the molecular terminal of benzodithiophene, and it may form many 2D RC salts, same as BEDT-TTF. In this paper, we report the structural and physical properties of the new RC salt (BEDT-BDT)AsF_6. In particular, we describe the correlation between the dihedral angle and transfer integral in the θ-mode arrangement, and discuss the characteristics of the molecular orbitals that are peculiar to BEDT-BDT, which are not found in BEDT-TTF.

2. Experimental Methods

2.1. Electrocrystallization

BEDT-BDT was synthesized according to a previous study [13]. Briefly, 4–5 mg of BEDT-BDT and 20−25 mg of $(n\text{-}Bu)_4NAsF_6$ were dissolved in distilled chlorobenzene (15 mL). Initially, a current of 0.1 μA was applied. The current was increased to 0.5 μA over a month. Very thin, reddish-blown, plate-like crystals were obtained.

2.2. X-ray Diffraction and Structural Analysis

X-ray diffraction data of the crystals were collected at 270 and 93 K using a Rigaku Mercury 70 diffractometer and multilayer mirror monochromated Mo-Kα radiation (λ = 0.71075 Å). The crystal structures were solved using a direct method (SIR2014) [14] and refined by full-matrix least squares on F^2 (SHELXL-2014/3) [15].

2.3. Calculation of Transfer Integrals and Fermi Surfaces

The intermolecular transfer integrals were estimated using the Amsterdam Density Functional (ADF) program at the PW91/TZP level [16]. Using these transfer integrals, the Fermi surfaces were calculated in the tight-binding approximation [17].

2.4. Resistivity Measurements

Transport measurements were performed on single crystals. Gold wires (10 μM in diameter) were attached to the conducting plane with carbon paste. The temperature dependence of the in-plane resistivity was measured using the conventional four-probe DC technique using a Keithley 6221/2182A Delta Mode System.

2.5. Magnetic-Susceptibility Measurements

Magnetic susceptibility was measured using a commercial superconducting quantum interface device magnetometer (Quantum Design, MPMS-XL5HG) in the temperature and magnetic-field range of 2–300 K and $-5 \leq H \leq 5$ T, respectively. The measurements were performed with polycrystalline samples (~3.32 mg). The background was subtracted from the data.

3. Results and Discussion

Figure 1 shows the crystal structure of the (BEDT-BDT)AsF_6. Half of the BEDT-BDT and AsF_6 anion are crystallographically independent. Because the donor: anion composition ratio was 1:1, the AsF_6 salt is considered to be a half-filled Mott insulator. The crystal system was monoclinic and the space group was C2/c. Since the Z value is 4, there are four molecules each of BEDT-BDT and AsF_6 in the unit cell. The AsF_6 salt has the same structure as the PF_6 salt reported in a previous study [12]; that is, the BEDT-BDT molecules form alternate layers and are arranged in θ-mode packing. However, the anion molecule AsF_6 is disordered. As the temperature factor of the fluorine atoms is large, we analyzed the crystallographic data obtained at 270 K, assuming the temperature factor of the fluorine atoms to be isotropic. The BEDT-BDT molecules form a tilted uniform stack along the *b*-axis. The interplanar distance is 3.955 Å, which is slightly larger than that in the PF_6 salt. There are many short contacts between the BEDT-BDT molecules. S···S short contacts (3.599 Å; blue dotted lines) and S···C short contacts (3.470 Å; red dotted lines) exist between the donor stacks (Figure 1a), whereas S···H short contacts (2.812 Å; green dotted lines) exist within the donor stack (Figure 1b). The intermolecular transfer integrals are $t_1 = -27$ meV and $t_2 = 22$ meV. The value of t_1 is slightly smaller than that of the PF_6 salt. This is consistent with the increased interplanar distance caused by the larger size of the anion. There is no structural phase transition down to 93 K. The anion molecules are disordered even at 93 K; however, the fluorine atoms could be analyzed as anisotropic (Table S1, see Supporting Information).

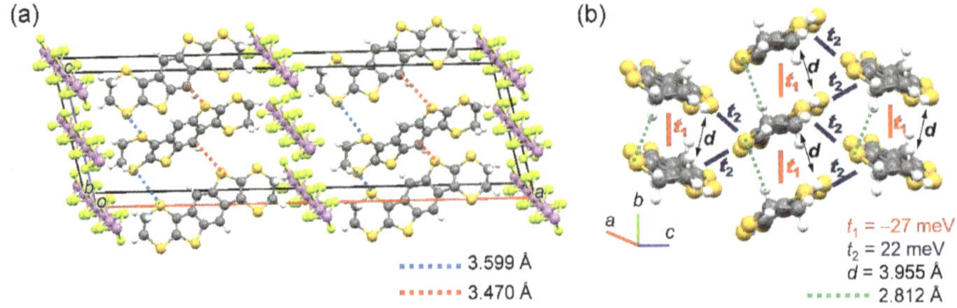

Figure 1. Crystal structures of (BEDT-BDT)AsF_6 at 270 K. (**a**) Molecular arrangements of the donor and anion layers. (**b**) Donor arrangement in θ-(BEDT-BDT)AsF_6 (the solid lines show the intermolecular transfer integrals; the AsF_6 anions are omitted for clarity).

Figure 2 shows the temperature dependence of the resistivity. The resistivity at 298 K is 3.4 Ω cm. Below 298 K, the resistivity gradually increases, and the sample exhibits semiconducting behavior (Figure 2a). The activation energy E_A is estimated to be 91 meV (Figure 2b). These values are not significantly different from those of the PF_6 salt (E_A = 87 meV). The difference in the temperature-dependent behavior of the resistivity is likely due to the difference in the measurement method (the PF_6 salt was measured using the two-terminal method). These results also support the conclusion that the AsF_6 salt is a Mott insulator with strong electron correlation.

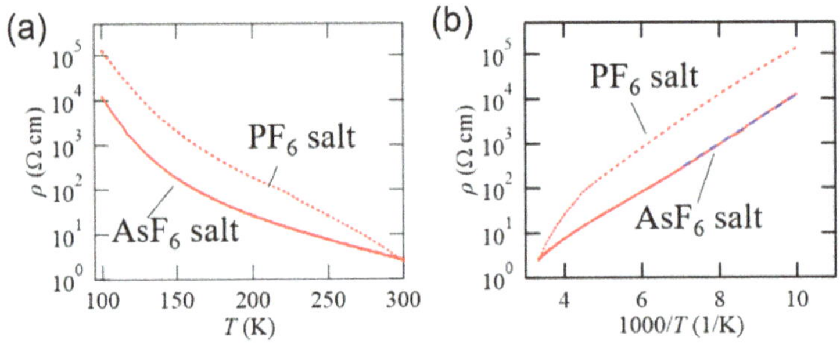

Figure 2. (**a**) Temperature dependence of the electrical resistivity ρ. (**b**) Arrhenius plot. The red dashed lines are the results for the PF_6 salt, which are taken from ref. [12]. The blue dashed line exhibits the fitting curve to estimate an activation energy.

The temperature dependence of the static magnetic susceptibility (χ) is shown in Figure 3a. The magnitude of χ is 3.96×10^{-4} emu/mol at 300 K, which is of the same order as that for typical molecular conductors. The temperature dependence of χ seems to follow the Curie–Weiss law $\chi = C/(T - \theta_W)$ for a wide temperature range above ~25 K. The $1/\chi$ plot provides a Curie constant C = 0.14 emu K/mol and a Weiss temperature $\theta_W = -5.97$ K, respectively (Figure 3b). The spin concentration calculated from C is 37.3% of the total spin localized at each BEDT-BDT site. Below 25 K, the spins change from a classical-property state to a quantum-property state. Therefore, the behavior of the magnetic susceptibility data follow the Heisenberg model in place of the Curie–Weiss law. It should be noted that the PF_6 salt from previous studies showed good agreement with the 2D triangular model [12], while the AsF_6 salt is consistent with the 1D chain model with the exchange coupling J = 7.4 K [18]. (Figure S3, see Supporting Information).

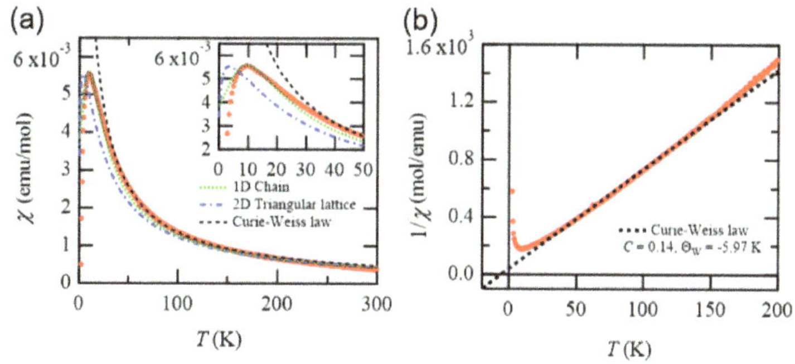

Figure 3. Temperature dependence of (**a**) magnetic susceptibility. The blue, green, and black dotted lines represent the fitting to the 2D triangular model, 1D chain model, and Curie–Weiss law, respectively. (**b**) $1/\chi$ plot.

The different behaviors observed at low temperature for the PF_6 and AsF_6 salts are discussed with the Fermi surface calculations in the next paragraph.

Figure 4 shows temperature dependence of the calculated Fermi surface. In this study, we performed a structural analysis of the PF_6 salt at 93 K and calculated the Fermi surface. The PF_6 salt has almost the same band structure at both 273 and 93 K, and the Fermi surface forms a type of 2D cylinder. At 270 K, the AsF_6 salt has a Fermi surface similar to that of the PF_6 salt. However, at 93 K, the band structure changes and the cylinder-type Fermi surface is significantly distorted. This is because the transfer integral t_1 in the stack direction becomes small. The temperature dependence of the Fermi surface supports the assumption that the 1D chain model is more appropriate that the 2D triangular model in the low-temperature region of magnetic susceptibility. However, this change in the Fermi surface of the AsF_6 salt is not due to a structural phase transition (Table S1, see Supporting Information).

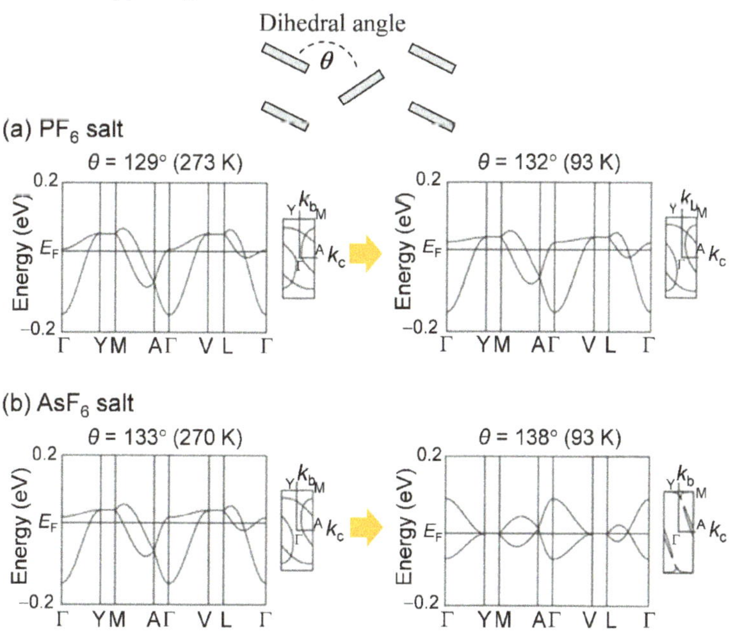

Figure 4. Temperature dependence of the band dispersions and Fermi surfaces of the (**a**) PF_6 and (**b**) AsF_6 salts.

Table 1 shows the HOMO energy, transfer integral, interplanar distance, and dihedral angle at each temperature. The value of the HOMO energy in one molecule of BEDT-BDT is ~−4.5 eV. This value does not depend greatly on either the difference between the counter anions or the temperature. Therefore, it is possible to compare the transfer integrals. The value of t_2 does not change between stacks; however, the value of $|t_1|$ within a stack is extremely small at 93 K for the AsF_6 salt. This change in $|t_1|$ depends more on the dihedral angle θ between the stacks than on the interplanar distance d between the donor molecules in the stack. Specifically, we confirm that $|t_1|$ tends to decrease as θ increases. Studies using molecular-orbital calculations have reported that the value of $|t_1|$ in the stack changes depending on the dihedral angle θ in a θ-type molecular arrangement [19]. Typical examples with θ-type arrangements are the BEDT-TTF salts [20,21]. In these BEDT-TTF salts, $|t_1|$ is smaller than $|t_2|$. This is because BEDT-TTF has multiple sulfur atoms aligned around the molecule. By contrast, the θ-type BEDT-BDT salts have $|t_1| > |t_2|$ as shown in Table 1.

Table 1. Temperature dependence of the HOMO energy, transfer integrals (t_1 and t_2), interplanar distance d, and dihedral angle θ of the BEDT-BDT salts.

	T (K)	HOMO (eV)	t_1 (meV)	t_2 (meV)	d (Å)	θ (°)
(BEDT-BDT)PF_6	273	−4.50	−33	22	3.947	128.6
-	93	−4.56	−27	23	3.901	131.5
(BEDT-BDT)AsF_6	270	−4.58	−27	22	3.955	133.1
-	93	−4.52	3.6	20	3.910	137.8

Figure 5 shows the dihedral-angle dependence of the transfer integrals of the BEDT-BDT and BEDT-TTF salts. As illustrated in Figure 5a, the overlap between molecules in the stack is generally expected to increase as θ increases; thus, t_1 increases. In the θ-type BEDT-TTF salts, θ changes depending on the size of the counter anion [20]. As θ increases, the absolute value of t_1, $|t_1|$, increases linearly (Figure 5b). By contrast, in the BEDT-BDT salts, $|t_1|$ decreases as θ increases. The change is not a monotonic linear change but exhibits a sharp curve with a local maximum at the point $\theta = 128°$ (the PF_6 salt at 273 K). The changes in t_2 are $94 < t_2 < 108$ meV for the BEDT-TTF salts and $20 < t_2 < 23$ meV for the BEDT-BDT salts. No significant changes are observed in either salt compared with t_1. The fact that t_1 in the stack of the BEDT-BDT salts is sensitive to θ is due to the shape of the molecular orbital.

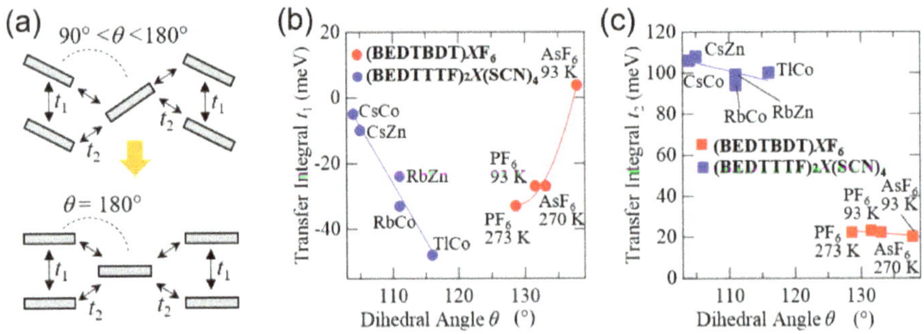

Figure 5. Dihedral-angle dependence of the (**a**) θ-type molecular arrangements and transfer integrals (**b**) t_1 and (**c**) t_2.

Figure 6 shows the phase (sign) symmetry of the HOMO for BEDT-TTF and BEDT-BDT. The HOMO phase of BEDT-TTF remains the same under the rotation operation. The shapes and phases of the HOMO are symmetrical with respect to both the long and short molecular axes. Therefore, the HOMO shape and phase have high symmetry. Regarding the HOMO, we focus on the sulfur atoms with d-orbitals that make substantial contributions to the intermolecular interactions. The HOMO of BEDT-TTF is localized at each sulfur atom, and the HOMO coefficient is particularly large at the sulfur atoms in the TTF skeleton [12]. Figure 6c demonstrates that the HOMO on the sulfur atoms at the TTF sites is larger than that on the carbon atoms. The HOMO above the sulfur atoms has the same phase, and the dominant HOMO phase in the molecular plane of BEDT-TTF is determined by the phase of the HOMO on the sulfur atoms in the TTF moiety. This property has been reported in previous studies [22,23]. This dominant HOMO phase is reversed with respect to the long plane of the molecule. For example, if the upper side of the molecular plane has a positive phase, the lower HOMO phase is negative. This is a characteristic of BEDT-TTF. However, the HOMO of BEDT-BDT has lower symmetry than that of BEDT-TTF because the only symmetry element is the rotation operation. The phase of the HOMO is inverted on the left and right sides of the central benzene ring. Because the positive and negative phases exist equally, there is no dominant phase throughout the molecular plane. That is, the sign of the HOMO is neutral on both the upper and lower sides of the molecular plane.

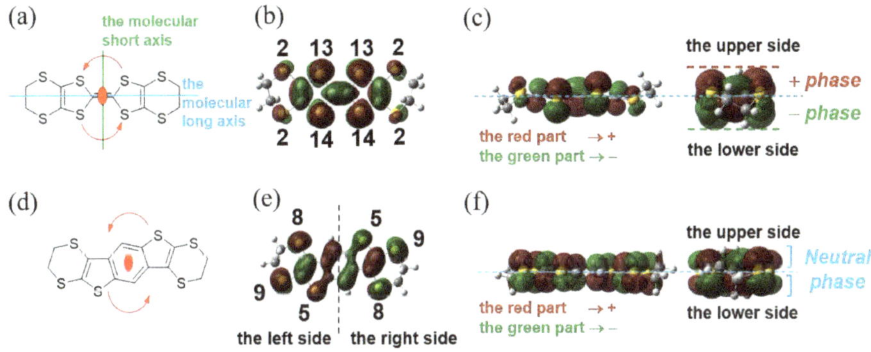

Figure 6. (**a**) Molecular structure and symmetry and (**b**,**c**) the HOMO distribution and shape of BEDT-TTF. The numbers represent the contributions (%) of the sulfur atoms to HOMO. (**d**) Molecular structure and symmetry and (**e**,**f**) the HOMO distribution and shape of BEDT-BDT.

The sign of t_1 in the BEDT-TTF salts switches periodically between positive and negative at intervals of $\theta - 60°$ due to the phase relationship of the HOMO [19]. By contrast, the sign of t_1 in the BEDT-BDT salts is much more sensitive to the dihedral angle θ than the sign of t_1 of the BEDT-TTF salts. The sign changes from negative to positive at intervals of $\theta = 10°$. That is, there is a state in which $t_1 = 0$ (the transfer integrals disappear) with only a slight change in the dihedral angle. This has a significant effect on the electronic structure of the BEDT-BDT-based RC salts. In fact, even a small variation such as thermal shrinkage due to a decrease in the temperature changes the dimensionality of the Fermi surface (Figure 4b). A slight change in the dihedral angle causes a significant change in the electronic structure, which is a characteristic of RC salts using BEDT-BDT because the HOMO does not have a dominant sign throughout the molecule. In other words, the BEDT-BDT salt is considered to be more pressure sensitive than conventional molecular conductors. This property is expected to lead to the development of new transistors whose conduction path changes due to the anisotropic pressure effects from the substrate [24–26].

4. Conclusions

In conclusion, we investigated the structural and physical properties of the new RC salt (BEDT-BDT)AsF_6. Moreover, we successfully performed the structural analysis of (BEDT-BDT)PF_6 at 93 K. The AsF_6 salt is a 2D Mott insulator with the same structure as the PF_6 salt. At temperatures below 300 K, the resistivity gradually increases and the salt exhibits semiconductor behavior with an activation energy of 91 meV. The static magnetic susceptibility essentially follows Curie–Weiss law but is consistent with a 1D chain model with J = 7.4 K below 25 K. When the temperature decreased from 270 to 93 K, the band dispersion and Fermi surface of the AsF_6 salt changed significantly. This is because the dihedral angle between the BEDT-BDT molecules increased due to thermal shrinkage. In a θ-type arrangement, the transfer integrals change depending on the dihedral angle, but the BEDT-BDT salts exhibit larger dihedral-angle dependence than the BEDT-TTF salts. This is because BEDT-BDT has only rotational symmetry and there is no dominant HOMO phase throughout the molecule. It is characteristic of the BEDT-BDT salts that the transfer integrals are sensitive to slight differences in the overlap between the molecules, and the dimensionality of the electronic structure changes more sensitively than that of conventional θ-type BEDT-TTF conductors.

Supplementary Materials: The following are available online at https://www.mdpi.com/article/10.3390/cryst11080868/s1, Figure S1: (a) Definition of label at each bond of BEDT-BDS molecule. (b) Relationship between bond length and molecular valence, Figure S2: (a) A plane definition of the BEDT-BDT and (b) a dihedral angle θ between the planes formed by two BEDT-BDT molecules on adjacent stacks, Figure S3: Exchange coupling J dependence of the calculation results for (a) 2D triangular lattice and (b) 1D chain model. Table S1: Crystallographic data of θ-(BEDT-BDT)AsF_6 and θ-(BEDT-BDT)PF_6, Table S2: List of bond lengths of BEDT-BDT and their charge-transfer salts.

Author Contributions: T.K. conceived and designed the experiments; T.K. developed all the molecular conductors; T.K., S.S., and T.S. participated in all the physical measurements; T.H. and K.T. (Keishiro Tahara) performed the DFT calculation. T.K., K.K., K.T. (Kazuo Takimiya), and J.-i.Y. wrote and/or reviewed the paper with contributions from all authors. All authors have read and agreed to the published version of the manuscript.

Funding: This research was partially supported by JSPS KAKENHI (Grant numbers: JP20K15356), the Kansai Research Foundation for Technology Promotion, Foundation of Kinoshita Memorial Enterprise, Masuyakinen basic research foundation, THE EBARA HATAKEYAMA MEMORIAL FOUNDATION (a public interest incorporated foundation), and Kato Foundation for Promotion of Science.

Data Availability Statement: Not applicable.

Conflicts of Interest: The authors declare no conflict of interest.

References

1. Ishiguro, T.; Yamaji, K.; Saito, G. *Organic Superconductors*, 2nd ed.; Springer: Berlin/Heidelberg, Germany, 1998.
2. Mori, T. *Electronic Properties of Organic Conductors*; Springer: Tokyo, Japan, 2016.
3. Mori, T. Principles that Govern Electronic Transport in Organic Conductors and Transistors. *Bull. Chem. Soc. Jpn.* **2016**, *89*, 973–986. [CrossRef]
4. Morawitz, H. Orientational Peierls Transition in Quasi One-Dimensional Organic Solids. *Phys. Rev. Lett.* **1975**, *34*, 1096–1099. [CrossRef]
5. Mori, H. Materials Viewpoint of Organic Superconductors. *J. Phys. Soc. Jpn.* **2006**, *75*, 051003. [CrossRef]
6. Saito, G.; Yoshida, Y. Development of Conductive Organic Molecular Assemblies: Organic Metals, Superconductors, and Exotic Functional Materials. *Bull. Chem. Soc. Jpn.* **2007**, *80*, 1–137. [CrossRef]
7. Shirahata, T.; Kohno, S.; Furuta, K.; Oka, Y.; Misaki, Y. Synthesis of New Electron Donor $ClMe_3$-TTP: Structures and Properties of $(ClMe_3\text{-TTP})_3X$ (X = PF_6 and AsF_6). *Bull. Chem. Soc. Jpn.* **2015**, *88*, 1086–1092. [CrossRef]
8. Kadoya, T.; Ashizawa, M.; Higashino, T.; Kawamoto, T.; Kumeta, S.; Matsumoto, H.; Mori, T. A Highly Conducting Organic Metal Derived from an Organic-Transistor Material: Benzothienobenzothiophene. *Phys. Chem. Chem. Phys.* **2013**, *15*, 17818–17822. [CrossRef] [PubMed]
9. Kiyota, Y.; Kadoya, T.; Yamamoto, K.; Iijima, K.; Higashino, T.; Kawamoto, T.; Takimiya, K.; Mori, T. Benzothienobenzothiophene-Based Molecular Conductors: High Conductivity, Large Thermoelectric Power Factor, and One-Dimensional Instability. *J. Am. Chem. Soc.* **2016**, *138*, 3920–3925. [CrossRef]
10. Higashino, T.; Kadoya, T.; Kumeta, S.; Kurata, K.; Kawamoto, T.; Mori, T. An Organic Metal Derived from a Selenium Analogue of Benzothienobenzothiophene. *Eur. J. Inorg. Chem.* **2014**, *24*, 3895–3898. [CrossRef]
11. Kadoya, T.; Oki, R.; Kiyota, Y.; Koyama, Y.; Higashino, T.; Kubo, K.; Mori, T.; Yamada, J. Transport Properties of Molecular Conductors $(BSBS)_2XF_6$ (X = As, Sb, and Ta): Investigation of Intermolecular Transfer Integrals in the Radical-Cationic State of BTBT-Type Semiconductors. *J. Phys. Chem. C* **2019**, *123*, 5216–5222. [CrossRef]
12. Kadoya, T.; Sugiura, S.; Tahara, K.; Higashino, T.; Kubo, K.; Sasaki, T.; Takimiya, K.; Yamada, J. Two-Dimensional Radical-Cationic Mott Insulator Based on an Electron Donor Containing Neither Tetrathiafulvalene nor Tetrathiapentalene Skeleton. *CrystEngComm* **2020**, *22*, 5949–5953. [CrossRef]

13. Wang, C.; Nakamura, H.; Sugino, H.; Takimiya, K. Thiacycle-fused benzo[1,2-*b*:4,5-*b*′]dithiophenes (BDTs): Synthesis, Packing, Molecular Orientation and Semiconducting Properties. *J. Mater. Chem. C* **2018**, *6*, 3604–3612. [CrossRef]
14. Burla, M.C.; Caliandro, R.; Carrozzini, B.; Cascarano, G.L.; Cuocci, C.; Giacovazzo, C.C.; Mallamo, M.; Mazzone, A.; Polidori, G. Crystal Structure Determination and Refinement via SIR2014. *J. Appl. Crystallogr.* **2015**, *48*, 306–309. [CrossRef]
15. Sheldrick, G.M. A short history of SHELX. *Acta. Cryst. A* **2008**, *64*, 112–122. [CrossRef] [PubMed]
16. *ADF: Powerful DFT Code for Modeling Molecules*; Scientific Computing and Modeling: Amsterdam, The Netherlands. Available online: http://www.scm.com/ADF/ (accessed on 30 June 2021).
17. Mori, T.; Kobayashi, A.; Sasaki, Y.; Kobayashi, H.; Saito, G.; Inokuchi, H. The Intermolecular Interaction of Tetrathiafulvalene and Bis(ethylenedithio)tetrathiafulvalene in Organic Metals. Calculation of Orbital Overlaps and Models of Energy-band Structures. *Bull. Chem. Soc. Jpn.* **1984**, *57*, 627–633. [CrossRef]
18. Ami, T.; Crawford, M.K.; Harlow, R.L.; Wang, Z.R.; Johnston, D.C.; Huang, Q.; Erwin, R.W. Magnetic Susceptibility and Low-Temperature Structure of the Linear Chain Cuprate Sr_2CuO_3. *Phys. Rev. B* **1995**, *51*, 5994. [CrossRef]
19. Kojima, H.; Mori, T. Dihedral Angle Dependence of Transfer Integrals in Organic Semiconductors with Herringbone Structures. *Bull. Chem. Soc. Jpn.* **2011**, *84*, 1049–1056. [CrossRef]
20. Mori, H.; Tanaka, S.; Mori, T.; Kobayashi, A.; Kobayashi, H. Crystal Structure and Physical Properties of M = Rb and Tl Salts (BEDT-TTF)2MM′(SCN)4 [M′ = Co, Zn]. *Bull. Chem. Soc. Jpn.* **1998**, *71*, 797–806. [CrossRef]
21. Mori, H.; Tanaka, S.; Mori, T. Systematic study of the electronic state in θ-type BEDT-TTF organic conductors by changing the electronic correlation. *Phys. Rev. B* **1998**, *57*, 12023. [CrossRef]
22. Mori, T. Structural Genealogy of BEDT-TTF-Based Organic Conductors I. Parallel Molecules: β and β″ Phases. *Bull. Chem. Soc. Jpn.* **1998**, *71*, 2509–2526. [CrossRef]
23. Mori, T.; Mori, H.; Tanaka, S. Structural Genealogy of BEDT-TTF-Based Organic Conductors II. Inclined Molecules: θ, α, and κ Phases. *Bull. Chem. Soc. Jpn.* **1999**, *72*, 179–197. [CrossRef]
24. Kawasugi, Y.; Yamamoto, H.M.; Hosoda, M.; Tajima, N.; Fukunaga, T.; Tsukagoshi, K.; Kato, R. Strain-Induced Superconductor/Insulator Transition and Field Effect in a Thin Single Crystal of Molecular Conductor. *Appl. Phys. Lett.* **2008**, *92*, 243508. [CrossRef]
25. Yamamoto, H.M.; Nakano, M.; Suda, M.; Iwasa, Y.; Kawasaki, M.; Kato, R. A Strained Organic Field-Effect Transistor with a Gate-Tunable Superconducting Channel. *Nat. Commun.* **2013**, *4*, 2379. [CrossRef] [PubMed]
26. Suda, M.; Kawasugi, Y.; Minari, T.; Tsukagoshi, K.; Kato, R.; Yamamoto, H.M. Strain-Tunable Superconducting Field-Effect Transistor with an Organic Strongly-Correlated Electron System. *Adv. Mater.* **2014**, *26*, 3490–3495. [CrossRef] [PubMed]

crystals

MDPI

Article

Electric Double Layer Doping of Charge-Ordered Insulators α-(BEDT-TTF)$_2$I$_3$ and α-(BETS)$_2$I$_3$

Yoshitaka Kawasugi [1,2,*], Hikaru Masuda [1], Jiang Pu [3], Taishi Takenobu [3], Hiroshi M. Yamamoto [2,4], Reizo Kato [1,2] and Naoya Tajima [1]

[1] Department of Physics, Toho University, Funabashi 274-8510, Chiba, Japan; 5416067m@st.toho-u.jp (H.M.); reizo@riken.jp (R.K.); naoya.tajima@sci.toho-u.ac.jp (N.T.)
[2] Condensed Molecular Materials Laboratory, RIKEN, Wako 351-0198, Saitama, Japan; yhiroshi@ims.ac.jp
[3] Department of Applied Physics, Nagoya University, Nagoya 464-8603, Aichi, Japan; jiang.pu@nagoya-u.jp (J.P.); takenobu@nagoya-u.jp (T.T.)
[4] Institute for Molecular Science, National Institutes of Natural Sciences, Okazaki 444-8585, Aichi, Japan
* Correspondence: yoshitaka.kawasugi@sci.toho-u.ac.jp

Abstract: Field-effect transistors based on strongly correlated insulators are an excellent platform for studying the electronic phase transition and simultaneously developing phase transition transistors. Molecular conductors are suitable for phase transition transistors owing to the high tunability of the electronic states. Molecular Mott transistors show field-induced phase transitions including superconducting transitions. However, their application to charge-ordered insulators is limited. In this study, we fabricated electric double layer transistors based on quarter-filled charge-ordered insulators α-(BEDT-TTF)$_2$I$_3$ and α-(BETS)$_2$I$_3$. We observed ambipolar field effects in both compounds where both electron and hole doping (up to the order of 10^{13} cm^{-2}) reduces the resistance by the band filling shift from the commensurate value. The maximum field-effect mobilities are approximately 10 and 55 cm^2/Vs, and the gate-induced conductivities are 0.96 and 3.6 e^2/h in α-(BEDT-TTF)$_2$I$_3$ and α-(BETS)$_2$I$_3$, respectively. However, gate-induced metallic conduction does not emerge. The gate voltage dependence of the activation energy in α-(BEDT-TTF)$_2$I$_3$ and the Hall resistance in α-(BETS)$_2$I$_3$ imply that the electric double layer doping in the present experimental setup induces hopping transport rather than band-like two-dimensional transport.

Keywords: organic conductor; charge-ordered insulator; electric double layer transistor; organic field-effect transistor

Citation: Kawasugi, Y.; Masuda, H.; Pu, J.; Takenobu, T.; Yamamoto, H.M.; Kato, R.; Tajima, N. Electric Double Layer Doping of Charge-Ordered Insulators α-(BEDT-TTF)$_2$I$_3$ and α-(BETS)$_2$I$_3$. *Crystals* **2021**, *11*, 791. https://doi.org/10.3390/cryst11070791

Academic Editor: Toshio Naito

Received: 17 June 2021
Accepted: 4 July 2021
Published: 7 July 2021

1. Introduction

Strongly correlated insulators at commensurate band fillings, such as the Mott insulator and the charge-ordered insulator, exhibit metal-insulator transitions by shifting the band filling [1]. Intriguing phenomena, such as high-temperature superconductivity and colossal magnetoresistance, emerge in the vicinity of the transition. Band filling control is generally accomplished by chemical substitution. Although this technique can change the band filling over a wide range, it cannot avoid the disorder caused by the introduction of dopants (impurities) and requires different samples for each band filling. Recently, electrostatic doping, which is based on the principle of field-effect transistors (FETs) and can avoid serious impurity effects that may occur in chemical doping, has found widespread use in the study of physical properties [2]. In particular, electric double layer transistors (EDLTs), which use ionic liquids as the gate electrolyte, have been widely adopted in the past decade because they allow a wider range of band filling control than FETs [3,4].

Molecular conductors are a suitable platform for the electrostatic doping of strongly correlated insulators. They contain various strongly correlated insulators and generally form single crystals with clean surfaces. Their lattice constants are generally larger than those of inorganic compounds, so that low electric fields can change large band fillings.

FETs and EDLTs based on molecular Mott insulators have been developed, and field-induced phase transitions have been investigated [5,6]. Studies of the doping effect on charge-ordered insulators have been limited. Yamamoto et al. [7,8] and Kimata et al. [9,10] fabricated FET devices using the charge-ordered insulator α-$(BEDT\text{-}TTF)_2I_3$ and observed decreases in the two-probe resistance between the source and drain electrodes by several tens of percent. However, the charge-ordered state was robust to field-effect doping, and the activation energy was essentially independent of the gate voltage. In this study, we fabricated EDLTs based on α-$(BEDT\text{-}TTF)_2I_3$ and α-$(BETS)_2I_3$ to investigate dense doping effects on the charge-ordered insulators compared to those in the FET measurements.

2. Materials and Methods

α-$(BEDT\text{-}TTF)_2I_3$ is a quasi-two-dimensional molecular conductor in which the conducting BEDT-TTF layer and the insulating I_3 layer are stacked alternately. Although α-$(BEDT\text{-}TTF)_2I_3$ is a semimetal according to band calculations, and the resistance decreases by cooling, it shows a metal-insulator transition at 135 K [11]. The insulating state is a charge-ordered state in which horizontal charge stripes are formed along the crystallographic b axis. The transition temperature is lowered when pressure is applied, and disappears at 1.5 GPa; at this pressure, the Dirac fermion phase emerges [12]. α-$(BETS)_2I_3$ is the selenium analog of α-$(BEDT\text{-}TTF)_2I_3$. It also shows metal–insulator transition but at a lower transition temperature (~ 50 K) [13]. Therefore, it is considered that the electronic state of α-$(BETS)_2I_3$ is similar to that of moderately pressurized α-$(BEDT\text{-}TTF)_2I_3$. However, recent X-ray diffraction and ^{13}C nuclear magnetic resonance experiments revealed that α-$(BETS)_2I_3$ maintains inversion symmetry below the transition temperature [14]. These results imply a different insulating mechanism from simple charge ordering (the spin-orbit interaction may play an important role). The mechanism is still under debate [14–16].

We fabricated EDLTs based on α-$(BEDT\text{-}TTF)_2I_3$ and α-$(BETS)_2I_3$ by laminating thin single crystals onto polyethylene terephthalate (PET) substrates where Au electrodes were pre-evaporated (Figure 1). The source, drain, and gate electrodes (18 nm thick Au) were patterned on the substrate using photolithography. We electrochemically synthesized thin (~100 nm) single crystals of α-$(BEDT\text{-}TTF)_2I_3$ (α-$(BETS)_2I_3$) from a chlorobenzene solution of BEDT-TTF (BETS) and tetrabutylammonium triiodide by applying 5 μA for 20 h. The thin crystal was transferred into 2-propanol with a pipette and guided onto the substrate. After the substrate was removed from the 2-propanol and dried, the crystal adhered to the substrate. Although the X-ray diffraction measurement was difficult, we were able to see through a polarizer that the crystal is a single crystal in which the two-dimensional conducting plane is parallel to the substrate (due to the polarizing property of I_3^-). As the gate electrolyte, ionic liquid 1-ethyl-3-methylimidazolium 2-(2-methoxyethoxy)ethyl sulfate was dropped to cover both the sample and the gate electrode. Lastly, the EDLT was covered by a 1.2 μm thick polyethylene naphthalate (PEN) film to reduce thermal stress at low temperatures by thinning the gate electrolyte. After completion, the EDLT was immediately cooled down in a cryostat, and the gate voltage at 220 K was changed to suppress the chemical reaction between the compounds and the ionic liquids. Charge displacement current measurements were performed by sweeping the gate voltage between ±0.5 V and measuring the gate current using a source-measure unit (KEITHLEY 2636B, Keithley Instruments, Cleveland, OH, USA). We derived the accumulated surface charge density from

$$p = \frac{Q}{eA} = \frac{\int I_G dV_G}{r_V eA} \quad (1)$$

where I_G, V_G, r_V, e, and A denote the gate current, gate voltage, sweep rate of the gate voltage, elementary charge, and area of the sample, respectively [17]. As for the temperature dependence of the resistance, we employed the standard four-probe method using a DC source (KEITHLEY 2400, Keithley Instruments, Cleveland, OH, USA) and a nano voltmeter (Agilent 34420A, Agilent Technologies, Santa Clara, CA, USA) under various gate voltages. The gate voltage was applied at 220 K in descending order from +0.4 V to −0.4 V in α-

(BEDT-TTF)$_2$I$_3$ and from +0.8 V to −0.6 V V in α-(BETS)$_2$I$_3$. Beyond those voltage ranges, the sample resistances tended to increase instead, probably because of degradation by the chemical reactions between the ionic liquid and the molecular conductors. The cooling rate was 0.75 K/min, and the data shown were captured during cooling. For α-(BETS)$_2$I$_3$, we also measured the Hall effect using a superconducting magnet that generates up to 8 T (TeslatronPT, Oxford Instruments, Abingdon, UK).

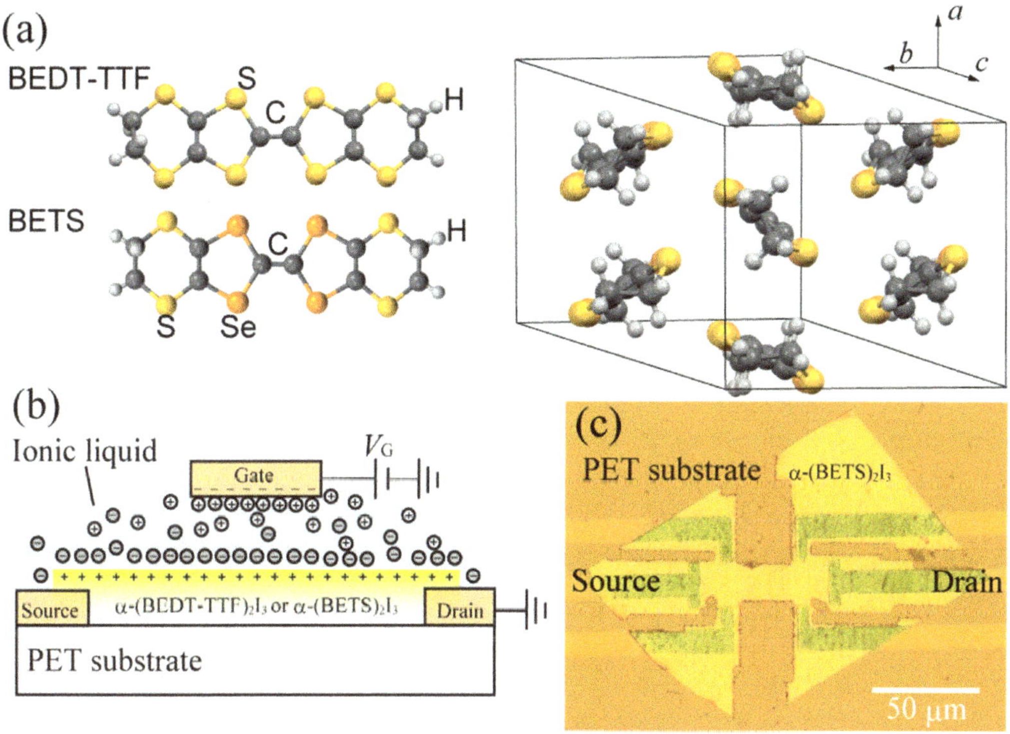

Figure 1. (**a**) BEDT-TTF and BETS molecules, and crystal structures of α-(BEDT-TTF)$_2$I$_3$ and α-(BETS)$_2$I$_3$ (I$_3$ is not shown). (**b**) Schematic side view and (**c**) optical top view of an EDLT device. The α-(BETS)$_2$I$_3$ crystal in (**c**) is laser-shaped into the Hall bar. The gate electrode (area: 800 × 800 cm^2) is patterned on the substrate, a few hundred micrometers away from the crystal in (**c**).

3. Results

3.1. Charge Displacement Current Measurement

First, we performed charge displacement current measurements at 220 K to confirm the accumulated charge density by electric double layer doping. Figure 2 shows the gate voltage V_G dependence of the charge density p, estimated from Equation (1). To the BEDT-TTF salt, the gate voltage of 0.5 V corresponds to approximately 10% doping (100% doping = 1 electron or hole per 2 BEDT-TTF molecules). Electrons are slightly more likely to be doped than holes at the same magnitude of gate voltage. In the case of the BETS salt, 0.5 V corresponds to only ~3% (hole) and ~5% (electron) doping. For reference, 10% doping corresponds to the gate voltage of approximately 250 V in an FET with a 300 nm thick SiO$_2$ film, which generally exceeds the withstand voltage of the SiO$_2$ film. In our previous study on the electric double layer doping of κ-(BEDT-TTF)$_2$Cu[N(CN)$_2$]Cl, the doping concentration at 0.5 V reached 20%, and the p–V_G curve was more linear [6].

3.2. Temperature Dependence of Resistance at Various Gate Voltages

3.2.1. α-(BEDT-TTF)$_2$I$_3$

Figure 3a shows the temperature dependence of the four-probe resistance in the EDLT device based on α-(BEDT-TTF)$_2$I$_3$ at various gate voltages. The device exhibits metal–insulator transitions at approximately 130 K, which is slightly lower than that for typical bulk α-(BEDT-TTF)$_2$I$_3$ crystals (135 K). The discrepancy is attributable to the biaxial strain effect that is due to the difference in the coefficient of thermal expansion between α-(BEDT-TTF)$_2$I$_3$ and the PET substrate. At low temperatures, the substrate shrinks and applies compressive strain to α-(BEDT-TTF)$_2$I$_3$, resulting in the reduction in the transition temperature [18]. The resistivity above the transition temperature is consistent with typical bulk crystals ($\sim 10^{-2}$ Ω cm) if we apply a typical thickness of the crystal (100 nm). Below the transition, the resistivity is lower than that of typical bulk crystals as the transition is moderate, probably because of the suppression of the lattice deformation.

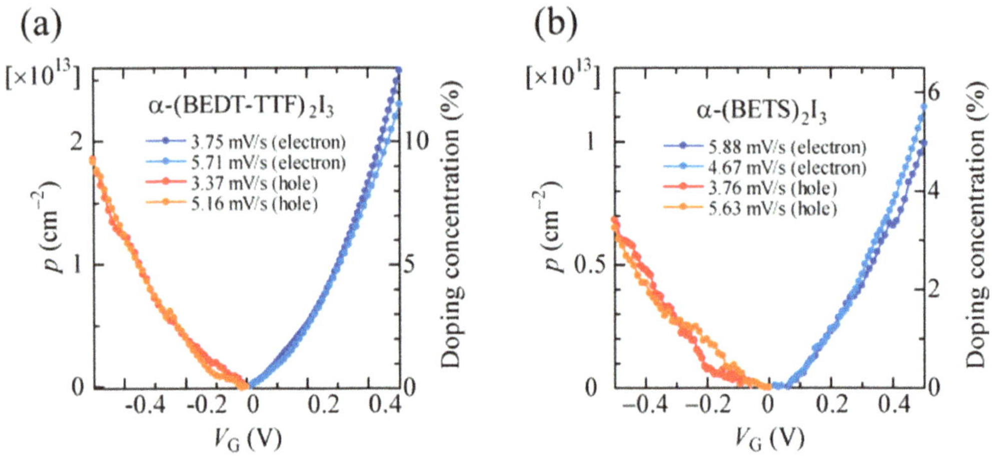

Figure 2. Gate voltage dependence of charge density p in (**a**) α-(BEDT-TTF)$_2$I$_3$ and (**b**) α-(BETS)$_2$I$_3$ at 220 K.

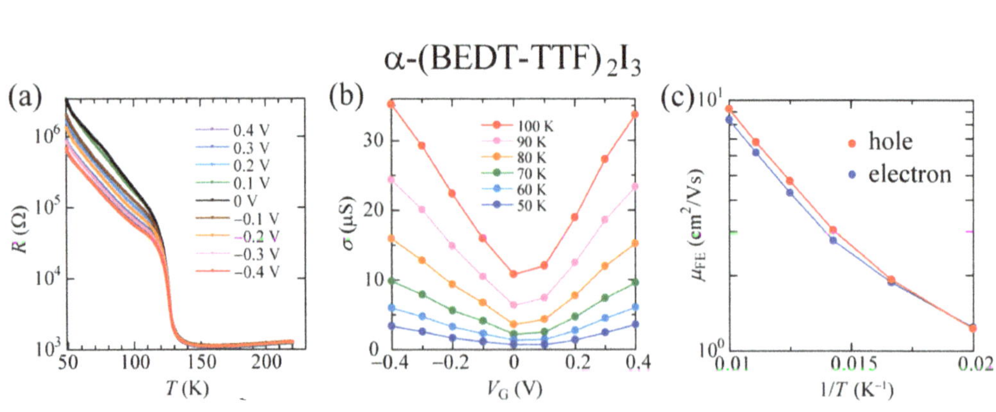

Figure 3. (**a**) Temperature dependence of the four-probe resistance at various gate voltages in α-(BEDT-TTF)$_2$I$_3$. (**b**) Gate voltage dependence of the sheet conductivity at 50–100 K. (**c**) Arrhenius plots of the field-effect mobility estimated in the range of $0.2\ \mathrm{V} < |V_G| < 0.4\ \mathrm{V}$.

Both positive and negative gate voltages enhance the conductance by up to a few times in the insulating region. The conductance enhancement is more significant than the case of FET and occurs at the channel (not only at the contacts) because we performed the four-probe measurements. We evaluated the field-effect mobility μ_{FE}, the effective mobility estimated from $\mu_{FE} = (1/C)\partial\sigma/\partial V_G$, where C is the capacitance per unit area of the electric double layer. We roughly estimated C from the slope (linear approximations for 0.2 V $< |V_G| <$ 0.4 V) in Figure 2a and obtained $\partial\sigma/\partial V_G$ from the gate voltage dependence of σ (Figure 3b). The derived μ_{FE} is approximately 10 cm^2/Vs at 100 K, which is three orders of magnitude higher than those in the FET measurements [9]. μ_{FE} shows similar thermal activation behaviors to the FETs (Figure 3c), although the values are similar for electron and hole in our EDLT. These values are not the actual carrier mobility values. However, they still indicate the switching performance of the resistance.

Nevertheless, as shown in Figures 3a and 4a, the metal–insulator transition temperature remains almost unchanged by the gate voltage, implying that the suppression of the charge ordering by doping, if any, is limited. Generally, the field effect is confined at the sample surface. Our sample consisted of several tens of conducting molecular layers, and the resistance of the bulk was not very high. Therefore, to determine the temperature dependence of the conductance at the doped surface, we had to extract the gate-induced conductance. Figure 4b shows the temperature dependence of the gate-induced sheet conductivity $\Delta\sigma = \frac{L}{W}\left(\frac{1}{R} - \frac{1}{R_{0V}}\right)$ at various gate voltages (where W is sample width, L is sample length). $\Delta\sigma$ monotonically decreased with cooling at all the gate voltages applied in this study, showing no emergence of metallic conduction, and gate-induced melting of charge ordering was unlikely. As shown in Figure 4a,c, neither the non-doped resistance ($V_G = 0$ V) nor $\Delta\sigma$ shows simple activation behaviors. Additionally, the variable range hopping mechanism does not describe $\Delta\sigma$ well either. According to Ivek and Čulo [19], the measured resistivity of α-(BEDT-TTF)$_2$I$_3$ can be decomposed as $1/\rho_{measured} = 1/\rho_{NNH} \mp 1/\rho_{remaining}$, where ρ_{NNH} is the resistivity of the nearest neighbor hopping (NNH) channel. NNH requires randomly distributed localized states, and ρ_{NNH} depends on the temperature-independent activation energy at low temperatures. However, $\rho_{remaining}$ depends on the mean-field-like activation energy originating from the charge ordering. In our EDLT device, the Arrhenius plots of $\Delta\sigma$ (Figure 4c) indicate that the activation energy is almost independent of V_G at relatively high temperatures, whereas it is dependent on V_G at low temperatures, as shown in Figure 4d. Assuming that the above two-channel conduction model by Ivek and Čulo [19] applies also to the doped case, we find that the activation energy for the NNH channel, which depends on the average energy difference between the localized states, decreases with gating. In contrast, the activation energy related to charge ordering is not markedly affected by V_G. The doped carriers seemingly fill the disorder-induced localized states and do not significantly prevent the formation of charge ordering.

3.2.2. α-(BETS)$_2$I$_3$

Figure 5a,b shows the temperature dependence of the four-probe resistance and the Arrhenius plots for α-(BETS)$_2$I$_3$ at various gate voltages. We observed large electron doping effects compared to α-(BEDT-TTF)$_2$I$_3$. However, the effects are highly asymmetric against the doping polarity (Figure 5c), where the hole doping slightly reduces the resistance only in the lowest temperature region. μ_{FE} under electron doping is 55 cm^2/Vs at 40 K and shows a thermal activation behavior that is similar to the α-(BEDT-TTF)$_2$I$_3$ device (Figure 5d). Under electron doping, the gate-induced conductivity $\Delta\sigma$ sufficiently exceeds the Mott–Ioffe–Regel conductivity limit in two dimensions (e^2/h, ~38.7 μS), around which the metallic conduction appears in Si-MOSFET [20] and devices based on κ-type BEDT-TTF salts [21,22]. However, $\Delta\sigma$ in α-(BETS)$_2$I$_3$ EDLT does not show metallic conduction below the transition temperature of $\sim$ 50 K, as shown in Figure 6a. The Arrhenius plots of $\Delta\sigma$ (Figure 6b) show that unlike in α-(BEDT-TTF)$_2$I$_3$, the activation energy near the transition temperature significantly decreases with increasing positive V_G. The mechanism is unclear because the insulating state of α-(BETS)$_2$I$_3$ is not a simple charge-ordered state, and the insulating mechanism is under debate [14–16]. However, the present results may provide clues to understanding the insulating phase of this compound.

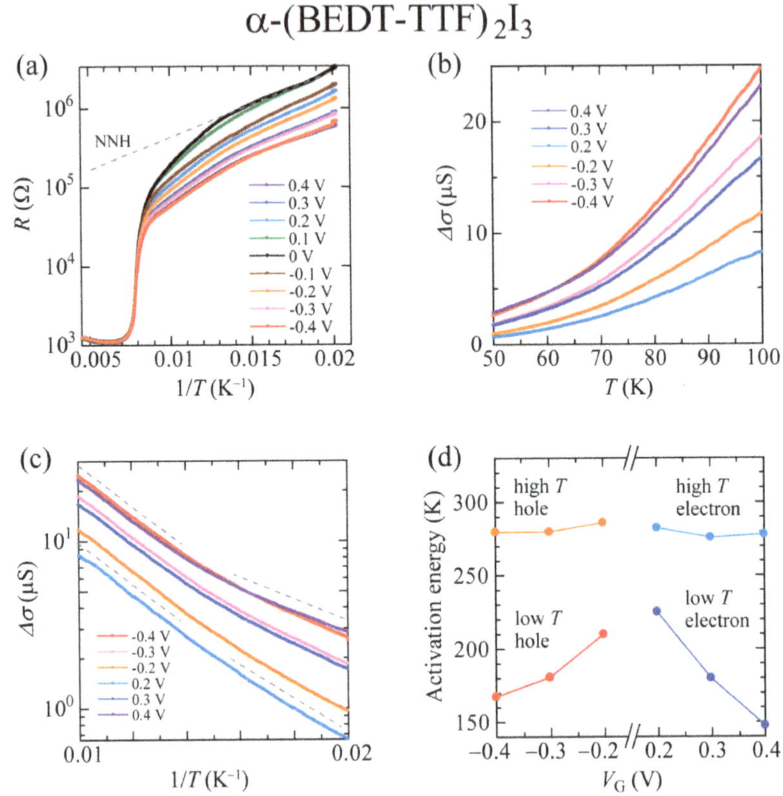

Figure 4. (**a**) Arrhenius plots of the four-probe resistance at various gate voltages in α-(BEDT-TTF)$_2$I$_3$. The dashed line is a guideline of the slope at low temperatures and is assigned to the NNH hopping conduction in the literature. (**b**) Temperature dependence of the gate-induced sheet conductivity. Note that the data at $|V_G| = 0.1$ V are not shown because $\Delta\sigma = \frac{L}{W}\left(\frac{1}{R} - \frac{1}{R_{0V}}\right)$ contains large errors when R is similar to R_{0V}. (**c**) Arrhenius plots of the gate-induced conductivity. Dashed lines are guides to the eye. (**d**) Gate voltage dependence of the activation energy estimated at relatively high (80–100 K) and low (50–60 K) temperatures.

We investigated the Hall effect in α-(BETS)$_2$I$_3$ at 10, 5, and 1.7 K, as shown in Figure 7. In order to eliminate any possible influence of magnetoresistance, the Hall resistance R_{xy} was determined as $R_{xy} = [R_{xy}(+B) - R_{xy}(-B)]/2$. Without gating, R_{xy} is positive and proportional to the magnetic field. The slope (= Hall coefficient R_H) increases with cooling, although R_{xy} is unmeasurable at 1.7 K due to the high resistance. R_{xy} decreases by $V_G = 0.2$ V, implying an electron conduction channel on the surface. However, by further electron doping, R_{xy} increases again and almost coincides with the ungated values. We explain below the complicated situation of the Hall effect in our devices [23].

First, R_{xy} of a bulk α-(BETS)$_2$I$_3$ crystal is negative. However, thin crystals on PET substrates show positive R_{xy} because hole carriers are doped by contact charging with the substrate. Therefore, two conduction channels (bulk and the α-(BETS)$_2$I$_3$/substrate interface) already exist without gating. The gate voltage would induce the third conduction channel (ionic liquid/α-(BETS)$_2$I$_3$ interface). It is difficult to distinguish the contribution of each conduction channel to R_{xy}. Nonetheless, as R_{xy} at $V_G = 0$ V is proportional to the magnetic field, we can regard the two conduction channels (bulk and bottom interface) as one effective V_G. -independent channel. Then, in principle, we can extract the mobility and the carrier density at the gated surface by fitting the data using the formula

$$R_{xy} = \frac{(\mu_0^2 n_0 + \mu_s^2 n_s) + (\mu_0 \mu_s B)^2 (n_0 + n_s)}{e\left[(\mu_0 |n_0| + \mu_s |n_s|)^2 + (\mu_0 \mu_s B)^2 (n_0 + n_s)^2\right]} B \tag{2}$$

where μ_0, n_0, μ_s, and n_s denote the effective mobility and the carrier density at $V_G = 0$ V, and the mobility and the carrier density at the gated surface, respectively. Using the values of μ_0 and n_0 derived from R_{xx} and R_{xy} at $V_G = 0$ V, and the constraint $R_{xx} = 1/e(\mu_0|n_0| + \mu_s|n_s|)$, we should obtain μ_s and n_s. However, we could not find any realistic solutions that reproduce the nonlinear R_{xy} at 0.2 V. Furthermore, assuming for simplicity that R_{xy} is proportional to B (the terms quadratic to B in Equation (2) can be ignored), we obtain hole carriers with $\mu_s = 8.9$ cm^2/Vs and $n_s = 6.4 \times 10^{12}$/cm^2. These values are obviously inconsistent with the charge injection by the positive gate voltage (Figure 2), implying a large discrepancy between the drift mobility and the Hall mobility. The discrepancy is probably due to hopping or one-dimensional (filamentary) conduction at the doped surface. A possible scenario is that as the gate voltage increases, the hopping or one-dimensionality is strengthened, resulting in R_{xy} becoming consistent with the ungated value as the Hall mobility at the doped surface approaches zero.

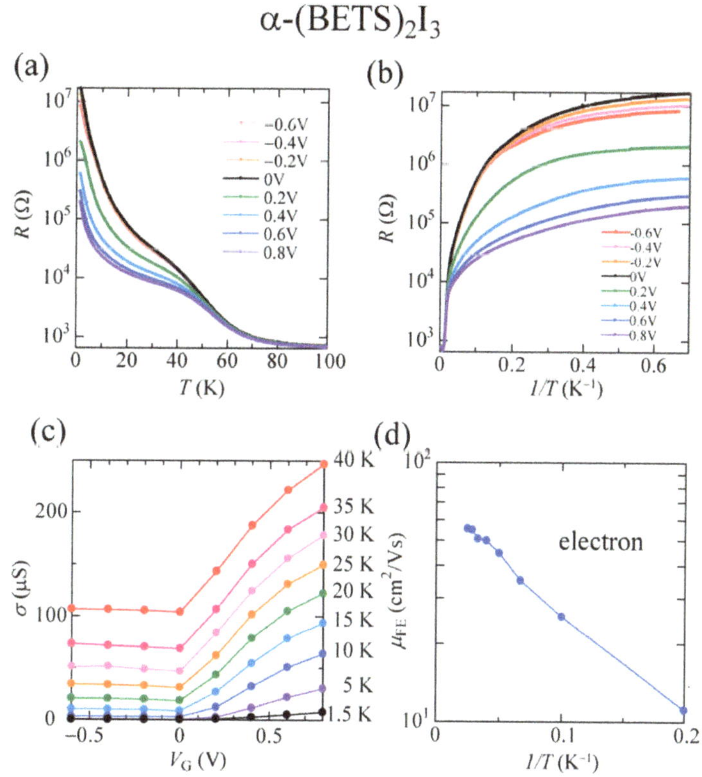

Figure 5. (**a**) Temperature dependence of the four-probe resistance at various gate voltages and (**b**) Arrhenius plots for α-(BETS)$_2$I$_3$. (**c**) Gate voltage dependence of the sheet conductivity at 1.5–40 K. (**d**) Arrhenius plots of the field-effect mobility estimated in the range of 0.2 V < V_G < 0.4 V.

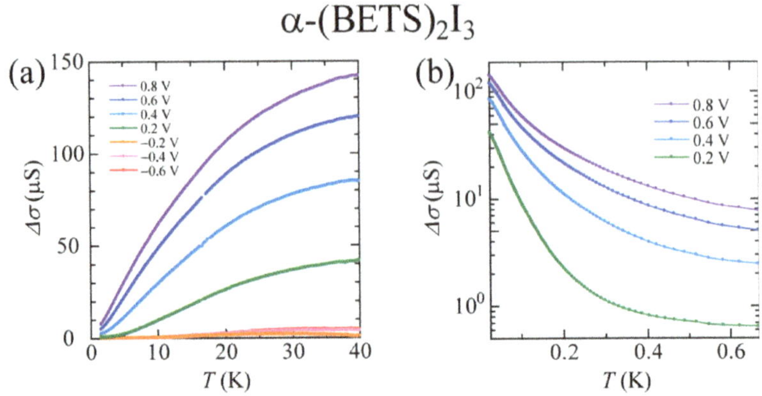

Figure 6. (**a**) Temperature dependence of the gate-induced conductivity in α-$(BETS)_2I_3$. (**b**) Arrhenius plots of the gate-induced conductivity.

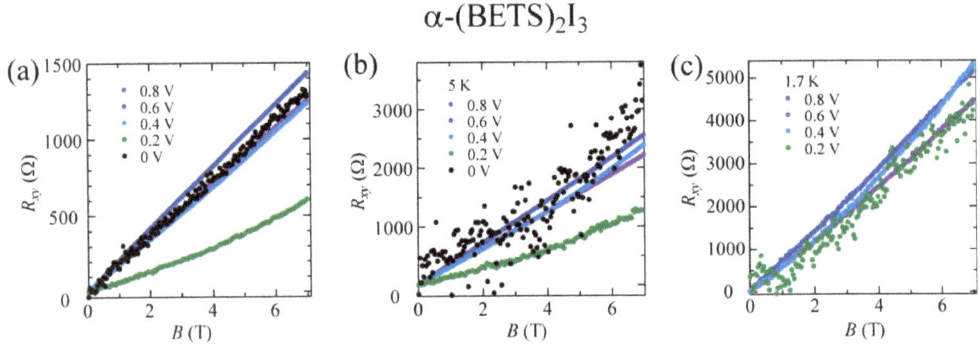

Figure 7. Magnetic field dependence of the Hall resistance under electron doping at (**a**) 10, (**b**) 5, and (**c**) 1.7 K.

4. Discussion and Conclusions

In both compounds, we achieved ambipolar gating effects on the four-probe resistance, indicating that the shift in the band filling in the charge-ordered (and related) insulating state reduces the resistance regardless of the doping polarity. However, we could not observe metallic conduction in the present experimental setup. In α-(BEDT-TTF)$_2$I$_3$, although the apparent activation energy is significantly decreased by the gate voltage, the extracted gate-induced conductivity implies that only the activation energy for the NNH conduction at low temperatures decreases. The doped carriers fill the randomly distributed localized states (due to the displacement of I_3^- anion chains, according to Ivek and Čulo [19]), and it seems difficult to suppress the charge ordering by further doping in the present device.

In α-(BETS)$_2$I$_3$, the gate-induced conductivity reaches ~140 μS, which largely exceeds the Mott–Ioffe–Regel conductivity limit in two dimensions (e^2/h, ~38.7 μS), around which metallic conduction emerges in various field-effect devices [20–22]. However, we could not observe metallic conduction either. At low electron doping (V_G = 0.2 V), the Hall resistance significantly decreases (by approximately 50% at 10 K), indicating that the electron doping induces partly two-dimensional, band-like transport at the doped surface. Nevertheless, with increasing gate voltage, the Hall resistance approaches the undoped values, suggesting that hopping or filamentary conduction that hardly contributes to the Hall effect is strengthened. Recently, we observed the Shubnikov–de Haas oscillations in thin single crystals of α-$(BETS)_2I_3$ doped by contact charging [23]. Although the doping concentration is comparable ($\sim 10^{12}$ cm^{-2}), we could not observe the oscillations in the present device, indicating that the present device has a more disordered surface.

Our results contrast those in Mott EDLTs based on κ-type BEDT-TTF salts, where gate-induced metal–insulator transitions and the Hall effect correspond to the reconstructed Fermi surface. However, the situation in this study may be somewhat typical for the electrolyte gating of organic molecular materials. In rubrene EDLTs, both the field-effect mobility and the Hall mobility generate peaks against the gate voltage [24]. Gate voltage application induces charge carriers at the surface. However, with increasing V_G, the accumulated ions start to form clusters with potential minima that trap the gate-induced carriers [25]. This phenomenon is considered unavoidable in materials with low dielectric constants and narrow bandwidths. To observe more intrinsic field effects, the combination with pressure may be effective because it generally increases the bandwidth. More essentially, the development of a surface treatment method for molecular conductors is necessary to reduce the unevenness of the potential at the gated surface.

Author Contributions: Conceptualization, Y.K. and H.M.Y.; methodology, J.P. and T.T.; investigation and analysis, Y.K., H.M. and N.T.; writing, Y.K., H.M. and N.T.; funding acquisition, Y.K. and R.K. All authors have read and agreed to the published version of the manuscript.

Funding: This research was funded by MEXT and JSPS KAKENHI, grant numbers JP16H06346, JP19K03730, JP19H00891, JP19K15383, JP19K22127, JP20H05664, JP20H05867, 20H05189, and 20H05862. J.P. also acknowledges support from KONDO-ZAIDAN.

Informed Consent Statement: Not applicable.

Data Availability Statement: All data needed to evaluate the conclusions in the paper are presented in the paper. Additional data related to this paper may be requested from the authors.

Acknowledgments: We would like to acknowledge Teijin DuPont Films Japan Limited for providing the PET films and Professor Yutaka Nishio for valuable discussions and advice.

Conflicts of Interest: The authors declare no conflict of interest.

References

1. Imada, M.; Fujimori, A.; Tokura, Y. Metal-insulator transitions. *Rev. Mod. Phys.* **1998**, *70*, 1039–1263. [CrossRef]
2. Ahn, C.H.; Bhattacharya, A.; Di Ventra, M.; Eckstein, J.N.; Frisbie, C.D.; Gershenson, M.E.; Goldman, A.M.; Inoue, I.H.; Mannhart, J.; Millis, A.J.; et al. Electrostatic modification of novel materials. *Rev. Mod. Phys.* **2006**, *78*, 1185–1212. [CrossRef]
3. Fujimoto, T.; Awaga, K. Electric-double-layer field-effect transistors with ionic liquids. *Phys. Chem. Chem. Phys.* **2013**, *15*, 8983–9006. [CrossRef] [PubMed]
4. Ueno, K.; Shimotani, H.; Yuan, H.; Ye, J.; Kawasaki, M.; Iwasa, Y. Field-induced superconductivity in electric double layer transistors. *J. Phys. Soc. Jpn.* **2014**, *83*, 032001. [CrossRef]
5. Yamamoto, H.M.; Nakano, M.; Suda, M.; Iwasa, Y.; Kawasaki, M.; Kato, R. A strained organic field-effect transistor with a gate-tunable superconducting channel. *Nat. Commun.* **2013**, *4*, 2379. [CrossRef]
6. Kawasugi, Y.; Seki, K.; Tajima, S.; Pu, J.; Takenobu, T.; Yunoki, S.; Yamamoto, H.M.; Kato, R. Two-dimensional ground-state mapping of a Mott-Hubbard system in a flexible field-effect device. *Sci. Adv.* **2019**, *5*, eaav7278. [CrossRef]
7. Yamamoto, H.M.; Ito, H.; Ikeda, M.; Kato, R.; Shigeto, K.; Tsukagoshi, K. Nano-size molecular conductors directly formed on silicon substrates. In *Multifunctional Conducting Molecular Materials*; Royal Society of Chemistry: London, UK, 2006; pp. 97–100.
8. Yamamoto, H.M.; Hosoda, M.; Kawasugi, Y.; Tsukagoshi, K.; Kato, R. Field effect on organic charge-ordered/Mott insulators. *Physica B* **2009**, *404*, 413–415. [CrossRef]
9. Kimata, M.; Ishihara, T.; Tajima, H. Electrostatic charge carrier injection into the charge-ordered organic material α-$(BEDT\text{-}TTF)_2I_3$. *J. Phys. Soc. Jpn.* **2012**, *81*, 073704. [CrossRef]
10. Kimata, M.; Ishihara, T.; Ueda, A.; Mori, H.; Tajima, H. Fabrication of a field effect transistor structure using charge-ordered organic materials α-$(BEDT\text{-}TTF)_2I_3$ and α'-$(BEDT\text{-}TTF)_2IBr_2$. *Synth. Met.* **2013**, *173*, 43–45. [CrossRef]
11. Bender, K.; Hennig, I.; Schweitzer, D.; Dietz, K.; Endres, H.; Keller, H.J. Synthesis, Structure and Physical Properties of a Two-Dimensional Organic Metal, Di[bis(ethylenedithiolo)tetrathiofulvalene] triiodide, $(BEDT\text{-}TTF)^{+}{}_2I^{-}{}_3$. *Mol. Cryst. Liq. Cryst.* **1984**, *108*, 359–371. [CrossRef]
12. Tajima, N.; Sugawara, S.; Tamura, M.; Nishio, Y.; Kajita, K. Electronic phases in an organic conductor α-$(BEDT\text{-}TTF)_2I_3$: Ultra narrow gap semiconductor, superconductor, metal, and charge-ordered insulator. *J. Phys. Soc. Jpn.* **2006**, *75*, 051010. [CrossRef]

13. Inokuchi, M.; Tajima, H.; Kobayashi, A.; Ohta, T.; Kuroda, H.; Kato, R.; Naito, T.; Kobayashi, H. Electrical and optical properties of α-$(BETS)_2I_3$ and α-$(BEDT\text{-}STF)_2I_3$. *Bull. Chem. Soc. Jpn.* **1995**, *68*, 547–553. [CrossRef]
14. Kitou, S.; Tsumuraya, T.; Sawahata, H.; Ishii, F.; Hiraki, K.; Nakamura, T.; Katayama, N.; Sawa, H. Ambient-pressure Dirac electron system in the quasi-two-dimensional molecular conductor α-$(BETS)_2I_3$. *Phys. Rev. B* **2021**, *103*, 035135. [CrossRef]
15. Ohki, D.; Yoshimi, K.; Kobayashi, A. Transport properties of the organic Dirac electron system α-$(BEDT\text{-}TSeF)_2I_3$. *Phys. Rev. B* **2021**, *102*, 235116. [CrossRef]
16. Tsumuraya, T.; Suzumura, Y. First-principles study of the effective Hamiltonian for Dirac fermions with spin-orbit coupling in two-dimensional molecular conductor α-$(BETS)_2I_3$. *Eur. Phys. J. B* **2021**, *94*, 17. [CrossRef]
17. Xie, W.; Frisbie, C.D. Organic electrical double layer transistors based on rubrene single crystals: Examining transport at high surface charge densities above 10^{13} cm^{-2}. *J. Phys. Chem. Rev.* **2011**, *115*, 14360–14368. [CrossRef]
18. Kawasugi, Y.; Yamamoto, H.M.; Hosoda, M.; Tajima, N.; Fukunaga, T.; Tsukagoshi, K.; Kato, R. Strain-induced superconductor/insulator transition and field effect in a thin single crystal of molecular conductor. *Appl. Phys. Lett.* **2008**, *92*, 243508. [CrossRef]
19. Ivek, T.; Čulo, M. Semimetallic and charge-ordered α-$(BEDT\text{-}TTF)_2I_3$: On the role of disorder in dc transport and dielectric properties. *Phys. Rev. B* **2017**, *96*, 075141. [CrossRef]
20. Kravchenko, S.V.; Kravchenko, G.V.; Furneaux, J.E.; Pudalov, V.M.; D'Iorio, M. Possible metal-insulator transition at B=0 in two dimensions. *Phys. Rev. B* **1994**, *50*, 8039–8042. [CrossRef]
21. Sato, Y.; Kawasugi, Y.; Suda, M.; Yamamoto, H.M.; Kato, R. Critical behavior in doping-driven metal–insulator transition on single-crystalline organic Mott-FET. *Nano Lett.* **2017**, *17*, 708–714. [CrossRef]
22. Kawasugi, Y.; Seki, K.; Pu, J.; Takenobu, T.; Yunoki, S.; Yamamoto, H.M.; Kato, R. Non-Fermi-liquid behavior and doping asymmetry in an organic Mott insulator interface. *Phys. Rev. B* **2019**, *100*, 115141. [CrossRef]
23. Kawasugi, Y.; Masuda, H.; Uebe, M.; Yamamoto, H.M.; Kato, R.; Nishio, Y.; Tajima, N. Pressure-induced phase switching of Shubnikov–de Haas oscillations in the molecular Dirac fermion system α-$(BETS)_2I_3$. *Phys. Rev. B* **2021**, *103*, 205140. [CrossRef]
24. Xia, Y.; Xie, W.; Ruden, P.P.; Frisbie, C.D. Carrier localization on surfaces of organic semiconductors gated with electrolytes. *Phys. Rev. Lett.* **2010**, *105*, 036802. [CrossRef] [PubMed]
25. Xie, W.; Liu, F.; Shi, S.; Ruden, P.P.; Frisbie, C.D. Charge density dependent two-channel conduction in organic electric double layer transistors (EDLTs). *Adv. Mater.* **2014**, *26*, 2527–2532. [CrossRef] [PubMed]

crystals

MDPI

Article

Absence of Superconductivity in the Hubbard Dimer Model for κ-(BEDT-TTF)$_2$X

Dipayan Roy [1], R. Torsten Clay [1,*] and Sumit Mazumdar [2]

[1] Department of Physics & Astronomy and HPC^2 Center for Computational Sciences, Mississippi State University, Starkville, MS 39762, USA; dr1413@msstate.edu
[2] Department of Physics, University of Arizona, Tucson, AZ 85721, USA; sumit@physics.arizona.edu
* Correspondence: r.t.clay@msstate.edu

Abstract: In the most studied family of organic superconductors κ-(BEDT-TTF)$_2$X, the BEDT-TTF molecules that make up the conducting planes are coupled as dimers. For some anions X, an antiferromagnetic insulator is found at low temperatures adjacent to superconductivity. With an average of one hole carrier per dimer, the BEDT-TTF band is effectively $\frac{1}{2}$-filled. Numerous theories have suggested that fluctuations of the magnetic order can drive superconducting pairing in these models, even as direct calculations of superconducting pairing in monomer $\frac{1}{2}$-filled band models find no superconductivity. Here, we present accurate zero-temperature Density Matrix Renormalization Group (DMRG) calculations of a dimerized lattice with one hole per dimer. While we do find an antiferromagnetic state in our results, we find no evidence for superconducting pairing. This further demonstrates that magnetic fluctuations in the effective $\frac{1}{2}$-filled band approach do not drive superconductivity in these and related materials.

Keywords: organic superconductors; correlated electron materials

Citation: Roy, D.; Clay, R.T.; Mazumdar, S. Absence of Superconductivity in the Hubbard Dimer Model for κ-(BEDT-TTF)$_2$X. *Crystals* **2021**, *11*, 580. https://doi.org/10.3390/cryst11060580

Academic Editor: Toshio Naito

Received: 29 March 2021
Accepted: 18 May 2021
Published: 22 May 2021

1. Introduction

Correlated-electron superconductivity (SC) continues to be one of the most challenging problems in condensed matter physics. Even as the large majority of investigators in this research area have focused on cuprates and iron-based superconductors (and more recently, ruthenates), it is now widely recognized that there exist many other unconventional or correlated-electron superconductors, although with lower superconducting critical temperature T_c [1–9]. Superconducting organic charge-transfer (CT) solids hold a special place in this context, as SC in CT solids was discovered significantly prior to the discovery of the phenomenon in the cuprates [10]. Despite intensive investigations over the past four decades, however, the mechanism of SC in CT solids remains a mystery, and organic SC continues to be an active research area, as illustrated by the appearance of many review articles on the subject over the past decade [11–15]. In the present paper, we report the results of our latest calculations that have probed a key question in the field of organic SC.

The common theme among many (though not all, see below) correlated-electron superconductors is the proximity of SC to spin-density wave (SDW) or antiferromagnetism (AFM) . This property is shared by organic superconductors based on the cation TMTSF as well as some κ-phase BEDT-TTF compounds. Unlike in the cuprates, though, transition from the insulating magnetic phase to the superconducting phase is driven by the application of moderate pressure as opposed to doping. Both TMTSF and BEDT-TTF based superconductors (hereafter (TMTSF)$_2$X and (BEDT-TTF)$_2$X) are characterized by cation:anion ratio of 2:1, with closed-shell monovalent anions X^-, the implication being that each individual cation monomer has charge $+\frac{1}{2}$. In both families, the cation lattice is highly dimerized, such that the overall charge on a dimer of cations is exactly +1. This has led to various effective half-filled band theories of organic CT superconductors, within which the dimer of cations is considered as a single site in a lattice described within anisotropic triangular lattice Hubbard models (we emphasize that these theories are therefore not applicable to nondimerized (BEDT-TTF)$_2$X families, such as those belonging to the θ-family, which also exhibit SC; we revisit this issue later). Within mean-field and dynamic mean field theory (DMFT) calculations on the half-filled band anisotropic triangular lattice Hubbard model, pressure increases the lattice frustration, causing thereby a bandwidth-driven transition from the AFM insulator to the superconducting state (or in the case of a lattice that is already strongly frustrated, as in κ-(BEDT-TTF)$_2$Cu$_2$(CN)$_3$, from a quantum spin liquid (QSL) to SC) [16–28]. Direct numerical calculations of superconducting pair-pair correlations within the $\frac{1}{2}$-filled Hubbard band on triangular lattices have, however, consistently found absence of SC [29–33]. Taken together with the very recent convincing demonstration of the absence of SC within the weakly doped square lattice two-dimensional (2D) Hubbard Hamiltonian [34], numerical calculations then cast serious doubt on the existing approximate spin fluctuation mechanisms of AFM-to-SC transition.

Conceivably, one possible origin of disagreement between the mean-field theories of organic SC and numerical calculations is that retention of the explicit dimeric structure of the Hubbard lattice sites is essential for superconducting correlations to dominate. That is, even as intra-dimer charge fluctuations are precluded within the effective $\frac{1}{2}$-filled band models (since that would make the effective $\frac{1}{2}$-filled description invalid and can even destroy the AFM in the weakly frustrated region of the Hamiltonian), the superconducting singlets should be defined such that they span over a pair of dimers, as opposed to a pair of effective single sites. Needless to say such a numerical calculation is far more involved than where dimers are replaced with single sites, as the required lattice sizes needed are twice that in the previous calculations [29,30,33]. We report precisely such a calculation here, which again finds absence of SC in such a dimer lattice.

2. Methods

We consider the Hubbard Hamiltonian,

$$H = -\sum_{\langle i,j \rangle,\sigma} t_{ij}(t^{\dagger}_{i,\sigma} t_{j,\sigma} + H.c.) + U \sum_{i} n_{i,\uparrow} n_{i,\downarrow}. \quad (1)$$

In Equation (1) $c^{\dagger}_{i,\sigma}$ creates an electron of spin σ on molecule i and $n_{i,\sigma} = c^{\dagger}_{i,\sigma} c_{i,\sigma}$. U is the onsite Hubbard interaction. It is important to note that this U is the Coulomb repulsion between carriers on each monomer site and is not the same as the effective dimer U_d. The lattice we consider is a rectangular lattice of dimers (see Figure 1). In this lattice there are three hopping integrals t_{ij}: t_d is the intra-dimer hopping, t the inter-dimer hopping, and t' a frustrating hopping (see Figure 1). We take the system to be $\frac{1}{4}$-filled with an average electron density of $\frac{1}{2}$ per monomer site. This lattice is expected to show Néel AFM order in the unfrustrated limit of small t'/t.

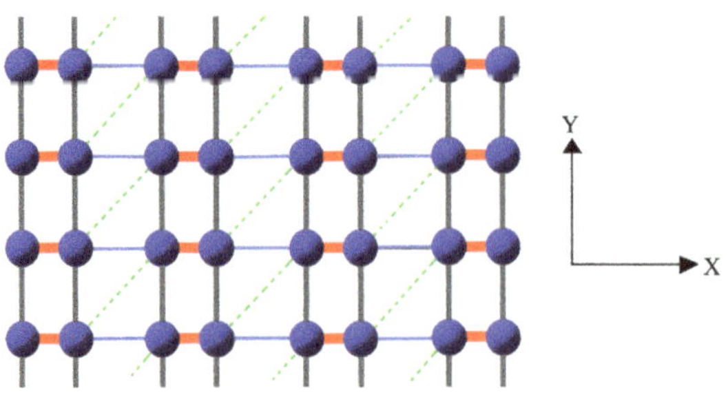

Figure 1. The dimer lattice we consider. Thick lines are the intra-dimer hopping t_d, thin lines are the inter-dimer hopping t, and dashed lines are the frustrating bond t'. Boundary conditions in our DMRG calculations are open in x and periodic in y.

To solve for the ground state of Equation (1) we employ the Density Matrix Renormalization Group (DMRG) method [35] using the ITensor library [36] with real-space parallelization [37]. DMRG can calculate essentially numerically exact correlation functions in quasi-one-dimensional systems. However, its primary limitation is an exponential scaling in the transverse size of systems that can be studied. Here, we consider lattices of dimension 4 sites in y and up to 32 sites in x (see Figure 1), corresponding to a 16 × 4 lattice in terms of dimers. We choose t_d = 1.5 and t = 0.5 such that the average hopping along the x axis is 1.0; our choice of $t_d \sim 3t$ is comparable to the dimerization found in the CT solid superconductors [14]. We take boundary conditions that are open (periodic) in x (y). The same boundary condition has been successfully used in several recent calculations on the Hubbard model [34,38], and has been tested against other approaches. We used a DMRG bond dimension m of up to 15,000 with minimum truncation error of 10^{-8}–10^{-7}; all results we show are extrapolated to zero truncation error.

3. Results

We measure dimer spin–spin and pair–pair correlation functions. The dimer spin operator is

$$n^d_{i,\sigma} = n_{i_1,\sigma} + n_{i_2,\sigma},$$

where the sites i_1 and i_2 make up the dimer i. We calculate the dimer spin–spin correlation function $S^d_z(i,j) = \langle (n^d_{i,\uparrow} - n^d_{i,\downarrow})(n^d_{j,\uparrow} - n^d_{j,\downarrow}) \rangle$. In Figure 2 for a 16 × 4 lattice we plot $\langle S^d_z(1,j) \rangle (-1)^j$ where dimer 1 is on the first chain and dimer j is on the nearest neighbor chain for $t' = 0.2$ and 0.4.

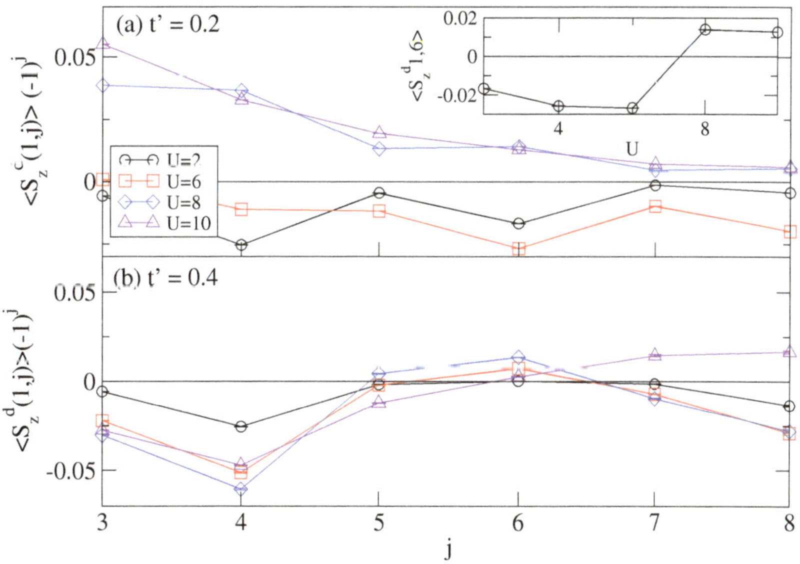

Figure 2. Dimer spin-spin correlations $\langle S^d_z(1,j) \rangle$ for a 16 × 4 lattice between the first dimer on chain 1 and dimer j on chain 2, multiplied by $(-1)^j$, the expected sign for Néel AFM. (**a**) $t' = 0.2$. The inset shows the U dependence of a single spin-spin correlation at long distance. Néel AFM order is present at t' for $U \gtrapprox 8$. (**b**) $t' = 0.4$. Here, we find no AFM order.

As shown in Figure 2a, for $t' = 0.2$ and $U \gtrapprox 8$ this quantity is positive for all spin correlations, indicating the presence of Néel AFM order. For $t' = 0.4$ (Figure 2b) we do not find Néel AFM order up to at least $U = 12$. These results show that Néel AFM order is only present in this system when the frustration is low, and for realistic U values AFM will not be present when $t'/t \gtrapprox 0.4$.

To define superconducting pair–pair correlations in the effective $\frac{1}{2}$-filled representation, we first define an operator creating an effective particle on dimer i with equal monomer populations, as required within effective $\frac{1}{2}$-filled band theories,

$$d^{\dagger}_{i,\sigma} = \frac{1}{\sqrt{2}}\left(c^{\dagger}_{i_1,\sigma} + c^{\dagger}_{i_2,\sigma}\right). \quad (2)$$

The $\Delta^{\dagger}_{i,i+\delta}$ operator creates a singlet pair between nearest-neighbor (n.n.) dimers separated by δ:

$$\Delta^{\dagger}_{i,i+\delta} = \frac{1}{\sqrt{2}}\left(d^{\dagger}_{i,\uparrow}d^{\dagger}_{i+\delta,\downarrow} - d^{\dagger}_{i,\downarrow}d^{\dagger}_{i+\delta,\uparrow}\right). \quad (3)$$

Note that Equation (3) involves *four* different lattice sites. We consider two kinds of n.n. pairs with δ taken as the n.n. distance between dimers in the x and y directions,

$$P(r = |\vec{r}_i - \vec{r}_j|)_{\delta,\delta'} = \langle \Delta^{\dagger}_{i,i+\delta}\Delta_{j,j+\delta'}\rangle. \quad (4)$$

In the effective dimer model pairing is expected to have d-wave symmetry. Because of the small transverse dimensions of our lattice we do not define full $d_{x^2-y^2}$ pairs involving four n.n. singlets, but instead calculate only single singlet-singlet correlations. To check if the pairing has the expected d-wave sign structure we calculate two different correlation functions, $P_{||}(r) \equiv P(r)_{\delta=x,\delta'=x}$ and $P_{\perp}(r) \equiv P(r)_{\delta=x,\delta'=y}$. $P_{||}(r)$ corresponds to correlations between two singlets both oriented along x, while $P_{\perp}(r)$ corresponds to one singlet oriented along x and one along y. For $d_{x^2-y^2}$ pairing $P_{\perp}(r)$ should be negative.

In a 2D superconductor long-range order should be present in $P(r)$ at zero temperature. Because of the quasi-one-dimensional nature of cylindrical geometry of the lattice that we can solve in DMRG long-range order in $P(r)$ is not possible even in the limit $x \to \infty$. If such a quasi-one-dimensional system showed a tendency to SC however, the system would behave as a Luther-Emery liquid [39] with $P(r)$ decaying as a power law $r^{-\alpha}$ with $\alpha < 1$. In addition to long-range order in $P(r)$, if SC in the system is driven by Coulomb interactions, one expects that $P(r, U) > P(r, U = 0)$ [29].

In Figure 3 we plot $P_{||}(r)$ as a function of r. In Figure 3a the indices i and j in Equation (4) are taken on the same chain, in Figure 3b they are on nearest neighbor chains, and in Figure 3c they are on second neighbor chains. In each panel r is the center-to-center distance between pairs in units of the dimer-dimer spacing. In Figure 3a we also plot the function r^{-1}. While the distances we have on the lattice are limited and we cannot determine if $P(r)$ decays with r as a power law or as an exponential, we find that on each chain $P(r)$ decays significantly faster than r^{-1}. We discuss the distance decay of $P(r)$ further below.

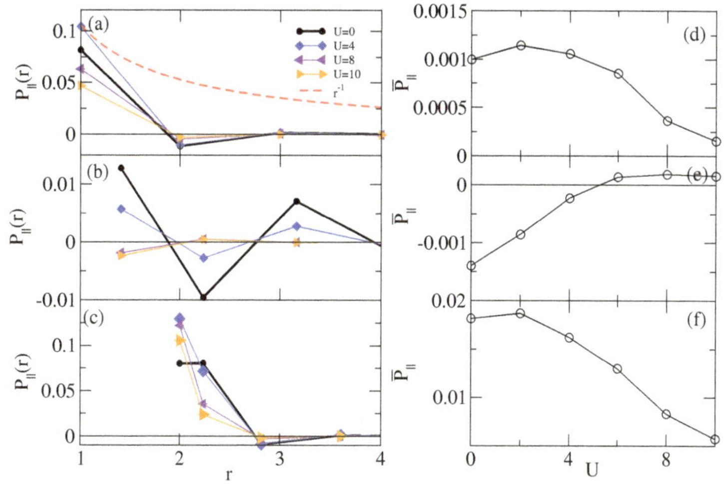

Figure 3. Pair–pair correlation for $t' = 0.2$ for parallel-oriented n.n. pairs on a 16×4 lattice. (**a**) Pair-pair correlation $P_{||}(r)$ for n.n. x-axis pairs on chain 1 (**b**) $P_{||}(r)$ for n.n. x-axis pairs on chains 1 and 2 (**c**) $P_{||}(r)$ for n.n. x-axis pairs on chains 1 and 3. In (**a–c**), r is the center-to-center pair distance. In (**a**) we show the function r^{-1} for comparison as the dashed curve. In panels (**d**) we plot the U dependence of the average long-range correlation $\bar{P}_{||}$ (see text) for the chain 1—chain 1 correlations in (**a**); panels (**e**,**f**) are similar for chain 1—chain 2 and chain 1—chain 3 correlations, respectively.

As seen in Figure 3a,c, only at very short range ($r \leq 2$) we find that $P(r, U)$ is enhanced very slightly over its value at $U = 0$. Similar enhancement at the shortest pair separations was previously found in calculations in the monomer $\frac{1}{2}$-filled Hubbard model on an anisotropic triangular lattice [29,30]. As noted in [40], at $r = 0$ the pair-pair correlation can be written as a linear combination of charge-charge and spin-spin correlations; the latter are enhanced by U [40]. To measure the effect of correlations on pairing we define the average long-range pair-pair correlation [41],

$$\bar{P} = \frac{1}{N_p} \sum_{2<|\vec{r}|<r_{\max}} P(r). \tag{5}$$

In Equation (5) N_p is the number of terms in the sum, and the sum is over r greater than 2 in dimer-dimer units but less than a maximum distance $r_{\max}$. The upper cutoff is necessary because of the open boundary conditions along x in our lattice and is chosen so that the maximum pair-pair distance along the chain is 4 (one half of the lattice length of 8 dimers). Figure 3d–f show $\bar{P}_{||}$ as a function of U. In each case the magnitude of $\bar{P}_{||}$ decreases with U apart from an insignificant increase at very small U. A large decrease in $\bar{P}_{||}$ also takes place for $U \gtrsim 8$ upon entering the AFM insulating phase. A similar decrease is seen in calculations within the monomer $\frac{1}{2}$-filled band [29,30].

Figure 4a–c shows perpendicular correlations $P_{\perp}(r)$ as a function of r. Like $P_{||}$, we find that $P_{\perp}$ along the chain direction decays faster than r^{-1}. These are negative in many cases (Figure 4a,c), which would be consistent with a $d_{x^2-y^2}$ type pairing symmetry. However, some $P_{\perp}(r)$ correlations are positive rather than negative (see Figure 4b). This apparent sign inconsistency with $d_{x^2-y^2}$ order may be a finite-size effect related to the small transverse dimensions of our lattice. It may also indicate that another type of pair symmetry besides $d_{x^2-y^2}$ is relevant, as found in other 4-chain calculations [38]. Figure 4d shows a clear decrease in magnitude of $P_{\perp}(r)$ with increasing U. The U dependence of Figure 4e,f are harder to interpret. We believe that the unusual U dependence here is due to shorter range correlations, for example as in Figure 4b where $P_{\perp}(r)$ changes discontinuously at small

U at $r \sim 2$ [42]. As in Reference [41], as distances $r < 3$ are excluded, we again find a continuous decrease of $\bar{P}_\perp$ with increasing U.

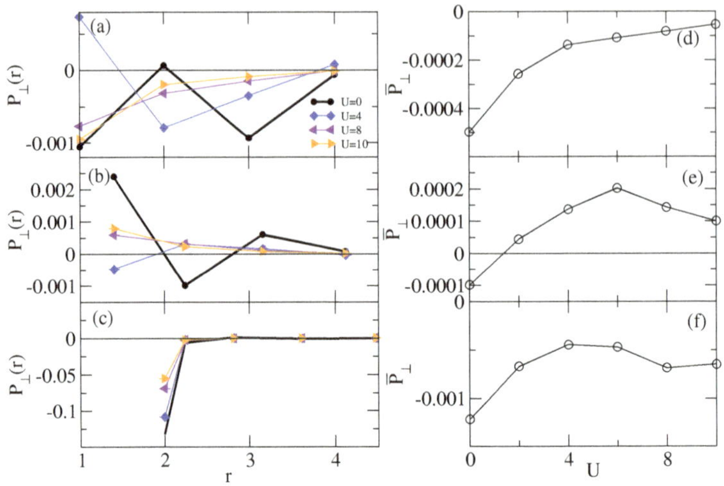

Figure 4. Panels (**a–f**) are the same as in Figure 3 except that the pair-pair correlation $P_\perp(r)$ is for the perpendicular orientation of pairs. Note that in (**d–f**), $\bar{P}_\perp$ is negative, with decreasing magnitude as U increases. $\bar{P}_\perp$ in (**e**) is strongly affected by shorter-range correlations which lead to the unusual U dependence (see text).

Figures 5 and 6 show the pair–pair correlations at $t' = 0.6$. Here AFM order is absent over the whole range of U we studied. We again find that $P_{||}(r)$ decreases faster than r^{-1} (see Figure 5a). As at smaller t' we again find that some pair-pair correlations are enhanced by U at short range ($r \leq 2$). However, here both $\bar{P}_{||}$ and $\bar{P}_\perp$ decrease continuously with U, again with the exception of one point at shorter r where there is a discontinuous change at small U (see Figure 6b,e).

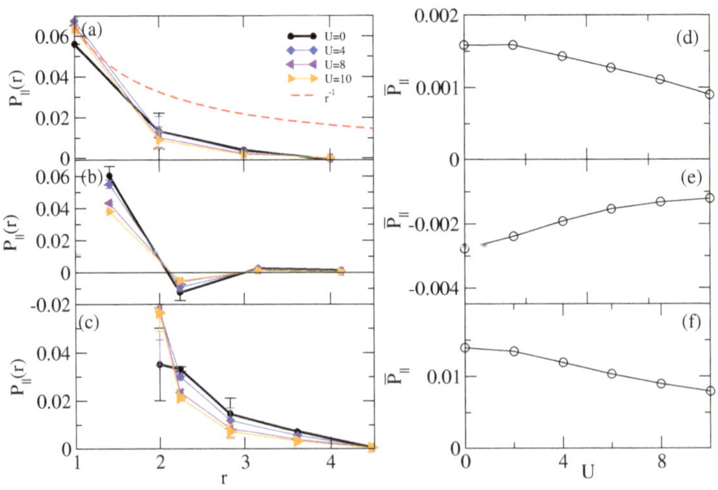

Figure 5. Panels (**a–f**) are the same as Figure 3 but with $t' = 0.6$.

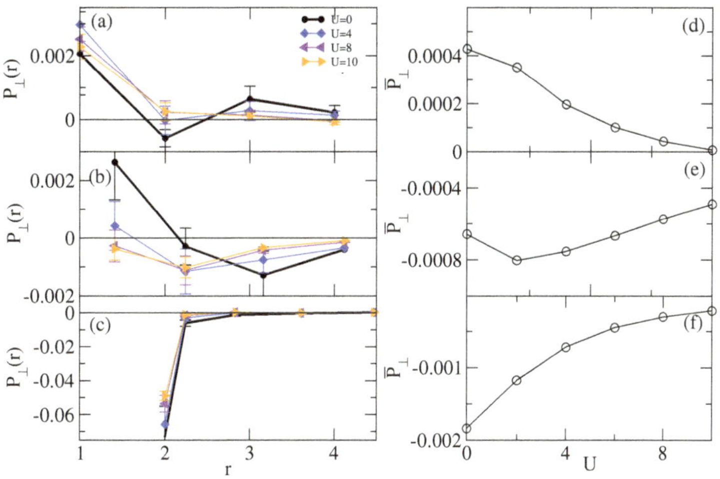

Figure 6. Panels (**a–f**) are the same as Figure 4 but with $t' = 0.6$.

A critical question is whether $P(r)$ decays as an exponential or power law with r [34]. In Figure 7 we compare $P_{||}(r)$ for the 16 × 4 lattice and the 32 × 4 lattice for intermediate $t' = 0.4$ where AFM order is not present. Figure 7a shows the distance dependence of $P_{||}(r)$ for pairs on the first chain of the lattice. We cannot definitively distinguish between a power law and an exponential decay with the system lengths available to us. However, as shown in Figure 7a, we find that if $P_{||}(r)$ decays with distance as $r^{-\alpha}$, then $\alpha \approx 2$. This decay is significantly faster with distance than found in quasi-1D superconducting systems like the weakly doped two-leg Hubbard ladder, where $\alpha < 1$ [43]. Additionally, in the 32 × 4 lattice we find that the magnitude of $P_{||}(r)$ decreases with U at long range (Figure 7a), which is also seen in DMRG calculations on the doped single-band Hubbard model (see [Reference [34], Figure 11]). Finally, in the larger lattice the average long-range pairing $\bar{P}_{||}$ also decreases monotonically with increasing U (see Figure 7b), which would preclude AFM-driven SC.

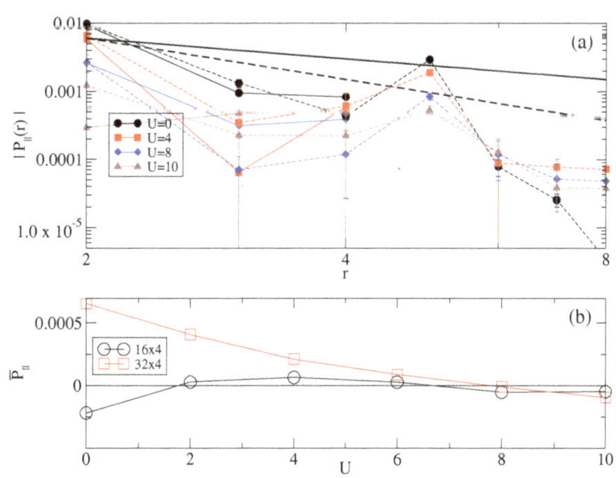

Figure 7. Pair–pair correlations for t' = 0.4 for parallel-oriented n.n. pairs on chain 1. In (**a**) points connected with solid (dashed) lines are for a 16 × 4 (32 × 4) lattice. For comparison we show the power law functions r^{-1} and r^{-2}. In (**b**), the U dependence of the average long-range correlation $\bar{P}_{||}$ for both lattices.

Summarizing our data, while we cannot prove the presence of long range AFM order without undertaking a full finite-size scaling, we do find strong AFM correlations for weak frustration and moderately strong U. We find no evidence that within the effective $\frac{1}{2}$-filled picture that the superconducting pair-pair correlations are enhanced over their value for non-interacting electrons, and on our cylindrical lattice the distance dependence of $P(r)$ in the x direction does not support the presence of long-range order.

4. Discussion

The conclusions of our work are: (i) SC is absent within the model; and (ii) the presence or absence of AFM order is not related to the tendency towards SC.

In κ-(BEDT-TTF)$_2$X with X = Cu[N(CN)$_2$]Cl where AFM order is found under ambient pressure, estimates for t'/t vary between 0.44 and 0.53 [14]. Similarly, no AFM order is found for X = $Cu_2(CN)_3$ where $0.80 \lessapprox t'/t \lessapprox 0.99$ [14]. In our present results, we find AFM order at $t' = 0.2$ ($t'/t = 0.4$) but no AFM order at $t'/t = 0.8$, which is roughly consistent with what is observed experimentally. SC is however found in both X = Cu[N(CN)$_2$]Cl and X = $Cu_2(CN)_3$, under pressures of 0.03 and 0.6 GPa, respectively. Moreover, there is no obvious connection between the superconducting T_c in κ-(BEDT-TTF)$_2$X and the value of t'/t [14]. For example SC is also found for $t'/t > 1$ ($t'/t = 1.3$ in X = CF_3SO_3 [44]). This suggests that SC in κ-(BEDT-TTF)$_2$X is not related to the proximity to an AFM or QSL phase. This is also consistent with our present results, where we find little difference in the pair–pair correlations between $t' = 0.2$ where AFM is present and $t' = 0.6$ where AFM is absent.

Our results on the absence of SC in the effective $\frac{1}{2}$-filled Hubbard dimer model is in complete agreement with our previous results on the $\frac{1}{2}$-filled Hubbard monomer model [29,30,33]. Independent of whether one considers the dimers of BEDT-TTF molecules as effective single sites with charge occupancy +1 (previous calculations) or as dimers with monomer charge occupancies of $\frac{1}{2}$ each (present calculations), the superconducting pair correlations are found to be suppressed by Hubbard U continuously from $U = 0$. This fundamental observation was arrived at from exact diagonalization [29,33], Path Integral Renormalization Group [30] and now DMRG on completely different lattices, thereby confirming that the result is not an artifact of the choice of a specific lattice or a particular computational technique. Taken together, these results very clearly indicate that spin fluctuations are not driving organic SC. With hindsight, this conclusion should not be surprising: as already pointed out in the above, AFM occurs in only three κ-(BEDT-TTF)$_2$X, X = Cu[N(CN)$_2$Cl, deuterated X = Cu[N(CN)$_2$Br and CF_3SO_3. Whether or not X = $Cu_2(CN)_3$ and $Ag_2(CN)_3$ are true QSLs continue to be debated, and both claims of a CO phase below 6 K [45] as well its absence [46] exist in the literature. What is perhaps very relevant in this context is that transition from CO to SC is far more common in CT solids, including in the β, β', β'' and θ-families [14]. Even among the κ-phase materials, CO has been detected in X = Hg(SCN)$_2$Cl [47–49] while in the related material X = Hg(SCN)$_2$Br there occurs a dipolar liquid phase [50] the formation of which is related to the mechanism of CO formation [51,52]. While these latter materials are also strongly correlated, CO requires an effective $\frac{1}{4}$-filled (or $\frac{3}{4}$-filled) description of CT solids, within which intra-dimer charge dimer degrees of freedom are explicitly taken into account [14,53–56]. The charge-charge correlations in the CO phase in these systems are different from that in Wigner crystals. The CO phase is a paired-electron crystal, with nearest neighbor spin-singlet charge-rich molecules separated by pairs of charge-poor molecules. Pressure increases frustration, causing the spin singlets to be mobile. Calculations on an anisotropic triangular lattice [41] as well as on κ-lattices [42,57] have indeed found that superconducting pair correlations are enhanced by the Hubbard U uniquely at $\frac{1}{4}$-filling. DMRG calculations on a width-three triangular lattice found evidence for p-wave and d-wave SC in the density range 0.40∼0.65 [58], which would appear to agree with the conclusion from the previous work [41]. Additionally, electron-electron and electron-bond phonon interactions act co-operatively at this bandfilling to enhance superconducting pair correlations, even as they compete with one another at all other fillings [59]. These results are encouraging and suggest further research along this direction.

Author Contributions: Conceptualization and methodology, R.T.C. and S.M.; software, R.T.C.; Validation, formal analysis, investigation, visualization, D.R., R.T.C. and S.M.; data curation, D.R. and R.T.C.; writing, R.T.C. and S.M.; Supervision, project administration, R.T.C. and S.M.; funding acquisition, S.M. All authors have read and agreed to the published version of the manuscript.

Funding: S.M. acknowledges support by the National Science Foundation grant NSF-CHE-1764152.

Data Availability Statement: Not applicable.

Conflicts of Interest: The authors declare no conflict of interest.

References

1. Mazumdar, S. A unified theoretical approach to superconductors with strong Coulomb correlations: the organics, $LiTi_2O_4$, electron- and hole-doped copper oxides and doped $BaBiO_3$. In *Interacting Electrons in Reduced Dimensions. Proceedings of a NATO Advanced Research Workshop*; Baeriswyl, D., Campbell, D.K., Eds.; Plenum: New York, NY, USA, 1989; pp. 315–329.
2. Uemura, Y.J.; Le, L.P.; Luke, G.M.; Sternlieb, B.J.; Wu, W.D.; Brewer, J.H.; Riseman, T.M.; Seaman, C.L.; Maple, M.B.; Ishikawa, M.; et al. Basic similarities among cuprate, bismuthate, organic, Chevrel-phase, and heavy-fermion superconductors shown by penetration-depth measurements. *Phys. Rev. Lett.* **1991**, *66*, 2665–2668. [CrossRef]
3. Fukuyama, H.; Hasegawa, Y. Superconductivity in organics and oxides: Similarity and dissimilarity. *Physica B+C* **1987**, *148*, 204–211. [CrossRef]
4. Fukuyama, H. On some organic conductors in the light of oxide superconductors. In *Physics and Chemistry of Organic Superconductors. Proceedings of the ISSP International Symposium*; Saito, G., Kagoshima, S., Eds.; Springer: Berlin, Germany, 1990; pp. 15–20.
5. McKenzie, R.H. Similarities between organic and cuprate superconductors. *Science* **1997**, *278*, 820–821. [CrossRef]
6. Iwasa, Y.; Takenobu, T. Superconductivity, Mott-Hubbard states, and molecular orbital order in intercalated fullerides. *J. Phys. Condens. Matter* **2003**, *15*, R495. [CrossRef]
7. Capone, M.; Fabrizio, M.; Castellani, C.; Tosatti, E. Colloquium: Modeling the unconventional superconducting properties of expanded A_3C_{60} fullerides. *Rev. Mod. Phys.* **2009**, *81*, 943–958. [CrossRef]
8. Mazumdar, S.; Clay, R.T. Is there a common theme behind the correlated-electron superconductivity in organic charge-transfer solids, cobaltates, spinels and fullerides? *Phys. Stat. Solidi* **2012**, *249*, 995–998. [CrossRef]
9. Baskaran, G. RVB states in doped band insulators from Coulomb forces: Theory and a case study of superconductivity in BiS_2 layers. *Supercond. Sci. Technol.* **2016**, *29*, 124002. [CrossRef]
10. Jérome, D.; Mazaud, A.; Ribault, M.; Bechgaard, K. Superconductivity in a synthetic organic conductor $(TMTSF)_2PF_6$. *J. Phys. (Paris) Lett.* **1980**, *41*, L95–L98. [CrossRef]
11. Kanoda, K.; Kato, R. Mott physics in organic conductors with triangular lattices. *Annu. Rev. Condens. Matter Phys.* **2011**, *2*, 167–188. [CrossRef]
12. Kato, R. Special issue: Molecular conductors. *Crystals* **2012**, *2012*, 56–1482.
13. Brown, S.E. Organic superconductors: The Bechgaard salts and relatives. *Physica C* **2015**, *514*, 279–289. [CrossRef]
14. Clay, R.T.; Mazumdar, S. From charge- and spin-ordering to superconductivity in the organic charge-transfer solids. *Phys. Rep.* **2019**, *788*, 1–89. [CrossRef]
15. Dressel, M.; Tomić, S. Molecular quantum materials: electronic phases and charge dynamics in two-dimensional organic solids. *Adv. Phys.* **2020**, *69*, 1–120. [CrossRef]
16. Kino, H.; Kontani, H. Phase diagram of superconductivity on the anisotropic triangular lattice Hubbard model: An effective model of κ-(BEDT-TTF) salts. *J. Phys. Soc. Jpn.* **1998**, *67*, 3691–3694. [CrossRef]
17. Schmalian, J. Pairing due to spin fluctuations in layered organic superconductors. *Phys. Rev. Lett.* **1998**, *81*, 4232–4235. [CrossRef]
18. Kondo, H.; Moriya, T. Spin Fluctuation-Induced Superconductivity in Organic Compounds. *J. Phys. Soc. Jpn.* **1998**, *67*, 3695–3698. [CrossRef]
19. Vojta, M.; Dagotto, E. Indications of unconventional superconductivity in doped and undoped triangular antiferromagnets. *Phys. Rev. B* **1999**, *59*, R713–R716. [CrossRef]
20. Baskaran, G. Mott Insulator to High T_c Superconductor via Pressure: Resonating Valence Bond Theory and Prediction of New Systems. *Phys. Rev. Lett.* **2003**, *90*, 197007. [CrossRef]
21. Liu, J.; Schmalian, J.; Trivedi, N. Pairing and Superconductivity Driven by Strong Quasiparticle Renormalization in Two-Dimensional Organic Charge Transfer Salts. *Phys. Rev. Lett.* **2005**, *94*, 127003. [CrossRef]
22. Kyung, B.; Tremblay, A.M.S. Mott transition, antiferromagnetism, and d-wave superconductivity in two-dimensional organic conductors. *Phys. Rev. Lett.* **2006**, *97*, 046402. [CrossRef]
23. Yokoyama, H.; Ogata, M.; Tanaka, Y. Mott Transitions and *d*-Wave Superconductivity in Half-Filled-Band Hubbard Model on Square Lattice with Geometric Frustration. *J. Phys. Soc. Jpn.* **2006**, *75*, 114706. [CrossRef]

24. Watanabe, T.; Yokoyama, H.; Tanaka, Y.; Inoue, J. Superconductivity and a Mott Transition in a Hubbard Model on an Anisotropic Triangular Lattice. *J. Phys. Soc. Jpn.* **2006**, *75*, 074707. [CrossRef]
25. Sahebsara, P.; Sénéchal, D. Antiferromagnetism and Superconductivity in Layered Organic Conductors: Variational Cluster Approach. *Phys. Rev. Lett.* **2006**, *97*, 257004. [CrossRef] [PubMed]
26. Nevidomskyy, A.H.; Scheiber, C.; Sénéchal, D.; Tremblay, A.M.S. Magnetism and d-wave superconductivity on the half-filled square lattice with frustration. *Phys. Rev. B* **2008**, *77*, 064427. [CrossRef]
27. Sentef, M.; Werner, P.; Gull, E.; Kampf, A.P. Superconducting Phase and Pairing Fluctuations in the Half-Filled Two-Dimensional Hubbard Model. *Phys. Rev. Lett.* **2011**, *107*, 126401. [CrossRef]
28. Hebert, C.D.; Semon, P.; Tremblay, A.M.S. Superconducting dome in doped quasi-two-dimensional organic Mott insulators: A paradigm for strongly correlated superconductivity. *Phys. Rev. B* **2015**, *92*, 195112. [CrossRef]
29. Clay, R.T.; Li, H.; Mazumdar, S. Absence of superconductivity in the half-filled band Hubbard model on the anisotropic triangular lattice. *Phys. Rev. Lett.* **2008**, *101*, 166403. [CrossRef]
30. Dayal, S.; Clay, R.T.; Mazumdar, S. Absence of long-range superconducting correlations in the frustrated $\frac{1}{2}$-filled band Hubbard model. *Phys. Rev. B* **2012**, *85*, 165141. [CrossRef]
31. Watanabe, T.; Yokoyama, H.; Tanaka, Y.; Inoue, J. Predominant magnetic states in the Hubbard model on anisotropic triangular lattices. *Phys. Rev. B* **2008**, *77*, 214505. [CrossRef]
32. Tocchio, L.F.; Parola, A.; Gros, C.; Becca, F. Spin-liquid and magnetic phases in the anisotropic triangular lattice: The case of κ-$(ET)_2X$. *Phys. Rev. B* **2009**, *80*, 064419. [CrossRef]
33. Gomes, N.; Clay, R.T.; Mazumdar, S. Absence of superconductivity and valence bond order in the Hubbard-Heisenberg model for organic charge-transfer solids. *J. Phys. Condens. Matter* **2013**, *25*, 385603. [CrossRef] [PubMed]
34. Qin, M.; Chung, C.M.; Shi, H.; Vitali, E.; Hubig, C.; Schollwöck, U.; White, S.R.; Zhang, S. Absence of superconductivity in the pure two-dimensional Hubbard model. *Phys. Rev. X* **2020**, *10*, 031016. [CrossRef]
35. White, S.R. Density matrix formulation for quantum renormalization groups. *Phys. Rev. Lett.* **1992**, *69*, 2863–2866. [CrossRef]
36. Fishman, M.; White, S.R.; Stoudenmire, E.M. The ITensor Software Library for Tensor Network Calculations. *arXiv* **2020**, arXiv:2007.14822.
37. Stoudenmire, E.M.; White, S.R. Real-space parallel density matrix renormalization group. *Phys. Rev. B* **2013**, *87*, 155137. [CrossRef]
38. Chung, C.M.; Qin, M.; Zhang, S.; Schollwöck, U.; White, S.R. Plaquette versus ordinary d-wave pairing in the t'-Hubbard model on a width 4 cylinder. *Phys. Rev. B* **2020**, *102*, 041106. [CrossRef]
39. Luther, A.; Emery, V.J. Backward Scattering in the One-Dimensional Electron Gas. *Phys. Rev. Lett.* **1974**, *33*, 589–592. [CrossRef]
40. Aimi, T.; Imada, M. Does Simple Two-Dimensional Hubbard Model Account for High-T_c Superconductivity in Copper Oxides? *J. Phys. Soc. Jpn.* **2007**, *76*, 113708. [CrossRef]
41. Gomes, N.; De Silva, W.W.; Dutta, T.; Clay, R.T.; Mazumdar, S. Coulomb Enhanced Superconducting Pair Correlations in the Frustrated Quarter-Filled Band. *Phys. Rev. B* **2016**, *93*, 165110. [CrossRef]
42. Clay, R.T.; Gomes, N.; Mazumdar, S. Theory of triangular lattice quasi-one-dimensional charge-transfer-solids. *Phys. Rev. B* **2019**, *100*, 115158. [CrossRef]
43. Dolfi, M.; Bauer, B.; Keller, S.; Troyer, M. Pair correlations in doped Hubbard ladders. *Phys. Rev. B* **2015**, *92*, 195139. [CrossRef]
44. Ito, H.; Asai, T.; Shimizu, Y.; Hayama, H.; Yoshida, Y.; Saito, G. Pressure-induced superconductivity in the antiferromagnet κ-$(ET)_2CF_2(SO)_3$. *Phys. Rev. B* **2016**, *94*, 020503(R). [CrossRef]
45. Kobayashi, T.; Ding, Q.P.; Taniguchi, H.; Satoh, K.; Kawamoto, A.; Furukawa, Y. Charge disproportionation in the spin-liquid candidate κ-$(ET)_2Cu_2(CN)_3$ at 6 K revealed by ^{63}Cu NQR measurements. *Phys. Rev. Res.* **2020**, *2*, 042023(R). [CrossRef]
46. Sedlmeier, K.; Elsässer, S.; Neubauer, D.; Beyer, R.; Wu, D.; Ivek, T.; Tomic, S.; Schlueter, J.A.; Dressel, M. Absence of charge order in the dimerized κ-phase BEDT-TTF salts. *Phys. Rev. B* **2012**, *86*, 245103. [CrossRef]
47. Drichko, N.; Beyer, R.; Rose, E.; Dressel, M.; Schlueter, J.A.; Turunova, S.A.; Zhilyaeva, E.I.; Lyubovskaya, R.N. Metallic state and charge-order metal-insulator transition in the quasi-two-dimensional conductor κ-$(BEDT\text{-}TTF)_2Hg(SCN)_2Cl$. *Phys. Rev. B* **2014**, *89*, 075133. [CrossRef]
48. Hassan, N.M.; Thirunavukkuarasu, K.; Lu, Z.; Smirnov, D.; Zhilyaeva, E.I., Torunova, S.; Lyubovskaya, R.N.; Drichko, N. Melting of charge order in the low-temperature state of an electronic ferroelectric-like system. *NPJ Quant. Mater.* **2020**, *5*, 15. [CrossRef]
49. Gati, E.; Fischer, J.K.; Lunkenheimer, P.; Zielke, D.; Köhler, S.; Kolb, F.; von Nidda, H.A.K.; Winter, S.M.; Schubert, H.; Schlueter, J.A.; et al. Evidence for Electronically Driven Ferroelectricity in a Strongly Correlated Dimerized BEDT-TTF Molecular Conductor. *Phys. Rev. Lett.* **2018**, *120*, 247601. [CrossRef]
50. Hassan, N.; Cunningham, S.; Mourigal, M.; Zhilyaeva, E.I.; Torunova, S.A.; Lyubovskaya, R.N.; Schlueter, J.A.; Drichko, N. Evidence for a quantum dipole liquid state in an organic quasi–two-dimensional material. *Science* **2018**, *360*, 1101–1104. [CrossRef]
51. Hotta, C. Quantum electric dipoles in spin-liquid dimer Mott insulator κ-$(ET)_2Cu_2(CN)_3$. *Phys. Rev. B* **2010**, *82*, 241104(R). [CrossRef]
52. Naka, M.; Ishihara, S. Electronic Ferroelectricity in a Dimer Mott Insulator. *J. Phys. Soc. Jpn.* **2010**, *79*, 063707. [CrossRef]
53. Li, H.; Clay, R.T.; Mazumdar, S. The paired-electron crystal in the two-dimensional frustrated quarter-filled band. *J. Phys. Condens. Matter* **2010**, *22*, 272201. [CrossRef] [PubMed]
54. Dayal, S.; Clay, R.T.; Li, H.; Mazumdar, S. Paired electron crystal: Order from frustration in the quarter-filled band. *Phys. Rev. B* **2011**, *83*, 245106. [CrossRef]

55. Seo, H. Charge Ordering in Organic ET Compounds. *J. Phys. Soc. Jpn.* **2000**, *69*, 805–820. [CrossRef]
56. Kaneko, R.; Tocchio, L.F.; Valentí, R.; Becca, F. Charge orders in organic charge-transfer salts. *New J. Phys.* **2017**, *19*, 103033. [CrossRef]
57. De Silva, W.W.; Gomes, N.; Mazumdar, S.; Clay, R.T. Coulomb enhancement of superconducting pair-pair correlations in a $\frac{3}{4}$-filled model for κ-(BEDT-TTF)$_2$X. *Phys. Rev. B* **2016**, *93*, 205111. [CrossRef]
58. Venderley, J.; Kim, E.A. Density matrix renormalization group study of superconductivity in the triangular lattice Hubbard model. *Phys. Rev. B* **2019**, *100*, 060506. [CrossRef]
59. Clay, R.T.; Roy, D. Superconductivity due to cooperation of electron-electron and electron-phonon interactions at quarter-filling. *Phys. Rev. Res.* **2020**, *2*, 023006. [CrossRef]

crystals

Article

High-Pressure Crystal Structure and Unusual Magnetoresistance of a Single-Component Molecular Conductor [Pd(dddt)$_2$] (dddt = 5,6-dihydro-1,4-dithiin-2,3-dithiolate)

Hengbo Cui [1,*], Hamish H.-M. Yeung [2,3,*], Yoshitaka Kawasugi [1], Takaaki Minamidate [1], Lucy K. Saunders [4] and Reizo Kato [1,*]

[1] Condensed Molecular Materials Laboratory RIKEN, 2-1 Hirosawa, Wako-shi, Saitama 351-0198, Japan; kawasugi@riken.jp (Y.K.); takaaki.minamidate@riken.jp (T.M.)
[2] School of Chemistry, University of Birmingham, Edgbaston, Birmingham B15 2TT, UK
[3] Inorganic Chemistry Laboratory, Oxford University, South Parks Road, Oxford OX1 3QR, UK
[4] Beamline I19, Diamond Light Source, Harwell Campus, Didcot OX11 0DE, UK; lucy.saunders@diamond.ac.uk
* Correspondence: hcui@riken.jp (H.C.); h.yeung@bham.ac.uk (H.H.-M.Y.); reizo@riken.jp (R.K.)

Abstract: A single-component molecular crystal [Pd(dddt)$_2$] has been shown to exhibit almost temperature-independent resistivity under high pressure, leading theoretical studies to propose it as a three-dimensional (3D) Dirac electron system. To obtain more experimental information about the high-pressure electronic states, detailed resistivity measurements were performed, which show temperature-independent behavior at 13 GPa and then an upturn in the low temperature region at higher pressures. High-pressure single-crystal structure analysis was also performed for the first time, revealing the presence of pressure-induced structural disorder, which is possibly related to the changes in resistivity in the higher-pressure region. Calculations based on the disordered structure reveal that the Dirac cone state and semiconducting state coexist, indicating that the electronic state at high pressure is not a simple Dirac electron system as previously believed. Finally, the first measurements of magnetoresistance on [Pd(dddt)$_2$] under high pressure are reported, revealing unusual behavior that seems to originate from the Dirac electron state.

Keywords: single-component molecular conductor; pressure effect; Dirac electron system; resistivity; magnetoresistance; synchrotron X-ray diffraction; band calculation

Citation: Cui, H.; Yeung, H.H.-M.; Kawasugi, Y.; Minamidate, T.; Saunders, L.K.; Kato, R. High-Pressure Crystal Structure and Unusual Magnetoresistance of a Single-Component Molecular Conductor [Pd(dddt)$_2$] (dddt = 5,6-dihydro-1,4-dithiin-2,3-dithiolate). *Crystals* **2021**, *11*, 534. https://doi.org/10.3390/cryst11050534

Academic Editor: Martin Dressel

Received: 29 March 2021
Accepted: 5 May 2021
Published: 11 May 2021

1. Introduction

Most single-component molecular crystals are semiconductors or insulators because of large HOMO-LUMO energy band gaps (HOMO = Highest Occupied Molecular Orbital, LUMO = Lowest Unoccupied Molecular Orbital). Compared to other simple organic molecular crystals, metal–dithiolene complexes have small HOMO-LUMO gaps (~0.5 eV) and are, therefore, considered to be important candidates in the search for conducting single-component molecular systems. Through the last two decades of research, it has been revealed that chemical modifications, such as increasing π-conjugation and ligand planarity, can reduce the HOMO-LUMO band gap. This method, however, reduces the solubility of source materials and makes crystal growth very difficult. So far, only three single-component molecular metals, [M(tmdt)$_2$] (M = Ni, Au; tmdt = trimethylenetetrathiafulvalenedithiolate) [1,2] and [Au(Me-thiazdt)$_2$] (Me-thiazdt = N-methyl-1,3-thiazoline-2-thione-4,5-dithiolate) [3], and one single-component Dirac electron candidate [Pt(dmdt)$_2$] (dmdt = dimethyltetrathiafulvalenedithiolate) [4], have been reported under ambient conditions. On the other hand, molecular crystals have soft lattices and so the bandwidths of HOMO and LUMO bands may be easily enlarged by the application of external pressure. Therefore, the application of pressure has been found to be an efficient way to discover new single-component molecular metals and superconductors. Indeed, three pressure-induced single-component molecular metals, $[Ni(ptdt)_2]$ (ptdt = propylenedithiotetrathiafulvalenedithiolate) [5], $[Cu(dmdt)_2]$ [6] and $[Ni(dmit)_2]$ (dmit = 1,3-dithiole-2-thione-4,5-dithiolate) [7], have been found at pressures of 19.4 GPa, 4.7 GPa, and 15.9 GPa, respectively. Furthermore, we also discovered the first pressure-induced single-component molecular superconductor, $[Ni(hfdt)_2]$ (hfdt = bis(trifluoromethyl)tetrathiafulvalenedithiolate), at 8 GPa (T_c = 5.5 K) [8]. Lorcy and coworkers reported that single-component molecular crystals $[Au(Et\text{-}thiazdt)_2]$ (Et-thiazdt = N-ethyl-1,3-thiazoline-2-thione-4,5-dithiolate) [9] and $[Au(Et\text{-}thiazds)_2]$ [10] exhibit metallic conduction at the even lower pressures of 1.3 GPa and 0.6 GPa, respectively.

Because of the multi-orbital character of single-component molecular conductors, it is also expected that under certain conditions, more exotic electronic states may occur. For example, the HOMO and LUMO bands may be induced to touch at a single point to form a Dirac electron system. Indeed, we have found that a single-component molecular crystal, $[Pd(dddt)_2]$ (dddt = 5,6-dihydro-1,4-dithiin-2,3-dithiolate), forms a Dirac electron system under high pressure. First-principles density functional theory (DFT) calculations indicated that Dirac cones emerge from the crossing between the HOMO and LUMO bands, which originate from crystallographically independent layers within the crystal [11]. In that work, the HOMO-LUMO energy gap was observed to be 0.65 eV at ambient pressure, and the room temperature resistivity was higher than the measurable range of the equipment. Pressure-dependent resistivity measurements revealed that the room temperature resistivity (ρ_{rt}) and activation energy (E_a) decrease sharply with increasing pressure above 4.2 GPa and increase slowly with increasing pressure above 12.6 GPa [12]; electrical properties in the lower pressure region are still unknown.

$[Pd(dddt)_2]$ belongs to the monoclinic crystal system with space group $P2_1/n$ (no. 14) [11]. The unit cell parameters at ambient pressure are as follows: a = 17.8698(82)) Å, b = 4.7281(19) Å, c = 18.4657(85) Å, β = 111.6313(63)°, V = 1450.3(1) Å^3, Z = 4. Figure 1 shows the molecular structure and crystal structure of $[Pd(dddt)_2]$. $[Pd(dddt)_2]$ molecules are uniformly stacked along the b axis, where Pd atoms are located at inversion centers. The Pd···Pd distance is 4.728 Å, which is identical to the magnitude of the b axis. The structure may further be thought of as crystallographically independent layers stacked in the a–c direction: Layer 1 contains molecules in which the Pd atoms are located on the 2b Wyckoff sites at (0, 0, 1/2) and (1/2, 1/2, 0), while Layer 2 contains molecules in which the Pd atoms are located on the 2a Wyckoff sites at (0, 0, 0) and (1/2, 1/2, 1/2). It is this layered structure that gives rise to the Dirac electron state at high pressure.

In other materials, the Dirac electron system shows interesting quantum magnetotransport phenomena, such as quantum Hall effects (QHEs) reported in graphene [13] and topological insulators [14], and linear relationships of magnetoresistance (MR = $[\rho(B) - \rho(0)]/\rho(0)$) with magnetic field in iron pnictide, $Ba(FeAs)_2$, single crystals, when the magnetic field is perpendicular to the current direction at low temperature [15]. Magnetoresistance (MR) measurements in a bulk two-dimensional Dirac electron system, α-$(BEDT\text{-}TTF)_2I_3$, showed an unusual negative interlayer MR in the low magnetic field region [16]. In the case of $[Pd(dddt)_2]$, no MR measurements have been reported to date. Theoretical calculations have shown that it is a nodal line semimetal, in which the Dirac points form a loop in three-dimensional reciprocal space [17,18], which may give rise to further interesting MR behavior.

To better understand the high-pressure behavior of $[Pd(dddt)_2]$, we report a series of new measurements and calculations that verify previous theoretical calculations and provide insight into the Dirac electron state. In particular, we report more detailed high-pressure resistivity measurements, the first experimental structure determinations performed at high pressure using single-crystal synchrotron X-ray diffraction, tight binding calculations performed using these data, and the first high-pressure MR measurements performed on single crystals of $[Pd(dddt)_2]$. We find that the new experimental crystal structures verify previous theoretical reports of the Dirac electron state and suggest that changes in resistivity may be affected by structural disorder. Tight binding calculations indicate that the Dirac electron state coexists with semiconducting behavior, also owing to the disorder. The new MR measurements are also indicative of exotic quantum transport behavior: negative MR is observed at low temperatures and the curvature of the field-dependent MR is heavily dependent on crystal orientation.

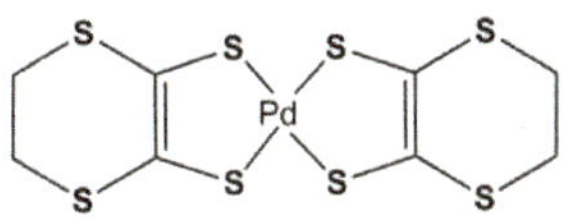

[Pd(dddt)2]

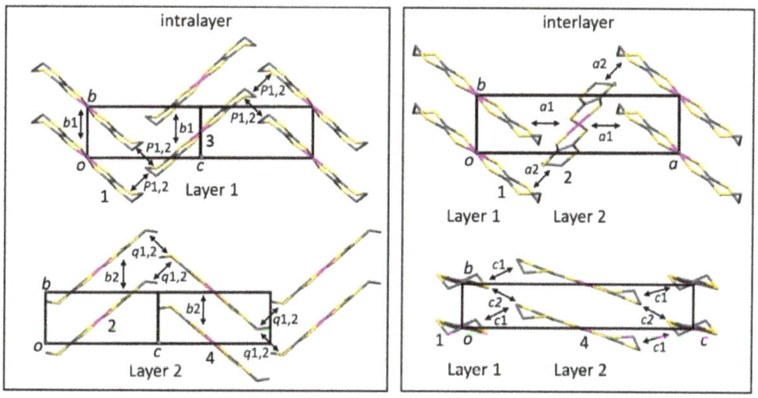

Figure 1. Molecular structure and crystal structure of [Pd(dddt)$_2$] at 5.9 GPa with symbols of intermolecular couplings. Palladium, sulfur and carbon atoms are shown in pink, yellow and grey, respectively. Hydrogen atoms are omitted for clarity.

2. Materials and Methods

2.1. High-Pressure Electrical Resistivity

Four-probe high-pressure resistivity measurements were performed by using the same procedure as previously performed for [Ni(ptdt)$_2$] [5]. A culet size of 0.56 mm DAC and Inconel 625 was used as the metal gasket. Four contacts were made from 5 μm gold wires and gold paint, and Daphne Oil 7373 was used as the pressure medium. The sample was encapsulated with a mixture of Araldite AR-S30 (Huntsman) and alumina powder (see Figure 2a). A small Ruby chip was attached to the center of the protective layer and pressure was determined by measuring the shift of Ruby fluorescence R1 lines at room temperature. Samples 5 and 6 were measured using the cryocooler helium compressor system (Sumitomo Heavy Industries, Ltd.) for cooling the DAC with a cooling rate of 1.5 K/min. Samples 7 and 8 were measured using a Quantum Design physical property measurement system (PPMS) with a cooling rate of 0.5 K/min.

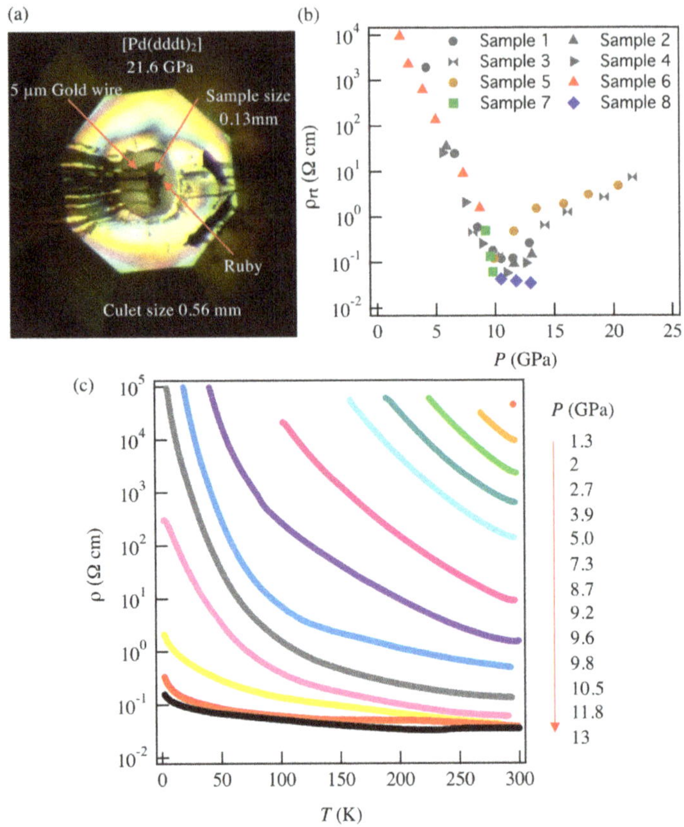

Figure 2. (**a**) Optical microscope image of the $[Pd(dddt)_2]$ sample at 21.6 GPa assembled in a Diamond Anvil Cell (DAC). (**b**) The pressure dependence of room temperature resistivity (samples 1~4 = data from previous work [11]). (**c**) Temperature-dependent resistivity ρ under various pressures from 1.3 GPa to 13 GPa.

2.2. High-Pressure Single Crystal Structure Determination

The high-pressure single-crystal structure measurements were performed at beamline I19-2, Diamond Light Source synchrotron, UK using 0.4859 Å radiation. A single crystal of $[Pd(dddt)_2]$ was coated in a thin layer of Araldite and placed in a DAC with culet size 0.6 mm and tungsten gasket and 4:1 methanol:ethanol was used as the pressure medium. The pressure was determined by the shift of Ruby fluorescence R1 lines at room temperature, before and after each measurement. Data were processed via in-house semi-automated routines using aimless [19], ccp4 [20], dials [21], pointless [22], and xia2 [23], and solved and refined using SHELX [24] within Olex2 [25].

The ambient pressure structure refinement was performed using riding H atoms. Owing to low completeness of the high-pressure datasets, only Pd atoms were refined with anisotropic ADPs, which were restrained using the ISOR command in the 5.9 GPa and 10.6 GPa structures. Structural disorder on the S/CH_2 moiety was modeled in two parts with refined occupancies, using two S positions. The C atom was modeled on the same position in both cases with sets of riding H atoms allowed to vary according to the respective S atom position. The S–C distance was restrained to 1.76(2) Å using the DFIX command, and the S atoms were constrained to have the same isotropic ADPs. In addition, terminal CH_2–CH_2 distances were restrained to 1.54(2) Å.

2.3. Tight-Binding Calculation

Calculations of molecular orbitals and intermolecular overlap integrals (S) between frontier molecular orbitals were carried out by the extended Hückel method. Reported sets of semi-empirical parameters for Slater-type atomic orbitals and valence shell ionization potentials for H [26], C [26], S [27], and Pd [28] were used for the calculations. Intermolecular transfer integrals, t (eV), were estimated using the equation $t = -10S$. Since the unit cell contains four [Pd(dddt)$_2$] molecules (molecules 1, 2, 3, 4: Figure 1), four HOMOs (H1, H2, H3, H4) and four LUMOs (L1, L2, L3, L4) are considered in the calculation. For later discussions on the electrical structure, we define a new cell as $a = -(a_o + c_o)$, $b = -b_o$, $c = c_o$, where a_o, b_o, and c_o are the original lattice vectors [11]. In the new unit cell, the a-axis is parallel to Layers 1 and 2. The band energies E (**k**) are obtained as eigenvalues of an 8×8 Hermite matrix **H** ($\boldsymbol{k}$), where a wavevector $\boldsymbol{k} = k_x\boldsymbol{a}^* + k_y\boldsymbol{b}^* + k_z\boldsymbol{c}^* = (k_x, k_y, k_z)$, and $\boldsymbol{ka} = 2\pi k_x$, $\boldsymbol{kb} = 2\pi k_y$, and $\boldsymbol{kc} = 2\pi k_z$. Matrix elements h_{mn} described in Appendix A correspond to the HOMO–HOMO couplings (m, n = 1–4), the LUMO–LUMO couplings (m, n = 5–8), and the HOMO–LUMO couplings (m = 1–4, n = 5–8 and m = 5–8, n = 1–4). Intermolecular transfer integrals between frontier molecular orbitals in h_{mn} are shown in Tables 1–5. The intralayer transfer integrals are given by $b1$ and p (1,2) for Layer 1, and $b2$ and q(1,2) for Layer 2. The interlayer transfer integrals are represented by a (1,2) and c (1,2). These transfer integrals are indexed by H (between HOMOs), L (between LUMOs), and HL (between HOMO and LUMO). In the case of the disordered structure, we examined two extreme crystal structures in which one of the possible conformations for molecule 2(4) is fully occupied (Structures A and B).

Table 1. HOMO–HOMO (H–H), LUMO–LUMO (L–L), and HOMO–LUMO (H–L) transfer integrals (meV) at 5.9 GPa.

	H–H	L–L	H–L	
$b1$	209.3	−1.9	−51.2	
$p1$ (p)	28.1	−12.4	19.9	Layer 1
$p2$	—	—	17.1	
$b2$	49.9	−80.4	−67.2	
$q1$ (q)	10.8	8.1	9.3	Layer 2
$q2$	—	—	9.2	
$a1$	−28.2	14.6	−20.1	
$a2$	2.2	1.3	−1.7	Interlayer
$c1$	15.4	12.7	14.1	
$c2$	−3.9	15.8	−11.8	

Table 2. HOMO–HOMO (H–H), LUMO–LUMO (L–L), and HOMO–LUMO (H–L) transfer integrals (meV) at 7.8 GPa (Structure A).

	H–H	L–L	H–L	
$b1$	190.2	53.2	−34.3	
$p1$ (p)	28.1	−17.0	23.1	Layer 1
$p2$	—	—	20.5	
$b2$	−0.9	−121.5	−71.9	
$q1$ (q)	7.3	7.2	8.5	Layer 2
$q2$	—	—	6.1	
$a1$	−37.7	19.0	−26.6	
$a2$	3.3	2.7	−3.1	Interlayer
$c1$	9.1	6.0	7.3	
$c2$	1.2	9.4	−5.6	

Table 3. HOMO–HOMO (H–H), LUMO–LUMO (L–L), and HOMO–LUMO (H–L) transfer integrals (meV) at 7.8 GPa (Structure B).

	H–H	L–L	H–L	
*b*1	190.2	53.2	−34.3	
*p*1 (*p*)	28.1	−17.0	23.1	Layer 1
*p*2	—	—	20.5	
*b*2	107.2	−5.4	−46.2	
*q*1 (*q*)	11.2	11.0	12.1	Layer 2
*q*2	—	—	10.2	
*a*1	−32.9	16.2	−22.9	
*a*2	2.0	1.9	−2.0	Interlayer
*c*1	−7.7	−15.2	−12.6	
*c*2	−9.3	12.4	−11.3	

Table 4. HOMO–HOMO (H–H), LUMO–LUMO (L–L), and HOMO–LUMO (H–L) transfer integrals (meV) at 10.6 GPa (Structure A).

	H–H	L–L	H–L	
*b*1	203.4	6.1	−67.5	
*p*1 (*p*)	24.7	−13.4	18.6	Layer 1
*p*2	—	—	17.7	
*b*2	−0.3	−135.0	−81.6	
*q*1 (*q*)	7.8	7.2	9.3	Layer 2
*q*2	—	—	5.9	
*a*1	−36.3	21.4	−28.4	
*a*2	4.9	4.1	−4.6	Interlayer
*c*1	7.6	4.8	6.0	
*c*2	−5.0	17.9	−12.5	

Table 5. HOMO–HOMO (H–H), LUMO–LUMO (L–L), and HOMO–LUMO (H–L) transfer integrals (meV) at 10.6 GPa (Structure B).

	H–H	L–L	H–L	
*b*1	203.4	6.1	−67.5	
*p*1 (*p*)	24.7	−13.4	18.6	Layer 1
*p*2	—	—	17.7	
*b*2	143.9	32.7	−49.8	
*q*1 (*q*)	13.9	12.9	14.8	Layer 2
*q*2	—	—	12.2	
*a*1	−31.6	21.1	−26.4	
*a*2	2.0	2.4	−2.4	Interlayer
*c*1	−6.4	−15.3	−11.7	
*c*2	−9.6	9.3	−9.3	

2.4. Magnetoresistance Measurements

Magnetoresistance measurements were performed with a Quantum Design physical property measurement system (PPMS) in a magnetic field range of −8 to 8 T in the range of 2–100 K. As shown in Figure 3a,b, the orientation of the four-probe attachment on samples was such that the magnetic field was either perpendicular or parallel to the current direction. As for samples 7 and 8, magnetoresistance was measured after the measurement of the temperature dependence of resistivity down to 2 K.

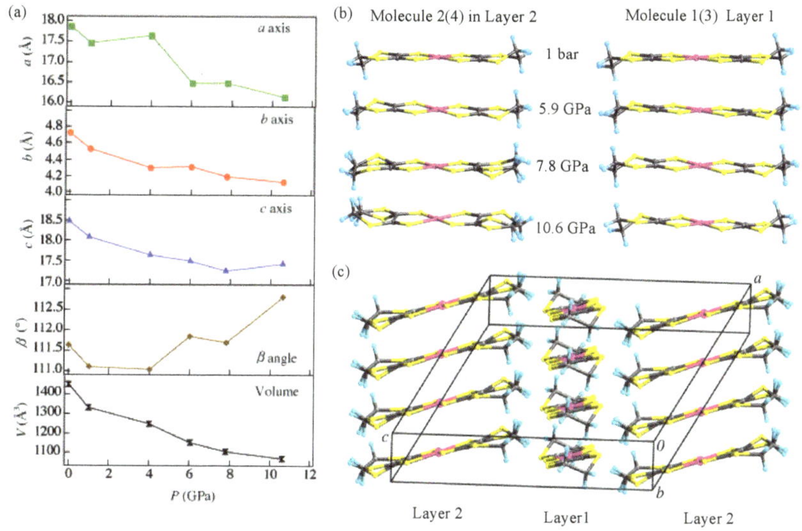

Figure 3. (**a**) Pressure dependence of cell parameters a, b, c, β and V. (**b**) Crystallographically independent [Pd(dddt)$_2$] molecules at different pressure in Layer 1 and Layer 2. (**c**) Crystal structure of [Pd(dddt)$_2$] at 7.8 GPa. Palladium, sulfur, carbon and hydrogen atoms are shown in pink, yellow, grey and cyan, respectively.

3. Results

3.1. High-Pressure Electrical Properties

A series of new high-pressure electrical resistivity measurements was performed using a diamond anvil cell (DAC) (Figure 2a), in which the pressure was carefully controlled for closer inspection of the material's behavior. The resistivity was measurable from 1.3 GPa, at which pressure the room temperature conductivity (σ_{rt}) was 2×10^{-5} S cm^{-1} (Figure 2b). The temperature dependence of the resistivity was measured from 2 GPa, at which pressure the activation energy (E_a) was 0.29 eV. The room temperature resistivity (ρ_{rt}) and E_a were found to initially decrease with increasing pressure then, at 13 GPa, a weak metal-like behavior was observed from room temperature to 230 K (see Figure 2c). At pressures higher than 13 GPa, ρ_{rt} and E_a were found to increase with increasing pressure, as observed previously [12]. Measurements performed on multiple samples (Figure 2b) and at multiple pressures (Figure 2c) show that the temperature dependence of the resistivity under various pressures is reproducible and consistent with previous results [11].

3.2. High-Pressure Single-Crystal Structure

To provide experimental structural information to explain the physical properties, high-pressure single-crystal X-ray diffraction measurements were performed. The single crystal of [Pd(dddt)$_2$] used was from the same batch of crystals that had been synthesized for high-pressure electrical property measurements. Unlike previously reported high-pressure single-crystal structure measurements of the isostructural system [Ni(dddt)$_2$] [29], crystals of [Pd(dddt)$_2$] suffered from twinning even at pressures below 1 GPa. To avoid twinning of [Pd(dddt)$_2$] crystals, the crystal was encapsulated with layer of Araldite epoxy resin, made thin to reduce background scattering. The improved hydrostatic conditioning effect of a mixture of alumina and araldite layers was demonstrated by previous DAC high-pressure resistivity measurements [5].

High-pressure diffraction data collection was performed at increasing pressures of 0.1 MPa (ambient pressure), 1.0 GPa, 4.0 GPa, 5.9 GPa, 7.8 GPa, 8.4 GPa, 10.6 GPa and 12.4 GPa (see Supplementary Materials). The monoclinic system and $P2_1/n$ space group of the ambient pressure structure were maintained at all measured pressures. Figure 3a shows the pressure dependence of unit cell parameters. The unit cell lengths and volume were found to gradually decrease with increasing pressure: at 10.6 GPa, the *a*, *b* and *c* axes reduce to 0.9, 0.87 and 0.94 times their ambient pressure values, respectively. On the other hand, the β angle decreases from 111.631° at 0 GPa to 110.04° at 4 GPa, then increases again to 112.8° at 10.6 GPa. At this pressure, the change in unit cell volume is reduced to 73.6% of its ambient pressure value, reflecting the soft nature of the organic crystal. Interestingly, at 4.0 GPa, coupling between the mechanical responses of the different axes results in decreases in *b* and *c*, alongside an increase in *a*. Unlike in well-understood winerack, honeycomb, helical or polyhedral framework structures [30], there is not an obvious structural motif that is responsible for this complex behavior in [Pd(dddt)$_2$]. However, it is not unusual for certain axes to expand upon compression (so-called negative linear compressibility) so long as the fundamental thermodynamic requirement for volume to decrease as pressure increases is obeyed, which is true in this case.

Crystal structures could be solved and refined at ambient pressure (0.1 MPa), 5.9 GPa, 7.8 GPa and 10.6 GPa. Unfortunately, degradation of crystal quality prevented structure refinement at 1.0 GPa, 4.0 GPa, 8.4 GPa and 12.4 GPa. At ambient pressure, there was no evidence of structural disorder. At 5.9 GPa, the crystal structure was also refined without disorder [31]. It should be mentioned that the data quality was not good enough to completely eliminate the existence of a disordered state. The crystal structure could be solved even with disorder, with similar refinement statistics to the ordered structure; however, when using anisotropic atomic displacement parameters (ADPs) for sulfur atoms, the disorder appears to be absent. We believe that better single-crystal structure measurements will be needed in the future to determine the structure with greater precision.

In all successful structure determinations at high pressure above 5.9 GPa, evidence was found in the residual electron density for disorder at one of the thin S atoms in molecule 2(4) within Layer 2 (Figure 1 or Figure 3b). Modeling this as 2-fold disorder resulted in a decrease in the residual factor R_1 of approximately 1% for each dataset. In addition, the atomic displacement parameter for the neighboring CH_2 unit was larger than average, indicating it may also be affected by the positional disorder. The cause of the disorder is likely to be increasing steric congestion at high pressure, which forces the molecules to adopt one of two conformations, giving rise to two distinct CH-S hydrogen bonds (Figure A1).

While the *b*-axis (i.e., the intermolecular Pd–Pd distance) decreases monotonously with increasing pressure, as shown in Figure 3a, changes in intramolecular distances are less smooth and vary depending on the molecule. At ambient pressure, all Pd–S distances are almost identical (2.26 Å). At 5.9 GPa the values decrease to within the range 2.21–2.24 Å, before diverging further at 7.8 GPa (2.21–2.26 Å) and 10.6 GPa (2.18–2.29 Å). In the latter case, Pd–S distances in Layer 2 are approximately 0.08 Å shorter that those found in Layer 1. This typifies the difference between the molecules that underlies their different contributions to the electronic structure at high pressure. C–C distances within the conjugated part of each molecule increase with increasing pressure, further suggesting subtle changes in electronic structure. At ambient pressure, they are 1.37 Å and 1.40 Å, increasing to 1.38/1.41 Å, 1.45/1.44 Å and 1.45/1.50 Å at 5.9 GPa, 7.8 GPa and 10.6 GPa, respectively.

The application of high pressure causes increasing distortion of the [Pd(dddt)$_2$] molecules. Bond angles, such as S–Pd–S angles, vary between the molecules in question; however, owing to low precision in atomic positions, there is insufficient evidence to indicate whether or not the application of high pressure causes any significant changes. On the other hand, with the exception of half of the CH_2 groups, the molecules are flat at ambient pressure but buckle significantly under pressure, owing to increasing steric congestion. This effect can be most clearly seen in the terminal CH_2-CH_2 groups, which bend out of the plane at 7.8 GPa (Figure 3b,c) and 10.6 GPa.

Structural disorder in Layer 2 (molecule 2(4)) also has a significant effect on the distance between adjacent molecules along the stacking direction. The structure, denoted as Structure A, in which all sulfur atoms, including the disordered atom (S8A), are almost coplanar, is descended from the low-pressure molecular structure. On the other hand, in the alternative structure, denoted as Structure B, the disordered sulfur atom (S8B) in the outer heteroring deviates significantly from the molecular plane (Figure 4). The occupancy of S8A and S8B (i.e., the probability of Structure A and Structure B existing in any part of the crystal) are 0.55 and 0.45 at 7.8 GPa, and 0.64 and 0.36 at 10.6 GPa, respectively. As shown in Table 6, this protruding sulfur atom provides a very short intermolecular S···S distance (*d*1) between molecules 2(4) in the stacking direction.

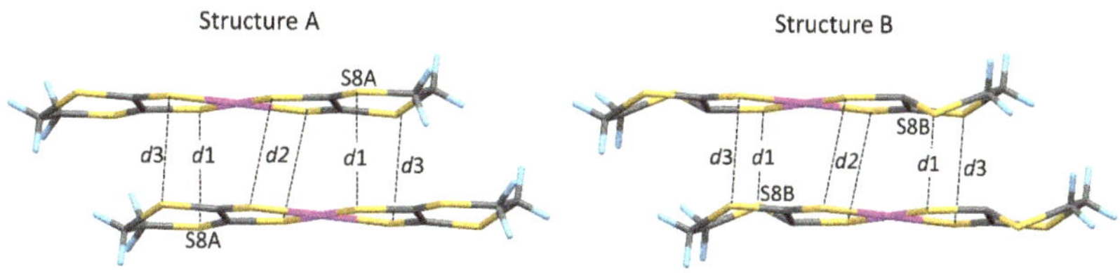

Figure 4. Two extreme crystal structures in which one of the possible conformations for molecule 2(4) is fully occupied (Structures A and B).

Table 6. Intermolecular S···S distances (Å) between Molecules 2(4) in Layer 2 along the stacking direction.

	Ambient Pressure	5.9 GPa	7.8 GPa		10.6 GPa	
			Structure A	Structure B	Structure A	Structure B
*d*1	3.953	3.612	3.780	3.199	3.707	3.028
*d*2	3.922	3.596	3.484	3.484	3.517	3.517
*d*3	3.850	3.547	3.453	3.453	3.352	3.352

3.3. Band Structure

To understand the high-pressure electronic states, the experimentally determined crystal structures were examined using a tight-binding band model based on the extended Hückel molecular orbital method [32]. Focusing on the 5.9 GPa structure, we reported details of the theoretical model and mechanism of the Dirac cone formation separately [31]. The measured cell volume at 5.9 GPa (1152.8 Å^3) is close to that obtained by the DFT calculation for the theoretically optimized 8 GPa structure (1147.5 Å) in the previous work [11]. The calculated transfer integrals for the 5.9 GPa structure (Table 1) provide considerable difference from the previous tight-binding model [31]. An important point is the existence of the direct interlayer HOMO–LUMO couplings between molecule 1(3) and molecule 2(4) (h_{16}, h_{18}, h_{36}, h_{38} and their complex conjugate elements) that were neglected in the previous model. The present model, however, reproduces the essential features of the DFT band calculations well. Figure 5 shows the electronic band structure at 5.9 GPa connecting TRIMs (time reversal invariant momenta) where the Dirac cone emerges on the Y–G line. In single-component molecular conductors, the Dirac cone can be formed at a point where the HOMO and LUMO bands cross without the HOMO–LUMO couplings [33]. As for the present system, the Dirac cone originates from the HOMO band in Layer 1 and the LUMO band in Layer 2 (Figure 6). The HOMO band is convex upward and the LUMO is convex downward, which leads to the HOMO-LUMO band crossing. This is because the major transfer integrals, $b1_H$ in Layer 1 and $b2_L$ in Layer 2, have opposite signs. The interlayer HOMO-LUMO couplings remove the degeneracy at the intersection of the HOMO and LUMO bands and open an energy gap. In this system, however, there is a surface on which the HOMO-LUMO interaction is zero and gap formation cannot occur. When this nodal surface of the HOMO-LUMO couplings meets a crossing plane of the HOMO and LUMO bands, the Dirac cones emerge in the *k*-space. The present model, revealed by the new X-ray diffraction data, includes both direct and indirect interlayer HOMO-LUMO couplings, and both of them play an important role in the Dirac cone formation [31]. The indirect HOMO–LUMO couplings are obtained by a second-order perturbation using the intralayer HOMO–LUMO and interlayer LUMO–LUMO or HOMO–HOMO couplings [11].

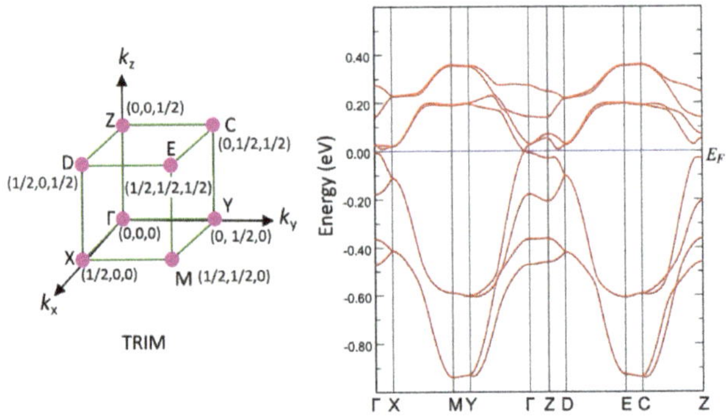

Figure 5. Electronic band structure of [Pd(dddt)$_2$] at 5.9 GPa connecting TRIMs.

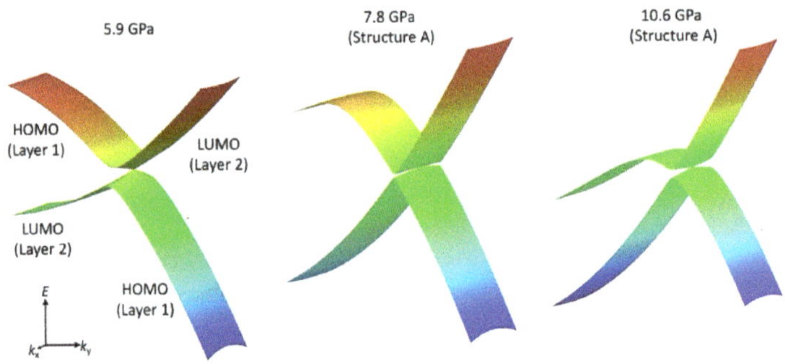

Figure 6. Dirac cone at $k_z = 0$ in [Pd(dddt)$_2$] at 5.9 GPa, 7.8 GPa and 10.6 GPa.

A unique feature of this Dirac electron system is that the Dirac point describes a closed line (loop) within the first Brillouin zone (Figure 7). The Dirac point in the single-component molecular Dirac electron system is located near the Fermi level, because charge carriers are generated by electron transfer from the fully occupied HOMO band to the completely empty LUMO band. The energy at the Dirac point E_D shows slight variations from the Fermi energy E_F on the loop, which gives hole and electron pockets (Figure 8). This means that the system is a nodal line semimetal. The deviation $\delta E = E_D - E_F$ is very small ($|\delta E| < 0.4$ meV) at 5.9 GPa.

In discussing the electronic structure at higher pressures (7.8 GPa and 10.6 GPa) with the disordered crystal structure, we start with Structure A, because Structure A relates more closely with the low-pressure structures including the 5.9 GPa structure. Indeed, the electronic band structures of Structure A at 7.8 and 10.6 GPa indicate that the system remains a nodal line semimetal (Figures 6, 7, 9 and 10). One effect of the higher pressure is enlargement of the nodal loop. This can be explained by an enhancement of the LUMO–LUMO coupling $b2_L$ (Tables 2 and 4), which gives wider LUMO bands and expanded contact of the HOMO and LUMO bands. Another effect of the increased pressure is an enhancement of the deviation from the Fermi level, which lies within the range -6.7 meV $< \delta E <$ 8.7 meV at 7.8 GPa and -3.6 meV $< \delta E <$ 5.2 meV at 10.6 GPa. This results in enlarged hole and electron pockets (Figure 8). In addition, it should be noted that an inversion of the hole and electron pockets occurs at 7.8 and 10.6 GPa (Figure 7).

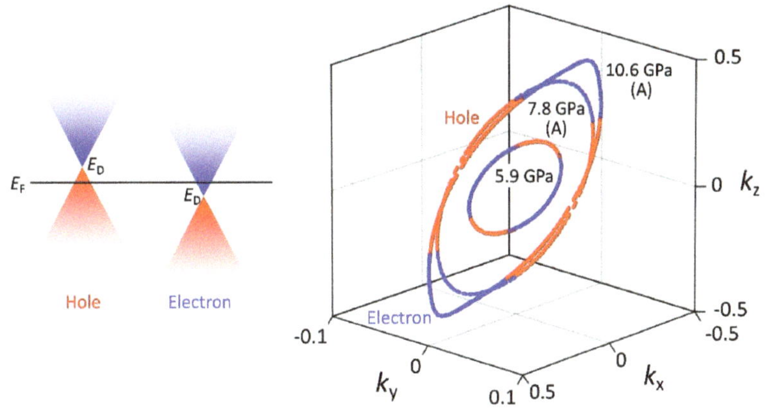

Figure 7. Nodal loops in [Pd(dddt)$_2$] at 5.9 GPa, 7.8 GPa and 10.6 GPa. Points with hole-like characteristics are indicated in red and points with electron-like characteristics in blue.

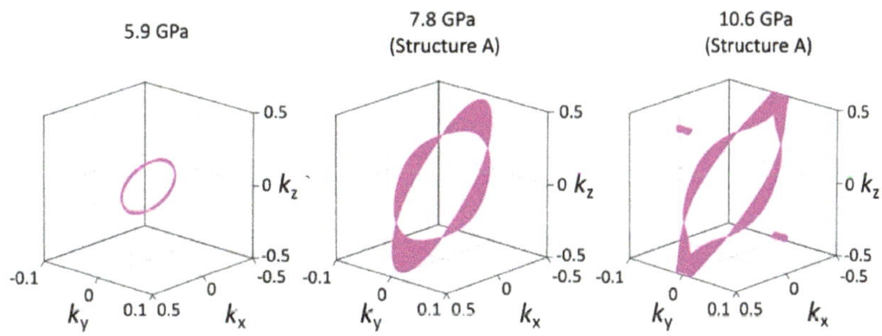

Figure 8. Fermi surfaces in [Pd(dddt)$_2$] at 5.9 GPa, 7.8 GPa and 10.6 GPa.

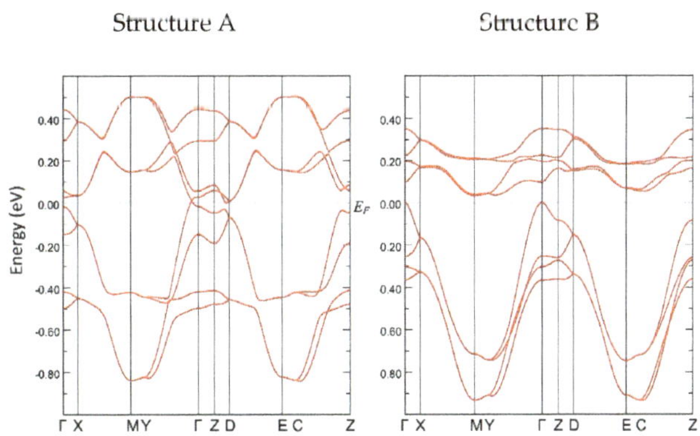

Figure 9. Electronic band structure of [Pd(dddt)$_2$] at 7.8 GPa connecting TRIMs.

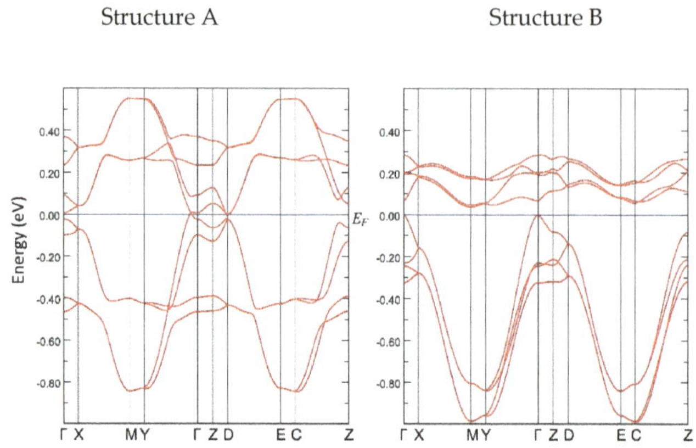

Figure 10. Electronic band structure of [Pd(dddt)$_2$] at 10.6 GPa connecting TRIMs.

Structure B is only found at higher pressures. Tables 3 and 5 indicate that the very short intermolecular S···S distance ($d1$) between molecules 2(4) in the stacking direction significantly affects intermolecular couplings $b2$ between frontier molecular orbitals ($b2_H$, $b2_L$, $b2_{HL}$). In particular, the change of $b2_L$ reduces the bandwidth of the LUMO bands and varies the curvature of the LUMO bands, which provides a semiconducting band structure with no Dirac cone (Figures 9 and 10). The drastic difference in the electronic state of structures A and B suggests that the system is not a simple Dirac electron system when disorder is present.

3.4. *Magnetoresistance*

Temperature-dependent anisotropic MR measurements under various pressures were performed on [Pd(dddt)$_2$] crystals by measuring the electrical response in orientations parallel to and perpendicular to the crystallographic b-axis. Figure 11a schematically shows the crystal shape, axes, and orientation of the four-probe attachment on the samples, as determined by single-crystal X-ray diffraction face-indexing. The first measurement, in which the magnetic field was perpendicular to the direction of electrical current, was performed at 11.8 GPa (Sample 8). At this pressure, the resistivity is almost temperature independent (Figure 2) and the system may be expected to be closest to the Dirac electron state. However, only a small signature of conventional semiconductor MR (i.e., a parabolic response) was observed below 10K (see Figure A2).

When the pressure was reduced to 10.5 GPa, a small negative MR is observed in the low magnetic field region ($B < 1$ T) at all measured temperatures, and linear MR appears in the low temperature and the high magnetic field region (Figures 11b and A3). The minimum value of the negative MR at 2 K was −0.86% and the maximum value was 9.14%. At higher temperatures, MR was reduced to less than 1%; however, negative MR at fields below 1 T and a linear relationship between the MR and magnetic field at higher fields was still observed up to 70 K. Interestingly, when the magnetic field and current directions were perpendicular, the MR effect increased at lower pressures. The second crystal (Sample 7) was mounted in the same orientation as shown in Figure 11a, i.e., with the magnetic field perpendicular to the direction of electrical current. Figure 11c shows temperature dependence of MR at 9.6 GPa. A relatively large MR was observed at 2 K: the maximum MR was 42.29% and the minimum negative MR was −2.41%.

The most interesting MR was observed when the magnetic field and current directions were parallel. Figure 11d shows a schematic of the crystal shape, axes and the orientation of the four-probe attachment on Sample 9, for which the temperature dependence of resistivity is shown in Figure A4. The σ_{rt} and E_a were 94 S cm^{-1}and 38 meV at 8.7 GPa, and 219 S cm^{-1} and 19 meV at 9.3 GPa, respectively. At 9.3 GPa, a relatively large negative MR was observed below 1 T over a wide temperature range from 2 K to 50 K and, in the B >1 T region, the MR showed a linear relationship with magnetic field (see Figures 11e and A5). At 2 K, the MR showed a maximum value of 6.56% and the negative MR decreased with increasing temperature up to 10 K, reaching −4.2%. At higher temperatures, both positive and negative MR decreased with increasing temperature, with no MR observed at 100 K. At 8.7 GPa, the linear relationship above 1 T and negative MR below 1 T was similar to that obtained at 9.3 GPa (Figure 11f). The maximum MR was 7.07% and minimum negative MR was −3.53%. Interestingly, at these two pressures, the MR at different temperatures was almost identical between 2 K and 20 K. At high temperatures, MR decreased and the negative to positive transition field remained constant.

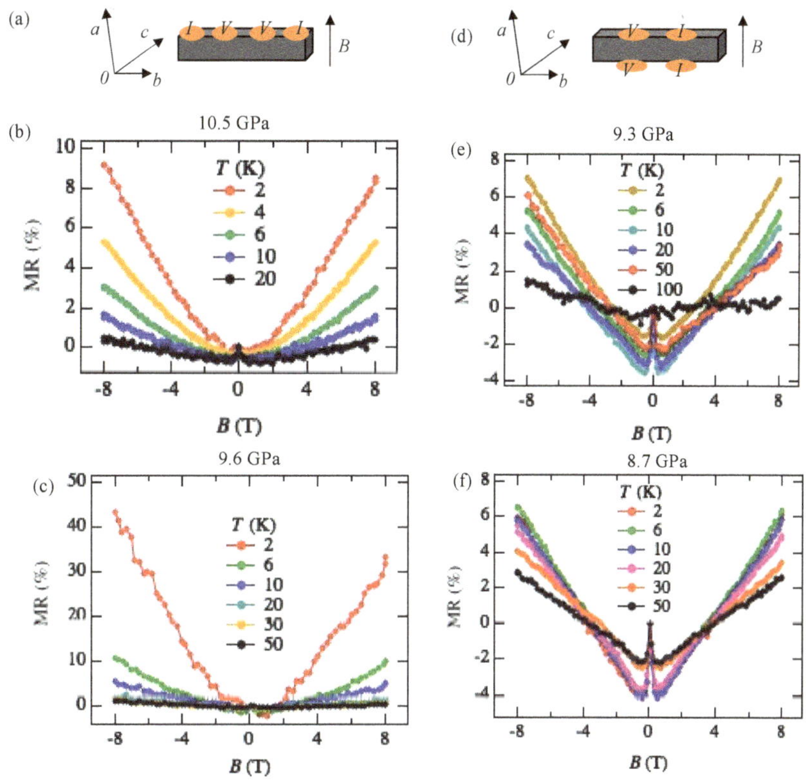

Figure 11. (**a**) Schematic of the crystal shape, axes and orientation of the four-probe attachment on Samples 7 and 8, in which the magnetic field (B) is perpendicular to the current direction. (**b**) Temperature (T) dependence of magnetoresistance (MR) as a function of B at 10.5 GPa (Sample 8). (**c**) Temperature dependence of MR as a function of B at 9.6 GPa (Sample 7). (**d**) Schematic of the crystal shape, axes and the orientation of the four-probe attachment on Sample 9, in which the magnetic field is parallel to the current direction. (**e**) MR as a function of B and T at 9.3 GPa (Sample 9). (**f**) MR as a function of B and T at 8.7 GPa (Sample 9).

4. Discussion

For isostructural single-component molecular conductors, the contribution of the central transition metal d-orbitals to the LUMO is often assumed to be similar. However, our experimental results show that the central metal substitution in [M(dddt)$_2$] (M = Ni, Pd), directly effects the high-pressure properties and structure. In contrast to the high-pressure structure of the Ni analogue [29], the Pd-Pd distance in [Pd(dddt)$_2$] decreases smoothly from 4.7281 Å at 1 bar to 4.117 Å at 10.6 GPa. Unlike [Ni(dddt)$_2$], which does not exhibit a structural transition up to 11.2 GPa [29], [Pd(dddt)$_2$] undergoes an order–disorder transition, which may be the cause of the increase in resistivity and coexistence of the Dirac cone state with semiconductivity in the higher-pressure region.

The tight-binding band structure calculations using the experimentally determined crystal structures show that the Dirac fermion state appears at least from 5.9 GPa. At 7.8 GPa and 10.6 GPa, the structural disorder causes coexistence of the Dirac cone state and a semiconducting state. This semiconductor gap state steadily appears in the resistivity data, manifesting as increasing resistivity.

The MR measurements show a linear relationship over wide temperature and pressure ranges. If the MR linear relationship is correlated with the Dirac electron, our results indicate that Dirac cones emerged in the pressure range from 8.7 GPa to 10.5 GPa at least. This result is consistent with the theoretical calculations for resistivity under pressure that indicates a nodal line semimetal state from 7.7 GPa to 8 GPa [34]. In this calculation, a weak metal-like behavior was also observed near the room temperature, which is in good agreement with the measured resistivity at 13 GPa. Previous studies have suggested that the temperature-independent resistivity at 12.6 GPa may be associated with the Dirac electron state [11]. However, in the case of Sample 7 under the same sample mounting conditions as the previous measurement, no linear MR results were observed above 12.5 GPa. It is possible that the sample mounting direction was not good for MR measurements or the disordered structure may not show linear MR. To fully understand the Dirac electronic state of this material, it is necessary to measure the Hall effect by applying electric current and magnetic field in different crystal directions. The topological property of wave function given by the Berry phase [35] is important for understanding Dirac nodal lines; it is predicted that the Hall effect may be observed when the Berry phase component is oriented parallel to the magnetic field [34]. To clarify these issues, additional MR measurements and Hall effect measurements are ongoing.

Regarding the negative MR below 1 T, such behavior has been previously observed in weakly localized semiconductor CrVTiAl thin films [36] or multilayer graphene [37]. Interestingly, in the absence of the Dirac cone, the negative MR quickly disappeared above 10 K [36]. In the multilayer graphene measurement, the magnetic field was parallel to the current direction [37]. MR measurements of few-layer graphene with the magnetic field perpendicular to current direction showed that increasing the number of layers enhances the weak localization effect [38]. In both of those experiments the negative MR appears in magnetic fields below 1 T and persists up to 50 K. By comparing these results with ours, we suggest that the negative MR observed in [Pd(dddt)$_2$] can be understood through the localization of the Dirac electron system.

Finally, we note that the critical pressure for observation of the Dirac electron state appears to be different depending on hydrostatic conditions in different pressure cells, pressure media and in silico pressurization methods. Our resistivity measurements show the smallest band gap at 13 GPa, while first principles DFT calculations showed that the Dirac cones appear at 8 GPa. However, the result of tight-binding calculations using the structure data from high-pressure crystal structure measurements indicates that Dirac cones appear at 5.9 GPa [31]. Several factors make the different methods challenging to compare directly. On one hand, solidification of pressure transmitting media, which occurs at different pressures depending on composition, can induce uniaxial effects in resistivity measurements. To mitigate this risk, the samples involved in both the structural and the electrical measurements were covered with a protective layer to improve the hydrostatic condition. On the other hand, the generalized gradient approximation (GGA) used in first principles calculations cannot describe long-range attractive van der Waals interactions (dispersion forces), which affects the estimation of lattice parameters in the lower pressure region. In addition, GGA calculations tend to underestimate electronic band gaps, which induces a difference between the calculated and experimental pressures [39].

5. Conclusions

By carefully controlling pressure and re-measuring the electrical properties under high pressure, we obtained a clearer understanding of the electrical behavior of [Pd(dddt)$_2$] at low pressure and reproduced the temperature-independent resistivity and increase in resistivity under high pressure. Furthermore, we successfully found a high-pressure-induced order–disorder transition in [Pd(dddt)$_2$] using synchrotron X-ray diffraction measurements. The unit cell lengths decrease as the pressure is increased, and crystal maintains monoclinic symmetry with the space group $P2_1/n$ at all pressures up to 12.4 GPa. To understand the high-pressure electronic state, tight-binding band structure calculations were performed using the experimentally obtained structures at 7.8 GPa and 10.6 GPa. Because of disorder at these two pressures, we examined two extreme crystal structures in which a single conformation of the disordered molecule 2(4) is fully occupied (Structures A and B). In Structure A, Dirac cones emerge in k-space; however, in Structure B, the system exhibits a semiconducting band structure. Therefore, it is suggested that the transition from the Dirac electron system to the semiconductor at high pressure is triggered by the structure transition. In addition, we have observed a linear relationship in the variable-pressure variable-temperature MR data as a function of the magnetic field and negative MR owing to the weak localization of the Dirac fermion system.

Supplementary Materials: Crystallographic information files are available from the CCDC, reference numbers 2072169-2072173. These data can be obtained free of charge via http://www.ccdc.cam.ac.uk/conts/retrieving.html.

Author Contributions: Conceptualization, H.C., H.H.-M.Y. and R.K.; Data Collection, H.C., T.M., H.H.-M.Y., Y.K. and L.K.S.; Writing—Original Draft Preparation, H.C., H.H.-M.Y. and R.K.; Writing—Review & Editing, H.C., H.H.-M.Y., and R.K. All authors have read and agreed to the published version of the manuscript.

Funding: This research was funded by JSPS Grant-in-Aid for Scientific Research (S) grant number JP16H06346, JSPS Grant-in-Aid for Scientific Research (C) grant number JP17K05850.

Informed Consent Statement: Not applicable.

Data Availability Statement: "MDPI Research Data Policies" at https://www.mdpi.com/ethics (accessed on 11 May 2021).

Acknowledgments: Research leading to these results was supported by Diamond Light Source (beamtime MT20934-1). We thank Ben Williams, Richard Gildea and Graeme Winters for continuing improvements to the xia2 software.

Conflicts of Interest: The authors declare no conflict of interest. The funders had no role in the design of the study; in the collection, analyses, or interpretation of data; in the writing of the manuscript, or in the decision to publish the results.

Appendix A

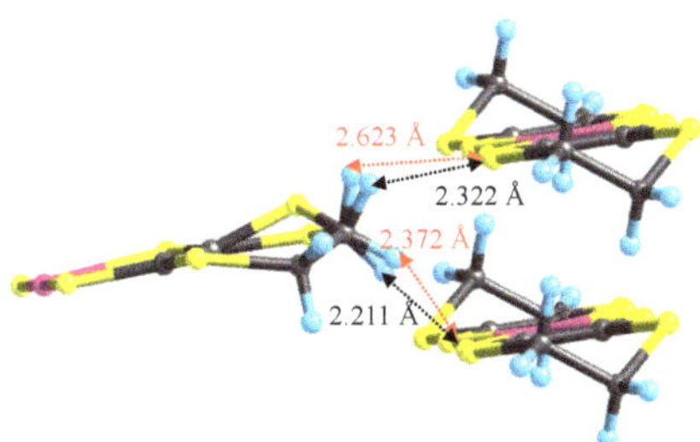

Figure A1. Two distinct CH-S hydrogen bonds by the positional disorder.

A1. Matrix elements of $\mathbf{H}(k)$

$h_{11} = b1_{\mathrm{H}}[\mathrm{e}^{ikb} + \mathrm{e}^{-ikb}]$
$h_{12} = a1_{\mathrm{H}}[\mathrm{e}^{ik(a+c)} + \mathrm{e}^{ikb}] + a2_{\mathrm{H}}[1 + \mathrm{e}^{ik(a+b+c)}]$
$h_{13} = p_{\mathrm{H}}[1 + \mathrm{e}^{ikb} + \mathrm{e}^{ika} + \mathrm{e}^{ik(a+b)}]$
$h_{14} = c1_{\mathrm{H}}[1 + \mathrm{e}^{-ikc}] + c2_{\mathrm{H}}[\mathrm{e}^{ikb} + \mathrm{e}^{ik(-b-c)}]$
$h_{15} = b1_{\mathrm{HL}}[\mathrm{e}^{-ikb} - \mathrm{e}^{ikb}]$
$h_{16} = a1_{\mathrm{HL}}[\mathrm{e}^{ikb} - \mathrm{e}^{ik(a+c)}] + a2_{\mathrm{HL}}[1 - \mathrm{e}^{ik(a+b+c)}]$
$h_{17} = p1_{\mathrm{HL}} + p2_{\mathrm{HL}}\mathrm{e}^{ikb} - p1_{\mathrm{HL}}\mathrm{e}^{ik(a+b)} - p2_{\mathrm{HL}}\mathrm{e}^{ika}$
$h_{18} = c1_{\mathrm{HL}}[1 - \mathrm{e}^{-ikc}] + c2_{\mathrm{HL}}[\mathrm{e}^{ikb} - \mathrm{e}^{ik(-b-c)}]$
$h_{22} = b2_{\mathrm{H}}[\mathrm{e}^{ikb} + \mathrm{e}^{-ikb}]$
$h_{23} = c1_{\mathrm{H}}[1 + \mathrm{e}^{-ikc}] + c2_{\mathrm{H}}[\mathrm{e}^{-ikb} + \mathrm{e}^{ik(b-c)}]$
$h_{24} = q_{\mathrm{H}}[\mathrm{e}^{ik(-a-c)} + \mathrm{e}^{ik(-b-c)} + \mathrm{e}^{ik(-a-b-c)} + \mathrm{e}^{-ikc}]$
$h_{25} = a1_{\mathrm{HL}}[\mathrm{e}^{ik(-a-c)} - \mathrm{e}^{-ikb}] + a2_{\mathrm{HL}}[1 - \mathrm{e}^{ik(-a-b-c)}]$
$h_{26} = b2_{\mathrm{HL}}[\mathrm{e}^{-ikb} - \mathrm{e}^{ikb}]$
$h_{27} = c1_{\mathrm{HL}}[\mathrm{e}^{-ikc} - 1] + c2_{\mathrm{HL}}[\mathrm{e}^{-ikb} - \mathrm{e}^{ik(b-c)}]$
$h_{28} = q_{1\mathrm{HL}}\mathrm{e}^{ik(-a-b-c)} + q_{2\mathrm{HL}}\mathrm{e}^{ik(-a-c)} - q_{1\mathrm{HL}}\mathrm{e}^{-ikc} - q_{2\mathrm{HL}}\mathrm{e}^{ik(-b-c)}$
$h_{33} = b1_{\mathrm{H}}[\mathrm{e}^{ikb} + \mathrm{e}^{-ikb}]$
$h_{34} = a1_{\mathrm{H}}[1 + \mathrm{e}^{ik(-a-b-c)}] + a2_{\mathrm{H}}[\mathrm{e}^{ik(-a-c)} + \mathrm{e}^{-ikb}]$
$h_{35} = -p2_{\mathrm{HL}} - p1_{\mathrm{HL}}\mathrm{e}^{-ikb} + p2_{\mathrm{HL}}\mathrm{e}^{ik(-a-b)} + p1_{\mathrm{HL}}\mathrm{e}^{-ika}$
$h_{36} = c1_{\mathrm{HL}}[1 - \mathrm{e}^{ikc}] + c2_{\mathrm{HL}}[\mathrm{e}^{ikb} - \mathrm{e}^{ik(-b+c)}]$
$h_{37} = b1_{\mathrm{HL}}[-\mathrm{e}^{-ikb} + \mathrm{e}^{ikb}]$
$h_{38} = a1_{\mathrm{HL}}[1 - \mathrm{e}^{ik(-a-b-c)}] + a2_{\mathrm{HL}}[\mathrm{e}^{-ikb} - \mathrm{e}^{ik(-a-c)}]$
$h_{44} = b2_{\mathrm{H}}[\mathrm{e}^{ikb} + \mathrm{e}^{-ikb}]$
$h_{45} = c1_{\mathrm{HL}}[1 - \mathrm{e}^{ikc}] + c2_{\mathrm{HL}}[\mathrm{e}^{ik(b+c)} - \mathrm{e}^{-ikb}]$
$h_{46} = q1_{\mathrm{HL}}\mathrm{e}^{ik(a+c)} + q2_{\mathrm{HL}}\mathrm{e}^{ik(a+b+c)} - q1_{\mathrm{HL}}\mathrm{e}^{ik(b+c)} - q2_{\mathrm{HL}}\mathrm{e}^{ikc}$
$h_{47} = a1_{\mathrm{HL}}[1 - \mathrm{e}^{ik(a+b+c)}] + a2_{\mathrm{HL}}[\mathrm{e}^{k(a+c)} - \mathrm{e}^{ikb}]$
$h_{48} = b2_{\mathrm{HL}}[\mathrm{e}^{-ikb} - \mathrm{e}^{ikb}]$
$h_{55} = \Delta E + b1_{\mathrm{L}}[\mathrm{e}^{ikb} + \mathrm{e}^{-ikb}]$
$h_{56} = a1_{\mathrm{L}}[\mathrm{e}^{ik(a+c)} + \mathrm{e}^{ikb}] + a2_{\mathrm{L}}[1 + \mathrm{e}^{ik(a+b+c)}]$
$h_{57} = p_{\mathrm{L}}[1 + \mathrm{e}^{ikb} + \mathrm{e}^{ika} + \mathrm{e}^{ik(a+b)}]$
$h_{58} = c1_{\mathrm{L}}[1 + \mathrm{e}^{-ikc}] + c2_{\mathrm{L}}[\mathrm{e}^{ikb} + \mathrm{e}^{ik(-b-c)}]$
$h_{66} = \Delta E + b2_{\mathrm{L}}[\mathrm{e}^{ikb} + \mathrm{e}^{-ikb}]$
$h_{67} = -c1_{\mathrm{L}}[1 + \mathrm{e}^{-ikc}] - c2_{\mathrm{L}}[\mathrm{e}^{-ikb} + \mathrm{e}^{ik(b-c)}]$
$h_{68} = q_{\mathrm{L}}[\mathrm{e}^{ik(-a-c)} + \mathrm{e}^{ik(-b-c)} + \mathrm{e}^{ik(-a-b-c)} + \mathrm{e}^{-ikc}]$
$h_{77} = \Delta E + b1_{\mathrm{L}}[\mathrm{e}^{ikb} + \mathrm{e}^{-ikb}]$
$h_{78} = -a1_{\mathrm{L}}[1 + \mathrm{e}^{ik(-a-b-c)}] - a2_{\mathrm{L}}[\mathrm{e}^{ik(-a-c)} + \mathrm{e}^{-ikb}]$
$h_{88} = \Delta E + b2_{\mathrm{L}}[\mathrm{e}^{ikb} + \mathrm{e}^{-ikb}]$

The energy difference between HOMO and LUMO ΔE is set to 0.696 eV to reproduce the energy band obtained by the DFT calculation.

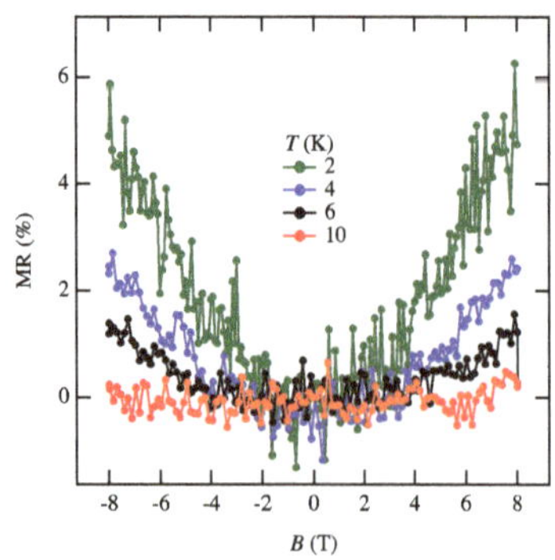

Figure A2. Temperature dependence of MR as a function of magnetic field at 11.8 GPa (sample 8).

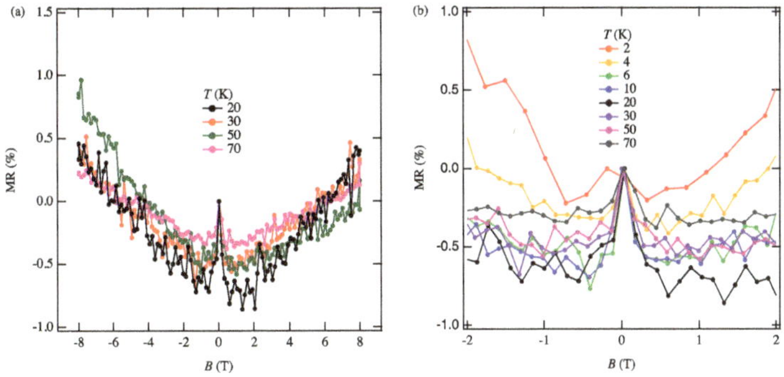

Figure A3. (**a**) Temperature dependence of MR as a function of magnetic field at 10.5 GPa (sample 8). (**b**) Enlarged view of low magnetic field region of MR as a function of magnetic field and temperature at 10.5 GPa.

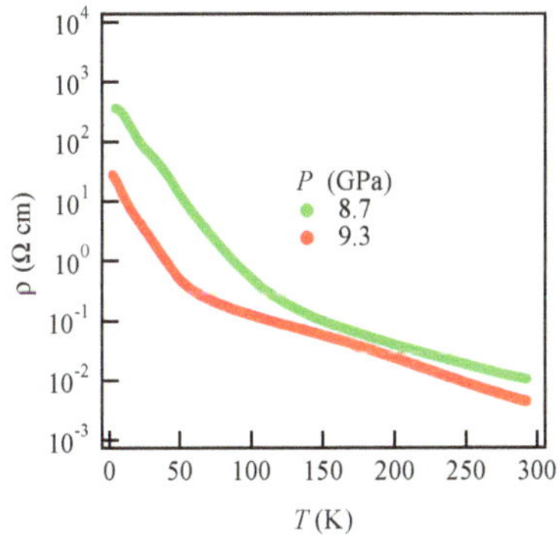

Figure A4. Temperature dependence of resistivity (sample 9).

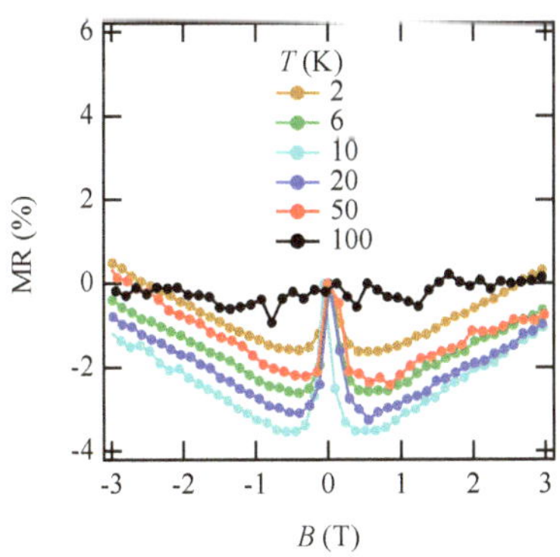

Figure A5. Enlarged view of Figure 11e in the low magnetic field region of MR as a function of magnetic field and temperature at 9.3 GPa.

References

1. Tanaka, H.; Okano, Y.; Kobayashi, H.; Suzuki, W.; Kobayashi, A. A three-dimensional synthetic metallic crystal composed of single-component molecules. *Science* **2001**, *291*, 285–287. [CrossRef] [PubMed]
2. Suzuki, W.; Fujiwara, E.; Kobayashi, A.; Fujishiro, Y.; Nishibori, E.; Takata, M.; Sakata, M.; Fujiwara, H.; Kobayashi, H. Highly conducting crystals based on single-component gold complexes with extended-TTF dithiolate ligands. *J. Am. Chem. Soc.* **2003**, *125*, 1486–1487. [CrossRef] [PubMed]
3. Le Gal, Y.; Roisnel, T.; Auban-Senzier, P.; Bellec, N.; Íñiguez, J.; Canadell, E.; Lorcy, D. Stable metallic state of a neutral-radical single-component conductor at ambient pressure. *J. Am. Chem. Soc.* **2018**, *140*, 6998–7004. [CrossRef] [PubMed]
4. Zhou, B.; Ishibashi, S.; Ishii, T.; Sekine, T.; Takehara, R.; Miyagawa, K.; Kanoda, K.; Nishibori, E.; Kobayashi, A. Single-component molecular conductor $[Pt(dmdt)_2]$—A three-dimensional ambient-pressure molecular Dirac electron system. *Chem. Commun.* **2019**, *55*, 3327–3330. [CrossRef]
5. Cui, H.; Brooks, J.S.; Kobayashi, A.; Kobayashi, H. Metallization of the single component molecular semiconductor $[Ni(ptdt)_2]$ under very high pressure. *J. Am. Chem. Soc.* **2009**, *131*, 6358–6359. [CrossRef]
6. Zhou, B.; Idobata, Y.; Kobayashi, A.; Cui, H.; Kato, R.; Takagi, R.; Miyagawa, K.; Kanoda, K.; Kobayashi, H. Single-component molecular conductor $[Cu(dmdt)_2]$ with three-dimensionally arranged magnetic moments exhibiting a coupled electric and magnetic transition. *J. Am. Chem. Soc.* **2012**, *134*, 12724–12731. [CrossRef]
7. Cui, H.; Tsumuraya, T.; Miyazaki, T.; Okano, Y.; Kato, R. Pressure-induced metallic conductivity in the single-component molecular crystal $[Ni(dmit)_2]$. *Eur. J. Inorg. Chem.* **2014**, *2014*, 3837–3840. [CrossRef]
8. Cui, H.; Kobayashi, H.; Ishibashi, S.; Sasa, M.; Iwase, F.; Kato, R.; Kobayashi, A. A single-component molecular superconductor. *J. Am. Chem. Soc.* **2014**, *136*, 7619–7622. [CrossRef]
9. Tenn, N.; Bellec, N.; Jeannin, O.; Piekara-Sady, L.; Auban-Senzier, P.; Íñiguez, J.; Canadell, E.; Lorcy, D. A single-component molecular metal based on a thiazole dithiolate gold complex. *J. Am. Chem. Soc.* **2009**, *131*, 16961–16967. [CrossRef]
10. Yzambart, G.; Bellec, N.; Nasser, G.; Jeannin, O.; Roisnel, T.; Fourmigué, M.; Auban-Senzier, P.; Íñiguez, J.; Canadell, E.; Lorcy, D. Anisotropic chemical pressure effects in single-component molecular metals based on radical dithiolene and diselenolene gold complexes. *J. Am. Chem. Soc.* **2012**, *134*, 17138–17148. [CrossRef]
11. Kato, R.; Cui, H.; Tsumuraya, T.; Miyazaki, T.; Suzumura, Y. Emergence of the Dirac electron system in a single-component molecular conductor under high pressure. *J. Am. Chem. Soc.* **2017**, *139*, 1770–1773. [CrossRef]
12. Cui, H.; Kato, R. Electrical properties of single-component molecular crystals under high pressure. *Rev. H. Pres. Sci. Tech.* **2018**, *28*, 217–224. [CrossRef]
13. Novoselov, K.; Geim, A.; Morozov, S.; Jiang, D.; Grigorieva, M.; Dubonos, S.; Firsov, A. Two-dimensional gas of massless Dirac fermions in graphene. *Nature* **2005**, *438*, 197–200. [CrossRef]
14. Hasan, M.; Kane, C. Colloquium: Topological insulators. *Rev. Mod. Phys.* **2010**, *82*, 3045–3067. [CrossRef]
15. Huynh, K.K.; Tanabe, Y.; Tanigaki, K. Both electron and hole Dirac cone states in $Ba(FeAs)_2$ confirmed by magnetoresistance. *Phys. Rev. Lett.* **2011**, *106*, 217004. [CrossRef]
16. Tajima, N.; Sato, M.; Sugawara, S.; Kato, R.; Nishio, Y.; Kajita, K. Spin and valley splittings in multilayered massless Dirac fermion system. *Phys. Rev. B* **2010**, *82*, 121420(R). [CrossRef]
17. Kato, R.; Suzumura, Y. Novel Dirac electron in single-component molecular conductor $[Pd(dddt)_2]$ (dddt =5,6-dihydro-1,4-dithiin-2,3-dithiolate). *J. Phys. Soc. Jpn.* **2017**, *86*, 064705. [CrossRef]
18. Tsumuraya, T.; Kato, R.; Suzumura, Y. Effective hamiltonian of topological nodal line semimetal in single-component molecular conductor $[Pd(dddt)_2]$ from first-principles. *J. Phys. Soc. Jpn.* **2018**, *87*, 113701. [CrossRef]
19. Evans, P.R.; Murshudov, G.N. How good are my data and what is the resolution? *Acta Crystallogr. Sect. D Biol. Crystallogr.* **2013**, *69*, 1204–1214. [CrossRef]
20. Winn, M.D.; Ballard, C.C.; Cowtan, K.D.; Dodson, E.J.; Emsley, P.; Evans, P.R.; Keegan, R.M.; Krissinel, E.B.; Leslie, A.G.W.; McCoy, A.; et al. Overview of the CCP4 suite and current developments. *Acta Crystallogr. Sect. D Biol. Crystallogr.* **2011**, *67*, 235–242. [CrossRef]
21. Winter, G.; Waterman, D.G.; Parkhurst, J.M.; Brewster, A.S.; Gildea, R.J.; Gerstel, M.; Fuentes-Montero, L.; Vollmar, M.; Michels-Clark, T.; Young, I.D.; et al. DIALS: Implementation and evaluation of a new integration package. *Acta Crystallogr. Sect. D Struct. Biol.* **2018**, *74*, 85–97. [CrossRef]
22. Evans, P. Scaling and assessment of data quality. *Acta Crystallogr. Sect. D Biol. Crystallogr.* **2006**, *62*, 72–82. [CrossRef]
23. Winter, G. xia2: An expert system for macromolecular crystallography data reduction. *J. Appl. Crystallogr.* **2010**, *43*, 186–190. [CrossRef]
24. Sheldrick, G.M. SHELXT—Integrated space-group and crystal-structure determination. *Acta Crystallogr. Sect. A Found. Adv.* **2015**, *71*, 3–8. [CrossRef]
25. Dolomanov, O.V.; Bourhis, L.J.; Gildea, R.J.; Howard, J.A.K.; Puschmann, H. OLEX2: A complete structure solution, refinement and analysis program. *J. Appl. Crystallogr.* **2009**, *42*, 339–341. [CrossRef]
26. Hoffmann, R. An Extended Hückel theory. I. hydrocarbons. *J. Chem. Phys.* **1963**, *39*, 1397. [CrossRef]
27. Clementi, E.; Roetti, C. Roothaan-Hartree-Fock atomic wavefunctions: Basis functions and their coefficients for ground and certain excited states of neutral and ionized atoms, $Z \leq 54$. *At. Data Nucl. Data Tables* **1974**, *14*, 177–478. [CrossRef]

28. QuantumATK R-2020.09 Documentation, Manual of Atomic data. Available online: https://docs.quantumatk.com/manual/AtomicData.html#sect3-atomicdata-huckel-muller (accessed on 11 May 2021).
29. Cui, H.; Tsumuraya, T.; Yeung, H.H.-M.; Coates, C.S.; Warren, M.R.; Kato, R. High pressure crystal structure and electrical properties of a single component molecular crystal [Ni(dddt)$_2$] (dddt = 5,6-dihydro-1,4-dithiin-2,3-dithiolate). *Molecules* **2019**, *24*, 1843. [CrossRef]
30. Cairns, A.B.; Goodwin, A.L. Negative linear compressibility. *Phys. Chem. Chem. Phys.* **2015**, *17*, 20449–20465. [CrossRef]
31. Kato, R.; Cui, H.; Minamidate, T.; Yeung, H.H.-M.; Suzumura, Y. Electronic Structure of a Single-Component Molecular Conductor [Pd(dddt)$_2$] (dddt = 5,6-dihydro-1,4-dithiin-2,3-dithiolate) under High Pressure. *J. Phys. Soc. Jpn.* **2020**, *89*, 124706. [CrossRef]
32. Mori, T.; Kobayashi, A.; Sasaki, Y.; Kobayashi, H.; Saito, G.; Inokuchi, H. The intermolecular interaction of tetrathiafulvalene and bis(ethylenedithio) tetrathiafulvalene in organic metals. Calculation of orbital overlaps and models of energy-band Structures. *Bull. Chem. Soc. Jpn.* **1984**, *57*, 627–633. [CrossRef]
33. Kato, R.; Suzumura, Y. A Tight-binding model of an ambient-pressure molecular Dirac electron system with open nodal lines. *J. Phys. Soc. Jpn.* **2020**, *89*, 044713. [CrossRef]
34. Suzumura, Y.; Cui, H.; Kato, R. Conductivity and resistivity of Dirac electrons in single-component molecular conductor [Pd(dddt)$_2$]. *J. Phys. Soc. Jpn.* **2018**, *87*, 084702. [CrossRef]
35. Berry, M.V. Quantal phase factors accompanying adiabatic changes. *Proc. R. Soc. Lond. Ser. A* **1984**, *392*, 45–57.
36. Stephen, G.M.; Lane, C.; Buda, G.; Graf, D.; Kaprzyk, S.; Barbiellini, B.; Bansil, A.; Heiman, D. Electrical and magnetic properties of thin films of the spin-filter material CrVTiAl. *Phys. Rev. B* **2019**, *99*, 224207. [CrossRef]
37. Bodepudi, S.C.; Singh, A.P.; Pramanik, S. Current perpendicular to plane magnetoresistance in chemical vapor deposition grown multilayer graphene. *Electronics* **2013**, *2*, 315–331. [CrossRef]
38. Liu, Y.; Lew, W.S.; Sun, L. Enhanced weak localization effect in few-layer graphene. *Phys. Chem. Chem. Phys.* **2011**, *13*, 20208–20214. [CrossRef]
39. Heyd, J.; Peralta, J.E.; Scuseria, G.E.; Martin, R.L. Energy band gaps and lattice parameters evaluated with the Heyd-Scuseria-Ernzerhof screened hybrid functional. *J. Chem. Phys.* **2005**, *123*, 174101. [CrossRef]

crystals MDPI

Article

Evolution of Shape and Volume Fraction of Superconducting Domains with Temperature and Anion Disorder in $(TMTSF)_2ClO_4$

Kaushal K. Kesharpu [1], Vladislav D. Kochev [1] and Pavel D. Grigoriev [1,2,3,*]

1 Department of Theoretical Physics and Quantum Technology, National University of Science and Technology "MISiS", 119049 Moscow, Russia; kesharpu.k@misis.ru (K.K.K.); vd.kochev@misis.ru (V.D.K.)
2 L.D. Landau Institute for Theoretical Physics, 142432 Chernogolovka, Russia
3 P.N. Lebedev Physical Institute of RAS, 119991 Moscow, Russia
* Correspondence: grigorev@itp.ac.ru

Abstract: In highly anisotropic organic superconductor $(TMTSF)_2ClO_4$, superconducting (SC) phase coexists with metallic and spin-density wave phases in the form of domains. Using the Maxwell-Garnett approximation (MGA), we calculate the volume ratio and estimate the shape of these embedded SC domains from resistivity data at various temperature and anion disorder, controlled by the cooling rate or annealing time of $(TMTSF)_2ClO_4$ samples. We found that the variation of cooling rate and of annealing time affect differently the shape of SC domains. In all cases the SC domains have oblate shape, being the shortest along the interlayer z-axis. This contradicts the widely assumed filamentary superconductivity along the z-axis, used to explain the anisotropic superconductivity onset. We show that anisotropic resistivity drop at the SC onset can be described by the analytical MGA theory with anisotropic background resistance, while the anisotropic T_c can be explained by considering a finite size and flat shape of the samples. Due to a flat/needle sample shape, the probability of percolation via SC domains is the highest along the shortest sample dimension (z-axis), and the lowest along the sample length (x-axis). Our theory can be applied to other heterogeneous superconductors, where the size d of SC domains is much larger than the SC coherence length ξ, e.g., cuprates, iron-based or organic superconductors. It is also applicable when the spin/charge-density wave domains are embedded inside a metallic background, or vice versa.

Keywords: organic superconductors; Beechgard salts; Maxwell-Garnett approximation; high-Tc

Citation: Kesharpu, K.K.; Kochev, V.D.; Grigoriev, P.D. Evolution of Shape and Volume Fraction of Superconducting Domains with Temperature and Anion Disorder in $(TMTSF)_2ClO_4$. *Crystals* **2021**, *11*, 72. https://doi.org/10.3390/cryst11010072

Received: 1 December 2020
Accepted: 13 January 2021
Published: 17 January 2021

1. Introduction

Raising the superconducting transition temperature (Tc) has been the goal of active research for a century. Compounds like cuprates [1,2], iron-based superconductors [3], organic superconductors (hereafter denoted as OrS) [4] are some major high-Tc superconductors at ambient pressure. These materials have several common properties: (i) layered crystal structure and, hence, high conductivity anisotropy; (ii) interplay between various types of electron ordering, i.e., between spin/charge-density wave and superconductivity; (iii) spatial inhomogeneity. Therefore, many effects and methods, both experimental and theoretical, are common for these materials. A recent and good review of all these systems is given by Stewart [5]. Our paper concerns the OrS and is devoted to two problems: (i) show that Maxwell-Garnett approximation can be used to estimate superconducting volume fraction in coexistence regime of superconducting, metallic and spin/charge-density wave phases; (ii) analysis of corresponding experimental data [6,7] in the organic superconductor $(TMTSF)_2ClO_4$ to study the effect of cooling rate and of disorder on the volume fraction, shape and size of SC domains.

$(TMTSF)_2X$ series belongs to quasi-1D OrS and has been widely studied for 40 years [4,8–10]. Many important effects have been discovered and investigated on these compounds, e.g., angular magnetoresistance oscillations (AMRO) in quasi-1D metals [9–11], field-induced spin-density waves (FISDW) [12,13], etc. [9,10]. An interesting and puzzling property of these materials related to our subject is that with the increase of pressure for $(TMTSF)_2PF_6$ or of anion ordering for $(TMTSF)_2ClO_4$, the superconductivity first appears along the least-conducting z-axis, while along the most conducting x-direction only in the last turn [6,7,14,15]. Recent experimental study on $(TMTSF)_2PF_6$ [14–17] and $(TMTSF)_2ClO_4$ [6,7] has shown that spin-density wave (SDW), superconducting (SC) and metallic phase coexist in form of segregated domains. With the increase in pressure (for $(TMTSF)_2PF_6$) [14,15,18] or in anion ordering (for $(TMTSF)_2ClO_4$) [6,7] the volume fraction of SC or metallic phase increases.

A few theories have been suggested to describe the coexistence regime in these materials, especially in $(TMTSF)_2PF_6$. One of them is the application of SO(4) symmetry that explains SDW and SC coexistence [19] but does not account for the observed hysteresis [16], for the strong enhancement of the upper critical field H_{c2} [20] and for the anisotropic SC onset [14,15,18] in the coexistence phase. Less exotic theories suggest a separation of SDW and SC in a coordinate [14–16,20–24] or momentum space [23]. The momentum SC-SDW separation assumes a semi-metallic state in a SDW phase, where, due to the imperfect nesting, small ungapped Fermi-surface pockets appear and become superconducting [23]. The H_{c2} enhancement can be explained in both scenarios [23,24], but the observed hysteresis suggests a spatial SDW/SC separation[16]. To explain the H_{c2} enhancement [20] the SC domain width must not exceed the in-plane SC penetration depth $\lambda_{ab} \equiv \lambda$. In $(TMTSF)_2ClO_4$ the in-plane penetration depth is [25–27] $\lambda_{ab}(T=0) \approx 1$ μm, and the out-of-plane penetration depth is [28] $\lambda_{bc}(T=0.19\text{ K}) \approx 40$ μm. One possible mechanism of the formation of such narrow domains in the SDW state could be the soliton phase [14,21,22,24,29]. It suggests that SDW order parameter becomes non-uniform with metallic domain appearing perpendicular to the highest conducting x-axis. However, the width of these soliton-wall domains, being of the order of SDW coherence length $\xi_{SDW} \sim 30$ nm, is too small to be consistent with the recent observation of AMRO and FISDW in $(TMTSF)_2PF_6$ [15] and in $(TMTSF)_2ClO_4$ [30], suggesting the domain size $d > 1$ μm. Moreover, the soliton-phase scenario accounts for the SC suppression along the most conducting x-axis only, but it could not explain why SC first appears along the least-conducting z-axis, because in this scenario the soliton walls are extended along both y and z-axes, which should result in SC along both these directions.

A probable explanation of this anisotropic SC onset was proposed recently [31]. It is based on two ideas. First, in anisotropic media the isolated SC islands increase conductivity much stronger along the least-conducting direction than along the others, as observed in FeSe [32,33] and described [32–34] using the Maxwell-Garnett approximation (MGA) [35] for small volume fraction ϕ of SC phase. However, this MGA theory cannot explain the anisotropic zero-resistance onset. For this we need the second idea, which takes into account the finite sample size L as compared to the size d of SC grains. If the sample shape is very anisotropic, e.g., a thin plate, then the current percolation via the SC grains, responsible for the zero-resistance onset, is most probable along the shortest sample dimension, [31] i.e., along the sample thickness, and least probable along the sample length. These two ideas were applied to explain [31] the experimental data[14,15,18] in PF6, taken on thin elongated samples with dimensions [14,18] $3 \times 0.2 \times 0.1$ mm^3. As $(TMTSF)_2ClO_4$ samples are usually flat shaped either [6,7], similar ideas can be applied to analyze the resistivity experimental data there too. In this paper, using the MGA we find superconducting volume ratio ϕ and shape of inclusions from available experimental data. We also investigate how disorder and cooling rate affect the inclusions' shape and size. This knowledge may help to better understand the microscopic structure and electronic properties of the SDW/SC coexistence phase in organic superconductors.

In Section 2 we briefly describe the important properties of $(TMTSF)_2ClO_4$ and argue that MGA can be applied to estimate temperature dependence of SC volume ratio (ϕ) in the presence of all 3 phases, i.e., metallic, SDW and SC. In Section 3 we describe the theoretical model in MGA. Furthermore, in Section 4 we apply our theoretical model to analyze the experimental data on $(TMTSF)_2ClO_4$. Using the experimental data from Ref. [7] we find out how cooling rate effects ϕ. Similarly, using the experiments of Ref. [6] we analyze the evolution of aspect ratio of SC inclusions with sample disorder. Finally, in Section 5 we discuss the main results of our investigation and their consequences.

2. Material and Method

2.1. Material

$(TMTSF)_2ClO_4$ is the only member of Beechgard salts which becomes superconducting at ambient pressure [36,37]. It is a quasi-1D superconductor in which cooper-pair ordering (SC) coexists with insulating Pierels ordering (SDW) (The analysis below equally applies for a charge-density wave (CDW) or SDW Pierels ordering. Our current study is mainly devoted to SDW/SC coexistence in $(TMTSF)_2ClO_4$, therefore we keep SDW notation for the insulating phase. However, it may equally be applied to other compounds with CDW/SC mixed phase.). This very competition between SC and SDW is the key to understand the unconventional superconductivity [38–41]. Among quasi-1D superconductors $(TMTSF)_2ClO_4$ is the only compound in which superconductivity can be controlled by the cooling rate of samples, which affects the disorderliness of ClO_4 anions. Slowly cooled $(TMTSF)_2ClO_4$ samples (relaxed state) undergo an SC transition at $T_c \approx 1.3$ K [42,43]. However, when cooled very fast (quenched state) $(TMTSF)_2ClO_4$ has an insulating SDW transition at $T_{SDW} \approx 4$–5 K [44–46]. This behavior can be ascribed to structural change that occurs at the anion-ordering temperature [47] $T_{AO} \approx 24.5$ K. For $T > T_{AO}$ the noncentrosymmetric tetrahedral ClO_4 anions, which are located at the inversion centers [48,49], preserve the inversion symmetry due to thermal motion of ClO_4 anions. ClO_4 anions randomly occupies one or other orientations, hence, on average the inversion symmetry is preserved [48]. For $T < T_{AO}$, if the sample is cooled fast enough then the randomness of orientation of ClO_4 anions is preserved [45]. However, if the sample is cooled slowly through T_{AO}, ClO_4 anions along a,c-axes are ordered uniformly; and along b-axis ordered alternatively [47,50,51]. The anion ordering introduces the new wave vector and the Fermi-surface folding. It disturbs the Fermi-surface nesting, preventing the SDW and favoring SC.

Recent experiments strongly support the presence of SC inclusions embedded in the background of metallic/SDW phase when $(TMTSF)_2ClO_4$ is cooled at intermediate rate [6,7]. This means granular superconductivity for partially ordered samples. From the crystallographic point of view, in this coexistence phase the domains of ordered ClO_4 anions are embedded inside disordered background. In these domains the anions have alternating ordering pattern, being disordered outside the domains [6,7,47]. The electronic state inside these domains is assumed to remain metallic even at $T < T_{AO}$, and at $T < T^*$ these ClO_4-ordered grains become superconducting [7]. Here T^* is the superconducting onset temperature. The volume fraction ϕ of superconducting phase increases with the decrease of temperature. For slow or intermediate cooling rate at $T = T_c$ the phase coherence of these SC islands establishes in the entire sample, leading to its zero resistance. In the temperature interval $T_c < T < T^*$ the effective medium model for highly anisotropic heterogeneous layered compounds [35], developed and applied by the authors to various anisotropic compounds in Refs. [32–34,52], can be used to estimate the volume fraction and the shape of SC inclusions inside the samples.

2.2. Method

To analyze the temperature dependence of resistivity at $T_c < T < T^*$ we use the Maxwell-Garnett approximation (MGA) [35], valid when the volume fraction of SC phase $\phi \ll 1$. Please note that the condition $\phi \ll 1$ is fulfilled in a wide range of parameters, because in 3D anisotropic samples even the SC percolation threshold ϕ_c, corresponding to the onset of nearly zero resistance, is considerably smaller than unity. This is illustrated in our recent work [31], where ϕ_c in needle-shaped flat $(TMTSF)_2PF_6$ samples from Refs. [14,18], typical for organic superconductors, was calculated numerically, assuming a rectangular sample shape and ellipsoidal SC inclusions of varying size at randomly distributed but fixed positions. This calculation has shown the strong anisotropy and small value of these ϕ_c (see Figures 3 and 4 of Ref. [31]). Thus, for SC domain size $d = 40$ µm the percolation threshold along the x-axis $\phi_c \approx 0.3$; for $d = 15$ µm $\phi_c \approx 0.15$ (see Figure 3 of Ref. [31]). For y- and z-axes ϕ_c is even smaller. In all these cases $\phi_c \ll 1$, suggesting a wide range where MGA can be applied. The $(TMTSF)_2ClO_4$ samples, studied in the present paper and in Refs. [6,7], are also flat and needle-shaped, similar to $(TMTSF)_2PF_6$ samples from Refs. [14,18].

In recent works [32–34,52] using MGA the superconducting volume ratio ϕ was found when SC inclusions were embedded inside metallic background. Similarly, here we also assume SC inclusions are embedded inside a background phase. However, here the background phase consists of SDW and metallic phases. This is permissible in MGA approximation as long as we know the effective conductivity of this mixed background phase. We make two more assumptions: (i) the SC inclusions are of ellipsoidal shape, which simplifies the calculations and allows deriving analytical formulas for conductivity; (ii) the size d and distance between SC inclusions l are much greater than the SC coherence length ξ, so that the SC proximity effects and the Josephson coupling between the SC grains do not change the results considerably. Both in $(TMTSF)_2PF_6$ and $(TMTSF)_2ClO_4$ the size $d \gtrsim 1$ µm of metal/SC inclusions is indeed much larger than the SC coherence length [7,53] $\xi = 70, 30$, and 2 nm along the a, b, and c axes, respectively. Such a large metal/SC grain size is demonstrated by the observation [15,30] of angular magnetoresistance oscillations (AMRO) in the mixed phase of these compounds. If $\phi \ll 1$, one can also take $l \gg \xi$.

Due to the proximity effect a shell of SC condensates around SC inclusions with thickness $\sim \xi$ gets created, which changes the effective size and shape of SC islands. Another quantum mechanical effect is the Josephson coupling between SC grains. The Josephson coupling energy E_J depends directly on Josephson junction current I_c, $E_J \equiv \hbar I_c/2e$, which exponentially decreases with the increase of distance l between neighboring SC grains: $I_c = I_0 \exp(-l/\xi)$. If $E_J \ll T$, the Josephson coupling can be disregarded [54]. As we assume $l, d \gg \xi$, both these effects can be neglected.

The conductance through N-S boundaries of a normal metal and SC may increase, maximum two times, due to the Andreev reflection (See Section 11.5.1 of Ref. [54] for the basic description of Andreev reflection). However, this increase of the interface conductance should not considerably change the effective conductivity of the whole sample because the main voltage drop (at a given current) comes not from the N-S interfaces, but from the large non-SC parts of the sample. Even if we take an infinite N-S interface conductance, this does not much affect the total sample resistance.

Considering the above facts, we neglect these small quantum mechanical effects, and treat conductivity due to embedded SC inclusions using the classical MGA approximation and percolation.

3. Theory

We use the Maxwell-Garnett approximation (MGA) [55,56] to find the SC volume ratio ψ. We derive a general formula for effective conductivity in anisotropic heterogeneous media, where unidirectional ellipsoidal SC (SDW) inclusions are embedded inside the background SDW/metallic (metallic) phase with anisotropic resistivity. We denote the diagonal effective conductivity tensor of this materials as $diag(\sigma_{xx}, \sigma_{yy}, \sigma_{zz})$. Similarly, the background conductivity tensor of this material is denoted as $diag(\sigma^b_{xx}, \sigma^b_{yy}, \sigma^b_{zz})$. Background phase consists of both metallic and SDW phase, or only of the metallic phase. Since the MGA in its standard form[35] can be applied only to an isotropic medium, we use the coordinate-space dilation to transform the initial problem with anisotropic conductivity into isotropic one [32–34] (see Appendix A for the details of this mapping and Equations (A4) and (A5) for the definition of dilation coefficients μ and η).

Before explaining the idea of MGA [35,55,56] we note that the stationary-current equation for electrostatic potential $V(\boldsymbol{r})$, coming from the continuity equation for the electric current J_i in heterogeneous media with coordinate-dependent diagonal conductivity $\sigma(\boldsymbol{r})$, (In Equations (1) and (2) the summation is assume over the repeated coordinate indices i and j.)

$$-\nabla_i J_i = \nabla_i\left[\sigma_{ij}(\boldsymbol{r})\nabla_j V(\boldsymbol{r})\right] = 0, \tag{1}$$

is completely equivalent to the electrostatic equation in heterogeneous media with coordinate-dependent dielectric constant $\varepsilon(\boldsymbol{r})$, as comes from the Maxwell's equations:

$$\nabla_i D_i \equiv \nabla_i(E_j\varepsilon_{ij}) = -\nabla_i\left[\varepsilon_{ij}(\boldsymbol{r})\nabla_j V(\boldsymbol{r})\right] = 0, \tag{2}$$

where E_i and D_i are the electric and displacement fields, and $\nabla_i \equiv \partial/\partial x_i$. Hence, the problem of an effective dielectric constant of such a medium with heterogeneous dielectric function $\varepsilon(\boldsymbol{r})$ is equivalent to the problem of effective conductivity of a heterogeneous medium with the same coordinate dependence of $\sigma(\boldsymbol{r})$.

To explain the idea of MGA (Originally, MGA was formulated in 1904 only for spherical inclusions in isotropic medium [57].) [35,55,56] of calculating the effective dielectric constant of heterogeneous isotropic media with a small volume fraction ϕ of inclusions of the second phase, we refer to Figure 1. The background phase is represented as "2", and the inclusion phase as "1" in the figure. We take a sphere of our heterogeneous media. Inside this sphere the inclusions are embedded as shown in Figure 1. Let this sphere be embedded in an infinite medium of background phase. An electric field E is applied to the infinite medium. Let $\mathcal{E}'$ be the electric field at a distant point P. $\mathcal{E}'$ includes polarization due to individual inclusions. We denote the inclusion volume ratio as ϕ and the background-phase volume ratio as $(1-\phi)$. We assume that the heterogeneous sphere with both inclusion and background phase can be substituted by a sphere of homogeneous phase with an effective dielectric constant ε_{eff}. Then the electric field at point P can be found in two ways: (i) By summing the polarization effect due to individual inclusions; the corresponding electric field at point P we denote as $\mathcal{E}'$. (ii) By taking the polarization effect at point P of homogeneous sphere with effective dielectric constant ε_{eff}; the corresponding electric field at point P we denote as $\mathcal{E}''$. The MGA assumes these fields to be the same, i.e., $\mathcal{E}' = \mathcal{E}''$, which gives an equation on ε_{eff} (See Section 18.1.1 of Ref. [35] for complete discussion on MGA). Using the well-known formula for the polarization of dielectric ellipsoid in isotropic medium, and replacing ε_{ii} by σ_{ii}, we find the following equation (see Equations (18.9) and (18.10) of Ref. [35]) for the effective conductivity σ_i^* along main axis i in the mapped space: (Equations (3), (5) and (8) assume that in the mapped space the conductivities of both phases, σ^{isl} and σ^b, are isotropic. For a generalization to anisotropic σ^{isl} or σ^h in the mapped space see Refs. [58,59]. In the case of superconducting inclusions, applied in Section IV to analyze the experimental data, $\sigma^{isl} = \infty$ is naturally isotropic.)

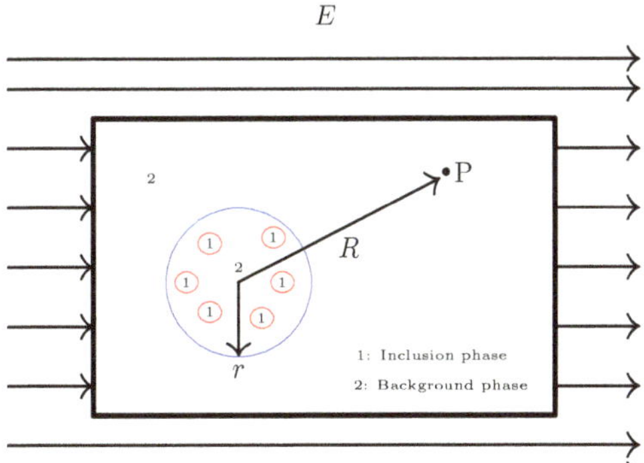

Figure 1. Schematic representation of Maxwell-Garnett approximation. The distance of point P from the center of sphere is very large compared to the sphere size, i.e., $r \ll R$. E is the applied electric field.

$$(1-\phi)(\sigma_i^* - \sigma^b) + \phi \left[\frac{\sigma^b \left(\sigma_i^* - \sigma^{isl} \right)}{\sigma^b + A_i \left(\sigma^{isl} - \sigma^b \right)} \right] = 0. \tag{3}$$

Here A_i are the depolarization factors of a dielectric ellipsoid with semiaxes a_i in the mapped space:

$$A_i = \prod_{n=1}^{3} a_n \int_0^\infty dt \Big/ 2(t + a_i^2) \sqrt{\prod_{n=1}^{3} (t + a_n^2)}. \tag{4}$$

The analytical solution of this integral can be found in terms of incomplete elliptic integrals of first and second kind (See Appendix B of Ref. [34]). If the inclusions have finite conductivity, then solving Equation (3) for σ_i^* we obtain

$$\sigma_i^* = \sigma_b \left[\frac{(A_i + (1 - A_i)\phi)(\sigma^{isl} - \sigma^b) + \sigma^b}{A_i(1-\phi)(\sigma^{isl} - \sigma^b) + \sigma^b} \right]. \tag{5}$$

If the inclusions are superconducting, we take $\sigma^{isl} \approx \infty$. Equation (3) in this case simplifies to

$$(1-\phi)(\sigma_i^* - \sigma_i^b) - \phi \frac{\sigma^b}{A_i} = 0 \tag{6}$$

Solving Equation (6) for σ_i^* we obtain

$$\sigma_i^* = \sigma^b \left[\frac{A_i + (1 - A_i)\phi}{A_i(1-\phi)} \right] \tag{7}$$

Both Equations (5) and (7) gives the effective conductivity σ^* in the mapped space. Although the background conductivity in the mapped space is isotropic, the effective conductivity σ^* is anisotropic because of the anisotropic ellipsoidal shape of inclusions. The effective conductivity of original anisotropic material in real space is found by the reverse mapping. It is done via multiplying the effective conductivity matrix in the mapped space $\sigma^* = diag(\sigma_{xx}^*, \sigma_{yy}^*, \sigma_{zz}^*)$ by the inverse mapping coefficient matrix $diag(1, \mu, \eta)$. Hence, multiplying Equation (5) by $diag(1, \mu, \eta)$, we find the effective conductivity in real space:

$$\begin{aligned} \frac{\sigma_{xx}}{\sigma_{xx}^b} &\equiv \frac{\sigma_{xx}}{\sigma^b} = \frac{(A_x + (1-A_x)\phi)(\sigma^{isl} - \sigma^b) + \sigma^b}{A_x(1-\phi)(\sigma^{isl} - \sigma^b) + \sigma^b}, \\ \frac{\sigma_{yy}}{\sigma_{yy}^b} &\equiv \frac{\sigma_{yy}}{\mu\sigma^b} = \frac{(A_y + (1-A_y)\phi)(\sigma^{isl} - \sigma^b) + \sigma^b}{A_y(1-\phi)(\sigma^{isl} - \sigma^b) + \sigma^b}, \\ \frac{\sigma_{zz}}{\sigma_{zz}^b} &\equiv \frac{\sigma_{zz}}{\eta\sigma^b} = \frac{(A_z + (1-A_z)\phi)(\sigma^{isl} - \sigma^b) + \sigma^b}{A_z(1-\phi)(\sigma^{isl} - \sigma^b) + \sigma^b}. \end{aligned} \tag{8}$$

Here σ_{ii}^b is the background conductivity along the axis i in real space. Similarly, multiplying Equation (7) by $diag(1, \mu, \eta)$, we find the effective conductivity of original inhomogeneous material with SC inclusions of volume fraction ϕ:

$$\frac{\sigma_{ii}}{\sigma_{ii}^b} = \frac{A_i + (1-A_i)\phi}{A_i(1-\phi)}. \tag{9}$$

From Equation (9) one can express the volume fraction ϕ via the effective σ_{ii} and background σ_{ii}^b conductivities and depolarization factor A_i along the same axis:

$$\phi = \frac{A_i(1 - \sigma_{ii}^b/\sigma_{ii})}{A_i + (1-A_i)\sigma_{ii}^b/\sigma_{ii}}. \tag{10}$$

Equation (8) is helpful when the conductivities of background and inclusion phases are both finite. Hence, it can be used to find the effective conductivity of heterogeneous material, when, e.g., SDW domains are embedded inside a metallic background, or vice versa. Equation (9) can be used when superconducting inclusions are embedded inside a background phase of finite conductivity. Below we use Equations (9) and (10) to analyze experimental resistivity data in $(TMTSF)_2ClO_4$ in the mixed SC/SDW state. Here the background phase is made up of metallic as well as SDW phases. Resistivity along the corresponding axis is found by taking the inverse of Equations (8) and (9).

4. Analysis of Experimental Data in $(TMTSF)_2ClO_4$

We consider partially ordered $(TMTSF)_2ClO_4$ samples. We denote T^* as superconducting onset temperature. In these compounds for $T > T^*$ there is no superconductivity. However, for $T < T^*$ the domains containing ordered ClO_4 anions partially transform to superconducting inclusions [7]. These ordered domains are embedded inside the phase of unordered ClO_4 anions, where SDW prevails but may coexist with metallic phase. Further cooling results in the increase of SC volume fraction ϕ and in the formation of coherent clusters of SC inclusions. At $T = T_c < T^*$ a complete SC channel gets opened [6,7], i.e., the SC phase coherence establishes in the whole sample. In Section 4.1, we use our theory to calculate the SC volume ratio ϕ. In Section 4.2, we study the influence of cooling rate on SC volume ratio. In the end, in Section 4.3, we find the approximate shape of SC inclusions in various disordered samples.

4.1. Application of MGA Theory to Describe Resistivity and to Find Superconducting Volume Ratio ϕ

To find a typical temperature dependence of SC volume ratio $\phi(T)$ in $(TMTSF)_2ClO_4$ at cooling rate $-dT/dt \leq 100$ K/min we choose the sample #4 in Figure 2 of Ref. [6] (This sample #4 was cooled at rate $-dT/dt = 100$ K/min and then annealed for some time at varying temperature between 15 and 23 K. Therefore, its disorder, presumably, corresponds to a slower cooling rate.) due to the availability of experimental resistivity data on $\rho_{xx}(T, H = 2T)$ in a magnetic field $H - H_z$, shown in Figure 2c of Ref. [6]. The magnetic field destroys superconductivity, and we can use these data to find the conductivity of background phase $\sigma^b_{xx}(T) = 1/(\rho_{xx}(T, H = 2T) - \Delta\rho_{xx})$ (see the inset in Figure 2), where the offset $\Delta\rho_{xx} = \rho_{xx}(T^*, H = 2T) - \rho_{xx}(T^*, H = 0)$ accounts for magnetoresistance of metallic phase at $H = 2T$. A magnetic field $H \gtrapprox 500$ Oe is usually enough to destroy superconductivity in $(TMTSF)_2ClO_4$ [60,61], but we take the data at $H = 2T$ where the SC effects can be safely ignored. Since the experimental data on $\rho_{zz}(T)$ under magnetic field for the same samples are absent, the background-phase conductivity σ^b_{zz} along the z-axis is found by extrapolating the metallic $\rho_{zz}(T)$ resistivity to low temperature by a second-order polynomial, similar to Ref. [61]. Here the second-order term comes from the electron-electron scattering at low temperature [61,62]. We take the x-axis as the reference axis for mapping to isotropic medium, i.e., σ^b_{xx} is taken as the background isotropic conductivity in the mapped space: $\sigma^b = \sigma^b_{xx}$. According to Equation (A4), the mapping coefficient along the z-axis is defined as $\eta = \sigma^b_{zz}/\sigma^b_{xx}$.

The SC volume ratio ϕ is found from Equation (10) for $i = x$. The calculated volume ratio as a function of temperature is plotted in Figure 2. Substituting ϕ found from Equation (10) for $i = x$ to Equation (9) for $i = z$, we predict the effective resistivity along the z-axis, $\sigma_{zz}(T)$. Its comparison with the experimental data from Figure 2b of Ref. [6] is shown in Figure 3. We see a rather good agreement. In this calculation we have one fitting parameter—the unknown ratio of the semiaxes a_x and a_z of ellipsoidal SC inclusions. We found that at high temperature $T \approx 1.2$ K, typical inclusions have the aspect ratio $a_z/a_x \approx 0.16$. At low temperature $T \approx 0.5$ K, the inclusions have aspect ratio $a_z/a_x \approx 0.85$. It means that with a decrease in temperature the SC inclusion becomes more isotropic along x and z-axes, i.e., $a_z/a_x \to 1$. It may indicate the formation of large and almost isotropic clusters of small SC inclusions. Please note that at any temperature the found aspect ratio a_z/a_x is much larger than the ratio of coherence lengths in $(TMTSF)_2ClO_4$, $\xi_z/\xi_x \approx 0.03$.

This supports the fact that in $(TMTSF)_2ClO_4$ the heterogeneity and SC islands originate from disorder and anion ordering rather than from usual SC fluctuations, because for SC fluctuations $a_z/a_x \sim \xi_z/\xi_x$.

Figure 3 and the inset in Figure 2 show the temperature dependence of resistivity for the same sample along the z and x axes correspondingly. From their comparison one observes that the resistivity drop near T_c for ρ_{zz} is much stronger than along for ρ_{xx}. This feature originates from the strong anisotropy of background-phase resistivity ρ_{ii}^b and is naturally described in generalized MGA theory [32–34]. The qualitative interpretation of this anisotropic resistivity drop due to SC onset is illustrated in Figure 1 of Refs. [32] or [33]. Because of high resistivity ρ_{zz}^b, the interlayer current mainly flows via the SC islands serving as shortcuts for this current direction. The effective resistivity ρ_{zz} is then determined by the much smaller intralayer resistivity and by the typical length of in-plane path between two close SC domains, which is inversely proportional to the SC volume fraction ϕ.

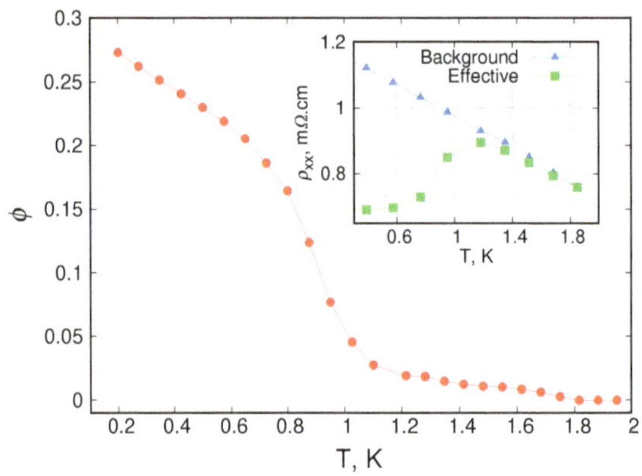

Figure 2. Dependence of SC volume ratio ϕ on temperature. ϕ (red,circle) is calculated using Equation (10) for $i = x$ and the experimental data on resistivity along the x-axis, taken from Figure 2c of Ref. [6] and shown in the inset. (inset) The effective medium resistivity ρ_{xx} containing SC, metallic and SDW phases (green squares), and the background resistivity ρ_{xx}^b in the absence of SC islands (blue triangles), taken from the data in Figure 2c of Ref. [6] in magnetic field.

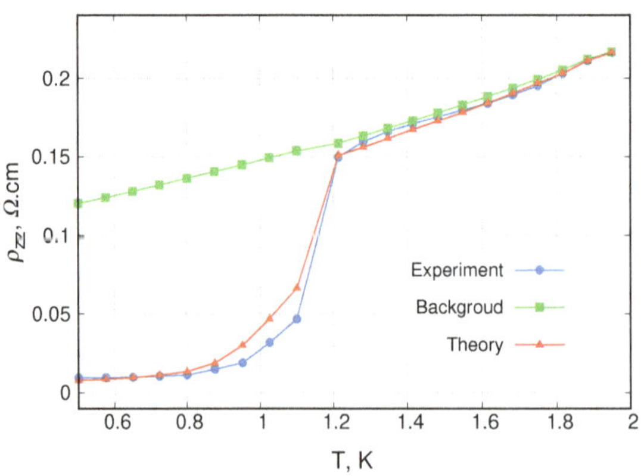

Figure 3. Temperature dependence of resistivity along z-axis. Experimental values (blue circles) are taken for sample #4 from Figure 2b of Ref. [6]. Background-phase metallic resistivity (green squares) is found by second-order approximation, i.e., $\rho_{zz}^b = 8.17 + 4.2\,T + 1.4\,T^2 \Omega{\cdot}\mathrm{cm}$. Theoretical values (red triangles) are found from z-axis resistivity in Equation (9).

4.2. Effect of Cooling Rate on Superconducting Volume Ratio

The cooling rate of $(TMTSF)_2ClO_4$ samples controls the fraction of ClO_4-ordered domains. At slow cooling, the ClO_4 anions have enough time to relax into ordered state. At fast cooling the thermal disorder remains in the samples, so that both ordered and disordered domains coexist. It was corroborated by resistivity [63], specific heat[64] and x-ray scattering [65] experiments.

The volume fraction ϕ_o of anion-ordered domains as a function of cooling rate has been studied using the X-ray scattering [65,66] and, recently, by resistivity and magnetic susceptibility [7] measurements. The corresponding results are compared in Figure 4 of Ref. [7] in the range of cooling rate $0 < -dT/dt < 20$ K/min. Several assumptions are made in extracting ϕ_o from these experiments[7,65]. First, (i) all these results assume that at the slowest cooling rate $-dT/dt|_{min} \sim 1$ K/min all ClO_4 are ordered at low temperature. It is not evident, as some degree of anion disorder may remain. Second, (ii) in the estimate of volume fraction ϕ_o from the resistivity measurements in the mixed SDW/metal phase in Ref. [7] the following equation (see Equation (1) of Ref. [7]) for the effective conductivity σ_{zz} along z-axis has been used:

$$\sigma_{zz} = \left[\phi\sigma_{min}^{1/3} + (1-\phi)\sigma_{max}^{1/3}\right]^3, \tag{11}$$

where $\rho_{min} = 1/\sigma_{min} = 0.03\ \Omega{\cdot}$cm is taken as a residual resistance of the sample with lowest cooling rate and $\rho_{max} = 1/\sigma_{max} = \rho_{min} + \Delta\rho_{c*} = 0.26\ \Omega{\cdot}$cm is determined assuming that the difference $\Delta\rho_{c*} = \rho_{max} - \rho_{min} = 0.23\ \Omega{\cdot}$cm is equal to the jump of resistivity at anion-ordering temperature $T_{AO} = 24.5$ K due to the scattering by anion disorder. In fact, at low temperature the anion disorder has much stronger effect on conductivity than just the electron scattering by this disorder itself, because it also favors the formation of insulating SDW state. Even if the fraction of insulating SDW domains is about one half, as in Figure 4 of Ref. [7], it may considerably affect the electron conductivity. In addition, (iii) Equation (11) does not take into account the conductivity anisotropy of $(TMTSF)_2ClO_4$, which strongly enhances the effect of metal/SC domains on resistivity along the least-conducting axis [32–34], as given by Equation (8). (iv) The extraction of ϕ_o from the magnetic susceptibility $\chi(T)$ data, especially at rapid cooling rate when the SC volume fraction $\phi \ll 1$, depends strongly on the size and shape of SC domains [34,54]; therefore the assumption [7] that $\phi_o = [\chi(T \to 0) - \chi(T_c)]/[\chi(T \to 0) - \chi(T_c)]_{dT/dt=0.02K/min}$ is not valid when the size of SC domains is smaller than the London penetration depth.

In this subsection we estimate the volume fraction ϕ of SC phase in $(TMTSF)_2ClO_4$ using the resistivity data from Ref. [7] and applying Equations (9) and (10) derived in MGA. Please note that the volume fractions ϕ of SC phase and ϕ_o of anion-ordered phase may differ, e.g., because the former depends on temperature. Since the MGA approximation is valid only at $\phi \ll 1$, it can only give $\phi(T)$ at $T > T_c$. Thus, our method of estimating ϕ works better for higher cooling rate when T_c is lower, therefore our results are rather complimentary to those in Ref. [7]. However, the cooling-rate dependence of $\phi(T > T_c)$ also gives the general tendency. Please note that at cooling rate $dT/dt = 100$ K/min and some annealing, by extrapolating the $\phi(T)$ curve in Figure 3 to $T = 0$ we obtain $\phi(T \to 0) \approx 0.3$, which is in a good agreement with other data in Figure 4 of Ref. [7].

To observe the influence of cooling rate on SC volume ratio $\phi(T)$, we use resistivity data from the inset of Figure 1d of Ref. [7]. These experimental values are taken as the effective resistivity ρ_{zz} along z-axis for different cooling rates. For background-phase resistivity ρ_{zz}^b along the z-axis we use the 2nd order polynomial fit of metallic resistivity at $T > T^*$. Substituting these $\rho_{zz}(T)$ and ρ_{zz}^b in Equation (10) we obtain $\phi(T)$ for various cooling rates. Unfortunately, in Ref. [7] there is no resistivity data along other two axes which would allow us to find the ellipsoid aspect ratio. Therefore, in Figure 4 we take the depolarization factor $A_i = 1/3$, i.e., the spherical inclusions in the mapped space instead of ellipsoidal. This choice corresponds to ellipsoid semiaxes $a_i \propto \sqrt{\sigma_{ii}(T^*)} \propto \xi_i$ in real space, as one expects for SC fluctuations. (The metallic conductivity $\sigma_{ii} \propto v_i^2 \propto \xi_i^2$, where v_i is the

Fermi velocity along axis i, and the BCS coherence length $\xi_i = \hbar v_i / \pi\Delta$.) The obtained $\phi(T)$ at cooling rates 0.02 K/min, 0.052 K/min, 2.5 K/min, 7.6 K/min and 18 K/min are shown in Figure 4. The curves in Figure 4 are similar to those in Figure 4 of Ref. [7], but the values of SC volume fraction ϕ are expectedly smaller than ϕ_o in Ref. [7], because at $T > T_c$ only a fraction of ClO_4-ordered domains becomes superconducting. However, ϕ increases with decreasing temperature and, probably, reaches ϕ_o at $T \to 0$.

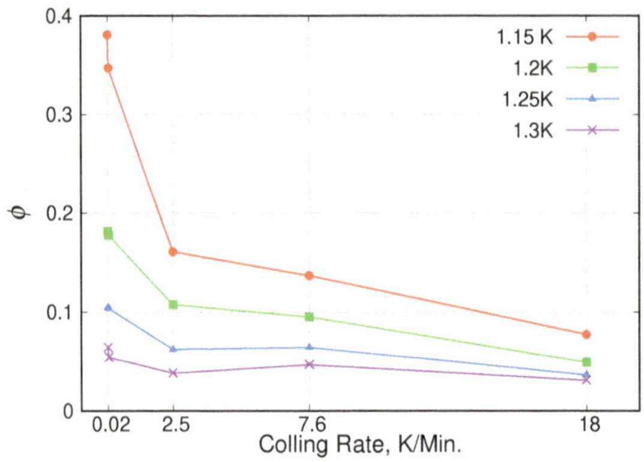

Figure 4. Dependence of SC volume ratio ϕ on cooling rate calculated using Equation (10) for different temperatures. For this calculation, the experimental data from the inset of Figure 1d in Ref. [7] are used.

4.3. *Effect of Disorder on the Shape of Superconducting Inclusions*

In Figure 4 we investigated the effect of cooling rate on ϕ. However, along with ϕ the shape of SC domains also plays an important role. Recent work [31] has shown that the probability of percolation along the shortest sample dimension, i.e., along sample thickness, is higher than along other directions. It was corroborated by the experiment on FeSe [33,52], where, by reducing the z-axis thickness of sample from 300 nm to ∼50 nm, one raised T_c from 8 K to 12 K [52]. The $(TMTSF)_2ClO_4$ samples are also usually flat. This effect of anisotropic superconductivity onset also depends on the shape of SC inclusions [31]. The knowledge of the shape of SC domains in $(TMTSF)_2ClO_4$ is also helpful to better understand the mechanism of their formation. In Section 4.1 we found a_z/a_x for sample # 4 in Ref. [6] at cooling rate 100 K/min. Below we find a_z/a_y and a_y/a_x for the samples cooled at rate 600 K/min with various times of subsequent annealing. This gives the effect of disorder on the shape of SC inclusions.

To study the evolution of aspect ratios a_z:a_y:a_x with disorder we use the experimental data from Figures 3 and 4 of Ref. [6]. Unfortunately, the curves with equal numbers in these two figures correspond to different samples. Thus, we do not have the data on resistivity along all three axes for the same sample and parameters, required to determine the full shape of ellipsoidal inclusions. However, we use the fact that the depolarization factors A_i in Equation (4) depend most strongly on the semiaxis a_i along the same direction, which allows us to vary only one parameter for each fit.

First we find the evolution of a_z/a_y with disorder. From the resistivity data along the z-axis, given in Figure 4a of Ref. [6], using Equation (10) we find $\phi(T)$ for different degrees of disorder. Using these $\phi(T)$ in Equation (9) we predicted resistivity along y-axis. From best fit values of predicted and experimental resistivity along y-axis, given in Figure 4b of Ref. [6], we find the ratio a_z/a_y for various degrees of disorder. The results are shown in Figure 5a and reveal that at low disorder, up to sample #9, the ratio a_z/a_y almost remains same. However, at higher disorder the ratio a_z/a_y increases.

Similarly, to find the evolution of a_y/a_x with disorder we use the data from Figure 3 of Ref. [6]. As before, from the resistivity data along y-axis using Equation (10) we find

$\phi(T)$ for different degrees of disorder. Using this $\phi(T)$ we predict resistivity along x-axis. We change the semiaxes of ellipsoids along y- and x-direction, so that the theoretical and experimental values of resistivity agree. Thus, the obtained ratio of a_y/a_x is shown in Figure 5b.

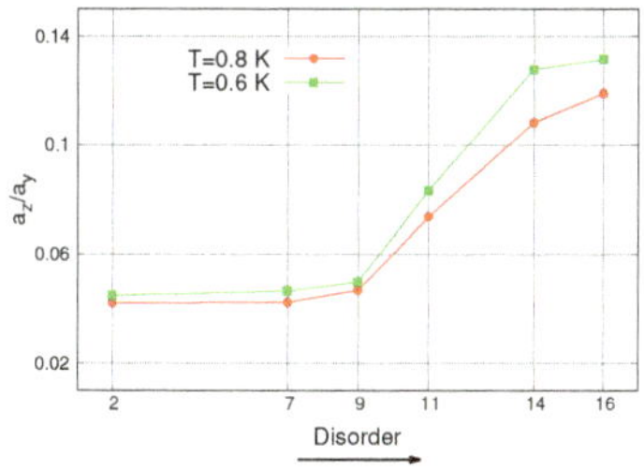

(a)

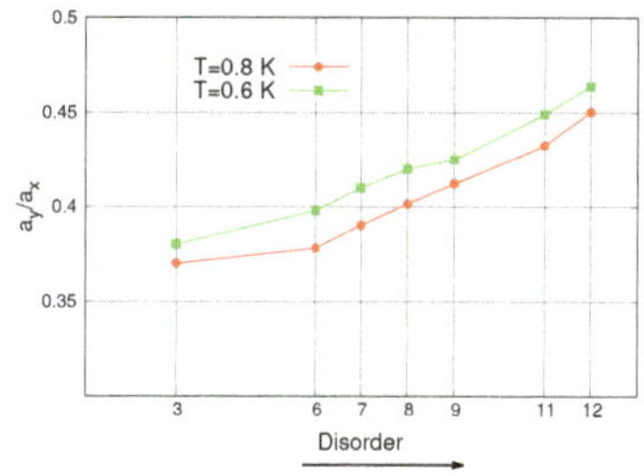

(b)

Figure 5. The dependence of aspect ratios a_z/a_y (**a**) and a_y/a_x (**b**) of superconducting domains in $(TMTSF)_2ClO_4$ on disorder at two temperatures $T = 0.8$ K and 0.6 K, calculated using Equations (4), (9) and (10) and resistivity data from Figures 3 and 4 of Ref. [6], taken at cooling rate 600 K/min and various annealing times. At longer annealing time, i.e., at weaker disorder, the shape of SC domains is more anisotropic.

5. Discussion

In this paper, we propose a method based on MGA to investigate the microscopic parameters of heterogeneous superconductors from resistivity data. We apply this method to the organic superconductor $(TMTSF)_2ClO_4$, where SC coexists with SDW in the form of isolated domains. Using our method we study the SC volume fraction ϕ and the shape of SC islands as a function of external parameters, such as temperature and ClO_4 anion disorder, which can be experimentally controlled by the cooling rate through the anion-ordering transition at $T_{AO} \approx 24.5$ K [7,65] or by the annealing of $(TMTSF)_2ClO_4$ samples [6].

For the best use of proposed method, one needs the following experimental data: (i) Temperature dependence of resistivity $\rho_{ii}(T)$ along each of non-equivalent main crystal axes (If the crystal has orthorhombic or lower symmetry, one needs the data along all three axes. If two or three crystal main axes are equivalent by symmetry, one only needs the data along two or one axes correspondingly.); (ii) $\rho_{ii}(T, H_0)$ in a magnetic field $H_0 > H_c$ destroying SC, to get the resistivity $\rho^b_{ii}(T)$ of the background homogeneous phase. In the absence of $\rho_{ii}(T, H_0)$ or in the case of non-SC inclusions, one needs to make an extrapolation of $\rho_{ii}(T)$ from $T > T^*$ to lower T to get $\rho^b_{ii}(T)$, which is less accurate. In the case of SDW/CDW inclusions one can also apply an external pressure destroying SDW/CDW to get $\rho^b_{ii}(T)$. If also (iii) magnetic susceptibility data are present, especially for all non-equivalent magnetic-field orientations, they help to independently check the obtained microscopic parameters and allow the estimate of the average size of SC inclusions as compared to the SC penetration depth [33,34]. Ideally, all these data are available for several values of external parameters that one is interested in, for example, at each studied cooling rate of $(TMTSF)_2ClO_4$. Unfortunately, despite an active experimental investigation of $(TMTSF)_2ClO_4$ by resistivity measurements, e.g., performed recently in Refs. [6,7], this full set of data is absent. Nevertheless, we have analyzed the available data from Refs. [6,7] to make some physical predictions concerning the mixed SC/SDW phase in $(TMTSF)_2ClO_4$.

In Figures 2 and 4 we show the obtained SC volume fraction ϕ for various cooling rates and temperatures $T_c < T < T^*$. Figure 3 illustrates how well the MGA model typically fits the experimental data. Figures 5 and 6 show the evolution of the shape of SC grains with the change of disorder by the annealing of rapidly cooled samples.

From Figure 5a we observe that for partially ordered samples #2, #7, #9 the aspect ratio $a_z/a_y \approx 0.05$ of SC domains at temperature $T \approx 0.6$–0.8 K depends weakly on disorder. The corresponding SC volume ratio $\phi \approx 0.1$ also weakly depends on disorder. At shorter annealing time, i.e., at larger disorder, the SC volume ratio decreases to $\phi \approx 0.03$, while the aspect ratio increases to $a_z/a_y \approx 0.13$. Another aspect ratio $a_y/a_x \approx 0.4$ depends much weaker on disorder, as shown in Figure 5b. Please note that the obtained ratios a_x:a_y:a_z at cooling rate 600 K/min are close to those of SC coherence length. In Figure 2 we found that for partially ordered sample #4 at cooling rate 100 K/min the aspect ratio of SC domains $a_z/a_x \approx 0.85$ at temperature $T \approx 0.5$ K, which differs considerably from what we get at 600 K/min even for long annealing time. This means that the decrease of cooling rate, as in Refs. [7,65] is not completely equivalent to the increase of annealing time at $T < T_{AO}$ used in Ref. [6]: they have similar effect on SC volume fraction ϕ but different effect on the shape of SC domains. Therefore, it would be very interesting to study their effect on the SC domain size, which can be extracted from the simultaneous magnetic susceptibility measurements as done for other compounds [33,34].

In organic superconductors $(TMTSF)_2ClO_4$ [6] and $(TMTSF)_2PF_6$ [14,15] superconductivity onsets anisotropically, i.e., first along the highest conducting z-axis and only in the end along the lowest-conducting x-axis. This behavior was first explained by assuming filamentary SC inclusions elongated along z-axis [14], but this hypothesis received neither theoretical nor experimental proof till now. Our analysis also shows that the SC domains are not elongated along z-axis but, on contrary, are oblate. Nevertheless, we predict much stronger decrease of resistivity along the least-conducting direction (compare Figure 3 and the inset in Figure 2), similar to experimental observations in $(TMTSF)_2ClO_4$ [6] and in many other heterogeneous anisotropic superconductors [32–34]. In our recent work [31] we proposed a simple model to explain the anisotropic zero-resistance onset also. We have shown [31] that the percolation probability along SC islands in needle or flat shaped samples is the highest along the shortest direction. A schematic illustration of this idea is given in Figure 6. The same idea can be applied to $(TMTSF)_2ClO_4$ to explain the anisotropic onset of superconductivity. Usually, the $(TMTSF)_2ClO_4$ samples are much shorter along the interlayer z-axis than along other two, e.g., the dimensions of samples in Ref. [6] are $3 \times 0.1 \times 0.03$ mm^3. Hence, the probability of percolation for the ellipsoidal inclusion, even with the obtained anisotropic aspect ratios a_z/a_x and a_z/a_y, will be the highest along the z-axis. Our preliminary calculations of percolation threshold for such flat elongated samples confirm this statement and show that the zero resistance, i.e., the percolation threshold, can be achieved even when the SC volume ratio $\phi = \phi_c \ll 1$. Hence, without invoking a filamentary SC one can easily explain the anisotropic onset of superconductivity in organic metals.

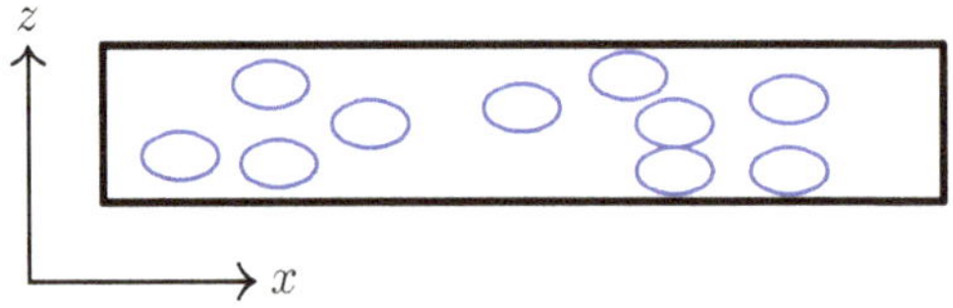

Figure 6. Schematic illustration of superconducting ellipsoid inclusions embedded inside a long thin conductor with dimensions $L_z \ll L_x$. It shows that the percolation probability along z is higher than along x-axis if $a_z/a_x > L_z/L_x$.

In our semi-classical theory based on MGA we neglect the dynamic fluctuations. Probably, the position and size of domains may fluctuate. However, in the studied organic metal the competition between SDW and superconductivity, leading to the domain formation, is governed by the disorder of ClO_4 anion orientation, which is frozen well below the anion-ordering transition temperature $T_{AO} \approx 24.5$ K. Hence, below $T_{SDW} \approx 4$–5 K we may consider the metal/SC and SDW domains as time-independent.

For all our calculation in Section 4 we have used Equations (9) and (10) because the inclusions are superconducting. However, the similar approach and Equation (8) can be used when the conductivity of inclusions is finite, e.g., when SDW inclusions are embedded inside a metallic background or vice versa. This is a usual occurrence in organic superconductors [4]. Thus, instead of using Equation (11) or similar phenomenological formulas to analyze the resistivity data in mixed metal/SDW or metal/CDW phases in organic metals, we recommend using MGA Formulas (8) and (4), which take into account the strong anisotropy of layered organic metals and the actual shape of domains. Apart from that, in several high-Tc superconductors as $Bi_2Sr_xCa_{3-x}Cu_2O_8$ [67], $La_{2-x}Sr_xCuO_4$ [68], $Ba(Fe_{1-x}Co_x)_2As_2$ [69–71], and in many other materials, e.g., $HfTe_3$ [72], there is also a spatially separated coexistence of metallic and SC or SDW/CDW phases. Hence in all these heterogeneous materials, provided the conductivity of both metallic and SDW/CDW phases is known and the domain size exceeds the coherence length, one can estimate the second-phase volume ratio and the domain shape from resistivity data using the above method.

6. Conclusions

Using the Maxwell-Garnett approximation we estimate the volume fraction and the shape of superconducting domains from resistivity measurements. We apply this method to investigate the heterogeneous electronic structure in organic superconductor $(TMTSF)_2ClO_4$, where superconductivity coexists with the spin-density wave in the form of isolated domains. This material is especially important to study such coexistence because it appears even at ambient pressure and can be easily controlled by changing the cooling rate or annealing time of the samples. From available resistivity data we study the evolution of the volume fraction and of the shape of superconducting domains in $(TMTSF)_2ClO_4$ with disorder and temperature. Our method applies not only to superconductors, but also when the density-wave or other-type domains are embedded inside a metallic background, or vice versa.

Author Contributions: P.D.G. conceptualized and developed the Methodology. Formal analysis, validation, manuscript writing and review was done by K.K.K., P.D.G. and V.D.K. All authors have read and agreed to the published version of the manuscript.

Funding: This article is partly supported by the Ministry of Science and Higher Education of the Russian Federation in the framework of Increase Competitiveness Program of MISiS, by RFBR grant No. 21-52-12027, and by the "Basis" Foundation for development of theoretical physics and mathematics. V. D. K. acknowledges the MISiS project No. K2-2020-001, and K. K. K. the MISiS support project for young research engineers and RFBR grants Nos. 19-32-90241 & 19-31-27001. P. D. G. acknowledges the State Assignment No. 0033-2019-0001 and RFBR grants Nos. 19-02-01000 & 21-52-12043.

Institutional Review Board Statement: Not applicable.

Informed Consent Statement: Not applicable.

Data Availability Statement: Data available on request.

Conflicts of Interest: The authors declare no conflict of interest.

Appendix A. Anisotropic Dilation of the Problem of Static Current Distribution

Anisotropic medium is converted to isotropic one by mapping the real space to a mapped space, where the solution is simpler. The mapping should satisfy the following conditions: (i) the conductivity of background phase in the mapped space should be isotropic, and (ii) the electrostatic continuity equation should be satisfied in the mapped space with the same solution.

In our notations, $\sigma^b_{xx}, \sigma^b_{yy}, \sigma^b_{zz}$ are the constant conductivity components of background phase in the original heterogeneous medium. Let J and V be the current density and the applied potential, respectively. The electrostatic continuity equation in the background phase is then written as

$$-\nabla J = \sigma^b_{xx}\frac{\partial^2 V}{\partial x^2} + \sigma^b_{yy}\frac{\partial^2 V}{\partial y^2} + \sigma^b_{zz}\frac{\partial^2 V}{\partial z^2} = 0. \tag{A1}$$

Heterogeneity is hidden in the boundary conditions on the surface of each grain and of the sample.

Let x', y' and z' be the axes in mapped space, where conductivity should be isotropic: $\sigma^b_{x'x'} = \sigma^b_{y'y'} = \sigma^b_{z'z'} = \sigma^b$. If J' and V' are the current density and electrostatic potential respectively in the mapped space, the continuity equation in the mapped space is written as

$$-\nabla' J' = \sigma^b \left(\frac{\partial^2 V'}{\partial x'^2} + \frac{\partial^2 V'}{\partial y'^2} + \frac{\partial^2 V'}{\partial z'^2} \right) = 0. \tag{A2}$$

The condition (ii) for our mapping means that the solution

$$V(x,y,z) = V'(x',y',z') \tag{A3}$$

of Equations (A1) and (A2) is the same. This solution completely determines the effective conductivity of heterogeneous medium, as it also gives the current density via $J_i = \sigma^b_{ii} \nabla_i V \neq J'_i = \sigma^b \nabla'_i V'$. Equations (A1)–(A3) are consistent if the mapping is the anisotropic scaling, e.g.,

$$x = x', \qquad y = \sqrt{\mu} y', \qquad z = \sqrt{\eta} z', \tag{A4}$$

with constant mapping coefficients μ and η determined by conductivity anisotropy in real space:

$$\mu = \frac{\sigma^b_{yy}}{\sigma^b_{xx}}, \qquad \eta = \frac{\sigma^b_{zz}}{\sigma^b_{xx}}. \tag{A5}$$

Instead of (A4) one could choose a product of the mapping (A4) and of any isotropic scaling by a factor α with the simultaneous change of $\sigma^b \to \alpha^2 \sigma^b$. We have chosen $\sigma^b = \sigma^b_{xx}$, so that $\mu, \nu < 1$.

The current components in the real space $r = (x,y,z)$ and in the mapped space $r' = (x',y',z')$ are related as

$$J_x(r) = J'_{x'}(r'), \qquad J_y(r) = \sqrt{\mu} J'_{y'}(r'), \qquad J_z(r) = \sqrt{\eta} J'_{z'}(r'). \tag{A6}$$

The shapes of inclusions are not preserved during this mapping procedure. For example, a sphere with radius a_x described by the equation $x^2/a_x^2 + y^2/a_x^2 + z^2/a_x^2 = 1$ in non-homogeneous medium will transform to an ellipsoid described by the equation $x'^2/a^2_{x'} + y'^2/a^2_{y'} + z'^2/a^2_{z'} = 1$ in mapped space, where the semiaxes are given by

$$a_{x'} = a_x, \quad a_{y'} = a_x/\sqrt{\mu}, \quad a_{z'} = a_x/\sqrt{\eta}. \tag{A7}$$

If z is the lowest-conducting axis, and the highest conducting axis is x, then a sphere in real space transforms to an ellipsoid elongated along z-axis. Due to the temperature dependence of conductivity anisotropy, the coefficients μ, η and the shape of inclusions change with temperature either. If in real space the inclusion is ellipsoid with semiaxes $a = a_x$, $b = \beta a_x$ and $c = \gamma a_x$, then it transforms to ellipsoid in mapped space with semiaxes

$$a_{x'} = a_x, \quad a_{y'} = a_x \beta/\sqrt{\mu}, \quad a_{z'} = a_x \gamma/\sqrt{\eta}. \tag{A8}$$

In MGA we take ellipsoidal inclusions with fixed aspect ratios β and γ, but varying size.

References

1. Shen, K.M.; Davis, J.S. Cuprate High-Tc Superconductors. *Mater. Today* **2008**, *11*, 14–21. [CrossRef]
2. Norman, M.R. Cuprates—An Overview. *J. Supercond. Nov. Magn.* **2012**, *25*, 2131–2134. [CrossRef]
3. Hosono, H.; Kuroki, K. Iron-Based Superconductors: Current Status of Materials and Pairing Mechanism. *Phys. C Supercond. Appl.* **2015**, *514*, 399–422. [CrossRef]

4. Jerome, D. Organic Conductors: From Charge Density Wave TTF-TCNQ to Superconducting $(TMTSF)_2PF_6$. *Chem. Rev.* **2004**, *104*, 5565–5591. [CrossRef] [PubMed]
5. Stewart, G.R. Unconventional Superconductivity. *Adv. Phys.* **2017**, *66*, 75–196. [CrossRef]
6. Gerasimenko, Y.A.; Sanduleanu, S.V.; Prudkoglyad, V.A.; Kornilov, A.V.; Yamada, J.; Qualls, J.S.; Pudalov, V.M. Coexistence of Superconductivity and Spin-Density Wave in $(TMTSF)_2$ ClO_4: Spatial Structure of the Two-Phase State. *Phys. Rev. B* **2014**, *89*, 054518. [CrossRef]
7. Yonezawa, S.; Marrache-Kikuchi, C.A.; Bechgaard, K.; Jérome, D. Crossover from Impurity-Controlled to Granular Superconductivity in $(TMTSF)_2ClO_4$. *Phys. Rev. B* **2018**, *97*, 014521. [CrossRef]
8. Jerome, D. Organic Superconductors: A Survey of Low Dimensional Phenomena. *Mol. Cryst. Liq. Cryst.* **1982**, *79*, 511–538. [CrossRef]
9. Ishiguro, T.; Yamaji, K.; Saito, G. *Organic Superconductors*; Springer Series in Solid-State Sciences; Springer: Berlin/Heidelberg, Germany, 1998; Volume 88. [CrossRef]
10. Lebed, A.; Hull, R.; Osgood, R.M.; Parisi, J.; Warlimon, H. (Eds.) *The Physics of Organic Superconductors and Conductors*; Springer Series in Materials Science; Springer: Berlin/Heidelberg, Germany, 2008; Volume 110. [CrossRef]
11. Kartsovnik, M.V. High Magnetic Fields: A Tool for Studying Electronic Properties of Layered Organic Metals. *Chem. Rev.* **2004**, *104*, 5737–5782. [CrossRef]
12. Chaikin, P.M. Magnetic-Field-Induced Transition in Quasi-Two-Dimensional Systems. *Phys. Rev. B* **1985**, *31*, 4770–4772. [CrossRef]
13. Gor'kov, L.; Lebed', A. On the Stability of the Quasi-Onedimensional Metallic Phase in Magnetic Fields against the Spin Density Wave Formation. *J. Phys. Lett.* **1984**, *45*, 433–440. [CrossRef]
14. Kang, N.; Salameh, B.; Auban-Senzier, P.; Jérome, D.; Pasquier, C.R.; Brazovskii, S. Domain Walls at the Spin-Density-Wave Endpoint of the Organic Superconductor $(TMTSF)_2PF_6$ under Pressure. *Phys. Rev. B* **2010**, *81*, 100509. [CrossRef]
15. Narayanan, A.; Kiswandhi, A.; Graf, D.; Brooks, J.; Chaikin, P. Coexistence of Spin Density Waves and Superconductivity in $(TMTSF)_2PF_6$. *Phys. Rev. Lett.* **2014**, *112*, 146402. [CrossRef] [PubMed]
16. Vuletić, T.; Auban-Senzier, P.; Pasquier, C.; Tomić, S.; Jérome, D.; Héritier, M.; Bechgaard, K. Coexistence of Superconductivity and Spin Density Wave Orderings in the Organic Superconductor $(TMTSF)_2PF_6$. *Eur. Phys. J. B* **2002**, *25*, 319–331. [CrossRef]
17. Kornilov, A.V.; Pudalov, V.M.; Kitaoka, Y.; Ishida, K.; Zheng, G.Q.; Mito, T.; Qualls, J.S. Macroscopically Inhomogeneous State at the Boundary between the Superconducting, Antiferromagnetic, and Metallic Phases in Quasi-One-Dimensional $(TMTSF)_2PF_6$. *Phys. Rev. B* **2004**, *69*, 224404. [CrossRef]
18. Pasquier, C.; Kang, N.; Salameh, B.; Auban-Senzier, P.; Jérome, D.; Brazovskii, S. Evolution of the Spin-Density Wave-Superconductivity Texture in the Organic Superconductor $(TMTSF)_2PF_6$ under Pressure. *Phys. B Condens. Matter* **2012**, *407*, 1806–1809. [CrossRef]
19. Podolsky, D.; Altman, E.; Rostunov, T.; Demler, E. SO(4) Theory of Antiferromagnetism and Superconductivity in Bechgaard Salts. *Phys. Rev. Lett.* **2004**, *93*, 246402. [CrossRef]
20. Lee, I.J.; Chaikin, P.M.; Naughton, M.J. Critical Field Enhancement near a Superconductor-Insulator Transition. *Phys. Rev. Lett.* **2002**, *88*, 207002. [CrossRef]
21. Gor'kov, L.P.; Grigoriev, P.D. Soliton Phase near Antiferromagnetic Quantum Critical Point in Q1D Conductors. *EPL Europhys. Lett.* **2005**, *71*, 425–430. [CrossRef]
22. Gor'kov, L.P.; Grigoriev, P.D. Nature of the Superconducting State in the New Phase in $(TMTSF)_2$ PF_6 under Pressure. *Phys. Rev. B* **2007**, *75*, 020507. [CrossRef]
23. Grigoriev, P.D. Properties of Superconductivity on a Density Wave Background with Small Ungapped Fermi Surface Parts. *Phys. Rev. B* **2008**, *77*, 224508. [CrossRef]
24. Grigoriev, P. Superconductivity on the Density-Wave Background with Soliton-Wall Structure. *Phys. B Condens. Matter* **2009**, *404*, 513–516. [CrossRef]
25. Dordevic, S.V.; Basov, D.N.; Homes, C.C. Do Organic and Other Exotic Superconductors Fail Universal Scaling Relations? *Sci. Rep.* **2013**, *3*, 1713. [CrossRef]
26. Greer, A.; Harshman, D.; Kossler, W.; Goonewardene, A.; Williams, D.; Koster, E.; Kang, W.; Kleiman, R.; Haddon, R. Muon Spin Rotation Study of the $(TMTSF)_2ClO_4$ System. *Phys. C Supercond.* **2003**, *400*, 59–64. [CrossRef]
27. Pratt, F.L.; Lancaster, T.; Blundell, S.J.; Baines, C. Low-Field Superconducting Phase of $(TMTSF)_2ClO_4$. *Phys. Rev. Lett.* **2013**, *110*, 107005. [CrossRef]
28. Schwenk, H.; Andres, K.; Wudl, F.; Aharon-Shalom, E. Direct Observation of a Large London Penetration Depth in an Organic Superconductor. *Solid State Commun.* **1983**, *45*, 767–769. [CrossRef]
29. Su, W.P.; Kivelson, S.; Schrieffer, J.R. Theory of Polymers Having Broken Symmetry Ground States. In *Physics in One Dimension*; Cardona, M., Fulde, P., Queisser, H.J., Bernasconi, J., Schneider, T., Eds.; Springer: Berlin/Heidelberg, Germany, 1981; Volume 23, pp. 201–211. [CrossRef]
30. Gerasimenko, Y.A.; Prudkoglyad, V.A.; Kornilov, A.V.; Sanduleanu, S.V.; Qualls, J.S.; Pudalov, V.M. Role of anion ordering in the coexistence of spin-density-wave and superconductivity in $(TMTSF)_2ClO_4$. *JETP Lett.* **2013**, *97*, 419–424. [CrossRef]
31. Kochev, V.D.; Kesharpu, K.K.; Grigoriev, P.D. Anisotropic Zero-Resistance Onset in Organic Superconductors. *arXiv* **2020**, arXiv:cond-mat/2007.14388.

32. Sinchenko, A.A.; Grigoriev, P.D.; Orlov, A.P.; Frolov, A.V.; Shakin, A.; Chareev, D.A.; Volkova, O.S.; Vasiliev, A.N. Gossamer High-Temperature Bulk Superconductivity in FeSe. *Phys. Rev. B* **2017**, *95*, 165120. [CrossRef]
33. Grigoriev, P.D.; Sinchenko, A.A.; Kesharpu, K.K.; Shakin, A.; Mogilyuk, T.I.; Orlov, A.P.; Frolov, A.V.; Lyubshin, D.S.; Chareev, D.A.; Volkova, O.S.; et al. Anisotropic Effect of Appearing Superconductivity on the Electron Transport in FeSe. *JETP Lett.* **2017**, *105*, 786–791. [CrossRef]
34. Seidov, S.S.; Kesharpu, K.K.; Karpov, P.I.; Grigoriev, P.D. Conductivity of Anisotropic Inhomogeneous Superconductors above the Critical Temperature. *Phys. Rev. B* **2018**, *98*, 014515. [CrossRef]
35. Torquato, S. *Random Heterogeneous Materials: Microstructure and Macroscopic Properties*; Interdisciplinary Applied Mathematics; Springer: New York, NY, USA, 2002.
36. Bechgaard, K.; Carneiro, K.; Olsen, M.; Rasmussen, F.B.; Jacobsen, C.S. Zero-Pressure Organic Superconductor: Di-(Tetramethyltetraselenafulvalenium)-Perchlorate $(TMTSF)_2ClO_4$. *Phys. Rev. Lett.* **1981**, *46*, 852–855. [CrossRef]
37. Bechgaard, K.; Carneiro, K.; Eg, O.; Olsen, M.; Rasmussen, F.B.; Jacobsen, C.; Rindorf, G. Superconductivity in $(TMTSF)_2ClO_4$ at Zero Pressure. *Mol. Cryst. Liq. Cryst.* **1982**, *79*, 627–632. [CrossRef]
38. Bychkov, Y.A.; Gor'kov, L.P.; Dzyaloshinskiĭ. Possibility of Superconductivity Type Phenomena in a One-Dimensional System. *JETP* **1966**, *23*, 489.
39. Sólyom, J. The Fermi Gas Model of One-Dimensional Conductors. *Adv. Phys.* **1979**, *28*, 201–303. [CrossRef]
40. Jérome, D.; Creuzet, F.; Bourbonnais, C. A Survey of the Physics of Organic Conductors and Superconductors. *Phys. Scr.* **1989**, *T27*, 130–135. [CrossRef]
41. Parkin, S.S.P.; Ribault, M.; Jerome, D.; Bechgaard, K. Superconductivity in the Family of Organic Salts Based on the Tetramethyltetraselenafulvalene (TMTSF) Molecule: $(TMTSF)_2X$ ($X=ClO_4$, PF_6, AsF_6, SbF_6, TaF_6). *J. Phys. C Solid State Phys.* **1981**, *14*, 5305–5326. [CrossRef]
42. Garoche, P.; Brusetti, R.; Jerome, D.; Bechgaard, K. Specific-Heat Measurements of Organic Superconductivity in $(TMTSF)_2ClO_4$. *J. Phys. Lett.* **1982**, *43*, L147–L152. [CrossRef]
43. Schwenk, H.; Neumaier, K.; Andres, K.; Wudl, F.; Aharon-Shalom, E. Meissner Anisotropy in Deuterated $(TMTSF)_2ClO_4$. *Mol. Cryst. Liq. Cryst.* **1982**, *79*, 633–638. [CrossRef]
44. Tomić, S.; Jérome, D.; Monod, P.; Bechgaard, K. EPR and Electrical Conductivity of the Organic Superconductor Di-Tetramethyltetraselenafulvalenium-Perchlorate, $(TMTSF)_2ClO_4$ and a Metastable Magnetic State Obtained by Fast Cooling. *J. Phys. Lett.* **1982**, *43*, 839–844. [CrossRef]
45. Takahashi, T.; Jérome, D.; Bechgaard, K. Observation of a Magnetic State in the Organic Superconductor $(TMTSF)_2ClO_4$: Influence of the Cooling Rate. *J. Phys. Lett.* **1982**, *43*, 565–573. [CrossRef]
46. Ishiguro, T.; Murata, K.; Kajimura, K.; Kinoshita, N.; Tokumoto, H.; Tokumoto, M.; Ukachi, T.; Anzai, H.; Saito, G. Superconductivity and Metal-Nonmetal Transitions in $(TMTSF)_2ClO_4$. *J. Phys. Colloq.* **1983**, *44*, C3–831–C3–838. [CrossRef]
47. Pouget, J.P. Structural Aspects of the Bechgaard and Fabre Salts: An Update. *Crystals* **2012**, *2*, 466–520. [CrossRef]
48. Bechgaard, K.; Carneiro, K.; Rasmussen, F.B.; Olsen, M.; Rindorf, G.; Jacobsen, C.S.; Pedersen, H.J.; Scott, J.C. Superconductivity in an Organic Solid. Synthesis, Structure, and Conductivity of Bis(Tetramethyltetraselenafulvalenium) Perchlorate, $(TMTSF)_2ClO_4$. *J. Am. Chem. Soc.* **1981**, *103*, 2440–2442. [CrossRef]
49. Rindorf, G.; Soling, H.; Thorup, N. The Structures of Di(2,3,6,7-Tetramethyl-1,4,5,8-Tetraselenafulvalenium) Perrhenate, (TMTSF)2ReO4, and Perchlorate, $(TMTSF)_2ClO_4$. *Acta Crystallogr. Sect. B Struct. Crystallogr. Cryst. Chem.* **1982**, *38*, 2805–2808. [CrossRef]
50. Pouget, J.P.; Shirane, G.; Bechgaard, K.; Fabre, J.M. X-Ray Evidence of a Structural Phase Transition in Di-Tetramethyltetraselenafulvalenium Perchlorate [$(TMTSF)_2ClO_4$], Pristine and Slightly Doped. *Phys. Rev. B* **1983**, *27*, 5203–5206. [CrossRef]
51. Le Pévelen, D.; Gaultier, J.; Barrans, Y.; Chasseau, D.; Castet, F.; Ducasse, L. Temperature and Pressure Dependencies of the Crystal Structure of the Organic Superconductor $(TMTSF)_2ClO_4$. *Eur. Phys. J. B* **2001**, *19*, 363–373. [CrossRef]
52. Mogilyuk, T.I.; Grigoriev, P.D.; Kesharpu, K.K.; Kolesnikov, I.A.; Sinchenko, A.A.; Frolov, A.V.; Orlov, A.P. Excess Conductivity of Anisotropic Inhomogeneous Superconductors Above the Critical Temperature. *Phys. Solid State* **2019**, *61*, 1549–1552. [CrossRef]
53. Murata, K.; Tokumoto, M.; Anzai, H.; Kajimura, K.; Ishiguro, T. Upper Critical Field of the Anisotropic Organic Superconductors, $(TMTSF)_2ClO_4$. *Jpn. J. Appl. Phys.* **1987**, *26*, 1367. [CrossRef]
54. Tinkham, M. *Introduction to Superconductivity*; Courier Corporation: Mineola, NY, USA, 2004.
55. Markel, V.A. Introduction to the Maxwell Garnett Approximation: Tutorial. *J. Opt. Soc. Am. A* **2016**, *33*, 1244. [CrossRef]
56. Markel, V.A. Maxwell Garnett Approximation (Advanced Topics): Tutorial. *J. Opt. Soc. Am. A* **2016**, *33*, 2237. [CrossRef] [PubMed]
57. Maxwell Garnett, J.C. Colours in Metal Glasses and in Metallic Films. *Philos. Trans. R. Soc. Lond. Ser. A* **1904**, *203*, 385–420.
58. Stroud, D. Generalized effective-medium approach to the conductivity of an inhomogeneous material. *Phys. Rev. B* **1975**, *12*, 3368–3373. [CrossRef]
59. Apresyan, L.A.; Vlasov, D.V. On depolarization factors of anisotropic ellipsoids in an anisotropic medium. *Tech. Phys.* **2014**, *59*, 1760–1765. [CrossRef]
60. Gubser, D.U.; Fuller, W.W.; Poehler, T.O.; Cowan, D.O.; Lee, M.; Potember, R.S.; Chiang, L.Y.; Bloch, A.N. Magnetic Susceptibility and Resistive Transitions of Superconducting $(TMTSF)_2ClO_4$: Critical Magnetic Fields. *Phys. Rev. B* **1981**, *24*, 478–480. [CrossRef]

61. Auban-Senzier, P.; Jérome, D.; Doiron-Leyraud, N.; René de Cotret, S.; Sedeki, A.; Bourbonnais, C.; Taillefer, L.; Alemany, P.; Canadell, E.; Bechgaard, K. The Metallic Transport of (TMTSF) $_2$ X Organic Conductors Close to the Superconducting Phase. *J. Physics: Condens. Matter* **2011**, *23*, 345702. [CrossRef]
62. Rice, M.J. Electron-Electron Scattering in Transition Metals. *Phys. Rev. Lett.* **1968**, *20*, 1439–1441. [CrossRef]
63. Schwenk, H.; Andres, K.; Wudl, F. Resistivity of the Organic Superconductor Ditetramethyltetraselenafulvalenium Perchlorate, $(TMTSF)_2ClO_4$, in Its Relaxed, Quenched, and Intermediate State. *Phys. Rev. B* **1984**, *29*, 500–502. [CrossRef]
64. Pesty, F.; Wang, K.; Garoche, P. Analysis of the Pair Breaking Effect of the Anion Disorder $(TMTSF)_2ClO_4$. *Synth. Met.* **1988**, *27*, 137–143. [CrossRef]
65. Pouget, J.P.; Kagoshima, S.; Tamegai, T.; Nogami, Y.; Kubo, K.; Nakajima, T.; Bechgaard, K. High Resolution X-ray Scattering Study of the Anion Ordering Phase Transition of $(TMTSF)_2ClO_4$. *J. Phys. Soc. Jpn.* **1990**, *59*, 2036–2053. [CrossRef]
66. Kagoshima, S.; Yasunaga, T.; Ishiguro, T.; Anzai, H.; Saito, G. Quenching effect of the anion ordering in the organic superconductor $(TMTSF)_2ClO_4$: An X-ray study. *Solid State Commun.* **1983**, *46*, 867–870. [CrossRef]
67. Martin, S.; Fiory, A.T.; Fleming, R.M.; Schneemeyer, L.F.; Waszczak, J.V. Temperature Dependence of the Resistivity Tensor in Superconducting $Bi_2Sr_{2.2}$ $Ca_{0.8}Cu_2O_8$ Crystals. *Phys. Rev. Lett.* **1988**, *60*, 2194–2197. [CrossRef] [PubMed]
68. Nakamura, Y.; Uchida, S. Anisotropic Transport-Properties of Single-Crystal $La_{2-x}Sr_xCuO_4$—Evidence for the Dimensional Crossover. *Phys. Rev. B* **1993**, *47*, 8369–8372. [CrossRef] [PubMed]
69. Ni, N.; Tillman, M.E.; Yan, J.Q.; Kracher, A.; Hannahs, S.T.; Bud'ko, S.L.; Canfield, P.C. Effects of Co Substitution on Thermodynamic and Transport Properties and Anisotropic H_{c2} in $Ba(Fe_{1-x}Co_x)_2$ As_2 Single Crystals. *Phys. Rev. B* **2008**, *78*, 214515. [CrossRef]
70. Wang, X.F.; Wu, T.; Wu, G.; Liu, R.H.; Chen, H.; Xie, Y.L.; Chen, X.H. The Peculiar Physical Properties and Phase Diagram of $BaFe_{2-x}Co_xAs_2$ Single Crystals. *New J. Phys.* **2009**, *11*, 045003. [CrossRef]
71. Tanatar, M.A.; Ni, N.; Thaler, A.; Bud'ko, S.L.; Canfield, P.C.; Prozorov, R. Pseudogap and Its Critical Point in the Heavily Doped $Ba(Fe_{1-x}Co_x)_2As_2$ from c-Axis Resistivity Measurements. *Phys. Rev. B* **2010**, *82*, 134528. [CrossRef]
72. Li, J.; Peng, J.; Zhang, S.; Chen, G. Anisotropic Multichain Nature and Filamentary Superconductivity in the Charge Density Wave System $HfTe_3$. *Phys. Rev. B* **2017**, *96*, 174510. [CrossRef]

crystals

MDPI

Article

Crystal Structures and Electrical Resistivity of Three Exotic TMTSF Salts with I_3^-: Determination of Valence by DFT and MP2 Calculations

Harukazu Yoshino [1,*], Yoshiki Iwasaki [2], Rika Tanaka [3], Yuka Tsujimoto [1] and Chiaki Matsuoka [1]

1 Graduate School of Science, Osaka City University, Osaka 558-8585, Japan; tsujimoto@e.sci.osaka-cu.ac.jp (Y.T.); matsuoka@e.sci.osaka-cu.ac.jp (C.M.)
2 Center for Advanced High Magnetic Field Science, Graduate School of Science, Osaka University, Toyonaka, Osaka 560-0043, Japan; iwasaki@ahmf.sci.osaka-u.ac.jp
3 Graduate School of Engineering, Osaka City University, Sumiyoshi-ku, Osaka 558-8585, Japan; g-tanaka@eng.osaka-cu.ac.jp
* Correspondence: yoshino@sci.osaka-cu.ac.jp; Tel.: +81-6-6605-3070

Received: 2 November 2020; Accepted: 5 December 2020; Published: 8 December 2020

Abstract: Three novel organic conductors $(TMTSF)_8(I_3)_5$, $(TMTSF)_5(I_3)_2$, and $(TMTSF)_4(I_3)_4 \cdot THF$ (THF = tetrahydrofuran) were synthesized and their crystal structures were characterized by X-ray diffraction analyses, where TMTSF denotes tetramethyltetraselenafulvalene. The crystals of both the $(TMTSF)_8(I_3)_5$ and $(TMTSF)_5(I_3)_2$ are composed of one-dimensional stacks of TMTSF trimers separated by TMTSF monomers. The crystal of the $(TMTSF)_4(I_3)_4 \cdot THF$ is composed of the TMTSF tetramers and I_3^- tetramers; and regarded as the elongated rock-salt structure. The electrical conductivity of the $(TMTSF)_8(I_3)_5$ and $(TMTSF)_5(I_3)_2$ is about 60 and 50 $S{\cdot}cm^{-1}$ at room temperature, respectively. The electrical resistivity of $(TMTSF)_8(I_3)_5$ is weakly metallic below room temperature and rapidly increases below 88 and 53 K on cooling suggesting two possible phase transitions. The electrical resistivity of $(TMTSF)_5(I_3)_2$ is semiconducting below room temperature but shows an anomaly around 190 K, below which the activation energy becomes small. The application of hydrostatic pressure up to 1.7 GPa do not change these behaviors of $(TMTSF)_8(I_3)_5$ and $(TMTSF)_5(I_3)_2$ very much. A method to evaluate the non-integer valence of crystallographically independent TMTSF molecules is developed by using the DFT (density-functional-theory) and MP2 (Hartree–Fock calculations followed by Møller–Plesset correlation energy calculations truncated at second order) calculations. It is shown that the method gives the valence of the TMTSF molecules of the I_3 salts consistent with their electrical properties.

Keywords: $(TMTSF)_8(I_3)_5$; $(TMTSF)_5(I_3)_2$; $(TMTSF)_4(I_3)_4 \cdot THF$; organic conductor; crystal structure; electrical resistivity; high pressure; DFT; MP2

1. Introduction

The organic donor TMTSF (= tetramethyltetraselenafulvalene.) is known to give the first organic superconductor $(TMTSF)_2PF_6$ (T_c = 0.9 K, $p_c \sim$ 1.2 GPa) [1] and first ambient-pressure organic superconductor $(TMTSF)_2ClO_4$ (T_c = 1.2 K) [2]. These superconductors are most studied members of Bechgaard salts, which are the 2:1 radical salts of TMTSF and monovalent anions (PF_6^-, AsF_6^-, SbF_6^-, TaF_6^-, BF_4^-, ClO_4^-, NO_3^-) [3,4]. In addition to the starting seven members, yet eight isostructural salts have been synthesized with NbF_6^- [5], ReO_4^- [6,7], BrO_4^- [8], FSO_3^- [8], $PF_2O_2^-$ [9], $CF_3SO_3^-$ [10,11], $H_2F_3^-$ [12], and SiF_5^- [13], respectively.

Their crystal structures are characterized by the weakly dimerized stacks of the planar TMTSF molecules along the most-conducting ***a***-axis. There are weak interaction between the neighboring stacks along the second-conducting ***b***-axis forming the TMTSF layers. Finally, each TMTSF layer is separated by an anion layer resulting in a sandwich structure.

The physical properties of the PF_6 and ClO_4 salts are so interesting that the studies of the TMTSF salts seem to have been mainly focused on the starting members of Bechgaard salts including the two. Many experimental and theoretical efforts have been devoted to reveal the nature of, for example, possible exotic superconductivity [14–18], spin-density-wave (SDW) transitions [19–27], field-induced SDW transitions [28–35], and a variety of angular-dependent magnetoresistance oscillations [36–44], all of which is closely related to the low dimensionality of electronic systems of Bechgaard salts.

On the other hand, much less attention has been paid to "exotic" TMTSF salts other than Bechgaard ones. To the best of the authors' knowledge, there are reports on the six charge-transfer complexes with acceptor molecules [45–50] as well as twenty five salts with inorganic or complex anions, where TMTSF molecules have the average valence of $(1/2)+$ [51–55], $(2/3)+$ [56–61], $(3/4)+$ [62,63], $(4/5)+$ [64], and $1+$ [65–71], respectively. Actually, the number of the exotic TMTSF salts is larger than that of the fifteen Bechgaard ones mentioned above. Although the superconductivity has never been reported, new phenomena to be explored will be probably provided by the variety of their exotic crystal structures giving different oxidation states and/or packing patterns of TMTSF molecules from that of Bechgaard salts.

In this paper, we present three novel exotic TMTSF salts with the same counter anion I_3^-, namely $(TMTSF)_8(I_3)_5$ (8:5 salt), $(TMTSF)_5(I_3)_2$ (5:2 salt), and $(TMTSF)_4(I_3)_4{\cdot}THF$ (4:4 salt, THF = tetrahydrofuran), respectively. Their crystal structures are of different types from those of any TMTSF salts ever reported. We carried out the X-ray crystal structure analyses; and the electrical resistivity measurements at ambient pressure as well as at pressures up to 1.7 GPa.

The information on the valence of donors and/or acceptors is helpful to understanding the electronic states of charge-transfer complexes with partial charge transfer. In some literature of the exotic TMTSF salts, they attempted to estimate the TMTSF valence on the basis of bond lengths within the TMTSF molecule [46,60,63,71]. It is, however, shown below that such an method does not give plausible estimates for the present I_3 salts.

Instead, we estimated the valence from the difference between calculated energies of the neutral and cationic TMTSF molecules using the conformations observed for the crystals. By comparing these energies with that obtained for the quantum mechanically optimized conformations, it is shown that one can obtain reasonably quantitative valence of TMTSF molecules in the crystals. This method is probably applicable to the other exotic TMTSF salts as well as to salts of other organic donors and/or acceptors.

2. Results

2.1. $(TMTSF)_8(I_3)_5$

2.1.1. Crystal Structure

The crystal structure of the 8:5 salt is shown in Figure 1; and its crystallographic and refinement data are summarized in Table 1. Crystallographic data files are available as Supplemental Materials. The appearance of single crystals are black thick cuboids reflecting the orthorhombic crystal system with the space group *Cmcm*.

There are four TMTSF stacks along the ***c***-axis in the unit cell with the period of six TMTSF molecules. There are two crystallographically independent TMTSF molecules (P and Q) in the stack, which is made of repetition of trimers (PQP). The normal to the molecular plane defined with the four Se atoms is tilted from the stacking direction ($\|\boldsymbol{a}$) by 2.23° for P, while that of Q is not tilted as its carbon and selenium atoms are on a mirror plane ($\|\boldsymbol{bc}$).

The stacks are separated by another independent TMTSF molecules (R) along the ***b***-axis; and by I_3^- chains along the ***a***-axis. Although there exist shorter Se$\cdots$Se contacts than the sum of van der Waals radii of two Se atoms (<4.0 Å) between Q and R as in Figure 2, such contacts do not form networks along the **b**-axis. Thus, the electronic system is probably quasi-one-dimensional along the *c*-axis. Here we adopted 2.0 Å as the van der Waals radius of Se by Pauling [72], while 1.9 Å by Bondi [73] is used in some literature [13,56,58,60]. The length and number of such short contacts between neighboring molecules are the primitive but important measure to find strong intermolecular interactions; namely the stronger interactions along a direction can give the wider band dispersion than that along the other directions.

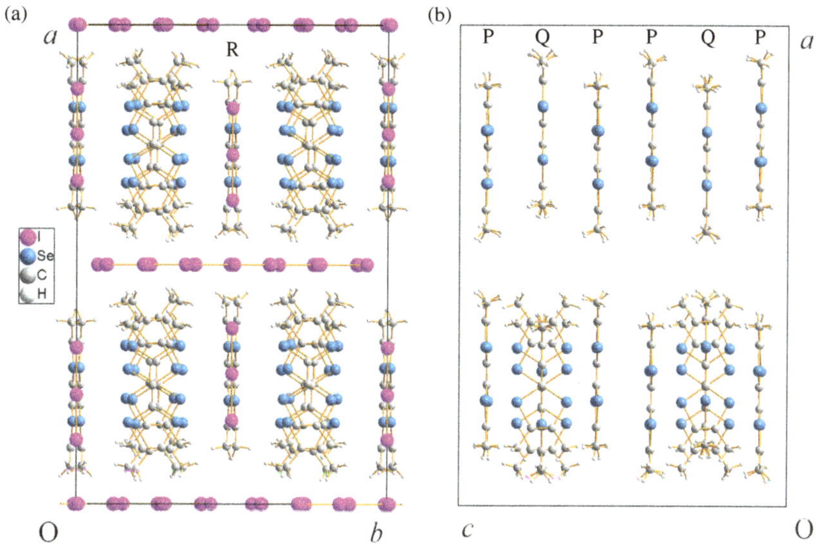

Figure 1. The crystal structure of $(TMTSF)_8(I_3)_5$ at room temperature viewed along (**a**) the **c** axis and (**b**) the **b**-axis. The crystallographically independent TMTSF molecules are labeled P, Q, and R, respectively.

The distance between the molecular planes, which is defined by the four Se atoms in each molecule, is almost the same for that between P and Q ($d_1 = 3.56(1)$ Å) and between P and P ($d_2 = 3.559(8)$ Å), respectively (Figure 2). On the other hand, the average short Se···Se contacts between P and Q and between P and P are 3.95 Å and 3.82 Å suggesting that the interaction between P and P is slightly stronger than that between P and Q.

While dimer units of donors are widely observed as building blocks of organic conductors such as Bechgaard salts, TMTSF trimers are also not so rare in the exotic TMTSF salts. For example, there are reports on $(TMTSF)_3[Ti_2F_8(C_2O_4)]$ [56], $(TMTSF)_3M(CN)_4$ (M = Pt, Ni) [57], $(TMTSF)_3W_6O_{19}(DMF)_2$ (DMF = *N*,*N*-dimethylformamide) [58], $(TMTSF)_3[Cr(NCS)_4(phen)]_2{\cdot}CH_2Cl_2$ (phen = 1,10-phenanthroline) [59], and $(TMTSF)_3(TFPB)_2$ (TFPB = tetrakis[3,5-bis(trifluoromethyl)phenyl]borate) [60], $(TMTSF)_3Ta_2F_{10}O$ [61], respectively. In most of these salts, the average valence of a TMTSF molecule is (2/3)+ suggesting the stability of the $[(TMTSF)_3]^{2+}$ unit.

Table 1. X-ray crystallographic and refinement data for $(TMTSF)_8(I_3)_5$, $(TMTSF)_5(I_3)_2$, and $(TMTSF)_4(I_3)_4\cdot THF$, respectively.

Compound	$(TMTSF)_8(I_3)_5$	$(TMTSF)_5(I_3)_2$	$(TMTSF)_4(I_3)_4\cdot THF$
Formula	$C_{80}H_{96}Se_{32}I_{15}$	$C_{50}H_{60}Se_8I_6$	$C_{44}H_{56}OSe_{16}I_{12}$
Formula weight	5488.21	2054.18	3387.28
Crystal system	Orthorhombic	Monoclinic	Orthorhombic
Space group	*Cmcm* (No.63)	$P2_1/n$ (No.14)	*Pnma* (No.62)
a (Å)	30.362(5)	11.263(9)	26.900(6)
b (Å)	20.389(3)	18.174(14)	9.914(2)
c (Å)	21.346(3)	18.166(14)	28.547(7)
α (°)	90	90	90
β (°)	90	90.355(9)	90
γ (°)	90	90	90
Volume (Å^3)	13,214.26	3718.40	7613.10
Z †	4	2	4
λ (Å)	0.71070	0.71070	0.71070
μ (mm^{-1})	12.364	12.328	12.562
Temperature (K)	296(2)	296(2)	296(2)
Number of reflections collected	7935	8512	9107
Number of reflections with $[F_0 > 2\sigma(F_0)]$	7216	5621	7309
Number of parameters refined	361	337	368
$(2\theta)_{max}$ (°)	25.11	25.36	25.08
final R indices $[I > 2\sigma(I)]$ R_1	0.0997	0.1018	0.0656
wR_2	0.2401	0.2559	0.1629
R indices (all data) R_1	0.1062	0.1376	0.0812
wR_2	0.2431	0.2918	0.1749

† Z = the number of chemical formula units per unit cell.

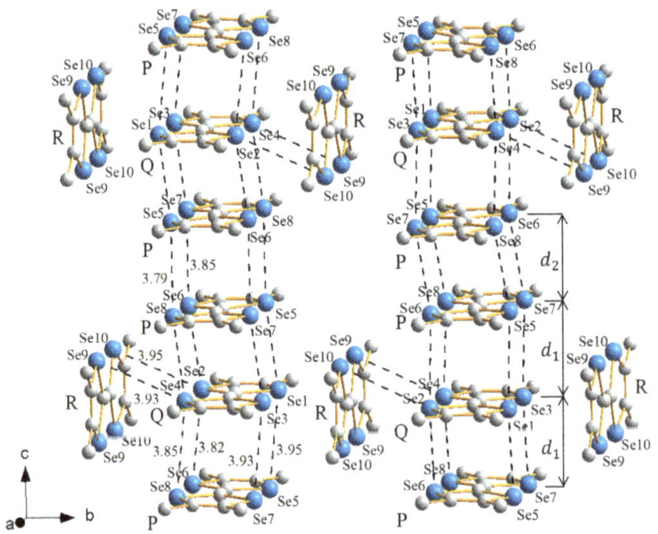

Figure 2. Donor arrangement in a half unit cell of $(TMTSF)_8(I_3)_5$. Hydrogen atoms are not shown for clarity. The crystallographically independent TMTSF molecules are labeled P, Q, and R, respectively. The broken lines show the Se··· Se contacts shorter than 4.0 Å, namely the sum of van der Waals radii of two Se atoms (see text). The interplanar distance between P and Q is $d_1 = 3.56(1)$ Å; and that between P and P is $d_2 = 3.559(8)$ Å, respectively.

On the other hand, the existence of the TMTSF monomer separating TMTSF stacks like R in the present 8:5 salt was reported for $(TMTSF)_3[Cr(NCS)_4(phen)]_2·CH_2Cl_2$ [59] and $(TMTSF)_4(TMTSF)Nb_6Cl_{18}·(CH_2Cl_2)_{0.5}$ [62,63]. This is also the case with the 5:2 salt shown below. When a rather strong interaction is expected between such a monomer and a donor stack, it is difficult to estimate the valence of TMTSF molecules because the crystallographically independent molecules naturally have different charges from one another and a TMTSF molecule easily possesses an irrational charge when it participates in forming energy bands. The estimation of the valence of TMTSF molecules in the present salts is discussed in Section 3.

There exist three types of disorders in the 8:5 salt at room temperature as shown in Figure 3. The first is a conformational disorder of three methyl groups of the TMTSF molecule Q. Each of the methyl groups at the carbon C5, C9, and C10 is disordered between two energetically favorable states.

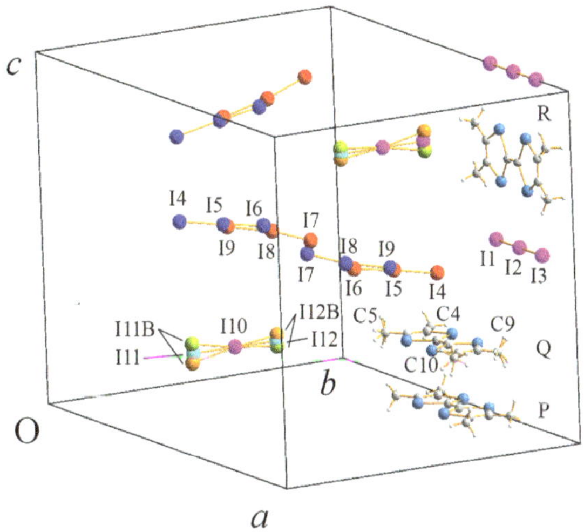

Figure 3. Some molecules in the unit cell of $(TMTSF)_8(I_3)_5$ selected to show the three types of disorders. (1) The conformational disorder of three methyl groups at the carbons C5, C9, and C10. (2) The positional disorder of the I_3^- anions of I4–I5–I6 and I7–I8–I9. (3) The orientational disorder of the I_3^- anions of I11–I10–I12 and I11B–I10–I12B.

The second is the positional disorder of the I_3^- anions on a mirror plane and they are labeled I4–I5–I6 and I7–I8–I9, respectively. The crystal structure analysis was carried out by assuming the same probability of 1/2 for each of the red and blue sites. The slight bending of I–I–I is probably spurious and caused by the overlap of the electron density of the disordered anions.

The third is the orientational disorder of the I_3^- anion whose center is labeled I10. The crystal structure was solved by assuming I11–I10–I12 (pink) has the probability of 1/2 and each of the two orientations of I11B–I10–I12B (lime green and orange) has 1/4.

In addition to the orthorhombic crystal system, these disorders and the large unit cell containing the thirty two TMTSF molecules and twenty I_3^- anions are characteristic to the 8:5 salt.

2.1.2. Electrical Resistivity

The electrical resistivity ρ of $(TMTSF)_8(I_3)_5$ was measured at ambient pressure and under hydrostatic pressures up to 1.73 GPa.

The single crystals are rather thick and plate-like. The dimensions are $0.5 \times 0.2 \times 0.07$ mm^3 on average for the seven samples. Although the correspondence between the crystal edges and the lattice vectors has not been determined, the anisotropy of ρ—those along the most and least grown directions are ~60 and ~2 S·cm^{-1}—suggests that the longest crystal edge is parallel to the stacking direction of TMTSF molecules (*c*-axis).

The ρ along the most-grown edge ($\boldsymbol{I} \parallel$ longest edge) is weakly metallic below room temperature, but resistance jumps make its intrinsic behavior very unclear and hysteretic. On the other hand, the reproducibility was good when the electrical current was applied along the least-grown edge, in other words perpendicular to the widest surface ($\boldsymbol{I} \perp$ plane). Such behavior is very similar to that observed for the Bechgaard salts. Thus, we measured the pressure dependence of the ρ with $\boldsymbol{I} \perp$ plane.

The result obtained for the sample #1908 is shown in Figure 4a. At ambient pressure, the ρ weakly increases below 300 K on cooling and becomes almost temperature independent below about 190 K. The ρ starts to increase very rapidly below 88 K suggesting a kind of phase transition or a change in the electronic state. After the increase in the ρ becomes gradual approaching 60 K, another rapid increase starts at 53 K.

The behavior of the ρ at higher pressures is similar to that observed at ambient pressure. The ρ becomes smaller and, at 1.7 GPa at room temperature, it reaches about 3.3% of that at ambient pressure. The temperatures of the steps also become lower at higher pressures.

One can see that the weak metallic behavior appears in the intermediate temperature region above 0.88 GPa. Please note that the hydrostatic pressure was applied using a clamped type pressure cell with Daphne 7373 oil as pressure medium as described in Section 4. Then the inside pressure decreases on cooling [74,75] resulting in the increase in the ρ, which sometimes cancels out the weak metallic behavior. Thus, the observation of the metallic behavior for the 8:5 salt suggests the existence of the small number of free charge carriers and a Fermi pocket, or a narrow band gap at Fermi level.

In the temperature dependence at ambient pressure, we see subtle hysteresis above 190 K as shown by the arrows (Figure 4a). Although the overall behavior under pressure is very similar to that at ambient pressure, no hysteresis was recognized at and above 0.30 GPa. It suggests that the apparent hysteresis has the same origin as the resistance jumps for $\boldsymbol{I} \parallel$ longest edge and is caused by an extrinsic effect like micro-cracks.

The steps below 88 K and 53 K are clearly seen also in the Arrhenius plot of ρ shown in Figure 4b. In addition to these steps, the Arrhenius plot reveals the existence of the third step above 200 K at each pressure. Figure 4c shows the numerical derivative of the Arrhenius plot. The calculations were carried out after smoothing the data. The rapid increases in the ρ in Figure 4a,b are recognized as the peaks in Figure 4c. Here we assume each peak corresponds to a phase transition or a change in the electronic state.

It should be noted that the pressure medium Daphne 7373 gradually solidifies around 200 K at ambient pressure; and the solidification temperature increases up to 300 K at 2.2 GPa [74,75]. The solidification results in rather rapid decrease in the inside pressure (~0.1 GPa) on cooling and can be detected as a subtle extrinsic anomaly in the ρ as indicated by the arrows in Figure 4c. This is, however, not the case with the result at ambient pressure as the sample was in vacuum. Therefore, the the third anomaly above 200 K is not that caused by the pressure change.

We can define the possible transition temperatures T_{c1} and T_{c2} as the peak tops in Figure 4c as well as that of the cross point in Figure 4a,b. The T_{c1} and T_{c2} defined by the cross point is 2–3 K higher than that at the peak top. On the other hand, it is difficult to define the T_{c3}, since the steps in Figure 4a,b, and the peaks in Figure 4c are incomplete. This shows that the step-like change in the high temperature region in the Arrhenius plot probably starts above 310 K.

Figure 4d shows the T–p phase diagram made by plotting the T_{c1} and T_{c2} defined as the cross points. The pressures at low temperatures were corrected by using the clamped pressure at room temperature and the temperature dependence of the pressure inside the clamped-type cell with Daphne 7373 as the pressure medium [74,75]. The nature of the phase transitions is unclear at present, but it is obvious that we need much higher pressure to suppress the semiconducting/insulating states (high-R 1 and high-R 2) than 1.7 GPa.

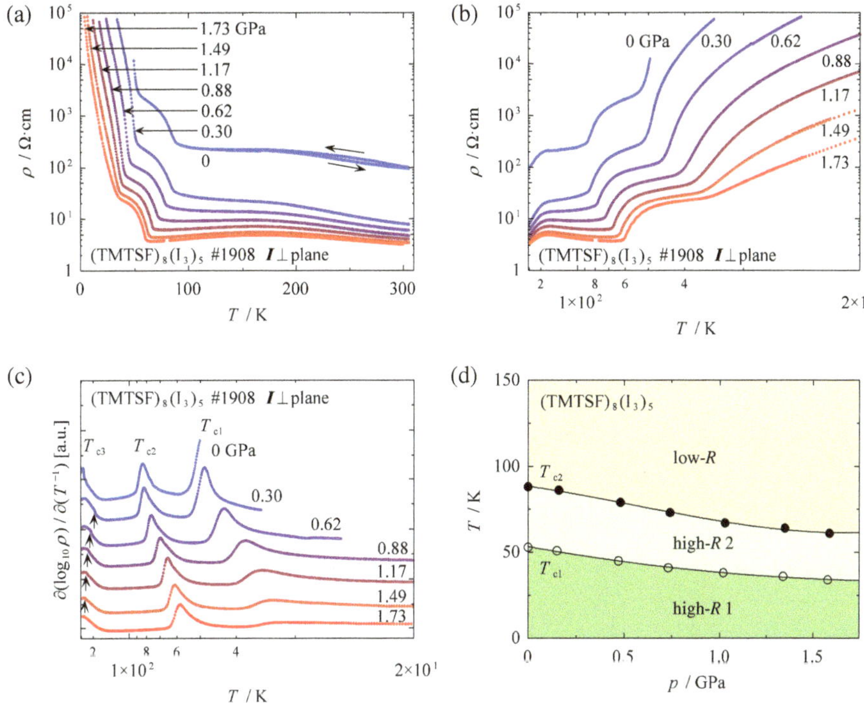

Figure 4. (**a**) Temperature dependence of the electrical resistivity ρ of $(TMTSF)_8(I_3)_5$ measured perpendicular to the most grown crystal surface. (**b**) Arrhenius plot of the same data in (**a**). Please note that the values in the horizontal axis is T but the scale is linear in T^{-1}. (**c**) Numerical derivative of the Arrhenius plot in (**b**) calculated after smoothing the data. A series of data at each pressure was shifted from one another for clarity. The arrows at high temperatures indicate the subtle anomalies caused by the solidification of the pressure medium (see text). (**d**) Temperature–pressure phase diagram of $(TMTSF)_8(I_3)_5$ below 150 K.

2.2. $(TMTSF)_5(I_3)_2$

2.2.1. Crystal Structure

The crystal structure of $(TMTSF)_5(I_3)_2$ at room temperature is shown in Figure 5; and its crystallographic and refinement data are summarized in Table 1. The crystal system is monoclinic with the space group $P2_1/n$, but $\beta = 90.355°$ is close to 90°.

There are two TMTSF stacks along the $\boldsymbol{a}$-axis in the unit cell. The stack at the center of the unit cell is crystallographically identical to that at the corners, which are related by the two-fold screw axis along the $\boldsymbol{b}$-axis to each other.

Each stack is made of the repetition of trimers, where the donors are labeled as B, A, and B, respectively. Please note that the center of the molecule A is located at an inversion center. The tilt angle of the normal to the molecular plane from the $\boldsymbol{a}$-axis is 24.26° and 23.79° for A and B, respectively. In this sense, the trimer stack of the 5:2 salt is rather different from that of the 8:5 salt, where the molecular plane is almost perpendicular to the stacking direction.

The TMTSF stacks are separated by the TMTSF monomers C and the I_3^- anions as in the 8:5 salt. The molecular plane of C is, however, not flat as the monomer R in the 8:5 salt. This suggests that C is in a different oxidation state from that of R.

Figure 6 shows the arrangement of the TMTSF molecules of the 5:2 salt. There are six Se$\cdots$Se short contacts between the neighboring TMTSF molecules in each stack. The interplanar distance between A and B is $d_1 = 3.45(3)$ Å; and that between B and B is $d_2 = 3.39(4)$ Å, respectively. Thus the much stronger interactions than that in the 8:5 salt are expected for the 5:2 salt. In addition, each monomer C shares four short contacts with one of neighboring stacks. Thus, we cannot simply conclude that C is neutral. The valence of each TMTSF molecule is estimated together with that in the other I_3 salts in Section 3.

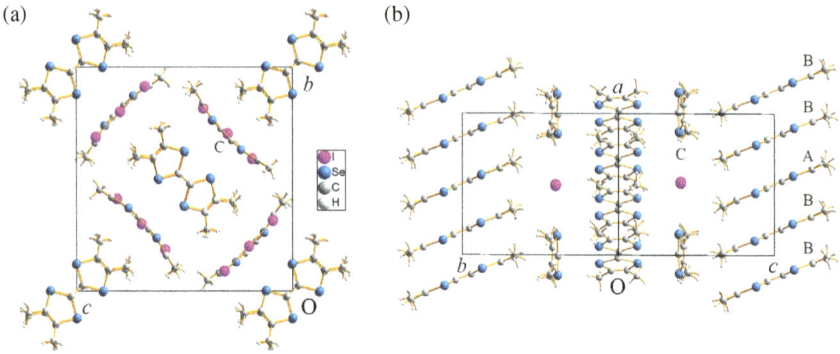

Figure 5. The crystal structure of $(TMTSF)_5(I_3)_2$ at room temperature viewed along (**a**) the ***a***-axis and (**b**) the (***b*** + ***c***)-direction, respectively. The crystallographically independent TMTSF molecules are labeled A, B, and C.

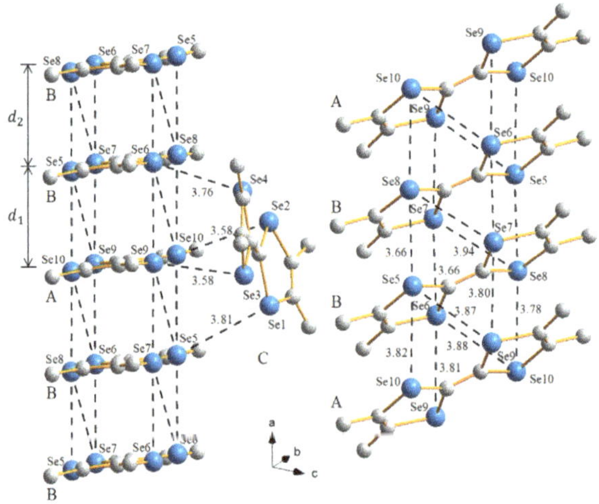

Figure 6. Donor arrangement in $(TMTSF)_5(I_3)_2$. Hydrogen atoms are not shown for clarity. The crystallographically independent TMTSF molecules are labeled A, B, and C, respectively. The broken lines show the Se$\cdots$Se contacts shorter than the sum of van der Waals radii of two Se atoms (4.0 Å). The interplanar distance between A and B is $d_1 = 3.45(3)$ Å; and that between B and B is $d_2 = 3.39(4)$ Å, respectively.

2.2.2. Electrical Resistivity

The ρ was measured along the most grown crystal axis. Although the correspondence between the crystal edges and the lattice vectors was not determined, it is natural to assume the crystal grows the best along the TMTSF stacks. The room temperature electrical conductivity is typically 50 S·cm^{-1} but scattered between 1 and 200 S·cm^{-1} among the seven single crystals measured.

The temperature dependence of the ρ is shown for the sample #1910 in Figure 7a. The ρ increases with decreasing temperature suggesting the existence of a band gap at Fermi level. Anomalous change in slope was observed around 190 K. It is clearly seen in the Arrhenius plot in Figure 7b. The slope becomes small below 190 K. Assuming the activation type temperature dependence, the tentative activation energy E_a above and below 190 K is estimated as 1.3×10^2 and 34 meV, respectively.

The application of hydrostatic pressure up to 1.73 GPa does not change the behavior very much. The room temperature resistivity was decreased to about 25% of that at ambient pressure. The tentative E_a was also reduced to 90 and 22 meV above and blow 185 K, respectively.

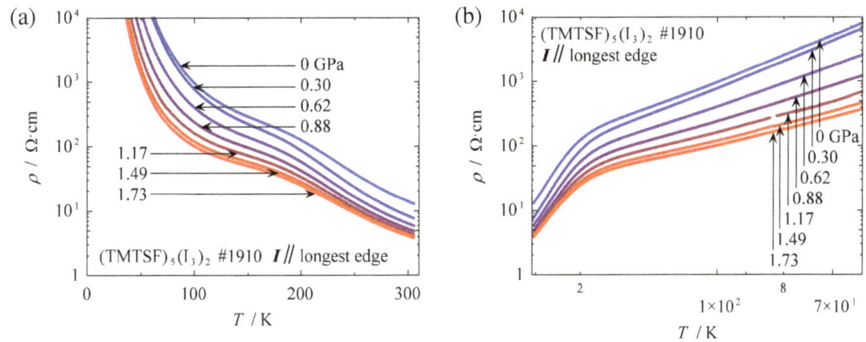

Figure 7. (**a**) Temperature dependence of the electrical resistivity ρ of $(TMTSF)_5(I_3)_2$ measured along the longest crystal edges. (**b**) Arrhenius plot of the same data in (**a**). Please note that the values in the horizontal axis is T but the scale is linear in T^{-1}.

2.3. *(TMTSF)$_4$(I$_3$)$_4$·THF*

The crystal structure of the 4:4 salt is shown in Figure 8; and its crystallographic and refinement data are summarized in Table 1. The crystal system is orthorhombic with the space group *Pnma*. The major building blocks of the crystal are the tetramers of TMTSF molecules and I_3^- anions in addition to the THF molecules coming from the crystal growth solvent. Since the composition of TMTSF : I_3^- is 1:1, the average valence of TMTSF should be 1+.

The TMTSF molecules in a tetramer are labeled K, L, M, and N as in Figure 8. They are not crystallographically identical to one another, while every molecule is on a mirror plane parallel to the *ac*-plane. Thus, it is natural to consider that the valence of these TMTSF molecules is not the same within the tetramer. This is discussed in Section 3.

The arrangement of the TMTSF molecules in the tetramer is shown in Figure 9. There are no short Se··· Se contacts between the nearest tetramers. The normal to the molecular plane defined with the four Se atoms is almost parallel to the stacking direction (*c*-axis) but slightly tilted. The angle between the normal to each molecular plane and the *c*-axis is 2.11°, 1.85°, 1.34°, and 2.04° for K, L, M, and N, respectively.

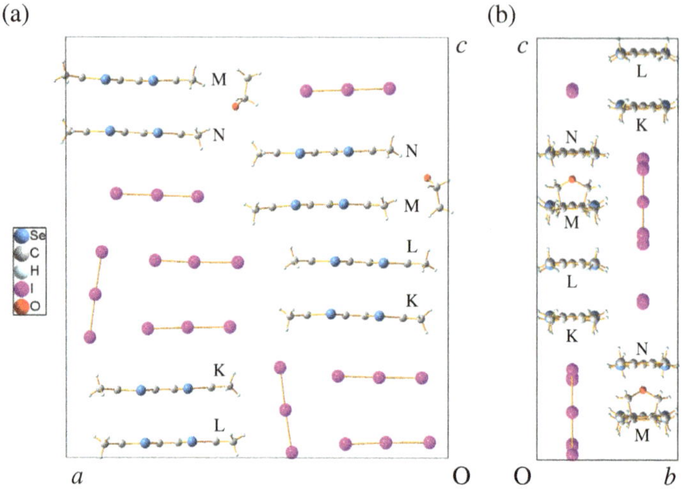

Figure 8. The crystal structure of $(TMTSF)_4(I_3)_4$·THF at room temperature viewed along (**a**) the ***b***-axis and (**b**) ***a***-axis, respectively. The crystallographically independent TMTSF molecules are labeled K, L, M, and N. The molecules in a half unit cell is shown in each of (**a**,**b**).

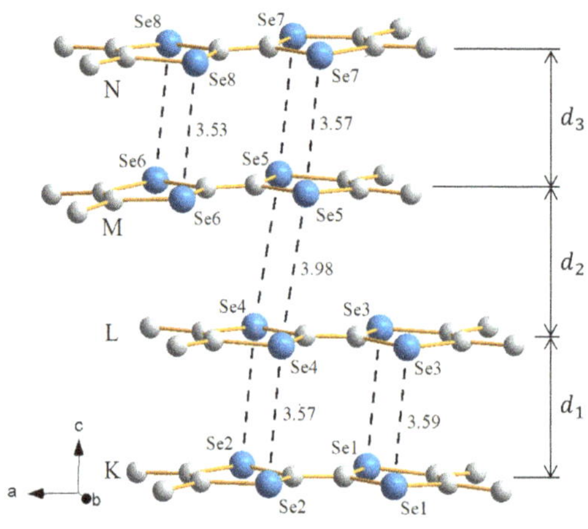

Figure 9. Donor arrangement in a tetramer of $(TMTSF)_4(I_3)_4$·THF. Hydrogen atoms are not shown for clarity. The crystallographically independent TMTSF molecules are labeled K, L, M and N, respectively. The broken lines show the Se$\cdots$Se contacts shorter than 4.0 Å. The interplanar distances are d_1 = 3.599(5) Å (K–L), d_2 = 3.92(7) Å (L–M), and d_3 = 3.53(2) Å (M–N), respectively.

The tetramer is regarded as made of two non-equivalent dimers K–L and M–N as is obvious in Figure 9. The intradimer distance d_1 = 3.60 Å (K–L) and d_3 = 3.63 Å (M–N) is much shorter than the interdimer distance d_2 = 3.92 Å (L–M). There are four short Se$\cdots$Se contacts within each of the dimers (K–L and M–N), while there are only two contacts between the dimers (L–M) and it is rather long (3.98 Å).

Each TMTSF tetramer is surrounded by six I_3^- tetramers and vice varsa. In this sense, the 4:4 salt is similar to the rock salt NaCl. In addition, there is no strong interaction between the nearest TMTSF tetramers. Thus, the system is probably insulating, though the electrical conductivity is not known due to the rarity of its single crystals.

Among the exotic TMTSF salts, $(TMTSF)_4(TMTSF)[Nb_6Cl_{18}]\cdot(CH_2Cl_2)_{0.5}$ also has a TMTSF tetramer as a building unit [62,63]. In this case, however, a one-dimensional TMTSF stacks are formed by the repetition of tetramers resulting in its moderate electrical conductivity ($0.5\ S\cdot cm^{-1}$) at room temperature. The orthorhombic rock-salt-like structure with the TMTSF and I_3^- tetramers of the 4:4 salt is unique even in the exotic TMTSF salts.

3. Discussion

3.1. Valence and Bond Lengths of TMTSF Molecules

The knowledge of the valence of donors in conducting salts is essential to understand their electronic states because it determines the band filling and, therefore, Fermi energy. Once the band structure is available, one can predict whether the system has a Fermi surface or not. In Bechgaard salts, where the ratio of the TMTSF molecule to the monovalent anion is 2:1, the valence of the TMTSF molecule is $(1/2)+$. In the exotic TMTSF salts, however, the evaluation is sometimes not so simple due to the existence of non-equivalent TMTSF molecules.

Some authors reported the attempts to estimate the valence of TMTSF molecules partially oxidized in salts on the basis of the intramolecular bond lengths [46,60,63,71]. Indeed, there seems to be a kind of correlation between the valence and the bond lengths of TMTSF molecules in the neutral crystal ($TMTSF^0$) [76]; the 2:1 radical salts ($TMTSF^{0.5+}$) [5,52,77–80]; and $(TMTSF)NO_3$ [69], $(TMTSF)[Cr(Br_4SQ)_2(Br_4Cat)$ (SQ = semiquinonate, Cat = catecholate) [70], and $(TMTSF)_3PW_{12}O_{40}$ [68] ($TMTSF^{1+}$) as tabulated in Table 2.

Here we label the bond lengths e, f_i, g_i, and h_i as in the inset of Table 2, where the subscript i distinguishes the chemically equivalent bonds from one another. For example, the length of the central and outer C=C bonds (e, h_1, and h_2) have rough tendency to increase when the molecular valence changes from 0 to 1+, while the C−Se single bonds (f_i and g_i) tend to become shorter.

Please note that we chose the BF_4, PF_6, and NbF_6 Bechgaard salts because they are representatives of the 2:1 radical salts with small (radius $r = 2.72$ Å), middle ($r = 2.95$ Å) [81] and large ($r = 3.20$ Å) [5] counter anions, respectively. $(TMTSF)_2NO_3$ is the last Bechgaard salt that we could find the atomic coordinate of hydrogen atoms for the quantum mechanical energy calculations. $(TMTSF)_2Ni(tds)_2$ (tds = [bis(trifluoromethyl)ethylene]diselenolato) [52] also has the 2:1 composition, but it is one of the exotic TMTSF salts with two crystallographically independent TMTSF molecules A and A′ at room temperature.

On the other hand, $(TMTSF)NO_3$, $(TMTSF)[Cr(Br_4SQ)_2(Br_4Cat)$, and $(TMTSF)_3PW_{12}O_{40}$ are examples of $TMTSF^{1+}$ simply from their chemical compositions.

We also carried out quantum mechanical calculations [82] on the basis of the density functional theory (DFT, B3LYP/6-31G(d,p)) as well as Hartee–Fock (HF) calculations followed by Møller–Plesset correlation energy calculations truncated at the second order (MP2, 6-31G(d,p)) for $TMTSF^0$, $TMTSF^{1+}$, and $TMTSF^{2+}$ to obtain each optimized structure for comparison. The calculated bond lengths as well as those of the TMTSF molecules in the I_3 salts are also summarized in Table 2.

The rough correlation noted above is shown by plotting the e; and the averages of f_i, g_i, and h_i ($\bar{f}$, $\bar{g}$, and $\bar{h}$) against the donor valence in Figure 10. The bond lengths observed in the I_3 salts are indicated by the arrows at the right-side vertical axis in each plot.

In case of the central C=C bond e (Figure 10a), the DFT and MP2 results coincide to each other showing the almost linear valence dependence. On the other hand, the DFT and MP2 calculations give 1–2% difference in the other bond lengths from each other, though they are still close. The experimental bond lengths are mostly around the lines connecting the calculated points, but we see exceptionally scattered points at the valence = 0.5 and 1.

Table 2. Comparison of bond lengths in TMTSF molecules in the crystals of neutral TMTSF and some salts. Each of $\overline{f}$, $\overline{g}$, and $\overline{h}$ is the average such as $\overline{f} = (f_1 + f_2 + f_3 + f_4)/4$.

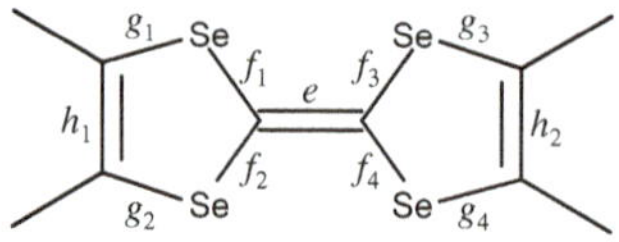

(Average) Valence	Material/Calculation	e/Å	$\overline{f}$/Å	$\overline{g}$/Å	$\overline{h}$/Å
0	neutral [1]	1.3526	1.8924	1.9062	1.3101
0.5+	BF_4 (A, 20 K) [2]	1.3720	1.8758	1.8914	1.3525
0.5+	BF_4 (B, 20 K) [2]	1.3589	1.8852	1.8993	1.3469
0.5+	PF_6 (20 K) [3]	1.3756	1.8821	1.8896	1.3796
0.5+	PF_6 [4]	1.3688	1.8755	1.8926	1.3295
0.5+	NO_3 (2:1, 125 K) [5]	1.4297	1.8641	1.9016	1.3271
0.5+	NbF_6 [6]	1.3501	1.8801	1.8908	1.3481
0.5+	Ni (A) [7]	1.3779	1.8653	1.8876	1.3297
0.5+	Ni (A') [7]	1.3939	1.8673	1.8908	1.3409
1+	NO_3 (1:1) [8]	1.3598	1.8646	1.8835	1.3235
1+	Cr [9]	1.4293	1.8437	1.8750	1.3550
1+	PWO (A) [10]	1.2951	1.8782	1.9094	1.2232
1+	PWO (B) [10]	1.4072	1.8513	1.8568	1.3392
	8:5 (P)	1.3786	1.8655	1.8855	1.3287
	8:5 (Q)	1.3513	1.8747	1.9016	1.3499
	8:5 (R)	1.3241	1.9014	1.8975	1.3747
	5:2 (A)	1.3801	1.8651	1.8815	1.3107
	5:2 (B)	1.3858	1.8629	1.8719	1.3568
	5:2 (C)	1.3244	1.8964	1.8950	1.3359
	4:4 (K)	1.3750	1.8636	1.8866	1.3522
	4:4 (L)	1.3864	1.8623	1.8834	1.3582
	4:4 (M)	1.3539	1.8685	1.8854	1.3404
	4:4 (N)	1.3676	1.8688	1.8967	1.3453
0	DFT [11]	1.3424	1.9055	1.9235	1.3409
1+	DFT [11]	1.3843	1.8710	1.8974	1.3522
2+	DFT [11]	1.4312	1.8439	1.8748	1.3720
0	MP2 [12]	1.3493	1.8869	1.9051	1.3474
1+	MP2 [12]	1.3848	1.8553	1.8709	1.3630
2+	MP2 [12]	1.4303	1.8376	1.8470	1.3859

[1] Neutral crystal [76]. [2] $(TMTSF)_2BF_4$ at the anion ordered state at 20 K with two crystallographically independent TMTSF molecules (A and B) [77]. [3] $(TMTSF)_2PF_6$ at 20 K with the neutron diffraction structural analysis [78]. [4] $(TMTSF)_2PF_6$ [79]. [5] $(TMTSF)_2NO_3$ at 125 K [80]. [6] $(TMTSF)_2NbF_6$ [5]. [7] $(TMTSF)_2Ni(tds)_2$ (tds = [bis(trifluoromethyl)ethylene]diselenolato) [52]. [8] $(TMTSF)NO_3$ [69]. [9] $(TMTSF)[Cr(Br_4SQ)_2(Br_4Cat)]\cdot(CH_2Cl_2)_2$ (SQ = semiquinonate, Cat = catecholate) [71]. [10] $(TMTSF)_3PW_{12}O_{40}$ [68]. [11] Optimized structures calculated by the density-functional theory method (B3LYP/6-31G(d,p)) [82]. [12] Optimized structures obtained by Hartree–Fock calculations (6-31G(d,p)) followed by Møller–Plesset correlation energy calculations truncated at second order [82].

The optimized structures are of the isolated molecules in vacuum and highly symmetric. On the other hand, the TMTSF molecules in the crystals are affected by their surroundings as well as by the formation of their own energy bands. Therefore, they are rather unsymmetric, even if the chemical structure of TMTSF is symmetric. This is the reason we compared the "average" bond lengths such as $\overline{f}$. In addition, the energy of TMTSF molecule also depends on bond angles, which are not considered here. Therefore, the scattering of the experimental data in Figure 10 is very natural.

Furthermore, the bond lengths observed for the crystallographically independent TMTSF molecules (PWO A and B) in $(TMTSF)_3PW_{12}O_{40}$ are anomalously far from the others at the valence = 1. The deviation is qualitatively understood assuming that A is close to neutral and B is more oxidized than 1+. This is consistent with the results of the quantum mechanical energy calculations discussed in Section 3.

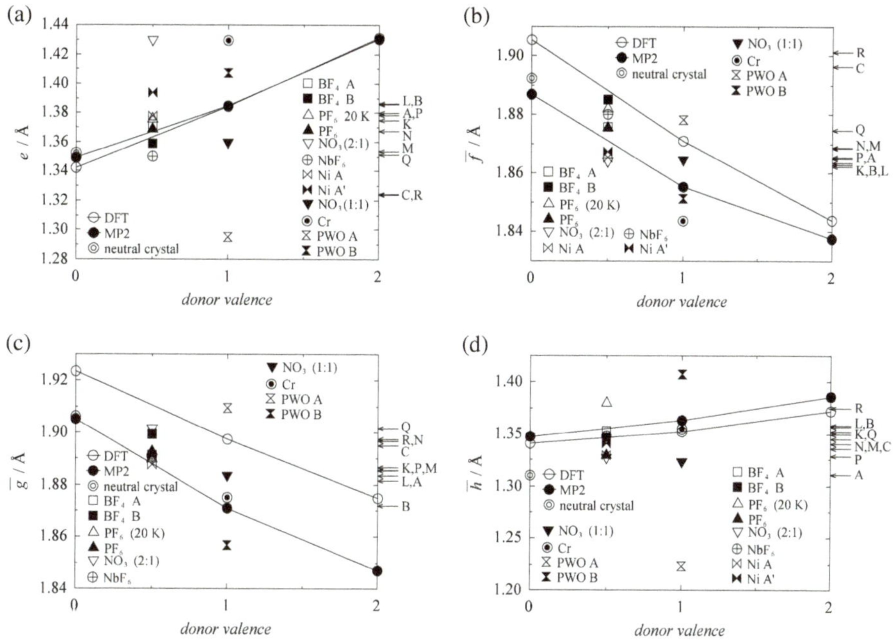

Figure 10. Dependence of the bond lengths (**a**) e, (**b**) $\bar{f}$, (**c**) $\bar{g}$, and (**d**) $\bar{h}$ of the TMTSF molecules in the crystals on the valence expected from their chemical compositions. The open and closed circles show the corresponding bond lengths in the optimized structures by the quantum mechanical calculations (see text and the caption in Table 2). The solid lines are guides to the eye.

A more important result we see here is, however, the unsystematic scattering of the arrows from A to Q at the right-side vertical axes. We can expect, for example, the monomers R and C in the 8:5 and 5:2 salts are probably less oxidized. Indeed, the C and R are far from the others in Figure 10a,b, but that is not the case in Figure 10c,d. Another difficulty is about K, L, M, and N in the 4:4 salt, where their average valence is most likely 1+ and higher than the others in the 8:5 and 5:2 salts. They are, however, inside the distribution.

It should be noted that Rosoka et al. [60] proposed an equation to estimate the effective valence by combining the bond lengths of the target molecule with those in the crystals of the neutral and cation radical salts. Although they succeeded in evaluating the TMTSF valence close to (2/3)+ in $(TMTSF)_3(TFPB)_2$ by the equation, we do not adopt their method in this study due to the low correlation in Figure 10. Instead, we carried out the quantum mechanical energy calculations, which consider all the information of each molecule not only the bond lengths but also bond angles, and even electron correlations.

3.2. Valence of TMTSF Molecule Estimated from Total Energy

3.2.1. Principles

The method we propose here to estimate the valence of TMTSF molecules in the crystals is as follows. This is base on the quantum mechanical calculations with the DFT or MP2 method used in the previous subsection.

1. The optimized structures and their total energies of TMTSF^0 ($E_0(0) = E_0$(optimized for 0)) and TMTSF^{1+} ($E_{1+}(1+) = E_{1+}$(optimized for 1+)) are calculated. Please note that the spin-unrestricted calculations are applied for cations; and all the electrons are considered in the MP2 calculations.
2. For example, for the TMTSF A in the 5:2 salt, the total energies of TMTSF^0 (E_0(A)) and TMTSF^{1+} (E_{1+}(A)) are calculated using the crystallographic data.
3. The energy differences $\Delta E_0 = E_0(\mathrm{A}) - E_0(0)$ and $\Delta E_{1+} = E_{1+}(\mathrm{A}) - E_{1+}(1+)$ are calculated. The ΔE_0 and ΔE_{1+} should be positive since the conformation of A is not optimum both in the neutral and cationic states.
4. The difference $\Delta E_0 - \Delta E_{1+}$ is used as a measure of the valence of A in the crystal.

The ΔE_0 and ΔE_{1+} are the energy increase of the TMTSF molecule from those at its neutral and cationic states with their optimized conformations, respectively. Therefore, the ΔE_0 will be smaller than the ΔE_{1+} when a TMTSF molecule in a crystal has a conformation which is more stable at a less oxidized state than at a more oxidized one. On the other hand, the ΔE_{1+} will become smaller than the ΔE_0 when a conformation approaches to what is stable at a valence close to 1+. In addition, if the effective valence of a molecule is (1/2)+ with a conformation in a crystal, the ΔE_0 and ΔE_{1+} will be comparable to each other. These are schematically shown in Figure 11a.

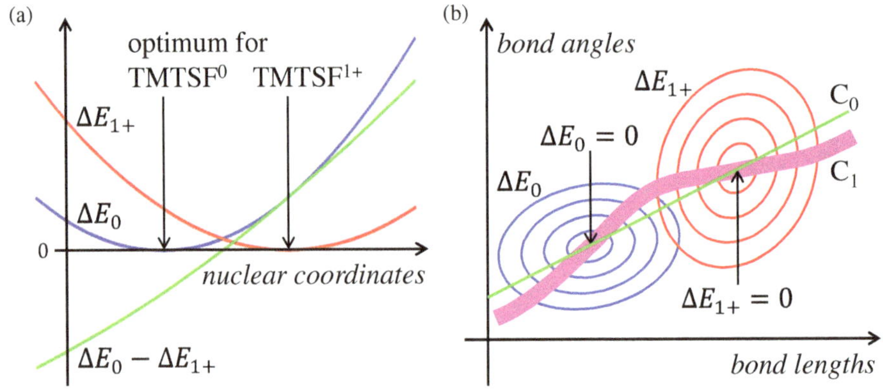

Figure 11. (**a**) Schematic image of ΔE_0 (red), ΔE_{1+} (blue), and $\Delta E_0 - \Delta E_{1+}$ (green) as functions of the nuclear coordinates of a TMTSF molecule in a crystal. (**b**) Simplified images of ΔE_0 and ΔE_{1+} as multidimensional paraboloids, which are the functions of bond lengths and bond angles of a TMTSF molecule in a crystal. The thin green line C_0 represents the path when the conformation changes continuously keeping the optimum one at each effective valence. In reality, the molecular conformations are not optimum in crystals, thus the molecules will be somewhere on the thick pink curve C_1 (see text).

These relations are simply expressed by the $\Delta E_0 - \Delta E_{1+}$. The energy increase ΔE_0 and ΔE_{1+} as functions of the nuclear coordinates/conformation are approximated by parabolas in the vicinity of their minima. Although the curvatures of the parabolas would be different from each other in general, the $\Delta E_0 - \Delta E_{1+}$ changes monotonically as shown by the green curve in Figure 11a as long as the difference between the curvatures is not so large.

The green curve ($\Delta E_0 - \Delta E_{1+}$) becomes negatively and positively large at the bottom of the blue (ΔE_0) and red (ΔE_{1+}) curves, respectively. In addition, the green curve becomes almost zero when the ΔE_0 and ΔE_{1+} are comparable to each other, where one can probably expect that the effective molecular valence is also in the middle between 0 and 1+. When the effective molecular valence changes continuously from 0 to 1+, the optimum conformation smoothly changes from that of TMTSF^0 to that of TMTSF^{1+} on the horizontal axis. Then the $\Delta E_0 - \Delta E_1$ will scale the molecular valence.

Actually the ΔE_0 and ΔE_{1+} are multivariable functions of the nuclear coordinates. Thus, each of the red and blue curves in Figure 11a is just one aspect of the multidimensional paraboloid, which is schematically shown in the two-dimensional space in Figure 11b, where the atomic coordinates in Figure 11a are separated into the bond lengths and bond angles. Each of the ΔE_0 and ΔE_{1+} is zero at a point in the multidimensional parametric space, where the molecule has the optimized conformation for each valence.

Using the $\Delta E_0 - \Delta E_{1+}$ as a gauge of the valence corresponds to use the green line C_0 in Figure 11b connecting the points of $\Delta E_0 = 0$ and $\Delta E_{1+} = 0$. It corresponds to the green curve $\Delta E_0 - \Delta E_{1+}$ in Figure 11a, when one adds the $\Delta E_0 - \Delta E_{1+}$ as the third axis in Figure 11b.

In reality, the "paraboloids" would have rather complicated contours in the multidimensional space; and the $\Delta E_0 - \Delta E_{1+}$ is not a curve but a multidimensional plane. In addition, even if the valence of a TMTSF molecule in a crystal is exactly 0 or 1+, its conformation is not the optimum one. Therefore each TMTSF molecule in different crystals or even at different temperatures is somewhere on a thick path connecting the points of $\Delta E_0 = 0$ and $\Delta E_{1+} = 0$ such as the pink curve C_1 in Figure 11b. The estimate of the effective molecular valence shown below will hold as long as the pink curve is not so thick and not far from the green line in the multidimensional space.

3.2.2. Energy Calculations

The results by the DFT and MP2 calculations are summarized in Tables 3 and 4, respectively. The E_0 and E_{1+} of the crystals are calculated by using the experimentally determined atomic coordinates whose significant figures are typically three or four. The change in the atomic positions from material to material results in the difference among the last four digits of the E_0 and E_{1+} in the tables. Therefore we estimated that the resulting $\Delta E_0 - \Delta E_{1+}$ would have three significant figures at most.

The energy differences ΔE_0, ΔE_{1+}, and $\Delta E_0 - \Delta E_{1+}$ are shown as histograms in Figure 12. The results by the DFT calculations (Figure 12a,b) are similar to those by the MP2 (Figure 12c,d), but the latter is considered to be more plausible in the absolute values as described below.

At first glance, both the ΔE_0 and ΔE_{1+} are almost the same for each TMTSF molecule in the crystals, thus one might feel that it does not tell anything. The large values (5–6 eV) in Figure 12a,c, just show the conformations of the TMTSF molecules are far from the optimized ones for the neutral and monocationic states.

It is, however, interesting that the results for the BF_4, PF_6, and NO_3 salts are very close to zero. This is probably because the crystal structures determined at low temperatures were used only for these salts, while the others are those at room temperature. It is reasonable because the optimized conformations by the calculations correspond to those at 0 K and no thermal oscillations are considered.

The relatively small size of the BF_4^-, PF_6^-, and NO_3^- anions may contribute to suppress the ΔE_0 and ΔE_{1+}. However, this is inconsistent with that the neutral crystal gives the large values. To find the conclusive results, we need the crystal structures of an identical material both at low and high temperatures, but the present authors could not find such reports. The atomic coordinate of hydrogen atoms is needed for this purpose, but it is often unavailable in early works.

Even though the thermal oscillations might affect the ΔE_0 and ΔE_{1+}, their effect seems to be canceled to an extent after calculating the $\Delta E_0 - \Delta E_{1+}$ as in Figure 12b,d. One can see a rough tendency that the $\Delta E_0 - \Delta E_{1+}$ increases from negative to positive values as the effective valence changes from 0 to 1+ for the materials reported previously. This suggests that the effects making the ΔE_0 and ΔE_{1+} large are regarded as systematic errors depending on materials. For example, the anharmonicity of the molecular vibrations always increase the bond lengths. This will increase both the ΔE_0 and ΔE_{1+} in a similar manner.

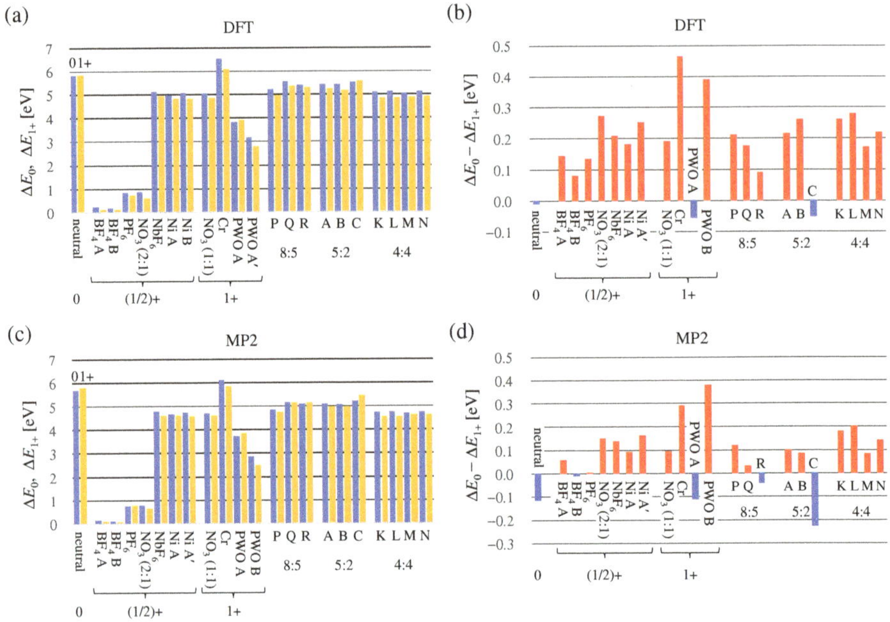

Figure 12. Visualization of (**a**) ΔE_0 and ΔE_{1+}; and (**b**) $\Delta E_0 - \Delta E_{1+}$ obtained by the DFT calculations in Table 3. (**c**,**d**) The correspondents by the HF energy calculations and its correlation-energy correction by the MP2 calculations in Table 4.

One significant exception in the tendency is the smaller value for $TMTSF^{1+}$ in (TMTSF)NO_3 than those for $TMTSF^{(1/2)+}$ of $(TMTSF)_2NO_3$, $(TMTSF)_2NbF_6$, and $(TMTSF)_2Ni(tds)_2$. The reason is unclear, but it would be the representative showing the limitation of the present method to determine the absolute values of the effective valence as discussed in the previous subsection.

On the other hand, the negative and large positive values of the $\Delta E_0 - \Delta E_{1+}$ for $(TMTSF)_3PW_{12}O_{40}$ (PWO A and B) are well understood as follows. The material has a "criss–cross" stacking of the monomer A and the dimer B–B along the ***a***-axis [68]. As is shown in Figure 10, the bond lengths observed in the material are anomalous as $TMTSF^{1+}$; and suggest that A is close to neutral and B is more oxidized than 1+. In Figure 12b,d, now we see that A and B show large negative and positive values of the $\Delta E_0 - \Delta E_{1+}$, respectively. If A is exactly neutral, the valence of B is (3/2)+ due to ${PW_{12}O_{40}}^{3-}$. This is consistent with the positively large $\Delta E_0 - \Delta E_{1+}$ assuming also that we can extrapolate the relation between the effective valence and $\Delta E_0 - \Delta E_{1+}$ to above 1+.

By considering the variation in the materials as well as in the methods to determine their crystal structures, the correlation between the effective valence of the TMTSF molecules and the value $\Delta E_0 - \Delta E_{1+}$ seems reasonably high. Despite the difficulty in determining the absolute value of the effective valence, it is probably possible to compare the degree of oxidation of the non-equivalent TMTSF molecules in the identical material as in $(TMTSF)_3PW_{12}O_{40}$.

3.2.3. $(TMTSF)_5(I_3)_2$

It should be noted that the crystal structure of $(TMTSF)_2PF_6$ used for the DFT and MP2 calculations was obtained by the neutron diffraction. It means that the C−H bond lengths are less precise and shorter in the materials other than $(TMTSF)_2PF_6$ (20 K) since neutrons are scattered by protons directly. It might change the ΔE_0 and ΔE_{1+} of the others slightly from the actual values, but the change will probably cancel out each other in $\Delta E_0 - \Delta E_{1+}$.

Finally, we can estimate the valence of TMTSF molecules in the present I_3 salts.

At a glance, the large negative value of the monomer C is noticeable. This is the strong evidence that the monomer C in the 5:2 salt is close to neutral. Assuming it is exactly neutral, the valence of the donors in the stacks is estimated as $(2/3)+$ on average because the 5:2 salt is regarded as a 3:2 salt by omitting C. We believe that each of A and B has almost the same valence $(2/3)+$ as the $\Delta E_0 - \Delta E_{1+}$ is almost the same.

Within the band picture, all of these seem consistent with the semiconducting behavior of the ρ. The one-dimensional stack made of the trimers will result in three separated energy bands. Then the lower two are filled, while the highest one remains empty at 0 K.

Another interesting possibility comes up when the upper two bands merge leaving the lowest third band apart. Then the upper band will be half-filled with highly one-dimensional nature. In such electronic systems, the Mott insulating state could be realized when the bandwidth is relatively small as compared with the on-site Coulomb energy [83]. The origin of the semiconducting state should be elucidated in a future work.

Table 3. The total energies of TMTSF molecules at neutral and cationic states (E_0 and E_{1+}); their relative values with respect to those calculated for the optimized structures ($\Delta E_0 = E_0 - E_0$(optimized for 0) and $\Delta E_{1+} = E_{1+} - E_{1+}$(optimized for 1+)); and their difference ($\Delta E_0 - \Delta E_{1+}$) as a measure of the valence of each conformation of TMTSF molecule. The calculations of E_0 and E_{1+} were carried out on the basis of the density-functional theory (B3LYP, 6-31G(d,p)) [82].

Conformation	E_0/Hartree [1]	E_{1+}/Hartree	ΔE_0/eV	ΔE_{1+}/eV	$(\Delta E_0 - \Delta E_{1+})$/eV
optimized for 0	−9985.7935	−9985.5699	0.000	0.279	−0.279
optimized for 1+	−9985.7884	−9985.5801	0.139	0.000	0.139
neutral [2]	−9985.5799	−9985.3662	5.813	5.822	−0.009
BF_4 (A, 20 K) [3]	−9985.7854	−9985.5773	0.221	0.077	0.144
BF_4 (B, 20 K) [3]	−9985.7877	−9985.5772	0.159	0.080	0.079
PF_6 (20 K) [4]	−9985.7636	−9985.5552	0.815	0.679	0.135
NO_3 (2:1) [5]	−9985.7622	−9985.5588	0.853	0.580	0.272
NbF_6 [6]	−9985.6057	−9985.3999	5.111	4.904	0.207
Ni (A) [7]	−9985.6104	−9985.4036	4.984	4.803	0.181
Ni (A′) [7]	−9985.6078	−9985.4036	5.055	4.805	0.250
NO_3 (1:1) [8]	−9985.6095	−9985.4029	5.009	4.824	0.185
Cr [9]	−9985.5534	−9985.3570	6.535	6.073	0.462
PWO (A) [10]	−9985.6532	−9985.4377	3.818	3.877	−0.058
PWO (B) [10]	−9985.6773	−9985.4782	3.164	2.775	0.388
8:5 (P)	−9985.6030	−9985.3973	5.185	4.976	0.209
8:5 (Q)	−9985.5909	−9985.3839	5.515	5.339	0.176
8:5 (R)	−9985.5961	−9985.3859	5.374	5.285	0.089
5:2 (A)	−9985.5943	−9985.3887	5.421	5.208	0.213
5:2 (B)	−9985.5945	−9985.3906	5.416	5.158	0.259
5:2 (C)	−9985.5916	−9985.3763	5.495	5.548	−0.053
4:4 (K)	−9985.6067	−9985.4028	5.084	4.825	0.259
4:4 (L)	−9985.6056	−9985.4024	5.114	4.836	0.278
4:4 (M)	−9985.6087	−9985.4016	5.030	4.860	0.170
4:4 (N)	−9985.6063	−9985.4010	5.095	4.876	0.219

[1] 1 hartree = 27.211386245988(53) eV. [2] Neutral crystal [76]. [3] $(TMTSF)_2BF_4$ at the anion ordered state at 20 K with two crystallographically independent TMTSF molecules (A and B) [77]. [4] $(TMTSF)_2PF_6$ at 20 K by the neutron diffraction structural analysis [78]. [5] $(TMTSF)_2NO_3$ at 125 K [80]. [6] $(TMTSF)_2NbF_6$ [5]. [7] $(TMTSF)_2Ni(tds)_2$ (tds = [bis(trifluoromethyl)ethylene]diselenolato) [52]. [8] $(TMTSF)NO_3$ [69]. [9] $(TMTSF)[Cr(Br_4SQ)_2(Br_4Cat)] \cdot (CH_2Cl_2)_2$ (SQ = semiquinonate, Cat = catecholate) [71]. [10] $(TMTSF)_3PW_{12}O_{40}$ [68].

Table 4. The total energies of TMTSF molecules at neutral and cationic states (E_0 and E_{1+}); their relative values with respect to those calculated for the optimized structures ($\Delta E_0 = E_0 - E_0$(optimized for 0) and $\Delta E_{1+} = E_{1+} - E_{1+}$(optimized for 1+)); and their difference ($\Delta E_0 - \Delta E_{1+}$) as a measure of the valence of each conformation of TMTSF molecule. The E_0 and E_{1+} were obtained by Hartree–Fock (HF) calculations (6-31G(d,p)) followed by Møller–Plesset correlation energy calculations truncated at second order (MP2) [82].

Conformation	E_0/Hartree [1]	E_{1+}/Hartree	ΔE_0/eV	ΔE_{1+}/eV	$(\Delta E_0 - \Delta E_{1+})$/eV
optimized for 0	−9977.9803	−9977.7594	0.000	0.361	−0.361
optimized for 1+	−9977.9745	−9977.7726	0.158	0.000	0.158
neutral [2]	−9977.7718	−9977.5598	5.675	5.792	−0.117
BF_4 (A, 20 K) [3]	−9977.9752	−9977.7696	0.141	0.082	0.059
BF_4 (B, 20 K) [3]	−9977.9761	−9977.7681	0.115	0.123	−0.009
PF_6 (20 K) [4]	−9977.9531	−9977.7455	0.741	0.739	0.002
NO_3 (2:1) [5]	−9977.9518	−9977.7495	0.778	0.629	0.148
NbF_6 [6]	−9977.8048	−9977.6021	4.777	4.641	0.136
Ni (A) [7]	−9977.8092	−9977.6046	4.657	4.571	0.085
Ni (A′) [7]	−9977.8070	−9977.6050	4.718	4.561	0.157
NO_3 (1:1) [8]	−9977.8086	−9977.6044	4.674	4.579	0.094
Cr [9]	−9977.7564	−9977.5594	6.094	5.804	0.290
PWO (A) [10]	−9977.8438	−9977.6319	3.716	3.830	−0.113
PWO (B) [10]	−9977.8760	−9977.6821	2.841	2.463	0.377
8:5 (P)	−9977.8026	−9977.5992	4.837	4.719	0.117
8:5 (Q)	−9977.7911	−9977.5846	5.150	5.118	0.032
8:5 (R)	−9977.7941	−9977.5848	5.068	5.112	−0.044
5:2 (A)	−9977.7942	−9977.5902	5.065	4.965	0.100
5:2 (B)	−9977.7951	−9977.5906	5.041	4.955	0.086
5:2 (C)	−9977.7888	−9977.5728	5.212	5.439	−0.227
4:4 (K)	−9977.8070	−9977.6059	4.717	4.537	0.180
4:4 (L)	−9977.8063	−9977.6060	4.736	4.535	0.201
4:4 (M)	−9977.8084	−9977.6037	4.679	4.596	0.083
4:4 (N)	−9977.8057	−9977.6032	4.752	4.612	0.140

[1] 1 hartree = 27.211386245988(53) eV. [2] Neutral crystal [76]. [3] $(TMTSF)_2BF_4$ at the anion ordered state at 20 K with two crystallographically independent TMTSF molecules (A and B) at 20 K [77]. [4] $(TMTSF)_2PF_6$ at 20 K by the neutron diffraction structural analysis [78]. [5] $(TMTSF)_2NO_3$ at 125 K [80]. [6] $(TMTSF)_2NbF_6$ [5]. [7] $(TMTSF)_2Ni(tds)_2$ (tds = [bis(trifluoromethyl)ethylene]diselenolato) [52]. [8] $(TMTSF)NO_3$ [69]. [9] $(TMTSF)[Cr(Br_4SQ)_2(Br_4Cat)] \cdot (CH_2Cl_2)_2$ (SQ = semiquinonate, Cat = catecholate) [71]. [10] $(TMTSF)_3PW_{12}O_{40}$ [68].

3.2.4. $(TMTSF)_8(I_3)_5$

A simple picture of the oxidation of the TMTSF molecules in the 8:5 salt is obtained by assuming that R is almost neutral as it is a monomer. Then the molecules 4P + 2Q will have the charge of +5 in total since the composition of the 8:5 salt is P:Q:R:I_3^- = 4:2:2:5. In other words, the average valence of P and Q is $(5/6)+$. This is, however, not reasonable since the $\Delta E_0 - \Delta E_{1+}$ of P and Q is smaller than that of the $TMTSF^{(1/2)+}$ of the 2:1 NO_3, NbF_6, and $Ni(tds)_2$ salts as in Figure 12d.

The monomer R has a smaller negative value than that of the monomer C in the 5:2 salt. This is not strange since the molecular plane of R is flat such as P and Q (Figure 1a), while C has a curved structure (Figure 5a). The small difference in the $\Delta E_0 - \Delta E_{1+}$ among P, Q, and R probably suggests that all the TMTSF molecules in the 8:5 salt accompany to form energy bands and their average valence is considered to be $(5/8)+$ as expected simply from the composition. On the other hand, P and R are probably the most and least oxidized, respectively, on the basis of the $\Delta E_0 - \Delta E_{1+}$ in Figure 12d.

Even though the knowledge of the valence of TMTSF is available, the highly symmetric but the rather complicated donor arrangement of the 8:5 salt makes the discussion of its band structure difficult without any band calculations. The sixteen HOMO's (highest occupied molecular orbitals) in one TMTSF layer will give the sixteen energy bands separated by narrow band gaps. When ten out of thirty two electrons in the HOMO's are removed (5/16-filled by holes in total), the Fermi level will be inside such a small band gap. This picture is consistent with the weak metallic behavior of the ρ for $\boldsymbol{I} \parallel$ longest edge below room temperature, and also with the semiconducting behavior for $\boldsymbol{I} \perp$ plane, respectively.

Or, in the real space picture, the possible difference in the effective valence of P, Q, and R probably implies that the electronic system is in a charge-disproportionated (or ordered) state in the 8:5 salt.

3.2.5. $(TMTSF)_4(I_3)_4 \cdot THF$

The average of $\Delta E_0 - \Delta E_{1+}$ of K, L, M, and L in the 4:4 salt is 0.151 eV and it is relatively high in Figure 12d. This is consistent with the average valence 1+ expected from the composition. On the other hand, we see that these crystallographically independent TMTSF molecules naturally give the relatively large distribution of $\Delta E_0 - \Delta E_{1+}$, namely 0.083 eV (M), 0.140 eV (N), 0.180 eV (K), and 0.201 eV (L), respectively.

The distribution poses an interesting possibility that some of the TMTSF molecules have the valence larger than a unity (>1+). One will easily understand that it is certainly possible by taking a look at the crystal structure in Figure 8a.

In a simplified picture, the 4:4 salt has an elongated rock-salt structure, but the cation and anion units are composed of not by single atoms such as Na^+ and Cl^- but by the tetramers of molecules. Therefore, the electrons do not have to distribute uniformly within each tetramer.

Indeed, the TMTSF molecules K and L are surrounded by more I_3^- anions than M and N. Especially, the molecule M is next to a neutral THF molecule instead of an I_3^- anion. Thus, the molecules K and L can possess more positive charge than M; and N will be in between them. Considering it like this, the order of the height of bars in Figure 12b,d is understood naturally.

3.2.6. Ionization Potentials

The difference between the total energy obtained for the optimized structure at "neutral" state (E_0(optimized for 0)) and that for the "cationic" state (E_{1+}(optimized for 1+)) corresponds to the adiabatic ionization potential I_a of the TMTSF molecule as,

$$I_a = E_{1+}(\text{optimized for } 1+) - E_0(\text{optimized for } 0). \tag{1}$$

On the other hand, the difference between E_0(optimized for 0) and the total energy of the cationic state with the structure optimized for the "neutral" state (E_{1+}(optimized for 0)) gives the vertical ionization energy I_v as,

$$I_v = E_{1+}(\text{optimized for } 0) - E_0(\text{optimized for } 0). \tag{2}$$

Comparison of the I_a and I_v values by the present calculations with the experimental ones will be a good test for knowing the reliability of the fundamental values of E_0 and E_{1+} in Tables 3 and 4.

By the present calculations, the values $I_a = 5.81$ eV (DFT) and 5.65 eV (MP2); and $I_v = 6.09$ eV (DFT) and 6.01 eV (MP2) are obtained for the isolated TMTSF molecule, while the values $I_a = 6.27$ eV and $I_v = 6.58$ eV were determined by the photoelectron spectroscopy for that in gas phase [84]. The calculated values are fairly close to the experimental ones, though the former is 7–9% smaller than the latter.

We confirmed that the agreement is much more improved (within ~1%), when Møller–Plesset correlation energy calculations, in combination with the basis set cc-pVTZ, is truncated at fourth order (MP4(SDQ)), as $I_a = 6.17$ eV and $I_v = 6.55$ eV, respectively. Therefore, the results in Figure 12c,d will be further refined by the MP4 calculations (in progress), though it takes much longer machine time than the DFT and MP2 ones.

4. Materials and Methods

4.1. Sample Preparation

All the present I_3 salts were grown by the electrochemical oxidation using stick-type platinum electrodes.

The single crystals of the 8:5 and 5:2 salts were obtained from the mixed solution of TMTSF (22.7 mg) and tetrabutylammonium triiodide (8.7 mg) with THF (98%, 30 mL) as the solvent. The constant DC electrical current of 2 μA was applied between electrodes for 143 h in air at 20 °C.

The crystals of both the 8:5 and 5:2 salts grew together on the anode. The crystals are black cuboids for the 8:5 salt and black needle-like for the 5:2 salt, respectively. After the crystals were removed from the electrode, they were washed by ethanol and dried in air.

The single crystals of the 4:4 salt were obtained from the mixed solution of TMTSF (22.8 mg) and tetrabutylammonium triiodide (9.0 mg) with THF (98%, 30 mL) as the solvent. The constant DC electrical current of 5 μA was applied between electrodes for 45 h in air at 20 °C.

The crystals of the 4:4 salt grew on the anode, but the major product was the 5:2 salt. The crystals of the 4:4 salt are black plate-like and most of them are much larger than those of the 5:2 salt.

4.2. X-ray Crystal Structure Analysis

Data collections for the present I_3 salts were performed using a MSC Mercury CCD X-ray system (Rigaku, Tokyo, Japan) equipped with MoKα radiation ($\lambda = 0.7107$ Å) at room temperature. All the structures were solved by the direct method using SHELXS-97 software package and were refined by the full-matrix least-squares method. The refinement data are summarized in Table 1.

4.3. Measurement of Electrical Resistivity

The electrical resistance was measured by a standard four-probe technique. Annealed gold wires (10 μm in diameter) were attached as electrodes with carbon paste (XC-12, Fujikura-kasei Co., Tokyo, Japan). The DC electrical current of 1–10 μA was applied by a current source (7651, Yokogawa, Tokyo, Japan) and the voltage signal was measured by a nano-volt/micro-ohm-meter (34420A, Keysight Technologies, Santa Rosa, CA, USA) or a digital multi-meter (34970A, Keysight Technologies, CA, USA). The polarity of the electrical current was switched to eliminate the parasitic voltage. Thus, the electrical resistance was determined as the slope of the voltage to the current. The electrical resistivity ρ was calculated using the size of each crystal assuming their shapes are rectangular.

The temperature dependence of the ρ at ambient pressure was measured in the vacuum chamber of a home-made cryostat. A clamped-type pressure cell made of BeCu was used for the measurement under pressure. We used Daphne 7373 (Idemitsu Kosan Co., Tokyo, Japan) as the pressure transmitting medium. The pressure inside was corrected by the clamped pressure at room temperature and the temperature dependence of the inside pressure in the previous reports [74,75].

After the measurement was finished, the lattice parameters of the samples #1908 and #1910 were determined by the X-ray diffraction pattern and they were identified as the 8:5 and 5:2 salts, respectively.

5. Conclusions

Three types of novel TMTSF salts $(TMTSF)_8(I_3)_5$, $(TMTSF)_5(I_3)_2$, and $(TMTSF)_4(I_3)_4 \cdot$THF were synthesized and their crystal structures were solved at room temperature. The electrical resistivity was measured for $(TMTSF)_8(I_3)_5$ and $(TMTSF)_5(I_3)_2$ at ambient pressure and under hydrostatic pressure up to 1.73 GPa.

The X-ray crystal structure analyses revealed that the 8:5 and 5:2 salts have one-dimensional TMTSF stacks made of TMTSF trimers. The TMTSF stacks are separated by TMTSF monomers in these salts. On the other hand, the crystal structure of the 4:4 salt is regarded as an elongated rock-salt type, where a TMTSF tetramer is surrounded by six I_3^- tetramers and vice versa.

The electrical resistivity of $(TMTSF)_8(I_3)_5$ is metallic for the electrical current along the highest conducting direction. The resistivity shows rapid increase below 88 and 53 K suggesting two phase transitions or changes in electronic states. The analysis of the ρ revealed the possibility of the third change in the electronic state above 310 K.

The electrical resistivity of $(TMTSF)_5(I_3)_2$ is semiconducting below room temperature. It shows the change in slope in the Arrhenius plot suggesting a kind of change in electronic state or a crossover from a wider-band-gap to a narrow-band-gap states around 190 K.

No large change in the behavior of the electrical resistivity was observed for the 8:5 and 5:2 salts by the application of hydrostatic pressures up to 1.73 GPa.

The valence of the crystallographically independent TMTSF molecules in the three I_3 salts has been estimated on the basis of the quantum mechanical calculations of the total energy. The method was found to give the plausible valence of each TMTSF molecule.

The large unit cell of the 8:5 salt is probably results in a number of energy bands separated by small band gaps. The estimated TMTSF valence ($(5/8)+$ on average) gives the band filled by $5/16$ with holes in total, though crystallographically independent TMTSF molecules P, Q, and R are probably less oxidized in this order.

The electronic state of the 5:2 salt is semiconducting and this is consistent with the trimer structures with $TMTSF^{(2/3)+}$ since the TMTSF monomer C is most likely neutral.

The average valence of TMTSF molecules is $1+$ in the 4:4 salt, but the quantum mechanical calculations and its crystal structure strongly suggest that the valence of two TMTSF molecules K and L in the tetramer is higher than $1+$.

It is concluded that the quantum mechanical energy calculations give relative change in the non-integer valence of TMTSF molecules in the I_3 salts, in spite of the limitation that the Gaussian package can accept only the integer valence as an input parameter for single molecule.

On the other hand, there exist more direct and powerful methods to determine not only valence but also the charge distribution within a molecule as well as a unit cell. For example, the electron density analysis based on the synchrotron X-ray diffraction has been successfully applied to investigate the charge-ordering transitions of α-$(BEDT\text{-}TTF)_2I_3$ (BEDT-TTF = bis(ethylenedithio)tetrathiafulvalene) [85] and $(TMTTF)_2PF_6$ (TMTTF = tetramethyltetrathiafulvalene) [86].

Please note that the Gaussian package provides a function to calculate the electron density map under the periodic boundary condition. In our calculation environment, however, the DFT calculations fail in case of, for example, the 4:4 salt due to the limitation of the number of atoms or basis sets. In addition, even though the electron density map has been obtained, there should be the problem in defining the borders between neighboring molecules to integrate the electron density numerically. This type of estimation of molecular valence in a crystal with a large unit cell will be another challenge.

The present method, however, provides a quick and convenient way to estimate the more precise valence of donors and acceptors than that just by comparing the bond lengths before trying a time-consuming method.

Author Contributions: Sample preparation, Y.I.; X-ray crystal structure analyses, R.T.; electrical resistivity measurements, H.Y., Y.I., Y.T. and C.M.; quantum-mechanical calculations, H.Y.; interpretation of results, H.Y.; writing—original draft preparation, H.Y.; All authors have read and agreed to the published version of the manuscript.

Funding: This work was supported by JSPS KAKENHI Grant Number JP25400380.

Conflicts of Interest: The authors declare no conflict of interest.

References

1. Jérome, D.; Mazaud, A.; Ribault, M.; Bechgaard, K. Superconductivity in a synthetic organic conductor $(TMTSF)_2PF_6$. *J. Phys. Lett.* **1980**, *41*, L95–L98. [CrossRef]
2. Bechgaard, K.; Carneiro, K.; Olsen, M.; Rasmussen, F.B. Zero-pressure organic superconductor: di-(tetramethyltetraselenafulvalenium)-perchlorate $[(TMTSF)_2ClO_4]$. *Phys. Rev. Lett.* **1981**, *46*, 852–855. [CrossRef]
3. Bechgaard, K.; Jacobsen, C.S.; Mortensen, K.; Pedersen, H.J.; Thorup, N. The properties of five highly conducting salts: $(TMTSF)_2X$, X = PF_6^-, AsF_6^-, SbF_6^-, BF_4^- and NO_3^-, derived from tetramethyltetraselenafulvalene (TMTSF). *Solid State Commun.* **1980**, *33*, 1119–1125. [CrossRef]
4. Parkin, S.S.S.; Ribault, M.; Jérome, D.; Bechgaard, K. Superconductivity in the family of organic salts based on the tetramethyltetraselenafulvalene (TMTSF) molecule: $(TMTSF)_2X$ (X = ClO_4, PF_6, AsF_6, SbF_6, TaF_6). *J. Phys. C: Solid State Phys.* **1981**, *14*, 5305–5326. [CrossRef]
5. Sakata, M.; Yoshida, Y.; Maesato, M.; Hagiwara, R. Preparation of superconducting $(TMTSF)_2NbF_6$ by electrooxidation of TMTSF using ionic liquid as electrolyte. *Mol. Cryst. Liq. Cryst.* **2006**, *452*, 103–112. [CrossRef]
6. Parkin, S.S.S.; Jérome, D.; Bechgaard, K. Pressure dependence of the metal–insulator and superconducting phase transitions in $(TMTSF)_2ReO_4$. *Mol. Cryst. Liq. Cryst.* **1981**, *79*, 213–224.
7. Rindorf, G.; Soling, H.; Thorup, N. The structures of di(2,3,6,7-tetramethyl-1,4,5,8-tetraselenafulvalenium) perrhenate, $(TMTSF)_2ReO_4$, and perchlorate, $(TMTSF)_2ClO_4$. *Acta Cryst. B* **1982**, *38*, 2805–2808. [CrossRef]
8. Williams, J.M.; Beno, M.A.; Appelman, E.H.; Capriotti, J.M.; Wudl, F.; Aharon-Shalom, E.; Nalewajek, D. New $(TMTSF)_2X$ derivatives: a change in the selenium network dimensionality derived from the molecular and crystal structures of $(TMTSF)_2(FSO_3)$ [T = 298K, 123K] and $(TMTSF)_2(BrO_4)$ T = 298K. *Mol. Cryst. Liq. Cryst.* **1982**, *79*, 319–326. [CrossRef]
9. Cox, S.; Boysel, R.M.; Moses, D.; Wudl, F.; Chen, J.; Ochsenbein, S.; Heeger, A.J.; Walsh,W.M., Jr.; Rupp, L.W., Jr. $(TMTSF)_2F_2PO_2$: An unusual member of the $(TMTSF)_2X$ family of organic metals. *Solid State Commun.* **1984**, *49*, 259–263. [CrossRef]
10. Soling, H.; Rindorf, G.; Thorup, N. Di(4,4′,5,5′-tetramethyl-$\Delta^{2,2'}$-bi-1,3-diselenolyliden)ium trifluoromethanesulfonate, $C_{21}H_{24}F_3O_3SSe_8$, $(TMTSF)_2CF_3SO_3$. *Acta Cryst. C* **1983**, *39*, 490–491. [CrossRef]
11. Thorup, N.; Rindorf, G.; Soling, H.; Johannsen, I.; Mortensen, K.; Bechgaard, K. Structural studies of some $(TMTSF)_2X$ compounds. *J. Phys.-Paris C* **1983**, *3*, 1017–1020. [CrossRef]
12. Beno, M.A.; Blackman, G.S.; Williams, J.M.; Bechgaard, K. Synthetic metals based on tetramethyltetraselenafulvalene (TMTSF): Synthesis, structure (T = 298 and 125 K), and novel properties of $(TMTSF)_2H_2F_3$. *Inorg. Chem.* **1982**, *21*, 3860–3862. [CrossRef]
13. Eriks, K.; Beno, M.A.; Bechgaard, K.; Williams, J.M. The structure of di(3,3′,4,4′-tetramethyl-2,2′,5,5′-tetraselenafulvalenium) pentafluorosilicate, $(C_{10}H_{11}Se_4)_2SiF_5$ at 293 and 125 K. *Acta Cryst. C* **1984**, *40*, 1715–1717. [CrossRef]
14. Nickel, J.C.; Duprat, R.; Bourbonnais, C.; Dupuis, N. Triplet superconducting pairing and density-wave instabilities in organic conductors. *Phys. Rev. Lett.* **2005**, *95*, 247001. [CrossRef]
15. Rozhkov, A.V. Superconductivity without attraction in a quasi-one-dimensional metal. *Phys. Rev. B* **2009**, *79*, 224520. [CrossRef]
16. Lebed, A.G. d-Wave-like nodal superconductivity in the organic conductor $(TMTSF)_2ClO_4$. *Physica B* **2012**, *404*, 1803–1805. [CrossRef]
17. Tanaka, Y.; Kuroki, K. Microscopic theory of spin-triplet f-wave pairing in quasi-one-dimensional organic superconductors. *Phys. Rev. B* **2004**, *70*, 060502R. [CrossRef]
18. Fuseya, Y.; Suzumura, Y. Superconductivity and density wave in the quasi-one-dimensional systems: renormalization group study. *J. Phys. Soc. Jpn.* **2005**, *74*, 1263–1269. [CrossRef]
19. Pouget, J.P.; Moret, R.; Comes, R.; Bechgaard, K.; Fabre, J.M.; Giral, L. X-ray diffuse scattering study of some $(TMTSF)_2X$ and $(TMTTF)_2X$ salts. *Mol. Cryst. Liq. Cryst.* **1982**, *79*, 129–143. [CrossRef]
20. Andrieux, A.; Jérome, D.; Bechgaard, K. Spin-density wave ground state in the one-dimensional conductor $(TMTSF)_2PF_6$: Microscopic evidence from ^{77}Se and ^{1}H NMR experiments. *J. Phys. Lett.* **1981**, *42*, L87–L90. [CrossRef]

21. Scott, J.C.; Pedersen, H.J.; Bechgaard, K. Proton NMR in the organic conductor tetramethyltetraselenafulvalinium hexafluorophosphate. *Phys. Rev. B* **1981**, *24*, 475–477. [CrossRef]
22. Mortensen, K.; Tomkiewicz, Y.; Bechgaard, K. Antiferromagnetism in the organic conductor bis-tetramethyltetraselenafulvalene hexafluoroarsenate [$(TMTSF)_2AsF_6$]: Static magnetic susceptibility. *Phys. Rev. B* **1982**, *25*, 3319–3325. [CrossRef]
23. Torrance, J.B.; Pedersen, H.J.; Bechgaard, K. Observation of antiferromagnetic resonance in an organic superconductor. *Phys. Rev. Lett.* **1982**, *49*, 881–884. [CrossRef]
24. Walsh, W.M., Jr.; Wudl, F.; Aharon-Shalom, E.; Rupp, L.W., Jr.; Vandenberg, J.M.; Andres, K.; Torrance, J.B. Itinerant-electron antiferromagnetism precursor to superconductivity in an organic conductor. *Phys. Rev. Lett.* **1982**, *49*, 885-888. [CrossRef]
25. Delrieu, J.M.; Roger, M.; Toffano, Z.; Moradpour, A.; Bechgaard, K. NMR proton lineshape in $(TMTSF)_2X$: Incommensurability of nesting vector and order parameter. *J. Phys.* **1986**, *47*, 839–861. [CrossRef]
26. Takahashi, T.; Maniwa, Y.; Kawamura, H.; Saito, G. Determination of SDW characteristics in $(TMTSF)_2PF_6$ by ^{1}H-NMR analysis. *J. Phys. Soc. Jpn.* **1986**, *55*, 1364–1373. [CrossRef]
27. Takahashi, T.; Maniwa, Y.; Kawamura, H.; Saito, G. Pressure dependence of SDW properties in the organic conductor, $(TMTSF)_2PF_6$. *J. Phys. Soc. Jpn.* **1989**, *58*, 703–709. [CrossRef]
28. Kwak, J.F.; Schirber, J.E.; Greene, R.L.; Engler, E.M. Magnetic quantum oscillations in tetramethyltetraselenafulvalenium hexafluorophosphate [$(TMTSF)_2PF_6$]. *Phys. Rev. Lett.* **1981**, *46*, 1296–1299. [CrossRef]
29. Bando, H.; Oshima, K.; Suzuki, M.; Kobayashi, H.; Saito, G. Shubnikov–de Haas effect in $(TMTSF)_2ClO_4$. *J. Phys. Soc. Jpn.* **1982**, *51*, 2711–2712. [CrossRef]
30. Chaikin, P.M.; Choi, M.-Y.; Kwak, J.F.; Brooks, J.S.; Martin, K.P.; Naughton, M.J.; Engler, E.M.; Greene, R.L. Tetramethyltetraselenafulvalenium perchlorate, $(TMTSF)_2ClO_4$, in high magnetic fields. *Phys. Rev. Lett.* **1983**, *51*, 2333–2336. [CrossRef]
31. Azevedo, L.J.; Williams, J.M.; Compton, S.J. Field-induced magnetic order in bis-tetramethyltetraselenafulvalenium perchlorate $(TMTSF)_2ClO_4$. *Phys. Rev. B* **1983**, *28*, 6600–6602. [CrossRef]
32. Takahashi, T.; Jérome, D.; Bechgaard, K. An NMR study of the organic superconductor $(TMTSF)_2ClO_4$. *J. Phys.* **1984**, *45*, 945–952. [CrossRef]
33. Montambaux, G.; Héritier, H.; Lederer, P. Spin susceptibility of the two-dimensional electron gas with open Fermi surface under magnetic field. *Phys. Rev. Lett.* **1985**, *55*, 2078–2081. [CrossRef] [PubMed]
34. Yamaji, K. Theory of field-induced spin-density-wave in Bechgaard salts. *Synth. Met.* **1986**, *13*, 29–43. [CrossRef]
35. Maki, K. Thermodynamics of field-induced spin-density-wave states in Bechgaard salts. *Phys. Rev. B* **1986**, *33*, 4826–4829. [CrossRef]
36. Lebed, A.G.; Bak, P. Theory of unusual anisotropy of magnetoresistance in organic superconductors. *Phys. Rev. Lett.* **1989**, *63*, 1315–1317. [CrossRef]
37. Boebinger, G.S.; Montambaux, G.; Kaplan, M.L.; Haddon, R.C.; Chichester, S.V.; Chiang, L.Y. Anomalous magnetoresistance anisotropy in metallic and spin-density-wave phases of the quasi-one-dimensional organic conductor $(TMTSF)_2ClO_4$. *Phys. Rev. Lett.* **1990**, *64*, 591–594. [CrossRef]
38. Osada, T.; Kawasumi, A.; Kagoshima, S.; Miura, N.; Saito, G. Commensurability effect of magnetoresistance anisotropy in the quasi-one-dimensional conductor tetramethyltetraselenafulvalenium perchlorate, $(TMTSF)_2ClO_4$. *Phys. Rev. Lett.* **1991**, *66*, 1525–1528. [CrossRef]
39. Osada, T.; Kagoshima, S.; Miura, N. Resonance effect in magnetotransport anisotropy of quasi-one-dimensional conductors. *Phys. Rev. B* **1992**, *46*, 1812–1815. [CrossRef]
40. Danner, G.M.; Kang, W.; Chaikin, P.M. Measuring the Fermi surface of quasi-one-dimensional metals. *Phys. Rev. Lett.* **1994**, *72*, 3714–3717. [CrossRef]
41. Lee, I.J.; Naughton, M.J. Effective electrons and angular oscillations in quasi-one-dimensional conductors. *Phys. Rev. B* **1998**, *57*, 7423–7426. [CrossRef]
42. Osada, T. Novel angular effect and electron orbital motion in quasi-one-dimensional conductors. *Synth. Met.* **1997**, *86*, 2143–2144. [CrossRef]
43. Lebed, A.G.; Bagmet, N.N. Nonanalytical magnetoresistance, the third angular effect, and a method to investigate Fermi surfaces in quasi-one-dimensional conductors. *Phys. Rev. B* **1997**, *55*, R8654–R8657. [CrossRef]

44. Yoshino, H.; Murata, K. Origin of the third angular effect of magnetoresistance of quasi-one-dimensional metals. *J. Phys. Soc. Jpn.* **1999**, *68*, 3027–3033. [CrossRef]
45. Bechgaard, K.; Kistenmacher, T.J.; Bloch, A.N.; Cowan, D.O. The crystal and molecular structure of an organic conductor from 4,4′,5,5′-tetramethyl-$\Delta^{2,2'}$-bis-1,3-diselenole and 7,7,8,8,-tetracyano-*p*-quinodimethane [TMTSF–TCNQ]. *Acta Cryst. B* **1977**, *33* 417–422. [CrossRef]
46. Kistenmacher, T.J.; Emge, T.J.; Bloch, A.N.; Cowan, D.O. Structure of the red, semiconducting form of 4,4′5,5′-tetramethyl-$\Delta^{2,2'}$-bi-1,3-deselenole-7,7,8,8,-tetracyano-*p*-quinodimethane, TMTSF–TCNQ. *Acta Cryst. B* **1982**, *38*, 1193–1199. [CrossRef]
47. Andersen, J.R.; Bechgaard, K.; Jacobsen, C.S.; Rindorf, G.; Soling, H.; Thorup, N. The crystal and molecular structure of the organic conductor 2,3,6,7-tetramethyl-1,3,5,8-tetraselenafulvalenium 2,5-dimethyl-7,7,8,8-tetracyano-*p*-quinodimethanide (TMTSF–DMTCNQ). *Acta Cryst. B* **1978**, *34*, 1901–1905. [CrossRef]
48. Stokes, J.P.; Emge, T.J.; Bryden, W.A.; Chappell, J.S.; Cowan, D.O.; Poehler, T.O.; Bloch, A.N.; Kistenmacher, T.J. $(TMTSF)_2(2,5\text{-}TCNQBr_2)$: Structure and physical properties. *Mol. Cryst. Liq. Cryst.* **1982**, *79*, 327–336. [CrossRef]
49. Konovalikhin, S.V.; D'yachenko, O.A.; Shilov, G.V.; Spitsyna, N.G.; Van, K.V.; Yagubskii, E.B. Crystal structure of a new molecular complex of fullerene with tetramethyltetraselenafulvalene: $C_{60}\cdot TMTSF\cdot 2CS_2$. *Russ. Chem. B+* **1997**, *46*, 1415–1420. [CrossRef]
50. Konovalikhin, S.V.; Dyachenko, O.A.; Shilov, G.V.; Spitsyna, N.G.; Yagubskii, E.B. Crystal and molecular structure of a new complex of [60]fullerene, $2(C_{60})\cdot 2(TMTSF)\cdot(C_6H_6)$. *Fullerene Sci. Techn.* **1998**, *6*, 563–575. [CrossRef]
51. Heuer, W.B.; Hoffman, B.M. A novel phase transition in conductive molecular crystals based on metal bis-diselenoalkene complexes: $[TMTSF]_2[M\{Se_2C_2(CF_3)_2\}_2]$; M = Ni, Pt; TMTSF = tetremethyltetraselenafulvalene. *J. Chem. Soc. Chem. Commun.* **1986**, 174–175. [CrossRef]
52. Heuer, W.B.; Squattrito, P.J.; Hoffman, B.M.; Ibers, J.A. Phase transition in the conductive molecular crystals bis(tetramethyltetraselenafulvalene)[bis(trifluoromethyl)ethylene]diselenolatonickel and -platinum, $[TMTSF]_2[M(tds)_2]$, M = Ni, Pt. Relation to the Copper Analogue. *J. Am. Chem. Soc.* **1988**, *110*, 792–803. [CrossRef]
53. Morgado, J.; Alcácer, L.; Almeida, M. Electrical resistivity and thermoelectric power of $(TMTSF)_2M(tds)_2$, M = Pt, Cu and Ni; evidence for the existence of two different phases. *Solid State Commun.* **1994**, *89*, 755–759. [CrossRef]
54. Kazheva, O.N.; Chekhlov, A.N.; Alexandrov, G.G.; Buravov, L.I.; Kravchenko, A.V.; Starodub, V.A.; Sivaev, I.B.; Bregadze, V.I.; Dyachenko, O.A. Synthesis, structure and electrical conductivity of fulvalenium salts of cobalt bis(dicarbollide) anion. *J. Organomet. Chem.* **2006**, *691*, 4225–4233. [CrossRef]
55. Kazheva, O.N.; Alexandrov, G.G.; Kravchenko, A.V.; Starodub, V.A.; Sivaev, I.B.; Lobanova, I.A.; Bregadze, V.I.; Bravov, L.I.; Dyachenko, O.A. New fulvalenium salts of bis(dicarbollide) cobalt and iron: synthesis, crystal structure and electrical conductivity. *J. Organomet. Chem.* **2007**, *692*, 5033–5043. [CrossRef]
56. Penicaud, A.; Batail, P.; Bechgaard, K.; Sala-Pala, J. A conductive tetramethyltetraselenafulvalenium salt based on a divalent oxalato-bridged bis-(tetrafluorometallate) complex: $(TMTSF)_3[Ti_2F_8(C_2O_4)]$. *Synth. Met.* **1988**, *22*, 201–207. [CrossRef]
57. Ouahab, L.; Paiou, J.; Grandjean, D.; Garrigou–Langrange, C.; Delhaes, P.; Bencharif, M. Charge transfer salts obtained with organic donors (TTF, TMTTF, TMTSF, and BEDT TTF) and tetracyanometallate planar dianions $[M(CN)_4^{2-}$, M = Pt^{II}, $Ni^{II}]$. *J. Chem. Soc. Chem. Commun.* **1998**, 1038–1041. [CrossRef]
58. Triki, S.; Ouahab, L.; Paidou, J.; Grandjean, D. The use of polyoxometallates as acceptors in charge transfer salts: preparation, X-ray crystal structures, and preliminary spectroscopic characterizations of $D_3M_6O_{19}$, D = TTF, TMTSF; M = Mo, W. *J. Chem. Soc. Chem. Commun.* **1998**, 1068–1070. [CrossRef]
59. Turner, S.S.; Le Pevelen, D.; Day, P.; Prout, K. TTF based charge transfer salts of $[Cr(NCS)_4(phen)]^-$: Bulk magnetic order and crystal structures of the TTF, TMTTF (tetramethyltetrathiafulvalene) and TMTSF (tetramethyltetraselenafulvalene) derivatives. *J. Chem. Soc. Dalton Trans.* **2000**, 2739–2744. [CrossRef]
60. Rosokha, S.V.; Stern, C.L.; Ritzert, J.T. π-bonded molecular wires: Self-assembly of mixed-valence cation-radical stacks within the nanochannels formed by inert tetrakis[3,5-bis(trifluoromethyl)phenyl]borate anions. *CrystEngComm* **2013**, *15*, 10638. [CrossRef]

61. Mroweh, N.; Mézière, C.; Allain, M.; Auban–Senzier, P.; Canadell, E.; Avarvari, N. Conservation of structural arrangements and 3:1 stoichiometry in a series of crystalline conductors of TMTTF, TMTSF, BEDT-TTF, and chiral DM-EDT-TTF with the oxo-bis[pentafluorotantalate(V)] dianion. *Chem. Sci.* **2020**, *11*, 10078–10091. [CrossRef]
62. Penicaud, A.; Batail, P.; Perrin, C.; Coulon, C.; Parkin, S.S.P.; Torrance, J.B. First cation radical mixed-valence hybrid salts of the paramagnetic octahedral cluster $Nb_6Cl_{18}{}^{3-}$. Preparation, crystal structures, and conducting and magnetic properties of pentakis[2,3,6,7-tetramethyl-1,4,5,8-tetra-selena- and -thia-fulvalenium) hexachloro(dodeca-μ_2-chloro-octahedro-hexaniobate). *J. Chem. Soc. Chem. Commun.* **1987**, 330–332. do:10.1039/C39870000330 [CrossRef]
63. Pénicaud, A.; Batail, P.; Coulon, C.; Canadell, E.; Perrin, C. Novel redox properties of the paramagnetic hexanuclear niobium cluster halide $Nb_6Cl_{18}{}^{3-}$ and the preparation, structures, and conducting and magnetic properties of its one-dimensional mixed-valence tetramethyltetra(selena and thia)fulvalenium salts: [TMTSF and TMTTF]$_5$[Nb_6Cl_{18}]·$(CH_2Cl_2)_{0.5}$. *Chem. Mater.* **1990**, *2*, 123–132.
64. Reinheimer, E.W.; Galán-Mascarós, J.R.; Cómez-García, C.J.; Zhao, H.; Fourmigué, M.; Dunbar, K.R. Radical salts of TTF derivatives with the metal–metal bonded $[Re_2Cl_8]^{2-}$ anion. *J. Mol. Struct.* **2008**, *890*, 81–89. [CrossRef]
65. Kobayashi, H.; Kobayashi, A.; Sasaki, Y.; Saito, G.; Inokuchi, H. The crystal structure of tetramethyltetraselenafulvalenium perrheneate, (TMTSF)$(ReO_4)(C_2H_3Cl_3)_{0.25}$. *B. Chem. Soc. Jpn.* **1983**, *56*, 2894–2898. [CrossRef]
66. Batail, P.; Ouahab, L.; Torrance, J.B.; Pylman, M.L.; Parkin, S.S.P. Cation radical salts with magnetic anions: preparation and characterization of $FeCl_4$ salts of TMTTF and TMTSF. *Solid State Commun.* **1985**, *55*, 597–600. [CrossRef]
67. Batail P.; Ouahab, L. Effect of anion volume on dimensionality of radical cation salts. *Mol. Cryst. Liq. Cryst.* **1985**, *125*, 205–214. [CrossRef]
68. Ouahab, L.; Grandjean, D. Structure of tris(3,3′,4,4′-tetramethyl-2,2′5,5′-tetraselenafulvalenium) phosphododecatungstate: $(TMTSF)_3PW_{12}O_{40}$. *Acta Cryst. C* **1991**, *47*, 2670–2672. [CrossRef]
69. Wu, L.Y.; Coppens, P.; Bu, X.H. Crystal structure of tetramethyltetraselenafulvalene nitrate, $(C_{10}H_{12}Se_4)NO_3$. *Z. Krist.-New Cryst. St.* **1997**, *212*, 101–102. [CrossRef]
70. Chang, H.-C.; Kitagawa, S.; Kondo, M.; Ishii, T. Structural, spectroscopic and magnetic properties of charge-transfer complex, (TMTSF)[Cr$(Cl_4SQ)_2(Cl_4Cat)$]·$0.5CH_2Cl_2$. *Mol. Cryst. Liq. Cryst.* **1999**, *335*, 183–192. [CrossRef]
71. Chang, H.-C.; Ishii, T.; Kondo, M.; Kitagawa, S. Synthesis, X-ray crystal structures and properties of chromium complexes with semiquinonate and catecholate. *J. Chem. Soc. Dalton Trans.* **1999**, 2467–2476. [CrossRef]
72. Pauling, L. *The Nature of the Chemical Bond*, 3rd ed.; Cornell Univ. Press: Ithaca, NY, USA, 1960; p. 260.
73. Bondi, A. van der Waals volumes and radii. *J. Phys. Chem.* **1964**, 441–448. [CrossRef]
74. Murata, K.; Yoshino, H.; Yadav, H.O.; Honda, Y.; Shirakawa, N. Pt resistor thermometry and pressure calibration in a clamped pressure cell with the medium, Daphne 7373. *Rev. Sci. Instrum.* **1997**, *68*, 2490–2493. [CrossRef]
75. Yokogawa, K.; Murata, K.; Yoshino, H.; Aoyama, S. Solidification of high-pressure medium Daphne 7373. *Jpn. J. Appl. Phys.* **2007**, *46*, 3636–3639. [CrossRef]
76. Kistenmacher, T.J.; Emge, T.J.; Shu, P.; Cowan, D.O. 4,4′,5,5′-tetramethyl-$\Delta^{2,2'}$-bis-1,3-diselenole, TMTSF. *Acta Cryst. B* **1979**, *35*, 772–775. [CrossRef]
77. Emge, T.J.; Wang, H.H.; Beno, M.A.; Williams, J.M.; Whangbo, M.-H.; Evain, E. Effect of anion ordering on the H-anion interactions and band electronic structure of $(TMTSF)_2BF_4$ at 20 K. *J. Am. Chem. Soc.* **1986**, *108*, 8215–8223. [CrossRef]
78. Gallois, B.; Gaultier, J.; Hauw, C.; Lamcharfi, T.D.; Filhol, A. Neutron low-temperature (4 and 20 K) and X-ray high-pressure (6.5×10^2 and 9.8×10^2 MPa) structures of the organic superconductor di(2,3,6.7-tetramethyl-1,4,5,8-tetraselenafulvalenium) hexafluorophosphate, $(TMTSF)_2PF_6$. *Acta Cryst. B* **1986**, *42*, 564–575. [CrossRef]
79. Thorup, N.; Rindorf, G.; Soling, H.; Bechgaard, K. The structure of di(2,3,6,7-tetramethyl-1,4,5,8-tetraselenafulvalenium) hexafluorophosphate, $(TMTSF)_2PF_6$, the first superconducting organic solid. *Acta Cryst. B* **1981**, *37*, 1236–1240. [CrossRef]

80. Emge,T.J.; Beno, M.A.; Daws, C.A.; Wang, H.H.; Williams, J.M. Novel structural features, and their relationship to the electrical properties, of the organic conductor $(TMTSF)_2NO_3$ at 298 K and 125 K. *Mol. Cryst. Liq. Cryst.* **1984**, *116*, 153–171. [CrossRef]
81. Kistenmacher, T.J. Anion symmetry and the separability of structural parameters for tetramethyltetraselenafulvalenium salts, $(TMTSF)_2X$. *Mol. Cryst. Liq. Cryst.* **1986**, *136*, 361–382. [CrossRef]
82. Frisch, M.J.; Trucks, G.W.; Schlegel, H.B.; Scuseria, G.E.; Robb, M.A.; Cheeseman, J.R.; Scalmani, G.; Barone, V.; Petersson, G.A.; Nakatsuji, H.; et al. *Gaussian 16, Revision C.01*; Gaussian, Inc.: Wallingford, CT, USA, 2019.
83. Emery, V.J.; Bruinsma, R.; Barišić, S., Electron-electron Umklapp scattering in organic superconductors. *Phys. Rev. Lett.* **1982**, *48*, 1039–1043. [CrossRef]
84. Gleiter, R.; Kobayashi, M.; Spanget-Larsen, J.; Ferraris, J.P.; Bloch, A.N.; Bechgaard, K.; Cowan, D.O. Photoelectron and electronic absorption spectra of tetrathiafulvalene and related compounds. *Ber. Bursenges. Phys. Chem.* **1975**, *79*, 1218. [CrossRef]
85. Kakiuchi, T.; Wakabayashi, Y.; Sawa, H.; Takahashi, T.; Nakamura, T. Charge ordering in α-$(BEDT\text{-}TTF)_2I_3$ by synchrotron X-ray diffraction. *J. Phys. Soc. Jpn.* **2007**, *76*, 113702. [CrossRef]
86. Kitou, S.; Fujii, T.; Kawamoto, T.; Katayama, N.; Maki, S.; Nishibori, E.; Sugimoto, K.; Takata, M.; Nakamura, T.; H. Sawa, Successive dimensional transition in $(TMTTF)_2PF_6$ revealed by synchrotron X-ray diffraction. *Phys. Rev. Lett.* **2017**, *119*, 065701. [CrossRef]

Publisher's Note: MDPI stays neutral with regard to jurisdictional claims in published maps and institutional affiliations.

crystals

MDPI

Article

Experimental and Modeling Study on the High-Performance p^{++}-GaAs/n^{++}-GaAs Tunnel Junctions with Silicon and Tellurium Co-Doped InGaAs Quantum Well Inserted

Yudan Gou [1,2], Jun Wang [1,2,*], Yang Cheng [2], Yintao Guo [2], Xiao Xiao [2], Heng Liu [2], Shaoyang Tan [2], Li Zhou [2], Huomu Yang [1], Guoliang Deng [1] and Shouhuan Zhou [1]

1 College of Electronics and Information Engineering, Sichuan University, Chengdu 610065, China; gouyudan@stu.scu.edu.cn (Y.G.); scuyhm@scu.edu.cn (H.Y.); gdeng@scu.edu.cn (G.D.); zhoush@scu.edu.cn (S.Z.)

2 Suzhou Everbright Photonics Co., Ltd., Suzhou 215000, China; yang.cheng@everbrightphotonics.com (Y.C.); yintao.guo@everbrightphotonics.com (Y.G.); xiao.xiao@everbrightphotonics.com (X.X.); heng.liu@everbrightphotonics.com (H.L.); shaoyang.tan@everbrightphotonics.com (S.T.); li.zhou@everbrightphotonics.com (L.Z.)

* Correspondence: wjdz@scu.edu.cn

Received: 26 October 2020; Accepted: 24 November 2020; Published: 28 November 2020

Abstract: The development of high-performance tunnel junctions is critical for achieving high efficiency in multi-junction solar cells (MJSC) that can operate at high concentrations. We investigate silicon and tellurium co-doping of InGaAs quantum well inserts in p^{++}-GaAs/n^{++}-GaAs tunnel junctions and report a peak current density as high as 5839 A cm^{-2} with a series resistance of 5.86×10^{-5} Ω cm^2. In addition, we discuss how device performance is affected by the growth temperature, thickness, and V/III ratio in the InGaAs layer. A simulation model indicates that the contribution of trap-assisted tunneling enhances carrier tunneling.

Keywords: tunnel junction; MOCVD; quantum well; co-doping; solar cells

1. Introduction

Solar energy is a renewable and environmentally friendly source of energy. Efforts to generate greater electric power from solar energy have benefited from the high efficiency of solar cell technology [1]. Tunnel junctions are an important component of multi-junction solar cells because they connect the subcells, where each subcell is designed to absorb a specific range within the solar spectrum. Therefore, as the number of subcells increases, the overall absorption of the solar spectrum is enhanced and the thermalization losses are reduced, resulting in a 6-junction solar cell with an efficiency as high as 47.1% [2]. The peak tunneling current density of tunnel junctions must be greater than the photocurrent density of the devices, and the tunnel junctions should have low electrical resistivity and high optical transparency [3].

The peak tunneling current is described below [4]:

$$J_{\text{peak}} \propto \exp \frac{-E_g^{3/2}}{\sqrt{N_{\text{eff}}}} \tag{1}$$

where E_g is the energy bandgap of the depletion region, and $N_{\text{eff}} = (N_{p++}N_{n++})/(N_{p++} + N_{n++})$ is the effective doping concentration, where N_{p++} and N_{n++} are the doping concentrations of the p++and n++ regions, respectively. Equation (1) implies that increased effective doping concentration or a narrower bandgap leads to increased peak tunneling current density. The tunnel-junction resistance is defined as the reciprocal of the initial slope of the *J-V* curve [4].

The usual way to fabricate tunnel junctions with high peak tunnel-current density and low electrical resistivity is to use highly doped material. Take the p^{++}-GaAs/n^{++}-GaAs tunnel junction as an example: obtaining heavy *p*-type doping is easily done by using carbon as the dopant. However, it is more complicated for *n*-type GaAs with a high doping level because, for common epitaxial growth conditions, the compensation and saturation issues of silicon-doped GaAs limits the performance of GaAs tunnel junctions with a doping level around 1×10^{19} cm^{-3} [5]. Therefore, the peak tunneling current density of silicon-doped GaAs tunnel junctions is around 25 A cm^{-2} [6]. However, the tunnel junction should be able to operate at current densities up to 90 A cm^{-2} in a high concentration photovoltaic (HCPV) system (HCPV > 6000 suns, CPV: 500–1000 suns) under non-uniform optical irradiation [7].

A promising way to improve the electrical performance of the tunnel junction is to insert a quantum well (QW) in the *p-n* interface [8,9]. The band structure of the tunnel junctions with and without quantum well were calculated by solving Poisson's equation, taking the bandgap narrow into account. As shown in Figure 1, the tunneling distance is shortened with the In0.07GaAs quantum well inserted due to the band offset, so a higher peak tunneling current density can be obtained. However, K. Louarn et al. obtained a peak tunneling current of 30 A cm^{-2} by using a silicon-doped InGaAs quantum well inserted in a GaAs tunnel junction, which does not satisfy the requirements for the HCPV system [10].

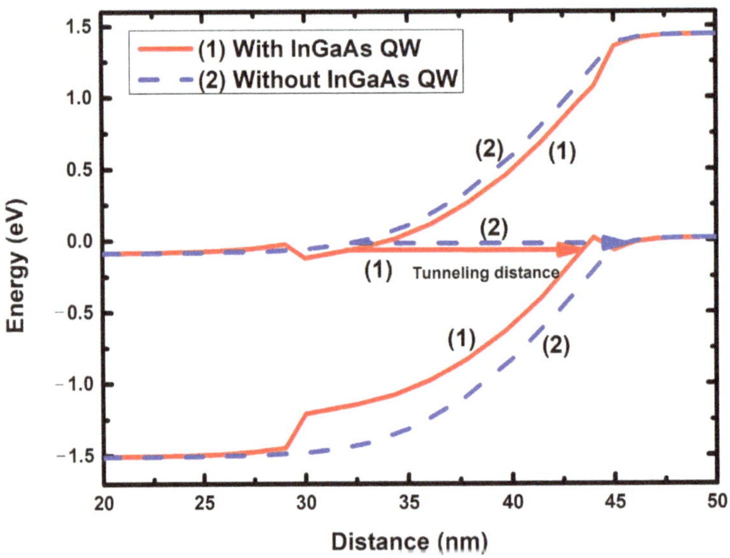

Figure 1. Band structure of (**1**) p^{++}-GaAs//n^{++}-$In_{0.07}$GaAs/n^{++}-GaAs (solid line) and (**2**) p^{++}-GaAs/n^{++}-GaAs (dashed line) tunnel junctions. The differences in band bending shorten the tunneling distance for the InGaAs quantum well inserted in the GaAs tunnel junction structure.

In this work, we proposed the silicon (Si) and tellurium (Te) co-doped InGaAs quantum well inserts in a p^{++}-GaAs/n^{++}-GaAs tunnel junction. Te likely acts as a surfactant that helps the incorporation of Si [11], and Si compensates for the delay time between the injection of DeTe into the reactor and the onset of Te incorporation into the epitaxial layer [12], resulting in a higher doping level in this layer.

The present study investigates the performance of the tunnel junction device, which is affected by the growth conditions of the co-doped InGaAs layer, including the temperature, thickness, and V/III ratio. A peak tunneling current density of 5839 A cm^{-2} with a resistance of 5.86×10^{-5} Ω cm^2 is achieved in this study. In addition, a simulation model was proposed to investigate the high peak tunneling current density.

2. Materials and Methods

By using metal-organic chemical vapor deposition (MOCVD), a series of tunnel junctions were grown on (100) *n*-type GaAs substrates misoriented 6° toward the <111> A direction. We used C (CBr_4 source) and Si (Si_2H_6 source) as *p*-and *n*-type dopants, respectively, and the InGaAs quantum well layer was doped with Si and Te (DeTe source). Figure 2 shows a typical tunnel junction structure, which consists of a 30 nm thick n^{++} (1×10^{19} cm^{-3}) GaAs layer and a 20 nm thick p^{++} (1×10^{20} cm^{-3}) GaAs layer, with an InGaAs quantum well layer embedded at the p^{++}-n^{++} junction interface. The doping level was assessed by using the electrical capacitance-voltage (ECV) profile and secondary ion mass spectroscopy (SIMS). The tunnel junction was surrounded by a 150 nm *n*-type (2×10^{18} cm^{-3}) GaAs buffer layer and a 100 nm *p*-type (5×10^{19} cm^{-3}) GaAs cap layer on the top to ensure good ohmic contact. The growth conditions of the InGaAs quantum well layer were varied, as summarized in Table 1.

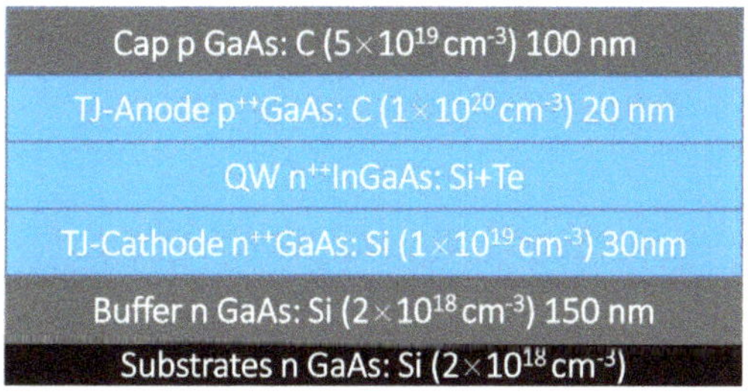

Figure 2. Schematic of tunnel-junction structure. The growth conditions of the InGaAs layer are summarized in Table 1.

Table 1. Summary of growth conditions for $In_{0.07}GaAs$ layer.

Sample No.	Thickness (nm)	V/III	Temperature (°C)
a-1	16	74	600
a-2	16	74	580
a-3	16	74	550
b-1	11	74	550
c-1	16	15	550
c-2	16	5	550

The tunnel junction devices were patterned to different sizes, and chemical etching was used to form square mesa structures of 50×50, 100×100, 200×200, 500×500, 1000×1000, and 1500×1500 μm^2. Figure 3 schematically illustrates the device structure. A AuGe/Ni/Au alloyed metal was sputtered onto the backside of the thinned substrates. The devices were fabricated by using conventional photolithographic and wet-etching techniques, and 100 nm SiO_x was deposited by plasma-enhanced chemical vapor deposition (PECVD) to isolate the mesa sidewalls. A Ti/Pt/Au top contact metal was deposited by thermal evaporation. To ensure accurate resistance measurements, the four-probe technique was used to measure the *J-V* curve.

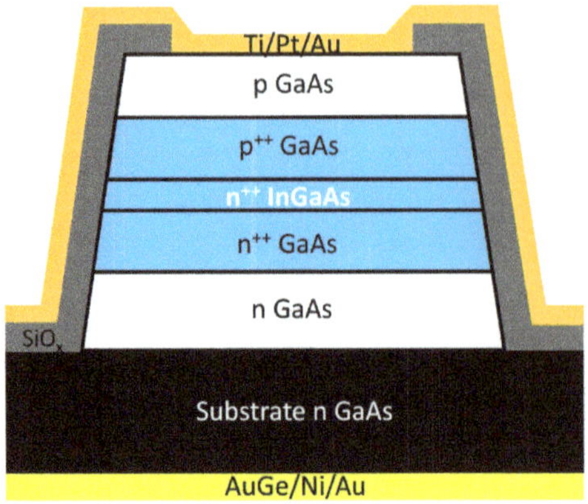

Figure 3. Schematic illustration of the tunnel-junction device structure.

3. Results and Discussion

3.1. Devices Performance of the Tunnel Junctions

We investigated how device performance is affected by growth temperature, thickness, and V/III ratio of the $In_{0.07}GaAs$ layer.

3.1.1. Influence of Growth Temperature

As shown in Figure 4, the *J-V* curve depends on the growth temperature, with both the peak current density and resistance changes. The results are summarized in rows a-1 to a-3 of Table 2. As the growth temperature decreases from 600 to 550 °C, the peak current density increases from 16 to 2130 A cm^{-2}, and the resistance decreases from 1.25×10^{-2} to 1.17×10^{-4} Ω cm^2.

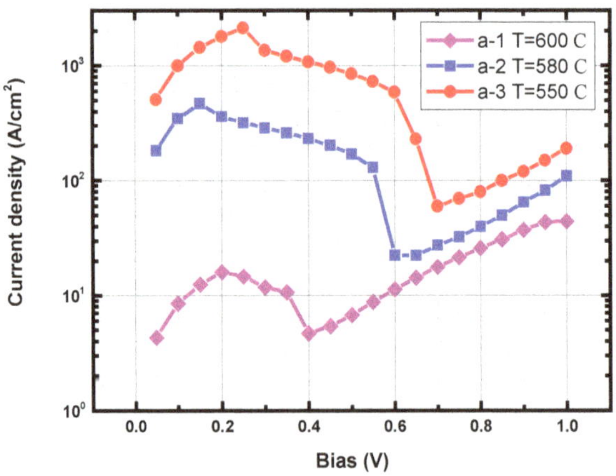

Figure 4. Experimental results of tunnel-junction devices under various growth temperature of the $In_{0.07}GaAs$ layer.

Table 2. Summary of experimental results.

Sample No.	Doping Level (cm^{-3})	Peak Current Density (A cm^{-2})	Resistance (Ω cm^2)
a-1	1.13×10^{19}	16	1.25×10^{-2}
a-2	2.00×10^{19}	472	3.17×10^{-4}
a-3	4.03×10^{19}	2130	1.17×10^{-4}
b-1	3.55×10^{19}	1222	1.94×10^{-4}
c-1	4.22×10^{19}	3320	9.00×10^{-5}
c-2	5.09×10^{19}	5839	5.86×10^{-5}

The growth temperature is a critical factor for Te incorporation, which occupies the arsenic sublattice, especially when aiming for high doping levels. The decrease in carrier concentration with increasing growth temperature can be explained as follows: As the growth temperature increases, the degree of thermal cracking of arsine increases, arsenic overpressure occurs, and the concentration of arsenic vacancies decreases, which reduces the concentration of substitutional vacancies for tellurium [13].

3.1.2. Influence of Thickness

Consider the results shown in Figure 5 and in rows b-1 and a-3 of Table 2. As the $In_{0.07}GaAs$ thickness increases from 11 to 16 nm, the peak current density increases from 1222 to 2130 A cm^{-2}, and the resistance decreases from 1.94×10^{-4} to 1.17×10^{-4} Ω cm^2.

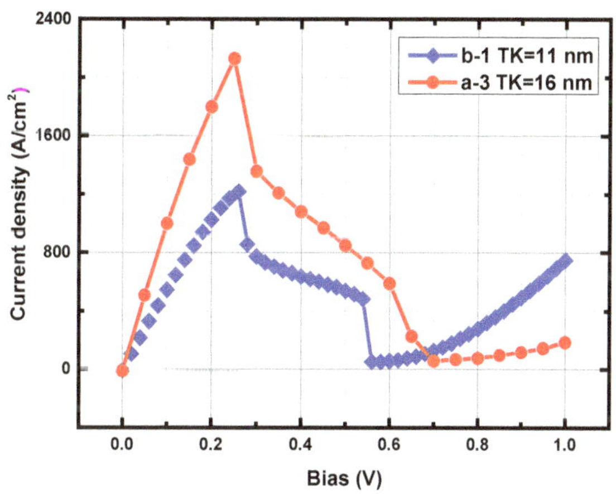

Figure 5. Experimental results of tunnel-junction devices under various thickness of the $In_{0.07}GaAs$ layer.

According to K. Louarn et al. [3], with a thick InGaAs layer, the depletion region expands because the *p*-side doping concentration exceeds that of the n side. Increasing the thickness of the *n*-side InGaAs layer reduces the band bending that extends up to the n-GaAs layer and increases the density of states of the direct band-to-band tunneling (DBBT) process [14], resulting in an increased peak tunneling current density because the tunneling probability is increased. With a thin InGaAs layer, a significant band offset occurs at the n^{++}-GaAs/n^{++}-InGaAs interface due to band misalignment, resulting in a quantized energy level due to the potential drop near the tunneling area. Quantum confinement reduces the density of states during the DBBT process, resulting in the reduction of the tunneling probability. Increasing the thickness of the InGaAs quantum well layer reduces the quantum confinement and weakens the discretization of states in the tunnel junction area.

3.1.3. Influence of V/III Ratio

Consider the results shown in Figure 6 and in rows c-1, c-2, and a-3 of Table 2. As the V/III ratio decreases from 74 to 5, the peak current density increases from 2130 to 5839 A cm^{-2}, and the resistance decreases from 1.17×10^{-4} to 5.86×10^{-5} Ω cm^2.

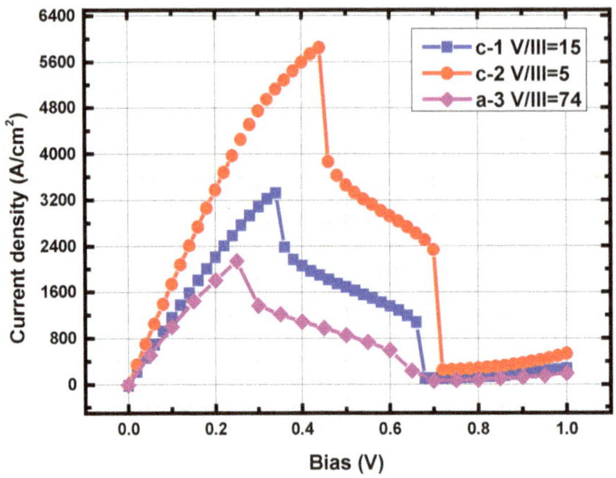

Figure 6. Experimental results of tunnel-junction devices under various V/III ratio for the $In_{0.07}GaAs$ layer.

Decreasing the V/III ratio in the arsine flow at a fixed dopant flow rate increases the probability of the Te substitution into arsenic vacancies due to the fact that Te atoms are n dopants. As shown in Table 2, a higher V/III ratio leads to a lower doping concentration in this layer, we see that lowest V/III ratio used (5), corresponding to a doping concentration up to 5.09×10^{19} cm^{-3} is reached, which is already above the maximum doping level of 1×10^{19} cm^{-3} achievable with Si dopant.

3.2. Role of Trap-Assisted Tunneling

The high peak tunneling current of this study cannot be explained as the usual carrier tunneling due to the doping levels and materials [15]. The contribution of the trap-assisted tunneling mechanism must be considered, which may enhance the carrier tunneling. The heavy doping of the InGaAs layer tends to create clusters [16], and the relaxation of the lattice mismatch between the InGaAs and GaAs layers promotes the formation of defects, which can act as traps in the trap-assisted tunneling (TAT) process [17].

The high peak current density is investigated by using a simulation model implemented in Crosslight APSYS software and that includes a direct band-to-band tunneling [18], bandgap narrowing, and trap-assisted tunneling. In the APSYS simulator, the tunneling probability is solved by Wenzel–Kramers–Brillouin (WKB) approximation and described as following [19]:

$$D = P_0 \exp(-E_\perp/\underline{E}) \quad (2)$$

$$J(E_\parallel) = \int_{x_1}^{x_2} \left| \left(\frac{2m^*}{\hbar^2}\right) \left[\left(\frac{E_g}{2}\right)^2 - (\varepsilon_c)^2 \right] / E_g + E_\perp \right|^{\frac{1}{2}} dx \quad (3)$$

$$P_0 = \exp\left[\frac{\pi m^{*\frac{1}{2}} (E_g)^{\frac{3}{2}}}{2(2)^{\frac{1}{2}} qF\hbar} \right] = \exp\left(-\frac{E_g}{4\underline{E}}\right) \quad (4)$$

$$\underline{E} = \frac{(2)^{\frac{1}{2}} qF\hbar}{2\pi m^{*\frac{1}{2}} \left(E_g\right)^{\frac{1}{2}}} \tag{5}$$

where $E_{\perp}$ and $E_{\|}$ are the electron kinetic energies in perpendicular and parallel to the tunneling direction, respectively, $\underline{E}$ is a measure of the significance of perpendicular momentum, P_0 is the tunneling probability with zero perpendicular momentum, m * is the effective tunneling mass.

With a highly doped InGaAs quantum well inserted in the GaAs tunnel junction, bandgap narrowing should be considered in the model [20]. The bandgap narrowing is expressed as

$$\Delta E_g = A\left\{ \ln \frac{N}{B} + \left[\left(\ln \frac{N}{B} \right)^2 + \frac{1}{2} \right]^{\frac{1}{2}} \right\} \tag{6}$$

where A and B are constants taken from Slotboom [21], and N is the dopant concentration.

The TAT model is based on the assumption that the traps are able to emit carriers and thus generate a current flux, which is expressed as [22]:

$$J = q\, N_{\text{trap}} \frac{d}{dF} (S_{\text{TAT}} f_T) \Delta V \tag{7}$$

where S_{TAT} is the emission rate, N_{trap} is the bulk trap density, F is the electrical field, f_T is a correction factor due to temperature, related to thermal activation of trapped carriers. The Poole-Frenkel model of field dependence is implemented [22], where the Poole-Frenkel shift of the trap level is expressed as:

$$\Delta E_{\text{PF}} = \sqrt{\frac{qF}{\pi \epsilon_0 \epsilon}} \tag{8}$$

To adapt the high peak tunneling current of the sample c-2, two free changeable parameters: the trap carrier lifetimes (t_n, t_p) and the Poole–Frenkel shift (ΔE_{PF}) are used to calibrate the simulation model. Figure 7 shows how the model be tuned to get a good agreement between the simulated and experimental results, including peak tunneling current values and the secondary peak values. With lower t_p and higher Poole-Frenkel shifts, the peak tunneling current values and secondary peak values increase. The simulation results are consistent with experimental data for trap carrier lifetimes of t_n = 0.45 × 10^{-4} s and t_p = 4 × 10^{-2} s, and with a Poole-Frenkel shift in the trap level of 1.41 eV, respectively. However, the simulation results are not consistent with the experimental data in the region of negative differential resistance (NDR) because of the instability of the secondary peak [23], which is probably caused by the "two-step" tunneling through the trap states [24]. However, Figure 8 shows that the peak tunneling current density decreases if the TAT effect is not taken into account, which indicates that the trap-assisted tunneling enhances the peak tunneling current values.

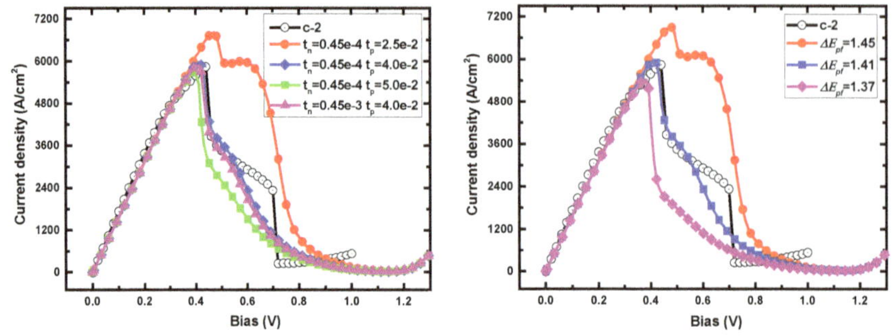

(a) (b)

Figure 7. Influence of (**a**) trap carrier lifetime and (**b**) Poole–Frenkel shift on simulated *J-V* curve. The experimental data are from sample c-2.

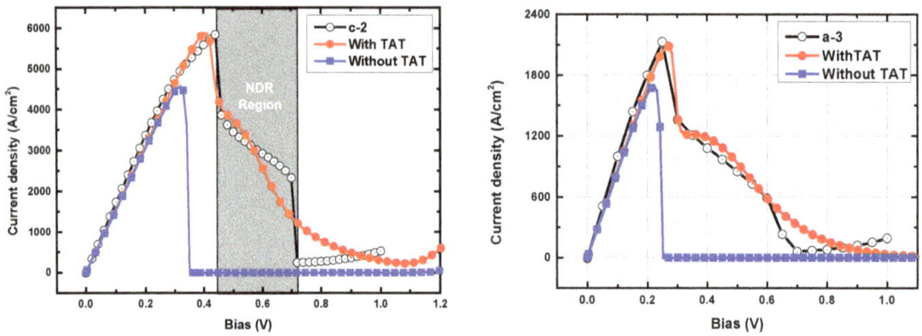

(a) (b)

Figure 8. Simulated *J-V* curve with and without using the trap-assisted tunneling (TAT) model. The experimental data is from the sample (**a**) c-2 and (**b**) a-3.

4. Conclusions

We implement Si and Te co-doped InGaAs quantum well inserts in a GaAs tunnel junction and obtain the result with a peak current density as high as 5839 A cm^{-2} and a series resistance of 5.86×10^{-5} Ω cm^2. The performance of the devices is investigated at various growth conditions of the InGaAs layer, including the growth temperature, thickness, and V/III ratio. A simulation model is used to investigate the performance of the devices, and the results indicate that the trap-assisted tunneling contributes to increasing the peak tunneling current.

Author Contributions: Conceptualization, Y.G. (Yudan Gou) and J.W.; methodology, Y.G. (Yudan Gou); software, Y.G. (Yudan Gou); validation, Y.G. and Y.G. (Yudan Gou); formal analysis, Y.G. (Yudan Gou); investigation, Y.G. (Yudan Gou), X.X., S.Z., L.Z. and J.W.; resources, J.W.; data curation, Y.G. (Yudan Gou); writing—original draft preparation, Y.G. (Yudan Gou); writing—review and editing, G.D., H.Y., H.L., S.T., Y.C., Y.G. (Yintao Guo), X.X., S.Z.; visualization, Y.G. (Yudan Gou) and J.W.; supervision, J.W.; project administration, Y.G. (Yudan Gou) and J.W. All authors have read and agreed to the published version of the manuscript.

Funding: This research received no external funding.

Acknowledgments: The authors thank Sheng Yang and Xia Changsheng from the Crosslight software Inc. for helpful discussions regarding this work.

Conflicts of Interest: The authors declare no conflict of interest.

References

1. Kabir, E.; Kumar, P.; Kumar, S.; Adelodun, A.A.; Kim, K.H. Solar energy: Potential and future prospects. *Renew. Sustain. Energy Rev.* **2018**, *82*, 894. [CrossRef]
2. Geisz, J.F.; France, R.M.; Schulte, K.L.; Steiner, M.A.; Norman, A.G.; Guthrey, H.L.; Young, M.R.; Song, T.; Moriarty, T. Six-junction III–V solar cells with 47.1% conversion efficiency under 143 Suns concentration. *Nat. Energy* **2020**, *5*, 326. [CrossRef]
3. Louarn, K.; Claveau, Y.; Fontaine, C.; Arnoult, A.; Marigo-Lombart, L.; Massiot, I.; Piquemal, F.; Bounouh, A.; Cavassilas, N.; Almuneau, G. Thickness Limitation of Band-to-Band Tunneling Process in GaAsSb/InGaAs Type-II Tunnel Junctions Designed for Multi-Junction Solar Cells. *ACS Appl. Energy Mater.* **2019**, *2*, 1149. [CrossRef]
4. Wheeldon, J.F.; Valdivia, C.E.; Walker, A.W.; Kolhatkar, G.; Jaouad, A.; Turala, A.; Riel, B.; Masson, D.; Puetz, N.; Fafard, S.; et al. Performance comparison of AlGaAs, GaAs and InGaP tunnel junctions for concentrated multijunction solar cells. *Prog. Photovolt.* **2011**, *19*, 442. [CrossRef]
5. Northrup, J.E.; Zhang, S.B. Dopant and defect energetics: Si in GaAs. *Phys. Rev. B* **1993**, *47*, 6791. [CrossRef] [PubMed]
6. Hermle, M.; Létay, G.; Philipps, S.P.; Bett, A.W. Numerical simulation of tunnel diodes for multi-junction solar cells. *Prog. Photovolt.* **2008**, *16*, 409. [CrossRef]
7. Bedair, S.; Carlin, C.Z.; Harmon, J.L.; Sayed, I.E.H.; Colter, P. High Performance Tunnel Junction with Resistance to Thermal Annealing. In Proceedings of the 12th International Conference on Concentrator Photovoltaic Systems (CPV-12), Freiburg, Germany, 25–27 April 2016; p. 020003.
8. Lumb, M.P.; Yakes, M.K.; Gonzalez, M.; Vurgaftman, I.; Bailey, C.G.; Hoheisel, R.; Walters, R.J. Double quantum-well tunnel junctions with high peak tunnel currents and low absorption for InP multi-junction solar cells. *Appl. Phys. Lett.* **2012**, *100*, 213907. [CrossRef]
9. Bedair, S.; Harmon, L.J.; Carlin, C.Z.; Hashem, I.E.; Colter, P.C. High performance as-grown and annealed high band gap tunnel junctions: Te behavior at the interface. *Appl. Phys. Lett* **2016**, *108*, 203903. [CrossRef]
10. Louarn, K.; Claveau, Y.; Marigo-Lombart, L.; Fontaine, C.; Arnoult, A.; Piquemal, F.; Bounouh, A.; Cavassilas, N.; Almuneau, G. Effect of low and staggered gap quantum wells inserted in GaAs tunnel junctions. *J. Phys. D Appl. Phys.* **2018**, *51*, 145107. [CrossRef]
11. Law, J.J.M.; Carter, A.D.; Lee, S.; Huang, C.Y.; Lu, H.; Rodwell, M.J.W.; Gossard, A.C. Co-doping of InxGa1-xAs with silicon and tellurium for improved ultra-low contact resistance. *J. Cryst. Growth* **2013**, *378*, 92. [CrossRef]
12. Nguyen, X.S.; Yadav, S.; Lee, K.H.; Kohen, D.; Kumar, A.; Made, R.I.; Lee, K.E.K.; Chua, S.J.; Gong, X.; Fitzgerald, E.A. MOCVD growth of high quality InGaAs HEMT layers on large scale Si wafers for heterogeneous integration with Si CMOS. *IEEE Trans. Semicond. Manuf.* **2017**, *30*, 456. [CrossRef]
13. Hamon, G.; Paillet, N.; Alvarez, J.; Larrue, A.; Decobert, J. Te doping of GaAs and GaInP using diisopropyl telluride (DIPTe) for tunnel junction applications. *J. Cryst. Growth* **2018**, *498*, 301. [CrossRef]
14. Louarn, K. Development of Tunnel Junctions Based on III-V Semiconductors Heterostructures for High Efficiency Multi-Junction Solar Cells. Ph.D. Thesis, Paul Sabatier University, Toulouse, France, 2018.
15. Jandieri, K.; Baranovskii, S.D.; Rubel, O.; Stolz, W.; Gebhard, F.; Guter, W.; Hermle, M.; Bett, A.W. Resonant electron tunneling through defects in GaAs tunnel diodes. *J. Appl. Phys.* **2008**, *104*, 094506. [CrossRef]
16. Park, Y.; Qian, W.; Skowronski, M. Precipitation of Tellurium-Rich Phase in Heavily Diethyltellurium-Doped GaAs during Organometallic Vapor Phase Epitaxy. *J. Electrochem. Soc.* **1995**, *142*, 4294. [CrossRef]
17. García, I.; Rey-Stolle, I.; Algora, C. Performance analysis of AlGaAs/GaAs tunnel junctions for ultra-high concentration photovoltaics. *J. Phys. D Appl. Phys.* **2012**, *45*, 045101.
18. Crosslight Software Inc., Canada. Available online: http://www.crosslight.com/ (accessed on 20 October 2019).
19. Li, Z.Q.; Xiao, Y.G.; Li, Z.M.S. Modeling of multi-junction solar cells by Crosslight APSYS. In *High and Low Concentration for Solar Electric Applications*; International Society for Optics and Photonics: San Diego, CA, USA, 2006; p. 633909.
20. Colter, P.; Hagar, B.; Bedair, S. Tunnel junctions for III-V multijunction solar cells review. *Crystals* **2018**, *8*, 445. [CrossRef]

21. Slotboom, J.W.; Degraaff, H.C. Measurements of bandgap narrowing in Si bipolar transistors. *Solid-State Electron.* **1976**, *19*, 857. [CrossRef]
22. Gehring, A. Simulation of Tunneling in Semiconductor Devices. Ph.D. Thesis, Technical University, Vienna, Austria, 2003.
23. Wheeldon, J.F.; Valdivia, C.E.; Walker, A.; Kolhatkar, G.; Hall, T.J.; Hinzer, K.; Masson, D.; Fafard, S.; Jaouad, A.; Turala, A. AlGaAs tunnel junction for high efficiency multi-junction solar cells: Simulation and measurement of temperature-dependent operation. In Proceedings of the 34th IEEE Photovoltaic Specialists Conference (PVSC), Philadelphia, PA, USA, 7–12 June 2009; pp. 106–111.
24. Narasimhan, V.K.; Yastrebova, N.; Valdivia, C.E.; Hall, T.J.; Hinzer, K.; Masson, D.; Fafard, S.; Jaouad, A.; Arès, R.; Aimez, V. Effect of parameter variations on the current-voltage behavior of AlGaAs tunnel junction models. In Proceedings of the 1st Microsystems and Nanoelectronics Research Conference, Ottawa, ON, Canada, 15 October 2008; p. 165.

crystals

MDPI

Article

Deuteration Effects on the Transport Properties of $(TMTTF)_2X$ Salts

Andrea Rohwer [1], Martin Dressel [1,*] and Toshikazu Nakamura [2]

1 1. Physikalisches Institut, Universität Stuttgart, Pfaffenwaldring 57, 70569 Stuttgart, Germany; andrea.rohwer@pi1.physik.uni-stuttgart.de

2 Institute for Molecular Science, Myodaiji, Okazaki 444-8585, Japan; t-nk@ims.ac.jp

* Correspondence: dressel@pi1.physik.uni-stuttgart.de; Tel.: +49-711-685-64946

Received: 20 October 2020; Accepted: 24 November 2020; Published: 27 November 2020

Abstract: The electronic properties in the quasi-one-dimensional Fabre salts are strongly affected by electronic correlations along the molecular stacks, but also by the interactions with the anions located in a cage that is formed by the methyl end groups. We systematically compare the charge transport in deuterated and protonated $(TMTTF)_2X$ salts with the anions X = Br, PF_6, SbF_6, and ClO_4, ranging from Mott and Efros–Shklovskii variable-range hopping to activated band transport with a temperature dependent energy gap. The strong dependence of charge localization and ordering on the anion size and deuteration confirms the subtle structural involvement of the anions in the charge transport along the TMTTF stack.

Keywords: charge-transfer salts; $(TMTTF)_2X$; deuteration; anions; charge transport

1. Introduction

Fascinated by the wealth of exciting physical phenomena summarized in the temperature-pressure phase landscape plotted in Figure 1, the Fabre salts $(TMTTF)_2X$ here TMTTF denotes tetramethyl-tetrathiafulvalene and X is a monovalent anion—have been intensely investigated since their discovery in the 1970s [1–3].

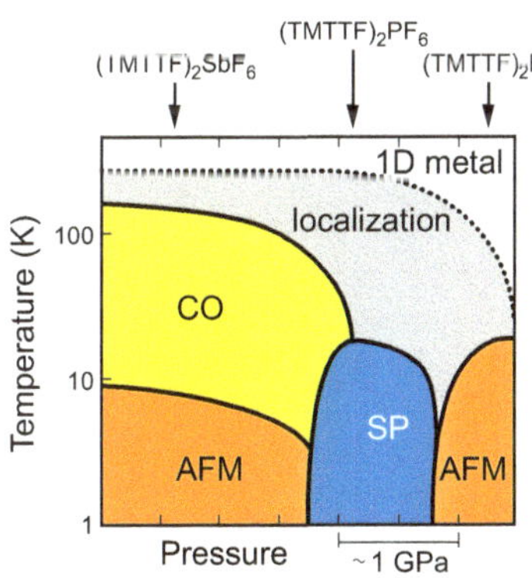

Figure 1. Phase diagram of the $(TMTTF)_2X$ family. The ambient pressure positions of the charge-transfer salts with centrosymmetric anions X = SbF_6, PF_6, and Br are indicated by arrows. 1D metal denotes the quasi-one-dimensional metallic properties, CO the charge-ordered phase, AFM stands for antiferromagnetism and SP for spin-Peierls state.

Under ambient conditions, these salts possess a quasi-one-dimensional metallic state, where the large anisotropy of their electronic properties is based in their stacked molecular structure.

With decreasing temperature, the localization of charges takes place over a broad temperature range—a crossover rather than a phase transition. Salts with larger anions, like X = SbF_6 and TaF_6, exhibit a charge-ordering phase transition at T_{CO} below the localization temperature T_ρ. At much lower temperatures $T_N < 10$ K, these compounds undergo a transition to an antiferromagnetic ground state. For salts with smaller anions, e.g., X = PF_6 and AsF_6, a weaker charge order is observed at temperatures between 100 and 50 K [4]. In addition, these compounds exhibit a spin-Peierls ground state below 20 K. For compounds with even smaller anions, such as, X = Br, no clear charge ordering can be observed; albeit there are indications that weak charge disproportionation takes place around 50 K [5,6]. Here again, antiferromagnetic ordering occurs at low temperatures. Salts containing tetrahedral anions, such as ClO_4, ReO_4 and BF_4, also exhibit some ordering of the anions; for that reason, they are not included in Figure 1. This structural phase transition normally occurs at temperatures T_{AO} well below the charge-ordering temperature [7–9]. In $(TMTTF)_2ClO_4$, however, no charge ordering is observed at all; but the anion transition T_{AO} occurs below T_ρ.

As indicated in the phase diagram (Figure 1), the transitions (going from left to right) can be induced by increasing the external pressure by a few GPa. A similar effect can be seen when 'internal pressure' is increased, i.e., by reducing the anion size. Why this is so can be better understood when considering the molecular structure of these salts depicted in Figure 2: The anions, which are segregated between the TMTTF stacks, are positioned in a cage formed by the methyl groups terminating the TMTTF molecules. In a first approximation, the size of the anions determines the separation of the molecular stacks. The degree of this anion confinement in the methyl end-group cage also affects the anion ordering in the tetrahedral anion salts. Even more important, the anion-stack interaction enhances the dimerization within the TMTTF stacks and consequently suppresses charge order [10]. In other words, how freely the anion can move within the confined space affects the anisotropic charge transport in the salt as well as how charges localize and order with decreasing temperature down to their antiferromagnetic and spin Peierls ground states [11].

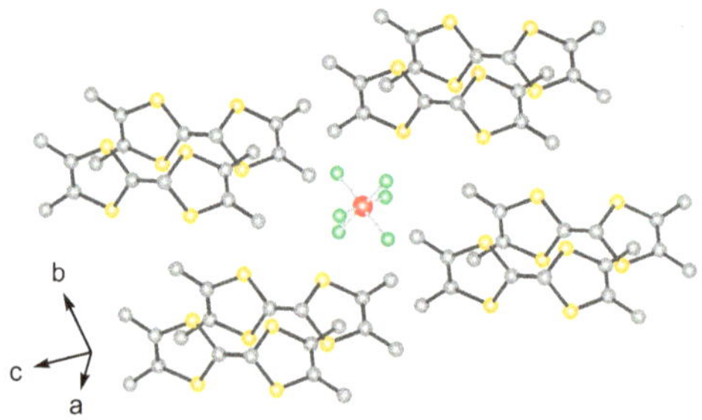

Figure 2. Molecular structure model of the $(TMTTF)_2X$ salt. As a representative example, an octahedral salt, e.g., $(TMTTF)_2PF_6$, is depicted here. All Fabre salts, containing octahedral, tetrahedral, or mono-atomic anions, have this triclinic crystal structure. The crystal structure is the same for undeuterated and deuterated samples investigated in this study.

In order to study these interactions—and more specifically, the effect of confinement of the anion within its methyl end-group cage—we compare protonated with deuterated Fabre salts: $(TMTTF)_2X$, for anions X = Br, PF_6, SbF_6, and ClO_4. When the hydrogen atoms in the CH_3 group are exchanged with the heavier deuterium isotope, the dynamics and interaction between the anion and its molecular neighbors are modified. A close comparison of the dc transport results obtained for deuterated and protonated samples should, therefore, provide valuable information on the electronic properties and elucidate further details of this intricate interplay.

In the present paper—after providing details of the experiments in Section 2—we present and discuss the results in Section 3. In the subsequent Section 4, the data are analyzed further with respect to various charge transport mechanisms.

2. Experimental Details

Single crystals of deuterated and protonated $(TMTTF)_2X$ with X = Br, PF_6, SbF_6, ClO_4 were grown electrochemically in an H-type glass cell under ambient conditions, as described in Reference [7]. Needle shaped crystals, ~1–2 mm long, ~100 µm wide, and less than 100 µm thick, were harvested after several weeks. Here, we denote the hydrogenated $(H_{12}C_{10}S_4)_2X$ as $(H_{12}\text{-}TMTTF)_2X$ and the deuterated analogue $(D_{12}C_{10}S_4)_2X$ is labelled $(D_{12}\text{-}TMTTF)_2X$.

Direct current (dc) resistivity measurements were performed across a temperature range between 300 K and 4 K, at a ramp rate of $|0.3\ \text{K/min}|$ in a glass cryostat using liquid nitrogen and helium for cooling, while the sample chamber was flooded with helium gas to ensure thermal contact between the sample and the cryogenic chambers. The dc resistivity was measured along the molecular stacking direction, i.e., the long needle axis, referred to as the a-axis of the crystal (see Figure 2). In order to perform four-point measurements, four gold wires were attached to the crystal needle with carbon paste. Samples were anchored on a sapphire plate to ensure good thermal contact and electrical insulation from the cryostat's sample chamber. Data were collected while cooling down as well as while warming up. A Keithley 2612B system source meter and a Keithley 2182 nanovoltmeter were used to perform resistivity measurements. In order to protect the sample from overheating, the current was set to a constant value; the voltage was ramped across four linearly spaced points around 0 V to check the samples' Ohmic behavior when recording data at a certain temperature.

While the anisotropy of the Fabre salts is interesting by itself, and was elaborated upon in the previous study by Köhler et al. [7], here we focus on measurements along the a-axis, where the data quality is superior and most of the relevant information can be extracted.

3. Results and Discussion

The recorded dc resistivity versus temperature, $\rho(T)$ is plotted in Figure 3 for $(TMTTF)_2X$, with the anions X = Br, PF_6, SbF_6, ClO_4, where in each of the panels, the respective protonated (black open squares) and deuterated (red solid diamonds) crystals are compared. All show an overall increase in resistivity upon cooling. The behavior can be typically traced up to the order of 10^6 Ωcm; beyond that bound, sample heating and non-Ohmic properties prevent reliable data acquisition. In the case of the rather good conductor $(TMTTF)_2Br$, data are recorded down to $T = 4$ K. For the salts with larger anions, the resistivity measurements become non-Ohmic already below 20 K or even higher temperatures. The resistivity curve of $(TMTTF)_2Br$ lies below its deuterated analogue, while for the other salts, the opposite behavior is observed. Arrows mark the charge order (CO) transition for the $(TMTTF)_2Br$, $(TMTTF)_2PF_6$ and $(TMTTF)_2SbF_6$ salts, and the anion order (AO) transition in the $(TMTTF)_2ClO_4$ samples (grey arrows for protonated, orange for deuterated crystals).

Table 1 summarizes all phase transition temperatures for the compounds under study. Upon being cooled down from room temperature, the samples exhibit metallic behavior, i.e., the resistivity decreases slightly with lowering temperature. Charges gradually localize and the metallic conductivity freezes: T_ρ is the localization temperature which occurs as a broad global minimum in the temperature dependent resistivity curves (Figure 3) and is defined as the temperature where the energy gap $\Delta(T) = 0$ in Figure 8 below. The change in localization temperature upon deuteration (see Table 2), i.e., $\Delta T_\rho = T_\rho^{\rm H} - T_\rho^{\rm D}$, ranges from tens to slightly more than one-hundred Kelvin. T_ρ increases upon deuteration for the Br, PF_6, and SbF_6 salts, but decreases in $(TMTTF)_2ClO_4$. Because this localization minimum is so shallow and broad, the uncertainties are rather large, as is reflected in the substantially differing T_ρ values reported in literature [7,12].

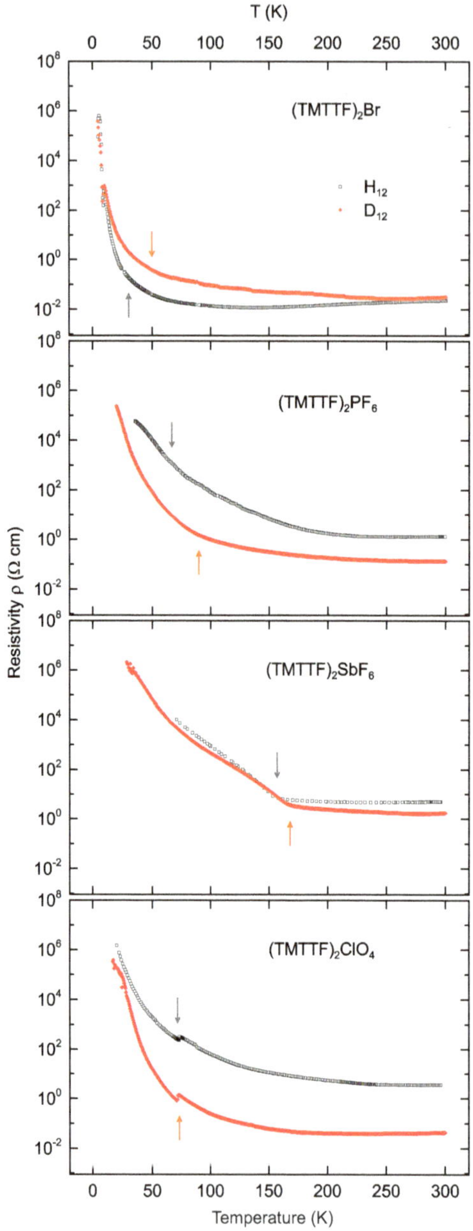

Figure 3. Temperature dependent dc resistivity $\rho(T)$ measured along the *a*-axis of deuterated (D_{12}) (red solid diamonds) and protonated (H_{12}) (black open squares) $(TMTTF)_2X$ salts (X = Br, PF_6, SbF_6, ClO_4). Note the logarithmic vertical axes. Arrows mark the charge ordering [for $(TMTTF)_2Br$, $(TMTTF)_2PF_6$, $(TMTTF)_2SbF_6$] and anion ordering [for $(TMTTF)_2ClO_4$] transitions for the deuterated (orange) and protonated (grey) samples.

Table 1. Phase transition temperatures T_ρ, T_{CO}, T_{AO}, T_{SP}, T_N, for H_{12}- and (D_{12}-TMTTF)$_2X$, with X = Br, PF_6, SbF_6, ClO_4, as determined from the dc resistivity data plotted in Figure 3. Note that some temperatures are better identified in the alternative presentation displayed in Figure 8. Since the low-temperature magnetic phase transitions for the Br, PF_6, and SbF_6 salts could not be resolved in our measurements, literature values are supplied for T_{SP} of H_{12}- and (D_{12}-TMTTF)$_2PF_6$ compounds from References [7,13], as well as for T_N of the (TMTTF)$_2$Br and (TMTTF)$_2SbF_6$ salts as they appear in the last column, from References [11,13–18].

Anions		Transition Temperatures (K)				
X		T_ρ	T_{CO}	T_{AO}	T_{SP}	T_N
Br	H_{12}	130	30			13
	D_{12}	260	50			19
PF_6	H_{12}	260	67		19	
	D_{12}	280	90		13	
SbF_6	H_{12}	240	157			6
	D_{12}	280	168			8
ClO_4	H_{12}	290		73.1		
	D_{12}	230		71.9		

Table 2. Changes observed in transition temperatures, ΔT, upon deuteration of the (TMTTF)$_2X$ salts, calculated from values given in Table 1 using $\Delta T = T_*^H - T_*^D$ for the respective quantities. In the case of resistivity minimum T_ρ, the error bar may reach ±25 K.

Anions	Change in Transition Temperature (K)				
X	ΔT_ρ	ΔT_{CO}	ΔT_{AO}	ΔT_{SP}	ΔT_N
Br	−130	−20			−6
PF_6	−20	−23		6	
SbF_6	−40	−11			2
ClO_4	60		1.2		

For most Fabre salts, a transition to a $4k_F$ charge ordered state occurs at T_{CO}. In our compounds containing centrosymmetric anions, this transition is visible as a steepening in the slope of the resistivity curve observed upon cooling, and is especially strong in (TMTTF)$_2SbF_6$, while being less pronounced in (TMTTF)$_2PF_6$ and (TMTTF)$_2$Br, see Figure 3. This observation is in agreement with the results previously reported [7]. For (TMTTF)$_2PF_6$ and (TMTTF)$_2$Br the transitions in deuterated and protonated crystals are difficult to infer from the raw data $\rho(T)$, but can be identified in the temperature dependent energy gap plotted in Figure 8. The CO transition temperatures, as marked in our data, are listed in Table 1 [6,7,12–14]. T_{CO} increases upon deuteration, as seen in Table 2. The enhancement of the CO transition temperature is largest for small anions [10]. This can be explained in terms of the competition between CO and intrastack dimerization: with the latter decreasing upon deuteration, the CO transition shifts to higher T_{CO} [15].

Non-centrosymmetric anions can arrange in a regular alternation pattern; in the case of (TMTTF)$_2ClO_4$ a clear anion ordering is observed at T_{AO} = 73 K. During this transition, a $2k_F$ superstructure forms along the stacking direction [12], which results from the orientation ordering of the tetrahedral anions in their TMTTF methyl cavity. Just to be clear, the anion order is mediated via the TMTTF stacks and not a direct interaction. For that reason, the coupling via the methyl groups is crucial. A zoomed-in view of the resistivity curves around this transition is presented in Figure 4. Here the cooling (solid symbols) and heating (open symbols) curves of the protonated (black symbols, top curve) and deuterated (red symbols, bottom curve) salts are plotted on a logarithmic resistivity and linear temperature axis. The hysteresis of this transition is more pronounced in the protonated sample ($T_{AO,H}^{warming} - T_{AO,H}^{cooling} = 1$ K) than in the deuterated sample ($T_{AO,D}^{warming} - T_{AO,D}^{cooling} = 0.4$ K). The hysteresis is well known from salts with non-centrosymmetric anions [7–9] and is an indicator of the first order nature of the AO phase transition. The broader hysteresis implies a stronger interaction of the anions with the caging methyl groups. In other words, if no hysteresis is present, there is basically no potential barrier preventing the phase transition. However, since the difference in hysteresis between the deuterated and protonated samples lies within the error margins of the experiment, it will not be further analysed here [19]. The other striking characteristic of the AO transition is the downward (upward) jump in resistivity upon cooling (heating). The tiny variation in T_{AO} brought about by deuteration ($\Delta T_{AO} = 1.2$ K)—as summarized in Table 2—is about the uncertainty of this measurement. This agrees with previous experiments that could not resolve any difference upon deuteration [10]. This observation indicates that the interaction between the anion and TMTTF stacks depends, to a lesser extent, on the hydrogen bonds of the methyl groups (indicated by the dotted lines and red left arrow in Figure 5), and must therefore depend more strongly on the Coulomb interaction between the anion ligands and the TMTTF molecules' sulphur atom (green downward arrow) [10,11].

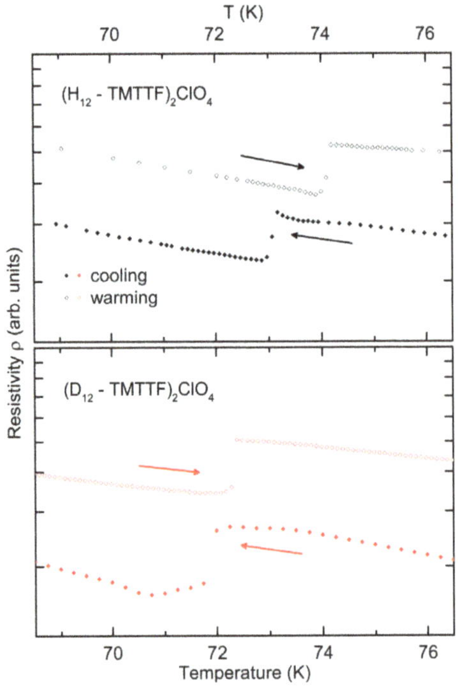

Figure 4. Anion transition of the protonated (black symbols, top panel) and deuterated (red symbols, bottom panel) $(TMTTF)_2ClO_4$ crystals. Cooling (solid symbols, indicated by left arrow) and warming (open symbols, indicated by right arrow) are plotted to show the hysteresis. Note the logarithmic resistivity axis and linear temperature axis.

The structural aspects of the charge and anion ordering transitions have been discussed in detail by Pouget [11,12,20]. Figure 5 illustrates the processes involved in stabilizing the states. Charge order in salts containing centrosymmetric anions (PF_6, SbF_6, and possibly Br) are most likely stabilized via the deformation of the methyl end group cavity—the process marked by the red left arrow. In this process, the displacement of the anion, X, polarizes the hydrogen bond network inside the cavity. This, in turn, causes a displacement of the charge on the σ-bonds connected to the hydrogen bonds, shifting the σ-electrons towards the center of the TMTTF molecules, which stabilizes the excess π-holes on the molecules (indicated by the grey ovals in Figure 5). This hypothesis is sustained by the fact that T_{CO} rises with larger anion size and upon deuteration—in both cases, the contact between the anion and its cavity periphery, i.e., the methyl end groups, increases. The second process proposed in reference [11], indicated by the green downward arrow in Figure 5, is that the anion moves towards the sulphur atom of the TMTTF molecule. This directly shortens the anion–sulphur contact distance and would also stabilize the enhanced π-hole density of the TMTTF molecule towards which the anion moves (different π-hole distribution than for the first process; not indicated in Figure 5). This second process is not substantiated experimentally for the charge-order transition, since structural refinement provides no evidence of a sizeable change in the anion-sulphur distance below T_{CO} for the centrosymmetric anion salts [11]. For the anion ordering transition, however, this second process seems to be a likely driving force in most tetrahedral anion salts, even though only very weak changes in the anion-sulphur distance have been reported below T_{AO} for $(TMTTF)_2ClO_4$ [11].

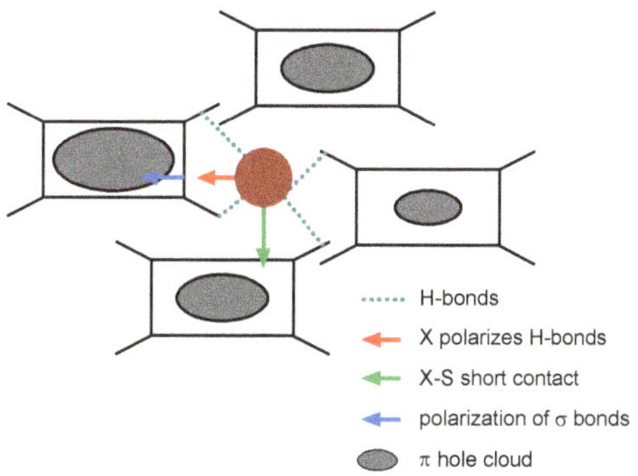

Figure 5. A schematic representation of the molecular structure of $(TMTTF)_2X$ salts. The rectangles represent the TMTTF molecules with their methyl end groups (four 'legs' sticking out). The latter form a cage between the organic molecules containing the anion, X (filled dark red sphere in the diagram). Depending on the size of the anion and the methyl end-groups (whether they are deuterated or protonated) this cage forms a spacious or a more confined trap for the anion to move about in. The blue dotted lines represent the hydrogen bonds forming between the anion and the methyl end groups. Figure adapted from Pouget [11].

Tables 1 and 2 list the transition temperatures and the difference resulting from deuteration for the spin-Peierls transitions T_{SP} in $(TMTTF)_2PF_6$ and the antiferromagnetic transitions T_N in $(TMTTF)_2SbF_6$ and $(TMTTF)_2Br$ samples as well, obtained by magnetic probes [18,21]. From our transport experiments on both protonated and deuterated $(TMTTF)_2Br$ down to $T = 4$ K, we can access the antiferromagnetic transitions located at $T_N = 13$ K and 19 K, respectively [11,14]. Figure 6 zooms into this low-temperature region of the warming (open symbols) and cooling (solid symbols) $\rho(T)$ curves of the protonated (black symbols, top panel) and deuterated (red symbols, bottom panel) Br-salts. T_N, as reported in Table 1, is marked by the grey (orange) arrow for the protonated (deuterated) crystal. The antiferromagnetic transition cannot be distinguished for the $(H_{12}\text{-TMTTF})_2Br$ compound, while there seems to be a slight bump in the warming curve of $(D_{12}\text{-TMTTF})_2Br$ at 19 K. This, however, does not exceed the experimental uncertainty sufficiently and should therefore not be interpreted as strong indication of the antiferromagnetic transition. Temperature dependent resistivity measurements performed by Tomić et al. have revealed anomalies at $T = 12.5$ K and 22 K in protonated $(TMTTF)_2Br$ [22]. Our cooling curves in Figure 6 show no clear evidence of an anomaly around 22 K. However, between 10 K and 7 K, the cooling curves of both crystals exhibit a pronounced dip with a minimum at 9 K (8 K) for the protonated (deuterated) salt. The underlying cause for this pronounced anomaly is as yet unclear.

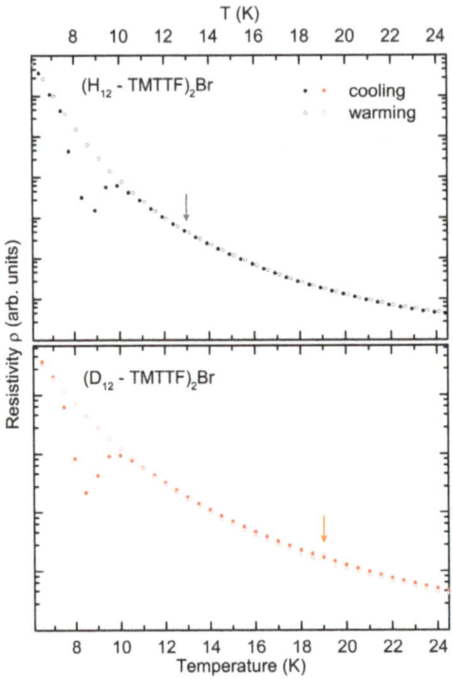

Figure 6. Resistivity $\rho(T)$ in the low-temperature regime of the protonated (black symbols, top panel) and deuterated (red symbols, bottom panel) of the $(TMTTF)_2Br$ compounds: cooling (solid symbols) and warming (open symbols). Arrows indicate the AFM transition temperatures obtained by magnetic probes.

4. Analysis of the Transport Mechanisms

In the following, we analyze and discuss the charge-transport mechanisms and the temperature dependent energy gap, respectively.

4.1. Charge Transport Mechanisms

There are a variety of models to describe the charge transport in organic compounds, ranging from the classical band transport to various hopping transport mechanisms. The underlying mechanism of band transport involves a delocalized molecular wavefunction over the entire volume of the crystalline sample with mean free paths larger than the inter-site distance—usually occurring in highly ordered systems. Hopping transport, on the other hand, is described by transitions between localized sites via tunneling or overcoming potential barriers, the mean free path being on the order of the inter-site distance, and mostly governs electronic properties in more disordered systems. Since the appropriate models depend on the concentration of localized states, carrier density, and electric field strength (among other factors), charge transport within the same sample may have to be described using different models, depending on the temperature range and applied field.

Band transport is applicable for high carrier densities and described by the Arrhenius model, based on an activated temperature dependence of the charge carrier concentration. A linear slope on a graph of logarithmic resistivity versus inverse temperature (left column in Figure 7) indicates band transport in the respective temperature range. Explicitly, the Arrhenuis law is given by

$$\rho(T) = \rho_0 \exp\left\{\frac{\Delta}{T}\right\}, \tag{1}$$

where the energy gap Δ is assumed to be temperature independent and ρ_0 is a constant prefactor [23]. This thermally activated transport was mostly observed along the c-axis of the crystal [7], i.e., perpendicular to the axis investigated here.

Variable range hopping is a phonon-activated process where charge carriers hop between localized states. This effect can be described in terms of a general formula proposed by Mott:

$$\rho(T) \propto \exp\left\{ \left(\frac{T_0}{T} \right)^{1/\gamma} \right\}, \tag{2}$$

where $\gamma = d + 1$ is related to the dimension d of the system [24]. Mott variable-range hopping in three dimensions (right column in Figure 7) assumes a constant density of states. Efros and Shklovskii showed that, at sufficiently low temperatures, the density of states near the Fermi energy is not constant, but vanishes linearly [25–27], thus accounting for the so-called Coulomb gap, and resulting in a resistivity dependence of Equation (2) with $\gamma = 2$ (central column in Figure 7) in all dimensions—this transport regime being referred to as Efros–Shklovskii variable-range hopping [7,28,29]. For some samples, the disorder may be so high that the Coulomb gap is dominant at all temperatures, leading to overall Efros–Shklovskii hopping. In samples displaying relatively low disorder, on the other hand, the energy scale is such that the carriers may have enough energy to overcome the Coulomb gap at all measurable temperatures, meaning that the density of states is practically constant, leading to overall Mott variable-range hopping [25]. At intermediate disorder, it may be possible to see a crossover from Mott to Efros–Shklovskii hopping with decreasing temperatures in the same samples [30]. Köhler et al. found three-dimensional Mott hopping transport to dominate below 60 K along the stacking axis [7], i.e., the orientation investigated in the present study.

For data analysis, the resistivity was plotted on a logarithmic y-axis versus the inverse temperature, i.e., T^{-1}, as well as $T^{-1/2}$, and $T^{-1/4}$, in Figure 7. A linear slope in the Arrhenius plot would mean that band transport is valid across the applicable temperature range. Similarly, a linear slope in the $\ln\{\rho\}$ versus $T^{-1/2}$ plot indicates Efros–Shklovskii variable-range hopping, while a linear slope in the $\ln\{\rho\}$ versus $T^{-1/4}$ plot is an indication of three-dimensional hopping. Figure 7 compares all three transport models (from left to right) for all four Fabre salts under study (from top to bottom) with protonated and deuterated molecules (open black and solid red symbols, respectively). The arrows indicate the transition temperatures: the localization temperature T_ρ being the left most arrows in each panel, and the ordering transitions T_{CO} or T_{AO} being the right arrows (all other transition temperatures were omitted for clarity). The bottom axes are linear in T^{-1}, $T^{-1/2}$, and $T^{-1/4}$, from left to right, while the top axes indicate the converted temperature in Kelvin. Linear fits to the slopes were performed for the temperature ranges 300 K to T_ρ, T_ρ to T_{CO} or T_{AO}, and temperatures below T_{CO} or T_{AO}. The results are summarized in Table A1 (see Appendix A).

The range above T_ρ shows metallic behavior, i.e., the resistivity decreases upon cooling, and none of the above models apply. Below the localization temperature, all samples show an upturn in resistivity, indicating semiconducting behavior, which can be described by activated band transport or variable-range hopping.

In the range between T_ρ and T_{AO}, the protonated $(TMTTF)_2Br$ salt exhibits a slight kink around $T = 60$ K. Above and below this kink, straight lines could be fitted over similar temperature ranges for all three transport models; hence, no firm conclusion can be drawn from the data about the dominant transport mechanism in this sample. For the deuterated compound of the Br-salt, the Efros–Shklovskii variable-range hopping model could be fitted across the largest temperature range, 255 K–54 K. This may be taken as an indication that the system is rather disordered, as mentioned in the discussion above. The deuterated compound shows a kink around 25 K, in the range between T_{CO} and T_N, which is close to the 22 K anomaly reported by Tomić et al. [22].

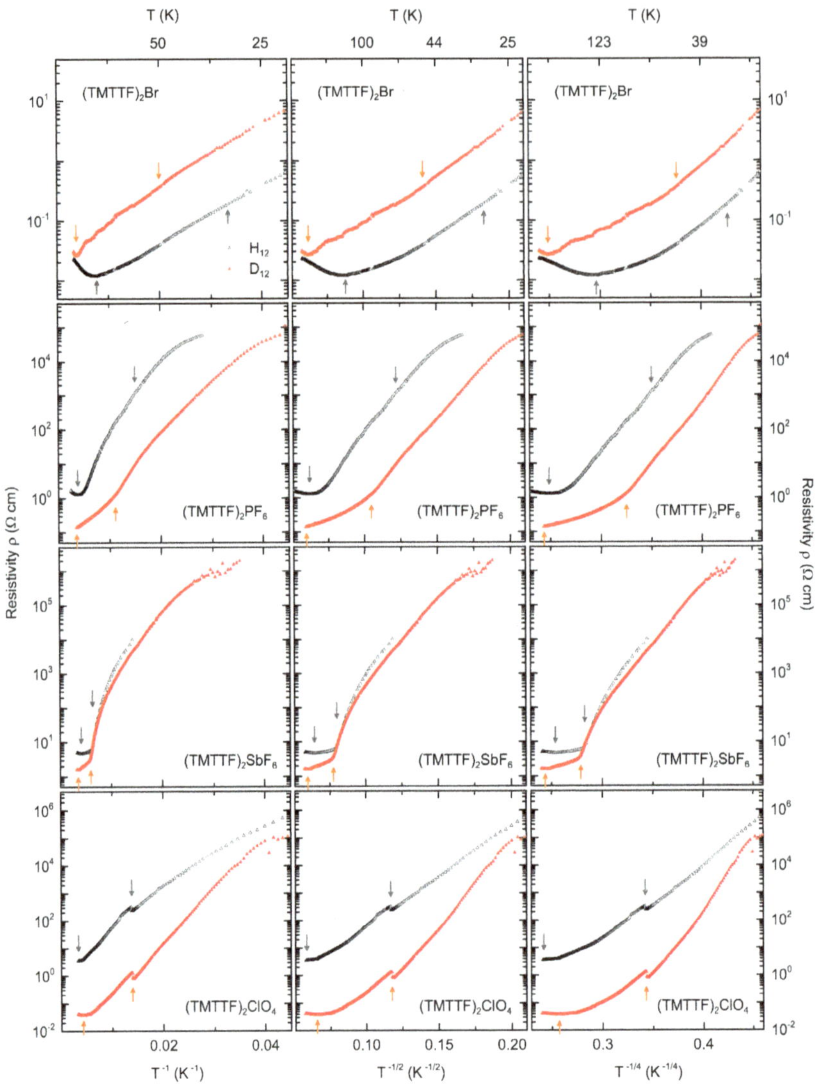

Figure 7. Comparison between three transport models, namely, Arrhenius (left column), three-dimensional variable-range hopping with Coulomb gap considered according to Efros and Shklovskii (central column) and three-dimensional Mott variable-range hopping without electronic correlations (right column), for all four protonated (shown by black symbols) and deuterated (red) salts (top to bottom). A given model is relevant in regions where the corresponding curves are linear. Supplementary data regarding the linear fits are provided in Table A1 (see Appendix A).

In $(H_{12}\text{-TMTTF})_2PF_6$, the resistivity follows a behavior corresponding to three-dimensional Mott variable-range hopping that extends beyond the T_ρ-T_{CO} range to low temperatures. The deuterated counterpart has a kink around $T = 160$ K and exhibits variable-range hopping across an extended range below T_{CO} as well. This indicates that the $(TMTTF)_2PF_6$ salts are more ordered systems than the deuterated Br-salt.

The other salt containing an octahedral anion, $(TMTTF)_2SbF_6$, exhibits three-dimensional variable-range hopping according to Mott from T_ρ to T_{CO} with a kink around $T = 190$ K in the protonated compound, extending to lower temperatures with another kink. The deuterated analogue, on the other hand, shows no clear match to any of the three models at higher temperatures. However, in the low-temperature range, it might exhibit some variable-range hopping.

$(TMTTF)_2ClO_4$ does not undergo any pure CO transition, but a pronounced AO transition slightly above 70 K. No significant change in slope is expected across a simple structural transition at T_{AO} when the transport mechanism remains the same; this was observed in other Fabre salts with tetrahedral anions [7,8], for instance. In the Arrhenius plot (left column of Figure 7), we find the slope before and after the AO transition to be rather similar only for the deuterated compound. In other words, $(D_{12}\text{-}TMTTF)_2ClO_4$ exhibits an activated behavior with the same energy gap Δ on both sides of the phase transition, as analyzed in Table A1 (see Appendix A). For the protonated sister compound, however, the slope in the Arrhenius plot differs before and after the AO transition. The hopping model in the central column of Figure 7, however, may indicate that the sample is so disordered that Efros–Shklovskii variable-range hopping is present across the entire temperature range.

To summarize our findings on the discussed transport mechanisms at play in the deuterated and protonated Fabre salts investigated—between room temperature and their localization crossover temperature, T_ρ– all compounds display metallic behavior. Below this, all samples show semiconducting behavior, which can be described best by an activated band transport model, or variable-range hopping. The resistivity of $(D_{12}\text{-}TMTTF)_2Br$ follows Efros–Shklovskii model across a broad temperature range, indicating it to be a more disordered system than the other compounds. $(H_{12}\text{-}TMTTF)_2SbF_6$ and both types of $(TMTTF)_2PF_6$ could be fitted with Mott variable-range hopping over a broad temperature range extending below their respective charge-order transition temperatures—an indication for ordered systems. This finding agrees with previous studies [7], where variable-range hopping transport was found to dominate below $T = 60$ K in octahedral $(TMTTF)_2X$šalts. $(D_{12}\text{-}TMTTF)_2ClO_4$ could be fitted by an Arrhenius model with the same slope above and below its anion transition, indicating band transport with a constant activation energy across the phase transition, while the behavior observed in the protonated sister compound is best fitted by Efros–Shklovskii variable-range hopping across the AO phase transition, indicating a temperature dependent energy gap in this system.

4.2. Energy Gap

The above discussion illustrates that band transport is not necessarily the best overall model applicable to these samples; certainly, the electrical resistivity $\rho(T)$ cannot be simply described by a temperature independent energy gap, like in Equation (1). In particular, when considering the charge ordering as a second-order phase transition, the energy gap should follow a mean-field temperature dependence [7]. In order to investigate this further, the temperature dependent gap

$$\Delta(T) = \ln\left\{\frac{\rho(T)}{\rho_0}\right\}T, \tag{3}$$

was deduced from Equation (1) and plotted in Figure 8 for all samples under investigation. Equation (3) assumes constant ρ_0, the values of which are listed in Table 3.

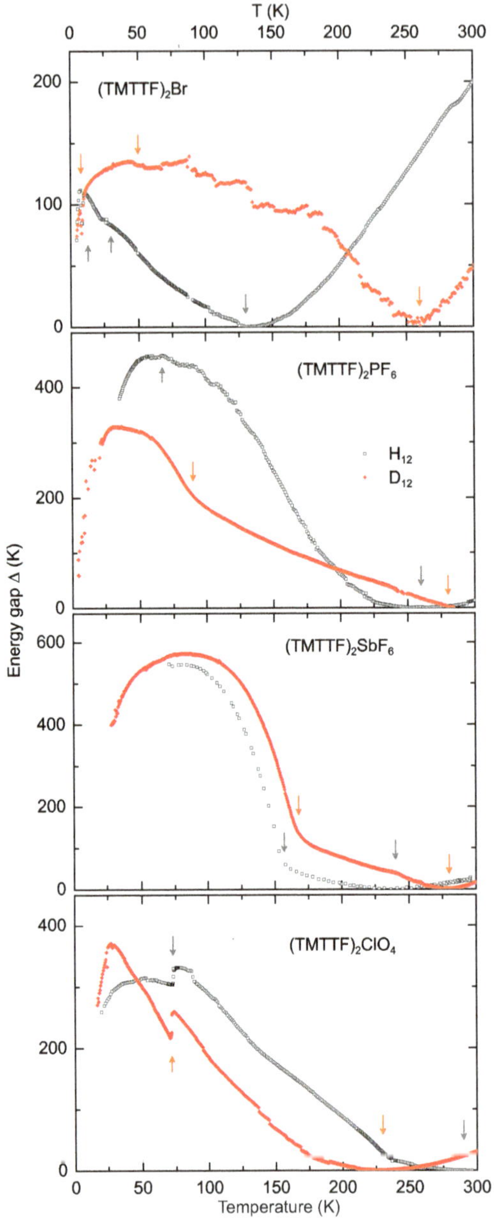

Figure 8. Temperature dependent energy gap $\Delta(T)$ for the protonated (black symbolds) and deuterated (red dots) Fabre salts $(TMTTF)_2X$ (X = Br, PF_6, SbF_6 and ClO_4) as calculated via Equation (3). The right most orange (grey) arrows indicate T_ρ of the deuterated (protonated) compounds. From right to left, the next set of arrows indicate the charge and anion ordering transitions T_{CO} or T_{AO}, while the left most set of arrows in the top panel [$(TMTTF)_2Br$ salts] indicate the AFM ordering at T_N.

Table 3. The constant resistivity ρ_0 from Equation (3) for the protonated and deuterated Fabre salts $(TMTTF)_2Br$, $(TMTTF)_2PF_6$, $(TMTTF)_2SbF_6$ and $(TMTTF)_2ClO_4$, as determined for Figure 8. The energy gap at the localization temperature T_ρ was set to zero by adjusting ρ_0.

Anions X		ρ_0 (Ωcm)
Br	H_{12}	0.01184
	D_{12}	0.02606
PF_6	H_{12}	1.286
	D_{12}	0.1401
SbF_6	H_{12}	4.638
	D_{12}	1.585
ClO_4	H_{12}	3.557
	D_{12}	0.03818

The presentation of the temperature dependent energy gap in Figure 8 makes all discussed phase transitions at T_ρ, T_{CO}, and T_{AO} clearly visible; one has to keep in mind, however, that the values in the metallic regime are not physically meaningful. Below T_ρ different regions of varying gap-behavior can be distinguished. For $(TMTTF)_2Br$ the energy gap Δ_0 is attributed to the bond dimerization resulting in unequal charge distribution between molecules [7]. Further cooling through T_{CO} causes stronger charge disproportionation between the molecules. Because of the second-order nature of this transition, the corresponding energy gap shows a BCS-like increase. The AO transition occuring in tetrahedral samples at lower temperatures was also found to contribute in a mean-field fashion. The total energy gap is given by [7]

$$\Delta(T) = \sqrt{\Delta_0^2 + [\Delta_{CO}(T) + \Delta_{AO}(T)]^2} \quad . \tag{4}$$

The gaps extracted from Figure 8 are summarized in Table 4, where further details are given. The localization energy gap, Δ_0, is highest in the $(TMTTF)_2PF_6$ and $(TMTTF)_2ClO_4$ compounds, where deuteration reduces the value. The opposite is true for $(TMTTF)_2Br$ and $(TMTTF)_2SbF_6$ salts, which show similar values for this gap. The charge-order gap, Δ_{CO}, exhibits a clear trend, increasing with anion size from just above 100 K for the protonated Br-salt to almost 600 K in the deuterated SbF_6-compound. The anion gap, Δ_{AO}, for $(TMTTF)_2ClO_4$ was read off directly from the plot. Since these compounds show no CO transition and the AO causes only little change of slope in the Arrhenuis plot, Δ_{AO} can be set to zero for this special case [7]. The abrupt changes in resistivity are attributed to the prefactor ρ_0 in Equation (1). The fact that the value of Δ_{AO} is smaller than Δ_0 for $(TMTTF)_2ClO_4$ seems to be unphysical. For the deuterated sample, however, a simple Arrhenius fit could be identified rather well (Figure 7), and correspondingly no discrepancies were found in the activation energy values. Hence, the ClO_4-salts seem to pose a challenge to the analysis proposed so far and further investigation is necessary to clarify this case. The total energy gap—last column in Table 4—also reveals a trend of increasing gap energy with increasing anion size. For all samples, except the $(TMTTF)_2PF_6$ salts, the deuterated compounds exhibit higher total gap energies than their protonated counterparts.

The increase in charge order gap energy and total gap energy with increasing anion size and upon deuteration (in general), indicates that the deformation of the methyl end-group cavity has a significant effect on the charge order transition—and thus confirms the deuteration and anion size effect shown in the trend of the change in T_{CO}. In a next step, local probes should be applied to monitor the charge distribution in these compounds, and in particular, how it varies with temperature and deuteration.

Table 4. Energy gaps for $(TMTTF)_2X$ (X = Br, PF_6, SbF_6 and ClO_4) as read off Figure 8. As suggested by Köhler et al., Δ_0 was extracted at T_{CO}, while Δ_{CO} and Δ_{AO} were read off at the highest point between T_{CO} or T_{AO} and $T \to 0$ K, respectively. Δ is the total gap energy as calculated according to Equation (4).

Anions X		Δ_0 (K)	Δ_{CO} (K)	Δ_{AO} (K)	Δ (K)
Br	H_{12}	90	110		140
	D_{12}	130	135		190
PF_6	H_{12}	460	460		650
	D_{12}	200	330		390
SbF_6	H_{12}	60	545		550
	D_{12}	140	570		585
ClO_4	H_{12}	330		315	450
	D_{12}	260		370	450

5. Summary

This comparative study between deuterated and protonated Fabre salts aimed to advance our understanding how the methyl end-groups effect the electronic properties in quasi-one-dimensional charge transfer salts. The different anions are dynamically confined by the terminal methyl groups and their interaction affects the electronic properties and phase transitions. The dc resistivity measurements performed along the stacking axis of the centrosymmetric and tetrahedral $(TMTTF)_2X$ salts and the data analysis, considering different transport models and a temperature dependent energy gap, have ascertained that the charge ordering transition is strongly affected by the methyl end-group cavity in which the anions are trapped: The change in T_ρ and T_{CO}, as well as the energy gap, increase upon deuteration and with increasing anion size. For the centrosymmetric anion salts, variable range hopping was confirmed to describe the charge transport in these samples best. $(TMTTF)_2ClO_4$ showed band transport behavior across its anion ordering phase transition. Overall, deuteration seems to have a similar effect on the system as an increase in anion size.

Author Contributions: Conceptualization and supervision, M.D.; synthesis of deuterated molecules, T.N.; investigation and analysis, A.R.; writing and editing, A.R. and M.D. All authors have read and agreed to the published version of the manuscript.

Funding: The project was supported by the Deutsche Forschungsgemeinschaft (DFG).

Acknowledgments: We acknowledge continuous crystal growth and thorough sample preparation by G. Untereiner; deuterated TMTTF crystals were grown by A. Moradpour. We acknowledge fruitful discussions with E. Rose and finally thank R. Rösslhuber and D. Liu for experimental support.

Conflicts of Interest: The authors declare no conflict of interest.

Appendix A

Table A1. Summary of linear fit parameters of the plots $\rho(T)$ vs. T^{-1}, $T^{-1/2}$, and $T^{-1/4}$ in Figure 7.

			#Linear Fit# over Entire# Temp Range	Best# Linear# Fit	Linear Fit# over En:ire# Temp Range	Best# Linear# Fit			Linear Fit# over Entire# Temp Range	Best# Linear# Fit	
$(TMTTF)_2X$			300 K-T_ρ			$T_\rho T_{CO}$ or T_{AO}				T_{AO} or $T_{CO}T_N$ or T_{SP} or 4 K	
H_{12}–Br	actual temperature range (K)		300–130			130–30				30–13	
	T^{-1}	best fit temp range (K)		300–180		130–60	60–30			24–13	
		slope	−89.595	−106.554	48.078	38.048	52.804		54.757	57.149	
		y-intercept	−1.352	−1.282	−2.349	−2.233	−2.457		−2.598	−2.741	
	$T^{-1/2}$	best fit temp range (K)		300–181		130–61	60–27			24–13	
		slope	−11.206	−14.082	12.462	8.118	16.354		24.579	26.979	
		y-intercept	−1.005	−0.819	−3.126	−2.662	−3.711		−5.317	−5.899	
	$T^{-1/4}$	best fit temp range (K)		300–167		130–63	49–27			22–10	
		slope	−6.044	−7.076	8.819	5.167	13.597		25.026	29.727	
		y-intercept	−0.193	0.069	−4.671	−3.482	−6.529		−11.669	−13.983	
D_{12}–Br	actual temperature range (K)		300–260			260–50				50–8	
	T^{-1}	best fit temp range (K)		295–260		157–174	173–89	89–48		50–25	25–10
		slope	−127.055	−119.421	71.181	141.631	74.232	54.111	40.052	52.862	42.037
		y-intercept	−1.096	−1.124	−1.789	−2.146	−1.789	−1.537	−1.050	−1.461	−1.002
	$T^{-1/2}$	best fit temp range (K)		295–260		255–54				50–25	25–10
		slope	−15.423	−14.362	14.053	13.994			18.514	17.777	21.234
		y-intercept	−0.628	−0.692	−2.443	−2.438			−3.047	−2.942	−3.640
	$T^{-1/4}$	best fit temp range (K)		295–261		257–177	186–67	67–48		50–25	25–10
		slope	−7.535	−6.995	8.638	10.224	8.547	11.621	17.357	14.510	21.126
		y-intercept	0.292	0.159	−3.751	−4.155	−3.730	−4.816	−7.047	−5.896	−8.875
H_{12}–PF_6	actual temperature range (K)		300–260			260–67				67–19 (data ends at 35)	
	T^{-1}	best fit temp range (K)		300–260		200–107	109–67			100–51	
		slope	−34.808	(not linear)	299.215	371.644	235.220		132.518	211.301	
		y-intercept	0.240		−1.166	−1.650	−0.419		1.274	−0.124	
	$T^{-1/2}$	best fit temp range (K)		300–260		181–65				124–47	
		slope	−4.125	(not linear)	53.106	56.448			38.392	50.922	
		y-intercept	0.363		−3.432	−3.758			−1.483	−3.168	
	$T^{-1/4}$	best fit temp range (K)		300–260		188–44				67–35	
		slope	−1.974	(not linear)	31.610	34.964			29.290	(see previous range)	
		y-intercept	0.598		−8.103	−9.142			−7.059	(see previous range)	
D_{12}–PF_6	actual temperature range (K)		300–280			280–90				90–13 (good data up to 20)	
	T^{-1}	best fit temp range (K)		300–280		239–119	157–90			59–30	
		slope	4.722	(too small)	124.022	115.341	139.517		148.637	151.236	
		y-intercept	−0.869		−1.325	−1.282	−1.458		−1.221	−1.113	
	$T^{-1/2}$	best fit temp range (K)		300–280		287–161	163–101			87–26	
		slope	0.532	(too small)	20.038	14.253	23.332		46.486	47.952	
		y-intercept	−0.883		−2.115	−1.710	−2.426		−4.693	−4.879	
	$T^{-1/4}$	best fit temp range (K)		300–280		280–155	163–103			89–23	
		slope	0.268	(too small)	11.213	7.761	13.707		36.144	36.353	
		y-intercept	−0.917		−3.675	−2.764	−4.438		−11.642	−11.717	

Table A1. *Cont.*

			#Linear Fit# over Entire# Temp Range	Best# Linear# Fit	Linear Fit# over Entire# Temp Range	Best# Linear# Fit					Linear Fit# over Entire# Temp Range	Best# Linear# Fit		
H_{12}–SbF_6	actual temperature range (K)		300–240		240–157						157–8 (good data ends at 70)			
	T^{-1}	best fit temp range (K)		296–251		228–165 (not linear)						110–70 (not linear)		
		slope	−47.048	−54.097	70.625	66.416					393.592	273.853		
		y-intercept	0.859	0.884	0.351	0.367					−1.177	0.196		
	$T^{-1/2}$	best fit temp range (K)		296–249		219–165 (not linear)						157–114	112–70	
		slope	−5.746	−6.532	10.123	10.124					78.505	121.771	59.678	
		y-intercept	1.034	1.081	−0.011	−0.018					−5.042	−8.825	−3.043	
	$T^{-1/4}$	best fit temp range (K)		297–252		234–182	199–157					157–114	116–71	
		slope	−2.734	−3.209	5.413	3.589	7.915				49.339	71.707	38.894	
		y-intercept	1.359	1.475	−0.734	−0.256	−1.421				−12.770	−19.378	−9.373	
D_{12}–SbF_6	actual temperature range (K)		300–280		280–166						166–8 (good data to 27)			
	T^{-1}	best fit temp range (K)		300–286		231–181						111–64 (not linear)		
		slope	−101.615	−126.631	134.304	128.026					203.142	247.267		
		y-intercept	0.560	0.644	−0.296	−0.274					0.336	0.176		
	$T^{-1/2}$	best fit temp range (K)		300–287		231–184						112–44		
		slope	−11.269	−14.926	18.243	17.713					52.762	52.129		
		y-intercept	0.871	1.084	−0.912	−0.886					−2.845	−2.557		
	$T^{-1/4}$	best fit temp range (K)		300–286		229–183						126–38		
		slope	−5.666	−7.150	9.511	9.482					37.198	36.190		
		y-intercept	1.582	1.940	−2.150	−2.154					−9.280	−8.818		
H_{12}–ClO_4	actual temperature range (K)		300–290		290–73					74–72	73–4 (good data ends at 20)			
	T^{-1}	best fit temp range (K)		300–290		232–73.2		87–73.2		73.17–72.87		72.5–52		53–29
		slope	4.567	(too small)	204.468	209.608		147.605		−2694.026	116.853	147.693		119.150
		y-intercept	0.536		−0.264	−0.316		0.495		39.318	0.838	0.334		0.903
	$T^{-1/2}$	best fit temp range (K)		300–290		233–131	130–88	87–73.0		73.17–72.82		72.6–54		72–24
		slope	0.533	(too small)	35.389	28.927	43.917	31.979		−630.607	37.248	36.931		37.704
		y-intercept	0.520		−1.723	−1.304	−2.591	−1.233		73.220	−2.002	−1.970		−2.063
	$T^{-1/4}$	best fit temp range (K)		300–290		237–136	138–89	87–73.1	135–73.1	73.17–72.87		72.8–53	53–22	72.6–22
		slope	0.257	(too small)	21.392	15.578	26.857	22.312	27.831	−308.947	29.329	26.114	30.649	29.310
		y-intercept	0.489		−4.929	−3.397	−6.691	−5.118	−6.988	108.098	−7.715	−6.583	−8.268	−7.709
D_{12}–ClO_4	actual temperature range (K)		300–230		230–73					73–70	73–4 (good data to 18)			
	T^{-1}	best fit temp range (K)		300–230		138–71.1		90–72.4		72.23–70.74		70.6–56		70.4–29
		slope	−44.010	(not linear)	172.713	197.157		193.418		−655.223	167.089	220.575		208.745
		y-intercept	−1.234		−2.309	−2.569		−2.514		0.137	−2.198	−3.206		−3.014
	$T^{-1/2}$	best fit temp range (K)		300–230		137–96		96–72.2		72.26–70.73		70.8–57		56–30
		slope	−5.544	(not linear)	30.793	33.654		43.408		−185.305	56.477	55.196		64.775
		y-intercept	−1.060		−3.641	−3.989		−4.948		21.896	−6.765	−6.656		−7.990
	$T^{-1/4}$	best fit temp range (K)		300–230		137–96		95–72.1		72.24–70.74		70.6–56	70.9–43	46–25
		slope	−2.778	(not linear)	18.215	20.594		28.769		−106.880	45.719	39.238	41.638	55.276
		y-intercept	−0.712		−6.314	−7.137		−9.713		36.728	−15.919	−13.627	−14.473	−19.754

References and Notes

1. Lebed, A. (Ed.) *The Physics of Organic Superconductors and Conductors*; Springer: Berlin, Germany, 2008
2. Dressel, M. Spin-charge separation in quasi one-dimensional organic conductors. *Naturwissenschaften* **2003**, 90, 337–344. [CrossRef] [PubMed]
3. Dressel, M. Ordering phenomena in quasi-one-dimensional organic conductors. *Naturwissenschaften* **2007**, *94*, 527–541. [CrossRef] [PubMed]
4. Dressel, M.; Dumm, M.; Knoblauch, T.; Masino, M. Comprehensive optical investigations of charge order in organic chain compounds $(TMTTF)_2X$. *Crystals* **2012**, *2*, 528–578. [CrossRef]
5. Nad, F.; Monceau, P.; Fabre, J.M. Low frequency dielectric permittivity of quasi-one-dimensional conductor $(TMTTF)_2Br$. *Eur. Phys. J. B* **1998**, *3*, 301–306. [CrossRef]
6. Coulon, C.; Lalet, G.; Pouget, J.P.; Foury-Leylekian, P.; Mouradpour, A.; Fabre, J.M. Anisotropic conductivity and charge ordering in $(TMTTF)_2X$ salts probed by ESR. *Phys. Rev. B* **2007**, *76*, 085126. [CrossRef]
7. Köhler, B.; Rose, E.; Dumm, M.; Untereiner, G.; Dressel, M. Comprehensive transport study of anisotropy and ordering phenomena in quasi-one-dimensional $(TMTTF)_2X$ salts ($X = PF_6$, AsF_6, SbF_6, BF_4, ClO_4, ReO_4). *Phys. Rev. B* **2011**, *84*, 035124. [CrossRef]
8. Rose, E.; Dressel, M. Coupling between molecular chains and anions in $(TMTTF)_2X$ salts. *Phys. B Condens. Matter* **2012**, *407*, 1787–1792. [CrossRef]
9. Rösslhuber, R.; Rose, E.; Ivek, T.; Pustogow, A; Breier, T.; Geiger, M.; Schrem, K.; Untereiner, G.; Dressel, M. Structural and Electronic Properties of $(TMTTF)_2X$ Salts with Tetrahedral Anions. *Crystals* **2018**, *8*, 121. [CrossRef]
10. Pustogow, A.; Peterseim, T.; Kolatschek, S.; Engel, L.; Dressel, M. Electronic correlations versus lattice interactions: Interplay of charge and anion orders in $(TMTTF)_2X$. *Phys. Rev. B* **2016**, *94*, 195125. [CrossRef]
11. Pouget, J.P. Structural Aspects of the Bechgaard and Fabre Salts: An Update. *Crystals* **2012**, *2*, 466–520. [CrossRef]
12. Pouget, J.P.; Alemany, P.; Canadell, E. Donor-anion interactions in quarter-filled low-dimensional organic conductors. *Mater. Horiz.* **2018**, *5*, 590–640. [CrossRef]
13. Pouget, J.P.; Foury-Leylekian, P.; Bolloc'h, D.; Hennion, B.; Ravy, S.; Coulon, C.; Cardoso, V.; Moradpour, A. Neutron-scattering evidence for a spin-peierls ground state in $(TMTTF)_2PF_6$. *J. Low Temp. Phys.* **2006**, *142*, 147–152. [CrossRef]
14. Asada, M.; Nakamura, T. Magnetic resonance investigation for a possible antiferromagnetic subphase in $(TMTTF)_2Br$. *Phys. Rev. B* **2017**, *96*, 125120. [CrossRef]
15. Furukawa, K.; Hara, T.; Nakamura, T. Deuteration effect and possible origin of the charge-ordering transition of $(TMTTF)_2X$. *J. Phys. Soc. Jpn.* **2005**, *74*, 3288. [CrossRef]
16. Nad, F.; Monceau, P.; Nakamura, T.; Furukawa, K. The effect of deuteration on the transition into a charge ordered state of $(TMTTF)_2X$ salts. *J. Phys. Condens. Matter* **2005**, *17*, L399–L406. [CrossRef]
17. Dumm, M.; Loidl, A.; Fravel, B.W.; Starkey, K.P.; Montgomery, L.K.; Dressel, M. Electron spin resonance studies on the organic linear-chain compounds $(TMTCF)_2X$(C = S, Se; $X = PF_6$, AsF_6, ClO_4, Br). *Phys. Rev. B* **2000**, *61*, 511–521. [CrossRef]
18. Salameh, B.; Yasin, S.; Dumm, M.; Untereiner, G.; Montgomery, L.; Dressel, M. Spin dynamics of the organic linear chain compounds $(TMTTF)_2X$ ($X = SbF_6$, AsF_6, BF_4, ReO_4, and SCN). *Phys. Rev. B* **2011**, *83*, 205126. [CrossRef]
19. We cannot guarantee that the experiments have been conducted under exactly the same conditions. Cooling rates as well as heat transfer rates may have differed between the measurements on the deuterated and protonated crystals, which may influence the hysteresis of this transition.
20. Pouget, J.P.; Ravy, S. Structural aspects of the Bechgaard salts and related compounds. *J. Phys. I* **1996**, *6*, 1501–1525. [CrossRef]
21. Dumm, M.; Loidl, A.; Alavi, B.; Starkey, K.P.; Montgomery, L.K.; Dressel, M. Comprehensive ESR-Study of the Antiferromagnetic Ground States in the One-Dimensional Spin Systems $(TMTSF)_2PF_6$, $(TMTSF)_2AsF_6$, and $(TMTTF)_2Br$. *Phys. Rev. B* **2000**, *62*, 6512–6520. [CrossRef]
22. Tomić, S.; Biskup, N.; Dolamski Babić, S.; Maki, K. Commensurate Spin-Density Wave State in $(TMTTF)_2Br$: Single-Particle and Collective Charge Dynamics. *Europhys. Lett.* **1994**, *26*, 295–301. [CrossRef]

23. Baranovski, S. *Charge Transport in Disordered Solids with Applications in Electronics*; John Wiley & Sons: West Sussex, UK, 2006.
24. Mott, N.F.; Davies, E. *Electronic Processes in Non-Crystalline Materials*, 2nd ed.; Clarendon Press: Oxford, UK, 1979.
25. Efros, A.L.; Shklovskii, B.I. Coulomb gap and low temperature conductivity of disordered systems. *J. Phys. C* **1975**, *8*, L49. [CrossRef]
26. Shklovskii, B.I.; Efros, A.L. *Electronic Properties of Doped Semiconductors*; Springer: Berlin, Germany, 1984.
27. Khondaker, S.I.; Shlimak, I.S.; Nicholls, J.T.; Pepper, M.; Ritchie, D.A. Crossover phenomenon for two-dimensional hopping conductivity and density-of-states near the Fermi level. *Solid State Commun.* **1999**, *109*, 751. [CrossRef]
28. Strictly speaking $\gamma = 2$ is also obtained in the case of one-dimensional transport ($d = 1$). The thorough study of anisotropic transport reveals that at low-temperatures one-dimensional models do not apply despite the appreciable anisotropy.
29. Dressel, M.; Petukhov, K.; Salameh, B.; Zornoza, P.; Giamarchi, T. Scaling Behavior of the Longitudinal and Transverse Transport in Quasi One-Dimensional Organic Conductors. *Phys. Rev. B* **2005**, *71*, 075104. [CrossRef]
30. Joung, D.; Khondaker, S.I. Efros-Shklovskii variable range hopping in reduced graphene oxide sheets of varying carbon sp^2 fraction. *Phys. Rev. B* **2012**, *86*, 235423. [CrossRef]

crystals

MDPI

Article

Synthesis, Structure and Physical Properties of (trans-TTF-py$_2$)$_{1.5}$(PF$_6$)·EtOH: A Molecular Conductor with Weak CH···N Hydrogen Bondings

Shohei Koyama [1], Morio Kawai [1], Shinya Takaishi [1], Masahiro Yamashita [1,2], Norihisa Hoshino [3], Tomoyuki Akutagawa [3], Manabu Kanno [1] and Hiroaki Iguchi [1,*]

1 Department of Chemistry, Graduate School of Science, Tohoku University, 6-3 Aramaki-Aza-Aoba, Aoba-ku, Sendai 980-8578, Japan; shohei.koyama.t3@dc.tohoku.ac.jp (S.K.); morio118ts@gmail.com (M.K.); shinya.takaishi.d8@tohoku.ac.jp (S.T.); yamasita.m@gmail.com (M.Y.); manabu.kanno.d2@tohoku.ac.jp (M.K.)

2 School of Materials Science and Engineering, Nankai University, Tianjin 300350, China

3 Institute of Multidisciplinary Research for Advanced Materials, Tohoku University, 2-1-1 Katahira, Aoba-ku, Sendai 980-8577, Japan; norihisa.hoshino.b5@tohoku.ac.jp (N.H.); akutagawa@tohoku.ac.jp (T.A.)

* Correspondence: h-iguchi@tohoku.ac.jp

Received: 31 October 2020; Accepted: 24 November 2020; Published: 26 November 2020

Abstract: The studies of crystal structures with hydrogen bonds have been actively pursued because of their moderate stabilization energy for constructing unique structures. In this study, we synthesized a molecular conductor based on 2,6-bis(4-pyridyl)-1,4,5,8-tetrathiafulvalene (trans-TTF-py$_2$). Two pyridyl groups were introduced into the TTF skeleton toward the structural exploration in TTF-based molecular conductors involved by hydrogen bonds. In the obtained molecular conductor, (trans-TTF-py$_2$)$_{1.5}$(PF$_6$)·EtOH, short contacts between the pyridyl group and the hydrogen atom of the TTF skeleton were observed, indicating that hydrogen bonding interactions were introduced in the crystal structure. Spectroscopic measurements and conductivity measurement revealed semiconducting behavior derived from π-stacked trans-TTF-py$_2$ radical in the crystal structure. Finally, these results are discussed with the quantified hydrogen bonding stabilization energy, and the band calculation of the crystal obtained from density functional theory calculation.

Keywords: tetrathiafulvalene; molecular conductor; hydrogen bonding

1. Introduction

Since the first metallic molecular conductor, TTF-TCNQ was reported [1], many researchers have been eagerly exploring TTF-based charge-transfer complexes [2,3], and exotic physical properties have been reported in well-designed molecular crystals in recent years [4,5]. In the progress of molecular conductors, modulating crystal structures has been one of the biggest challenges because the physical properties of molecular conductors heavily depend on their structures. Up to now, there have been several attempts to manipulate crystal structures by introducing supramolecular interactions into the TTF skeleton, such as intermolecular hydrogen bond [6–12] and halogen bond [6,13–16]. Among them, utilizing the hydrogen bond has been gathering attention because of its ability to form various unique structures owing to the moderate bond-dissociation energy (~40 kcal/mol) [17]. Although developments of TTF-based molecular crystals with hydroxy, amide, and cyano groups have been reported so far [6–12], TTF-based molecular conductors with pyridyl groups have been scarcely reported [18,19]. Furthermore, the pyridyl group in the TTF skeleton has often been used for metal coordination [20,21] but not for hydrogen bonding. The hydrogen bonding via pyridyl groups has many advantages in the arrangement of crystal structures because of its well-directed interaction derived from the rigid structure of the pyridyl group. In addition, the pyridyl group is easy to be introduced into π-conjugated molecular backbones by applying a hetero coupling reaction [22]. To investigate a novel crystal structure of TTF-based molecular conductors, we chose 2,6-bis(4-pyridyl)-1,4,5,8-tetrathiafulvalene (trans-TTF-py_2) as a starting material. This molecule has two pyridyl groups on the TTF skeleton (Scheme 1), and a unique structure based on a hydrogen bonding via the pyridyl groups can be expected. Although TTF-py_2 has been reported as a neutral molecular crystal [23] and as a ligand in coordination polymers [24–28], there is no report of trans-TTF-py_2-based conductive crystal. Even in trans-bis-substituted TTF molecules, the crystal structure of the molecular conductor was reported in only one paper to the best of our knowledge [29]. In our paper, a molecular conductor, (trans-TTF-py_2)$_{1.5}$(PF_6)·EtOH (**TTF-py_2_PF_6**), was prepared through electrochemical crystallization. Single crystal X-ray diffraction (SXRD) revealed a one-dimensional π-stacking structure of oxidized trans-TTF-py_2, where the nitrogen atoms of the pyridyl groups have short contacts with hydrogen atoms of the adjacent TTF-py_2 molecules, suggesting the existence of significant hydrogen bondings. Spectroscopic analyses and the measurement of electrical conductivity of **TTF-py_2_PF_6** were carried out. Its band structure and the stabilization energy of hydrogen bondings calculated by density functional theory (DFT) are also discussed.

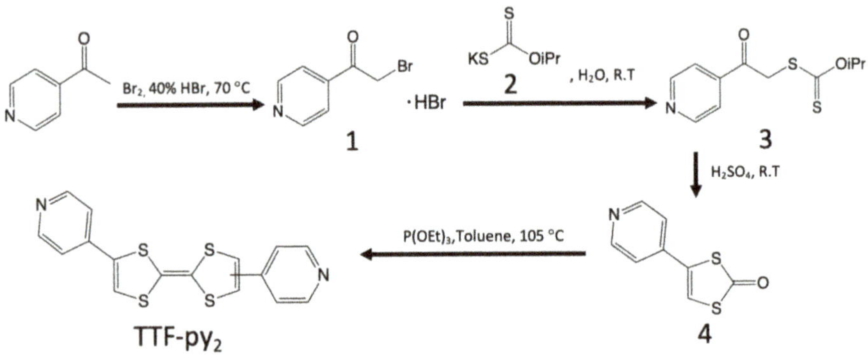

Scheme 1. Synthesis of TTF-py_2.

2. Materials and Methods

2.1. Methods

All solvents and reagents used in the syntheses were obtained from commercial sources without further purification. IR spectrum was recorded as KBr pellets on a FT/IR-4200 spectrometer of JASCO, Tokyo, Japan at room temperature (RT). UV-Vis-NIR absorption spectrum was measured as KBr pellets on a V-670 spectrophotometer of JASCO, Tokyo, Japan at RT. Both spectra were connected to represent the solid-state absorption spectrum in a wide range (Figure 2b). The ESR spectra were acquired by using a JES-FA100 of JEOL, Tokyo, Japan. Microscopic Raman Spectrum was measured by using a LabRAM HR-800 of HORIBA, Kyoto, Japan at RT.

^{1}H NMR measurements were performed on Bruker AV500 of Bruker Japan, Kanagawa, Japan at RT. The temperature dependence of the electrical conductivity was measured in a liquid He cryostat of a Quantum Design Physical Property Measuring System (PPMS) MODEL 6000 of Quantum Design Japan, Tokyo, Japan by using the two-probe method in direct current (DC) mode with Keithley sourcemeter model 2611 of Tektronix, Beaverton, Oregon, United States. The cooling rate was 2 K/min. The electrical leads (15 μmφ gold wires) were attached to a single crystal with carbon paste (Dotite XC-12 in diethyl succinate).

2.1.1. Single X-ray Diffraction

The diffraction data for **TTF-py$_2$_PF$_6$** were collected on a XtaLAB AFC10 diffractometer with a HyPix-6000HE hybrid pixel array detector, graphite monochromated Mo Kα radiation (λ = 0.7107 Å) and a cryogenic equipment GN-2D/S of Rigaku, Tokyo, Japan. The crystal structure was solved using direct methods (SHELXT) followed by Fourier synthesis. Structure refinement was performed using full matrix least-squares procedures with SHELXL [30,31] on *F* [2], where *F* is the crystal structure factor, in the Olex2 software. [32]

2.1.2. Computational Methods

DFT calculations were performed using the gaussian 16 package [33] for estimation of hydrogen bonding with the counterpoise method [34]. The B3LYP functional [35–37] and the cc-pVTZ basis sets [38] were used for the calculation of hydrogen bonding, because the B3LYP functional has been validated in previous studies for the calculation of hydrogen bonds in several models such as clathrate, radical, and cation-anion systems [39–42]. The atomic coordination of trans-TTF-py$_2$ ligands was extracted from the cif files of **TTF-py$_2$_PF$_6$** reported herein and used for the calculation without any optimizations. The molecular orbital energies were presented against the vacuum level standard.

Amsterdam Modeling Suite (AMS) packages were applied for the calculations of charge transfer integrals [43,44] and band structure [45] with tight-binding approximation. Charge transfer integrals between adjacent trans-TTF-py$_2$ molecules and band structure of **TTF-py$_2$_PF$_6$** were investigated by the B3LYP/TZP method, and third-order density-functional-based tight binding (DFTB3) model, respectively, without structural optimization.

2.2. *Syntheses*

TTF-py$_2$ was synthesized by following already described procedures [46] (Scheme 1) with a little modification (e.g., reaction temperature and the solvent used for washing in the synthesis of **1**, reaction temperature and the addition of the extraction process in the synthesis of **2**, reaction temperature in the reaction in the synthesis of **3**, and recrystallization process in the synthesis of **4**).

2.2.1. Synthesis of 4-(2-bromoacetyl)pyridine Hydrobromide (1)

Bromine (20.00 g, 0.13 mol) was added dropwise to 4-acetylpyridine (15.36 g, 0.13 mol) dissolved in 40 mL of HBr aq.(47–49% wt.) at 70 °C. Transparent crystalline solid was immediately precipitated in the solution after the addition of bromine. The precipitate was filtered, washed with acetonitrile, and dried in a desiccator to obtain 4-(2-bromoacetyl)pyridine hydrobromide. Yield: 74.8% ^{1}H NMR (500 MHz, D_2O, 298 K, ppm) δ 8.78 (d, 2H), 8.19 (d, 2H), 3.72 (s, 2H).

2.2.2. Synthesis of Potassium Isopropylxanthate (2)

Potassium hydroxide (14.03 g, 0.25 mol) were dissolved in 100 mL isopropanol at 80 °C and then cooled to 40 °C. Pale pinkish suspension was obtained after carbon disulfide (19.04 g, 0.25 mol) were slowly added dropwise to the solution. The precipitate was filtered and extracted with acetone to obtain a bright yellow crystalline solid. Yield: 76.3%. ^{1}H NMR (500 MHz, $CDCl_3$, 298 K, ppm) δ 5.33 (sep, 1H), 1.35 (d, 6H).

2.2.3. Synthesis of O-(1-methylethyl)S-[2-oxo-2-(4-pyridinyl)ethyl]carbonodithioate (3)

1 (7.02 g, 25 mmol) dissolved in 75 mL H_2O and **2** (1.30 g, 38 mmol) dissolved in 50 mL H_2O were mixed with stirring. Formed transparent crystals were filtered, washed with H_2O, and dried in a desiccator overnight to give **3**. Yield: 77.8% ^{1}H NMR (500 MHz, $CDCl_3$, 298 K, ppm) δ 8.86 (d, 2H), 7.80 (d, 2H), 5.70(sep, 1H), 4.59 (s, 2H), 1.38 (d, 6H).

2.2.4. Synthesis of 4-(4-pyridyl)-1,3-dithiol-2-one (**4**)

3 (4.9 g,19 mmol) was dissolved in 7.5 mL conc. H_2SO_4, and the resulting black solution was heated to 50 °C for 30 min. The solution was poured into 200 mL of H_2O, neutralized by $NaHCO_3$, and extracted with dichloromethane. The solvent was dried with $MgSO_4$ and evaporated to obtain a brown solid. This solid was recrystallized with EtOH to get a brown crystalline solid. ^{1}H NMR (500 MHz, $CDCl_3$, 298 K, ppm) δ 8.70 (d, 2H), 7.33 (d, 2H), 7.18 (s, 1H).

2.2.5. Synthesis of TTF-py$_2$ (the Mixture of cis and trans Isomers)

4 (2.0 g,10 mmol) was suspended in 10 mL dehydrated toluene and 10 mL dehydrated $P(OEt)_3$ under N_2 atmosphere. This suspension was heated to 105 °C for 4 h. Red precipitate was filtered, washed with MeOH, and recrystallized with nitromethane to acquire red crystalline solid. ^{1}H NMR (500 MHz, $CDCl_3$, 298 K, ppm) δ 8.63 (2H), 7.77 (1H), 7.47 (2H). Since two structural isomers are produced in the synthesis of TTF-py$_2$ (cis-TTF-py$_2$ and trans-TTF-py$_2$), we had difficulty in assigning the correct coupling from this NMR spectrum. Anal. Calcd for $C_{16}H_{10}N_2S_4$: C, 53.60; H, 2.81; N, 7.81; S, 35.77. Found: C, 53.21; H, 2.85; N, 7.99; S, 35.44.

2.2.6. Synthesis of (trans-TTF-py$_2$)$_{1.5}$(PF_6)·EtOH (TTF-py$_2$_PF$_6$)

TTF-py$_2$, which is the mixture of cis and trans isomers (10 mg, 28 μmol), and TBAPF$_6$ (40 mg, 104 μmol) were dissolved in 10 mL of $CHCl_3$/EtOH (7:3, *v*/*v*) solution. A constant current of 2.5 μA was applied to the solution. After a week, black needle crystals were isolated from the anode, and subsequently washed by $CHCl_3$ and EtOH followed by drying in air. Anal. Calcd for $C_{26}H_{21}F_6N_3OPS_6$: C, 42.85; H, 2.90; N, 5.77; S, 26.40; Cl, 0.00. Found: C, 42.32; H, 2.74; N, 5.85; S, 26.46; Cl, 0.06.

3. Results

3.1. Crystal Structure

The trans-TTF-py$_2$-based molecular conductor, **TTF-py$_2$_PF$_6$**, was obtained through electrochemical crystallization in a $CHCl_3$/EtOH (7:3, *v*/*v*) solution dissolving TTF-py$_2$ and TBAPF$_6$ [47]. The similar conditions with TBAClO$_4$, TBABF$_4$, or TBAX (X = Cl and Br) did not give any crystals. Single-crystal X-ray structure analysis shows two hexafluorophosphates for every three trans-TTF-py$_2$ molecules, indicating that the average charge of trans-TTF-py$_2$ is +0.66 (Figure 1). The cis isomer was not incorporated into the crystal. Two crystallographically independent trans-TTF-py$_2$ (**A** and **B**) molecules exist in the crystal structure (Figure 1a), repeating the **A**–**B**–**B** order to form a one-dimensional columnar structure by π–π stacking. The closest intermolecular S···S distances between adjacent trans-TTF-py$_2$ molecules along the π-stack axis are summarized in Table 1. This table indicates that the S···S distances between **A**–**B** are shorter than the sum of van der Waals radius of the sulfur atoms (3.60 Å) [48], whereas those between **B**–**B** are longer than that value. In fact, the side view of the π-stacking structure (Figure 1c) clearly shows that the shift of π-stacking between **B** and **B** is larger than that between **A** and **B**. Focusing on the environment of π-stack columnar structures (Figure 1d,e), one π-stack column is surrounded by four other columns. Figure 1d,e highlight the interactions of **A** and **B** with neighboring molecules, respectively. It is noteworthy that there are short contacts between the nitrogen atoms of the pyridyl groups and the hydrogen atoms of TTF cores, emphasized with orange broken lines in Figure 1d and e. **A** is surrounded by four **B** molecules and has short contacts with two of them. **B**, on the other hand, is surrounded by two each of **A** and **B**, and it interacts with all but one **A**. The intermolecular N···H distances are 2.365 Å (between **A** and **B**) and 2.626 Å (between **B** and **B**), which are shorter than the sum of the van der Waals radii of hydrogen and nitrogen atoms [48], suggesting the construction of C–H···N hydrogen bondings. To obtain a deeper insight into the hydrogen bondings connecting trans-TTF-py$_2$ molecules, a DFT calculation was applied for estimating stabilization energies of the hydrogen bondings between **A** and **B** and **B** and **B** depicted in Figure 1d,e. The calculation indicated that the stabilization energies are −0.74 kcal/mol for the hydrogen bonding between **A** and **B** and −1.72 kcal/mol for that between **B** and **B**. Hence, the sum of the stabilization energies of the hydrogen bondings around **A** is −1.48 kcal/mol (= 2 × (−0.74) kcal/mol) and that around **B** is −4.18 kcal/mol (= 2 × (−1.72) −0.74 kcal/mol). Calculated effective transfer integrals (V_{eff}) between neighboring trans-TTF-py$_2$ molecules along the columnar structure are V_{eff}(A–B) = 390.8 meV for **A–B** and V_{eff}(B–B) = 222.9 meV for **B–B** (Figure 1c), reflecting the difference in the stacking feature. On the other hand, V_{eff} of trans-TTF-py$_2$ molecules between adjacent columns are less than 10 meV in all combinations. This much difference of V_{eff} obviously demonstrates the character of one-dimensional electron conductor in **TTF-py$_2$_PF$_6$**.

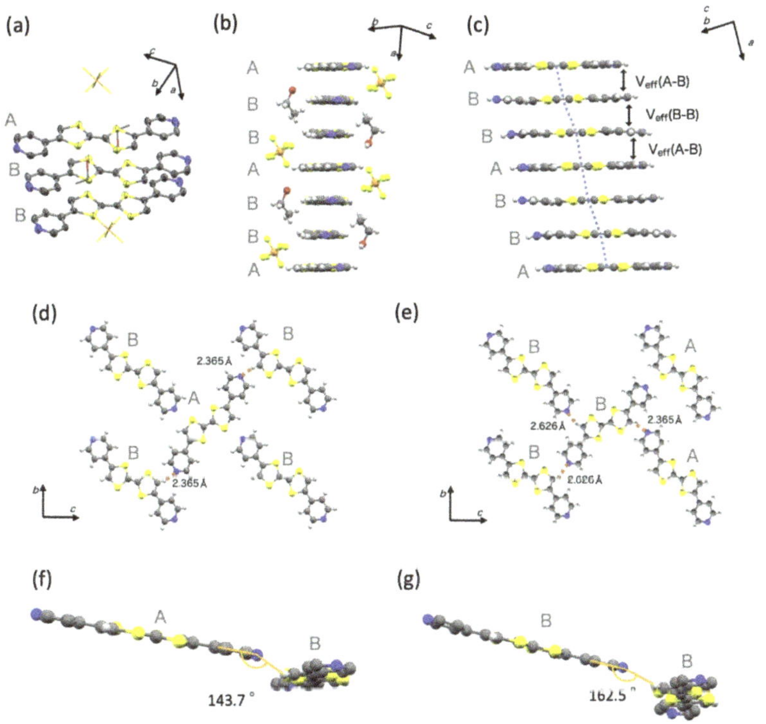

Figure 1. Crystal structure of **TTF-py$_2$_PF$_6$**. (**a**) Thermal ellipsoid plot of **TTF-py$_2$_PF$_6$**. Counteranion (PF_6^-) and crystal solvent (EtOH) were pictured in capped sticks, and hydrogen atoms are omitted for clarity. (**b**,**c**) π-stacked TTF-py$_2$ columns viewed from different sides. Blue broken lines are written to represent the shift of π-stacks. Counteranion and crystal solvent were omitted for clarity in (**c**). (**d**,**e**) Perspective views of **TTF-py$_2$_PF$_6$** along the *a* axis with placing **A** or **B** at the center, respectively. Short contacts with neighboring columns are represented as broken orange lines. (**f**,**g**) View of the angle of the hydrogen bonding site between **A** and **B** (**f**) and **B** and **B** (**g**). Hydrogen atoms of pyridyl groups were omitted for clarity. Yellow, S; Orange, P; Light green, F; Red, O; Blue, N; Gray, C; White, H.

Table 1. The distances between the nearest sulfur atoms along the π-stack axis.

	A–B	B–B
Intracolumnar S···S distance/Å	3.584 3.481 3.493 3.408	3.701 3.666

3.2. Electronic States and Conductivity of TTF-py2_PF6

To understand the electronic state of **TTF-py2_PF6**, microscopic Raman spectrum, solid-state absorption spectrum, and ESR spectrum of **TTF-py2_PF6** were recorded and shown in Figure 2. Microscopic Raman spectrum of **TTF-py2_PF6** shows a strong peak around 1450 cm^{-1} and weak peaks around 1420 and 1520 cm^{-1}. It is well known that there is a correlation between the valence of the TTF molecule and Raman frequency [49], and the valence corresponding to 1450 cm^{-1} is found to be about +0.6 based on previous studies [49]. This value is in approximate agreement with the valence obtained from the formula, +0.66. The absorption spectrum consists of major absorption bands above 1.7 eV, a broad band around 0.4 eV, and a minor band around 1.5 eV (Figure 2b). Each band was assigned by following previous reports [50–52]. Major bands above 1.7 eV correspond to the mixture of the π–π^* transition for the radical monomer trans-TTF-$py_2^{\bullet+}$ and radical dimer (trans-TTF-$py_2^{\bullet+})_2$. A broad band around 0.4 eV and a minor band around 1.5 eV are characteristics of intermolecular charge transfer from trans-TTF-$py_2^{\bullet+}$ to trans-TTF-py_2^0 (forming trans-TTF-py_2^0 and trans-TTF-$py_2^{\bullet+}$) and from trans-TTF-$py_2^{\bullet+}$ to trans-TTF-$py_2^{\bullet+}$ (forming trans-TTF-py_2^0 and trans-TTF-py_2^{2+}), respectively [53]. The existence of trans-TTF-$py_2^{\bullet+}$ was also supported by the results of the ESR spectrum (Figure 2c), which shows a singlet peak with $g = 2.003$ derived from a TTF radical.

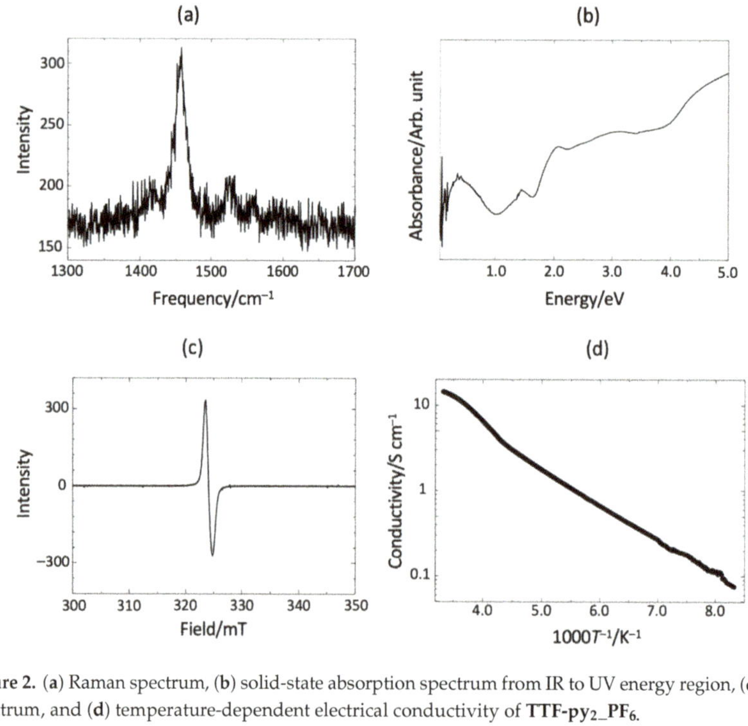

Figure 2. (**a**) Raman spectrum, (**b**) solid-state absorption spectrum from IR to UV energy region, (**c**) ESR spectrum, and (**d**) temperature-dependent electrical conductivity of **TTF-py2_PF6.**

A single crystal direct-current conductivity (= σ) of **TTF-py2_PF6** along the π-stack direction is shown in Figure 2d. The electrical conductivity of **TTF-py2_PF6** is 12 S cm^{-1} at RT, and the decreases with cooling temperature, suggesting semiconducting behavior. The slope of the Arrhenius plot is constant with $1000T^{-1}$ above 4.5 K^{-1}. From the fitting of the Arrhenius plot above 4.5 K^{-1} with an Arrhenius dependence of $\sigma = \sigma_0 \exp[-(E_g/2kT)]$ where σ_0 is a constant, E_g is the bandgap and k is Boltzmann constant [54], it shows that the E_g of the carriers is estimated to be 162 meV.

4. Discussion

The structural study shows short contacts by the hydrogen bondings, and the DFT calculation with the counterpoise correction shows the significant stabilization energy. It is noteworthy that although the distance of the hydrogen bonding between **A** and **B** (2.365 Å) is shorter than that between **B** and **B** (2.626 Å), the stabilization energy of the hydrogen bonding between **A** and **B** (= −0.74 kcal/mol) is less than that between **B** and **B** (= −1.72 kcal/mol). This tendency did not change even when the exchange correlation functional was changed from B3LYP to B3PW91 or PBEPBE in the DFT calculations. The nature of the hydrogen bond is complicated by the fact that the hydrogen bond has multiple parameters such as angles, distances and chemical species (elements of hydrogen donor and acceptor atoms). [17,55]. In the present case, the N···H distance seems to be less important to determine the energy of the hydrogen bond. As shown in Figure 1f,g, the angles between the axis of the pyridyl groups and the hydrogen atoms of the TTF core (∠C(position 4)···N···H) are 143.7° for **A** and **B** and 162.5° for **B** and **B**. Because the latter angle is closer to 180°, where the lone pair of nitrogen atom is headed to the hydrogen atom, the hydrogen bonding between **B** and **B** is stronger than that between **A** and **B**. Wood et al. examined the distance- and the angle-dependence of hydrogen bondings with pyridine in terms of computational chemistry [55]. In the paper, the structural dependence of the stabilization energy varies with chemical species, and the dependence of the energy on the interatomic distance is found to be relatively weak for the weak hydrogen bondings such as those with benzene. Even in this system, the hydrogen bondings are also weak C-H···N bonds, thus the dependence of the energy on the N···H distance is small, and the change of the angle is more likely to be involved in the stabilization energy.

The temperature-dependent electrical conductivity of **TTF-py$_2$_PF$_6$** shows semiconducting behavior along the *a* axis. Effective charge transfer integrals of adjacent TTF-py$_2$ clarify the nature of one-dimensional electron conductivity along the *a* axis, which is also denoted from the band structure. Given the difference of the effective charge transfer integral between **A** and **B** (V_{eff} = 390.8 meV), and **B** and **B** (V_{eff} = 220.9 meV), the semiconducting behavior of **TTF-py$_2$_PF$_6$** is probably due to the localization of carriers in the columnar structure. In fact, band calculation by using tight-binding approximation without structural optimization (Figure 3) shows an inherent bandgap, E_g = 140 meV, and hence a significant localization of carrier can be assumed from the crystal structure. The Calculated bandgap is almost consistent with the bandgap acquired from temperature-dependent conductivity (E_g = 162 meV). The carriers in the dominant part of the electron conduction were located where the path of the band structure is along a^* axis, the direction of π–π stacking of TTF-py$_2$. Hence, not only the effective charge transfer integrals but also the band structure show the one-dimensional conducting character of **TTF-py$_2$_PF$_6$**.

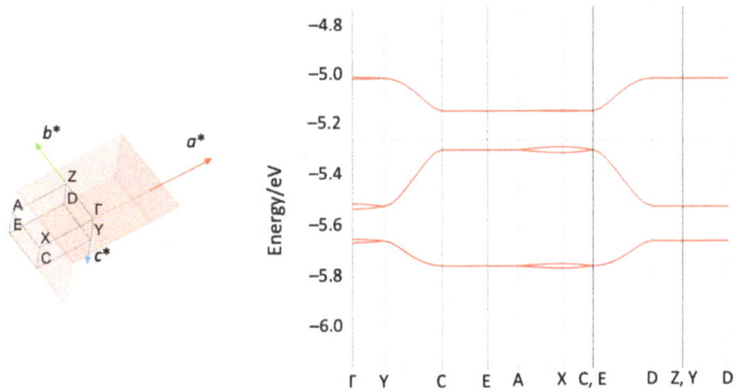

Figure 3. Brillouin zone of TTF-py2_PF6 (**left**) showing the path corresponding to the band structure of TTF-py2_PF6 (**right**).

5. Conclusions

In this paper, we discussed the crystal structure and the physical properties of the molecular conductor **TTF-py$_2$_PF$_6$**. It is the first molecular conductor containing trans-TTF-py$_2$ molecules and is the second one with trans-bis-substituted TTF molecules, to the best of our knowledge. Although **TTF-py$_2$_PF$_6$** has one dimensional electron transport properties, which is typical in molecular conductors, the hydrogen bondings between the pyridyl groups and the hydrogen atoms of the TTF skeleton were successfully introduced in the crystal structure. We believe that this result provides a potential for further structural explorations and physical properties of TTF-based molecular conductors with a substitution group with hydrogen bondings such as pyridyl and other moieties. Additionally, TTF-py$_2$ is a promising ligand for constructing both conductive π-stacked arrays and coordination networks in a crystal, such as porous molecular conductors [56,57] and conductive π-stacked metal-organic frameworks [58–61].

CCDC-2040487 contains the supplementary crystallographic data for this paper. Crystal structure information is available online at the Cambridge Crystallographic Data Centre (CCDC) database via www.ccdc.cam.ac.uk/data_request/cif.

Author Contributions: Conceptualization, S.K. and H.I.; methodology, S.K.; software, S.K., H.I.; validation, H.I., S.T. and M.K. (Manabu Kanno); formal analysis, S.K.; investigation, S.K., M.K. (Morio Kawai); resources, H.I., S.T., N.H., T.A.; data curation, S.K.; writing—original draft preparation, S.K.; writing—review and editing, H.I., S.T. and M.Y.; visualization, S.K.; supervision, H.I., and M.Y.; project administration, H.I. and S.T.; funding acquisition, S.K., H.I., and S.T. All authors have read and agreed to the published version of the manuscript.

Funding: This work was partly supported by JSPS KAKENHI Grant Numbers JP18H04498 (H.I.) for "Soft Crystals", JP18K14233 (H.I.), JP19H05631 (H.I., S.T. and M.Y.) and JP20J22404 (S.K.), by the Toyota Riken Scholar Program (H.I.) and by the Kato Foundation for Promotion of Science KJ-2916 (H.I.).

Acknowledgments: Authors acknowledge the Research and Analytical Center for Giant Molecules, Graduate School of Science, Tohoku University, for the measurement of elemental analysis and common equipment unit of Advanced Institute of Material Research, Tohoku University, for the measurement of microscopic Raman spectra.

Conflicts of Interest: The authors declare no conflict of interest.

References

1. Ferraris, J.; Cowan, D.O.; Walatka, V.; Perlstein, J.H. Electron transfer in a new highly conducting donor-acceptor complex. *J. Am. Chem. Soc.* **1973**, *95*, 948–949. [CrossRef]
2. Frère, P.; Skabara, P.J. Salts of extended tetrathiafulvalene analogues: Relationships between molecular structure, electrochemical properties and solid state organisation. *Chem. Soc. Rev.* **2005**, *34*, 69–98. [CrossRef] [PubMed]
3. Otsubo, T.; Takimiya, K. Recent Synthetic Advances of Tetrathiafulvalene-Based Organic Conductors. *Bull. Chem. Soc. Jpn.* **2004**, *77*, 43–58. [CrossRef]
4. Yamochi, H.; Koshihara, S. Organic metal (EDO-TTF)2PF6 with multi-instability. *Sci. Technol. Adv. Mater.* **2009**, *10*, 024305. [CrossRef] [PubMed]
5. Ueda, A. Development of Novel Functional Organic Crystals by Utilizing Proton- and π-Electron Donating/Accepting Abilities. *Bull. Chem. Soc. Jpn.* **2017**, *90*, 1181–1188. [CrossRef]
6. Fourmigué, M.; Batail, P. Activation of Hydrogen- and Halogen-Bonding Interactions in Tetrathiafulvalene-Based Crystalline Molecular Conductors. *Chem. Rev.* **2004**, *104*, 5379–5418. [CrossRef]
7. Mroweh, N.; Pop, F.; Mézière, C.; Allain, M.; Auban-Senzier, P.; Vanthuyne, N.; Alemany, P.; Canadell, E.; Avarvari, N. Combining Chirality and Hydrogen Bonding in Methylated Ethylenedithio-Tetrathiafulvalene Primary Diamide Precursors and Radical Cation Salts. *Cryst. Growth Des.* **2020**, *20*, 2516–2526. [CrossRef]
8. Sahadevan, S.A.; Monni, N.; Abhervé, A.; Auban-Senzier, P.; Canadell, E.; Mercuri, M.L.; Avarvari, N. Synthesis and Physical Properties of Purely Organic BEDT-TTF-Based Conductors Containing Hetero-/Homosubstituted Cl/CN-Anilate Derivatives. *Inorg. Chem.* **2017**, *56*, 12564–12571. [CrossRef]

9. Lopes, G.; Gama, V.; Belo, D.; Simão, D.; Santos, I.C.; Almeida, M.; Rabaça, S. Cyanobenzene–Ethylenedithio–Tetrathiafulvalene Salts with ClO4–: Bilayer Polymorphs and Different Stoichiometries. *Cryst. Growth Des.* **2017**, *17*, 2801–2808. [CrossRef]
10. Baudron, S.A.; Batail, P.; Coulon, C.; Clérac, R.; Canadell, E.; Laukhin, V.; Melzi, R.; Wzietek, P.; Jérome, D.; Auban-Senzier, P.; et al. (EDT-TTF-CONH$_2$)6[Re6Se8(CN)6], a Metallic Kagome-Type Organic–Inorganic Hybrid Compound: Electronic Instability, Molecular Motion, and Charge Localization. *J. Am. Chem. Soc.* **2005**, *127*, 11785–11797. [CrossRef]
11. Rabaça, S.; Oliveira, S.; Santos, I.C.; Gama, V.; Belo, D.; Lopes, E.B.; Canadell, E.; Almeida, M. Polymorphism and Superconductivity in Bilayer Molecular Metals (CNB-EDT-TTF)4I3. *Inorg. Chem.* **2016**, *55*, 10343–10350. [CrossRef] [PubMed]
12. Baudron, S.A.; Avarvari, N.; Batail, P.; Coulon, C.; Clérac, R.; Canadell, E.; Auban-Senzier, P. Singular Crystalline β′-Layered Topologies Directed by Ribbons of Self-Complementary Amide···Amide Ring Motifs in [EDT-TTF-(CONH$_2$)$_2$]$_2$X (X = HSO_4^-, ClO_4^-, ReO_4^-, AsF_6^-): Coupled Activation of Ribbon Curvature, Electron Interactions, and Magnetic Susceptibility. *J. Am. Chem. Soc.* **2003**, *125*, 11583–11590. [CrossRef] [PubMed]
13. Lieffrig, J.; Le Pennec, R.; Jeannin, O.; Auban-Senzier, P.; Fourmigué, M. Toward chiral conductors: Combining halogen bonding ability and chirality within a single tetrathiafulvalene molecule. *CrystEngComm* **2013**, *15*, 4408–4412. [CrossRef]
14. Shin, K.S.; Brezgunova, M.; Jeannin, O.; Roisnel, T.; Camerel, F.; Auban-Senzier, P.; Fourmigué, M. Strong Iodine···Oxygen Interactions in Molecular Conductors Incorporating Sulfonate Anions. *Cryst. Growth Des.* **2011**, *11*, 5337–5345. [CrossRef]
15. Nakano, Y.; Takahashi, Y.; Ishida, K.; Ishikawa, M.; Yamochi, H.; Uruichi, M. Crystal structure and physical properties of radical cation salt based on 4,5-ethylenedioxy-4′-iodotetrathiafulvalene (EDO-TTF-I) with iodine bonding ability. *Mater. Chem. Front.* **2018**, *2*, 752–759. [CrossRef]
16. Shin, K.-S.; Jeannin, O.; Brezgunova, M.; Dahaoui, S.; Aubert, E.; Espinosa, E.; Auban-Senzier, P.; Świetlik, R.; Frąckowiak, A.; Fourmigué, M. Inter-layer charge disproportionation in the dual-layer organic metal (tTTF-I)2ClO4with unsymmetrical I···O halogen bond interactions. *Dalton Trans.* **2014**, *43*, 5280–5291. [CrossRef]
17. Steiner, T. The hydrogen bond in solid state. *Angew. Chem. Int. Ed.* **2002**, *41*, 48–76. [CrossRef]
18. Kanetou, T.; Tsunashima, R.; Hoshino, N.; Akutagawa, T. Solid-state structure and electronic states of hydrogen-bonded dimer of pyridyl-substituted tetrathiafulvalene salted with PF 6? *RSC Adv.* **2017**, *7*, 6236–6241. [CrossRef]
19. Lee, S.C.; Ueda, A.; Kamo, H.; Takahashi, K.; Uruichi, M.; Yamamoto, K.; Yakushi, K.; Nakao, A.; Kumai, R.; Kobayashi, K.; et al. Charge-order driven proton arrangement in a hydrogen-bonded charge-transfer complex based on a pyridyl-substituted TTF derivative. *Chem. Commun.* **2012**, *48*, 8673–8675. [CrossRef]
20. Liu, S.-X.; Dolder, S.; Franz, P.; Neels, A.; Stoeckli-Evans, H.; Decurtins, S. Structural studies of transition metal complexes with 4,5-*Bis*(2-pyridylmethylsulfanyl)-4′,5′-ethylenedithiotetrathiafulvalene: Probing their potential for the construction of multifunctional molecular assemblies. *Inorg. Chem.* **2003**, *42*, 4801–4803. [CrossRef]
21. Jia, C.; Liu, S.-X.; Ambrus, C.; Neels, A.; Labat, G.; Decurtins, S. One-Dimensional μ-Chloromanganese(II)−Tetrathiafulvalene (TTF) Coordination Compound. *Inorg. Chem.* **2006**, *45*, 3152–3154. [CrossRef] [PubMed]
22. Mitamura, Y.; Yorimitsu, H.; Oshima, K.; Osuka, A. Straightforward access to aryl-substituted tetrathiafulvalenes by palladium-catalysed direct C–H arylation and their photophysical and electrochemical properties. *Chem. Sci.* **2011**, *2*, 2017–2021. [CrossRef]
23. Han, Y.-F.; Zhang, J.-S.; Lin, Y.-J.; Dai, J.; Jin, G.-X. Synthesis and characterization of half-sandwich iridium complexes containing 2,6(7)-bis(4-pyridyl)-1,4,5,8-tetrathiafulvalene and ancillary ortho-carborane-1,2-dichalcogenolato ligands. *J. Organomet. Chem.* **2007**, *692*, 4545–4550. [CrossRef]
24. Wang, R.; Kang, L.-C.; Xiong, J.; Dou, X.; Chen, X.-Y.; Zuo, J.-L.; You, X.-Z. Structures and physical properties of oligomeric and polymeric metal complexes based on bis(pyridyl)-substituted TTF ligands and an inorganic analogue. *Dalton Trans.* **2011**, *40*, 919–926. [CrossRef] [PubMed]

25. Tang, Z.-Z.; Weng, Y.-G.; Yin, W.-Y.; Jiang, M.; Zhu, Q.-Y.; Dai, J. A Potential Hybrid Hole-Transport Material Incorporating a Redox-Active Tetrathiafulvalene Derivative with CuSCN. *Inorg. Chem.* **2019**, *58*, 15824–15831. [CrossRef] [PubMed]
26. Yin, Z.-N.; Li, Y.-H.; Sun, Y.-G.; Chen, T.; Xu, J.; Zhu, Q.-Y.; Dai, J. 3D Copper Tetrathiafulvalene Redox-Active Network with 8-Fold Interpenetrating Diamond-like Topology. *Inorg. Chem.* **2016**, *55*, 9154–9157. [CrossRef] [PubMed]
27. Zappe, L.; Schönfeld, S.; Hoerner, G.; Zenere, K.A.; Leong, C.F.; Kepert, C.J.; D'Alessandro, D.M.; Weber, B.; Neville, S.M. Spin crossover modulation in a coordination polymer with the redox-active bis-pyridyltetrathiafulvalene (py2TTF) ligand. *Chem. Commun.* **2020**, *56*, 10469–10472. [CrossRef]
28. Sherman, D.A.; Murase, R.; Duyker, S.G.; Gu, Q.; Lewis, W.; Lu, T.; Liu, Y.; D'Alessandro, D.M. Reversible single crystal-to-single crystal double [2+2] cycloaddition induces multifunctional photo-mechano-electrochemical properties in framework materials. *Nat. Commun.* **2020**, *11*, 1–10. [CrossRef]
29. Cerrada, E.; Laguna, M.; Bartolomé, J.; Campo, J.; Orera, V.; Jones, P.G. Cation-radical salts with organometallic gold anions. X-ray structure of [TTFPh]2[Au(C6F5)2]. *Synth. Met.* **1998**, *92*, 245–251. [CrossRef]
30. Sheldrick, G. SHELXT– Integrated space-group and crystal-structure determination. *Acta Crystallogr. Sect. A Found. Adv.* **2015**, *71*, 3–8. [CrossRef]
31. Sheldrick, G.M. Crystal structure refinement with SHELXL. *Acta Crystallogr. Sect. C Struct. Chem.* **2015**, *C71*, 3–8. [CrossRef] [PubMed]
32. Dolomanov, O.V.; Bourhis, L.J.; Gildea, R.J.; Howard, J.A.K.; Puschmann, H. OLEX2: A complete structure solution, refinement and analysis program. *J. Appl. Cryst.* **2009**, *42*, 339–341. [CrossRef]
33. *Gaussian 16, Revision C.01*; Gaussian, Inc.: Wallingford, CT, USA, 2016.
34. Boys, S.; Bernardi, F. The calculation of small molecular interactions by the differences of separate total energies. Some procedures with reduced errors. *Mol. Phys.* **1970**, *19*, 553–566. [CrossRef]
35. Becke, A.D. A new mixing of Hartree–Fock and local density—Functional theories. *J. Chem. Phys.* **1993**, *98*, 1372–1377. [CrossRef]
36. Lee, C.; Yang, W.; Parr, R.G. Development of the Colle-Salvetti correlation-energy formula into a functional of the electron density. *Phys. Rev. B* **1988**, *37*, 785–789. [CrossRef]
37. Miehlich, B.; Savin, A.; Stoll, H.; Preuss, H. Results obtained with the correlation energy density functionals of becke and Lee, Yang and Parr. *Chem. Phys. Lett.* **1989**, *157*, 200–206. [CrossRef]
38. Kendall, R.A.; Dunning, T.H., Jr. Electron affinities of the first-row atoms revisited. Systematic basis sets and wave functions. *J. Chem. Phys.* **1992**, *96*, 6796–6806. [CrossRef]
39. Del Bene, J.E.; Person, W.B.; Szczepaniak, K. Properties of Hydrogen-Bonded Complexes Obtained from the B3LYP Functional with 6-31G(d,p) and 6-31+G(d,p) Basis Sets: Comparison with MP2/6-31+G(d,p) Results and Experimental Data. *J. Phys. Chem.* **1995**, *99*, 10705–10707. [CrossRef]
40. Shultz, M.J.; Vu, T.H. Hydrogen Bonding between Water and Tetrahydrofuran Relevant to Clathrate Formation. *J. Phys. Chem. B* **2015**, *119*, 9167–9172. [CrossRef]
41. Joshi, R.; Ghanty, T.K. Hydrogen bonding interaction between HO2 radical and selected organic acids, RCOOH (R=CH3, H, Cl and F). *Chem. Phys. Lett.* **2013**, *584*, 43–48. [CrossRef]
42. Gao, H.; Zhang, Y.; Wang, H.-J.; Liu, J.; Chen, J. Theoretical Study on the Structure and Cation–Anion Interaction of Amino Acid Cation Based Amino Acid Ionic Liquid [Pro]+[NO3]$^-$. *J. Phys. Chem. A* **2010**, *114*, 10243–10252. [CrossRef] [PubMed]
43. Velde, G.; Bickelhaupt, F.M.; Baerends, E.J.; Guerra, C.F.; van Gisbergen, S.J.A.; Snijders, J.G.; Ziegler, T. Chemistry with ADF. *J. Comput. Chem.* **2001**, *22*, 931–967. [CrossRef]
44. *ADF 2019.3, SCM, Theoretical Chemistry*; Vrije Universiteit: Amsterdam, The Netherlands. Available online: https://www.scm.com/product/adf/ (accessed on 26 November 2020).
45. *AMS DFTB 2019.3, SCM, Theoretical Chemistry*; Vrije Universiteit: Amsterdam, The Netherlands. Available online: https://www.scm.com/product/ams/ (accessed on 26 November 2020).
46. Levi, O.P.-T.; Becker, J.Y.; Ellern, A.; Khodorkovsky, V. Synthesis of 2,3-dimethylthio-6-pyridyl tetrathiafulvalene: A precursor for a new system involving a direct linkage between a strong donor (D) and a strong acceptor (A). *Tetrahedron Lett.* **2001**, *42*, 1571–1573. [CrossRef]

47. Crystallographic data for TTF-py2_PF6: C52H42F12N6O2P2S12, M = 1457.57, T = 120 K, Monoclinic, Space group P21/c, Z = 2, a = 10.5747(10) Å, b = 13.6826(7) Å, c = 19.9351(11) Å, β = 90.256(6)°, V = 2884.4(3) Å3, Dc = 1.678 mg cm–3, 16917 reflections collected, final R1 and wR2 (I > 2σ(I)) = 0.0910 and 0.2264. CCDC deposition number 2040487. Available online: https://www.ccdc.cam.ac.uk/ (accessed on 26 November 2020).
48. Bondi, A. van der Waals Volumes and Radii. *J. Phys. Chem.* **1964**, *68*, 441–451. [CrossRef]
49. Matsuzaki, S.; Moriyama, T.; Toyoda, K. Raman spectra of mixed valent TTF salts; relation between Raman frequency and formal charge. *Solid State Commun.* **1980**, *34*, 857–859. [CrossRef]
50. Rosokha, S.V.; Kochi, J.K. Molecular and Electronic Structures of the Long-Bonded π-Dimers of Tetrathiafulvalene Cation-Radical in Intermolecular Electron Transfer and in (Solid-State) Conductivity. *J. Am. Chem. Soc.* **2007**, *129*, 828–838. [CrossRef]
51. Scott, B.A.; La Placa, S.J.; Torrance, J.B.; Silverman, B.D.; Welber, B. The crystal chemistry of organic metals. Composition, structure, and stability in the tetrathiafulvalinium-halide systems. *J. Am. Chem. Soc.* **1977**, *99*, 6631–6639. [CrossRef]
52. Torrance, J.B.; Scott, B.A.; Welber, B.; Kaufman, F.B.; Seiden, P.E. Optical properties of the radical cation tetrathiafulvalenium (TT F+) in its mixed-valence and monovalence halide salts. *Phys. Rev. B* **1979**, *19*, 730–741. [CrossRef]
53. Jacobsen, C.S. *Optical Properties in Semiconductors and Semimetals*; Conwell, E., Ed.; Academic Press: Boston, MA, USA, 1988; Volume 27, pp. 293–384.
54. Kittel, C. *Introduction to Solid State Physics*; Wiley: Hoboken, NJ, USA, 2004.
55. Wood, P.A.; Allen, F.H.; Pidcock, E. Hydrogen-bond directionality at the donor H atom—Analysis of interaction energies and database statistics. *CrystEngComm* **2009**, *11*, 1563–1571. [CrossRef]
56. Qu, L.; Iguchi, H.; Takaishi, S.; Habib, F.; Leong, C.F.; D'Alessandro, D.M.; Yoshida, T.; Abe, H.; Nishibori, E.; Yamashita, M. Porous Molecular Conductor: Electrochemical Fabrication of Through-Space Conduction Pathways among Linear Coordination Polymers. *J. Am. Chem. Soc.* **2019**, *141*, 6802–6806. [CrossRef] [PubMed]
57. Koyama, S.; Tanabe, T.; Takaishi, S.; Yamashita, M.; Iguchi, H. Preliminary chemical reduction for synthesizing a stable porous molecular conductor with neutral metal nodes. *Chem. Commun.* **2020**, *56*, 13109–13112. [CrossRef] [PubMed]
58. Koo, J.Y.; Yakiyama, Y.; Lee, G.R.; Lee, J.; Choi, H.C.; Morita, Y.; Kawano, M. Selective Formation of Conductive Network by Radical-Induced Oxidation. *J. Am. Chem. Soc.* **2016**, *138*, 1776–1779. [CrossRef] [PubMed]
59. Chen, D.; Xing, H.; Su, Z.-M.; Wang, C. Electrical conductivity and electroluminescence of a new anthracene based metal–organic framework with π-conjugated zigzag chains. *Chem. Commun.* **2016**, *52*, 2019–2022. [CrossRef] [PubMed]
60. Park, S.S.; Hontz, E.R.; Sun, L.; Hendon, C.H.; Walsh, A.; Van Voorhis, T.; Dincă, M. Cation-Dependent Intrinsic Electrical Conductivity in Isostructural Tetrathiafulvalene-Based Microporous Metal–Organic Frameworks. *J. Am. Chem. Soc.* **2015**, *137*, 1774–1777. [CrossRef] [PubMed]
61. Skorupskii, G.; Trump, B.A.; Kasel, T.W.; Brown, C.M.; Hendon, C.H.; Dincă, M. Efficient and tunable one-dimensional charge transport in layered lanthanide metal-organic framworks. *Nat. Chem.* **2020**, *12*, 131–136. [CrossRef]

crystals

MDPI

Article

Chiral Conducting Me-EDT-TTF and Et-EDT-TTF-Based Radical Cation Salts with the Perchlorate Anion

Nabil Mroweh [1], Pascale Auban-Senzier [2], Nicolas Vanthuyne [3], Elsa B. Lopes [4], Manuel Almeida [4], Enric Canadell [5] and Narcis Avarvari [1,*]

1 MOLTECH-Anjou, UMR 6200, CNRS, UNIV Angers, 2 bd Lavoisier, CEDEX, 49045 Angers, France; nabil.mroweh@cea.fr
2 Laboratoire de Physique des Solides, Université Paris-Saclay CNRS UMR 8502, Bât. 510, 91405 Orsay, France; pascale.senzier@universite-paris-saclay.fr
3 CNRS, Centrale Marseille, iSm2, Aix Marseille Université, 13397 Marseille, France; nicolas.vanthuyne@univ-amu.fr
4 Centro de Ciencias e Tecnologias Nucleares (C^{2}TN) and Departmento de Engenharia e Ciencias Nucleares (DECN), Instituto Superior Técnico (IST), Universidade de Lisboa, E.N. 10, 2695-066 Bobadela LRS, Portugal; eblopes@ctn.tecnico.ulisboa.pt (E.B.L.); malmeida@ctn.tecnico.ulisboa.pt (M.A.)
5 Campus de la UAB, Institut de Ciència de Materials de Barcelona, ICMAB-CSIC, 08193 Bellaterra, Spain; canadell@icmab.es
* Correspondence: narcis.avarvari@univ-angers.fr; Tel.: +33-(24)-173-5084

Received: 10 November 2020; Accepted: 19 November 2020; Published: 23 November 2020

Abstract: Introduction of chirality in the field of molecular conductors has received increasing interest in recent years in the frame of modulation of the crystal packing, and hence conducting properties, with the number of stereogenic centers and absolute configuration, e.g., racemic or enantiopure forms. Here, we describe the preparation by electrocrystallization of chiral radical cation salts, based on the donors methyl-ethylenedithio-tetrathiafulvalene (Me-EDT-TTF) **1** and ethyl-ethylenedithio-tetrathiafulvalene (Et-EDT-TTF) **2** containing one stereogenic center, with the perchlorate anion. Donor **1** provided the series of crystalline materials [(*rac*)-**1**]ClO_4, [(*S*)-**1**]$_2ClO_4$ and [(*R*)-**1**]$_2ClO_4$, while for donor **2** only the 1:1 salts [(*rac*)-**2**]ClO_4 and [(*R*)-**2**]ClO_4 could be prepared as suitable single crystals for X-ray analysis. The enantiopure salts of **1** show β-type packing and metallic conductivity in the high temperature regime, with room temperature conductivity values of 5–10 S cm^{-1}, whereas compound [(*rac*)-**2**]ClO_4 is a very poor semiconductor. Tight-binding extended Hückel band structure calculations support the metallic conductivity of the enantiopure salts of **1** and suggest that small structural changes, possibly induced by thermal contraction or pressure, could lead to a pseudo-elliptic closed Fermi surface, typical for a 2D metal.

Keywords: organic conductors; chirality; tetrathiafulvalene; crystal structures; electrical resistivity; band structure calculations

1. Introduction

In the frame of the increasingly developing field of chiral molecular materials [1], we recently introduced methyl-ethylenedithio-tetrathiafulvalene (Me-EDT-TTF) **1** and ethyl-ethylenedithio-tetrathiafulvalene (Et-EDT-TTF) **2** (Scheme 1) as valuable precursors for chiral molecular conductors [2,3], through a series of radical cation salts with the PF_6^- anion, in which modulation of the donor:anion stoichiometry, crystal packing and conducting properties with the absolute configuration and steric bulkiness of the substituent was observed [4]. Interesting differences were observed, for example, between the enantiopure PF_6^- salts of **1** and those of the dimethylated donor DM-EDT-TTF

(Scheme 1), since the former crystallize in the triclinic space group $P1$ and show metallic behaviour in the high-temperature regime, while the latter crystallize in the monoclinic space group $P2_1$ and are bandgap semiconductors [5]. These results point out the importance of the number of stereogenic centres in the modulation of intermolecular interactions with the anion, through the establishment of C–H···F hydrogen bonding, donor packing, and thus electron transport properties. Moreover, the (*rac*), (*S*) and (*R*) $(\mathbf{1})_2PF_6$ salts are isostructural with the $(DM\text{-}EDT\text{-}TTF)_2XF_6$ (X = As, Sb) series [6], showing that the size of the anion, in conjunction with the number of stereogenic centres, can also play a paramount role in the modulation of the intermolecular interactions. Enantiopure monoalkylated EDT-TTF **1** and **2** and derivatives [7] have afforded so far only the PF_6^- series of radical cation salts. Comparatively, since the first report of an enantiopure TTF precursor, namely (*S,S,S,S*)-TM-BEDT-TTF [8], enantiopure dimethylated EDT-TTF (DM-EDT-TTF) [5,6,9] and BEDT-TTF (DM-BEDT-TTF) [10–12], together with the tetramethylated BEDT-TTF (TM-BEDT-TTF) [13–16] (Scheme 1), provided several series of radical cation salts, culminating in the first observation of the electrical magnetochiral anisotropy effect (eMChA) in a bulk crystalline conductor [17]. Note that diethyl BEDT-TTF (DEt-BEDT-TTF) and tetraethyl BEDT-TTF (TEt-BEDT-TTF) (Scheme 1) precursors have been reported only in the racemic form [18]. Besides the hexafluorophosphate, another anion very often encountered in TTF based radical cation salts, including in the chiral ones, is the perchlorate. Indeed, enantiopure DM-EDT-TTF provided enantiomorphic mixed-valence salts $(DM\text{-}EDT\text{-}TTF)_2ClO_4$ showing the eMChA effect [17], the use of (*S,S,S,S*)-TM-BEDT-TTF afforded a 3:2 salt with metallic conductivity [13], while an orthorhombic semiconducting phase [11] and a metallic monoclinic phase [19], both presenting a 2:1 stoichiometry, have been reported for (*R,R*)-DM-BEDT-TTF and (*S,S*)-DM-BEDT-TTF, respectively. Moreover, a superconducting transition was suggested for the latter [19], yet this assumption was recently invalidated by more complete physical measurements on both enantiomeric salts [20].

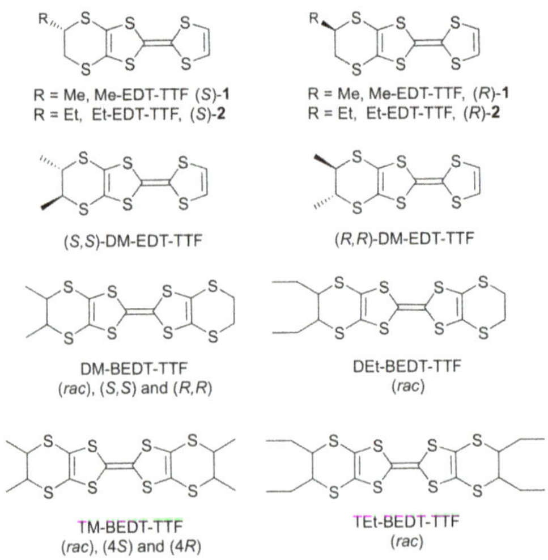

Scheme 1. Methyl and ethyl substituted ethylenedithio-tetrathiafulvalene (EDT-TTF) and bis(ethylenedithio)-tetrathiafulvalene (BEDT-TTF) donors.

In the continuation of our research line dedicated to chiral molecular conductors, we describe herein the synthesis, structural characterization and electron transport properties of racemic and enantiopure radical cation salts of Me-EDT-TTF **1** and Et-EDT-TTF **2** with the perchlorate anion, together with band structure calculations, highlighting some peculiar features of their electronic structures.

2. Materials and Methods

Donors **1** and **2** were prepared as recently described [4]. For each of the three electrocrystallization experiments with **1**, $[NBu_4]ClO_4$ (20 mg) was dissolved in 6 mL THF and the solution was poured into the cathodic compartment of an electrocrystallization cell. The anodic chamber was filled respectively with 5 mg of (*rac*)-**1**, (*S*)-**1** and (*R*)-**1** dissolved in 6 mL THF. Crystals of each salt grown at 3 °C over a period of one week on a platinum wire electrode, by applying a constant current of 1 µA. Black plates were obtained on the electrode for [(*rac*)-**1**]ClO_4 and brown thin narrow plates for [(*S*)-**1**]$_2ClO_4$ and [(*R*)-**1**]$_2ClO_4$. Similar experimental conditions were applied for the electrocrystallization of (*rac*)-**2**, (*S*)-**2** and (*R*)-**2** in the presence of $[NBu_4]ClO_4$. Black crystalline needles were obtained on the electrode for [(*rac*)-**2**]ClO_4, while for (*R*)-**2** a mixture of small green thin plates, formulated as [(*R*)-**2**]$_2ClO_4$, and dark violet thick plates, corresponding to [(*R*)-**2**]ClO_4, were collected after several trials. The quality of the former was not sufficient to allow a high quality crystal structure, yet it allowed to clearly identify the components and to determine the packing of the donors. Unfortunately, no salt with (*S*)-**2** crystallized under these conditions.

Single crystals of the compounds were mounted on glass fibre loops using a viscous hydrocarbon oil to coat the crystal and then transferred directly to cold nitrogen stream for data collection. Data collection was performed on an Agilent Supernova diffractometer with CuKα (λ = 1.54184 Å). The structures were solved by direct methods with the SIR92 program and refined against all F^2 values with the SHELXL-97 program using the WinGX graphical user interface. All non-H atoms were refined anisotropically. Hydrogen atoms were introduced at calculated positions (riding model), included in structure factor calculations but not refined. Further details about data collection and solution refinement are given in Table 1; Table 2.

Crystallographic data for the five structures were deposited with the Cambridge Crystallographic Data Centre, deposition numbers CCDC 2043591 ([(*rac*)-**1**]ClO_4), 2043592 ([(*R*)-**1**]$_2ClO_4$), 2043593 ([(*S*)-**1**]$_2ClO_4$), 2043594 ([(*rac*)-**2**]ClO_4), 2043595 ([(*R*)-**2**]ClO_4). These data can be obtained free of charge from CCDC, 12 Union road, Cambridge CB2 1EZ, UK (e-mail: deposit@ccdc.cam.ac.uk or http://www.ccdc.cam.ac.uk).

Table 1. Crystal Data for compounds [(*rac*)-**1**]ClO_4, [(*R*)-**1**]$_2ClO_4$ and [(*S*)-**1**]$_2ClO_4$.

Compound	[(*rac*)-1]ClO_4	[(*R*)-1]$_2ClO_4$	[(*S*)-1]$_2ClO_4$
formula	$C_9H_8ClO_4S_6$	$C_{18}H_{16}ClO_4S_{12}$	$C_{18}H_{16}ClO_4S_{12}$
M (g mol^{-1})	407.96	716.48	716.48
(T)	150.00(10)	155.00(7)	170.58(10)
crystal system	Triclinic	Monoclinic	Monoclinic
space group	*P*-1	C2	C2
a (Å)	6.4126(4)	15.1135(5)	15.1561 (5)
b (Å)	9.4439(6)	6.5681(2)	6.5791(2)
c (Å)	12.2713(8)	26.6686(9)	26.6154(9)
α (°)	97.293(5)	90	90
β (°)	94.927(5)	091.898(3)	91.828(3)
γ (°)	96.828(5)	90	90
V (Å^3)	728.05(8)	2645.86(15)	2652.56(15)
Z	2	4	4
ρ_{calcd} (g cm^{-3})	1.861	1.799	1.794
μ (mm^{-1})	10.464	10.390	10.364
Flack parameter	-	0.06(2)	0.04(4)
Completeness (%)	99.1%	99.5%	99.8%
goodness-of-fit on F^2	1.061	1.081	1.040
final R1/wR2 [I > 2σ(I)]	R1 = 0.0676, wR2 = 0.1781	R1 = 0.0611, wR2 = 0.1583	R1 = 0.0515, wR2 = 0.1331
R1/wR2 (all data)	R1 = 0.0809, wR2 = 0.1903	R1 = 0.0629, wR2 = 0.1597	R1 = 0.0560, wR2 = 0.1361

Table 2. Crystal Data for compounds [(*rac*)-**2**]ClO_4 and [(*R*)-**2**]ClO_4.

Compound	[(*rac*)-2]ClO_4	[(*R*)-2]ClO_4
formula	$C_{10}H_{10}ClO_4S_6$	$C_{10}H_{10}ClO_4S_6$
M (g mol^{-1})	421.99	421.99
T (K)	150.00(10)	150.00(10)
crystal system	Triclinic	Triclinic
space group	P-1	$P1$
a (Å)	7.5501(9)	7.5729(5)
b (Å)	9.3157(10)	9.1781(6)
c (Å)	11.7417(11)	11.8048(9)
α (°)	80.593(9)	80.392(6)
β (°)	75.196(11)	76.155(6)
γ (°)	79.910(9)	81.022(6)
V (Å^3)	779.95(14)	779.68(9)
Z	2	2
ρ_{calcd} (g cm^{-3})	1.797	1.797
μ (mm^{-1})	9.791	9.794
Flack parameter	-	0.10(6)
Completeness (%)	99.5%	97.1%
goodness-of-fit on F^2	1.047	1.049
final R1/wR2 [I > 2σ(I)]	R1 = 0.0662, wR2 = 0.1751	R1 = 0.0531, wR2 = 0.1423
R1/wR2 (all data)	R1 = 0.0795, wR2 = 0.1981	R1 = 0.0574, wR2 = 0.1487

Electrical conductivity and the thermoelectric power measurements for [(*S*)-**1**]$_2ClO_4$ and [(*R*)-**1**]$_2ClO_4$ single crystals were made along the long axis of the crystals (*a* axis) in the temperature range of 20/50–310 K, using a measurement cell attached to the cold stage of a closed-cycle helium refrigerator. The thermopower was measured by using a slow AC (ca. 10^2 Hz) technique [21], by attaching two Ø = 25 μm diameter 99.99% pure Au wires (Goodfellow, UK), thermally anchored to two quartz blocks, with Pt paint (Demetron, Germany, 308A) to the extremities of an elongated sample as in a previously described apparatus [22], controlled by a computer [23]. The oscillating thermal gradient was kept below 1 K and was measured with a differential Au 0.05 at. % Fe vs. chromel thermocouple of the same type. The absolute thermoelectric power of the samples was obtained after correction for the absolute thermopower of the Au leads, by using the data of Huebner [24]. Electrical resistivity measurements were done both in a four-in-line contact configuration, using a low-frequency AC method (77 Hz) with a 5–10 μA current, the voltage is measured with an SRS (Stanford Research Systems, California, USA) Model SR83 lock-in amplifier, and in two points configuration. For resistivity measurements with two contacts, gold wires were glued with silver paste directly on both ends of the crystals and low temperature was provided by a homemade cryostat equipped with a 4 K pulse-tube. For [(*S*)-**1**]$_2ClO_4$ and [(*R*)-**1**]$_2ClO_4$ single crystals, we applied a DC current (0.1 μA) and measured the voltage with a Keithley 2401 microvoltmeter (Tektronix, Inc Beaverton, OR 97077 United States). Note that, in this configuration, the resistance of the two contacts is added to the resistance of the sample which makes it difficult to evidence a metallic behavior when cooling down. For measuring much higher resistance values in [(*rac*)-**2**]ClO_4 and [(*R*)-**2**]ClO_4 single crystals, we used a different technique always in two points configuration. We applied a constant voltage (10–15 V) and measured the current using a Keithley 6487 Picoammeter/Voltage Source (Tektronix, Inc Beaverton, OR 97077 United States). However, [(*R*)-**2**]ClO_4 single crystals could not be cooled down because the limit in current detection (around 50 pA) was nearly reached at room temperature.

The tight-binding band structure calculations were of the extended Hückel type [25]. A modified Wolfsberg–Helmholtz formula was used to calculate the non-diagonal $H_{\mu\nu}$ values [26]. All valence electrons were taken into account in the calculations and the basis set consisted of Slater-type orbitals of double-ζ quality for C 2s and 2p, S 3s and 3p and of single-ζ quality for H 1s. The ionization potentials, contraction coefficients and exponents were taken from previous work [27].

3. Results and Discussion

Donors **1** and **2** were prepared as racemic mixtures according to the protocol we recently reported, followed by chiral HPLC separation to afford the pure enantiomers [4]. The racemic and enantiopure forms of both donors were engaged in electrocrystallization experiments in the presence of $[NBu_4]ClO_4$ as a supporting electrolyte. Suitable single crystals for X-ray diffraction analysis of radical cation salts with ClO_4^- were obtained for (*rac*)-**1**, (*R*)-**1**, (*S*)-**1**, (*rac*)-**2** and (*R*)-**2**.

3.1. Solid-State Structures of the Radical Cation Salts

3.1.1. [(*rac*)-1]ClO_4, [(R)-1]$_2ClO_4$ and [(S)-1]$_2ClO_4$

Black thick crystalline plates of [(*rac*)-**1**]ClO_4 were collected on the electrode. The compound crystallizes in the triclinic centrosymmetric space group *P*–1, with one independent donor molecule and one anion in the asymmetric unit (Figure 1a). The methyl substituent lies in the axial position and the oxygen atom O1 is disordered over two positions O1A (s.o.f. 0.28) and O1B (s.o.f. 0.72).

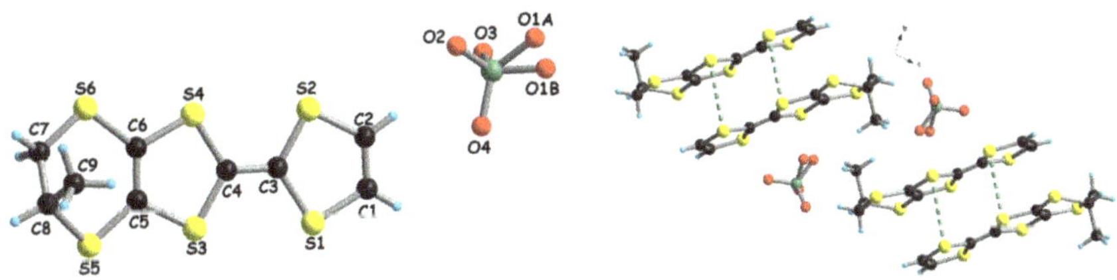

Figure 1. (**a**) Molecular structure of [(*rac*)-**1**]ClO_4 (O1 atom is disordered). Only the (*S*) enantiomer of the donor is shown; (**b**) Packing diagram of [(*rac*)-**1**]ClO_4 in the *bc* plane, with an emphasis on the intradimer S···S short contacts (3.38 and 3.47 Å, green dotted lines).

The donors form very stable centrosymmetric head-to-tail dimers (*vide infra*) through the establishment of very short S···S intermolecular contacts of 3.385 and 3.466 Å (Figure 1b). These dimers interact especially laterally along the *a*-direction as denoted by the short S···S intermolecular contacts ranging between 3.39 and 3.87 Å (Figure S1). The stoichiometry 1:1 is in favor of a +1 oxidation state of the donors, in agreement with the values of the central C=C and internal C–S bonds (Table 3).

Table 3. Selected C=C and C–S bond lengths for [(*rac*)-**1**]ClO_4, [(*R*)-**1**]$_2ClO_4$ and [(*S*)-**1**]$_2ClO_4$.

	Bond Lengths (Å)				
	compound	[(*R*)-1]$_2ClO_4$	[(*S*)-1]$_2ClO_4$	[(*rac*)-1]ClO_4	
A	C3A—C4A	1.385(13)	1.372(8)	C3—C4	1.390(8)
	S1A—C3A	1.738(10)	1.746(6)	S1—C3	1.721(6)
	S2A—C3A	1.731(10)	1.729(6)	S2—C3	1.715(6)
	S3A—C4A	1.720(9)	1.749(6)	S3—C4	1.710(6)
	S4A—C4A	1.730(10)	1.730(6)	S4—C4	1.722(5)
B	C3B—C4B	1.368(13)	1.392(7)	-	-
	S1B—C3B	1.744(9)	1.734(6)	-	-
	S2B—C3B	1.736(10)	1.733(6)	-	-
	S3B—C4B	1.750(10)	1.711(5)	-	-
	S4B—C4B	1.733(10)	1.726(6)	-	-

As already extensively discussed for the $(\mathbf{1})_2PF_6$ series of salts [4], the anion is also involved here in several hydrogen-bonding interactions between the oxygen atoms and protons of different types (CH_{vinyl}, CH_{Me}, CH_2 and CH_3) (Figure S2).

The enantiopure salts $[(R)\text{-}\mathbf{1}]_2ClO_4$ and $[(S)\text{-}\mathbf{1}]_2ClO_4$ are isostructural and crystallize in the monoclinic non-centrosymmetric space group *C*2, the asymmetric unit containing two independent donor molecules and two anion halves located on C_2 symmetry axes. Unlike the racemic salt, now the methyl substituent is located in an equatorial position in both independent molecules (Figure 2a for the (*R*) enantiomer). Since the two compounds are isostructural, only $[(R)\text{-}\mathbf{1}]_2ClO_4$ will be detailed hereafter. Parallel columns of donors oriented in a head-to-tail manner are present in the organic layer, an arrangement reminiscent of the β-type packing [28]. Intrastack and interstack short S···S contacts of 3.52–3.54 Å and 3.62–3.63 Å, respectively, are observed (Figure 2b).

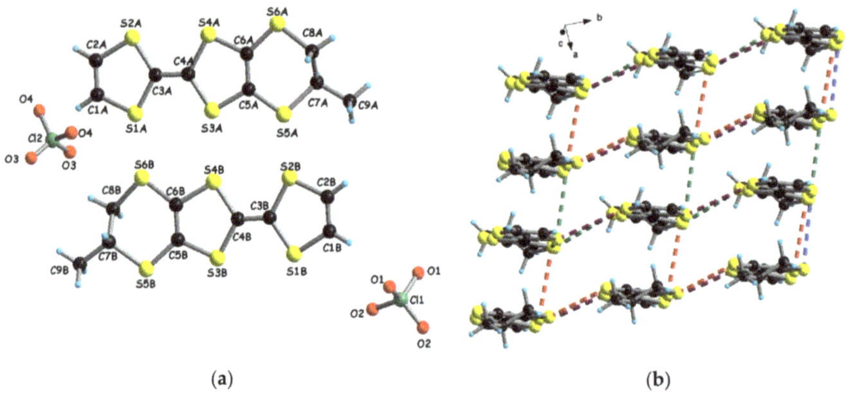

Figure 2. (**a**) Molecular structure of $[(R)\text{-}\mathbf{1}]_2ClO_4$ along with the atom numbering scheme; (**b**) Donor layer with an emphasis on the S···S short contacts: red dotted lines (3.52–3.54 Å), blue dotted lines (3.81 Å), green dotted lines (3.62–3.63 Å) and violet dotted lines (4.02 Å).

The analysis of the central C=C and internal C–S bond lengths (Table 3) suggests that the charge is equally distributed over the two independent donor molecules which are thus in a mixed-valence state. Again, the anions and the donors engage in a set of hydrogen bond interactions, such as CH_{vinyl}···O (2.43–2.51 Å), H_{CH2}··· O (2.46–2.63 Å) and H_{Me}···O (2.62 Å) (Figure 3).

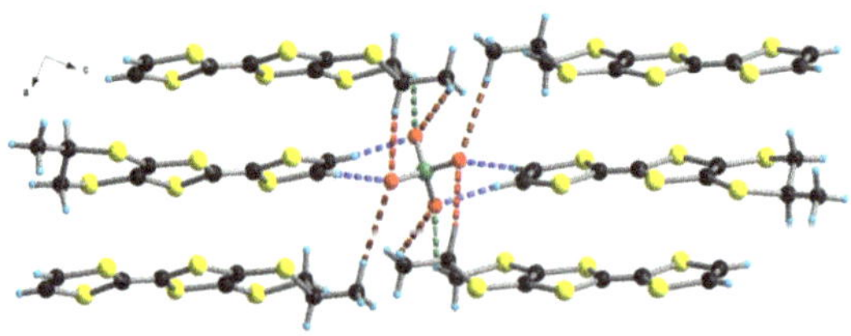

Figure 3. Solid-state structure of $[(R)\text{-}\mathbf{1}]_2ClO_4$, with an emphasis on the C–H···O short contacts: blue dotted lines for CH_{vinyl} (2.43–2.51 Å), green dotted lines for CH_2 (2.46–2.63 Å), brown dotted lines for Me (2.62 Å) and red dotted line for CH_{Me} (2.99 Å).

The crystal structures of $[(R)\text{-}\mathbf{1}]_2ClO_4$ and $[(S)\text{-}\mathbf{1}]_2ClO_4$ are thus completely different from those of the analogous salts with the dimethylated donor DM-EDT-TTF, i.e., $[(R,R)\text{-DM-EDT-TTF}]_2ClO_4$ and $[(S,S)\text{-DM-EDT-TTF}]_2ClO_4$, which crystallized in hexagonal enantiomorphic space groups [17], pointing out the importance of the number of stereogenic centers in the modulation of the packing.

3.1.2. [(rac)-2]ClO_4 and [(R)-2]ClO_4

Electrocrystallization of donor **2** provided crystalline materials only for the racemic and (*R*) enantiomer, with two different phases of 1:1 and 2:1 stoichiometry for the latter. However, the crystals of the [(*R*)-**2**]$_2ClO_4$ phase were not of sufficient quality to allow an accurate description of the structure, yet the packing of the donors is similar to the one in [(*R*)-**1**]$_2ClO_4$ (*vide supra*). [(*rac*)-**2**]ClO_4 crystallized in the triclinic system, *P*–1 space group, with one independent donor and one anion in the asymmetric unit (Figure 4a), while [(*R*)-**2**]ClO_4, which is isostructural to the racemic counterpart, crystallized in the non-centrosymmetric space group *P*1 of the triclinic system, with two donors and two anions in the asymmetric unit (Figure S3).

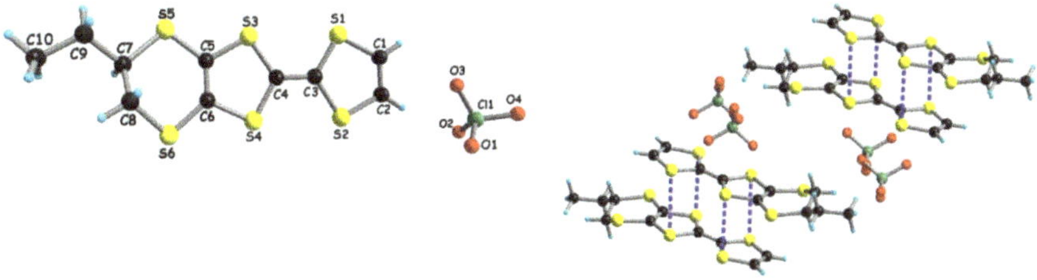

(**a**) (**b**)

Figure 4. (**a**) Molecular structure of [(*rac*)-**2**]ClO_4 along with the atom numbering scheme. Only the (*R*) enantiomer of the donor is shown; (**b**) Packing diagram of [(*rac*)-**2**]ClO_4, with an emphasis on the intradimer S···S short contacts (3.37 and 3.47 Å, blue dotted lines).

According to the 1:1 stoichiometry the donors should bear a +1 charge, in agreement with the central C=C and internal C–S bond lengths (Table S1). In both structures the ethyl substituents are located in the equatorial position, although the donors arrange in head-to-tail dimers as in [(*rac*)-**1**]ClO_4 where the methyl substituent is axial, with very short intradimer S···S distances of 3.37–3.47 Å for [(*rac*)-**2**]ClO_4 (Figure 4b) and 3.35–3.37 Å for [(*R*)-**2**]ClO_4 (Figure S4). This strong dimerization should lead, very likely, to rather poor electron transport properties (*vide infra*). As in the case of donor **1**, a complex set of hydrogen bond interactions between the oxygen atoms of the anion and the hydrogen atoms of the donors can be disclosed (Figure S5 for [(*rac*)-**2**]ClO_4 and Figure S6 for [(*R*)-**2**]ClO_4), emphasizing once again the template role of the anion in the structural disposition of the donors.

3.2. *Single Crystal Conductivity Measurements*

In the (**1**)$_2ClO_4$ series only the crystals of the enantiopure phases, namely [(*R*)-**1**]$_2ClO_4$ and [(*S*)-**1**]$_2ClO_4$, were of suitable dimensions for two- and four-contact single-crystal resistivity measurements, although great care had to be taken because of the fragility of the crystalline plates. The measured room temperature conductivity values range between 5–10 S cm^{-1} depending on the quality of the sample and of the contacts. However, the temperature dependence of the resistivity suggests metal-like conductivity in the high-temperature range (partially masked in the two points measurements), followed upon cooling by a localized regime with a very low activation energy of 29–47 meV (340–540 K) (Figure 5a). The thermoelectric power measurements are also indicative of metallic behavior in the high-temperature regime (Figure 5b) when considering the very small positive values decreasing towards zero upon cooling.

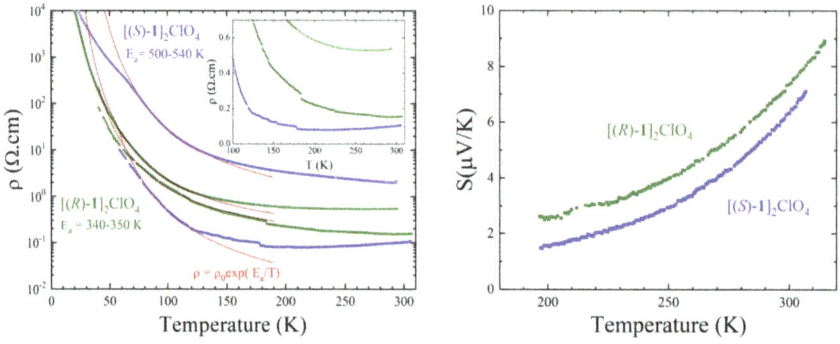

(**a**) (**b**)

Figure 5. (**a**) Temperature dependence of the electrical resistivity ρ for single crystals of [(*S*)-**1**]$_2$ClO$_4$ (blue curves) and [(*R*)-**1**]$_2$ClO$_4$ (green curves) measured using either four in-line contacts (two lower curves, full symbols) or two contacts (two upper curves, empty symbols). The red lines are the fit to the activation law $\rho = \rho_0 \exp(E_a/T)$ in the 80–130 K temperature range. The inset shows the same resistivity data on a linear scale in order to emphasize the high temperature conducting regime in the three lower curves. (**b**) Temperature dependence of the thermoelectric power for a single crystal of [(*S*)-**1**]$_2$ClO$_4$ (blue curve) and [(*R*)-**1**]$_2$ClO$_4$ (green curve).

The 2:1 stoichiometry and *β*-type packing of the donors in [(*R*)-**1**]$_2$ClO$_4$ and [(*S*)-**1**]$_2$ClO$_4$ are clearly in favor of high electrical conductivity in the organic layer, as suggested by the single crystal electrical resistivity measurements. The situation is drastically different in the case of (**2**)ClO$_4$ salts. Suitable crystals for temperature-dependent resistivity measurements could be obtained only for [(*rac*)-**2**]ClO$_4$. As expected when considering the strong dimerization of the donors in the crystal structure, the room temperature value of the resistivity of 10^7 Ω cm and activation energy E_a = 5100 K clearly indicate a very poor semiconducting, almost insulating, behavior of this material (Figure 6 and Figure S7). Moreover, a value of $\rho = 2 \times 10^8$ Ω cm room temperature resistivity could be measured on small crystals of [(*R*)-**2**]ClO$_4$, which are isostructural with the racemic form.

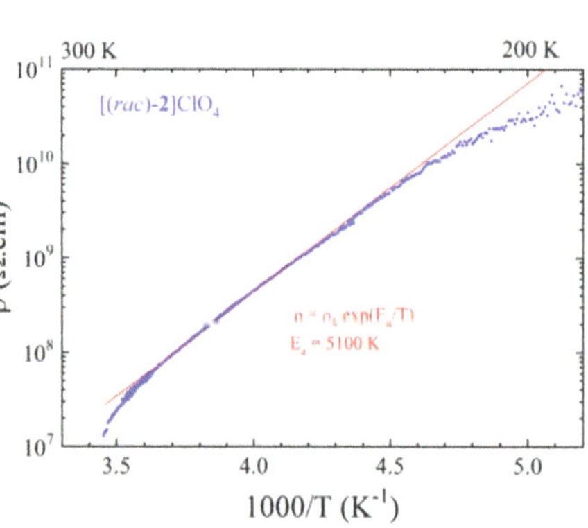

Figure 6. Temperature dependence of the electrical resistivity, ρ, plotted as log ρ versus 1000/T for a single crystal of [(*rac*)-**2**]ClO$_4$ measured using two contacts. The red line is the linear fit to the data giving the activation energy.

Since the two racemic salts [(*rac*)-**1**]ClO_4 and [(*rac*)-**2**]ClO_4 are structurally very similar it can be inferred that the former is also a poor semiconductor.

3.3. *Band Structure Calculations*

3.3.1. [(*rac*)-**1**]ClO_4 and [(*rac*)-**2**]ClO_4

Although the samples of [(*rac*)-**1**]ClO_4 did not allow single-crystal resistivity measurements, unlike those of [(*rac*)-**2**]ClO_4, we performed band structure calculations on both racemic salts for comparison purpose. The donor lattice of the [(*rac*)-**1**]ClO_4 salt is shown in Figure 7a, highlighting the presence of dimers. Although there are quite short contacts associated with all interactions (Table 4), the very good σ-type overlap associated with the interaction I should make this interaction largely dominant. Because of the head-to-tail overlap mode the S···S contacts along *b*- and *c*- are long. However, note that whereas the methyl substituents practically cut any interaction along *c*, the donors face each other through the non-substituted side of the molecule along *b*, providing a better situation for possible inter-dimer interaction. The S···S contacts shorter than 3.9 Å as well as the associated $\beta_{HOMO\text{-}HOMO}$ values are reported in Table 4. Interaction I is almost two orders of magnitude larger than the other interactions, so that this salt must be considered as made of chains, along the *a*-direction, of very stable $(Me\text{-}EDT\text{-}TTF_2)^{2+}$ dimers.

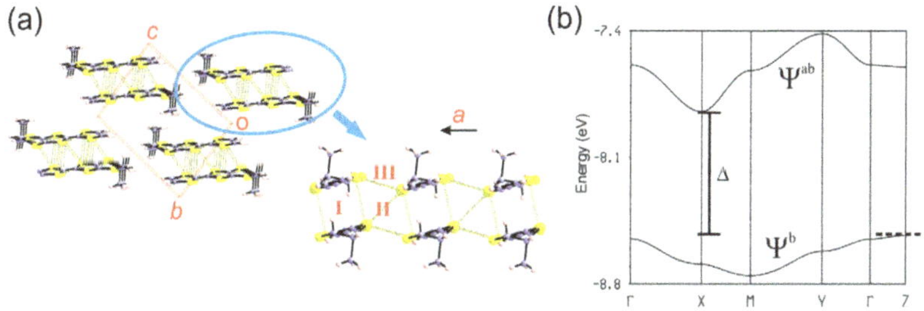

Figure 7. (**a**) Donor lattice of the [(*rac*)-**1**]ClO_4 salt where the chains and the main intermolecular interactions are labeled; (**b**) Extended Hückel band structure for the donor lattice of [(*rac*)-**1**]ClO_4. The dashed line refers to the highest occupied level and Γ = (0, 0, 0), X = (a^*/2, 0, 0), Y = (0, b^*/2, 0), M = (a^*/2, b^*/2, 0) and Z = (0, 0, c^*/2).

Table 4. Intermolecular S···S contacts shorter than 3.9 Å and absolute values of the $\beta_{HOMO\text{-}HOMO}$ interaction energies (eV) [[29]] for the different donor···donor interactions in the [(*rac*)-**1**]ClO_4 salt.

Interaction	S···S (Å)	$\lvert\beta_{HOMO\text{-}HOMO}\rvert$ (eV)
I	3.385 (×2), 3.466 (×2)	0.9401
II	3.518 (×2), 3.793	0.0207
II	3.394, 3.497, 3.839, 3.849, 3.877	0.0401

The calculated band structure is shown in Figure 7b where an indirect bandgap (from Z to X) of 0.67 eV clearly separates the two HOMO bands (i.e., only the lower band, built from the bonding combination of the two HOMOs, Ψ^b, is filled). Note that whereas the dispersion of the lower band is weak, that of the upper one is quite considerable and of the same order along the directions *a* and *b*. Thus the effective mass of the electron carriers should be considerably larger than that of the hole carriers and may confer a considerably 2D character to the activated conductivity. The larger dispersion associated with the upper band is a consequence of the antibonding nature of the upper Ψ^{ab} orbital of the dimer for two reasons: (i) because of the inclination of the molecules with respect to the *a* axis the inter-dimer overlap along this direction is better, and (ii) because of the intra-dimer antibonding character the Ψ^{ab} orbital somewhat hybridizes with molecular σ levels so as to shift the wave function towards the outer region of the dimer, thus slightly decreasing the antibonding nature of Ψ^{ab} and somewhat favoring the inter-dimer overlap along *b*.

The calculated band structure for [(*rac*)-**2**]ClO_4 (see Figure S8) leads to an equivalent description and an indirect bandgap of 0.8 eV from Γ to X, in excellent agreement with our conductivity measurements.

3.3.2. [(*R*)-**1**]$_2$$ClO_4$ and [(*S*)-**1**]$_2$$ClO_4$

The donor layers of the [(*R*)-**1**]$_2$$ClO_4$ and [(*S*)-**1**]$_2$$ClO_4$ salts contain two different donors and six different intermolecular interactions (see Figure 8). The layer can be described as a series of parallel chains of AB dimers (interaction I) running along the (*a* + *b*)-direction. Every donor is implicated in two different interactions along the chain and four inter-chain interactions. To understand the electronic structure and transport properties of these salts we need to have a hint on the strength of the HOMO···HOMO interactions which may be done by looking at the absolute value of the so-called HOMO···HOMO interaction energy, $\beta_{HOMO\text{-}HOMO}$, [29] associated with each interaction. The calculated values for the six interactions in the two different layers of the two pure enantiomeric salts are reported in Table 5.

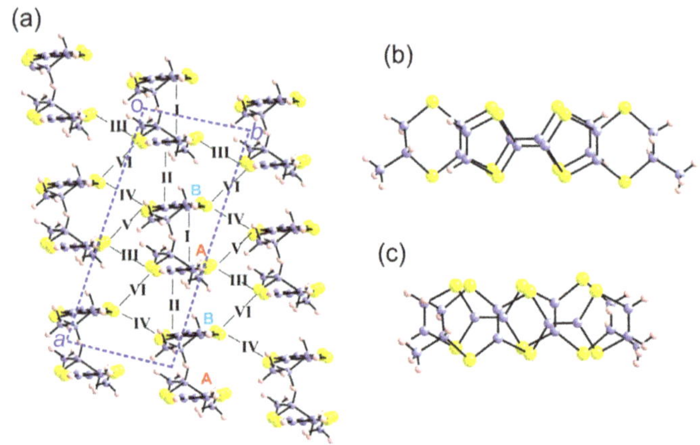

Figure 8. (**a**) Donor layer of the [(*R*)-**1**]$_2$$ClO_4$ salt where the different donors and intermolecular interactions are labeled; (**b**) Top view of the intra-dimer interaction I; (**c**) Top view of the inter-dimer interaction II.

Table 5. Absolute values of the $\beta_{HOMO\text{-}HOMO}$ interaction energies (eV) for the different donor···donor interactions in the [(*R*)-**1**]$_2$$ClO_4$ and [(*S*)-**1**]$_2$$ClO_4$ salts.

Interaction	[(*R*)-1]$_2$$ClO_4$	[(*S*)-1]$_2$$ClO_4$
I (A-B)	0.7880	0.7865
II (A-B)	0.6366	0.6285
III (A-A)	0.0303	0.0290
IV (B-B)	0.0370	0.0375
V (A-B)	0.0774	0.0765
VI (A-B)	0.1012	0.0945

The results for both enantiomers are practically identical. The more salient observation is that the interactions along the chain (I and II) are strong whereas the inter-chain interactions (III to VI) are practically one order of magnitude weaker. This should confer a clear one-dimensional (1D) character to the system. Note that the two intra-chain interactions, despite originating from different overlap modes (Figure 8b,c), are associated with similar interaction energy values, showing that a better orientation of the HOMOs (electronic effect) may perfectly compensate for a larger set of short contacts (metric effect). Because of this fact, the chains along the diagonal ($a + b$)-direction are quite uniform as far as the HOMO···HOMO interactions are concerned. Note that the lateral interactions although weaker are by no means negligible so that they should provide a substantial coupling between these quite uniform chains.

The calculated band structure and Fermi surface of the [(*R*)-**1**]$_2$ClO$_4$ salt are shown in Figure 9 (those for [(*S*)-**1**]$_2$ClO$_4$ are practically identical as shown in Figure S9). The diagram in Figure 9a contains four HOMO bands because there are four donors per repeat unit. Because of the stoichiometry, there are two holes in these bands so that two partially filled and very dispersive bands are generated, in agreement with the metallic character of these salts. The calculated Fermi surface may appear complex at a first sight although really it is not. The repeat unit of the layer contains two identical AB pairs so that the calculation could have been performed using a centered rectangular lattice with just two molecules in the unit cell. In that case, the calculated Fermi surface would simply be the red part of Figure 9b (and the area of the Brillouin zone would be twice larger). Using the crystallographic repeat unit, which is twice larger, two identical Fermi surfaces are folded into the new, smaller Brillouin zone shown in Figure 9b. The warped red lines of Figure 9b are perpendicular to the ($a + b$)-direction as it must be if the system is built from chains of HOMOs along the ($a + b$)-direction. The warping of the red lines is substantial because of the above noted lateral interactions; in fact the two lines almost touch at the S point so that it is possible that small structural changes brought about by thermal contraction or pressure could maybe close the lines and lead to a pseudo-elliptic closed Fermi surface (i.e., a 2D metal). Note that the red Fermi surface is almost identical to those reported for the (DM-EDT-TTF)XF$_6$ (X = P, As, Sb) salts [5,6], where the donor possesses two stereogenic centers, as well as for (**1**)$_2$PF$_6$ [4], in which, as in the present salts, the donor has a single stereogenic center, thus denoting a very similar organization of the layers in all these salts. We direct the reader to these references for further discussion of the electronic structure of these layers. However, we note that even if some parts of the red Fermi surface of Figure 9b are nested, such nesting is incomplete. Consequently, if metal to insulator low-temperature transitions occur in these salts, as for some of the above-mentioned ones, it is not expected that they may originate from the charge or spin density wave instabilities so frequent in low dimensional systems but from electronic or structural localization.

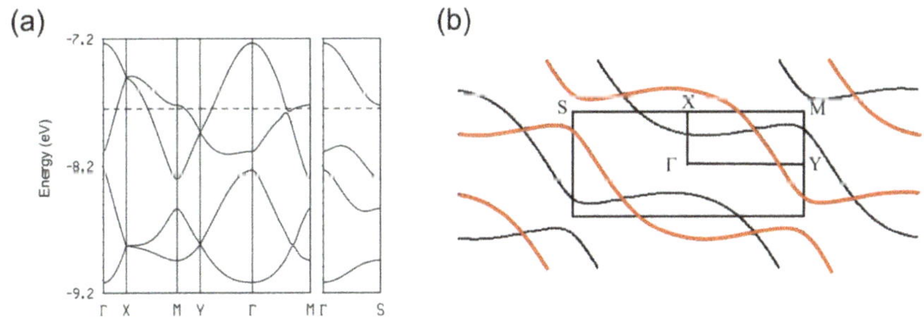

Figure 9. (**a**) Extended Hückel band structure and (**b**) Fermi surface for the donor layers of [(*R*)-**1**]$_2$ClO$_4$. The dashed line refers to the calculated Fermi level and Γ = (0, 0), X = (a^*/2, 0), Y = (0, b^*/2), M = (a^*/2, b^*/2) and S = (a^*/2, −b^*/2).

4. Conclusions

Electrocrystallization of the chiral EDT-TTF derivatives **1** and **2** provided a complete series of radical cation salts for the former and racemic and enantiopure (*R*) salts for the latter, with the same perchlorate anion. Interestingly, unlike the series with the PF$_6^-$ anion previously described [4], where the racemic and enantiopure salts were isostructural, here, the perchlorate ion discriminates between the racemic and enantiopure forms (Table 6).

Table 6. Comparison between the radical cation salts of donors **1** and **2** with the ClO_4^- and PF_6^- anions.

Radical Cation Salt	Stoichiometry (Donor:Anion)	Crystal System	Conductivity (High Temperature)
[(*rac*)-**1**]ClO_4	1:1	Triclinic	Poor semiconductor
[(*R*)-**1**]$_2ClO_4$	2:1	Monoclinic	Metal
[(*S*)-**1**]$_2ClO_4$	2:1	Monoclinic	Metal
[(*rac*)-**1**]$_2PF_6$	2:1	Triclinic	Metal
[(*R*)-**1**]$_2PF_6$	2:1	Triclinic	Metal
[(*S*)-**1**]$_2PF_6$	2:1	Triclinic	Metal
[(*rac*)-**2**]ClO_4	1:1	Triclinic	Poor semiconductor
[(*R*)-**2**]ClO_4	1:1	Triclinic	Poor semiconductor
[(*rac*)-**2**]PF_6•(C_4H_8O)	1:1	Triclinic	Poor semiconductor
[(*R*)-**2**]$_2PF_6$	2:1	Triclinic	Semiconductor
[(*S*)-**2**]$_2PF_6$	2:1	Triclinic	Semiconductor

In the [(*rac*)-**1**]ClO_4 salt, which should be a poor semiconductor with an indirect gap, the donors are fully oxidized and arrange in centrosymmetric dimers. In sharp contrast, the donors form a β-type packing in the mixed-valence enantiopure salts [(*R*)-**1**]$_2ClO_4$ and [(*S*)-**1**]$_2ClO_4$ salts, which show metal-like behavior in the high-temperature regime, in agreement with extended Hückel band structure calculations. From a structural, electron transport properties and electronic structure point of view, these enantiopure salts resemble the (**1**)$_2PF_6$ and (DM-EDT-TTF)$_2XF_6$ (X = As, Sb) series, but not to the enantiopure (DM-EDT-TTF)$_2ClO_4$ compounds [17], pointing out that the variation of the number of stereogenic centers and the use of racemic or enantiopure forms in combination with various anions is a simple mean of reaching a large collection of chiral molecular conductors with original electronic structures. Indeed, both donors **1** and DM-EDT-TTF provided mixed-valence salts of 2:1 stoichiometry with the ClO_4^- anion, showing metal-like conductivity, yet their crystal structures are drastically different, i.e., monoclinic space group for [(*R*)-**1**]$_2ClO_4$ and [(*S*)-**1**]$_2ClO_4$ and enantiomorphic hexagonal space groups for [(*R,R*)-DM-EDT-TTF]$_2ClO_4$ and [(*S,S*)-DM-EDT-TTF]$_2ClO_4$ [17]. This is a clear consequence of the different number of stereogenic centers and symmetry of the donor, impacting the intermolecular hydrogen bonding interactions with the anion and the overlap interactions between the donors. Further work in these families of materials will be devoted to conductivity measurements under pressure and under a magnetic field, and to the use of larger octahedral or tetrahedral anions.

Supplementary Materials: The following are available online at http://www.mdpi.com/2073-4352/10/11/1069/s1, Figure S1: Packing diagram of [(*rac*)-**1**]ClO_4 in the *ac* plane with an emphasis on the short lateral S···S contacts, Figure S2: Solid state structure of [(*rac*)-**1**]ClO_4 with an emphasis on the C–H···O short contacts, Figure S3: Molecular structure of [(*R*)-**2**]ClO_4 along with the atom numbering scheme, Figure S4: Packing diagram of [(*R*)-**2**]ClO_4, with an emphasis on the intradimer S···S short contacts, Figure S5: Solid state structure of [(*rac*)-**2**]ClO_4 with an emphasis on the C–H···O short contacts, Figure S6: Solid state structure of [(*R*)-**2**]ClO_4 with an emphasis on the C–H···O short contacts, Figure S7: Temperature dependence of the electrical resistivity ρ for a single crystal of [(*rac*)-**2**]ClO_4 measured using two contacts, Figure S8: Extended Hückel band structure for the donor lattice of [(*rac*)-**2**]ClO_4, Figure S9: Extended Hückel band structure (**a**) and Fermi surface (**b**) for the donor layers of [(*S*)-**1**]$_2ClO_4$, Table S1: Selected C=C and C–S bond lengths for [(*rac*)-**2**]ClO_4 and [(*R*)-**2**]ClO_4.

Author Contributions: N.A. conceived and designed the experiments; N.M. synthesized and characterized the materials; N.V. performed the chiral HPLC separation of the precursors; P.A.-S., E.B.L. and M.A. investigated the electron transport properties; E.C. undertook the theoretical study; N.A. and E.C. wrote and/or reviewed the manuscript with contributions from all authors. All authors have read and agreed to the published version of the manuscript.

Funding: This research was partially funded in France by the National Agency for Research (ANR), Project 15-CE29-0006-01 ChiraMolCo, in Spain by the Spanish MICIU through Grant PGC2018-096955-B-C44 and the Severo Ochoa FUNFUTURE (CEX2019-000917-S) Excellence Center distinction, as well as by Generalitat de Catalunya (2017SGR1506) and in Portugal by FCT under contracts UIDB/04349/2020 and LISBOA-01-0145-FEDER-029666.

Acknowledgments: This work was supported in France by the CNRS and the University of Angers. The collaboration between the Portuguese and French team members was also supported by a FCT–French Ministry of Foreign Affairs bilateral action FCT/PHC-PESSOA 2020-21 (Project 44647UB).

Conflicts of Interest: The authors declare no conflict of interest.

References

1. Pop, F.; Zigon, N.; Avarvari, N. Main-Group-Based Electro- and Photoactive Chiral Materials. *Chem. Rev.* **2019**, *119*, 8435–8478. [CrossRef] [PubMed]
2. Avarvari, N.; Wallis, J.D. Strategies towards Chiral Molecular Conductors. *J. Mater. Chem.* **2009**, *19*, 4061–4076. [CrossRef]
3. Pop, F.; Avarvari, N. Chiral Metal-Dithiolene Complexes. *Coord. Chem. Rev.* **2017**, *346*, 20–31. [CrossRef]
4. Mroweh, N.; Auban-Senzier, P.; Vanthuyne, N.; Canadell, E.; Avarvari, N. Chiral EDT-TTF precursors with one stereogenic centre: Substituent size modulation of the conducting properties in the (R-EDT-TTF)$_2$PF$_6$ (R = Me or Et) series. *J. Mater. Chem. C* **2019**, *7*, 12664–12673. [CrossRef]
5. Pop, F.; Auban-Senzier, P.; Frąckowiak, A.; Ptaszyński, K.; Olejniczak, I.; Wallis, J.D.; Canadell, E.; Avarvari, N. Chirality Driven Metallic versus Semiconducting Behavior in a Complete Series of Radical Cation Salts Based on Dimethyl-Ethylenedithio-Tetrathiafulvalene (DM-EDT-TTF). *J. Am. Chem. Soc.* **2013**, *135*, 17176–17186. [CrossRef]
6. Pop, F.; Auban-Senzier, P.; Canadell, E.; Avarvari, N. Anion size control of the packing in the metallic versus semiconducting chiral radical cation salts (DM-EDT-TTF)$_2$XF$_6$ (X = P, As, Sb). *Chem. Commun.* **2016**, *52*, 12438–12441. [CrossRef]
7. Mroweh, N.; Pop, F.; Mézière, C.; Allain, M.; Auban-Senzier, P.; Vanthuyne, N.; Alemany, P.; Canadell, E.; Avarvari, N. Combining chirality and hydrogen bonding in methylated ethylenedithio-tetrathiafulvalene primary diamide precursors and radical cation salts. *Cryst. Growth Des.* **2020**, *20*, 2516–2526. [CrossRef]
8. Wallis, J.D.; Karrer, A.; Dunitz, J.D. Chiral metals? A chiral substrate for organic conductors and superconductors. *Helv. Chim. Acta* **1986**, *69*, 69–70. [CrossRef]
9. Mroweh, N.; Mézière, C.; Allain, M.; Auban-Senzier, P.; Canadell, E.; Avarvari, N. Conservation of structural arrangements and 3:1 stoichiometry in a series of crystalline conductors of TMTTF, TMTSF, BEDT-TTF, and chiral DM-EDT-TTF with the oxo-bis[pentafluorotantalate(V)] dianion. *Chem. Sci.* **2020**, *11*, 10078–10091. [CrossRef]
10. Matsumiya, S.; Izuoka, A.; Sugawara, T.; Taruishi, T.; Kawada, Y. Effect of Methyl Substitution on Conformation and Molecular Arrangement of BEDT-TTF Derivatives in the Crystalline Environment. *Bull. Chem. Soc. Jpn.* **1993**, *66*, 513–522. [CrossRef]
11. Matsumiya, S.; Izuoka, A.; Sugawara, T.; Taruishi, T.; Kawada, Y.; Tokumoto, M. Crystal Structure and Conductivity of Chiral Radical Ion Salts (Me$_2$ET)$_2$X. *Bull. Chem. Soc. Jpn.* **1993**, *66*, 1949–1954. [CrossRef]
12. Pop, F.; Allain, M.; Auban-Senzier, P.; Martínez-Lillo, J.; Lloret, F.; Julve, M.; Canadell, E.; Avarvari, N. Enantiopure Conducting Salts of Dimethylbis(ethylenedithio)tetrathiafulvalene (DM-BEDTTTF) with the Hexachlororhenate(IV) Anion. *Eur. J. Inorg. Chem.* **2014**, *2014*, 3855–3862. [CrossRef]
13. Karrer, A.; Wallis, J.D.; Dunitz, J.D.; Hilti, B.; Mayer, C.W.; Bürkle, M.; Pfeiffer, J. Structures and Electrical Properties of Some New Organic Conductors Derived from the Donor Molecule TMET (*S,S,S,S*-Bis(dimethylethylenedithio) tetrathiafulvalene). *Helv. Chim. Acta* **1987**, *70*, 942–953. [CrossRef]
14. Pop, F.; Laroussi, S.; Cauchy, T.; Gómez-García, C.J.; Wallis, J.D.; Avarvari, N. Tetramethyl-Bis(ethylenedithio)-Tetrathiafulvalene (TM-BEDT-TTF) Revisited: Crystal Structures, Chiroptical Properties, Theoretical Calculations, and a Complete Series of Conducting Radical Cation Salts. *Chirality* **2013**, *25*, 466–474. [CrossRef] [PubMed]
15. Yang, S.; Pop, F.; Melan, C.; Brooks, A.C.; Martin, L.; Horton, P.; Auban-Senzier, P.; Rikken, G.L.J.A.; Avarvari, N.; Wallis, J.D. Charge transfer complexes and radical cation salts of chiral methylated organosulfur donors. *CrystEngComm* **2014**, *16*, 3906–3916. [CrossRef]
16. Pop, F.; Batail, P.; Avarvari, N. Enantiopure Radical Cation Salt Based on Tetramethyl-bis(ethylenedithio)-tetrathiafulvalene and Hexanuclear Rhenium Cluster. *Crystals* **2016**, *6*, 8. [CrossRef]
17. Pop, F.; Auban-Senzier, P.; Canadell, E.; Rikken, G.L.J.A.; Avarvari, N. Electrical magneto-chiral anisotropy in a bulk chiral molecular conductor. *Nat. Commun.* **2014**, *5*, 3757. [CrossRef]
18. Kini, A.M.; Parakka, J.P.; Geiser, U.; Wang, H.-H.; Rivas, F.; DiNino, E.; Thomas, S.; Dudek, J.D.; Williams, J.M. Tetraalkyl- and dialkyl-substituted BEDT-TTF derivatives and their cation-radical salts: Synthesis, structure, and properties. *J. Mater. Chem.* **1999**, *9*, 883–892. [CrossRef]

19. Zambounis, J.S.; Mayer, C.W.; Hauenstein, K.; Hilti, B.; Hofherr, W.; Pfeiffer, J.; Bürkle, M.; Rihs, G. Crystal Structure and Electrical Properties of κ-((*S*,*S*)-DMBEDT-TTF)$_2$ClO$_4$. *Adv. Mater.* **1992**, *4*, 33–35. [CrossRef]
20. Mroweh, N.; Mézière, C.; Pop, F.; Auban-Senzier, P.; Alemany, P.; Canadell, E.; Avarvari, N. In Search of Chiral Molecular Superconductors: κ-[(*S*,*S*)-DM-BEDT-TTF]$_2$ClO$_4$ Revisited. *Adv. Mater.* **2020**, *32*, 2002811. [CrossRef]
21. Chaikin, P.M.; Kwak, J.F. Apparatus for thermopower measurements on organic conductors. *Rev. Sci. Instrum.* **1975**, *46*, 218–220. [CrossRef]
22. Almeida, M.; Alcácer, L.; Oostra, S. Anisotropy of thermopower in N-methyl-N-ethylmorpholinium bistetracyanoquinodimethane, MEM(TCNQ)$_2$, in the region of the high-temperature phase transitions. *Phys. Rev. B* **1984**, *30*, 2839–2844. [CrossRef]
23. Lopes, E.B. *INETI-Sacavém*; Internal Report; INETI Press: Sacavém, Portugal, 1991.
24. Huebener, R.P. Thermoelectric Power of Lattice Vacancies in Gold. *Phys. Rev.* **1964**, *135*, A1281–A1291. [CrossRef]
25. Whangbo, M.-H.; Hoffmann, R. The Band Structure of the Tetracyanoplatinate Chain. *J. Am. Chem. Soc.* **1978**, *100*, 6093–6098. [CrossRef]
26. Ammeter, J.H.; Bürgi, H.-B.; Thibeault, J.; Hoffmann, R. Counterintuitive Orbital Mixing in Semiempirical and ab Initio Molecular Orbital Calculations. *J. Am. Chem. Soc.* **1978**, *100*, 3686–3692. [CrossRef]
27. Pénicaud, A.; Boubekeur, K.; Batail, P.; Canadell, E.; Auban-Senzier, P.; Jérome, D. Hydrogen-Bond Tuning of Macroscopic Transport Properties from the Neutral Molecular Component Site along the Series of Metallic Organic-Inorganic Solvates (BEDT-TTF)$_4$Re$_6$Se$_5$C1$_9$·[guest], [guest = DMF, THF, dioxane]. *J. Am. Chem. Soc.* **1993**, *115*, 4101–4112. [CrossRef]
28. Mori, T. Structural Genealogy of BEDT-TTF-Based Organic Conductors I. Parallel Molecules: β and β'' Phases. *Bull. Chem. Soc. Jpn.* **1998**, *71*, 2509–2526. [CrossRef]
29. Whangbo, M.-H.; Williams, J.M.; Leung, P.C.W.; Beno, M.A.; Emge, T.J.; Wang, H.H. Role of the Intermolecular Interactions in the Two-Dimensional Ambient-Pressure Organic Superconductors β-(ET)$_2$I$_3$ and β-(ET)$_2$IBr$_2$. *Inorg. Chem.* **1985**, *24*, 3500–3502. [CrossRef]

Publisher's Note: MDPI stays neutral with regard to jurisdictional claims in published maps and institutional affiliations.

crystals

MDPI

Article

Single Crystal Heat Capacity Measurement of Charge Glass Compound θ-(BEDT-TTF)$_2$CsZn(SCN)$_4$ Performed under Current and Voltage Application

Kosei Hino, Tetsuya Nomoto, Satoshi Yamashita and Yasuhiro Nakazawa *

Department of Chemistry, Graduate School of Science, Osaka University, Machikaneyama 1-1, Toyonaka, Osaka 560-0043, Japan; hino.kosei.ma@m-ep.co.jp (K.H.); nomotot15@chem.sci.osaka-u.ac.jp (T.N.); sayamash@chem.sci.osaka-u.ac.jp (S.Y.)

* Correspondence: nakazawa@chem.sci.osaka-u.ac.jp; Tel.: +81-6-6850-5396

Received: 30 October 2020; Accepted: 18 November 2020; Published: 21 November 2020

Abstract: Heat capacity measurements of θ-(BEDT-TTF)$_2$CsZn(SCN)$_4$ in its non-equilibrium electronic states induced by applying electric currents and voltages were performed by a modified relaxation calorimetry technique. We developed a single crystal heat capacity measurements system by which the Joule heating produced in samples by external currents and voltages can be balanced with the cooling power to make a non-equilibrium steady state. Although temperature versus time profiles in the relaxation process in calorimetry can be obtained as exponential curves as in the usual relaxation technique, we found that the change of resistivity that occurs during the heating and relaxation process should be taken into account in analyzing the data. By correcting this factor in the analyses, we succeeded in evaluating absolute values of $C_p(I)$ and $C_p(V)$ in these non-equilibrium states. The experiments up to 150 μA and the constant voltage of 20 mV do not induce visible change in the structure of the Boson peak in C_pT^{-3} vs. T suggestive of the glassy ground state of phonons. Although the suppression of the short-range fluctuations of the charge density has been reported, it does not seriously affect the glassy phonons in this current range.

Keywords: charge glass; heat capacity; electric current; electric voltage; Boson peak

1. Introduction

The θ-type charge transfer complexes consisting of an organic donor molecule, BEDT-TTF (bis(ethylenedithio)tetrathiafulvalene), and monovalent anions are attracting extensive interest in the area of electron correlation science related to the charge and lattice degrees of freedom [1,2]. Peculiar charge transport features, such as non-linear conductivity, large magnetoresistance [3], thyristor effect, etc. [4,5], and the formation of non-equilibrium charge glass (CG) states [6–8] are considered as manifestations of charge dynamics induced by inter-molecular Coulomb interactions in the frustrated lattice. These compounds have a layered structure of alternatively stacked π-electron donor sheets and insulating counter anion (X^{-1}) sheets and the chemical composition is expressed as (BEDT-TTF)$_2$X. The θ-type arrangement of donor molecules is classified as a typical non-dimeric structure and the electron filling of the HOMO band determined by the inter-molecules' transfer is exactly 3/4 in this case [9,10]. Although the crystal structure is orthorhombic, the neighboring molecules form a triangle lattice structure as shown in Figure 1a–c. Geometric frustration and inter-site Coulomb interaction to make charge disproportionation result in the Fermi liquid state becoming unstable at low temperatures and induce various ground states [11–14]. Although they show conductive behavior near room temperature reflecting the third-quarter band filling, charge ordering transitions accompanied by structural phase transitions occur in θ-(BEDT-TTF)$_2$RbZn(SCN)$_4$ at 195 K and in θ-(BEDT-TTF)$_2$RbCo(SCN)$_4$ at 190 K in the two-dimensional layer when the samples are cooled down slowly across these temperatures [9,15].

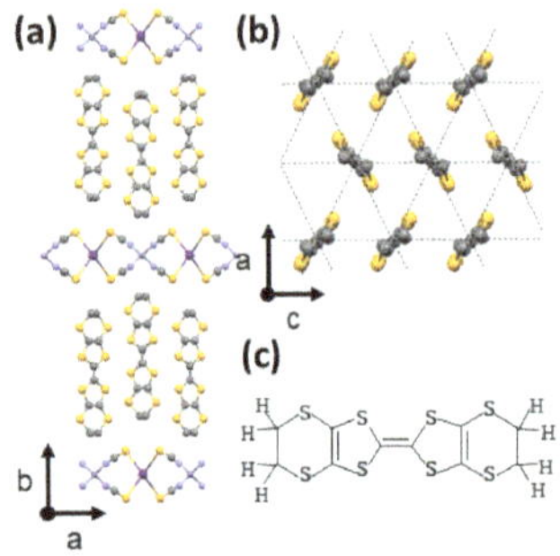

Figure 1. (**a**) Crystal structure of θ-(BEDT-TTF)$_2$CsZn(SCN)$_4$. The donor molecules and counter anions stack in a segregated structure. (**b**) The arrangement of the donor molecule of BEDT-TTF in the two-dimensional layer. The dashed line shows the triangle lattice in this arrangement. (**c**) The molecular structure of BEDT-TTF.

On the other hand, the rapidly cooled θ-(BEDT-TTF)$_2$RbZn(SCN)$_4$ and θ-(BEDT-TTF)$_2$CsZn(SCN)$_4$ are reported to freeze into the glassy state in the charge distribution (CG state) [16–18]. From the X-ray diffraction [16,17] and IR and Raman experiments [19], the competition of several types of short-range charge fluctuations coexists in them. In the case of θ-(BEDT-TTF)$_2$CsZn(SCN)$_4$, two kinds of short-range charge disproportionation modes with the wave vectors, q_1 = (2/3, k, 1/3) and q_2 = (0, k, 1/2), coexist at low temperatures. The former tends to form a three-hold periodicity, while the latter has a two-hold one. Sawano et al. observed that the application of electric currents during the X-ray analysis process suppresses the short-range fluctuations, meaning that a possible melting of the short-range correlations was induced by the currents [5].

Since the charge-lattice coupling is strong in the charge fluctuation compounds, the application of electronic currents that affects the short-range charge fluctuations may induce some changes in low-energy phonons in θ-(BEDT-TTF)$_2$CsZn(SCN)$_4$ [20]. Heat capacity measurements with electric currents for conductive materials have never been performed, although those with the application of large electric fields for perfectly insulating samples have been performed by the adiabatic method [21–23] and the relaxation method by Lashly et al. [24]. There are no reports up to now for semi-conductive and metallic materials due to the serious Joule heating produced by external currents and voltages. Here, we report the promising method for the measurement and discuss the thermodynamic properties of θ-(BEDT-TTF)$_2$CsZn(SCN)$_4$ in the non-equilibrium condition produced by external currents and voltages for the first time.

2. Experimental

2.1. Sample Preparation

The crystal used for this work was grown by the electrochemical oxidation method. The BEDT-TTF molecules were solved in 1,1,2-trichroloethane with CsSCN, Zn(SCN)$_2$, and 18-crown-6-ether. The electric current of 1 μA was applied to the solution using H-type cells for about three weeks to yield single crystals suitable for heat capacity measurements. The crystal used for this experiment was 616.4 μg with 0.40 × 0.45 × 2.0 mm^3.

2.2. Experimental Set Up of the Calorimetry Cell

The measurements of the heat capacity for the single crystal sample with applied currents were performed by a modified thermal relaxation technique. We used a calorimetry cell constructed for measuring single crystal samples in an order of 10^{1-3} μg [25,26]. To apply constant currents/voltages to the single crystal to suppress short-range charge disproportionation, we attached four thin lead wires by using a small amount of Ag paste on the gold-deposited portions on the surface of the crystal. We used constantan wires with a diameter of φ25 μm for electric leads. The same wires were also used for the leads of the Cernox sensor (CX1050, LakeShore, Carson, CA, USA) and the heater (1 kΩ strain gauge) of the calorimetry cell. The schematic view of the sample stage is shown in Figure 2a. The electric leads are connected to the anchored ports at the heat sink of the calorimeter of which the temperature is accurately controlled within ±5 mK. The applied currents induce Joule heating in the sample and give rise to temperature discrepancy between the sample part and the heat sink. This discrepancy depends on the magnitude of the external currents. However, it is possible to realize a steady state through the balance of the Joule heating and the cooling power. This discrepancy can be kept constant if the sample heater is not operated. Therefore, it is possible to get heating and relaxing curves by using the sample heater. Good thermal contact between the sample and the thermometer is required to stabilize the steady state. Therefore, we covered the bottom surface and side surfaces of the single crystal using Apiezon N grease on the sample holder. A thermal gradient occurs inside the sample part shown in Figure 2a as the sample heating is considered negligible, since the thermal conductivity between the sample and the Cernox sensor was evaluated as 50–150 μWK^{-1}, which is nearly two orders of magnitude larger than the κ value between the heat sink and sample stage. We confirmed that the base temperature during the current application can be stabilized within $\pm 5 \times 10^{-6}$ K s^{-1}.

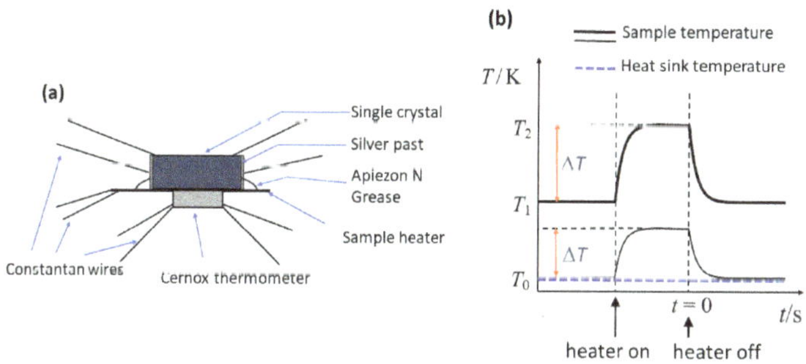

Figure 2. (**a**) Schematic illustration of the sample part of the relaxation calorimetry cell. Four lead wires are attached on the surfaces of the single crystal for applying current and voltage. (**b**) The temperature profiles against time for the standard and the modified relaxation calorimetry technique. The lower curve shows the heating and relaxation curve for the standard relaxation calorimetry. The upper one shows the case with constant currents or voltages applied for the crystal.

2.3. Analytic Method

Here, we discuss the method to determine the absolute values of the heat capacity. In the relaxation calorimetry system, the temperature profile of the sample cell is recorded as a function of time. We show the schematic view in the typical temperature change of the sample part in the normal condition and that under the applied current in Figure 2b. In the relaxation calorimetry, the absolute values of the heat capacity are determined by the analysis of the relaxation curve using Formula (1). If the temperature dependence of κ is small, it can be simplified as (2) as is used for the normal relaxation technique. The ΔT for the present experiments were set in a range between 60 and 100 mK. The typical temperature profile in the normal condition without currents is shown as the lower curve in Figure 2b:

$$P_{\text{heater}} = C_P \frac{\mathrm{d}T}{\mathrm{d}t} + \int_{T_0}^{T} \kappa(T)\mathrm{d}T, \tag{1}$$

$$P_{\text{heater}} = C_P\frac{dT}{dt} + \kappa(T - T_0). \tag{2}$$

When the constant current is applied to the sample, the sample itself also works as a heater to make constant deviation of the temperature from the heat sink. The constant current produces an increase of temperature from the base temperature, T_0, and makes a non-equilibrium state with a temperature, T_1. However, it is possible to consider it as a steady state and the thermal profile can be evaluated by Equations (3) and (4):

$$P_{\text{sample}}(T_1) = \int_{T_0}^{T_1} \kappa(T)dT, \tag{3}$$

$$P_{\text{heater}} + P_{\text{sample}}(T) - P_{\text{sample}}(T_1) = C_P\frac{dT}{dt} + \int_{T_1}^{T} \kappa(T)dT. \tag{4}$$

The possible temperature profile in the relaxation calorimetry performed in this process is shown in Figure 2b as the upper curve.

To evaluate the absolute values of the heat capacity realized in this current-induced non-equilibrium state, the evaluation of $P_{\text{sample}}(T) - P_{\text{sample}}(T_1)$ produced by the current is required. It should be pointed out that the constant deviation from the sample stage becomes large, especially in the low-temperature region, since the heat capacity of the sample becomes smaller while the electric resistivity becomes larger with the decrease of the temperature. Therefore, the minimum temperature we can attain shifts to a higher temperature. During the heating and relaxation process of the measurements, we can keep the condition of $\Delta T << (T_1 - T_0)$ and the temperature of the heat sink can be kept constant. In this case, we can reasonably consider that $P_{\text{sample}}(T_1)$ is kept constant. The temperature dependence of the electric resistivity of θ-(BEDT-TTF)$_2$CsZn(SCN)$_4$ is semi-conductive and shows an increase in the low-temperature region. Therefore, the change of the resistivity, namely the change of the Joule heating, $P_{\text{sample}}(T)$, during the heating and relaxation processes should be taken into account in each point. If we assume that the ΔT is small, in the range between 60 and 100 mK, and the resistivity change in this narrow temperature range can be expressed as a linear function as $R(T) = r_1T + r_0$, the effect of the Joule heating can be assumed as Equations (5) and (6) when the applied currents are constant and applied voltages are constant. In the case of semi-conductive temperature dependence in the present temperature range below 11 K, r_1 is always negative and has the larger absolute value:

$$P_{\text{heater}} = C_P\frac{dT}{dt} + \kappa(T - T_1) - I^2r_1(T - T_1) = C_P\frac{dT}{dt} + \kappa'(T - T_1), \tag{5}$$

$$P_{\text{heater}} = C_P\frac{dT}{dt} + \kappa(T - T_1) + V^2\frac{r_1}{{r_0}^2}(T - T_1) = C_P\frac{dT}{dt} + \kappa'(T - T_1). \tag{6}$$

The heat balance equation can be simplified by using the κ' term and similar relaxation fitting as the normal method can be used. The κ' can be described as $\kappa' = \kappa - I^2r_1$ and $\kappa' = \kappa + V^2\, r_1/{r_0}^2$ for the case of constant current I and constant voltage V being applied, respectively. The total heat capacity C_p of the sample part at various currents can be determined from the time constant of the relaxation curve and the κ' as in the usual relaxation method. The contribution of the Joule heating term to κ' for the constant current case gives a positive contribution, while it is negative for the constant voltage case.

3. Results and Discussion

In Figure 3, we show the temperature dependence of the total heat capacity of the sample part, including the contribution of the single crystal sample and those of the addenda, by the red squares. The data plotted by the blue triangles demonstrate the blank heat capacity measured without any sample and grease. The green circles show the contribution of the addenda part, including the contributions of Apiezon N grease, Au coating, and the Ag past, etc. The data shown by green circles were used as the background contribution for deriving the sample heat capacity, $C_p(I)$ and

$C_p(V)$. We can consider that the current dependence of the background heat capacity is negligible here. Figure 4 shows typical temperature relaxation curves obtained in steady state with I = 19 μA and V = 20 mV, together with the normal conduction obtained at nearly the same temperatures. Although the relaxation process can be given as single exponential curves, their time constants are different depending on the conditions of the steady states. This difference originates from the difference of the Joule heating that occurs during the relaxation processes. To evaluate this contribution quantitatively, we compared κ and κ' for several conditions in Figure 5a–d. We show the temperature dependence of κ obtained in the normal condition without any current and voltage by the red squares in Figure 5a–d. κ is mainly determined by the lead wires, which thermally terminate the sample part and the heat sink. Therefore, its temperature dependence can be described by the thermal conductivity of the constantan wires. In fact, those at 6 and 9 K can be calculated as 1.43 and 2.69 μWK^{-1}, respectively, for the case of 12 wires of φ25 μm in diameter and 5 mm in length in the present cell [27]. These values are consistent with the experimental values. Figure 5a–c also displays the experimentally determined κ' of constant current I = 19, 80, and 150 μA. respectively. Since the κ' includes the Joule heating contribution as shown in Formula (5), the temperature dependences of the κ' show large deviation from the κ values, especially in the low-temperature region. We also show the temperature dependence of κ' calculated using the experimental values of κ and resistivity values for the cases of I = 19, 80, 150 μA by blue curves in Figure 5a–c. We can confirm there is good agreement between the experimental values of κ' and the calculated ones. This result demonstrates that the analyses based on Formula (5) are reliable to determine $C_p(I)$ with various currents. We also display the temperature dependence of κ' obtained for the case of heating by applying a constant voltage of V = 20 mV in Figure 5d. Although it is difficult to detect an accurate sample current in this experiment to evaluate resistivity, usage of the measured κ' for the analysis according to Formula (6) is also available to determine $C_p(V)$.

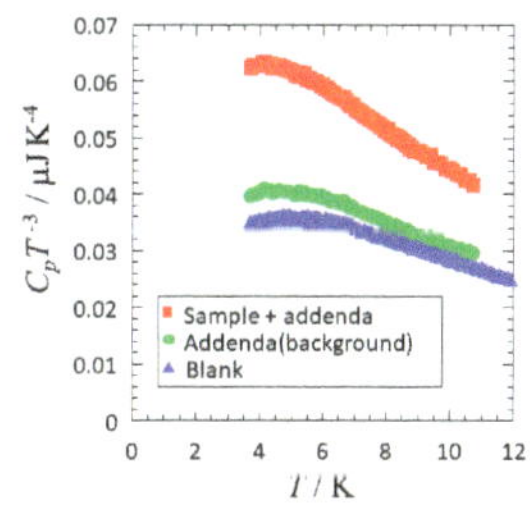

Figure 3. Temperature dependences of the total heat capacity of the sample part (red squares), addenda heat capacity (green circles), and blank heat capacity (blue triangles) shown in C_pT^{-3} vs. T plot.

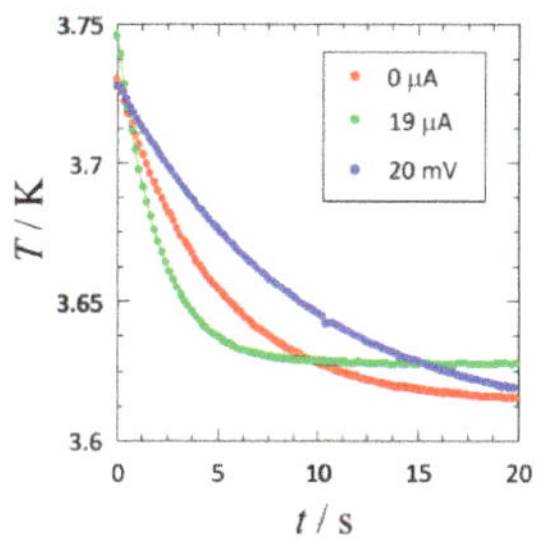

Figure 4. Typical thermal relaxation curves obtained in the present set up. The red circles show the standard case without current and voltage. The green circles denote the relaxation curve for the constant current of I = 19 μA and the blue ones show the curve for constant voltage with V = 20 mV.

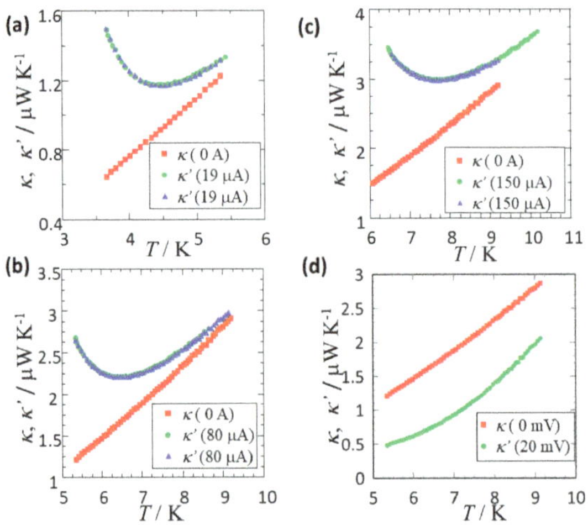

Figure 5. Temperature dependences of the experimentally determined κ (red square) and κ' (green circle) for (**a**) I = 19, (**b**) I = 80, (**c**) I = 150 μA, and (**d**) V = 20 mV. The blue circles show the calculated values from κ and P_{sample} obtained by the resistivity value of the sample and applied currents.

Since the charge disproportionation and fluctuations affect the structure of low-energy phonons through the electron-lattice coupling, the low-temperature heat capacity changes drastically depending on the ground states of the charge degrees of freedom [20,28]. A peculiar feature was observed in the heat capacity of θ-(BEDT-TTF)$_2$CsZn(SCN)$_4$ by Yoshimoto et al. in the normal condition [20]. The temperature dependence of the lattice heat capacity divided by T^3 shows a large Boson peak structure around 2–3 K, which is a typical feature of molecular glasses and amorphous compounds [20,28,29]. On the other hand, such a peak structure is absent in the charge-ordered (CO) state of slowly cooled θ-(BEDT-TTF)$_2$RbZn(SCN)$_4$ and metallic θ-(BEDT-TTF)$_2$I$_3$. This feature is consistent with the glass-like thermal conductivities of θ-(BEDT-TTF)$_2$CsZn(SCN)$_4$ and rapidly cooled θ-(BEDT-TTF)$_2$RbZn(SCN)$_4$ reported recently [30]. The Boson peak structure for the present sample was also detected by the heat capacity measurement in the normal state as shown in Figure 6a.

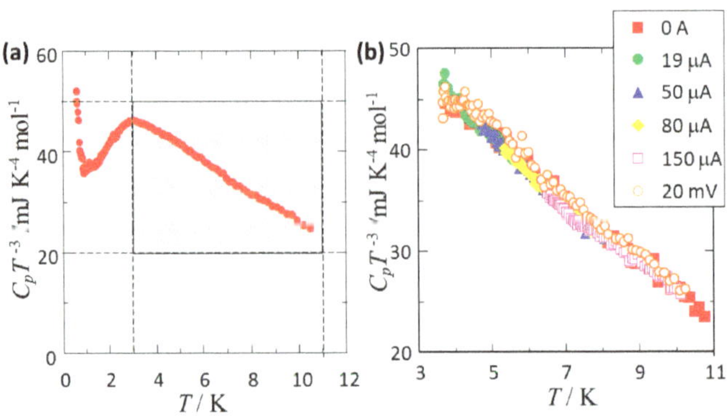

Figure 6. (**a**) Temperature dependence of the heat capacity of θ-(BEDT-TTF)$_2$CsZn(SCN)$_4$ for the single crystal sample used for the present experiment obtained by the standard relaxation technique. (**b**) Those obtained with constant current of I = 0, 19, 50, 80, 150 μA and V = 20 mV are shown in C_pT^{-3} vs. T plot. The rectangle area shown by the light blue color in (**a**) corresponding to the whole area shown in (**b**).

The molar heat capacity of θ-$(BEDT\text{-}TTF)_2CsZn(SCN)_4$ with the applied currents up to 150 μA and with the constant voltage of 20 mV obtained by using the κ' for each steady state and the addenda heat capacity data in Figure 3 is shown in Figure 6b in the C_pT^{-3} vs. T plot. Although the temperature region we can measure in the applied currents and voltage is limited, and we could not cover the whole shape of the Boson peak due to the shift of the base temperature by the larger Joule heating in the low-temperature region, we compare them with the data marked by a light blue color in Figure 5a. The agreement of the data, including the absolute values of molar heat capacity, with those of Figure 6a in the upper side region of the Boson peak is good, which means successful measurements in the applied currents and voltage. From the figure, we can confirm that the application of electric currents up to 150 μA and the application of constant voltage of 20 mV do not induce change in the heat capacity. In the previous X-ray experiments given by Sawano et al., suppression of the short-range peak denoted as q_2 = (0, k, 1/2) occurs with the application of an external current of 70 μA, while the short-range structure of q_1 = (2/3, k, 1/3) is not affected even by the current of 260 μA [5]. Since the contribution of the Boson peak is induced by the disordered phonons, the influence of the partial charge melting by applying current seems to be smaller in the present temperature and current range. It is suggested that the disordered phonons responsible for the Boson peak have the character of local phonons without long-range coherency as in acoustic phonons. This is consistent with the step-like decrease of thermal conductivity with decreasing temperature that occurs around the similar temperature region [30]. The local disorders, for example, produced by hydrogen bonding through ethylene conformations, may be a possible scenario as was reported in [31,32]. The relation with donors and counter anions should be taken into account in discussing the thermal properties of these charge-disproportionation phenomena. Probably, experiments with a much larger electric current or electric field, which can melt the glass-like charge state more drastically, can derive further information of the phonon-glass dynamics occurring in the low-energy region in θ-$(BEDT\text{-}TTF)_2CsZn(SCN)_4$. The effect of external currents and voltages for the charge disproportionation systems are quite interesting for charge-lattice coupling systems and detection by thermodynamic measurements in such non-equilibrium states should be developed in a wider temperature region.

4. Summary

In summary, we performed heat capacity measurements of θ-$(BEDT\text{-}TTF)_2CsZn(SCN)_4$ with applied current and voltage conditions. The Joule heating by the static currents was balanced with the cooling power from the heat sink to make a steady state. We also reported the method of data analyses to evaluate the accurate heat capacity with applied currents in detail. The experiments up to 150 μA and the constant bias of 20 mV did not change the Boson peak structure in the heat capacity, although some suppression effect was reported by the X-ray diffraction. We found that the suppression of the short-range fluctuations did not seriously affect the disorder in the phonon structure in this experimental range. The measurements of heat capacity under currents can give a new possibility to investigate phase transitions of electronic and lattice origin.

Author Contributions: K.H., T.N., and S.Y. worked on the synthesis and characterizations of the samples. K.H., T.N. and S.Y. conducted the low temperature heat capacity measurements with external currents and voltages. K.H. and Y.N. designed the research plan and all authors worked on the overall discussions throughout the work. K.H., T.N. and Y.N. mainly wrote the draft of the manuscript and all authors commented on the manuscript. All authors have read and agreed to the published version of the manuscript.

Funding: This research was financially supported by JSPS KAKENHI Grant Numbers JP19K22169 and 20H01862.

Acknowledgments: The authors thank T. Yamamoto (Ehime University) for valuable discussion in terms of charge disproportionation.

Conflicts of Interest: The authors declare no conflict of interest.

References

1. Mori, H. Materials Viewpoint of Organic Superconductors. *J. Phys. Soc. Jpn.* **2006**, *75*, 1–15. [CrossRef]
2. Mori, T.; Kawamoto, T. Organic conductors—From fundamentals to nonlinear conductivity. *Annu. Rep. Prog. Chem. Sect. C Phys. Chem.* **2007**, *103*, 134–172. [CrossRef]
3. Takahide, Y.; Konoike, T.; Enomoto, K.; Nishimura, M.; Terashima, T.; Uji, S.; Yamamoto, H.M. Large positive magnetoresistance of insulating organic crystals in the non-ohmic region. *Phys. Rev. Lett.* **2007**, *98*, 1–4. [CrossRef] [PubMed]
4. Inagaki, K.; Terasaki, I.; Mori, H.; Mori, T. Large Dielectric Constant and Giant Nonlinear Conduction in the Organic Conductor θ-$(BEDT\text{-}TTF)_2CsZn(SCN)_4$. *J. Phys. Soc. Jpn.* **2004**, *73*, 3364–3369. [CrossRef]
5. Sawano, F.; Terasaki, I.; Mori, H.; Mori, T.; Watanabe, M.; Ikeda, N.; Nogami, Y.; Noda, Y. An organic thyristor. *Nature* **2005**, *437*, 522–524. [CrossRef]
6. Kagawa, F.; Sato, T.; Miyagawa, K.; Kanoda, K.; Tokura, Y.; Kobayashi, K.; Kumai, R.; Murakami, Y. Charge-cluster glass in an organic conductor. *Nat. Phys.* **2013**, *9*, 419–422. [CrossRef]
7. Sasaki, S.; Hashimoto, K.; Kobayashi, R.; Itoh, K.; Iguchi, S.; Nishio, Y.; Ikemoto, Y.; Moriwaki, T.; Yoneyama, N.; Watanabe, M.; et al. Crystallization and vitrification of electrons in a glass-forming charge liquid. *Science* **2017**, *357*, 1381–1385. [CrossRef]
8. Sato, T.; Miyagawa, K.; Kanoda, K. Electronic crystal growth. *Science* **2017**, *357*, 1378–1381. [CrossRef]
9. Mori, H.; Tanaka, S.; Mori, T. Systematic study of the electronic state in θ-type BEDT-TTF organic conductors by changing the electronic correlation. *Phys. Rev. B* **1998**, *57*, 12023–12029. [CrossRef]
10. Williams, J.M.; Ferraro, J.R.; Thorn, R.J.; Carlson, K.D.; Geiser, U.; Wang, H.H.; Kini, A.M.; Whangbo, N.-H. *Organic Superconductors (Including Fullerenes): Synthesis, Structure, Properties, and Theory*; Prentice Hall: Englewood Cliffs, NJ, USA, 1992; pp. 92–93.
11. Mori, T. Non-Stripe Charge Order in the θ-Phase Organic Conductors. *J. Phys. Soc. Jpn.* **2003**, *72*, 1469–1475. [CrossRef]
12. Seo, H. Charge Ordering in Organic ET Compounds. *J. Phys. Soc. Jpn.* **2000**, *69*, 805–820. [CrossRef]
13. Clay, R.T.; Mazumdar, S.K.; Campbell, D. Charge Ordering in θ-$(BEDT\text{-}TTF)_2X$ Materials. *J. Phys. Soc. Jpn.* **2002**, *71*, 1816–1819. [CrossRef]
14. McKenzie, R.H.; Merino, J.; Marston, J.B.; Sushkov, O.P. Charge ordering and antiferromagnetic exchange in layered molecular crystals of the type. *Phys. Rev. B* **2005**, *64*, 085109. [CrossRef]
15. Mori, H.; Tanaka, S.; Mori, T. Magnetic Properties of Coexistent System of Itinerant and Localized Electrons, $(BEDT\text{-}TTF)_2MCo(SCN)_4$ (M = K, Rb, Cs). *J. Phys. I* **1996**, *6*, 1987–1996.
16. Nogami, Y.; Pouget, J.P.; Watanabe, M.; Oshima, K.; Mori, H.; Tanaka, S.; Mori, T. Structural modulation in θ-$(BEDT\text{-}TTF)_2CsM'(SCN)_4$[M′=Co, Zn]. *Synth. Met.* **1999**, *103*, 1911. [CrossRef]
17. Nogami, Y.; Hanasaki, N.; Watanabe, M.; Yamamoto, K.; Ito, T.; Ikeda, N.; Ohsumi, H.; Toyokawa, H.; Noda, Y.; Terasaki, I.; et al. Charge Order Competition Leading to Nonlinearity in Organic Thyristor Family. *J. Phys. Soc. Jpn.* **2010**, *79*, 1–5. [CrossRef]
18. Sato, T.; Kagawa, F.; Kobayashi, K.; Ueda, A.; Mori, H.; Miyagawa, K.; Kanoda, K.; Kumai, R.; Murakami, Y.; Tokura, Y. Systematic Variations in the Charge-Glass-Forming Ability of Geometrically Frustrated θ-$(BEDT\text{-}TTF)_2X$ Organic Conductors. *J. Phys. Soc. Jpn.* **2014**, *83*, 1–4. [CrossRef]
19. Suzuki, K.; Yamamoto, K.; Yakushi, K.; Kawamoto, A. Infrared and Raman Studies of θ-$(BEDT\text{-}TTF)_2$ $CsZn(SCN)_4$: Comparison with the Frozen State of θ-$(BEDT\text{-}TTF)_2RbZn(SCN)_4$. *J. Phys. Soc. Jpn.* **2005**, *74*, 2631–2639. [CrossRef]
20. Yoshimoto, R.; Takane, Y.; Hino, K.; Yamashita, S.; Nakazawa, Y. Coupling of charge and lattice degrees of freedoms in θ-type BEDT-TTF compound probed by low-temperature heat capacity measurements. *Physica B* **2014**, *449*, 19–24. [CrossRef]
21. Camnasio, J.A.; Gonzalo, A.J. Comparative Study of the Ferroelectric Specific Heat in TGS and DTGS. *J. Phys. Soc. Jpn.* **1975**, *39*, 451–459. [CrossRef]
22. Ramos, S.; del Cerro, J.; Zamora, M. Specific heat of triglycine sulfate at several applied electric fields near the critical temperature. *Phys. Status Solidi (a)* **1980**, *61*, 307–313. [CrossRef]
23. Del Cerro, J.; Martin, J.M.; Ramos, S. Specific heat measurements under non-equilibrium conditions. *J. Therm. Anal.* **1996**, *47*, 1691–1700. [CrossRef]

24. Lashley, J.C.; Hundley, M.F.; Mihaila, B.; Smith, J.L.; Opeil, C.P.; Finlayson, T.R.; Fisher, R.A.; Hur, N. Heat capacity in magnetic and electric fields near the ferroelectric transition in triglycine sulfate. *Appl. Phys. Lett.* **2007**, *90*, 1–3. [CrossRef]
25. Imajo, S.; Fukuoka, S.; Yamashita, S.; Nakazawa, Y. Construction of relaxation calorimetry for 10^{1-2} μg samples and heat capacity measurements of organic complexes. *J. Therm. Anal. Calorim.* **2015**, *123*, 1871–1876. [CrossRef]
26. Fukuoka, S.; Horie, Y.; Yamashita, S.; Nakazawa, Y. Development of heat capacity measurement system for single crystals of molecule-based compounds. *J. Therm. Anal. Calorim.* **2013**, *113*, 1303–1308. [CrossRef]
27. Duthil, P. Material Properties at Low Temperature. *arXiv* **2015**, arXiv:1501.07100.
28. Nishio, Y.; Nihei, Y.; Tamura, M.; Kajita, K.; Nakamura, T.; Takahashi, T. Specific heat and metal-insulator transition of $(BEDT\text{-}TTF)_2MZn(SCN)_4$(M = Cs, Rb). *Synth. Met.* **1999**, *103*, 1907–1908. [CrossRef]
29. Nomoto, T.; Yesil, E.; Yamashita, S.; Akutsu, H.; Nakazawa, Y. Thermodynamic properties of glassy phonon states induced by strong electron correlations in θ-type organic charge transfer salts. *Mod. Phys. Lett. B* **2020**, *34*, 1–7. [CrossRef]
30. Nomoto, T.; Yamashita, S.; Akutsu, H.; Nakazawa, Y.; Krivchikov, A.I. Phonon Glass Induced by Electron Correlation. *J. Phys. Soc. Jpn.* **2019**, *88*, 1–4. [CrossRef]
31. Pouget, J.-P.; Alemany, P.; Canadell, E. Donor-anion interactions in quarter-filled low-dimensional organic conductors. *Mat. Hori.* **2018**, *5*, 590–640. [CrossRef]
32. Alemany, P.; Pouget, J.-P.; Canadell, E. Structural and electronic control of the metal to insulator transition and local orderings in the θ-$(BEDT\text{-}TTF)_2X$ organic conductors. *J. Phys. Condens. Mat.* **2015**, *27*, 1–16. [CrossRef] [PubMed]

crystals

MDPI

Article

Microscopic Observation of π Spin Polarization by *d* Localized Spin in λ Type BETS Based Organic Superconductors

Ko-ichi Hiraki [1,*], Toshihiro Takahashi [2], Hiroshi Akiba [3], Yutaka Nishio [4] and Biao Zhou [5]

[1] Department of Natural Sciences, Fukushima Medical University, Fukushima 960-1295, Japan
[2] Department of Physics, Gakushuin University, Tokyo 171-8588, Japan; toshi.takahashi@gakushuin.ac.jp
[3] Institute for Solid State Physics, The University of Tokyo, Chiba 277-858, Japan; h-akiba@issp.u-tokyo.ac.jp
[4] Department of Physics, Toho University, Chiba 274-8510, Japan; nishio@ph.sci.toho-u.ac.jp
[5] Department of Humanities and Sciences, Nihon University, Tokyo 101-0061, Japan; zhou@chs.nihon-u.ac.jp
* Correspondence: hiraki@fmu.ac.jp

Received: 26 October 2020; Accepted: 18 November 2020; Published: 20 November 2020

Abstract: In this study, ^{77}Se NMR measurements were carried out to detect the π spin polarization of the organic BETS (BETS = Bis(ethylenedithio)tetraselenafulvalene) molecule of the field induced superconductor, λ-$(BETS)_2Fe_{1-x}Ga_xCl_4$, which shows a superconducting transition at relatively low magnetic field compared to the non-doped λ-$(BETS)_2FeCl_4$. From the analysis of the NMR spectrum at low temperature, it was clarified that the exchange interaction between π and 3*d* spins in the Ga doping system is smaller than that in the Fe salt. It is also clarified that the conduction π spins feel the "averaged" exchange field from the localized 3*d* spins at the dilute Fe sites.

Keywords: $(BETS)_2Fe_{1-x}Ga_xCl_4$; π-*d* interaction; NMR

1. Introduction

The organic charge transfer salt λ-$(BETS)_2FeCl_4$ has been attracting much physical interest due to the strong interaction between conducting π spins on the two-dimensional organic sheet and $S = 5/2$ 3*d* localized spins at the Fe site [1]. The system shows a metal–insulator transition (MIT) with antiferromagnetic ordering at 8 K under zero magnetic field [1–4]. Application of a strong magnetic field decreases the MIT temperature. A magnetic field stronger than 11 T kills the MIT and the system remains metallic down to low temperatures. The MIT of the system has been considered as a cooperative transition between organic π and *d* spins [4]. However, another scenario was proposed as a result of the specific heat measurements presented by Akiba et al. [5]. The mechanism of the MIT and the details of the antiferromagnetic ordered state are still unclear [6]. The most drastic phenomenon of this material is magnetic field-induced superconductivity (FISC) [7,8]. The explanation of this unusual phenomenon is the compensation mechanism [9]: polarized 3*d* electrons produce a strong magnetic field through the exchange interaction with π conduction electrons, which cancels the applied field. The magnitude of the exchange field by localized *d* spin was evaluated experimentally and the compensation mechanism was evidenced by several studies [5,10,11] including our NMR studies [12,13]. The details of the FISC of the system were reviewed by Uji and Brooks [14].

On the other hand, an isostructural λ-$(BETS)_2GaCl_4$ (Ga salt, hereafter) is a pure π electronic system because there is no *d* spin on the anion site [15]. The Ga salt shows superconductivity around 5 K at zero magnetic field and the superconducting state is killed by a magnetic field (H_{C2} ~ 12 T) as in a conventional superconductor [16].

The π-*d* interaction of the system can be tuned by making alloy compounds λ-$(BETS)_2Fe_{1-x}Ga_xCl_4$ [16,17]. The structural disorder effect is considered to be small because the size of the anions $FeCl_4$ and $GaCl_4$ is similar. In fact, in the Ga-doped system, the transition temperature, T_{MI}, decreases continuously with increasing x continuously [17] and the insulating phase disappears when x is larger than 0.25. The Ga doping also changes the field-induced phase. Uji et al. performed precise and systematic electrical resistivity measurements under various magnetic fields for different Ga contents and they proposed a global temperature-field phase diagram [17]. A schematic drawing of the temperature-field phase diagram is given in Figure 1. The dome-shaped FISC phase shifts towards the lower field side without broadening and the antiferromagnetic insulating (AFI) phase shrinks with Ga doping. Phenomenologically, the FISC phase of Ga-doped salt seems to be very similar to that of the Fe salt. The gravity center of the FISC dome, H_{max}, the magnetic field that gives the highest transition temperature, was observed at H_{max}~$32 \times (1 - x)$ in Tesla. The area of the FISC is not changed significantly by doping. The π-d exchange field seems to be homogeneous and reduced continuously by Ga doping. Several studies on the Ga-doped system have been carried out. Oshima et al. performed combined measurements of ESR and electrical resistance in the field-induced superconducting state of Ga 40% system [18]. They observed increases in the resistance in the superconducting state when the Fe spin polarization was disturbed by ESR [18]. This suggests that the π-d exchange field was modified by the ESR transition and, consequently, FISC was destroyed. Akiba et al. performed systematic heat capacity measurements on various samples with different Ga content x and evaluated the strength of the π-d interaction from the entropy analysis [19]. They reported that the π-d interaction decreased systematically with x as $J_{\pi d}$~J_0 $(1 - x)$, where $J_{\pi d}$ and J_0 are π-d interaction and that of the Fe salt, respectively. These reports suggest that the conduction π spins feel the "uniform" exchange field from the dilute Fe site. In the present study, we performed Se NMR measurements on a Ga-doped system, λ-$(BETS)_2(Fe_{1-x}Ga_x)Cl_4$, to investigate how the π-d interaction changes by the partial replacement of Fe with Ga, through detecting the π spin polarization quantitatively and microscopically.

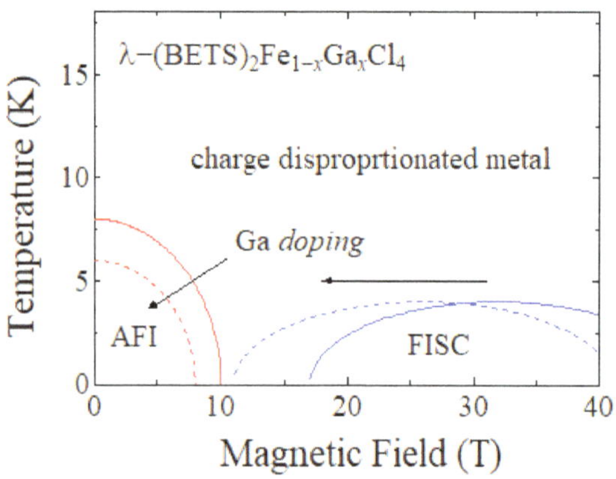

Figure 1. Schematic field-temperature phase diagram of λ-$(BETS)_2Fe_{1-x}Ga_xCl_4$, showing the effect of Ga doping.

2. Experimental Details

Single crystal samples were prepared by Zhou et al. at Nihon University by the electrochemical method using a mixture of Ga and Fe electrolytes. The Ga content is evaluated by the molar content of Ga and Fe electrolytes as $M_{Ga}/(M_{Ga} + M_{Fe})$, where M_{Ga} and M_{Fe} are molecular contents of Ga and Fe electrolytes, respectively. More precise estimations of the Ga content were obtained by the following two methods: (i) the electron probe microanalysis (EPMA) measurements and (ii) the heat capacity measurements. Akiba et al. reported that there is a clear relationship between Ga content and heat capacity [19]. The values of the Ga content thus agreed well with those in preparation.

A single crystal sample from the same batch which was characterized by the heat capacity measurements was used in the present study. The dimensions of the sample were ~0.1 × ~0.05 × ~3 mm^3. The natural abundance of NMR-active ^{77}Se isotopes contribute to the signal at a value of 7%.

The ^{77}Se NMR measurements were performed by a conventional pulsed NMR spectrometer with 9 T superconducting magnet. NMR spectrum was obtained by the fast Fourier transformation of the spin echo signal at a fixed magnetic field of 9 T. The typical pulse width was ~0.5 μs. In most of the present measurements, the *rf* pulse duration was short enough to cover the spectrum. The NMR frequency shift and the linewidth were defined as the first moment and the square root of the second moment of the spectrum, respectively. The sample and the NMR coil were mounted on a goniometer with the rotation axis along the needle axis (crystallographic *c*): the geometry of the experiments and analysis was almost the same as in our previous studies [13].

The NMR shift of electronic origin, δf, should be expressed as

$$\delta f(\theta, T) = \gamma A(\theta)\chi^{\pi}(H_0 + JM(T)) \tag{1}$$

where $\gamma, A(\theta), \chi^{\pi}, H_0, J$ and $M(T)$ are gyromagnetic ratio of ^{77}Se nucleus (8.127296 MHz/T) [13] hyperfine coupling constant, local susceptibility of the π spin system, external field, exchange interaction and Fe moment, respectively. The hyperfine coupling constant for ^{77}Se nuclei in the present system is known to be uniaxial, written as $A(\theta) = A^{iso} + A^{ax}(3\cos^2(\theta - \theta_0)\cos^2\phi_0 - 1)$, where A^{iso}/A^{ax}, θ, θ_0 and ϕ_0 are isotropic/axial components of the hyperfine coupling, the rotating angle in the a^*b^*-plane and the orientation angles of the Se $4p_z$ orbital with respect to the a^*b^*-plane (θ_0 is 17.3°) and the angle between the π orbital of the Se and rotating plane (13.6°), respectively. The geometrical situation is shown in the inset of the Figure 2. In our previous reports on Fe and Ga salts, the values of hyperfine parameters were already obtained [13,20]. We used the Brillouin function of $S = 5/2$ for the temperature dependence of the Fe moment, $M(T)$, in Equation (1) as in the previous work [13].

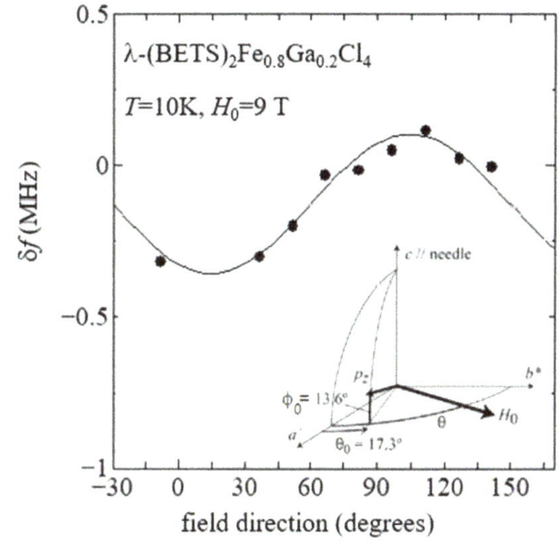

Figure 2. Angular dependence of the central shift, δf. The solid line is the calculated angular dependence assuming the exchange interaction $JM(10K) = -20$ T. The geometrical situation of the measurements is shown in the inset.

3. Results and Discussion

3.1. Evaluation of the Exchange Interaction

The NMR shift δf is a linear function of H_0 or M(T) under the fixed field or temperature, as shown in Equation (1), so that the temperature dependence of δf at a fixed magnetic field or the field dependence at a fixed temperature enables us to determine the exchange interaction J independently. The analysis of the temperature and field dependences of the NMR shift on the Fe salt gave the value of the low temperature (LT) exchange field JM(LT) as −32 T (J~5.4 T/μ_B) in our previous work [13]. We tried to perform similar measurements in the present study on the Ga-doped sample to evaluate the exchange interaction, J. However, the signal–noise ratio of the spectrum was poor due to the broadening, especially at low temperatures. Therefore, we decided to measure the angular dependence of the NMR spectrum at 10 K, where the linewidth was relatively narrow.

Figure 2 shows the angular dependence of δf for the Ga-doped sample, which exhibits a qualitatively similar sinusoidal angular dependence with opposite sign to the Ga salt and the same sign as the Fe salt [13,20]. The crystal structure and the hyperfine symmetry of the present Ga-doped sample are expected to be the same as the non-doped systems and we have chosen the same experimental geometry as those in our previous studies [13,20]. The observed NMR shift clearly indicates that the effective field at the Se nuclear site of the present system is opposite to the external field due to the negative exchange field produced by the localized d spin as observed in the pure Fe salt.

The solid line in Figure 2 is the calculated frequency shift assuming that the exchange interaction, J, is isotropic because we did not observe the angular dependence of the exchange interaction in the Fe salt. We also assumed that the hyperfine parameter, $A(\theta)\chi^{\pi}$, for the Ga-doped system should be given by a molar average of those for the pure salts as $A(\theta)\chi^{\pi}$(G20) = 0.8 $A(\theta)\chi^{\pi}$(Fe) + 0.2 $A(\theta)\chi^{\pi}$(Ga), where χ^{π}(G20), χ^{π}(Fe) and χ^{π}(Ga) are the π spin susceptibility of the Ga 20%, Fe and Ga salts, respectively. The hyperfine parameters of Fe and Ga salts obtained from Refs [13] and [20] are listed in Table 1. The best fit was obtained by $H^{\mathrm{eff}} = H_0 + JM$(10K)~ −11 T. The strength of the applied field in the present study was H_0 = 9 T. Therefore, the exchange interaction at 10 K was obtained as JM(10K) = −20 T. Considering the temperature dependence of the Fe moment given by the Brillouin function, the exchange field at low temperatures was estimated as JM(0K) = −25 T (J~5 T/μ_B), which is 20% smaller than that of Fe salt, as expected.

Table 1. NMR parameters of the Ga and the Fe salts obtained by Refs [13,20].

Material	$A^{iso}\chi^{\pi}$ (%)	$A^{ax}\chi^{\pi}$ (%)	$\Delta\chi^{\pi}/\chi^{\pi}$	Reference
Ga salt	0.1	0.28	0.3	[20]
Fe salt	0.08	0.19	0.7	[13]

We also measured the temperature dependence of the NMR shift, δf, in the temperature range between 10 and 100 K at a fixed field orientation (θ~30 deg) as in the earlier report (Figure 10 of [13]). The results are plotted as a function of Fe moment in Figure 3. In this geometry, the $A(\theta)\chi^{\pi}$(G20) was estimated to be 0.8$A(\theta)\chi^{\pi}$(Fe) + 0.2 $A(\theta)\chi^{\pi}$(Ga) = 0.29%. The solid line indicates the calculated values using the Equation (1) with J = −5 T/μ_B, mentioned above. The experimental data agree well with the calculation.

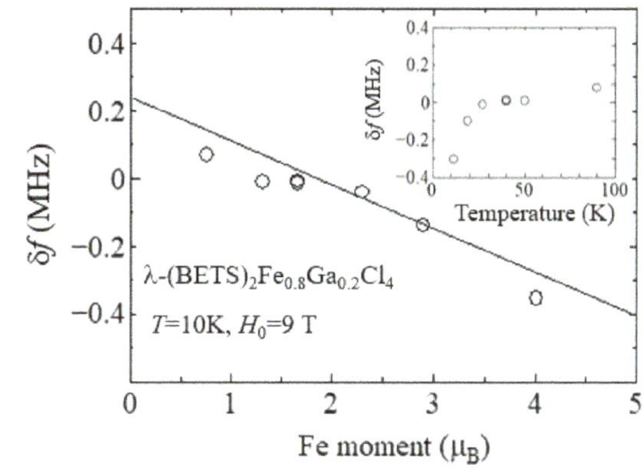

Figure 3. Temperature dependence of the NMR shift, δf obtained at the field orientation of 30 degrees; the horizontal axis is taken as the magnitude of the Fe moment at each temperature. The solid line indicates the calculated values taking J~5 T/μ_B. The inset shows the NMR shift as a function of temperature.

From the angular and temperature dependences of the NMR shift, it was clarified that the exchange interaction between π conduction spin and Fe moment in the Ga-doped salt is homogeneously reduced as Ga doping. The amount of the reduction of the J value of 20% is consistent with the Ga content $x = 0.2$. Although there is a broadening of the spectrum at low temperatures below ~5 K, no appreciable change in the spectrum such as a peak split due to the local inhomogeneity was observed. This indicates that the conduction π spins are polarized uniformly by the "averaged" exchange field from the localized 3d spins at the dilute Fe sites.

3.2. Linewidth Analysis: Charge Disproportionation

We reported an unexpected broadening of the ^{77}Se-NMR spectrum at low temperatures in the pure Fe and Ga salts [20]. From the analysis of the angular dependence of the linewidth, we concluded that the broadening is due to charge disproportionation (CD) in the BETS layer. Since the donor arrangement of the conducting layers is quite similar to the non-doped systems, the CD is also expected in the present system. In fact, we observed that the linewidth of the present salt shows a similar angular dependence. Results at 10 K are shown in Figure 4.

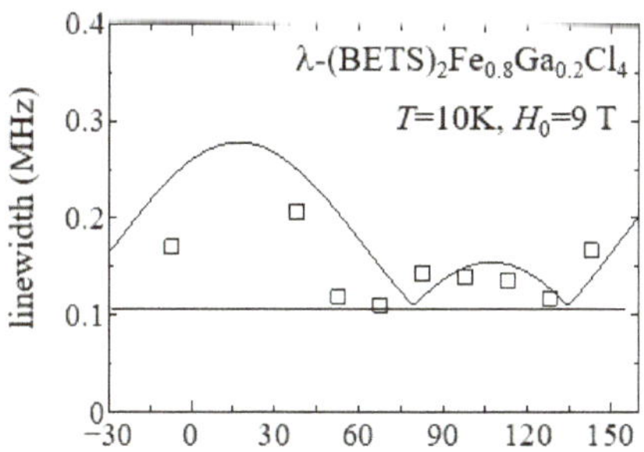

Figure 4. Angular dependence of the line width. The horizontal dashed line corresponds to the width of non-electronic origin. The solid line is the calculated value assuming charge disproportionation in the two-dimensional BETS conducting sheet. Details are given in the text.

To examine the angular dependence of the linewidth, we calculated the line width with the following assumptions. (i) We considered two independent contributions of angular dependent hyperfine component of the electronic origin, Δf_{el}, and of angular independent non-electronic component, Δf_{ne}, as was the case in the pure salts. (ii) The former is expressed as $\Delta f_{\mathrm{el}} = \gamma A(\theta)\Delta\chi^{\pi}$. Here, $\Delta\chi^{\pi}$ is the degree of the spatial inhomogeneity of the spin susceptibility and it should be proportional to the degree of CD. The ratios, $\Delta\chi^{\pi}/\chi^{\pi}$, for the Fe and Ga salts were obtained as 0.7 and 0.3, respectively, at low temperatures. For the present alloy system, we simply took this value as an adjusting parameter. (iii) We took the minimal value of the linewidth observed at 80 and 130 degrees as the latter contribution, Δf_{ne}. (iv) We calculated the linewidth as

$$\Delta f_{\mathrm{obs}} = \sqrt{\Delta f_{\mathrm{el}}^2 + \Delta f_{\mathrm{ne}}^2} \qquad (2)$$

The solid curve in Figure 4 indicates the calculated angular dependence with $\Delta\chi^{\pi}/\chi^{\pi} = 0.4$. The basic features were well reproduced while the experimental data were rather scattered.

There are some remarks on the value of $\Delta\chi^{\pi}/\chi^{\pi}$. (i) As mentioned above, the ratio $\Delta\chi^{\pi}/\chi^{\pi}$ should be proportional to the degree of CD on the BETS layer. The CD in the Fe salt was much more enhanced compared with that in the Ga salt, the mechanism of which was not clarified. We do not have any reasonable estimate of the degree of CD in the alloy system. (ii) The degree of CD in the pure salts was temperature-dependent; the large broadening was observed at low temperatures. (iii) The expected total line width at around $\theta \sim 30$ degrees is comparable to (or larger than) the *rf* excitation pulse width. In such a case, the *rf* pulse might not cover the whole spectrum and it was difficult to detect the linewidth with high accuracy. From the present measurements, we only claim that the existence of CD was confirmed in the alloy system as in the pure Fe and Ga salts. The temperature dependence of the CD and the quantitative comparison with the pure salts will be addressed in a future investigation.

4. Conclusions

In this study, ^{77}Se NMR measurements of λ-(BETS)$_2$Fe$_{1-x}$Ga$_x$Cl$_4$ with 20% Ga doping ($x = 0.2$) were performed. From the analysis of the angular dependence of Se NMR spectra, relatively small π-d exchange interaction was obtained as −25 T, which is 20% smaller than that of Fe salt. The amount of reduction of J is consistent with the Ga content. It was confirmed microscopically that the itinerant conduction spins feel the effectively uniform exchange field from the dilute Fe sites. We also observed charge disproportionation in the two-dimensional BETS conducting sheet in the Ga-doped system.

Author Contributions: K.-i.H. and T.T. designed this study. The single crystal samples were prepared by B.Z.; H.A. and Y.N. evaluated the Ga content, *x*, by the heat capacity analysis. K.-i.H. performed NMR experiments and analyzed the data. K.-i.H. and T.T. wrote the manuscript. All authors have read and agreed to the published version of the manuscript.

Funding: This work was partially supported by Grant-in-Aid for scientific research (C) (General) No. 23540423 and on Innovative Areas No. 20110002 from the Ministry of Education, Culture, Sports, Science and Technology, Japan.

Acknowledgments: The authors thank A. Kobayashi and H. Kobayashi (Nihon University) for EPMA measurements and fruitful discussions.

Conflicts of Interest: The authors declare no conflict of interest.

References

1. Kobayashi, H.; Cui, H.-B.; Kobayashi, A. Organic Metals and Superconductors Based on BETS (BETS = Bis(ethylenedithio)tetraselenafulvalene). *Chem. Rev.* **2004**, *104*, 5265–5288. [CrossRef] [PubMed]
2. Kobayashi, H.; Tomita, H.; Naito, T.; Kobayashi, A.; Sakai, F.; Watanabe, T.; Cassoux, P. New BETS Conductors with Magnetic Anions (BETS = bis(ethylenedithio)tetraselenafulvalene). *J. Am. Chem. Soc.* **1996**, *118*, 368–377. [CrossRef]
3. Tokumoto, M.; Naito, T.; Kobayashi, H.; Kobayashi, A.; Laukhin, V.N.; Brossard, L.; Cassoux, P. Magnetic anisotropy of organic conductor λ-(BETS)$_2$FeCl$_4$. *Synth. Met.* **1997**, *86*, 2161–2162. [CrossRef]

4. Brossard, L.; Clerac, R.; Coulon, C.; Tokumoto, M.; Ziman, T.D.; Petrov, K.; Laukhin, V.N.; Naughton, M.J.; Audouard, A.; Goze, F.; et al. Interplay between chains of $S = 5/2$ localised spins and two-dimensional sheets of organic donors in the synthetically built magnetic multilayer $\lambda-(BETS)_2FeCl_4$. *Eur. Phys. J.* **1998**, *1*, 439–452. [CrossRef]
5. Akiba, H.; Nakano, S.; Nishio, Y.; Kajita, K.; Zhou, B.; Kobayashi, A.; Kobayashi, H. Mysterious Paramagnetic States of Fe 3*d* Spin in Antiferromagnetic Insulator of λ-$BETS_2FeCl_4$ System. *J. Phys. Soc. Jpn.* **2009**, *78*, 033601. [CrossRef]
6. Oshima, Y.; Cui, H.-B.; Kato, R. Antiferromagnetic Insulating Ground State of Molecular π-d System λ-$(BETS)_2FeCl_4$ (BETS = Bis(ethylenedithio)tetraselenafulvalene): A Theoretical and Experimental Review. *Magnetochemistry* **2017**, *3*, 10. [CrossRef]
7. Uji, S.; Shinagawa, H.; Terashima, T.; Yakabe, T.; Terai, Y.; Tokumoto, M.; Kobayashi, A.; Tanaka, H.; Kobayashi, H. Magnetic-field-induced superconductivity in a two-dimensional organic conductor. *Nature* **2001**, *410*, 908–910. [CrossRef]
8. Balicas, L.; Brooks, L.S.; Storr, K.; Uji, S.; Tokumoto, M.; Tanaka, H.; Kobayashi, H.; Kobayashi, A.; Barzykin, V.; Gor'kov, L.P. Superconductivity in an Organic Insulator at Very High Magnetic Fields. *Phys. Rev. Lett.* **2001**, *87*, 067002. [CrossRef]
9. Jaccarino, V.; Peter, M. Ultra High Field Superconductivity. *Phys. Rev. Lett.* **1962**, *9*, 290–292. [CrossRef]
10. Uji, S.; Terakura, C.; Terashima, T.; Yakabe, T.; Terai, Y.; Tokumoto, M.; Kobayashi, A.; Sakai, F.; Tanaka, H.; Kobayashi, H. Fermi surface and internal magnetic field of the organic conductors λ-$(BETS)_2FexGa_{1-x}Cl_4$. *Phys. Rev. B* **2002**, *65*, 113101. [CrossRef]
11. Wu, G.; Clark, W.G.; Brown, S.E.; Brooks, J.S.; Kobayashi, A.; Kobayashi, H. ^{77}Se NMR measurements of the π-d exchange field in the organic conductor λ-$(BETS)_2FeCl_4$. *Phys. Rev. B* **2007**, *76*, 132510. [CrossRef]
12. Hiraki, K.; Mayaffre, H.; Horvati, M.C.; Berthier, C.; Tanaka, H.; Kobayashi, A.; Kobayashi, H.; Takahashi, T. π-d Interaction in the Field Induced Superconductor λ-$(BETS)_2FeCl_4$: Studied by ^{77}Se NMR. *J. Low Temp. Phys.* **2006**, *142*, 185–190. [CrossRef]
13. Hiraki, K.; Mayaffre, H.; Horvati, M.C.; Berthier, C.; Yamaguchi, T.; Uji, S.; Tanaka, H.; Kobayashi, A.; Kobayashi, H.; Takahashi, T. ^{77}Se NMR evidence for the Jaccarino-Peter mechanism in the field induced superconductor, $\lambda(BETS)_2FeCl_4$. *J. Phys. Soc. Jpn.* **2007**, *76*, 124708. [CrossRef]
14. Uji, S.; Brooks, J.S. Magnetic-Field-Induced Superconductivity in Organic Conductors. *J. Phys. Soc. Jpn.* **2006**, *75*, 051014. [CrossRef]
15. Tanaka, H.; Kobayashi, A.; Sato, A.; Akutsu, H.; Kobayashi, H. Chemical Control of Electrical Properties and Phase Diagram of a Series of λ-Type BETS Superconductors, λ-$(BETS)_2GaBr_xCl_{4-x}$. *J. Am. Chem. Soc.* **1999**, *121*, 760–768. [CrossRef]
16. Tanatar, M.A.; Ishiguro, T.; Tanaka, H.; Kobayashi, A.; Kobayashi, H. Anisotropy of the Upper Critical Field of the Organic Superconductor λ-$(BETS)_2GaCl_4$. *J. Supercond.* **1997**, *12*, 511–514. [CrossRef]
17. Uji, S.; Terashima, T.; Terakura, C.; Yakabe, T.; Terai, Y.; Yasuzuka, S.; Imanaka, Y.; Tokumoto, M.; Kobayashi, A.; Sakai, F.; et al. Global Phase Diagram of the Magnetic Field Induced Organic Superconductors λ-$(BETS)_2Fe_xGa_{1-x}Cl_4$. *J. Phys. Soc. Jpn.* **2003**, *72*, 369–373. [CrossRef]
18. Oshima, Y.; Nojiri, H.; Uji, S.; Brooks, J.S.; Tokumoto, T.; Cui, H.; Kato, R. Breakdown of the field-induced superconductivity by dynamical spin reversal. *Phys. Rev. B* **2012**, *86*, 024525. [CrossRef]
19. Akiba, H.; Sugawara, H.; Nobori, K.; Shimada, K.; Tajima, N.; Nishio, Y.; Kajita, K.; Zhou, B.; Kobayashi, A.; Kobayashi, H. Paramagnetic Metal–Antiferromagnetic Insulator Transition of λ-$BETS_2Fe_xGa_{1-x}Cl_4$ System. *J. Phys. Soc. Jpn.* **2012**, *81*, 053601. [CrossRef]
20. Hiraki, K.; Kitahara, M.; Takahashi, T.; Mayaffre, H.; Horvatic, M.; Berthier, C.; Uji, S.; Tanaka, H.; Zhou, B.; Kobayashi, A.; et al. Evidence of Charge Disproportionation in λ Type BETS Based Organic Superconductors. *J. Phys. Soc. Jpn.* **2010**, *79*, 074711. [CrossRef]

Publisher's Note: MDPI stays neutral with regard to jurisdictional claims in published maps and institutional affiliations.

crystals

MDPI

Article

Syntheses, Structures, and Physical Properties of Neutral Gold Dithiolate Complex, $[Au(etdt)_2]{\cdot}THF$

Kazuha Sakaguchi, Biao Zhou *, Yuki Idobata, Hajime Kamebuchi and Akiko Kobayashi

Department of Chemistry, College of Humanities and Sciences, Nihon University, Sakurajosui 3-25-40, Setagaya-Ku, Tokyo 156-8550, Japan; sakakazu0309@gmail.com (K.S.); 6sided.currencystone@ezweb.ne.jp (Y.I.); hkame@chs.nihon-u.ac.jp (H.K.); akoba@chs.nihon-u.ac.jp (A.K.)

* Correspondence: zhou@chs.nihon-u.ac.jp; Tel.: +81-3-5317-9364

Received: 13 October 2020; Accepted: 3 November 2020; Published: 4 November 2020

Abstract: In order to develop new types of single-component molecular conductors with novel electronic structures and physical properties, the neutral gold dithiolate complex with an etdt (= ethylenedithiotetrathiafulvalenedithiolate) ligand, $[Au(etdt)_2]$ was prepared. However, unlike the reported single-component molecular metals, the neutral gold complex $[Au(etdt)_2]{\cdot}THF$ (**2**) contains a solvent molecule of tetrahydrofuran (THF). The crystals of **2** form a two-dimensional conducting layer structure, which are separated by the terminal ethylene groups and THF molecules. The fairly high room-temperature conductivity of 0.2 S/cm and semiconducting behavior with a low activation energy of 0.1 eV of **2**, is consistent with the result of the density functional theory band structure calculations. The observed non-magnetic behavior of **2** is caused from the dimeric structure of $[Au(etdt)_2]$ molecules.

Keywords: single-component molecular conductors; extended-TTF dithiolate ligands; gold dithiolate complexes

1. Introduction

Since the discovery of the first molecular metal, TTF-TCNQ (TTF = tetrathiafulvalene, TCNQ = tetracyanoquinodimethane) charge transfer complex in 1973 [1], many various types of molecular conductors have been reported, including famous examples such as $(TMTSF)_2PF_6$ (TMTSF = tetramethyltetraselenafulvalene)—the first molecular organic superconductor [2], and λ-$(BETS)_2FeCl_4$ (BETS = bis(ethylenedithio)tetraselenafulvalene)—a field induced superconductor [3,4]. These traditional charge transfer-based molecular conductors inherently consist of more than one kind of molecule, a donor as well as an acceptor. However, the first single-component molecular metal $[Ni(tmdt)_2]$ (tmdt = trimethylenetetrathiafulvalenedithiolate) developed in 2001 has opened a new field of conducting materials [5,6]. $[Ni(tmdt)_2]$ consists of only one kind of neutral molecule, and exhibits metallic behavior down to 0.6 K. The observation of de Haas-van Alphen oscillations, at very high magnetic fields and low temperatures, showed the existence of the three-dimensional Fermi surfaces [7], which was also proved by ab initio band structure calculations [8]. After that, a number of single-component molecular conductors have been reported, such as $[Au(Me\text{-}thiazdt)_2]$ (Me-thiazdt = *N*-methyl-1,3-thiazoline-2-thione-4,5-dithiolate)—a single-component molecular metal without any TTF unit [9], and TED (= tetrathiafulvalene-extended dicarboxylate radical)—a single-component pure organic metal [10].

For single-component molecular conductor $[M(L)_2]$ systems, an important characteristic is that its electronic structure can be widely tuned by exchanging the central transition metal atom (M) for another transition metal atom, even among isostructural systems. The series of isostructural systems, $[M(tmdt)_2]$ (M = Ni, Cu, Pd, Pt, Au) is a typical case. $[Ni(tmdt)_2]$ and $[Pt(tmdt)_2]$ exhibit very high conductivity and metallic behavior down to extremely low temperatures [11,12]. $[Au(tmdt)_2]$ is a hybrid antiferromagnetic metal, and undertakes an antiferromagnetic transition at around 110 K as well as retaining its metallic state [13,14]. [Cu(tmdt)$_2$] is a hybrid Mott insulator, and shows a one-dimensional antiferromagnetic Heisenberg behavior with magnetic ordering at 13 K [15,16]. However, [Pd(tmdt)$_2$] with an even number of total electrons, is an antiferromagnetic semiconductor, and takes out a magnetic ordering onset exceeding 100 K due to strong electron correlation [17,18].

On the other hand, the difference of the ligands has an important effect or influence on the electronic structures and band structures of single-component molecular conductors. Actually, a variety of electronic structures have been realized by using similar extended-TTF dithiolate ligands with different terminal groups (as shown in Scheme 1). For example, unlike the isostructural [M(tmdt)$_2$] systems with a tight three-dimensional molecular packing, [M(ptdt)$_2$] (M = Ni, Au; ptdt = propylenedithiotetrathiafulvalenedithiolate) [19,20] and [M(hfdt)$_2$] (M = Ni, Au, Pd; hfdt = bis(trifluoromethyl)tetrathiafulvalenedithiolate) [21,22] crystallize in a layered two-dimensional (2D) molecular packing. Especially, [Ni(hfdt)$_2$] is a single-component molecular superconductor with transition temperatures at 5.5 K under high pressures around 8 GPa [23]. Recently, [Pt(dmdt)$_2$] has been found to host strongly correlated massless Dirac electrons with nodal lines at ambient pressure [24].

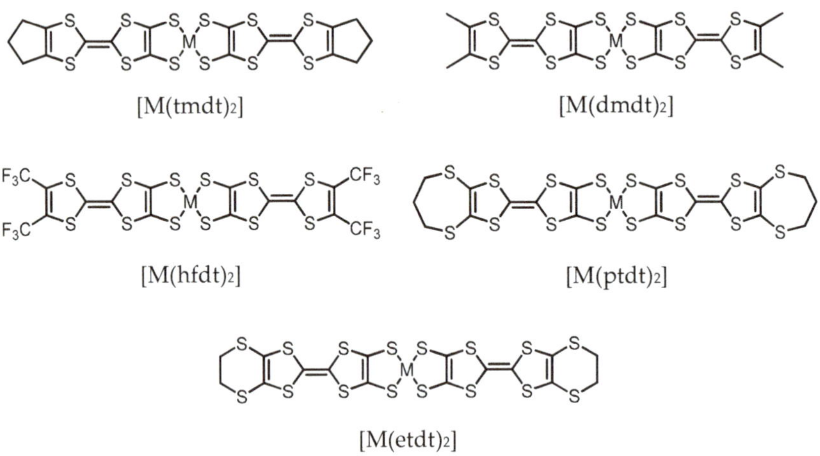

Scheme 1. Chemical structure of single-component molecular conductor [M(L)$_2$] systems mentioned in this paper.

It is well known that, for BEDT-TTF (= bis(ethylenedithio)tetrathiafulvalene) charge transfer complexes, the conformational flexibility of the terminal ethylene groups yields a variety of the crystal structures and the resultant electronic structures. For example, α-(BEDT-TTF)$_2$I$_3$ is a molecular Dirac electron system under high pressure [25], as well as β-, κ-, and θ-(BEDT-TTF)$_2$I$_3$ are ambient pressure molecular superconductors [26,27]. To develop new types of single-component molecular conductors with novel electronic structures and physical properties, we have tried to prepare a series of dithiolate complexes, with etdt (= ethylenedithiotetrathiafulvalenedithiolate) ligand, which contain the same terminal ethylene group to that of BEDT-TTF molecule. Although, some similar dithiolate complexes had been reported by G. Matsubayashi et al. [28,29], their oxidized neutral species have not been studied. We report here, syntheses, crystal structures, and physicals properties of new neutral gold dithiolate complexes, [Au(etdt)$_2$]·THF.

2. Materials and Methods

2.1. General Methods

All the syntheses were carried out under argon atmosphere using the Schlenk technique because the anionic state of metal complexes with extended-TTF dithiolate ligands are quite sensitive to oxygen. The etdt ligand [30] and the gold source, tetra-n-butylammonium tetrachloroaurate(III) ($^nBu_4N{\cdot}[AuCl_4]$) [31], were prepared according to procedures reported in literature. Tetrahydrofuran (THF) was distilled over Na metal and benzophenone. Methanol (MeOH) was distilled over Mg metal activated with I_2. The supporting electrolytes tetra-n-butylammonium hexafluorophosphate ($^nBu_4N{\cdot}PF_6$) used in the electrocrystallization was recrystallized three times with ethylacetate and dried in vacuo. All other reagents were used as purchased without any further purification.

2.2. Synthesis of $^nBu_4N{\cdot}[Au(etdt)_2]$ (**1**)

The etdt ligands (139 mg; 0.30 mmol) was hydrolyzed with a 25-wt% MeOH solution of tetramethylammonium hydroxide ($Me_4N{\cdot}OH$) (440 mg; 1.2 mmol) in dry THF at room temperature. The solution was stirred for 30 min, and a reddish-orange intermediate precipitate was obtained as the reaction proceeded. After cooling to −78 °C in a dry ice/MeOH bath, a THF solution of $^nBu_4N{\cdot}[AuCl_4]$ (92 mg; 0.16 mmol) was added dropwise to the reaction mixture. Then, the reaction mixture was warmed to room temperature overnight. The resulting microcrystals were collected by filtration and crystallized from a THF solution of $^nBu_4N{\cdot}PF_6$ at room temperature to afford dark-violet single crystals of **1**.

2.3. Electrochemical Synthesis of $[Au(etdt)_2]{\cdot}THF$ (**2**)

A mixture of **1** (~15 mg; 0.013 mmol) and $^nBu_4N{\cdot}PF_6$ (100 mg; 0.26 mmol) as a supporting electrolyte was poured into H-shaped glass cells containing Pt electrodes and dissolved in dry THF (20.0 mL). When a constant current of 0.2 μA was applied at room temperature, air-stable black needle-like single crystals with sizes less than 80 μm of **2** grew on the Pt electrode within approximately 3 weeks.

2.4. Crystal Structure Determination of **1** and **2**

The single-crystal X-ray diffraction data for compounds **1** and **2** were collected on a Rigaku Micro7HFM-VariMax Saturn 724R CCD system equipped with graphite monochromated Mo Kα radiation (λ = 0.71073 Å) and a confocal X-ray mirror. The crystal structures were solved using direct methods (SHELXT) [32] and refined by full-matrix least-squares (SHELXL) [33] under the Olex2 graphical interface [34]. Anisotropic temperature factors were applied for the non-hydrogen atoms. The calculated positions of the hydrogen atoms were not refined but included in the final calculations. The crystal data and experimental details of the crystal structure determination are listed in Table 1.

Table 1. X-ray crystallographic data of the gold complexes **1** and **2**.

	1	**2**
Empirical Formula	$C_{32}H_{44}AuNS_{16}$	$C_{20}H_{16}AuOS_{16}$
Formula Weight	1152.61	982.25
Crystal System	monoclinic	triclinic
Space Group	$P2_1/c$ (#14)	P-1 (#2)
a/Å	20.7276 (5)	6.4340 (3)
b/Å	16.6751 (4)	8.6985 (4)
c/Å	12.7080 (3)	28.8338 (13)
α/°	90	84.332 (4)
β/°	90.421 (2)	87.668 (4)
γ/°	90	69.599 (4)
V/Å^3	4392.21 (18)	1505.07 (13)
Z	4	2
D_{calc}/g·cm^{-3}	1.743	2.167
Temp./K	173	298 (R.T.)
F (000)	2312	958
μ/mm^{-1}	4.137	6.018
Reflections Collected	100,597	11,123
Independent Reflections	10,156	5934
Parameters	455	318
R_1 (I > 2σ) [1]	0.037	0.051
wR_2 (all data) [2]	0.074	0.138
GOF	1.013	1.104
CCDC	2,036,560	2,036,561

[1] $R_1 = \Sigma||F_o| - |F_c||/\Sigma|F_o|$. [2] $wR_2 = [\Sigma\omega(|F_o| - |F_c|)^2/\Sigma\omega F_o{}^2]^{1/2}$.

2.5. *Electrical Resistivity Measurements of* **2**

Four-probe resistivity measurements were performed on compressed pellets of polycrystalline samples of **2** cooled by the liquid helium using a HECS 994C-1 multichannel four-terminal conductometer. Annealed gold wires (15 μm in diameter) bonded to the sample by gold paint were used as leads.

2.6. *Magnetic Measurements of* **2**

Magnetic measurements of **2** were performed with a Quantum Design MPMS-7XL superconducting quantum interference device (SQUID) magnetometer in the temperature range of 2.0–300 K. The applied magnetic field was 5000 Oe. The samples were wrapped in clean aluminum foil whose magnetic susceptibility was separately measured and subtracted. The diamagnetic contribution was estimated from Pascal's constants.

2.7. *Molecular Orbital, Band Structure, and Density of State Calculations of* **2**

Molecular orbital (MO), band structure, and density of state (DOS) calculations were performed by density functional theory (DFT) and the general gradient approximation (GGA) method using the DMol^3 module [35,36] as implemented in Materials Studio v5.5 (Accelrys, San Diego, CA, USA). Becke–Lee–Yang–Parr (BLYP) functional [37] and double numeric plus polarization (DNP) basis set were used in the calculations.

3. Results and Discussion

3.1. *Crystal Structure of Monoanion Gold Complex* **1**

Monoanion complex **1** crystallizes in monoclinic system with space group of $P2_1/c$. One tetra-n-butylammonium cation and one $[Au(etdt)_2]^-$ anion are crystallographically independent in the unit cell. As shown in Figure 1a,b, the Au(III) atom in $[Au(etdt)_2]^-$ anion shows a square-planar coordination geometry, with Au-S distance of 2.316–2.326 Å, and an average S-Au-S angle of 91.30°, which is similar to those of the other reported gold dithiolate complexes [13,21]. The geometry of the two etdt ligands in $[Au(etdt)_2]^-$ anion is not symmetrical: One of the ligands is almost planar, while the other ligand is bent at the positions of the outermost S atoms with a dihedral angle of 50.9°.

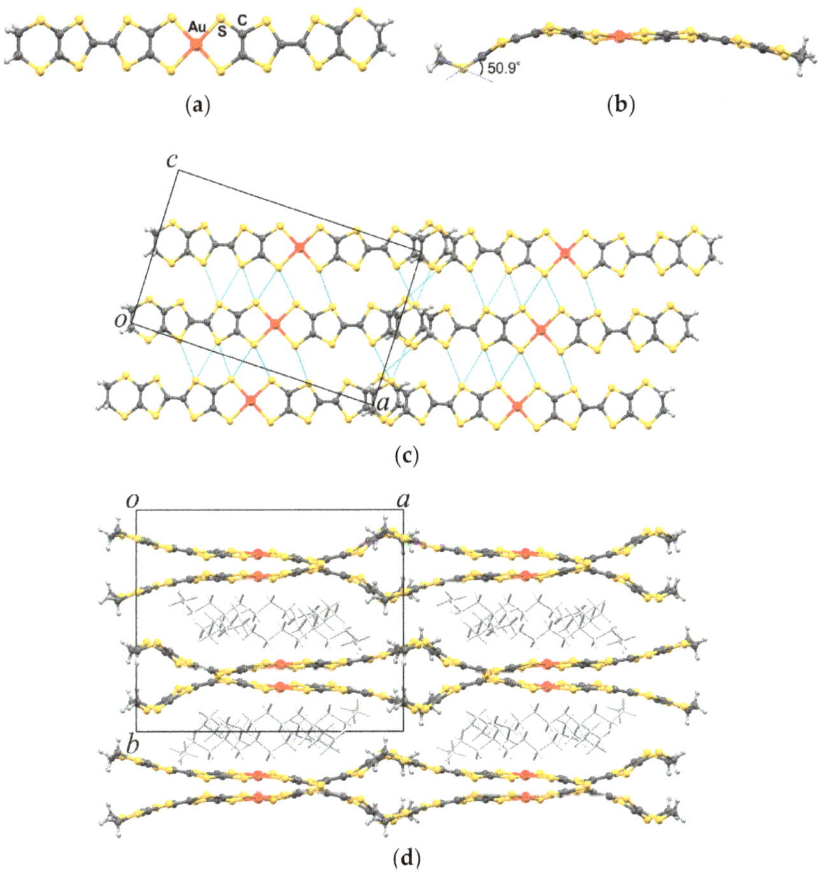

Figure 1. (**a**) Top view of the $[Au(etdt)_2]^-$ anion. (**b**) Side view of the $[Au(etdt)_2]^-$ anion. (**c**) Crystal structure of **1** viewed along the *b*-axis. The short S···S contacts (<3.7 Å) are shown as dotted line. The bulky tetra-n-butylammonium cations are omitted for clarity; (**d**) Crystal structure of **1** viewed along the *c*-axis.

As shown in Figure 1c, the $[Au(etdt)_2]^-$ anions form side-by-side arrays along the *c*-axis with several short S···S contacts less than the sum of the van der Waals radii (<3.7 Å), with a shortest intermolecular S···S distance of 3.352 Å, indicating strong intermolecular interaction along the *c*-axis. On the other hand, the $[Au(etdt)_2]^-$ anions are overlapped only on the terminal ethylene group along the long axis, to form a molecular layer parallel to the *ac* plane. As shown in Figure 1d, these molecular layers are separated by the bulky tetra-n-butylammonium cations along the *b*-axis. There is no short S···S contact between adjacent layers.

3.2. *Crystal Structure of Neutral Gold Complex* **2**

Although the single crystal size is very small and thin, the crystal structure of neutral complex **2** has been successfully determined by single crystal X-ray structure analysis. The neutral complex **2** crystallizes into the triclinic system with space group of *P*-1. One neutral [Au(etdt)$_2$] molecule and one unexpected THF molecule are crystallographically independent in the unit cell. In traditional donor-acceptor molecular conductor systems, there are several examples reported which contain solvent molecules, such as β''-(BEDT-TTF)$_4$[(H$_3$O)Fe(C$_2$O$_4$)$_3$]·PhCN—the first paramagnetic molecular superconductor [38], (Me$_4$N)[Ni(ptdt)$_2$]·Me$_2$CO—a monoanion dithiolate nickel complex [19]. However, to the best of our knowledge, **2** is a rare case of solvent-containing neutral dithiolate complexes with extended-TTF ligands. Since **2** exhibits a distance of 3.240 Å between the oxygen atom of THF and ethylene groups of the etdt ligand (O···H-C), the THF molecules may be stabilized by the weak intramolecular hydrogen bonding. In general, solvent molecules do not contribute to the electronic structure, hence the neutral gold complex **2** is still single-component molecular conductor.

The molecular structure and the packing diagram viewed along the different axes of **2** are shown in Figure 2. The Au(III) atom in the neutral [Au(etdt)$_2$] molecule also show a square-planar coordination geometry, with Au-S distance of 2.318–2.336 Å, and an average S-Au-S angle of 91.10°, which is similar to that of **1**. The oxidized extended-TTF ligands in single-component molecular conductors usually became planar. However, as shown in Figure 2b, one of the etdt ligands in the neutral [Au(etdt)$_2$] molecule is still bent, which might be due to the space steric hindrance effect caused by the presence of THF molecules. On the other hand, the C=C distances in the TTF unit of the planar ligand are 1.340–1.360 Å, which is longer than that of **1**. Consequently, similar to that of reported single-component molecular conductors, the electrochemical oxidization was mainly carried out at the extended-TTF ligands.

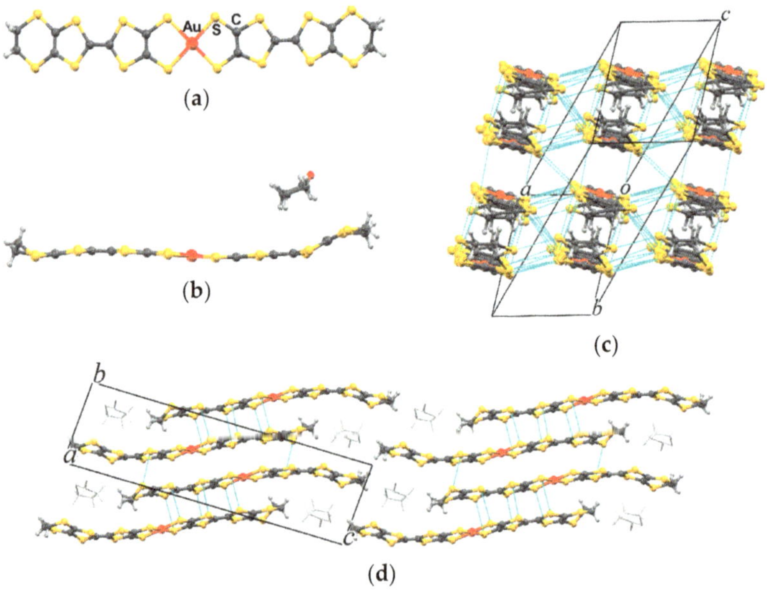

Figure 2. (**a**) Top view of the neutral [Au(etdt)$_2$] molecule without THF molecule. (**b**) Side view of the neutral [Au(etdt)$_2$] molecule with THF molecule. (**c**) Crystal structure of **2** viewed along the long axes of molecules. The short S···S contacts (<3.7 Å) are shown as dotted line. THF molecules are omitted for clarity. (**d**) Crystal structure of **2** viewed along the *a*-axis.

As shown in Figure 2c, the neutral [Au(etdt)$_2$] molecules are stacked face-to-face to form a dimeric column along the *b*-axis, with interplanar distances of 3.472 and 3.906 Å, respectively. The dimeric columns are arranged side-by-side along the *a*-axis to form a conduction layer parallel to the *ab* plane. There are many intermolecular S···S short contacts are observed in the molecular layer. Especially along the *a*-axis, a shortest intermolecular S···S distance of 3.394 Å suggests that **2** would be a satisfactory single-component molecular conductor with the relatively high conductivity. As shown in Figure 2d, the [Au(etdt)$_2$] conducting layers are strongly separated by the terminal ethylene groups and THF molecules along the *c*–axis, forming a 2D electronic structure. As mentioned before, such similar 2D electronic structure has also been observed in single-component molecular conductors with bulky ligands, [M(ptdt)$_2$] and [M(hfdt)$_2$].

3.3. *Electrical Properties of Neutral Gold Complex* 2

Since the single crystal size of **2** was very small, resistivity measurements were performed by the standard four-probe method using compressed pellets of polycrystalline samples down to 40 K cooled by the liquid helium. The room-temperature conductivity (σ_{RT}) of **2** is about 0.2 S/cm, which is somewhat high for compressed pellet sample of 2D molecular conductor. As shown in Figure 3, the resistivity increases with decreasing temperature, and exhibits a semiconducting behavior with an activation energy (E_a) of about 0.1 eV in the temperature range of 200–300 K. Considering that the measurements were carried out on compressed pellets, **2** should be a fairly good conductor in the single crystalline state.

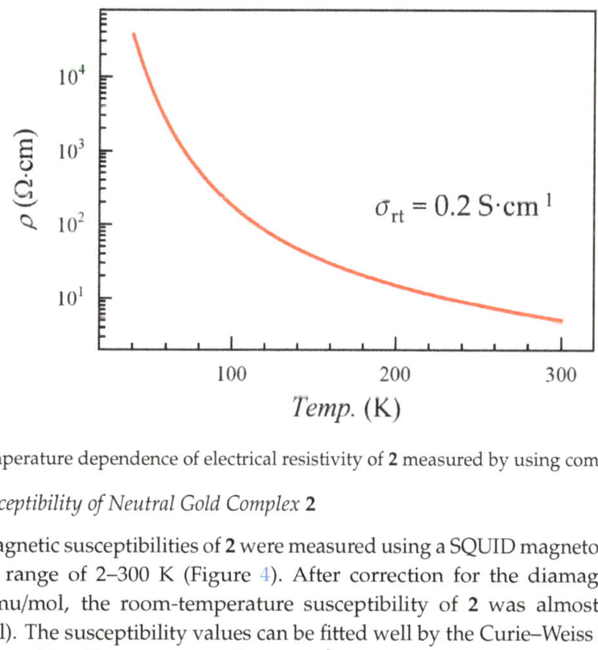

Figure 3. Temperature dependence of electrical resistivity of **2** measured by using compressed pellets.

3.4. *Magnetic Susceptibility of Neutral Gold Complex* 2

The static magnetic susceptibilities of **2** were measured using a SQUID magnetometer at 5000 Oe in the temperature range of 2–300 K (Figure 4). After correction for the diamagnetic contribution of -4.0×10^{-4} emu/mol, the room-temperature susceptibility of **2** was almost zero (small than 2×10^{-5} emu/mol). The susceptibility values can be fitted well by the Curie–Weiss law over the entire temperature range, with a Curie constant of 2.2×10^{-3} K·emu/mol and a very small Weiss temperature of −0.15 K, which usually correspond to paramagnetic ($S_{1/2}$) impurities of 0.6%. Consequently, the magnetic susceptibility measurements suggest that **2** is essentially non-magnetic, which is consistent with the dimeric structure, and its semiconducting nature.

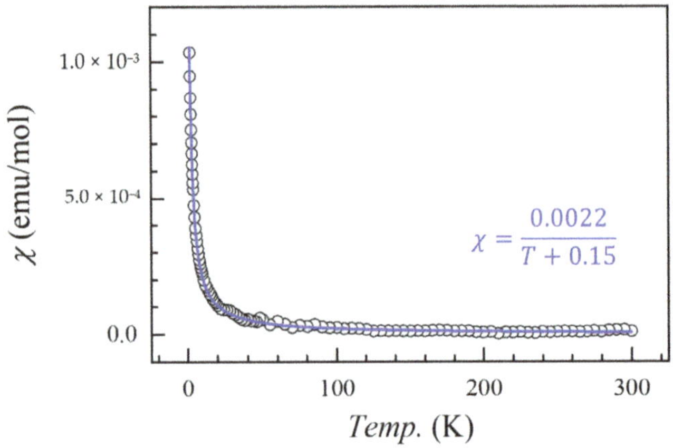

Figure 4. Temperature dependence of magnetic susceptibility of **2** at the field of 5000 Oe. The blue line is the Curie–Weiss fitting curve as described in the text.

3.5. Electronic Structures and Band Structure Calculations of Neutral Gold Complex 2

The MO, band structure, and DOS calculations were performed by the DFT method. The spin polarized molecular orbitals and the energy levels of the neutral [Au(etdt)$_2$] molecule are shown in Figure 5a. The frontier orbitals near the Fermi level are very similar to those of reported gold dithiolate complexes with extended-TTF ligands, such as [Au(tmdt)$_2$]. The singly occupied molecular orbital (SOMO) of [Au(etdt)$_2$] is composed of an anti-symmetric combination of the left- and right-ligand π orbitals and a small contribution of the d orbital of the Au(III) atom. As a result, the spin density distribution of the [Au(etdt)$_2$] molecule shown in Figure 5b is mainly distributed on the ligand. As compared to the neutral [Au(tmdt)$_2$], which becomes a magnetic metal exhibiting antiferromagnetic transition at 110 K, the neutral [Au(etdt)$_2$] becomes a non-magnetic semiconductor owing to the dimeric structure.

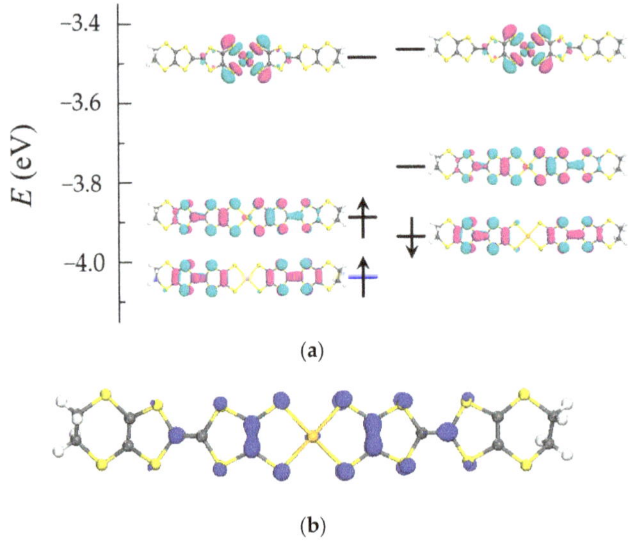

Figure 5. (**a**) Spin polarized molecular orbitals and the energy levels of the neutral [Au(etdt)$_2$] molecule. (**b**) Spin density distributions of the neutral [Au(etdt)$_2$] molecule.

The band energy dispersion curve and DOS of **2** are shown in Figure 6. The energy dispersion is very small along the c^* direction, but exhibits a considerable energy dispersion along the a^* and b^* directions, indicating the 2D nature of the system. The calculated DOS also give a band gap (ΔE) of about 0.20 eV, which is consistent with the semiconducting behavior and E_a ($\Delta E \approx 2E_a$) for the resistivity measurements.

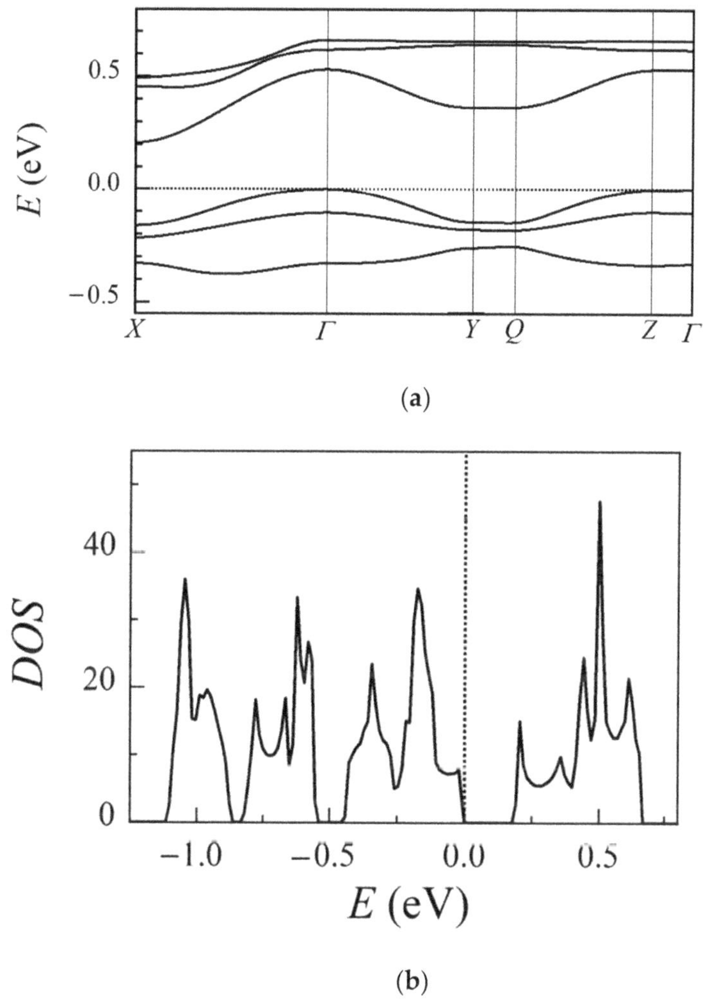

Figure 6. (**a**) The band energy dispersion curve of **2**. The symbols Γ, *X*, *Y*, *Z*, and *Q* represent the following positions in the reciprocal space: *Γ* (0,0,0), *X* (1/2,0,0), *Y* (0,1/2,0), *Z* (0,0,1/2), and *Q* (0,1/2,1/2). (**b**) The density of states (DOS) of **2**.

4. Conclusions

In conclusion, a new neutral gold dithiolate complex with an extended-TTF ligand, $[Au(etdt)_2](THF)$ (**2**), was prepared. Unlike the reported single-component molecular metals, **2** is a rare case of a solvent-containing single-component molecular conductor. The crystals of **2** are composed of 2D conducting layers of $[Au(etdt)_2]$ molecules, which are strongly separated by the terminal ethylene groups and THF molecules. The resistivity measurements performed on the compressed pellets of samples of **2** exhibit fairly high room-temperature conductivity of 0.2 S/cm and a low activation energy of 0.1 eV, which are consistent with the result of the DFT band structure calculations. The observed non-magnetic behavior of **2** is consistent with the dimeric structure of [Au(etdt)$_2$] molecules, and its semiconducting nature. Such results confirm that the crystal structures and electronic structures of the single-component molecular conductor [M(L)$_2$] system can be tuned by adopting various combinations of M and L.

Author Contributions: Conceptualization, B.Z. and A.K.; investigation, K.S., Y.I., B.Z. and A.K.; data curation, K.S., Y.I. and B.Z.; writing—original draft preparation, B.Z., H.K. and A.K.; writing—review and editing, B.Z., H.K. and A.K.; All authors have read and agreed to the published version of the manuscript.

Funding: This research was funded by "JSPS KAKENHI, grant number 17K05846".

Acknowledgments: The authors would like to thank Nanotechnology Platform Program (Molecule and Material Synthesis) of MEXT, Japan.

Conflicts of Interest: The authors declare no conflict of interest.

References

1. Ferraris, J.; Cowan, D.O.; Walatka, V.; Perlstein, J.H. Electron transfer in a new highly conducting donor-acceptor complex. *J. Am. Chem. Soc.* **1973**, *95*, 948–949. [CrossRef]
2. Jérome, D.; Mazaud, A.; Ribault, M.; Bechgaard, K.J. Superconductivity in a synthetic organic conductor(TMTSF)^{2}PF 6. *Phys. Lett.* **1980**, *41*, 95–98.
3. Kobayashi, H.; Tomita, H.; Naito, T.; Kobayashi, A.; Sakai, F.; Watanabe, T.; Cassoux, P. New BETS Conductors with Magnetic Anions (BETS = bis(ethylenedithio)tetraselenafulvalene). *J. Am. Chem. Soc.* **1996**, *118*, 368–377. [CrossRef]
4. Uji, S.; Shinagawa, H.; Terashima, T.; Yakabe, T.; Terai, Y.; Tokumoto, M.; Kobayashi, A.; Tanaka, H.; Kobayashi, H. Magnetic-field-induced superconductivity in a two-dimensional organic conductor. *Nature* **2001**, *410*, 908–910. [CrossRef]
5. Tanaka, H.; Okano, Y.; Kobayashi, H.; Suzuki, W.; Kobayashi, A. Three-dimenstional synthetic metallic crystal composed of single-component molecules. *Science* **2001**, *291*, 285–287. [CrossRef]
6. Kobayashi, A.; Fujiwara, E.; Kobayashi, H. Single-Component Molecular Metals with Extended-TTF Dithiolate Ligands. *Chem. Rev.* **2004**, *104*, 5243–5264. [CrossRef]
7. Tanaka, H.; Tokumoto, M.; Ishibashi, S.; Graf, D.; Choi, E.S.; Brooks, J.S.; Yasuzuka, S.; Okano, Y.; Kobayashi, H.; Kobayashi, A. Observation of Three-Dimensional Fermi Surfaces in a Single-Component Molecular Metal, [Ni(tmdt)$_2$]. *J. Am. Chem. Soc.* **2004**, *126*, 10518–10519. [CrossRef]
8. Rovira, C.; Novoa, J.J.; Mozos, J.-L.; Ordejón, P.; Canadell, E. First-principles study of the neutral molecular metalNi(tmdt)$_2$. *Phys. Rev. B* **2002**, *65*, 081104. [CrossRef]
9. Le Gal, Y.; Roisnel, T.; Auban-Senzier, P.; Bellec, N.; Íñiguez, J.; Canadell, E.; Lorcy, D. Stable Metallic State of a Neutral-Radical Single-Component Conductor at Ambient Pressure. *J. Am. Chem. Soc.* **2018**, *140*, 6998–7004. [CrossRef]
10. Kobayashi, Y.; Terauchi, T.; Sumi, S.; Matsushita, Y. Carrier generation and electronic properties of a single-component pure organic metal. *Nat. Mater.* **2017**, *16*, 109–114. [CrossRef]
11. Zhou, B.; Kobayashi, A.; Okano, Y.; Nakashima, T.; Aoyagi, S.; Nishibori, E.; Sakata, M.; Tokumoto, M.; Kobayashi, H. Single-Component Molecular Conductor [Pt(tmdt)$_2$] (tmdt = trimethylenetetrathiafulvalenedithiolate)—An Advanced Molecular Metal Exhibiting High Metallicity. *Adv. Mater.* **2009**, *21*, 3596–3600. [CrossRef]
12. Takagi, R.; Miyagawa, K.; Yoshimura, M.; Gangi, H.; Kanoda, K.; Zhou, B.; Idobata, Y.; Kobayashi, A. Magnetochiral nonreciprocity of volume spin wave propagation in chiral-lattice ferromagnets. *Phys. Rev. B* **2016**, *93*, 024403. [CrossRef]
13. Suzuki, W.; Fujiwara, E.; Kobayashi, A.; Fujishiro, Y.; Nishibori, E.; Takata, M.; Sakata, M.; Fujiwara, H.; Kobayashi, H. Highly Conducting Crystals Based on Single-Component Gold Complexes with Extended-TTF Dithiolate Ligands. *J. Am. Chem. Soc.* **2003**, *125*, 1486–1487. [CrossRef] [PubMed]

14. Hara, Y.; Miyagawa, K.; Kanoda, K.; Shimamura, M.; Zhou, B.; Kobayashi, A.; Kobayashi, H. NMR Evidence for Antiferromagnetic Transition in the Single-Component Molecular Conductor, [Au(tmdt)$_2$] at 110 K. *J. Phys. Soc. Jpn.* **2008**, *77*, 053706. [CrossRef]
15. Zhou, B.; Yajima, H.; Kobayashi, A.; Okano, Y.; Tanaka, H.; Kumashiro, T.; Nishibori, E.; Sawa, H.; Kobayashi, H. Single-Component Molecular Conductor [Cu(tmdt)$_2$] Containing an Antiferromagnetic Heisenberg Chain. *Inorg. Chem.* **2010**, *49*, 6740–6747. [CrossRef]
16. Takagi, R.; Hamai, T.; Gangi, H.; Miyagawa, K.; Zhou, B.; Kobayashi, A.; Kanoda, K. Single-component molecular material hosting antiferromagnetic and spin-gapped Mott subsystems. *Phys. Rev. B* **2017**, *95*, 094420. [CrossRef]
17. Ogura, S.; Idobata, Y.; Zhou, B.; Kobayashi, A.; Takagi, R.; Miyagawa, K.; Kanoda, K.; Kasai, H.; Nishibori, E.; Satoko, C.; et al. Antiferromagnetic Ordering in the Single-Component Molecular Conductor [Pd(tmdt)$_2$]. *Inorg. Chem.* **2016**, *55*, 7709–7716. [CrossRef]
18. Takagi, R.; Sari, D.P.; Mohd-Tajudin, S.S.; Ashi, R.; Watanabe, I.; Ishibashi, S.; Miyagawa, K.; Ogura, S.; Zhou, B.; Kobayashi, A.; et al. Antiferromagnetic Mott insulating state in the single-component molecular material Pd(tmdt)$_2$. *Phys. Rev. B* **2017**, *96*, 214432. [CrossRef]
19. Kobayashi, A.; Tanaka, H.; Kumasaki, M.; Torii, H.; Narymbetov, B.; Adachi, T. Origin of the High Electrical Conductivity of Neutral [Ni(ptdt)2] (ptdt^{2-}= propylenedithiotetrathiafulvalenedithiolate): A Route to Neutral Molecular Metal. *J. Am. Chem. Soc.* **1999**, *121*, 10763–10771. [CrossRef]
20. Zhou, B.; Yajima, H.; Idobata, Y.; Kobayashi, A.; Kobayashi, T.; Nishibori, E.; Sawa, H.; Kobayashi, H. Single-component Layered Molecular Conductor, [Au(ptdt)$_2$]. *Chem. Lett.* **2012**, *41*, 154–156. [CrossRef]
21. Sasa, M.; Fujiwara, E.; Kobayashi, A.; Ishibashi, S.; Terakura, K.; Okano, Y.; Fujiwara, H.; Kobayashi, H. Crystal structures and physical properties of single-component molecular conductors consisting of nickel and gold complexes with bis(trifluoromethyl)tetrathiafulvalenedithiolate ligands. *J. Mater. Chem.* **2005**, *15*, 155–163. [CrossRef]
22. Zhou, B.; Ogura, S.; Liu, Q.Z.; Kasai, H.; Nishibori, E.; Kobayashi, A. A Single-component Molecular Conductor with Metal–Metal Bonding, [Pd(hfdt)$_2$] (hfdt= bis(trifluoromethyl)tetrathiafulvalenedithiolate). *Chem. Lett.* **2016**, *45*, 303–305. [CrossRef]
23. Cui, H.; Kobayashi, H.; Ishibashi, S.; Sasa, M.; Iwase, F.; Kato, R.; Kobayashi, A. A Single-Component Molecular Superconductor. *J. Am. Chem. Soc.* **2014**, *136*, 7619–7622. [CrossRef]
24. Zhou, B.; Ishibashi, S.; Ishii, T.; Sekine, T.; Takehara, R.; Miyagawa, K.; Kanoda, K.; Nishibori, E.; Kobayashi, A. Single-component molecular conductor [Pt(dmdt)$_2$]—A three-dimensional ambient-pressure molecular Dirac electron system. *Chem. Commun.* **2019**, *55*, 3327–3330. [CrossRef]
25. Kajita, K.; Nishio, Y.; Tajima, N.; Suzumura, Y.; Kobayashi, A. Molecular Dirac Fermion Systems—Theoretical and Experimental Approaches—. *J. Phys. Soc. Jpn.* **2014**, *83*, 072002. [CrossRef]
26. Kajita, K.; Nishio, Y.; Moriyama, S.; Sasaki, W.; Kato, R.; Kobayashi, H.; Kobayashi, A. New organic superconductors Κ- and θ-(BEDT-TTF)2I3: Transport property. *Solid State Commun.* **1987**, *64*, 1279–1284. [CrossRef]
27. Yagubskiĭ, É.B.; Shchegolev, I.F.; Laukhin, V.N.; Kononovich, P.A.; Karatsovnik, M.V.; Zvarykina, A.V.; Buravov, L.I. Coexistence of different superconducting phases with transition temperatures between 1.5 and 7 K in the (BEDT-TTF)- $I^-{}_3$ system. *J. Exp. Theor. Phys.* **1984**, *39*, 12–16.
28. Nakano, M.; Kuroda, A.; Matsubayashi, G.-E. Extended bisdithiolene metal complexes: Preparation and electrical conductivities of [M($C_8H_4S_8$)$_2$] anion complexes (M = Ni(II), Pt(II), Au(III)). *Inorg. Chim. Acta* **1997**, *254*, 189–193. [CrossRef]
29. Kubo, K.; Nakano, M.; Tamura, H.; Matsubayashi, G.-E.; Nakamoto, M. Preparation and oxidation of polarized Au(III) complexes having both the C-deprotonated-2-phenylpyridine (ppy) and a sulfur-rich dithiolate ligand and X-ray crystal structure of [Au(η_2-C,N-ppy)(η_2-S,S-$C_8H_4S_8$)]·0.5DMF. *J. Organomet. Chem.* **2003**, *669*, 141–148. [CrossRef]
30. Binet, L.; Fabre, J.M.; Montginoul, C.; Simonsen, K.B.; Becher, J. Preparation and chemistry of new unsymmetrically substituted tetrachalcogenofulvalenes bearing CN(CH_2)$_2$X and HO(CH_2)$_2$X groups (X = S or Se). *J. Chem. Soc. Perkin Trans. 1* **1996**, 783–788. [CrossRef]
31. Braunstein, P.; Clark, R.J.H. The preparation, properties, and vibrational spectra of complexes containing the AuCl^{2-}, AuBr^{2-}, and AuI^{2-} ions. *J. Chem. Soc. Dalton Trans.* **1973**, 1845–1848. [CrossRef]
32. Sheldrick, G. A short history ofSHELX. *Acta Crystallogr. A* **2008**, *64*, 112–122. [CrossRef]

33. Sheldrick, G. SHELXT—Integrated space-group and crystal-structure determination. *Acta Crystallogr. C* **2015**, *71*, 3–8. [CrossRef]
34. Dolomanov, O.V.; Bourhis, L.J.; Gildea, R.J.; Howard, J.A.K.; Puschmann, H. OLEX2: A complete structure solution, refinement and analysis program. *J. Appl. Crystallogr.* **2009**, *42*, 339–341. [CrossRef]
35. Delley, B. An all-electron numerical method for solving the local density functional for polyatomic molecules. *J. Chem. Phys.* **1990**, *92*, 508–517. [CrossRef]
36. Delley, B. From molecules to solids with the DMol3 approach. *J. Chem. Phys.* **2000**, *113*, 7756–7764. [CrossRef]
37. Perdew, J.P.; Burke, K.; Ernzerhof, M. Generalized Gradient Approximation Made Simple. *Phys. Rev. Lett.* **1996**, *77*, 3865–3868. [CrossRef]
38. Kurmoo, M.; Graham, A.W.; Day, P.; Coles, S.J.; Hursthouse, M.B.; Caulfield, J.L.; Singleton, J.; Pratt, F.L.; Hayes, W.; Ducasse, L.; et al. Superconducting and Semiconducting Magnetic Charge Transfer Salts: (BEDT-TTF)$_4$AFe(C_2O_4)$_3$.cntdot.C_6H_5CN (A = H_2O, K, NH_4). *J. Am. Chem. Soc.* **1995**, *117*, 12209–12217. [CrossRef]

crystals

MDPI

Article

Direct Observation of Molecular Orbitals Using Synchrotron X-ray Diffraction

Shunsuke Kitou [1,*], Yuto Hosogi [1], Ryo Kitaura [2], Toshio Naito [3,4,5], Toshikazu Nakamura [6] and Hiroshi Sawa [1,*]

1 Department of Applied Physics, Nagoya University, Nagoya 464-8603, Japan; hosogi.yuto@g.mbox.nagoya-u.ac.jp
2 Department of Chemistry, Nagoya University, Nagoya 464-8602, Japan; r.kitaura@nagoya-u.jp
3 Graduate School of Science and Engineering, Ehime University, Matsuyama 790-8577, Japan; naito.toshio.mu@ehime-u.ac.jp
4 Research Unit for Development of Organic Superconductors, Ehime University, Matsuyama 790-8577, Japan
5 Geodynamics Research Center (GRC), Ehime University, Matsuyama 790-8577, Japan
6 Institute for Molecular Science, Myodaiji, Okazaki 444-8585, Japan; t-nk@ims.ac.jp
* Correspondence: kitou.shunsuke@h.mbox.nagoya-u.ac.jp (S.K.); hiroshi.sawa@cc.nagoya-u.ac.jp (H.S.)

Received: 15 October 2020; Accepted: 30 October 2020; Published: 3 November 2020

Abstract: The physical properties of molecular crystals are governed by the frontier orbitals of molecules. A molecular orbital, which is formed by superposing the atomic orbitals of constituent elements, has complicated degrees of freedom in the crystal because of the influence of electron correlation and crystal field. Therefore, in general, it is difficult to experimentally observe the whole picture of a frontier orbital. Here, we introduce a new method called "core differential Fourier synthesis" (CDFS) using synchrotron X-ray diffraction to observe the valence electron density in materials. By observing the valence electrons occupied in molecular orbitals, the orbital state can be directly determined in a real space. In this study, we applied the CDFS method to molecular materials such as diamond, C_{60} fullerene, $(MV)I_2$, and $(TMTTF)_2X$. Our results not only demonstrate the typical orbital states in some materials, but also provide a new method for studying intramolecular degrees of freedom.

Keywords: X-ray diffraction; single crystal; electron density; molecular orbital

1. Introduction

The degree of freedom of electron orbitals in materials is closely related to physical properties, such as electrical conductivity, magnetic, dielectric, and optical properties, and crystal structure [1–3]. An electron orbital corresponds to the existence probability of electrons in a real space. In other words, the orbital state can be correctly understood by observing the spatial distribution of electrons in materials. Among the three degrees of freedom in electrons (i.e., charge, spin, and orbital), because the charge and spin can readily respond to external electric and/or magnetic fields, their properties are relatively easy to measure. On the other hand, several experimental methods, such as electron spin resonance (ESR), nuclear magnetic resonance (NMR), high harmonics generated from femtosecond laser pulses [4], ultraviolet angle-resolved photoelectron spectroscopy [5], polarized neutron diffraction [6], resonant X-ray scattering [7], and multipole analysis [8] using X-ray diffraction (XRD), are known to extract orbital information that control the anisotropy of the physical properties. However, in these methods, the measurable substances are limited, and the results may depend on the applied model. Furthermore, as the molecular substance formed by the bonding of multiple atoms has several independent parameters that describe the crystal structure, it is very difficult to estimate the frontier orbital state of the entire system from quantum chemical or first-principles calculations.

The spatial distribution of electrons observed experimentally will be useful for understanding the orbital states in materials. In this sense, XRD is one of the most powerful experimental probes to observe electrons around atoms in a solid. XRD is based on the scattering phenomenon of electrons in a crystal, where the periodic structure of electrons and diffraction intensities are linked by a Fourier transform. In principle, a complete electron density (ED) distribution can be reproduced by performing the inverse Fourier transform on the 'infinite' number of diffraction intensities corresponding to the Fourier series. However, in reality, because the number of diffraction reflections obtained by the XRD experiment is 'finite', it is impossible to reproduce the complete ED distribution owing to the mathematical truncation effects of the Fourier synthesis. Moreover, of all the electrons around an atom, only valence electrons, which are very small in number compared to the total number of electrons in the crystal, are responsible for their physical properties. Therefore, to extract the frontier orbital information from the ED, only the valence electrons need to be extracted. Due to these difficulties, instead of observing the ED, the crystal structure is usually determined from XRD experiments by refining the internal parameters (e.g., atomic positions, atomic displacement parameters, and site occupancies) of the assumed crystal structural model through the least squares method to match the diffraction intensities. This method is called a crystal structural analysis. To access the orbital state (i.e., the distribution of electrons) with high accuracy in materials from the XRD, a new method of analysis is required.

In this paper, we introduce the core differential Fourier synthesis (CDFS) method [9,10] for directly observing the valence electron density (VED) in a crystal from XRD data. The details of the method are also explained in [10]. CDFS can efficiently extract only the valence electron information, which corresponds to the physical properties. By using this method, the orbital state that governs the anisotropy of the physical properties can be directly observed in a real space. Here, we show the results of applying the CDFS method to diamond, C_{60} fullerene, $(MV)I_2$ (MV = methyl viologen), and $(TMTTF)_2X$ (TMTTF = tetramethyl tetrathiafulvalene; X = PF_6, and AsF_6).

2. Methods

The XRD experiments using single crystals were performed on the BL02B1 beamline at the synchrotron radiation facility SPring-8 in Japan [11]. A helium-gas-blowing device (Japan Thermal Engineering Co., Ltd., Japan) was employed to cool the sample. A two-dimensional (2D) imaging plate (FUJIFILM Co., Japan) and a CdTe PILATUS (DECTRIS Ltd., Switzerland) detectors, which had a dynamic range of ~10^6, were installed in the diffractometer (Rigaku Co., Japan). Diffraction intensity averaging and a structural analysis were performed using SORTAV [12] and Jana2006 [13], respectively. The crystal structure and ED distribution figures were visualized using VESTA [14].

The diffraction intensity $I(\boldsymbol{K})$ obtained from the XRD experiment can be described as a Fourier transform of the ED $\rho(\boldsymbol{r})$ using the scattering vector $\boldsymbol{K}$ (Equation (1)).

$$I(\boldsymbol{K}) = S\left|\int_{\text{all}} \rho(\boldsymbol{r})e^{-i\boldsymbol{K}\boldsymbol{r}}d\boldsymbol{r}\right|^2 \tag{1}$$

Here, S is a scaling factor. The integration range ($\int_{\text{all}}$ in Equation (1)) corresponds to the range in which X-rays interfere in the crystal. In the case of a crystal, $I(\boldsymbol{K})$ can be described as Equation (2) using Laue functions $L(s)$, $L(t)$, and $L(u)$, where $\boldsymbol{K} = s\boldsymbol{a}^* + t\boldsymbol{b}^* + u\boldsymbol{c}^*$.

$$\begin{aligned} I(\boldsymbol{K}) &= S\,L(s)L(t)L(u)\left|\int_{\text{unit cell}} \rho(\boldsymbol{r})e^{-i\boldsymbol{K}\boldsymbol{r}}d\boldsymbol{r}\right|^2 \\ &\propto \left|F_{\text{obs}}(\boldsymbol{K})\right|^2 \end{aligned} \tag{2}$$

Here, $\left|F_{\text{obs}}(\boldsymbol{K})\right|$ is the absolute value of the experimentally observed crystal structure factor. Generally, if the number of unit cells N in a crystal is sufficiently large, then the Laue function becomes

N^2 only when all of the s, t, u are integers; otherwise, $I(\boldsymbol{K}) = 0$. In principle, the ED distribution can be reproduced by the inverse Fourier transform of the diffraction intensity according to

$$\rho(\boldsymbol{r}) = \frac{1}{V}\sum_{\boldsymbol{K}}^{\infty} F_{\text{obs}}(\boldsymbol{K})e^{i\boldsymbol{K}\boldsymbol{r}}, \tag{3}$$

if the infinite diffraction data were observed. However, the calculation of the ED generally has three problems. (i) To extract the VED with anisotropic information, a sufficiently wide dynamic range of intensity is required. (ii) $|F_{\text{obs}}(\boldsymbol{K})|$ is obtained from the experimentally diffraction intensity because of the relationship of $I(\boldsymbol{K}) \propto |F_{\text{obs}}(\boldsymbol{K})|^2$. In this case, $|F_{\text{obs}}(\boldsymbol{K})|$ does not include information of the phase term $P = e^{i\phi(\boldsymbol{K})}$ as $F_{\text{obs}}(\boldsymbol{K}) = |F_{\text{obs}}(\boldsymbol{K})|P = |F_{\text{obs}}(\boldsymbol{K})|e^{i\phi(\boldsymbol{K})}$, which is necessary for the calculation of $\rho(\boldsymbol{r})$. (iii) Because the number of the $F_{\text{obs}}(\boldsymbol{K})$ data is finite, the mathematical truncation effect seriously affects the ED distribution.

With regard to (i), the relationship between the number of focused electrons and the number of electrons in the unit cell is important. In the case of a quasi-one-dimensional (1D) molecular conductor $(TMTTF)_2X$, a 3/4-filled system, because X^- is a monovalent anion, such as PF_6^-, AsF_6^-, SbF_6^-, Br^-, or ClO_4^-, there is one hole for two TMTTF molecules forming a dimer. The number of electrons in the $(TMTTF)_2X$ unit is approximately 350. To estimate the amount of charge transfer of less than $1e$ in the TMTTF dimer in the charge ordering state, a dynamic range that can accurately extract $1^2/350^2 = 1/122500 \sim 10^{-5}$ of the measured maximum intensity is necessary. Here, the 10^{-5} signal information should be observed with a sufficient signal-to-noise (S/N) ratio. This requirement is sufficiently fulfilled using the current synchrotron radiation facility, in which the dynamic range of 10^6 is guaranteed.

With regard to (ii), because the phase term can be assigned from the calculated crystal structure factor

$$F_{\text{cal}}(\boldsymbol{K}) = \sum_j f_j T_j e^{-i\boldsymbol{K}\boldsymbol{r}_j} \tag{4}$$

as $P = F_{\text{cal}}(\boldsymbol{K})/|F_{\text{cal}}(\boldsymbol{K})|$, Equation (3) can be rewritten as

$$\rho(\boldsymbol{r}) = \frac{1}{V}\sum_{\boldsymbol{K}}^{\infty} |F_{\text{obs}}(\boldsymbol{K})| P e^{i\boldsymbol{K}\boldsymbol{r}}. \tag{5}$$

Here, $\boldsymbol{r}_j$ is the jth atomic position, T_j is the jth atomic displacement parameter, and f_j is the jth atomic scattering factor, which is described as

$$f_j(\boldsymbol{K}) = \int_{\text{atom}} \rho_j(\boldsymbol{r}) e^{-i\boldsymbol{K}\boldsymbol{r}} d\boldsymbol{r}. \tag{6}$$

Figure 1 shows the atomic scattering factor of carbon as a function of $\sin\theta/\lambda$ [15]. Here, λ is the wavelength of the incident X-ray, and θ is the XRD angle. While the contribution of the core electrons extends to the high-angle region, the contribution of the valence electrons exists only in the low-angle region ($\sin\theta/\lambda \leq 0.5$ Å^{-1}). When calculating the ED, it is necessary to determine $F_{\text{cal}}(\boldsymbol{K})$ with high accuracy to obtain the correct phase term P. For this purpose, the structural refinement is performed using only the high-angle $I(\boldsymbol{K})$ (i.e., high-angle analysis), where the contribution of the spatially spread valence electron is very small.

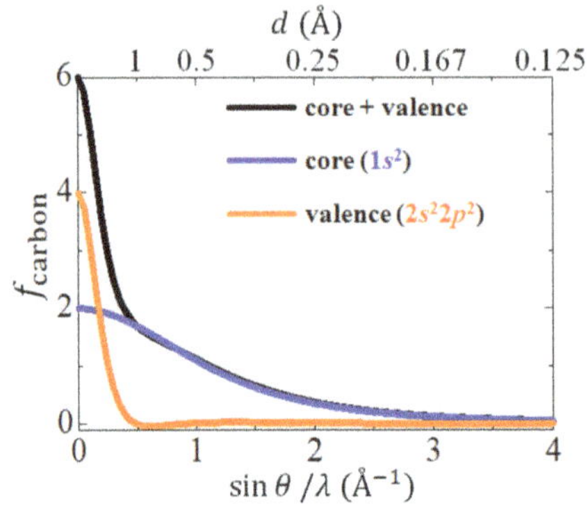

Figure 1. Atomic scattering factor of carbon [15]. Black, blue, and orange lines indicate the contribution of the total, core, and valence electrons, respectively.

With regard to (iii), the truncation effect of the Fourier synthesis can be suppressed using the CDFS method. The equation of the inverse Fourier transform by the CDFS method is described as

$$\rho_v(\boldsymbol{r}) = \frac{1}{V}\sum_{\boldsymbol{K}}\left[\left(|F_{\text{obs}}(\boldsymbol{K})|P - \left|\sum_j f_j^{\text{core}} T_j e^{-i\boldsymbol{K}\boldsymbol{r}_j}\right| P^{\text{core}}\right)e^{i\boldsymbol{K}\boldsymbol{r}}\right] + \frac{n_v}{V}. \tag{7}$$

Here, f_j^{core} is the jth atomic scattering factor with only the core electron contribution, which corresponds to the blue line in Figure 1. P^{core} is the phase term with only the core electron contribution calculated as $P = F_{\text{cal}}^{\text{core}}(\boldsymbol{K})/\left|F_{\text{cal}}^{\text{core}}(\boldsymbol{K})\right|$. n_v is the total number of valence electrons contained in the unit cell. The 0 0 0 Bragg reflection intensity cannot be observed experimentally. When ignoring the second term n_v/V in Equation (7), the total number of electrons in the unit cell becomes zero. Therefore, the VED distribution data are corrected by adding the n_v/V term. In general, $(\sin\theta/\lambda)_{\text{max}}$ measured experimentally is ~ 2.5Å^{-1} (d_{min} ~ 0.2Å) even when using the short-wavelength X-ray obtained at the synchrotron radiation facility. In Figure 1, a nonzero value remains in $f_{\text{carbon}}\left(= f_{\text{carbon}}^{\text{core}}\right)$ at $\sin\theta/\lambda = 2.5$ Å^{-1}. Therefore, even if a short-wavelength X-ray is used, the mathematical truncation effect cannot be avoided from the calculation using Equation (5). Meanwhile, in Equation (7), the inverse Fourier transform is performed on the term obtained by subtracting the core electron contribution from $\left|F_{\text{obs}}(\boldsymbol{K})\right|$, which contains the contribution of all electrons. Because most of the contribution of the remaining $f_{\text{carbon}}^{\text{valence}}$ exists only in $\sin\theta/\lambda \leq 0.5$ Å^{-1}, the VED can be extracted with a little truncation effect of the Fourier synthesis by the CDFS analysis.

3. Results and Discussion

3.1. Diamond

First, we describe the VED of diamond, which is a typical compound that forms sp^3 hybridized orbitals. Diamond forms a crystal structure in which C atoms are arranged in three dimensional space (space group: $Fd\bar{3}m$) (Figure 2a). Each C atom has six electrons, where the inner 1s orbital is occupied by two electrons, and the outer 2s and 2p orbitals are occupied by two electrons. In general, the valence electrons occupying the outer shell orbitals contribute to the physical properties. In the case of diamond, the four valence electrons of each C atom occupy the sp^3 hybridized orbitals, forming a very strong bond state called a covalent bond.

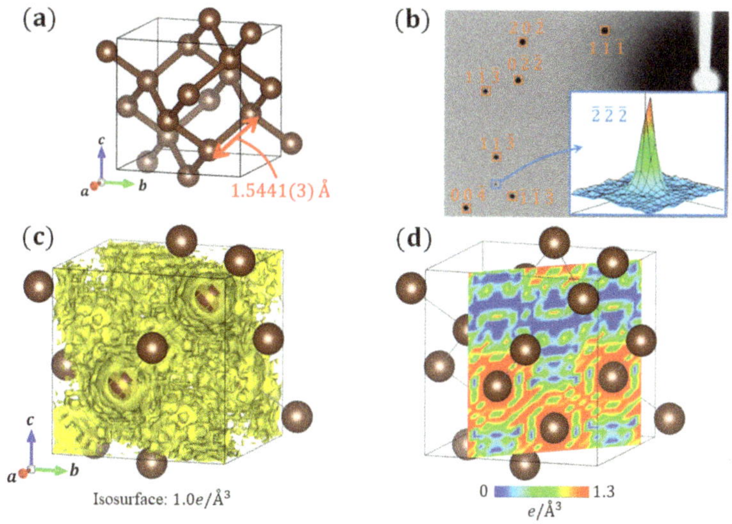

Figure 2. (**a**) Crystal structure of diamond; (**b**) Synchrotron X-ray diffraction (XRD) data of a diamond crystal at 30 K. (**c**,**d**) Total electron density (ED) distribution of diamond calculated by the general inverse Fourier transform of diffraction intensity using Equation (5); (**c**) Surface plot; (**d**) Sectional view of the (4 4 0) plane.

Because the d-glide ($h + k + l \neq 4n$ reflections disappear) exists in the space group $Fd\bar{3}m$, the diffraction intensity of a $\bar{2}\,\bar{2}\,\bar{2}$ reflection is originally zero. However, the $\bar{2}\,\bar{2}\,\bar{2}$ (and its equivalent reflections) intensity appears in the data obtained by the synchrotron XRD experiment at 30 K (Figure 2b). This finding corresponds to the fact that the extinction rule of *d*-glide is violated by the bonding electrons existing between C atoms. Using the high-resolution diffraction data obtained by the short-wavelength X-rays of synchrotron radiation, an ED analysis based on the usual inverse Fourier transform equation (Equation (5)) is performed, where reflections that satisfy the *d*-glide extinction rule are also included. Because the obtained total ED distribution of diamond is greatly disturbed (Figure 2c,d), the VED distribution corresponding to the covalent bonds cannot be confirmed. This is the mathematical truncation effect of the Fourier synthesis mentioned above.

Figure 3 shows the VED distribution of diamond obtained by the CDFS analysis using Equation (7). A very smooth ED distribution is observed, and the VED corresponding to the C–C covalent bonds is clearly visible. The density at the center point of the C–C covalent bonds is $\sim 1.3e/\text{Å}^3$. Compared with the VED distribution obtained by the first-principles calculation [16], this result not only qualitatively reproduces its shape, but also quantitatively reproduces its density well. As shown here, the VED distribution can be extracted with high accuracy through the CDFS analysis. In diamond, valence electrons contributed only to covalent bonds. In the following, we will introduce some systems in which the valence electrons play more roles. In particular, beginning with a small, simple molecule with a rather localized electronic system, we will progressively go into the details of more complicated molecular systems with delocalized electrons, concluding with discussion of those capable of phase transitions.

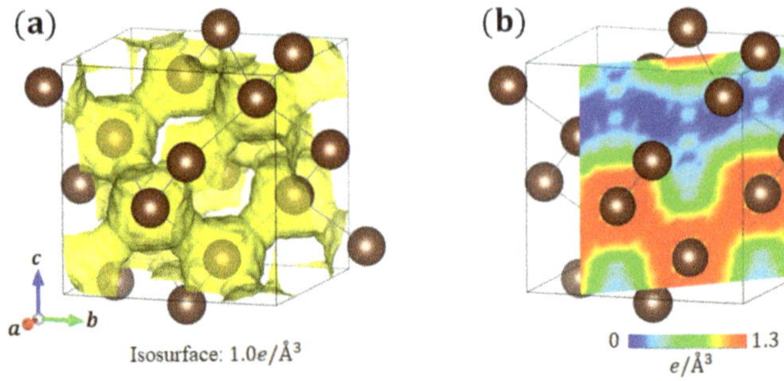

Figure 3. Valence electron density (VED) distribution of diamond calculated by the CDFS method using Equation (7): (**a**) Surface plot, and (**b**) Sectional view of the (4 4 0) plane.

3.2. C_{60} Fullerene

Next, we show the VED of C_{60} fullerene [17,18], which has two types of features of the sp^2 and sp^3 hybridizations. Many structural studies of C_{60} have been reported so far [19–24]. The C_{60} cluster consists of 20 hexagons and 12 pentagons. The former involves the sp^2 hybridization, and the latter sp^3 hybridization. There are two different types of C–C bonds; one is on the regular pentagon (single bond), and the other is shared by two hexagons (double bond). In the high-temperature phase of a C_{60} crystal, the C_{60} molecules are completely orientationally disordered and continue to randomly rotate, where the space group is $Fm\overline{3}m$. At 260 K, C_{60} exhibits a first-order phase transition; below that temperature, there is a discontinuous reorientation principally around the [111] direction, where the space group is $Pa\overline{3}$. Below 90 K, the discontinuous rotational motion is frozen, although a small amount of static disorder still exists [19,23].

In this study, we focus on the low-temperature phase at 30 K without a dynamical motion of C_{60} molecules. The lattice parameter in $Pa\overline{3}$ is $a = 14.0279(6)$ Å at 30 K. The detail of the structural analysis results is shown in Appendix A (Table A1). As reported in previous structural studies [22,23], there may be two types of domains in the low-temperature phase. One is the merohedral domain described in [22], in which $I(hkl)$ and $I(khl)$ are superimposed, and the other is the rotational domain around the [111] direction described in [23], in which two kinds of C_{60} configurations exist. Our structural analysis results show that the ratio of the merohedral domain is 0.4995(8):0.5005, and the ratio of the rotational domain is 0.818(3):0.182. When observing the VED distribution, the contributions of the former domain can be separated analytically. In the latter domain, the two components appear independently, where the VED in two directions coexist in a real space. However, in this case, because the ratio of the rotational domain was quite different, the VED was observed as almost one domain.

Figure 4a shows the crystal structure and VED of a C_{60} molecule, and Figure 4b shows a sectional view of the $\left(1\ 1\ \overline{1}\right)$ plane. The (i) and (iii) bonds, ~ 1.45 Å, correspond to the single bonds, and the (ii) bond, ~ 1.39 Å, corresponds to the double bond. Certainly, the double bond forms a stronger bond than the single bond. The VED distribution is clearly observed along the shape of the C_{60} molecule, which is very similar to the results of previous density functional theory calculations [25]. Furthermore, the difference in the ED is confirmed between the single (shown as (i) and (iii)) and double (shown as (ii)) bonds (Figure 4b). The ED at the double bond is higher than that at the single bond, which corresponds to the strength of the C–C covalent bond. We succeeded in quantitatively evaluating the covalent bond strength in the C_{60} molecule from the VED.

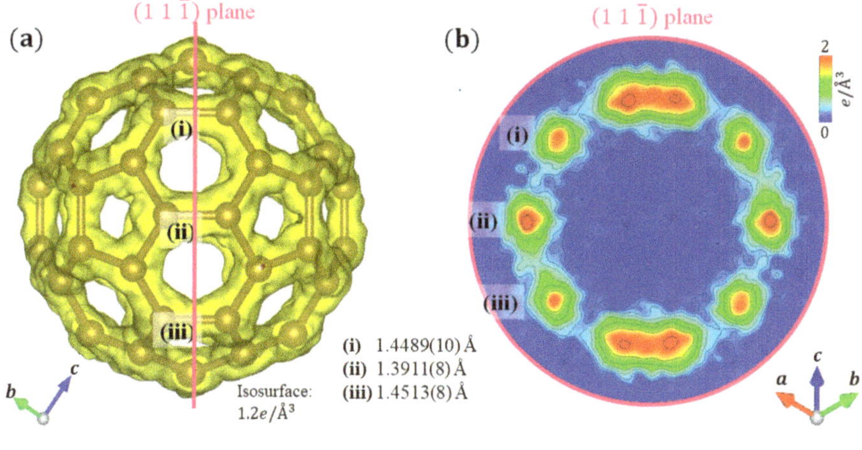

Figure 4. VED distribution of C_{60} fullerene: (**a**) Surface plot; (**b**) Sectional view of the $(1\,1\,\bar{1})$ plane. The (i) and (iii) bonds correspond to the single bonds, which are on the regular pentagons. The (ii) bond corresponds to the double bond, which is shared by two hexagons.

3.3. $(MV)I_2$

In $(MV)I_2$ (space group: $P2_1/c$) [26], the valence of an MV molecule is 2+ because of two I^- ions. The crystal structure of $(MV)I_2$ and the molecular structure of MV^{2+} are shown in Figure 5a,b, respectively. We performed the structural analysis and CDFS analysis of $(MV)I_2$ at 100 K. Figure 5d shows the VED of an MV^{2+} molecule, in which H, C, and N atoms have $1s^1$, $2s^22p^2$, and $2s^22p^3$ valence electrons, respectively. Although C and N^+ have the same number of valence electrons, the distribution states of valence electrons on the C and N atoms in the MV^{2+} molecule are quite different from each other. The ED around C is spatially spread, whereas that around N^+ is localized on the atom. This result corresponds to the difference in electronegativities between the C and N atoms, where the N atoms are more electronegative than the C atoms. Furthermore, there is a difference in the EDs around the two C atoms bonded to N in the six-membered ring of the MV^{2+} molecule. The ED around the C(1) atoms close to the I^- anions (surrounded by red dotted circles) on the MV^{2+} molecular plane, which is in the direction of the hydrogen bond, is lower than that of the C(2) atoms located at the opposite side. This result suggests that the anisotropic VED distribution on the MV^{2+} molecule is realized by the anisotropic electrostatic interactions with the surrounding the I^- ions. A CDFS analysis directly visualized the distribution state of complicated molecular orbitals composed of different types of atoms.

3.4. $(TMTTF)_2X$

Finally, we applied the CDFS analysis to quasi-1D molecular conductors $(TMTTF)_2X$ [27], which shows various electronic properties in the pressure–temperature phase diagram [28–31]. Figure 6a,b shows the crystal structure of $(TMTTF)_2PF_6$ and the molecular structure of TMTTF, respectively. For the charge ordering phase transition of $(TMTTF)_2PF_6$ (T_{CO} ~ 67 K) at ambient pressure, although responses associated with the charge ordering are confirmed by the dielectric constant [32,33], NMR [34,35], ESR [36], infrared, and Raman spectroscopies [37,38], no evidence of charge ordering has been observed from the crystal structure [39,40]. Thus, this transition from the dimer Mott phase to the charge ordering phase (Figure 6c), which is associated with the lack of an inversion center, has been called a mysterious 'structureless transition' [41–43]. We investigated the crystal structure and VED distribution of the charge ordering phase in $(TMTTF)_2PF_6$ using synchrotron XRD [9].

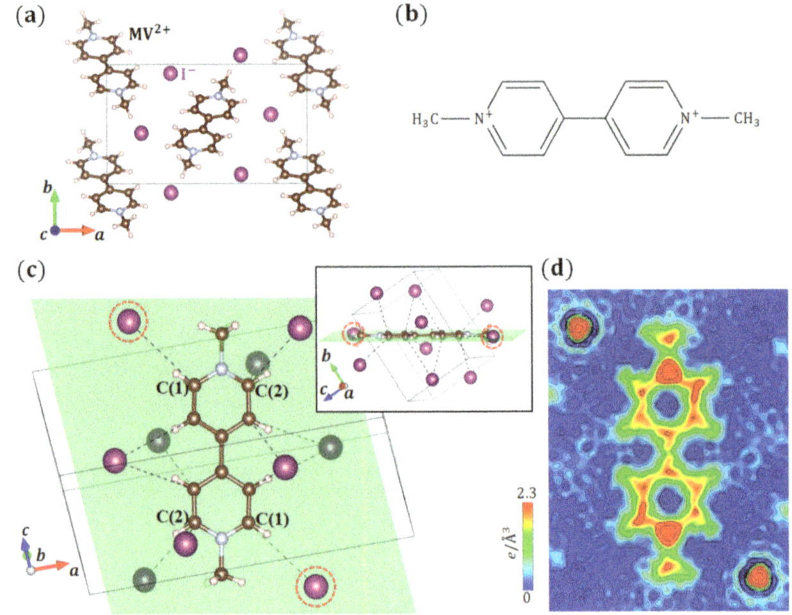

Figure 5. (**a**) Crystal structure of (MV)I_2; (**b**) Molecular structure of MV^{2+}; (**c**) Relationship between a MV^{2+} molecule and surrounding I^- ions. There is an inversion center at the central C–C bond in the MV^{2+} molecule. The least square (LS) plane, on which the central six C atoms are located, is shown in green. The I^- ions surrounded by the red dotted circles exist near the green plane. (**d**) VED distribution of the MV^{2+} molecule. This density plane corresponds to the green plane in (**c**). The orientations of the MV^{2+} molecules are the same between (**c**,**d**).

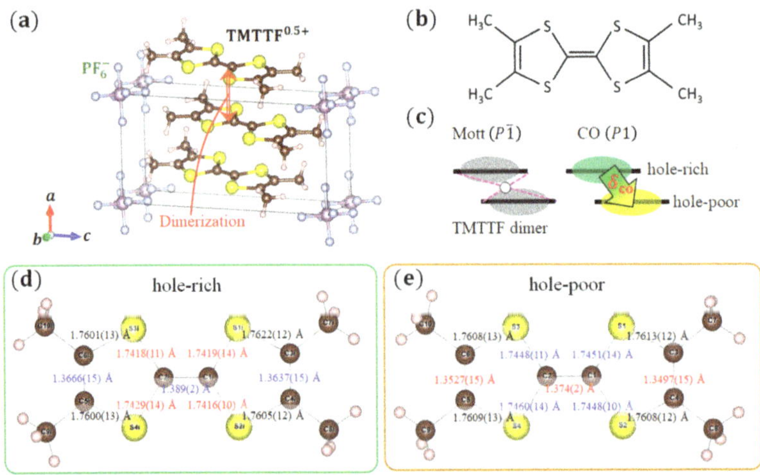

Figure 6. (**a**) Crystal structure of $(TMTTF)_2PF_6$; (**b**) Molecular structure of a TMTTF; (**c**) Schematic configuration of the dimer Mott and charge ordering (CO) phases; (**d**,**e**) Bond length in the (**d**) hole-rich and (**e**) hole-poor TMTTF molecules at 30 K in the charge ordering phase of $(TMTTF)_2PF_6$. The red (blue) values indicate that the bonds are shorter (longer) than the others at the same positions. The black values indicate that bonds at the same position are consistent within the error range.

The precise structural parameters at 30 K in the charge ordering phase are obtained by a high-angle analysis. Figure 6d,e shows the bond length in the hole-rich and hole-poor TMTTF molecules at 30 K, respectively. The C=C bonds at the center and both sides in the hole-rich molecule are longer than those in the hole-poor molecule. The central C–S bonds in the hole-rich molecule are shorter than those in the hole-poor molecule. These results correctly reflect the tendency of the charge ordering state in the TMTTF dimer. The amount of charge transfer δ_{co} is estimated from the bond length in the TMTTF molecule. Two types of formulas, $q = -15.55 + 20.42r$ (set 1) and $q = -26.88 + 34.98r$ (set 2), are given for empirically calculating the valence of a TMTTF molecule in [44]. Here, q is the valence of the TMTTF, and $r(= a/b)$ is a ratio of the central (a) C=C bond and (b) C–S bonds length. The amount of charge transfer δ_{co} in the TMTTF dimer is calculated as $\delta_{co} = \left(q_{\text{hole-poor}} - q_{\text{hole-rich}}\right)/2$. By using these formulas, $\delta_{co} = 0.10e$ (set 1) and $0.17e$ (set 2) are obtained from our structural analysis results at 30 K. The difference of the highest occupied molecular orbital levels between the hole-poor and hole-rich TMTTF molecules, $\sim$ 21.7 meV, was confirmed by the extended Hückel calculation [45]. These results show that our crystal structural analysis revealed the charge ordering state in $(TMTTF)_2PF_6$.

Figure 7a,b shows the VED of the hole-rich and hole-poor TMTTF molecules in the charge ordering phase of $(TMTTF)_2PF_6$ obtained from the CDFS analysis. In this case, the valence electrons of H, C, and S constituting the TMTTF molecule correspond to $1s^1$, $2s^2 2p^2$, and $3s^2 3p^4$, respectively. The ED corresponding to the bonding orbital on the C=C bonds and the node of ED corresponding to the antibonding orbital on the C–S bonds are clearly shown. Moreover, the ED reflecting the isotropic s orbitals is observed on the S and C atoms of the methyl groups. Almost no difference is observed between the appearance in the two VED distributions (Figure 7a,b), because the amount of charge transfer ($\delta_{co} \leq 0.5e$) is too small compared to the number of valence electrons of a TMTTF molecule ($64e$). Therefore, we compared the number of valence electrons between the two molecules in the dimer. By comparing the number of valence electrons in the atomic basin of the respective atoms in the TMTTF molecule calculated by Bader's topological analysis [46], the amount of charge transfer was determined as $\delta_{co} = 0.10e$ [9], which is consistent with the estimation from the bond length ($\delta_{co} = 0.10e$) (set 1), from the Raman ($\delta_{co} = 0.055e$ [38]), from the infrared ($\delta_{co} = 0.075e$ [31]), and from the NMR spectroscopies ($\delta_{co} = 0.14e$ [35]).

To investigate the intramolecular degrees of freedom, we focus on $(TMTTF)_2AsF_6$, which has the common crystal structure $(TMTTF)_2PF_6$ and undergoes a charge ordering phase transition at $T_{CO} \sim 100$ K [47]. A larger charge transfer in the dimer than that of $(TMTTF)_2PF_6$ was confirmed in the charge ordering phase of $(TMTTF)_2AsF_6$ by the Raman ($\delta_{co} = 0.09e$ [38]), infrared ($\delta_{co} = 0.105e$ [31]), and NMR spectroscopies ($\delta_{co} = 0.25e$ [48], $\delta_{co} = 0.17e$ [49]). We investigated whether the difference in the amount of charge transfer affects the VED distribution of the TMTTF molecule. As a result of the high-angle analysis of $(TMTTF)_2AsF_6$ at 30 K, $\delta_{co} = 0.34e$ (set 1), which is larger than that of $X = PF_6$, is determined from the bond lengths in the TMTTF molecules using the formulas in [44]. This tendency is consistent with those described in previous reports of $X = PF_6$ and AsF_6 [31,35,38,48,49].

Figure 7c,d shows the VED of the hole-rich and hole-poor TMTTF molecules in the charge ordering phase of $(TMTTF)_2AsF_6$ obtained from the CDFS analysis, respectively. Almost no difference is observed in the appearance between the hole-rich and hole-poor TMTTF molecules in $X = AsF_6$, and the VED distributions of $X = AsF_6$ are also approximately identical to those of $X = PF_6$. In this case, the difference in the magnitude of the absolute value of the ED between $X = PF_6$ and AsF_6 is of little significance because the CDFS analysis is based on the inverse Fourier transform of finite data (Equation (7)). However, when calculating the amount of the charge transfer in the dimer from the VED of $X = AsF_6$, the charge transfer $\delta_{co} = 0.43e$ is obtained, which is larger than that of $X = PF_6$. This tendency is consistent with that in our structural analysis.

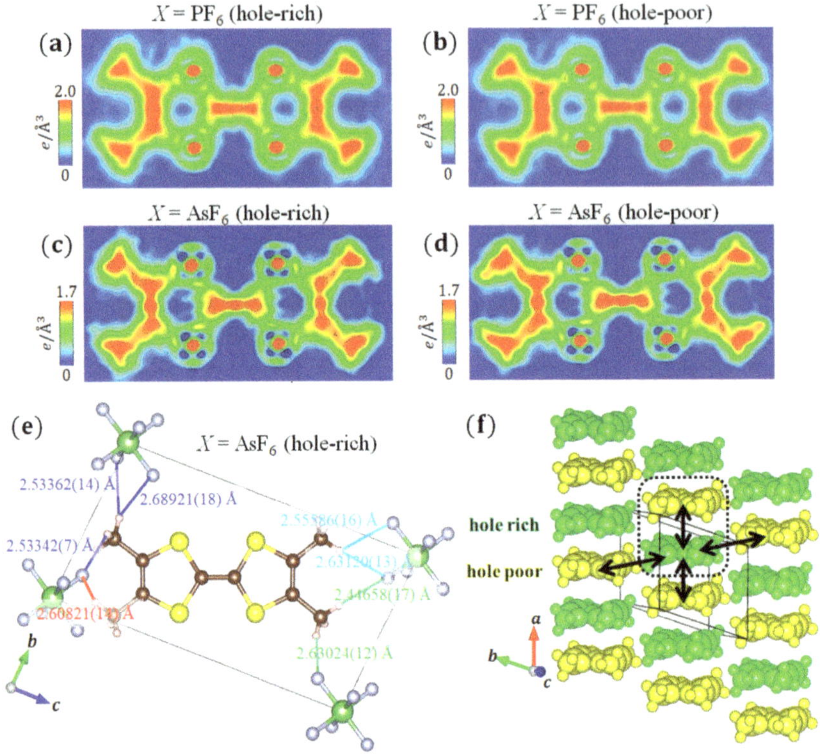

Figure 7. (**a**–**d**) VED distribution of hole-rich and hole-poor TMTTF molecules of $(TMTTF)_2PF_6$ and $(TMTTF)_2AsF_6$ on LS planes, respectively. The LS planes are defined by the central two C and four S atoms of the TMTTF molecules. (**e**) H–F distances of 2.7 Å or less in the hole-rich TMTTF molecule of $(TMTTF)_2AsF_6$. The orientations of the TMTTF molecules are the same between (**c**,**e**). (**f**) Charge ordering patterns of hole-rich and hole-poor TMTTF molecules in $(TMTTF)_2X$ (X = PF_6 and AsF_6), which indicate a 2D Wigner crystal state. The region surrounded by the dotted square shows a TMTTF molecule dimer.

The importance of H–F interactions between the methyl group in the TMTTF molecule and X anions is pointed out in this system [50]. Thus, we investigated the distance between hydrogen in the methyl group and fluorine in the anion. Figure 7e shows the H–F distances of 2.7 Å or less in the hole-rich TMTTF molecule. In this regard, we focused on the anisotropy of the VED distribution in the TMTTF molecule in X = AsF_6 (Figure 7c). The VED is concentrated around the methyl groups, but there seems to be no correlation between the distribution state and H–F distances. This tendency is the same for the hole-poor TMTTF in X = AsF_6 (Figure 7d) and for the hole-rich and hole-poor TMTTF in X = PF_6 (Figure 7a,b). Therefore, no clear effect of closed-shell anions was confirmed from the VED distribution of TMTTF.

From the above results, we directly revealed that the spatial charge ordering pattern formed a 2D Wigner crystal state (Figure 7f) from the molecular structure and VED distribution in the charge ordering phase in $(TMTTF)_2X$ (X = PF_6 and AsF_6). This pattern is also consistent with previous ESR experiment [36] and theoretical expectations [51–53].

4. Conclusions

We succeeded in directly observing the VED distribution of several molecular materials, i.e., diamond, C_{60} fullerene, $(MV)I_2$, and $(TMTTF)_2X$, using synchrotron XRD and the CDFS method. When a molecule is formed by bonds between atoms, various interactions work depending on the shapes and energy levels of the hybridized orbitals. As a result, even molecules with relatively simple structures produce complex molecular orbitals. Therefore, it is difficult to observe the whole picture of molecular orbitals with existing experimental methods. On the other hand, because the CDFS method using synchrotron XRD can directly observe the distribution state of valence electrons occupying the reconstructed molecular orbitals in a real space, it is possible to take an approach which differs from the existing methods for the study of molecular substances. Furthermore, the VED distribution obtained from the CDFS analysis corresponds to the square of the wave function, which provides essential information for quantum chemical and first-principles calculations. In fact, with complementary study of the CDFS analysis and first-principles calculations in transition metal oxides, we succeeded in clarifying the whole orbital state formed by the localized 3*d* orbitals on an atom and the metal–ligand hybridized orbitals [10]. The research methods we propose may signal a breakthrough in the study of the orbital states in materials.

Author Contributions: S.K. and H.S. designed and coordinated this study. R.K., T.N. (Toshio Naito) and T.N. (Toshikazu Nakamura) synthesized the samples. S.K., Y.H. and H.S. performed the XRD experiment; S.K. and Y.H. analyzed the XRD data. S.K. and H.S. wrote the manuscript. All authors have read and agreed to the published version of the manuscript.

Funding: This work was supported by a Grant-in-Aid for Scientific Research (No. JP19J11697) from JSPS. The synchrotron radiation experiments were performed at SPring-8 with the approval of the Japan Synchrotron Radiation Research Institute (JASRI) (Proposal No. 2011B0083, No. 2019A0070, and No. 2019B0070).

Acknowledgments: We thank Ken Niwa for providing materials used in the experiments, Takeshi Hara, Keita Kojima, Taishun Manjo, Naoyuki Katayama, and Kunihisa Sugimoto for their support in the synchrotron XRD experiments, Mariano De Souza, and Jean-Paul Pouget for fruitful discussions.

Conflicts of Interest: The authors declare no conflict of interest.

Appendix A

CCDC 2036797 contains the supplemental crystallographic data of C_{60} fullerene at 30 K. The data is provided free of charge by The Cambridge Crystallographic Data Centre [54].

Table A1. Summary of the crystallographic data of C_{60}.

CCDC Deposit Number	2036797
Temperature (K)	30
Wavelength (Å)	0.35044
Space group	$Pa\bar{3}$
a (Å)	14.0279(6)
V (Å^3)	2760.4(2)
Z	24
F (000)	1440
$(\sin\theta/\lambda)_{Max}$ (Å^{-1})	1.11
$N_{Total,obs}$	65,092
$N_{Unique,obs}$	4920
Average redundancy	11.4
Completeness (%)	100
# of reflections ($I > 1.5\sigma$)	5424
R ($I > 1.5\sigma$)	4.36
wR ($I > 1.5\sigma$)	3.73
GOF ($I > 1.5\sigma$)	1.21

References

1. Albright, T.A.; Burdett, J.K.; Whangbo, M.H. *Orbital Interactions in Chemistry*; Wiley: Hoboken, NJ, USA, 2013.
2. Beveridge, D.L. Approximate Molecular Orbital Theory of Nuclear and Electron Magnetic Resonance Parameters. In *Semiempirical Methods of Electronic Structure Calculation. Modern Theoretical Chemistry*; Segal, G.A., Ed.; Springer: Boston, MA, USA, 1977; Volume 8, pp. 163–214. [CrossRef]
3. Tokura, Y.; Nagaosa, N. Orbital Physics in Transition-Metal Oxides. *Science* **2000**, *288*, 462–468. [CrossRef] [PubMed]
4. Itatani, J.; Levesque, J.; Zeidler, D.L.; Niikura, H.; Pépin, H.; Kieffer, J.C.; Corkum, P.B.; Villeneuve, D.M. Tomographic imaging of molecular orbitals. *Nature* **2004**, *432*, 867–871. [CrossRef] [PubMed]
5. Puschnig, P.; Berkebile, S.; Fleming, A.J.; Koller, G.; Emtsev, K.; Seyller, T.; Riley, J.D.; Ambrosch-Draxl, C.; Netzer, F.P.; Ramsey, M.G. Reconstruction of Molecular Orbital Densities from Photoemission Data. *Science* **2009**, *326*, 702–706. [CrossRef] [PubMed]
6. Ito, Y.; Akimitsu, J. Observation of Orbital Ordering in K_2CuF_4. *J. Phys. Soc. Jpn.* **1976**, *40*, 1333–1338. [CrossRef]
7. Murakami, Y.; Hill, J.P.; Gibbs, D.; Blume, M.; Koyama, I.; Tanaka, M.; Kawata, H.; Arima, T.H.; Tokura, Y.; Hirota, K.; et al. Resonant X-ray Scattering from Orbital Ordering in $LaMnO_3$. *Phys. Rev. Lett.* **1998**, *81*, 582–585. [CrossRef]
8. Hansen, N.K.; Coppens, P. Testing aspherical atom refinements on small-molecule data sets. *Acta Crystallogr. Sect. A Cryst. Phys. Diffraction Theor. Gen. Crystallogr.* **1978**, *34*, 909–921. [CrossRef]
9. Kitou, S.; Fujii, T.; Kawamoto, T.; Katayama, N.; Maki, S.; Nishibori, E.; Sugimoto, K.; Takata, M.; Nakamura, T.; Sawa, H. Successive Dimensional Transition in $(TMTTF)_2PF_6$ Revealed by Synchrotron X-ray Diffraction. *Phys. Rev. Lett.* **2017**, *119*, 065701. [CrossRef] [PubMed]
10. Kitou, S.; Manjo, T.; Katayama, N.; Shishidou, T.; Arima, T.H.; Taguchi, Y.; Tokura, Y.; Nakamura, T.; Yokoyama, T.; Sugimoto, K.; et al. Collapse of the simple localized $3d^1$ orbital picture in Mott insulator. *Phys. Rev. Res.* **2020**, *2*, 033503. [CrossRef]
11. Sugimoto, K.; Ohsumi, H.; Aoyagi, S.; Nishibori, E.; Moriyoshi, C.; Kuroiwa, Y.; Sawa, H.; Takata, M. Extremely High Resolution Single Crystal Diffractometry for Orbital Resolution using High Energy Synchrotron Radiation at SPring-8. *AIP Conf. Proc.* **2010**, *1234*, 887. [CrossRef]
12. Blessing, R.H. Data Reduction and Error Analysis for Accurate Signal Crystal Diffraction Intensities. *Crystallogr. Rev.* **1987**, *1*, 3–58. [CrossRef]
13. Petříček, V.; Dušek, M.; Palatinus, L. Crystallographic Computing System JANA2006: General features. *Z. Krist. Cryst. Mater.* **2014**, *229*, 345. [CrossRef]
14. Momma, K.; Izumi, F. VESTA 3 for three-dimensional visualization of crystal, volumetric and morphology data. *J. Appl. Crystallogr.* **2011**, *44*, 1272–1276. [CrossRef]
15. Su, Z.; Coppens, P. Relativistic X-ray Elastic Scattering Factors for Neutral Atoms Z= 1–54 from Multiconfiguration Dirac–Fock Wavefunctions in the 0–12Å^{-1} sinθ/λ Range, and Six-Gaussian Analytical Expressions in the 0–6Å^{-1} Range. *Acta Crystallogr. Sect. A Found. Crystallogr.* **1997**, *53*, 749–762. [CrossRef]
16. Pavone, P.; Karch, K.; Schütt, O.; Strauch, D.; Windl, W.; Giannozzi, P.; Baroni, S. Ab initiolattice dynamics of diamond. *Phys. Rev. B* **1993**, *48*, 3156–3163. [CrossRef]
17. Kroto, H.W.; Heath, J.R.; O Brien, S.C.; Curl, R.F.; Smalley, R.E. C_{60}: Buckminsterfullerene. *Nature* **1985**, *318*, 162–163. [CrossRef]
18. Krätschmer, W.; Lamb, L.D.; Fostiropoulos, K.; Huffman, D.R. Solid C_{60}: A new form of carbon. *Nature* **1990**, *347*, 354–358. [CrossRef]
19. David, W.I.F.; Ibberson, R.M.; Matthewman, J.C.; Prassides, K.; Dennis, T.J.S.; Hare, J.P.; Kroto, H.W.; Taylor, R.; Walton, D.R.M. Crystal structure and bonding of ordered C_{60}. *Nature* **1991**, *353*, 147–149. [CrossRef]
20. Fischer, J.E.; Heiney, P.A.; McGhie, A.R.; Romanow, W.J.; Denenstein, A.M.; McCauley, J.P.; Smith, A.B. Compressibility of Solid C_{60}. *Science* **1991**, *252*, 1288–1290. [CrossRef] [PubMed]
21. Heiney, P.A.; Fischer, J.E.; McGhie, A.R.; Romanow, W.J.; Denenstein, A.M.; McCauley, J.P., Jr.; Smith, A.B.; Cox, D.E. Orientational ordering transition in solid C_{60}. *Phys. Rev. Lett.* **1991**, *66*, 2911–2914. [CrossRef]
22. Liu, S.; Lu, Y.J.; Kappes, M.M.; Ibers, J.A. The Structure of the C_{60} Molecule: X-ray Crystal Structure Determination of a Twin at 110 K. *Science* **1991**, *254*, 408–410. [CrossRef]

23. David, W.I.F.; Ibberson, R.M.; Dennis, T.J.S.; Hare, J.P.; Prassides, K. Structural Phase Transitions in the Fullerene C_{60}. *Europhysics Lett.* **1992**, *18*, 219–225. [CrossRef]
24. Bürgi, H.B.; Blanc, E.; Schwarzenbach, D.; Liu, S.; Lu, Y.J.; Kappes, M.M.; Ibers, J.A. The Structure of C_{60}: Orientational Disorder in the Low-Temperature Modification of C_{60}. *Angew. Chem. Int. Ed.* **1992**, *31*, 640–643. [CrossRef]
25. Saito, S.; Oshiyama, A. Cohesive mechanism and energy bands of solid C_{60}. *Phys. Rev. Lett.* **1991**, *66*, 2637–2640. [CrossRef] [PubMed]
26. Russell, J.H.; Wallwork, S.C. The crystal structures of the dichloride and isomorphous dibromide and diiodide of the N,N′-dimethyl-4,4′-bipyridylium ion. *Acta Crystallogr. Sect. B Struct. Crystallogr. Cryst. Chem.* **1972**, *28*, 1527–1533. [CrossRef]
27. Galigné, J.L.; Liautard, B.; Peytavin, S.; Brun, G.; Fabre, J.M.; Torreilles, E.; Giral, L. Etude structurale du bromure de tétraméthyltétrathiafulvalène, $(TMTTF)_2Br$. *Acta Crystallogr. Sect. B Struct. Crystallogr. Cryst. Chem.* **1978**, *34*, 620–624. [CrossRef]
28. Jérome, D.; Schulz, H.J. Organic conductors and superconductors. *Adv. Phys.* **1982**, *31*, 299–490. [CrossRef]
29. Jérome, D. The Physics of Organic Superconductors. *Science* **1991**, *252*, 1509–1514. [CrossRef] [PubMed]
30. Pouget, J.P.; Ravy, S. Structural Aspects of the Bechgaard Salts and Related Compounds. *J. Phys. I* **1996**, *6*, 1501–1525. [CrossRef]
31. Dressel, M.; Dumm, M.; Knoblauch, T.; Masino, M. Comprehensive Optical Investigations of Charge Order in Organic Chain Compounds $(TMTTF)_2X$. *Crystals* **2012**, *2*, 528–578. [CrossRef]
32. Nad, F.; Monceau, P.; Carcel, C.; Fabre, J.M. Dielectric response of the charge-induced correlated state in the quasi-one-dimensional conductor $(TMTTF)_2PF_6$. *Phys. Rev. B* **2000**, *62*, 1753–1756. [CrossRef]
33. Monceau, P.; Nad, F.Y.; Brazovskii, S. Ferroelectric Mott-Hubbard Phase of Organic $(TMTTF)_2X$ Conductors. *Phys. Rev. Lett.* **2001**, *86*, 4080. [CrossRef] [PubMed]
34. Chow, D.S.; Zamborszky, F.; Alavi, B.; Tantillo, D.J.; Baur, A.; Merlic, C.A.; Brown, S.E. Charge Ordering in the TMTTF Family of Molecular Conductors. *Phys. Rev. Lett.* **2000**, *85*, 1698–1701. [CrossRef]
35. Nakamura, M.; Furukawa, K.; Hara, T. Redistribution of Charge in the Proximity of the Spin-Peierls Transition: ^{13}C NMR Investigation of $(TMTTF)_2PF_6$. *J. Phys. Soc. Jpn.* **2007**, *76*, 64715. [CrossRef]
36. Nakamura, M. Possible Charge Ordering Patterns of the Paramagnetic Insulating States in $(TMTTF)_2X$. *J. Phys. Soc. Jpn.* **2003**, *72*, 213–216. [CrossRef]
37. Jacobsen, C.S.; Tanner, D.B.; Bechgaard, K. Optical and infrared properties of tetramethyltetraselenafulvalene $[(TMTSF)_2X]$ and tetramethyltetrathiafulvalene $[(TMTTF)_2X]$ compounds. *Phys. Rev. B* **1983**, *28*, 7019–7032. [CrossRef]
38. Świetlik, R.; Barszcz, B.; Pustogow, A.; Dressel, M. Raman spectroscopy evidence of domain walls in the organic electronic ferroelectrics $(TMTTF)_2X$ (X=SbF_6, AsF_6, PF_6). *Phys. Rev. B* **2017**, *95*, 085205. [CrossRef]
39. Pouget, J.P.; Moret, R.; Comes, R.; Bechgaard, K.; Fabre, J.M.; Giral, L. X-ray Diffuse Scattering Study of Some $(TMTSF)_2X$ and $(TMTTF)_2X$ Salts. *Mol. Cryst. Liq. Cryst.* **1982**, *79*, 485–499. [CrossRef]
40. Foury-Leylekian, P.; Le Bolloc'h, D.; Hennion, B.; Ravy, S.; Moradpour, A.; Pouget, J.P. Neutron-scattering evidence for a spin-Peierls ground state in $(TMTTF)_2PF_6$. *Phys. Rev. B* **2004**, *70*, 180–405. [CrossRef]
41. Coulon, C.; Parkin, S.S.P.; Laversanne, R. Structureless transition and strong localization effects in bis-tetramethyltetrathiafulvalenium salts $[(TMTTF)_2X]$. *Phys. Rev. B* **1985**, *31*, 3583–3587. [CrossRef]
42. De Souza, M.; Foury-Leylekian, P.; Moradpour, A.; Pouget, J.P.; Lang, M. Evidence for Lattice Effects at the Charge-Ordering Transition in $(TMTTF)_2X$. *Phys. Rev. Lett.* **2008**, *101*, 216–403. [CrossRef]
43. Foury-Leylekian, P.; Petit, S.; Andre, G.; Moradpour, A.; Pouget, J.P. Neutron scattering evidence for a lattice displacement at the charge ordering transition of $(TMTTF)_2PF_6$. *Phys. B Condens. Matter* **2010**, *405B*, S95–S97. [CrossRef]
44. Umland, T.C.; Allie, S.; Kuhlmann, T.; Coppens, P. Relation between geometry and charge transfer in low-dimensional organic salts. *J. Phys. Chem.* **1988**, *92*, 6456–6460. [CrossRef]
45. Mori, T.; Kobayashi, A.; Sasaki, Y.; Kobayashi, H.; Saito, G.; Inokuchi, H. The Intermolecular Interaction of Tetrathiafulvalene and Bis(ethylenedithio)tetrathiafulvalene in Organic Metals. Calculation of Orbital Overlaps and Models of Energy-band Structures. *Bull. Chem. Soc. Jpn.* **1984**, *57*, 627–633. [CrossRef]
46. A Computer Program Package for Multipole Refinement, Topological Analysis of Charge Densities and Evaluation of Intermolecular Energies from Experimental and Theoretical Structure Factors. Available online: https://www.chem.gla.ac.uk/~{}louis/xd-home/ (accessed on 3 November 2020).

47. Nad, F.; Monceau, P.; Carcel, C.; Fabre, J.M. Charge ordering phase transition in the quasi-one-dimensional conductor $(TMTTF)_2AsF_6$. *J. Phys. Condens. Matter* **2000**, *12*, L435–L440. [CrossRef]
48. Zámborszky, F.; Yu, W.; Raas, W.; Brown, S.E.; Alavi, B.; Merlic, C.A.; Baur, A. Competition and coexistence of bond and charge orders in $(TMTTF)_2AsF_6$. *Phys. Rev. B* **2002**, *66*, 081103. [CrossRef]
49. Fujiyama, S.; Nakamura, M. Redistribution of Electronic Charges in Spin-Peierls State in $(TMTTF)_2AsF_6$ Observed by ^{13}C NMR. *J. Phys. Soc. Jpn.* **2006**, *75*, 14705. [CrossRef]
50. Pouget, J.P. Structural Aspects of the Bechgaard and Fabre Salts: An Update. *Crystals* **2012**, *2*, 466–520. [CrossRef]
51. Seo, H.; Fukuyama, H. Antiferromagnetic Phases of One-Dimensional Quarter-Filled Organic Conductors. *J. Phys. Soc. Jpn.* **1997**, *66*, 1249–1252. [CrossRef]
52. Mazumdar, S.; Clay, R.T.; Campbell, D.K. Bond-order and charge-density waves in the isotropic interacting two-dimensional quarter-filled band and the insulating state proximate to organic superconductivity. *Phys. Rev. B* **2000**, *62*, 13400–13425. [CrossRef]
53. Shibata, Y.; Nishimoto, S.; Ohta, Y. Charge ordering in the one-dimensional extended Hubbard model: Implication to the TMTTF family of organic conductors. *Phys. Rev. B* **2001**, *64*, 235107. [CrossRef]
54. The Cambridge Crystallographic Data Centre. Available online: https://www.ccdc.cam.ac.uk/structures/ (accessed on 3 November 2020).

Publisher's Note: MDPI stays neutral with regard to jurisdictional claims in published maps and institutional affiliations.

crystals

MDPI

Article

One-Dimensional Alternating Extended Hubbard Model at Quarter-Filling and Its Applications to Structural Instabilities of Organic Conductors

M. Ménard and C. Bourbonnais *

Regroupement Québécois sur les Matériaux de Pointe, Départememt de physique, Université de Sherbrooke, Sherbrooke, QC J1K 2R1, Canada; menarmar@gmail.com

* Correspondence: Claude.Bourbonnais@Usherbrooke.ca

Received: 17 September 2020; Accepted: 6 October 2020; Published: 16 October 2020

Abstract: The one-dimensional extended Hubbard model with lattice dimerization and alternated site potentials is analyzed using the renormalization group method. The coupling of electrons to structural degrees of freedom such as the anion lattice and acoustic phonons is investigated to obtain the possible instabilities against the formation of lattice superstructures. Applications of the theory to anionic and spin-Peierls instabilities in the Fabre and Bechgaard salts series of organic conductors and ordered alloys are presented and discussed.

Keywords: interacting electrons in one dimension; electronic and lattice instabilities; renormalization group method; organic conductors

1. Introduction

The discovery of the amazing variety of electronic and structural phases in the Bechgaard and Fabre charge transfer salts series $(TMTSF)_2X$ and $(TMTTF)_2X$ over the last four decades or so continues to arouse great interest in the field of organic conductors [1–13]. This research paper is devoted to explore some theoretical aspects about the role played by lattice commensurability in the origin and the development of these phases.

The part played by lattice commensurability was very early suspected to be among the determinant factors behind the occurence of electronic and lattice instabilities found in the phase diagram of these charge transfer salts. This is the case of the importance attributed to the weak but finite dimerization of the organic stacks. It is characterized by a lattice periodicity of wave vector $4k_F$ (k_F being the Fermi wave vector along the stacks) that coincides with the one of the anion lattice X [14,15]. This superstructure superimposes to the basic quarter-filled commensurability of the hole band structure, which is fixed by the complete charge transfer and stoichiometry of the salts. This introduces half-filling type of Coulomb Umklapp scattering along the organic stacks, known to have a strong impact on low energy electron correlations in one spatial dimension (1D) [16–18]. Its magnitude was shown to be a key element in controlling the strength of magnetism and Mott insulating behavior across the whole phase diagram of both series of materials [15,19–21].

Another source of lattice periodicity for charge carriers of organic stacks comes from degrees of freedom associated to the anion X lattice, whose $4k_F$ periodicity is intrinsically linked to that of dimerization [15]. Non-centro-symmetrical anions like $X = ReO_4^-$, ClO_4^-, BF_4^-, ... and to a certain extend centrosymmetric ones like $X = PF_6^-$, AsF_6^- ... can order and form lattice superstructures also called anion orderings (AO) [9,10,22–24]. The coupling between anion degrees of freedom and charge carriers makes AO inevitably entangled to various electronic instabilities of organic stacks [10,25].

The interplay between superimposed lattice periodicities and electronic instabilities has been further revealed by chemistry. A notable example that we will focus on in the present work combines both families of compounds in the form of the alloys $[(TMTSF)_{1-x}(TMTTF)_x]_2ReO_4$ with the asymmetrical anion ReO_4^-. Remarkably, in the case where x∼0.5, the alloy is found to be ordered [26]. The alternation of different organic molecules along the stacks then acts as a distinct site potential with the same $4k_F$ periodicity found for the bond centered dimerization [26–29]. Transport and structural studies have showed that these superimposed types of commensurability have a strong impact on both the Mott insulating state and the staggered ReO_4^- AO found in the pure limits [26].

An alternating site periodicity can also be generated dynamically in pure systems of the $(TMTTF)_2X$ salts. This occurs for the charge ordering (CO) phase transition [30–32], which introduces a charge disproportionation in the dimerized unit cell that breaks the inversion symmetry and gives rise to a ferroelectric state [29]. The importance of Coulomb interaction between carriers of different molecular units in the unit cell has pointed to the key role of underlying quarter-filling commensurability component of Umklapp scattering in the electronic origin of this phase [33–36].

The existence of a CO state in the $(TMTTF)_2X$ series has been found to interfere with the formation of another ordered state involving both spin and lattice degrees of freedom. This the case of the spin-Peierls (SP) instability which corresponds to a $2k_F$ bond like tetramerization of the organic stacks that opens a gap in the spin sector [37,38]. On experimental grounds [39], the presence of the CO state is found to particularly weaken the strength of the SP order, an effect as we will see that results from the interplay between different types of commensurability [35,40–43].

In the present paper we shall address theoretically the combined influence of different types of lattice commensurability in the framework of a generalized extended Hubbard model. This will be examined in the 1D regime that characterizes both series of materials at relatively high temperature. The combined influence of site and bond dimerization lattice potentials together with the respective modulation of Coulomb interaction terms are considered. Their impact on excitation gaps and singular correlations of the phase diagram is analyzed using a two-loop renormalization group (RG) method in the electron gas limit. The $2k_F$ bond and site density-wave correlations with their phase relation to the underlying alternated lattice are determined. The model is integrated to the one proposed by Riera and Poilblanc for the mechanism of anion ordering in the $(TMTSF)_2X$ and $(TMTTF)_2X$ salts. The coupling of anions displacement to charge carriers in systems like the alloys $[(TMTSF)_{1-x}(TMTTF)_x]_2ReO_4$ is derived and the reduction (magnification) of the staggered anion ordering (Mott temperature scale) is obtained and shown to compare favorably with previous experiments performed on these materials [26]. Finally, the model is also applied to the influence of CO state on the pressure profile of the spin-Peierls instability occurring in $(TMTTF)_2X$ with centrosymmetric anions. By modeling CO in terms of a site lattice potential, it is found that lattice coupling responsible for the SP tetramerization is strongly altered by the presence of CO which mainly governs the variation of the temperature scale of the SP instability observed under pressure.

2. Alternating Extended Hubbard Model

The Model and Its Continuum Limit

We consider a 1D model of interacting electrons on a bipartite lattice. The Hamiltonian $H = H_0 + H_I$ consists of the one-body term

$$H_0 = -\sum_{r,\sigma}\Big[(t+\delta t)a^\dagger_{r,\sigma}b_{r,\sigma} + (t-\delta t)a^\dagger_{r,\sigma}b_{r-1,\sigma} + \mathrm{H.c.}\Big] \\ + \epsilon_0\sum_{r,\sigma}(m_{r,\sigma} - n_{r,\sigma}) \tag{1}$$

and the interaction term,

$$
\begin{aligned}
H_I = \frac{1}{2}\sum_{r,\sigma}\Big[(U+\delta U)m_{r,\sigma}m_{r,-\sigma} + (U-\delta U)n_{r,\sigma}n_{r,-\sigma}\Big] \\
+ \sum_{r,\sigma_{1,2}} (V+\delta V)m_{r,\sigma_1}n_{r,\sigma_2} + (V-\delta V)m_{r,\sigma_1}n_{r-1,\sigma_2}.
\end{aligned} \tag{2}
$$

Here, $a^{(\dagger)}_{r\sigma}$ and $b^{(\dagger)}_{r\sigma}$ are the annihilation (creation) operators of electrons of spin σ at even and odd sites of a bipartite lattice; $m_{r\sigma} = a^\dagger_{r\sigma}a_{r\sigma}$ and $n_{r\sigma} = b^\dagger_{r\sigma}b_{r\sigma}$ stand for the number operators on the corresponding sites. At quarter-filling, we have $\langle m_{r\sigma}\rangle = \langle n_{r\sigma}\rangle = \frac{1}{2}$. The standard extended Hubbard part of H is described by the hopping integral t, and the intra- and inter-site interaction terms U and V. The model includes the influence of two different alternating potentials. We have first a small lattice dimerization which modulates the hopping integral by δt [14,15] and the nearest-neighbor interaction V by a corresponding positive δV, whether the electrons interact on shorter or longer bonds as expected from quantum chemistry calculations on dimerized chains materials like the Fabre and the Bechgaard salts [44]. The second potential to be examined is an alternating site potential of amplitude ϵ_0 [27]. The latter can be found in practice in quasi-1D organic ordered alloys in which the molecular species alternates from site to site [26], or as a result of charge ordering which in systems like the Fabre salts introduces a charge disproportionation modulated along the stacks [31,32]. Besides the modulation of site energy, a molecular alternation is expected to modify Coulomb integral U from site to site, which is taken into account by adding (substracting) a negative δU when the site energy is increased (reduced) by ϵ_0 (see Figure 1).

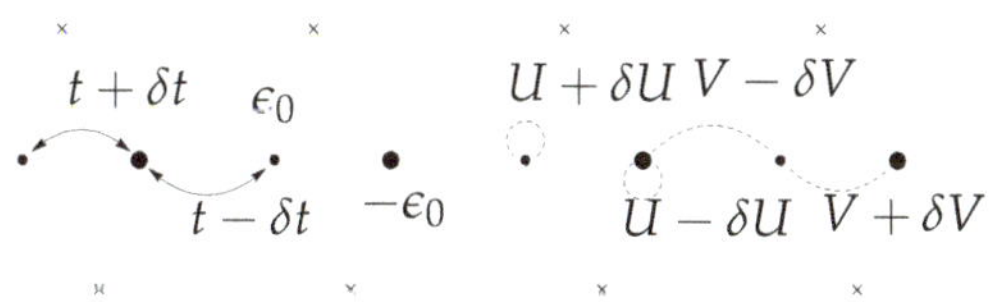

Figure 1. One dimensional alternating extended Hubbard model. The small and big full circles depict higher and lower potential energies, and crosses refer to the positions of anions in systems like $(TMTSF)_2X$, $(TMTTF)_2X$ and their alloys.

Using the Fourier transforms of each sublattice operator,

$$
a_{r,\sigma} = \sqrt{\frac{2}{N}}\sum_k a_{k\sigma}e^{ikr}, \tag{3}
$$

$$
b_{r,\sigma} = \sqrt{\frac{2}{N}}\sum_k b_{k\sigma}e^{ikr}, \tag{4}
$$

the one-body part of the Hamiltonian is diagonalized in the form

$$
H_0 = -\sum_{k,\sigma} E_k d^\dagger_{k,\sigma}d_{k,\sigma} + \sum_{k,\sigma} E_k f^\dagger_{k,\sigma}f_{k,\sigma}, \tag{5}
$$

which is written in terms of the new operators $d^{(\dagger)}_{k,\sigma}$ and $f^{(\dagger)}_{k,\sigma}$ for the lower and upper bands following the transformations

$$
a^\dagger_{k,\sigma} = e^{-i\left(\frac{k}{4}-\frac{\nu_k}{2}\right)}\left(\sin\frac{\gamma_k}{2}d^\dagger_{k,\sigma} + \cos\frac{\gamma_k}{2}f^\dagger_{k,\sigma}\right), \tag{6}
$$

$$
b^\dagger_{k,\sigma} = e^{i\left(\frac{k}{4}-\frac{\nu_k}{2}\right)}\left(\cos\frac{\gamma_k}{2}d^\dagger_{k,\sigma} - \sin\frac{\gamma_k}{2}f^\dagger_{k,\sigma}\right), \tag{7}
$$

where the phase factors obey the relations

$$\epsilon_0 = E_k \cos \gamma_k, \tag{8}$$

$$2t \cos \frac{k}{2} = E_k \sin \gamma_k \cos \nu_k, \tag{9}$$

$$2\delta t \sin \frac{k}{2} = E_k \sin \gamma_k \sin \nu_k. \tag{10}$$

The amplitude of the spectrum for each band is given by

$$E_k = 2\sqrt{t^2 \cos^2 \frac{k}{2} + \delta t^2 \sin^2 \frac{k}{2} + \left(\frac{\epsilon_0}{2}\right)^2}. \tag{11}$$

At quarter filling the lower band is occupied up to the Fermi points $\pm k_F = \pm\pi/2$, corresponding to the Fermi level $-E_{k_F} \equiv -E_F$ where

$$E_F = \sqrt{2t^2 + 2\delta t^2 + \epsilon_0^2}, \tag{12}$$

is the Fermi energy. The lower and upper bands in Figure 2 are separated by a gap $\Delta = 2E_\pi = 2\sqrt{4\delta t^2 + \epsilon_0^2}$.

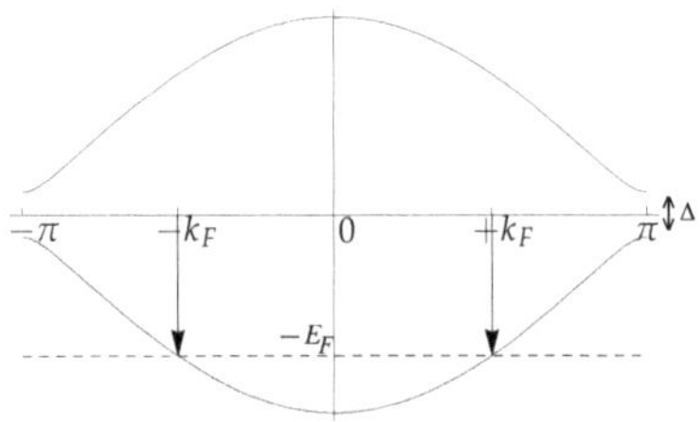

Figure 2. Electron spectrum of the one-dimensional extended alternated Hubbard model.

For the interacting part of the Hamiltonian, we shall focus on the weak coupling case by considering only low energy scattering processes taking place within the lower 'd' band (perturbative corrections coming from the upper f band at higher energy have been considered by Penc and Mila [27]). This yields

$$H_I = \frac{1}{2N} \sum_{\{k,\sigma\}} g(k_1, k_2, k_3, k_4) \delta_{k_4+k_3,k_1+k_2+G} \times d^\dagger_{k_4,\sigma_4} d^\dagger_{k_3,\sigma_3} d_{k_2,\sigma_2} d_{k_1,\sigma_1} + \ldots . \tag{13}$$

where $G = 0$ ($G = \pm 2\pi$) stands for normal (Umklapp) scattering processes. The interaction amplitude reads

$$\begin{aligned} g(k_1, k_2, k_3, k_4) &= (U + \delta U) e^{i(2\nu_{[3]} - \frac{G}{4})} \prod_{i=1}^{4} \sin \frac{\gamma_{k_i}}{2} \\ &+ (U - \delta U) e^{-i(2\nu_{[3]} - \frac{G}{4})} \prod_{i=1}^{4} \cos \frac{\gamma_{k_i}}{2} \\ &+ (2V \cos \frac{k_4 - k_1}{2} - 2i\delta V \sin \frac{k_4 - k_1}{2}) e^{i(\frac{G}{4} + 2\nu_{[1]})} \sin \frac{\gamma_{k_4}}{2} \cos \frac{\gamma_{k_3}}{2} \cos \frac{\gamma_{k_2}}{2} \sin \frac{\gamma_{k_1}}{2} \\ &+ (2V \cos \frac{k_4 - k_1}{2} + 2i\delta V \sin \frac{k_4 - k_1}{2}) e^{-i(\frac{G}{4} + 2\nu_{[1]})} \cos \frac{\gamma_{k_4}}{2} \sin \frac{\gamma_{k_3}}{2} \sin \frac{\gamma_{k_2}}{2} \cos \frac{\gamma_{k_1}}{2}, \end{aligned} \tag{14}$$

where

$$v_{[1,3]} \quad = \quad \frac{v_{k_4} \mp v_{k_3} \pm v_{k_2} - v_{k_1}}{4}. \tag{15}$$

In weak coupling, the lattice model can be transposed into the continuum limit of the electron gas model. This standard procedure first consists in the linearization of the spectrum around the Fermi points $pk_F = \pm k_F$,

$$E_F - E_k \approx E_p(k) = v_F(pk - k_F), \tag{16}$$

where

$$v_F = \frac{t^2 - \delta t^2}{\sqrt{2t^2 + 2\delta t^2 + \epsilon_0^2}}, \tag{17}$$

is the Fermi velocity ($\hbar = 1$ throughout). The operators for left and right moving electrons become $d_{\pm,k,\sigma}$ ($d^\dagger_{\pm,k,\sigma}$). The linearized spectrum is limited by an energy cut-off $E_0/2 = E_F$ on either side of the Fermi level, where E_0 is the total width of the lower band. The interaction is defined with respect to the Fermi points giving rise to four possible coupling constants, commonly noted as $g_{i=1-4}$ [16,17,45]. Together with the one-electron part, this leads to the usual form for the Hamiltonian of the electron gas model. For the lower band, one gets

$$\begin{aligned} H = \sum_{k,p,\sigma} E_p(k) d^\dagger_{p,k\sigma} d_{p,k\sigma} + \frac{1}{N} \sum_{\{k,q,\sigma\}} & g_1\, d^\dagger_{+,k_1+q,\sigma_1} d^\dagger_{-,k_2-q,\sigma_2} d_{+,k_2,\sigma_2} d_{-,k_1,\sigma_1} \\ & + g_2\, d^\dagger_{+,k_1+q,\sigma_1} d^\dagger_{-,k_2-q,\sigma_2} d_{-,k_2,\sigma_2} d_{+,k_1,\sigma_1} \\ & + \frac{1}{2} (g_3^+ d^\dagger_{+,k_1+q,\sigma_1} d^\dagger_{+,k_2-q+G,\sigma_2} d_{-,k_2,\sigma_2} d_{-,k_1,\sigma_1} + \text{H.c.}) \\ & + \frac{1}{2} g_4\, d^\dagger_{+,k_1+q,\sigma_1} d^\dagger_{+,k_2-q,\sigma_2} d_{+,k_2,\sigma_2} d_{+,k_1,\sigma_1} \\ & + \frac{1}{2} g_4\, d^\dagger_{-,k_1+q,\sigma_1} d^\dagger_{-,k_2-q,\sigma_2} d_{-,k_2,\sigma_2} d_{-,k_1,\sigma_1}, \end{aligned} \tag{18}$$

where the scattering amplitudes in the standard terminology [16,17], are $g_1 \equiv g(+k_F, -k_F, +k_F, -k_F)$, $g_2 \equiv g(+k_F, -k_F, -k_F, +k_F)$, $g_3^\pm \equiv g(\pm k_F, \pm k_F, \mp k_F, \mp k_F)$ and $g_4 \equiv g(\pm k_F, \pm k_F, \pm k_F, \pm k_F)$, which in order stand for backward, forward, half-filling Umklapp, and inner branch forward scattering processes. According to the expression (14) for the lattice model, the bare g_i amplitudes are given by

$$g_1 = U(1 + \epsilon_0^2/E_F^2) - 2\delta U \epsilon_0/E_F + 4t\delta V \delta t/E_F^2 \tag{19}$$

$$g_2 = U(1 + \epsilon_0^2/E_F^2) - 2\delta U \epsilon_0/E_F + 2V(1 - \epsilon_0^2/E_F^2) \tag{20}$$

$$\begin{aligned} g_3^p &= [U(1 + \epsilon_0^2/E_F^2) - 2\delta U \epsilon_0/E_F] \frac{2t\delta t}{t^2 + \delta t^2} \\ &+ 2\delta V(1 - \epsilon_0^2/E_F^2) \\ &+ ip[2U\epsilon_0/E_F - \delta U(1 + \epsilon_0^2/E_F^2)] \frac{t^2 - \delta t^2}{t^2 + \delta t^2} \\ &\equiv |g_3| e^{ip\theta} \end{aligned} \tag{21}$$

$$g_4 = g_2. \tag{22}$$

From these expressions, we first observe that alongside the bond alternation of hopping δt and the modulation of on-site energy ϵ_0, both local ($\delta U < 0$) and intersite ($\delta V > 0$) modulations of interactions contribute to an increase of backscattering and half-filling Umklapp. Furthermore, the local components of the modulation contribute to an imaginary part of $g_3^p = |g_3|e^{ip\theta}$, Ref. [28] whose argument θ plays an important role for the relative phase of charge/spin density-wave correlations with respect to the lattice. This will be analyzed in Section 3.2.

3. Renormalization Group Results

3.1. Formulation and Coupling Constants

We apply the renormalization group approach to our effective electron gas model described above. In the following, we sketch out only the main steps of the procedure that will be useful later on for applications [17,46,47]. We follow Ref. [46] and write the partition function in the functional integral form,

$$Z = \iint \mathfrak{D}\psi^* \mathfrak{D}\psi \, e^{S[\psi^*,\psi]}, \tag{23}$$

over a set anticommuting fermion variables $\{\psi^*, \psi\}$, where the measure is

$$\mathfrak{D}\psi^*\mathfrak{D}\psi = \prod_{\bar{k},p,\sigma} d\psi^*_{p,\sigma}(\bar{k}) d\psi_{p,\sigma}(\bar{k}),$$

and $\bar{k} = (k, \omega_n = (2n+1)\pi T)$ ($k_B = 1$ throughout). The action $S = S_0 + S_I$ splits into a free-quadratic-(S_0) and an interacting-quartic-(S_I) parts,

$$\begin{aligned}
S[\psi^*,\psi] = & \sum_{\bar{k},p,\sigma} z^{-1}[G^0_p(\bar{k})]^{-1} \psi^*_{p,\sigma}(\bar{k})\psi_{p,\sigma}(\bar{k}) \\
& - \frac{T}{N} \sum_{\bar{k}_1,\bar{k}'_1,\bar{k}_2,\bar{k}'_2} \sum_{\sigma,\sigma'} \delta_{\bar{k}_1+\bar{k}_2,\bar{k}'_1+\bar{k}'_2(\pm\bar{G})} \\
& \times \Big\{ z_1 g_1 \psi^*_{+,\sigma}(\bar{k}'_1)\psi^*_{-,\sigma'}(\bar{k}'_2)\psi_{+,\sigma'}(\bar{k}_2)\psi_{-,\sigma}(\bar{k}_1) \\
& + z_2 g_2 \psi^*_{+,\sigma}(\bar{k}'_1)\psi^*_{-,\sigma'}(\bar{k}'_2)\psi_{-,\sigma'}(\bar{k}_2)\psi_{+,\sigma}(\bar{k}_1) \\
& + \frac{1}{2} z_3 [g_3^+ \psi^*_{+,\sigma}(\bar{k}'_1)\psi^*_{+,\sigma'}(\bar{k}'_2)\psi_{-,\sigma'}(\bar{k}_2)\psi_{-,\sigma}(\bar{k}_1) + \text{"c.c."}] \\
& + \frac{1}{2} g_4 \psi^*_{+,\sigma}(\bar{k}'_1)\psi^*_{+,\sigma'}(\bar{k}'_2)\psi_{+,\sigma'}(\bar{k}_2)\psi_{+,\sigma}(\bar{k}_1) \\
& + \frac{1}{2} g_4 \psi^*_{-,\sigma}(\bar{k}'_1)\psi^*_{-,\sigma'}(\bar{k}'_2)\psi_{-,\sigma'}(\bar{k}_2)\psi_{-,\sigma}(\bar{k}_1) \Big\},
\end{aligned} \tag{24}$$

where z and $z_{1,2,3}$ are in order the renormalization factors for the one-particle propagator

$$G^0_p(\bar{k}) = [i\omega_n - E_p(k)]^{-1} \tag{25}$$

and the four-points electron-electron vertices $\Gamma_{1,2,3}$. The z_i are combined to z to give the renormalization factors $z^2 z_i$ for each coupling $g_{i=1,2,3}$. At the bare level, the couplings g_i are defined at the band edge energy cutoff $\Lambda_0 (\equiv E_F)$ above and below the Fermi level where both z and the z_i's equal unity.

The RG transformation is standard and consists in the succesive integrations of electronic degrees of freedom, denoted by $\bar{\psi}^{(*)}$, in outer energy shell of thickness $\Lambda(\ell)d\ell$ on both sides of the Femi level of the lower band, where $\Lambda(\ell) = \Lambda_0 e^{-\ell}$ is the cutoff at the step ℓ of the RG procedure. The integration of degrees of freedom from step ℓ to $\ell + d\ell$ is achieved perturbatively. This recursive transformation can be written in the form

$$\begin{aligned}
Z & \sim \iint [\mathfrak{D}\psi^*\mathfrak{D}\psi]_{\ell+d\ell} \, e^{S[\psi^*,\psi]_\ell} \iint [\mathfrak{D}\bar{\psi}^*\mathfrak{D}\bar{\psi}]_{d\ell} \, e^{S_0[\bar{\psi}^*,\bar{\psi}]_{d\ell}} \\
& \qquad \times \, e^{S_I[\psi^*,\psi,\bar{\psi}^*,\bar{\psi}]_{d\ell}} \\
& \sim \iint [\mathfrak{D}\psi^*\mathfrak{D}\psi]_{\ell+d\ell} e^{S[\psi^*,\psi]_\ell} \, e^{\sum_n \frac{1}{n!} \langle S^n_I[\psi^*,\psi,\bar{\psi}^*,\bar{\psi}]_{d\ell}\rangle_{0,d\ell}} \\
& \propto \iint [\mathfrak{D}\psi^*\mathfrak{D}\psi]_{\ell+d\ell} e^{S[\psi^*,\psi]_{\ell+d\ell}},
\end{aligned} \tag{26}$$

where $\langle S_I^n\rangle_{0,d\ell}$ are outer shell free (loop) averages with external fermion legs in the inner energy shells at $\Lambda \leq \Lambda(\ell + d\ell)$. The effective-renormalized-action $S[\psi^*, \psi]_{\ell+d\ell}$ at $\ell + d\ell$ leads to the recursion transformation for the $z's$.

Thus for the one-particle propagator, $z(\ell + d\ell) = z(\ell)z(d\ell)$, which leads to the familiar result at the two-loop level [17,46,47],

$$\frac{d\ln z}{d\ell} = -\frac{1}{16}(\pi v_F)^{-2}[(2g_2 - g_1)^2 + 3g_1^2 + 2g_3^+ g_3^-], \tag{27}$$

which is independent of the phase of the Umklapp term. The recursion relations $g_i(\ell + d\ell) = z_i(d\ell)z^2(d\ell)g_i(\ell)$ for the coupling constants lead to the two-loop flow equations

$$\frac{d}{d\ell}\bar{g}_1 = -\bar{g}_1^2 - \frac{1}{2}\bar{g}_1^3, \tag{28}$$

$$\frac{d}{d\ell}(2\bar{g}_2 - \bar{g}_1) = \bar{g}_3^+\bar{g}_3^- - \frac{1}{2}\bar{g}_3^+\bar{g}_3^-(2\bar{g}_2 - \bar{g}_1), \tag{29}$$

$$\frac{d}{d\ell}\bar{g}_3^p = \bar{g}_3^p(2\bar{g}_2 - \bar{g}_1) - \frac{1}{4}\bar{g}_3^p[(2\bar{g}_2 - \bar{g}_1)^2 + \bar{g}_3^+\bar{g}_3^-], \tag{30}$$

$$\frac{d\theta}{d\ell} = 0. \tag{31}$$

The first equation for $\bar{g}_1$ is connected to spin degrees of freedom and is decoupled from $(2\bar{g}_2 - \bar{g}_1, \bar{g}_3^p)$, which is connected to the charge. These extend the known flow equations of the electron gas model [17,46,47] to the case of a complex g_3^p. Note that only the amplitude of Umklapp $|g_3|$ renormalizes, whereas its phase θ remains scale invariant $[\mathrm{Im}(d_\ell \ln g_3^p) = 0]$ and is then fixed at the bare level by the expression (21). The renormalization of g_4 is here neglected. However the influence of this coupling has been incorporated through the normalization $\bar{g}_1 = g_1/\pi v_\sigma$ and $(2\bar{g}_2 - \bar{g}_1, \bar{g}_3^p) = (2g_2 - g_1, g_3^p)/\pi v_\rho$ for the decoupled spin (σ) and charge (ρ) interactions, respectively [17], where $v_{\sigma,\rho} = v_F \mp g_4/2\pi$ are the spin and charge velocities.

The properties of the above flow equations are standard [17,47] and can be summarized as follows. In the spin sector for instance, the negative (positive) sign of g_1 determines the conditions for the flow to strong (weak) attractive coupling, $\bar{g}_1^* \to -2$ ($\bar{g}_1^* \to 0$), as $\ell \to \infty$. In the attractive case, this indicates the emergence of a spin gap Δ_σ, whose scale is of the order of the cutoff energy $2\Lambda(\ell_\sigma)$ at which the flow of $\bar{g}_1$ in (28) becomes singular at the one-loop $\mathcal{O}(\bar{g}_1^2)$ level, namely $\Delta_\sigma \sim 2E_F e^{-1/|g_1|}$.

If we now consider the charge sector, the magnitude of $2\bar{g}_2 - \bar{g}_1$ with respect to $|\bar{g}_3|$ at the bare level determines the conditions for strong coupling or a charge gap Δ_ρ. Thus for $\bar{g}_1 - 2\bar{g}_2 \geq |\bar{g}_3|$, charge degrees of freedom remain gapless since $|\bar{g}_3^*| \to 0$ and $\bar{g}_1^* - 2\bar{g}_2^*$ is non universal as $\ell \to \infty$. In the whole region where $\bar{g}_1 - 2\bar{g}_2 < |\bar{g}_3|$, both $2\bar{g}_2^* - \bar{g}_1^* \to 2$ and $|\bar{g}_3^*| \to 2$ are marginally relevant and scale to strong coupling when $\ell \to \infty$. An order of magnitude for the charge gap Δ_ρ can be readily given by the singularities encountered at a finite ℓ_ρ in (29) and (30) at the one-loop, $\mathcal{O}(\bar{g}^2)$, level [16]. For $-|\bar{g}_3| < \bar{g}_1 - 2\bar{g}_2 < |\bar{g}_3|$, one has $\Delta_\rho \sim 2\Lambda(\ell_\rho) = 2E_F\exp[-c/\sqrt{|\bar{g}_3|^2 - (2\bar{g}_2 - \bar{g}_1)^2}]$, where $c = \arccos[(2\bar{g}_2 - \bar{g}_1)/|\bar{g}_3|]$; for $\bar{g}_1 - 2\bar{g}_2 = -|\bar{g}_3|$, $\Delta_\rho = 2E_F\exp[-(1/|\bar{g}_3|)]$; and finally for $\bar{g}_1 - 2\bar{g}_2 < -|\bar{g}_3|$, one has $\Delta_\rho = 2E_F\exp[-c'/\sqrt{(2\bar{g}_2 - \bar{g}_1)^2 - |\bar{g}_3|^2}]$, where $c' = \cosh^{-1}[(2\bar{g}_2 - \bar{g}_1)/|\bar{g}_3|]$.

We display in Figure 3 the contour plot of the scale for the charge gap $e^{-\ell_\rho} = \Delta_\rho/2E_F$ at the one-loop level in the $(\epsilon_0, \delta t)$ plane of alternating potentials and for repulsive (U, V) interactions and smaller modulations $(\delta U, \delta V)$. In the first quadrant where both δt and ϵ_0 are positive, the variation of Δ_ρ is not monotonous; it first increases with δt and ϵ_0 and then undergoes a smooth decreases. According to (21), both the real and imaginary parts of Umklapp increase at relatively small δt and ϵ_0; this is responsible for the increase of Δ_ρ. At sufficiently large δt, however, a reduction of the imaginary part of g_3^p becomes apparent and leads to the decrease of Δ_ρ. A similar variation of charge gap has been obtained in the bosonization approach to the alternating Hubbard model with positive U [28].

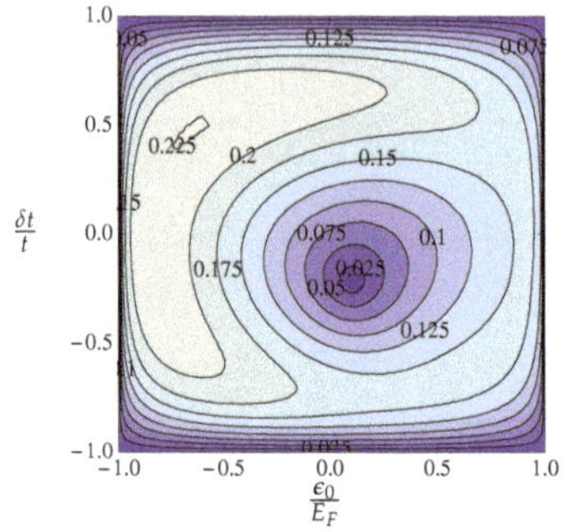

Figure 3. Contour plot of the normalized one-loop charge gap $\Delta_\rho/2E_F(\equiv e^{-\ell_\rho})$ as a function of the alternating site (ϵ_0) and bond (δt) potentials. The calculations are done for repulsive interactions $U/t = V/t = 0.5$ and $\delta U/t = \delta V/t = 0.1$.

Interestingly, if we broaden the analysis situation where positive $\delta U > 0$ and $\delta V > 0$ are considered, the competition between a positive ϵ_0 and negative δt can bring both the real and imaginary parts of g_3^p and in turn Δ_ρ to zero, as shown in the second quadrant of Figure 3. Around this point, the behavior of Equations (29) and (30), as $\ell \to \ell_\rho$ at the one-loop level shows that the gap vanishes following the power law $\Delta_\rho/2E_F \sim [\frac{1}{2}|\tilde{g}_3|/(2\tilde{g}_2 - \tilde{g}_1)]^{1/(2\tilde{g}_2-\tilde{g}_1)}$.

3.2. Response Functions and Phase Diagram

In order to analyze the nature of correlations and the possible phases of the above model, we proceed to the calculation of susceptibilities. To do so, we follow Ref. [46] and add to the action a linear coupling to a set of infinitesimal source fields $\{h\}$ to fermion pair fields. These are associated to susceptibilities that can become singular in the $2k_F$ density-wave and superconducting channels. As infinitesimal terms, they can be combined at the bare level to the interaction term S_I and treated as a perturbation. The action becomes

$$S[\psi^*, \psi, \{h\}] = S_0[\psi^*, \psi] + S_I[\psi^*, \psi] + S_h[\psi^*, \psi, \{h\}], \tag{32}$$

where

$$\begin{aligned} S_h[\psi^*, \psi, \{h\}] = &\sum_{\bar{q}} \Big[\sum_{\mu} z_\mu^{\rm s} h_\mu^{\rm s*}(\bar{q}) \Delta_\mu(\bar{q}) \\ &+ \sum_{\mu=+} z_{\theta_\mu}^{\rm c} h_\mu^{\rm c*}(\bar{q}) O_{\theta_\mu}(\bar{q}) \\ &+ \sum_{\mu=\pm} z_{\theta_\mu}^{\sigma} \vec{h}_\mu^{\sigma*}(\bar{q}) \cdot \vec{S}_{\theta_\mu}(\bar{q}) + \text{"c.c."} \Big]. \end{aligned} \tag{33}$$

In the superconducting channel, the pair fields are

$$\Delta_\mu(\bar{q}) = \sqrt{\frac{T}{N}} \sum_{\bar{k},\alpha,\beta} \alpha \psi_{-,-\alpha}(-\bar{k}) \sigma_\mu^{\alpha\beta} \psi_{+,\beta}(\bar{k} + \bar{q}) \tag{34}$$

for $\mu = 0$ singlet (SS) and $\mu = 1, 2, 3$ triplet (TS) superconductivity. Here $\sigma_0 = \mathbf{1}$, $\sigma_{1,2,3}$ are the Pauli matrices, and $\bar{q} = (q, \omega_m = 2\pi m T)$. The initial pair renormalization factors at $\ell = 0$ are $z_\mu^{\rm s} = 1$.

In the $2k_F$ density-wave channel, the presence of a complex Umklapp interaction term g_3^p in (21), which can be written as $g_3^{\pm} = \pm|g_3|e^{i\theta_\pm}$, where $\theta_+ = \theta$ and $\theta_- = \theta - \pi$, introduces spin and charge density-wave correlations with a particular phase relation with respect to the lattice. For $2k_F$ charge-density-wave (CDW), the pair field can be written in terms of two independent stationary waves,

$$O_{\theta_\pm}(\bar{q}) = e^{i\theta_\pm}O^*(\bar{q}) + O(\bar{q}), \tag{35}$$

where

$$O(\bar{q}) = \sqrt{\frac{T}{N}}\sum_{\bar{k},\alpha}\psi^*_{-,\alpha}(\bar{k}-\bar{q})\psi_{+,\alpha}(\bar{k})$$

for $\bar{q} = (q, \omega_m)$. The phase relation of CDW$_{\theta_\pm}$ maxima and minima with respect to the lattice is shown in Figure 4 at $q = 2k_F$. In the absence of ϵ_0 and δU, for instance, the imaginary part of g_3^p vanishes and $\theta_- = \pi$ and CDW$_{\theta_-}$ correlations are centered on bonds between dimers, whereas $\theta_+ = 0$ refers to CDW$_{\theta_+}$ whose maxima are centered on dimers. In the presence of a finite site potential ϵ_0 and/or δU, the inversion symmetry within the dimers is broken and the position of maxima for CDW$_{\theta_\pm}$ move accordingly (see Figure 4).

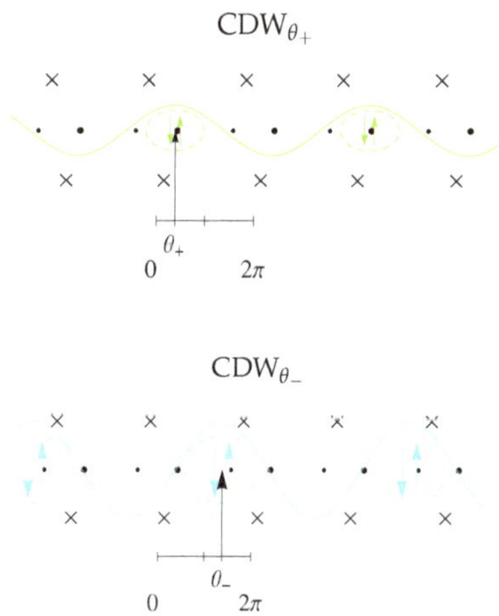

Figure 4. Site (+) and bond (−) $2k_F$ charge-density-wave and their respective phase ($\theta_\pm$) relative to the alternated lattice.

A similar decomposition can be made for $2k_F$ spin-density-wave (SDW$_{\theta_\pm}$) by introducing

$$\vec{S}_{\theta_\pm}(\bar{q}) = e^{i\theta_\pm}\vec{S}^*(\bar{q}) + \vec{S}(\bar{q}) \tag{36}$$

for SDW$_{\theta_\pm}$, where

$$\vec{S}(\bar{q}) = \sqrt{\frac{T}{N}}\sum_{\bar{k},\alpha,\beta}\psi^*_{-,\alpha}(\bar{k}-\bar{q})\vec{\sigma}^{\alpha\beta}\psi_{+,\alpha}(\bar{k}) \tag{37}$$

is the spin field at $\bar{q} = (q, \omega_m)$. When ϵ_0 and δU are absent, g_3^p is real and $\theta_\pm = 0(\pi)$, so that $\vec{S}_{\theta_\pm}$ describe $2k_F$ SDW with spin maxima centered on (between) the dimers, as shown in Figure 5. In the same Figure, for finite and positive ϵ_0 and/or δU, $\theta_\pm$ moves away from $0(\pi)$ alongside the maxima of spin density that move in (between) the unit cell.

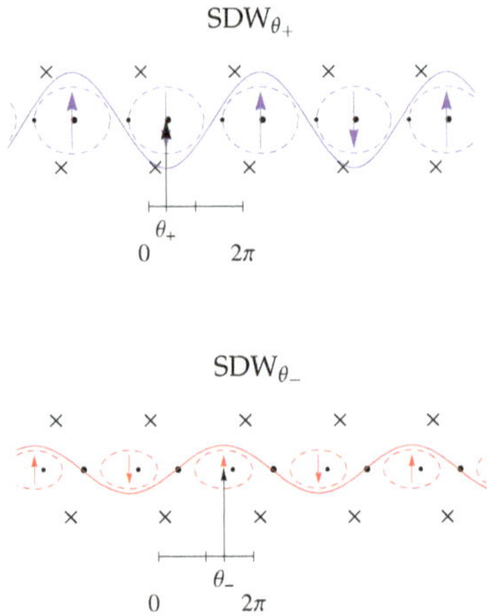

Figure 5. Site (+) and bond (−) $2k_F$ spin-density-wave and their respective phase ($\theta_\pm$) relative to the alternated lattice.

Making the substitution $S_I \to S_I + S_h$ in the RG transformation (26), the renormalized action at $\Lambda(\ell)$ reads

$$S[\psi^*, \psi, \{h\}]_\ell = S_0[\psi^*, \psi]_\ell + S_I[\psi^*, \psi]_\ell + S_h[\psi^*, \psi, \{h\}]_\ell \\ + \sum_{\mu,r} \chi^r_\mu(\bar{q}_\mu, \ell) h^{r*}_\mu(\bar{q}_\mu) h^r_\mu(\bar{q}_\mu) + \dots \; . \tag{38}$$

Here, the flows of renormalization factors in S_0 and S_I coincide with those obtained previously in (27)–(30), whereas the z^r_μ's associated to the pair vertices in (33) are governed at the two-loop level by an equation of the form [17,46,47]

$$\frac{d \ln z^r_\mu}{d\ell} = \frac{1}{2\pi v_F} g^r_\mu - \frac{1}{4\pi^2 v_F^2}[g_1^2 + g_2^2 - g_1 g_2 + g_3^+ g_3^- /2], \tag{39}$$

where for superconducting correlations ($r = \mathrm{s}$) the combinations of couplings g^r_μ are $g^{\mathrm{s}}_{\mathrm{SS}} = -g_1 - g_2$ and $g^{\mathrm{s}}_{\mathrm{TS}} = g_1 - g_2$ for singlet and triplet superconductivity; for density-wave correlations, one has $g^{\mathrm{c}}_{\theta_\pm} = g_2 - 2g_1 \mp |g_3|$ for CDW$_{\theta_\pm}$ in the charge sector ($r = \mathrm{c}$); and $g^{\sigma}_{\theta_\pm} = g_2 \pm |g_3|$ for SDW$_{\theta_\pm}$ in the spin sector ($r = \sigma$). The second term of (39), which is common to all pair vertices, refers to the self-energy corrections of Equation (27). According to (39), the behavior of z^r_μ is well known and follows the power law

$$z^r_\mu \sim [\Lambda(\ell)]^{-\frac{1}{2}\gamma^{r*}_\mu}, \tag{40}$$

at large ℓ. It signals a singularity when the exponent $\gamma^{r*}_\mu > 0$. The expression for $\frac{1}{2}\gamma^{r*}_\mu$ coincides with the right side expression of (39) evaluated at the fixed points values of scaling Equations (28)–(30). A singular behavior will also be found in the corresponding expressions for susceptibilities, which are given by the quadratic field terms of (38). These are generated by the RG transformation (26) and take the form,

$$\chi^r_\mu(\bar{q}_\mu) = (\pi v_F)^{-1} \int_0^\ell [z^r_\mu(\ell)]^2 d\ell, \tag{41}$$

which are defined positive and evaluated in the static limit $\bar{q}_\mu = (q_\mu, 0)$, for $q_{\rm SS,TS} = 0$ and $q_{\rm CDW,SDW} = 2k_F$. At large ℓ, the susceptibilities will be governed by a power law

$$\pi v_F \chi^r_\mu(\bar{q}_\mu) \approx A^r_\mu(\ell)[\Lambda(\ell)]^{-\gamma^{r*}_\mu} + c^r_\mu. \tag{42}$$

where c^r_μ is a positive constant.

The phase diagram determined by the dominant and subdominant singularities in the susceptibilities χ^r_μ is shown in Figure 6, as a function of initial g_i. Its structure necessarily presents many similarities with the known two-loop RG results of the electron gas model [17,47], but also some differences due to the presence of a complex g^p_3. In Figure 6 the massive ($\Delta_\rho \neq 0$) charge sector, delimited by the separatrix $g_1 - 2g_2 = |g_3|$, is enlarged with δt, ϵ_0, δV and negative δU, which is detrimental to the region of singular superconducting correlations on the left of this line. At $g_1 > 0$, this is also concomitant with the strengthening of dominant dimer or site like SDW$_{\theta_+}$, and subdominant interdimer or bond like CDW$_{\theta_-}$ singular correlations. In the attractive region where $g_1 < 0$, only the reinforcement of CDW$_{\theta_-}$ singular correlations is found on the right-hand side of the separatrix, where a gap in both spin and charge degrees of freedom occurs.

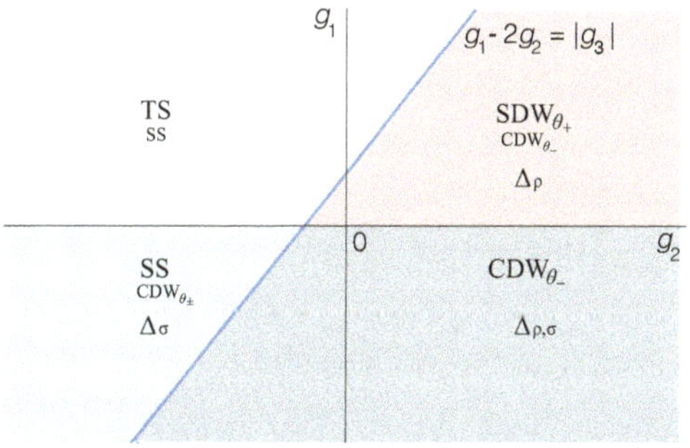

Figure 6. Phase diagram of the 1D alternated extended Hubbard model in the continuum electron gas limit.

4. Applications

4.1. Anion Ordering in [(TMTSF)$_{1-x}$TMTTF$_x$]ReO$_4$

As a first application of the above model, we analyze the role of CDW correlations in the mechanism of anion ordering (AO) in the concrete case of the ReO$_4$ sulfur-selenide alloys [(TMTSF)$_{1-x}$TMTTF$_x$]ReO$_4$.

4.1.1. Experimental Features

In both (TMTTF)$_2X$ and (TMTSF)$_2X$ series of hole 1/4-filled and dimerized compounds, with non centrosymmetric anions like X = ReO$_4$, NO$_3$, BF$_4$ ..., the possibility arises for the anion lattice to form superstructures below the ordering temperature $T^0_{\mathbf{q}_A}$ [22,23]. These superstructures superimpose displacement and orientation of the anions [10]. In the high-1D-temperature domain, the most frequent staggered anion ordering wave vectors are $\mathbf{q}_{A_1} = (\frac{1}{2}, \frac{1}{2})$ and $\mathbf{q}_{A_2} = (\frac{1}{2}, 0)$, expressed in units of the reciprocal lattice vector (a^*, b^*) in the ab plane of the materials (here the longitudinal part $\frac{1}{2}a^* = 2k_F$).

In the case of the Bechgaard salt (TMTSF)$_2$ReO$_4$, for instance, $(\frac{1}{2}, \frac{1}{2})$ AO takes place at $T^0_{\frac{1}{2},\frac{1}{2}} \simeq 177$ K [22], which also coincides with a metal-insulator transition [48]. A similar AO superstructure is found for the Fabre salt (TMTTF)$_2$ReO$_4$ at $T^0_{\frac{1}{2},\frac{1}{2}} \simeq 154$ K, a slightly lower temperature that falls within the Mott insulating state found below $T_\rho \simeq 230$ K [30,49].

An interesting situation is found for the alloys [(TMTSF)$_{1-x}$(TMTTF)$_x$]$_2$ReO$_4$ in solid solution [26]. At $x = 0.55$, the salt is ordered with TMTSF and TMTTF molecules alternating along the stacking a axis.

Remarkably, the AO transition temperature drops and reaches the minimum, $T^0_{\frac{1}{2},\frac{1}{2}} \simeq 82$ K, a value significantly lower than the pure limits at $x = 0$ and $x = 1$, as shown in Figure 7 [26]. By contrast, the Mott scale reaches instead a maximum at $T_\rho > 325$ K in the $x = 0.55$ alloy [26], a value significantly higher than the two pure limits. It is worthwhile to note that the variation of AO temperature in the alloys differs from the one found in the hybrid salt (TMDTDSF)$_2$ReO$_4$, in which each organic molecule is composed of two sulfur and two selenium atoms [50–52]. For the latter, $T^0_{\frac{1}{2},\frac{1}{2}} \simeq 165$ K, an ordering temperature which for this salt essentially falls on the midpoint between the two limits. As to the Mott scale $T_\rho \simeq 210$ K, it lies below the one found in the pure sulfur limit ($x = 1$).

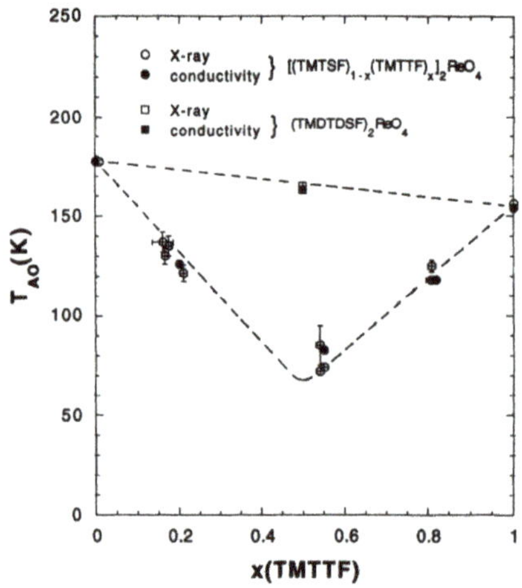

Figure 7. Anion ordering critical temperature for the ReO$_4$ salt in different families of organic conductors and their alloys. After Ilakovac et al. [26] and references there cited.

4.1.2. Electron–Anion Interaction

To examine the interplay between electronic and translational anionic degrees of freedom, we add an electron–anion interaction H_A to the purely electron part (18). Following Riera and Poilblanc [25], this interaction can be written in the form

$$H_a = \frac{1}{2}\sum_{r,j} K_a \delta^2_{r,j} + \sum_{r,j,\sigma} \big[(\delta_{r,j} - \lambda\delta_{r,j-1})m_{r,\sigma} + (\lambda\delta_{r,j} - \delta_{r,j-1})n_{r,\sigma}\big], \tag{43}$$

where the electron charge on even and odd sites in the dimer at r and chain j is coupled to the anionic displacement $\delta_{r,j}$, as shown in Figure 8. The constant λ takes into account the two inequivalent distances between the anions and the molecular sites in the strength of the coupling strength ($0 \leq \lambda \leq 1$). Here the anion displacements are treated in the classical harmonic approximation for which the spring constant K_a has been rescaled $K_a/g^2 \to K_a$ in order to incorporate the anion-electron coupling strength g. By using the transformations (6) and (7), the expression of H_a for the lower band electrons, when coupled to the relevant anionic distortions $\delta_{\frac{1}{2},q_\perp}$ in Fourier space, is given by

$$H_a = \frac{1}{2}\sum_{q_\perp} K_a\, \delta^2_{\frac{1}{2},q_\perp} + \frac{1}{\sqrt{N}} \sum_{p,k,q_\perp,\sigma} \xi^p_{q_\perp}\, \delta_{\frac{1}{2},q_\perp}\, d^\dagger_{p,k,\sigma} d_{-p,k-2pk_F,\sigma}. \tag{44}$$

The form factors for the electron–anion interaction read

$$\xi_0^p = (\lambda - 1)\frac{(t+\delta t)\epsilon_0/E_F + ip(t-\delta t)}{\sqrt{2t^2+2\delta t^2}}, \tag{45}$$

and

$$\xi_{\frac{1}{2}}^p = (\lambda + 1)\frac{(t+\delta t) + ip(t-\delta t)\epsilon_0/E_F}{\sqrt{2t^2+2\delta t^2}}, \tag{46}$$

for the $(\frac{1}{2},0)$ and $(\frac{1}{2},\frac{1}{2})$ anion orderings. One observes that due to the parity of the electron–anion coupling and the site potential, both factors have a real and an imaginary parts. This will then introduce a phase dependent coupling to CDW.

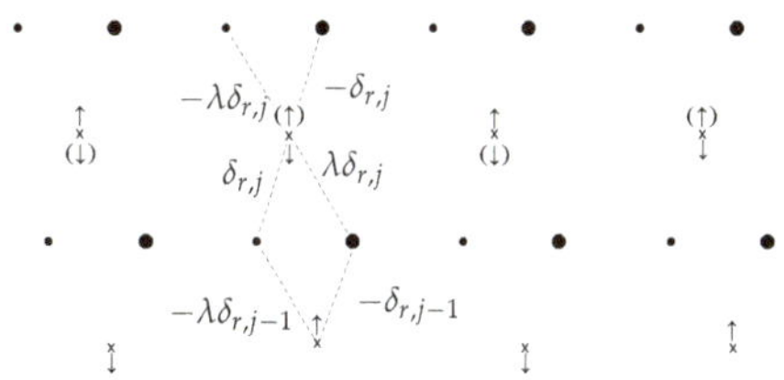

Figure 8. Electron–anion (×) interaction of the Riera-Poilblanc model [25] in systems like $(TMTSF)_2X$, $(TMTTF)_2X$ and their alloys. The arrows depict anion displacements for the $(\frac{1}{2},\frac{1}{2})$ $[(\frac{1}{2},0)]$ ordering.

When the anionic part of the Hamiltonian is incorporated into the action S, it becomes

$$S = S_0 + S_I - \frac{\beta}{2}\sum_{q_\perp}{}' K_a\, \delta^2_{\frac{1}{2},q_\perp} + S_{ea}. \tag{47}$$

The electron–anion coupling can be recast in the form

$$S_{ea} = -\sqrt{\beta}\sum_{q_\perp}\delta_{\frac{1}{2},q_\perp} e^{-i\frac{\theta_+}{2}}\Big\{z^c_{\theta_+}O_{\theta_+}\mathrm{Re}\big[\xi^+_{q_\perp}e^{-i\frac{\theta_\perp}{2}}\big] + iz^c_{\theta_-}O_{\theta_-}\mathrm{Re}\big[\xi^+_{q_\perp}e^{-i\frac{\theta_-}{2}}\big]\Big\} \tag{48}$$

where the renormalization factors $z_{\theta_\pm} = 1$ at $\ell = 0$.

Therefore in the presence of a site potential ϵ_0, which breaks the inversion symmetry within the dimer, the anion order parameter $\delta_{\frac{1}{2},q_\perp}$ is coupled to the two independent $CDW_{\theta_\perp}$ of the electron gas, as shown in Figure 4. However, there is a definite form factor for each CDW which in the end plays an important role in the type of AO stabilized. This is examined next.

4.1.3. Anion Ordering

The linear coupling (48) between the anion order parameter and the $CDW_{\theta_\pm}$ composite field is similar to the one encountered in Section 3.2 for the coupling of electrons to external source fields in calculation of susceptibilities χ_μ. This can be exploited to generate a Landau free energy expansion of the anion order parameter. Thus considering S_{ea} as a weak perturbation besides S_I, the successive partial integrations of electron degrees of freedom by the RG yield the effective action at the scale $\Lambda(\ell)$:

$$S[\psi^*,\psi]_\ell = S_0[\psi^*,\psi]_\ell + S_I[\psi^*,\psi]_\ell + S_{ea}[\psi^*,\psi]_\ell - \beta\mathfrak{F}[\delta]_\ell. \tag{49}$$

This procedure then generates the Landau free energy $\mathfrak{F}[\delta]_\ell$ of the anionic order parameter. Up to the quadratic level it takes the form

$$\mathfrak{F}[\delta]_\ell = \sum_{q_\perp} [K_a/2 - \chi_{\frac{1}{2},q_\perp}(\ell)]\delta^2_{\frac{1}{2},q_\perp} + \mathcal{O}(\delta^4), \tag{50}$$

where

$$\chi_{\frac{1}{2},q_\perp}(\ell) = \sum_{\mu=\pm} \mathrm{Re}[\xi^+_{q_\perp} e^{-i\frac{\theta_\mu}{2}}]\chi^c_{\theta_\mu}(\ell) \tag{51}$$

is the effective susceptibility involved in the AO at $(\frac{1}{2}, q_\perp)$. Relating the loop variable $\ell = \ln E_F/T$ to temperature, the zero of the quadratic coefficient leads to the following condition for the AO critical temperature $T^0_{\frac{1}{2},q_\perp}$,

$$\frac{K_a}{2} = \chi_{\frac{1}{2},q_\perp}(T^0_{\frac{1}{2},q_\perp}). \tag{52}$$

The strongest anion susceptibility will lead to the highest $T^0_{\frac{1}{2},q_\perp}$. For repulsive interactions $U = 2V > 0$ and non zero K_a, the regions of stability of both $(\frac{1}{2}, \frac{1}{2})$ and $(\frac{1}{2}, 0)$ AO in the $(\delta t, \epsilon_0)$ plane are shown in Figure 9 for different values of λ.

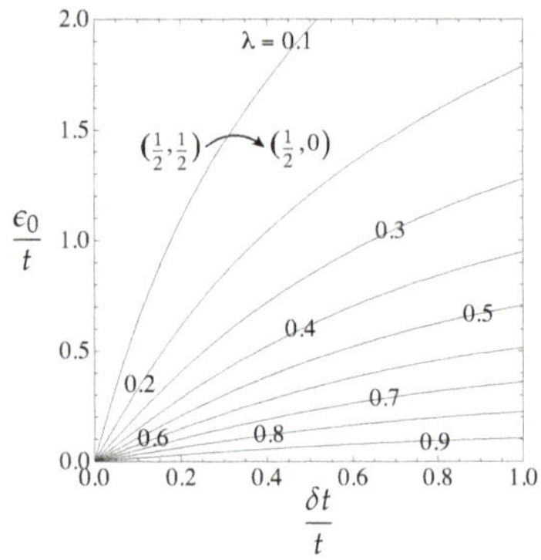

Figure 9. Iso$-\lambda$ phase boundaries between $(\frac{1}{2}, \frac{1}{2})$ and $(\frac{1}{2}, 0)$ anion orderings, as a function of the normalized site potential and dimerization.

It is useful to first consider limiting cases for the asymmetry parameter λ related to the position of the anions in the unit cell. Thus for $\lambda \to 1$, each anion is aligned with the center of a dimer and $\xi^+_0 \to 0$, so that only the $(\frac{1}{2}, \frac{1}{2})$ ordering is stabilized through a coupling to both $\mathrm{CDW}_{\theta_\pm}$. Interestingly, the most important coupling is to the CDW_{θ_+} correlations which according to Figure 6 are not singular, at variance with CDW_{θ_-}. This indicates that the driving force of the $(\frac{1}{2}, \frac{1}{2})$ AO is not the result of a $2k_F$ instability of the electron gas, as it is for the ordinary Peierls mechanism.

In the opposite decentered limit, where $\lambda \to 0$, both anion couplings are finite. At $\delta t = 0$, they becomes equal in amplitude and couple identically to both $\mathrm{CDW}_{\theta_\pm}$. The phase boundary then merges with the ϵ_0 axis at $\delta t = 0$, where $\theta_+ = \pi/2$ and $\theta_- = -\pi/2$ (or $3\pi/2$) and anions are coupled to completely site centered CDW (see Figure 4). It follows that for such a λ, the $(\frac{1}{2}, 0)$ phase is stable over the whole $(\delta t > 0, \epsilon_0)$ plane. In this case, the singular CDW_{θ_-} correlations are the main driving force of AO.

When λ grows from zero in Figure 9, the region of stability of $(\frac{1}{2}, \frac{1}{2})$ phase starts to increase against that of $(\frac{1}{2}, 0)$. According to (45) and (46), this is the consequence of a reinforcement (weakening) of the effective anion coupling $\mathrm{Re}\,[\xi^+_{\frac{1}{2}} e^{-i\theta_\pm/2}]$ ($\mathrm{Re}\,[\xi^+_0 e^{-i\theta_\pm/2}]$) to the $\mathrm{CDW}_{\theta_\pm}$. For the whole range of λ, the $(\frac{1}{2},\frac{1}{2})$ AO is essentially due to the coupling to the non singular CDW_{θ_+}, whereas the $(\frac{1}{2},0)$ AO is mainly coupled to the singular CDW_{θ_-}.

4.1.4. Theory and Experiment

One can proceed to the application of the above model to the variation of anion ordering $T_{\frac{1}{2},\frac{1}{2}}$ and Mott insulating scale T_ρ in the alloys $[(TMTTF)_{1-x}TMTSF_x]ReO_4$. First, we fix the various parameters of the model. The average longitudinal hopping t of the model can be set from the results of band calculations [53,54], $t = [364(1-x) + 200x]$ meV, interpolating between the pure Fabre ($x = 1$) and the Bechgaard ($x = 0$) limits. From these results, one can also determine the modulation of the hopping, $\delta t = [26(1-x) + 14.5x]$ meV. From photoemission experiments [55], the difference in ionization energy between the TMTSF and TMTTF organic molecules leads to a site modulation energy of $\epsilon_0 \simeq 200$ meV, in order to use $\epsilon_0(x) = 4x(1-x)200$ meV, as the effective alternating site potential as a function of x (this amplitude for the site potential ϵ_0 does not include the contribution coming from charge ordering, which is observed in $(TMTTF)_2ReO_4$ at $x = 1$ [30,49] and should be present at finite x. Although this contribution is not known in the alloy, its input in the calculations would further suppress anion ordering transition in Figure 10). Regarding interactions, we shall take $U = [200(1-x) + 268x]$ meV giving a stronger (weaker) site repulsive in the Fabre (Bechgaard) case. According to quantum chemistry calculations [44], the value of the nearest-neighbor interaction can be fixed at $V = U/2$ for all x. Finally, we shall link the small modulations of interactions, $\delta U = -\frac{\epsilon_0}{E_F}\frac{U}{3}$ and $\delta V = \frac{t\delta t}{E_F^2}\frac{U}{3}$, to those of ϵ_0, U and δt. In this way for instance, the values $U \pm \delta U$ and $V \pm \delta V$ at $x = 0.5$ corresponds to the interaction values in the Bechgaard (+) and Fabre (−) cases.

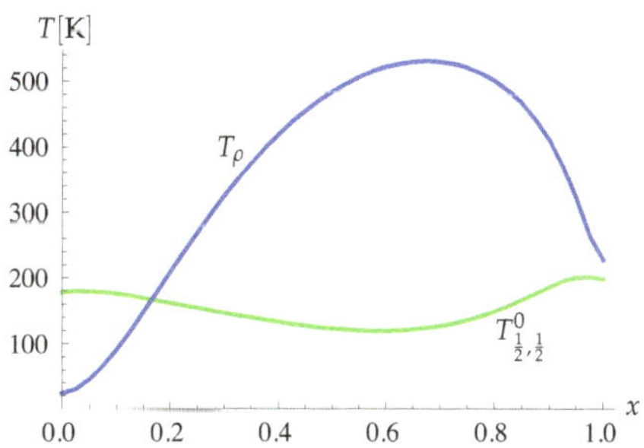

Figure 10. Calculated critical temperature for the $(\frac{1}{2},\frac{1}{2})$ anion ordering (**green**) and Mott scale (**blue**), as a function of TMTTF concentration x in $[(TMTSF)_{1-x}(TMTTF)_x]_2ReO_4$ alloys.

With the above figures, the $(\frac{1}{2},\frac{1}{2})$ AO will dominate for not too small λ and essentially arbitrary values of anion lattice stiffness K_a. We shall fix $K_a/[2(1+\lambda)^2] = 1.75\ (\text{eV})^{-1}$ to give from (52), the experimental value $T^0_{\frac{1}{2},\frac{1}{2}} \simeq 180$ K in the Bechgaard $x = 0$ limit [22]. Thus when the site potential $\epsilon_0(x)$ grows with x, alongside the effect of dimerization $\delta t/t$, $T^0_{\frac{1}{2},\frac{1}{2}}$ decreases due to the weakening of the coupling to non singular CDW_{θ_+} correlations in (51) and (52); it reaches a minimum at $x = 0.5$, namely where ϵ_0 is maximum, in qualitative agreement with the results of Figure 7. As x grows further, ϵ_0 goes down and $T^0_{\frac{1}{2},\frac{1}{2}}$ starts to increase and evolve toward the limiting value of 190 K in the $x = 1$ Fabre case. This value is nearly the same than for $x = 0$, but higher than the experimental value [30,49,56]. The minimum value $T^0_{\frac{1}{2},\frac{1}{2}} \simeq 105$ K at $x = 0.5$ corresponds to a decrease of 40% or so from the pure $x = 0$ value, compared with nearly 55% for experiments in Figure 7, showing a qualitative agreement (see the above note in Section 4.1.4).

In contrast to $T^0_{\frac{1}{2},\frac{1}{2}}$, the value of the Mott scale T_ρ in Figure 10 reaches a maximum at more than twice its value found in the $x = 1$ Fabre limit. This is a consequence of positive ϵ_0 and $\delta t/t$

alternating potentials, whose influences add in quadrature in the g_3 expression in (19) and the value of T_ρ. This conclusion for T_ρ agrees with those previous works [26–29].

4.2. Interplay between the Spin-Peierls and Charge Ordered States

As a second application of our model, we examine the influence of charge ordering on the spin-Peierls instability of weakly localized 1D Mott insulators of the Fabre salts series.

4.2.1. Experiments

The spin-Peierls instability is found in some members of the Fabre $(TMTTF)_2X$ series with $X = PF_6$ and AsF_6. These quarter-filled band but weakly dimerized systems show a 1D Mott insulating behavior below some temperature scale $T_\rho \simeq 220$ K [57,58]. In $(TMTTF)_2AsF_6$, for instance, the Mott insulating behavior is followed at lower temperature by a continuous transition toward the formation of charge ordered (CO) state at $T_{CO} \simeq 103$ K [29,31]. Below T_{CO}, there is a charge disproportionation in the dimer unit cell leading to a finite static alternated site potential ϵ_0. Within the CO state, there is an additional instability that involves both spins and lattice degrees of freedom. According to x-ray diffuse scattering [37,39], the lattice becomes unstable with the onset of 1D lattice fluctuations at the wave vector $2k_F$ below the characteristic temperature scale T^0_{SP}. For the AsF_6 compound, $T^0_{SP} \simeq 40$ K [10,37]. Occurring well below the Mott scale T_ρ, T^0_{SP} takes place in the presence of strong antiferromagnetic correlations and then refers to a spin-Peierls (SP) instability. A true 3D SP ordering, however, occurs only at a much lower temperature, namely, $T_{SP} \simeq 11$ K [37,39]. The latter obeys the empirical rule $T_{SP} \sim T^0_{SP}/3$ for the reduction of the ordering temperature by 1D fluctuations in weakly coupled Peierls and spin-Peierls chains.

From NMR experiments under pressure [39], T_{SP} is found to increase while T_{CO} is steadily decreasing. This indicates that a reduction of the charge disproportionation, that is the site alternated potential ϵ_0, enhances the SP ordering. The increase of T_{SP} with pressure carries on until its amplitude reaches T_{CO}, defining a critical pressure above which T_{SP} undergoes a steady decrease (see Figure 11).

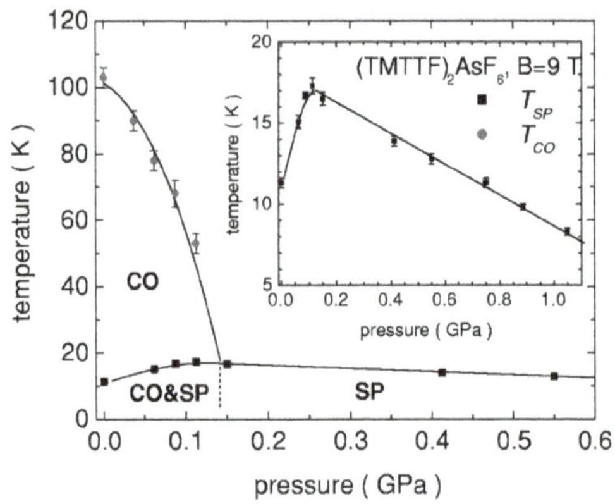

Figure 11. Critical temperatures of spin-Peierls and charge ordered states as a function of applied pressure in $(TMTTF)_2AsF_6$, as determined from NMR experiments. From Zamborszky et al. [39].

4.2.2. Electron–Lattice Coupling

The formation of a bond superstructure in the dimerized chains results from a coupling of the lattice degrees of freedom to bond charge density-wave correlations of electrons. By analogy with a Peierls electron–lattice coupling in a tight-binding band [59,60], the coupling develops from the modulation of the interdimer hopping integral $t_2 = t - \delta t$ by the displacement ϕ_r of the dimer from its equilibrium position. The expansion of $t_2(\phi)$ with respect to ϕ leads to an additional electron–lattice part to the Hamiltonian (5) which is of the form,

$$H_{\rm ep} + H_{\rm p}^0 = \sum_{r,\sigma} t_2'[a^\dagger_{r,\sigma} b_{r-1,\sigma}(\phi_{r-1} - \phi_r) + {\rm H.c.}] + \frac{K_D}{2}\sum_r (\phi_{r-1} - \phi_r)^2. \tag{53}$$

Here $t_2' = dt_2/d\phi$ and K_D is the spring constant of the harmonic potential energy, $H_{\rm p}^0$, of interdimer lattice modes in the static limit. From the canonical transformation (6) and (7), the Hamiltonian for the coupling of the tetramerization order parameter, ϕ_{2k_F}, to electrons close to Fermi level is given by

$$H_{\rm ep} + H_{\rm p}^0 = \frac{1}{\sqrt{N}}\sum_{k,\sigma}[g_{\rm ep}(\epsilon_0)\, d^\dagger_{+,k,\sigma} d_{-,k-2k_F,\sigma}\phi_{2k_F} + {\rm H.c.}] + 2K_D|\phi_{2k_F}|^2 \tag{54}$$

where

$$g_{\rm ep}(\epsilon_0) = -2i\, t_2' \sin\gamma_{k_F} = -2i\, t_2'\sqrt{1 - \frac{\epsilon_0^2}{E_F^2}} \tag{55}$$

is the electron–lattice coupling constant showing an explicit dependence on the potential amplitude ϵ_0 of CO, which reduces the strength of the coupling.

Transposing this term into the action allows to write at $\ell = 0$,

$$S = S_0 + S_I - 2\beta K_D|\phi_{2k_F}|^2 + S_{\rm ep} \tag{56}$$

where

$$S_{\rm ep} = -\sqrt{\beta}\, g_{\rm ep}(\epsilon_0)\big[(z^c_{\theta_+} O_{\theta_+} \sin\tfrac{1}{2}\theta_+ - iz^c_{\theta_-} O_{\theta_-} \sin\tfrac{1}{2}\theta_-)\phi_{2k_F} + {\rm c.c.}\big], \tag{57}$$

in which we have decomposed the BOW composite field in terms of the two independent CDW at $\theta_\pm = \theta_{g_3}$ $(\theta_{g_3} \pm \pi)$, and for $q_0 = (2k_F, 0)$. Here the initial conditions for the pair vertex renormalization factors at $\ell = 0$ are $z^c_{0_\pm} = 1$.

4.2.3. Spin-Peierls Instability

The presence of the SP order parameter ϕ_{2k_F} then linearly couples as an external field to the two independent CDW for a complex g_3. In weak coupling $S_{\rm ep}$ can be taken as a perturbative term alongside S_I. Thus the successive partial integrations of the RG transformation following (26), down to the energy distance $\Lambda_0(\ell)$ from the Fermi level will lead to the effective action at step ℓ:

$$S[\psi^*,\psi]_\ell = S_0[\psi^*,\psi]_\ell + S_I[\psi^*,\psi]_\ell + S_{\rm ep}[\psi^*,\psi,\phi]_\ell - \beta\mathfrak{F}[\phi]_\ell. \tag{58}$$

The RG then generates a ℓ dependent Landau free energy $\mathfrak{F}[\phi]_\ell$ of the SP order parameter. At the quadratic level, it takes the form

$$\mathfrak{F}[\phi]_\ell = \big[2K_D - |g_{\rm ep}(\epsilon_0)|^2 \sum_{\mu=\pm}(\sin^2\tfrac{1}{2}\theta_\mu)\chi^c_{\theta_\mu}(q_0,\ell)\big]|\phi_{2k_F}|^2 + \mathcal{O}(\phi^4). \tag{59}$$

The mean field criteria for the SP 'transition' temperature at $T^0_{\rm SP}$ is obtained from the zero of the quadratic term at $\ell^0_{\rm SP} = \ln E_F/T^0_{\rm SP}$. This leads to

$$\frac{2K_D}{|g_{\rm ep}(\epsilon_0)|^2} = \sum_{\mu=\pm} (\sin^2 \tfrac{1}{2}\theta_\mu)\chi^c_{\theta_\mu}(q_0, T^0_{\rm SP}). \tag{60}$$

The right-hand-side of this expression is essentially dominated by the CDW$_{\theta_-}$ susceptibility $\chi^c_{\theta_-}$ which is more centered on bonds. It presents a power law divergence with decreasing temperature, while $\chi^c_{\theta_+}$, which is more site-centered saturates at a small value. According to the results of Section 3.2, the bond susceptibility takes the form

$$\chi^c_{\theta_-}(q_0, T) \sim (\pi v_F)^{-1} C_{\theta_-}(T/T_\rho)^{-\gamma^*},$$

where $\gamma^* = 3/2$ for the fixed point behavior below the Mott scale T_ρ [47]. Here C_{θ_-} is a positive constant that gives the contribution to the susceptibility from all energy scales above T_ρ.

By singling out the dominant θ_- part, we obtain the approximate result for the mean field SP temperature

$$T^0_{\rm SP} \approx [C_{\theta_-}(\sin^2 \tfrac{1}{2}\theta_-)|\bar{g}_{\rm ep}(\epsilon_0)|^2/2K_D]^{\frac{2}{3}} T_\rho, \tag{61}$$

where $|\bar{g}_{\rm ep}(\epsilon_0)|^2 = |g_{\rm ep}(\epsilon_0)|^2/\pi v_F$. Although from (21) an increase of the site potential ϵ_0 raises the value of $|g_3|$ and then T_ρ, ϵ_0 is a major source of reduction of the electron-phonon matrix element $\bar{g}_{\rm ep}(\epsilon_0)$, which together with the shift of CDW$_{\theta_-}$ off the bonds, leads to an overall decrease of $T^0_{\rm SP}$ with ϵ_0. This is shown in Figure 12 (top). The charge imbalance between sites of neighbouring dimers is therefore acting as the main source of reduction of the lattice coupling to $2k_F$ bond correlations.

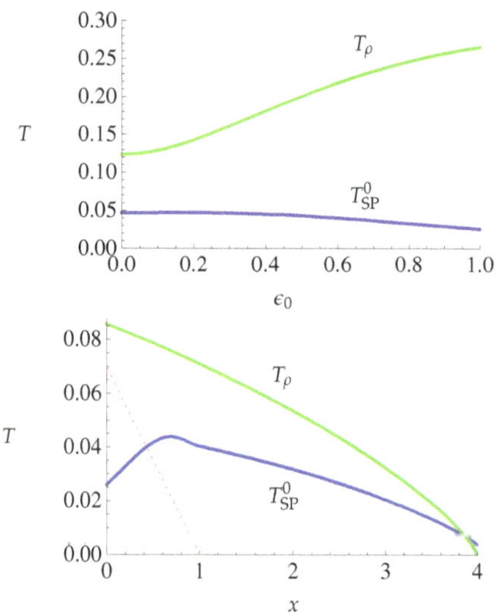

Figure 12. (**top**): Calculated variation of the mean-field SP ordering temperature $T^0_{\rm SP}$ and of the Mott scale T_ρ with the amplitude of site potential ϵ_0 due to charge ordering; (**bottom**): Calculated variations of the mean-field 1D (3D) spin-Peierls temperature $T^0_{\rm SP}$ ($T_{\rm SP} \sim T^0_{\rm SP}/3$), as a function of the tuning (pressure) parameter x. The dashed red line is a linear parametrization of the charge ordering temperature $T_{\rm CO} = \epsilon_0(x)/2$. All temperature scales are normalized by the average hopping t along the stacks.

It is worth noticing that the mean-field prediction $T^0_{\rm SP}$ does not coincide with a real transition temperature, but refers to the temperature scale for the onset of strong $2k_F$ 1D lattice SP fluctuations [61,62], such as those seen for instance in diffuse x-ray scattering [37]. For a true 3D transition to take place, interchain coupling $V_\perp$ (e.g., electron–lattice, Coulomb, ...) is necessary to stabilize long-range ordering for a one-component SP order parameter. The generalization of the above approach to the case of weakly coupled Landau-Ginzburg chains is well known [63]; it leads to the relation

$$T_{\rm SP} \simeq \frac{T^0_{\rm SP}}{\ln(T^0_{\rm SP}/V_\perp)} \sim T^0_{\rm SP}/\alpha, \tag{62}$$

between the $T^0_{\rm SP}$ and $T_{\rm SP}$ scales. For $V_\perp \ll T^0_{\rm SP}$, fluctuations reduce $T^0_{\rm SP}$ by a factor α, which is typically around 3.

4.2.4. Theory and Experiment

We are now in a position to apply our model to the evolution of the SP instability in the presence of CO for a compound like $(TMTTF)_2AsF_6$ under pressure (top of Figure 12). We consider the hopping modulation $\delta t(x) = 0.2 - 0.5x$ (normalized by $t = 1$), tuned downward by the pressure parameter x. Here, 0.2 is a typical band calculation value of δt for a compound like $(TMTTF)_2AsF_6$ at ambient pressure ($x = 0$) [64]. We fix the interaction parameters $U = 1$ and $V = 0.5$ in order to obtain a Mott temperature scale, $T_\rho \sim 100$ K ($T_\rho \sim 0.09$ in units of t), consistent with the range found in experiments for a value of $t = 1300$ K at ambient pressure [64]. Charge ordering is responsible for the onset of a static site potential, which we parametrize by the linear profile $\epsilon_0(x) = \epsilon_0(0)(1 - x/x_c)$. Here we shall use the relation $\epsilon_0(0) = \Delta_{\rm CO} \sim 2T_{\rm CO}$ between the CO gap and the observed critical temperature $T_{\rm CO} \approx 100$ K at ambient pressure. Finally, we let the normalized small variations $\delta U = -\epsilon_0(x)/E_F$ and $\delta V = t\delta t(x)/E_F^2$ evolving under pressure following $\epsilon_0(x)$ and $\delta t(x)$, respectively.

The solution of (60) for $T^0_{\rm SP}$ from the use of (42) and (55) as a function of the tuning parameter x, is displayed in the bottom panel of Figure 12. We see that to the fall of $\epsilon_0(x)$ under pressure corresponds an increase of $T^0_{\rm SP}$, the latter being a consequence of the boost in the electron lattice-coupling $g_{\rm ph}(\epsilon_0)$ that overcomes the impact of the reduction of T_ρ on $T^0_{\rm SP}$ under pressure. A maximum of $T^0_{\rm SP}$ is reached at x_c where $\epsilon_0(x_c)$ vanishes, leading to near 50% of increase of $T^0_{\rm SP}$ from its ambient pressure value. Beyond x_c, $\epsilon_0(x > x_c) = 0$ and T^0_{SP} undergoes a monotonic decreases with x, which according to (60) and (61) is governed by that of T_ρ under pressure—when only bond order wave correlations with $\theta_- = \pi$ couple to the lattice (at sufficiently high but intermediate pressure, interchain antiferromagnetic exchange coupling, which is present but not considered in the present high temperature calculations, enters into play and introduces a competition between the SP and the magnetic Néel states, from which the latter state ends up to be favored [20,21,61]. This competition is also documented on experimental grounds in the similar compound $(TMTTF)_2PF_6$ – see for instance Refs. [65,66]. In the very high pressure range, it is the turn of the Mott scale T_ρ to become irrelevant in $(TMTTF)_2X$, with the Néel state merging into an itinerant SDW state when interchain exchange evolves towards coherent interchain hopping, as found in $(TMTSF)_2X$ at low pressure [1,4]. In these conditions the absence of a tetramerization-Peierls-instability, despite the absence of charge ordering, may take its origin from the relatively weak amplitude of Umklapp scattering. This suppresses T_ρ and strongly reduces the growth of bond CDW correlations for the Peierls scale T^0_P in (60). The long-range Peierls ordering is likely to be preempted when conflicting sources of interchain coupling are added to the model such as single electron hopping and Coulomb interactions, along with the emergence of SDW long-range order [10]).

The results of the above model for $T^0_{\rm SP}$ and by extension $T_{\rm SP}$ fairly bear the comparison with the experiments shown in Figure 11. The present model could also offer an explanation as to the absence of a spin-Peierls transition in a compound like $(TMTTF)_2SbF_6$, which rather exhibits a Néel transition around 7K [67]. The SbF_6 Fabre salt is known to develop charge ordering at a temperature $T_{\rm CO} \approx 160$ K [29,30,68], which is sizably higher than for other SP salts like PF_6 and

AsF_6. Following (55), a significantly larger T_{CO} or ϵ_0, together with a relatively weaker value of T_ρ for SbF_6 [58], would further suppress the electron-phonon matrix element $g_{ep}(\epsilon_0)$ and from (61) the scales T^0_{SP} and T_{SP}. Including the influence of interchain exchange interaction between spins, which is present in practice but has been disregarded in this work (see for instance Refs. [20,21,61]), will favor the competition between the SP and the Néel states and may be responsible for the absence of long-range SP ordering and the relatively small Néel temperature for this salt (see also the note in the preceding paragraph). The observation of a reentrant SP state in $(TMTTF)_2SbF_6$ at higher pressure is consistent with that view [67]. These results obtained above are also compatible with those using the bosonization technique [40], and numerics [35,40].

5. Concluding Remarks

In the work developed above, we inquired into the properties of the one-dimensional extended Hubbard model at quarter-filling with superimposed dimerization, site and anion alternating lattice potentials, as they can be found in practice in low-dimensional charge transfer salts with a 2:1 stochiometry. The renormalization group method was applied to the continuum or electron gas limit of the model in order to determine the influence of dimerization and site commensurability potentials on low energy interactions, in particular effective half-filling type of Umklapp scattering that emerges from the lattice. The impacts of Umklapp on singular correlations and different excitation gaps of the model have been determined, along with spin and charge density-wave correlations and their specific phase relations to the underlying alternated lattice.

By coupling electrons to anionic displacements in systems like the $(TMTTF)_2X$ and $(TMTSF)_2X$ charge transfer salts, the mechanism of anion ordering in these materials has been investigated. It was found that a site alternated potential can significantly reduce the coupling of charge density wave correlations to anion displacements and in turn the critical temperature of staggered anion ordering. This occurs while the Mott insulating character of electrons is on the contrary strongly enhanced by site and bond commensurabilities. These opposed effects were found to be consistent with observations about anionic order and Mott insulating state in $[(TMTSF)_{1-x}(TMTTF)_x]_2ReO_4$ alloys.

The influence of the charge ordered state on the spin-Peierls ordering has also been investigated for members of the Fabre salts series at low applied pressure. Acting as alternate site commensurability, charge order was found to mainly reduce the inter-dimer tight-binding electron-phonon matrix element. The coupling of the lattice to bond density-wave correlations is then lessened by charge disproportionation and a competition between the spin-Peierls and the charge ordered states takes place. The interplay gives rise to a characteristic dome of the spin-Peierls ordering temperature as a function of the suppression of charge order under pressure, a result congruent with observations made in the Fabre salts series.

Allowing for anionic displacements besides lattice degrees of freedom would be an interesting possible extension of the latter calculations. This could provide the opportunity to check if the spin-Peierls lattice distortion is accompanied by staggered anion ordering, as suspected on experimental grounds for spin-Peierls systems of the Fabre salts series [11].

It is worth mentioning another straightforward application of the model of anion ordering which was not considered above. This concerns the anion displacement that goes with the charge ordering transition in the $(TMTTF)_2X$ series. Such anion displacement is known to be uniform in character [10]. The corresponding order parameter $\delta_{0,0}$ will thus be linearly coupled to uniform charge density of the stacks, in a way similar to the expression given in (48). Since uniform charge susceptibility is a quantity proportional to the dielectric constant, as the latter is being singular at the charge ordering transition [32,68], it will drive a collective shift of the anionic position as apparently found experimentally [9,68,69].

Author Contributions: M.M. performed the calculations and the comparisons with experiments in the framework of a PhD thesis at Université de Sherbrooke (2017). C.B. wrote the paper. Both authors contributed to the discussion of the results. All authors have read and agreed to the published version of the manuscript.

Funding: This research was funded by the National Science and Engineering Research Council of Canada (NSERC) and the Réseau Québcois des Matériaux de Pointe (RQMP).

Acknowledgments: The authors thank G. Chitov for useful comments on many aspects of this work.

Conflicts of Interest: The authors declare no conflict of interest.

References

1. Jérome, D.; Schulz, H.J. Organic conductors and superconductors. *Adv. Phys.* **1982**, *31*, 299. [CrossRef]
2. Ishiguro, T.; Yamaji, K. *Organic Superconductors*; Springer-Verlag Series in Solid-State Science; Springer: Berlin/Heidelberg, Germeny, 1990; Volume 88.
3. Bourbonnais, C.; Jérome, D. The normal phase of quasi-one-dimensional organic superconductors. In *Advances in Synthetic Metals, Twenty Years of Progress in Science and Technology*; Bernier, P., Lefrant, S., Bidan, G., Eds.; Elsevier: New York, NY, USA, 1999; pp. 206–261.
4. Bourbonnais, C.; Jérome, D. Interacting electrons In quasi-one-Dimensional organic superconductors. In *The Physics of Organic Superconductors and Conductors*; Springer Series in Materials Science; Lebed, A., Ed.; Springer: Heidelberg, Germany, 2008; Volume 110, p. 357.
5. Brown, S.E.; Chaikin, P.M.; Naughton, M.J. La tour des sels de Bechgaard. In *The Physics of Organic Superconductors and Conductors*; Springer Series in Materials Science; Lebed, A., Ed.; Springer: Heidelberg, Germany, 2008; Volume 110, p. 49.
6. Seo, H.; Hotta, C.; Fukuyama, H. Towards systematic understanding of diversity of electronic properties in low-dimensional molecular solids. *Chem. Rev.* **2004**, *104*, 5005. [CrossRef] [PubMed]
7. Brown, S.E. The Bechgaard salts and relatives. *Physica C* **2015**, *514*, 279. [CrossRef]
8. Dressel, M. Ordering phenomena in quasi-one-dimensional organic conductors. *Naturwissenschaften* **2007**, *94*, 527. [CrossRef] [PubMed]
9. Dressel, M.; Dumm, M.; Knoblauch, T.; Masino, M. Comprehensive optical investigations of charge order in organic chain compounds $(TMTTF)_2X$. *Crystals* **2012**, *2*, 528. [CrossRef]
10. Pouget, J.P. Interplay between electronic and structural degrees of freedom in quarter-filled low dimensional organic compounds. *Physica B* **2015**, *460*, 45. [CrossRef]
11. Pouget, J.P.; Alemany, P.; Canadell, E. Donor-anion interactions in quarter-filled low-dimensional organic conductors. *Mater. Horizons* **2018**, *5*, 590. [CrossRef]
12. Rosslhuber, R.; Rose, E.; Ivek, T.; Pustogow, A.; Breier, T.; Geiger, M.; Schrem, K.; Untereiner, G.; Dressel, M. Structural and electronic properties of $(TMTTF)_2X$ salts with tetrahedral anions. *Crystals* **2018**, *8*, 121. [CrossRef]
13. Clay, R.T.; Mazumdar, S. From charge- and spin-ordering to superconductivity in the organic charge-transfer solids. *Phys. Rep.* **2019**, *788*, 1. [CrossRef]
14. Barisic, S.; Brazovskii, S. *Recent Developments in Condensed Matter Physics*; Devreese, J.T., Ed.; Plenum: New York, NY, USA, 1981; Volume 1, p. 327.
15. Emery, V.J.; Bruinsma, R.; Barisic, S. Electron-electron umklapp scattering in organic superconductors. *Phys. Rev. Lett.* **1982**, *48*, 1039. [CrossRef]
16. Dzyaloshinskii, I.E.; Larkin, A.I. Possible states of quasi-unidimensional systems. *Sov. Phys. JETP* **1972**, *34*, 422.
17. Solyom, J. The Fermi gas model of one-dimensional conductors. *Adv. Phys.* **1979**, *28*, 201. [CrossRef]
18. Emery, V.J. *Highly Conducting One-Dimensional Solids*; Devreese, J.T., Evrard, R.E., van Doren, V.E., Eds.; Plenum: New York, NY, USA, 1979; p. 247.
19. Brazovskii, S.; Yakovenko, Y. On the theory of organic superconducting materials. *Sov. Phys. JETP* **1985**, *62*, 1340.
20. Caron, L.G.; Bourbonnais, C. Importance of one-dimensional correlations in the phase diagram of the $(TMTTF)_2$-$(TMTSF)_2X$ salts. *Phys. B + C* **1986**, *143*, 453. [CrossRef]
21. Bourbonnais, C.; Caron, L.G. New mechanisms of phase transition in quasi-one dimensional conductors. *Europhys. Lett.* **1988**, *5*, 209. [CrossRef]
22. Moret, R.; Pouget, J.P.; Comès, R.; Bechgaard, K. X-ray scattering evidence for anion ordering in the low-temperature phase of Di(Tetramethyltetraselanafulvalenium) Perrhenate. *Phys. Rev. Lett.* **1982**, *49*, 1008–1012. [CrossRef]

23. Moret, R.; Pouget, J.P.; Comes, R.; Bechgaard, K. Structural phase transitions in $(TMTSF)_2X$ and $(TMTTF)_2X$: A survey with some new results. *J. Phys. Colloq.* **1983**, *44*, C3-957. [CrossRef]
24. Foury-Leylekian, P.; Petit, S.; Andre, G.; Moradpour, A.; Pouget, J.P. Neutron scattering evidence for a lattice displacement at the charge ordering transition of $(TMTTF)_2PF_6$. *Physica B* **2010**, *405*, S95. [CrossRef]
25. Riera, J.; Poilblanc, D. Influence of anion potential on the charge ordering in the charge-transfer salts. *Phys. Rev. B* **2001**, *63*, R241102. [CrossRef]
26. Ilakovac, V.; Ravy, S.; Pouget, J.P.; Lenoir, C.; Boubekur, K.; Batail, P.; Babic, S.D.; Biskup, N.; Korin-Hamzic, B.; Tomic, S.; et al. Enhanced charge localization in the organic alloys $[(TMTSF)_{1-x}TMTTF_x]_2ReO_4$.*Phys. Rev. B* **1994**, *50*, 7136. [CrossRef] [PubMed]
27. Penc, K.; Mila, F. Charge gap in the one-dimensional Hubbard model at quarter-filling. *Phys. Rev. B* **1994**, *50*, 11429. [CrossRef] [PubMed]
28. Tsuchiizu, M.; Yoshioka, H.; Suzumura, Y. Theoretical study of the charge gap of organic conductors Bechgaard salts. *J. Phys. Soc. Jpn.* **1999**, *68*, 1809. [CrossRef]
29. Monceau, P.; Nad, F.; Brazovskii, S. Ferroelectric Mott-Hubbard phase of organic $(TMTTF)_2X$ conductors. *Phys. Rev. Lett.* **2001**, *86*, 4080. [CrossRef] [PubMed]
30. Coulon, C.; Parkin, S.S.P.; Laversanne, R. Structureless transition and strong localization effects in bis-tetramethyltetrathiafulvalenium salts $[(TMTTF)_2X]$. *Phys. Rev. B* **1985**, *31*, 3583. [CrossRef] [PubMed]
31. Chow, D.S.; Zamborszky, F.; Alavi, B.; Tantillo, D.; Baur, A.; Merlic, C.A.; Brown, S.E. Charge Ordering in the TMTTF Family of Molecular Conductors. *Phys. Rev. Lett.* **2000**, *85*, 1698. [CrossRef] [PubMed]
32. Nad, F.; Monceau, P.; Carcel, C.; Fabre, J.M. Charge ordering phase transition in the quasi-one-dimensional conductor $(TMTTF)_2AsF_6$. *J. Phys. Condens. Matter* **2000**, *12*, L435. [CrossRef]
33. Seo, H.; Fukuyama, H. Antiferromagnetic phases of one-dimensional quarter-felled organic conductors. *J. Phys. Soc. Jpn.* **1997**, *66*, 1249. [CrossRef]
34. Mazumdar, S.; Clay, R.T.; Campbell, D.K. Bond-order and charge-density waves in the isotropic interacting two-dimensional quarter-filled band and the insulating state proximate to organic superconductivity. *Phys. Rev. B* **2000**, *62*, 13400. [CrossRef]
35. Riera, J.; Poilblanc, D. Coexistence of charge-density waves, bond-order waves, and spin-density waves in quasi-one-dimensional charge-transfer salts. *Phys. Rev. B* **2000**, *62*, R16243. [CrossRef]
36. Giamarchi, T. Mott transition in one dimension. *Physica* **1997**, *B230-232*, 975. [CrossRef]
37. Pouget, J.; Moret, R.; Comes, R.; Bechgaard, K.; Fabre, J.M.; Giral, L. X-ray diffuse scattering of some $(TMTSF)_2X$ and $(TMTTF)_2X$ salts. *Mol. Cryst. Liq. Cryst.* **1982**, *79*, 129. [CrossRef]
38. Creuzet, F.; Bourbonnais, C.; Caron, L.G.; Jérome, D.; Bechgaard, K. A ^{13}C NMR study of the interplay between the spin-Peierls and antiferromagnetic ground states in $(TMTTF)_2PF_6$ under pressure. *Synth. Met.* **1987**, *19*, 289. [CrossRef]
39. Zamborszky, F.; Yu, W.; Raas, W.; Brown, S.E.; Alavi, B.; Merlic, C.A.; Baur, A. Competition and coexistence of bond and charge orders in $(TMTTF)_2AsF_6$. *Phys. Rev. B* **2002**, *66*, 081103(R). [CrossRef]
40. Sugiura, M.; Suzumura, Y. Competition of dimerization and charge ordering in the pin-Peierls state of organic conductors. *J. Phys. Soc. Jpn.* **2003**, *72*, 1458. [CrossRef]
41. Kuwabara, M.; Seo, H.; Ogata, M. Coexistence of charge order and spin-Peierls lattice distortion in one-dimensional organic conductors. *J. Phys. Soc. Jpn.* **2003**, *72*, 225. [CrossRef]
42. Tsuchiizu, M.; Yoshioka, H.; Suzumura, Y. Crossover from quarter-filling to half-filling in a one-dimensional electron system with a dimerized and quarter-filled band. *J. Phys. Soc. Jpn.* **2001**, *70*, 1460. [CrossRef]
43. Clay, R.T.; Mazumdar, S.; Campbell, D.K. Pattern of charge ordering in quasi-one-dimensional organic charge-transfer solids. *Phys. Rev. B* **2003**, *67*, 115121. [CrossRef]
44. Castet, F.; Fritsch, A.; Ducasse, L. Determination of coulombic interaction parameters of the extended Hubbard model in the organic conductors. *J. Phys. I (France)* **1996**, *6*, 583. [CrossRef]
45. Bychkov, Y.A.; Gorkov, L.P.; Dzyaloshinskii, I. Possibility of superconductivity type phenomena in a one-dimensional system. *Sov. Phys. JETP* **1966**, *23*, 489.
46. Bourbonnais, C.; Caron, L.G. Renormalization group approach to quasi-one-dimensional conductors. *Int. J. Mod. Phys. B* **1991**, *05*, 1033. [CrossRef]
47. Kimura, M. Possible phases and some properties of the one metal with the half-filled band. *Prog. Theor. Phys.* **1975**, *63*, 955. [CrossRef]

48. Jacobsen, C.S.; Pedersen, H.J.; Mortensen, K.; Rindorf, G.; Thorup, N.; Torrance, J.B.; Bechgaard, K. An unusual metal-insulator transition: bis(tetramethyltetraselenafulvalenium)-perrhenate $(TMTSF)_2ReO_4$. *J. Phys. C Solid State Phys.* **1982**, *15*, 2651. [CrossRef]
49. Pustogow, A.; Petersein, T.; Kolatschek, S.; Engel, L.; Dressel, M. Electronic correlations versus lattice interactions: Interplay of charge and anion orders in $(TMTTF)_2X$. *Phys. Rev. B* **2016**, *94*, 195125. [CrossRef]
50. Liu, Q.; Ravy, S.; Senhaji, A.; Pouget, J.P.; Johanssen, I.; Bechgaard, K. Structural investigation of the $(TMDTDSF)_2X$ series of organic conductors. *Synth. Met.* **1991**, *42*, 1711. [CrossRef]
51. Auban, P.; Jérome, D.; Johannsen, I.; Bechgaard, K. Physical properties of the new organic conductor series di(tetramethyl-dithiadiselenafulvalene) X; X = BF_4, ClO_4, ReO_4, PF_6, AsF_6, SbF_6. *Synth. Met.* **1991**, *41*, 2285. [CrossRef]
52. Gotschy, B.; Auban-Senzier, P.; Farrall, A.; Bourbonnais, C.; Jerome, D.; Canadell, E.; Henriques, R.T.; Johanssen, I.; Bechgaard, K. One-dimensional physics in organic conductors $(TMDTDSF)_2X$, X = PF_6, ReO_4:^{77}Se-NMR experiments. *J. Phys. I (France)* **1992**, *2*, 677. [CrossRef]
53. Grant, P.M. Electronic structure of the 2:1 charge transfer salts of tmtcf. *J. Phys. Colloq.* **1983**, *44*, C3 847. [CrossRef]
54. Jacko, A.C.; Feldner, H.; Rose, E.; Lissner, F.; Dressel, M.; Valenti, R.; Jeschke, H.O. Electronic properties of Fabre charge-transfer salts under various temperature and pressure conditions. *Phys. Rev. B* **2013**, *87*, 155139. [CrossRef]
55. Sato, N.; Saito, G.; Inokuchi, H. Ionization potentials and polarization energies of tetraselenafulvalene (TSF) derivatives determined from ultraviolet photoelectron spectroscopy. *Chem. Phys.* **1983**, *76*, 79. [CrossRef]
56. Parkin, S.S.P.; Mayerle, J.J.; Engler, E.M. Anion ordering in $(TMTTF)_2ReO_4$. *J. Phys. (Paris) Coll.* **1983**, *44*, C3-1105.
57. Coulon, C.; Delhaes, P.; Flandrois, S.; Lagnier, R.; Bonjour, E.; Fabre, J. A new survey of the physical properties of the $(TMTTF)_2X$ series. Role of the counterion ordering. *J. Phys. (Paris)* **1982**, *43*, 1059. [CrossRef]
58. Laversanne, R.; Coulon, C.; Gallois, B.; Pouget, J.P.; Moret, R. Structural and electrical properties of $(TMTTF)_2MF_6$ salts (M = P, As, Sb). Role of the anions. *J. Phys. (Paris) Lett.* **1984**, *45*, L393. [CrossRef]
59. Su, W.P.; Schrieffer, J.R.; Heeger, A.J. Solitons in polyaecetylene. *Phys. Rev. Lett.* **1979**, *42*, 1698. [CrossRef]
60. Barišić, S.; Labbé, J.; Friedel, J. Tight-binding and transition-metal superconductivity. *Phys. Rev. Lett.* **1970**, *25*, 919. [CrossRef]
61. Caron, L.G.; Bourbonnais, C., Creuzet, F.; Jerome, D. Evolution of the spin-Peierls transition in the $(TMTTF)_2X$ salts (X = PF_6, PF_6 Br) under pressure. *Synth. Met.* **1987**, *19*, 69. [CrossRef]
62. Bourbonnais, C.; Dumoulin, B. Theory of lattice and electronic fluctuations in weakly localized spin-Peierls systems. *J. Phys. I (France)* **1996**, *6*, 1727. [CrossRef]
63. Scalapino, D.J.; Imry, Y.; Pincus, P. Generalized Ginzburg-Landau theory of pseudo-one dimensional systems. *Phys. Rev. B* **1975**, *11*, 2042. [CrossRef]
64. Ducasse, L.; Abderrabba, A.; Hoarau, J.; Pesquer, M.; Gallois, B.; Gaultier, J. Temperature dependence of the transfer integrals in the $(TMTSF)_2X$ and $(TMTTF)_2X$ families. *J. Phys. C* **1986**, *19*, 3805. [CrossRef]
65. Caron, L.G.; Creuzet, F.; Butaud, P.; Bourbonnais, C.; Jérome, D.; Bechgaard, K. Spin-Peierls and antiferromagnetic transition in $(TMTTF)_2PF_6$: EPR and NMR measurements and theory. *Synth. Met.* **1988**, *27B*, 123. [CrossRef]
66. Chow, D.; Wzietek, P.; Foglatti, D.; Alavi, B.; Tantillo, D.J.; Merlic, C.A.; Brown, S.E. Singular behavior in the pressure-tuned competition between spin-Peierls and antiferromagnetic ground states of $(TMTTF)_2PF_6$. *Phys. Rev. Lett.* **1998**, *81*, 3984. [CrossRef]
67. Yu, W.; Zhang, F.; Zamborszky, F.; Alavi, B.; Baur, A.; Merlic, C.A.; Brown, S.E. Electron-lattice coupling and broken symmetries of the molecular salt ?$(TMTTF)_2SbF_6$. *Phys. Rev. B* **2004**, *70*, R121101. [CrossRef]
68. De Souza, M.; Squillante, L.; Sônego, C.; Menegasso, P.; Foury-Leylekian, P.; Pouget, J.P. Probing the ionic dielectric constant contribution in the ferroelectric phase of the Fabre salts. *Phys. Rev. B* **2018**, *97*, 045122. [CrossRef]

69. De Souza, M.; Pouget, J.P. Charge-ordering transition in $(TMTTF)_2X$ explored via dilatometry. *J. Phys. Condens. Matter* **2013**, *25*, 343201. [CrossRef] [PubMed]

Publisher's Note: MDPI stays neutral with regard to jurisdictional claims in published maps and institutional affiliations.

crystals

MDPI

Article

Electric Transport of Nodal Line Semimetals in Single-Component Molecular Conductors

Yoshikazu Suzumura [1,*], Reizo Kato [2] and Masao Ogata [3]

1 Department of Physics, Nagoya University, Nagoya 464-8602, Japan
2 RIKEN, 2-1 Hirosawa, Wako-shi, Saitama 351-0198, Japan; reizo@riken.jp
3 Department of Physics, University of Tokyo, Bunkyo, Tokyo 113-0033, Japan; ogata@phys.s.u-tokyo.ac.jp
* Correspondence: suzumura@s.phys.nagoya-u.ac.jp

Received: 31 August 2020; Accepted: 21 September 2020; Published: 24 September 2020

Abstract: We examine an effect of acoustic phonon scattering on the electric conductivity of a single-component molecular conductor [Pd(dddt)$_2$] (dddt = 5,6-dihydro-1,4-dithiin-2,3-dithiolate) with a half-filled band by applying the previous calculation in a two-dimensional model with Dirac cone [Phys. Rev. B. 98, 161205 (2018)], wherethe electric transport by the impurity scattering exhibits a noticeable interplay of the Dirac cone and the phonon scattering, resulting in maximum of the conductivity with increasing temperature. The conductor shows a nodal line semimetal, where the band crossing of HOMO (Highest Occupied Molecular Orbital) and LUMO (Lowest Unoccupied Molecular Orbital) provides a loop of Dirac points located close to the Fermi energy followed by the density of states (DOS) similar to that of a two-dimensional Dirac cone. Using a tight-binding (TB) model [arXiv:2008.09277], which was obtained using the crystal structure observed from a recent X ray diffraction experiment under pressure, it is shown that the obtained conductivity explains reasonably the anomalous behavior in [Pd(dddt)$_2$] exhibiting temperature-independent resistivity at finite temperatures. This paper demonstrates a crucial role of the acoustic phonon scattering at finite temperatures in the electric conductivity of Dirac electrons. The present theoretical results of conductivity are compared with those of the experiments.

Keywords: nodal line semimetal; single-component molecular conductor; conductivity; DOS; tight-binding model

1. Introduction

In molecular solids, various electronic properties from metallic to insulating states have been examined due to the interplay of transfer energies and mutual interactions between molecules [1]. However, there is another electronic state, where the temperature dependence of the resistivity does not show metallic or insulating behavior. It originates from a band structure of a Dirac electron [2,3], where the conduction and valence bands cross at a certain momentum in the Brillouin zone. Such a state has been discovered in the two-dimensional materials of graphene [4] and organic conductors [5,6]. This state has been extensively studied in three-dimensional systems to show a nodal line semimetal i.e., a loop of Dirac points in inorganic conductors [7–9] and in molecular conductor [10].

The Dirac electron in organic conductors with a zero-gap state was found in two-dimensional α-(BEDT-TTF)$_2$I$_3$(BEDT-TTF = bis(ethylenedithio)tetrathiafulvalene) [5,6], using a tight-binding (TB) model with transfer energies estimated by the extended Hückel method [11]. It should be noted that the TB model [12,13] describes the electronic states of molecular conductors. These Dirac electrons are studied experimentally and theoretically to comprehend physical properties in the bulk system. Noticeable temperature dependence of Hall coefficient [14,15] NMR [16–18], anisotropic conductivity [19] and the Nernst coefficient [20] have been obtained since the Dirac point is located close to the chemical potential.

Another molecular Dirac electron system was found in a single-component molecular conductor, $[Pd(dddt)_2]$ (dddt = 5,6-dihydro-1,4-dithiin-2,3-dithiolate) [10]. The application of pressure is useful for the modification of conducting properties in single-component molecular crystals with soft lattices [21–25]. $[Pd(dddt)_2]$ under high pressure exhibits almost temperature independent resistivity [26]. First-principles calculations indicate that this material belongs to the three-dimensional Dirac electron system [27] consisting of HOMO (Highest Occupied Molecular Orbital) and LUMO (Lowest Unoccupied Molecular Orbital) bands, and that a TB model exhibits a loop of Dirac points called a nodal line semimetal [28]. There are several studies on the effective Hamiltonian, where a general two-band model is introduced [29] and the explicit calculation is performed for the nodal line semimetal [30].

However, the conductivity is not yet clearly understood although the temperature-independent conductivity is believed to be an evidence of Dirac electrons. From the theoretical point of view, the conductivity at the zero doping shows a universal conductance at absolute temperature [4] but increases linearly with increasing temperature due to the linear increase of the density of states [14,31], when the temperature is larger than the energy of the damping by the impurity scattering. Such an increase is incompatible with the experiment of the almost constant resistivity in two-dimensions [32]. The problem comes from the fact that the constant behavior can be obtained only when the damping energy being larger than the temperature, which is not the case of the molecular conductors. Recently, to comprehend the almost constant conductivity, a possible mechanism has been proposed for two-dimensional Dirac electrons, where the scattering by an acoustic phonon plays a crucial role [33].

The conductivity of $[Pd(dddt)_2]$ showing almost temperature-independent behavior [10] was examined theoretically [34,35] using a TB model of the Hückel calculation for a crystal structure obtained by first-principles calculations under pressure. The DOS exhibits a linear dependence close to the chemical potential but the region for the relevant energy is narrow compared with that for the constant resistivity. The conductivity in terms of the above TB model shows a large anisotropy and the almost constant behavior at high temperature, since the reduction of the DOS for larger energy suppresses the increase of the conductivity. In this case, the TB model suggests the almost constant resistivity in $[Pd(dddt)_2]$ due to the decrease of the DOS, but not by the property relevant to the Dirac cone. Recently, a TB model was re-examined using the crystal structure, which was obtained under high pressure [36]. Surprisingly, this band calculation shows the DOS, which depends linearly on a wide region of the energy being compatible with temperature of almost constant conductivity. Thus, we re-examine the almost constant resistivity in $[Pd(dddt)_2]$ using such a newly found TB model and by taking acoustic phonon scattering. The present paper demonstrates that such a mechanism does exist as the evidence of Dirac electrons for single-component molecular conductors. Note that a three-dimensional system with nodal line of Dirac points is important compared with two-dimensional systems with a Dirac point, since the almost constant conductivity is obtained by the combined effect of the nodal line and the acoustic phonon.

In Section 2, based on the recently obtained TB model under pressure, the nodal line and the DOS are calculated to find a wide region for the linear dependence of the DOS. A formulation for the conductivity is given by taking account of both the impurity and electron–phonon (e–p) scatterings. In Section 3, by calculating the chemical potential for the half-filled band, we examine the temperature dependence of the anisotropic conductivity and show the almost constant conductivity for reasonable choices of the e–p coupling constant. The corresponding resistivity is compared with that of the experiment. In Section 4, a summary and discussion are given.

2. Model and Formulation

2.1. TB Model

Figure 1 displays the crystal structure of $[Pd(dddt)_2]$ of a three-dimensional system with eight molecular orbitals, which consist of four molecules (1, 2, 3, and 4) with HOMO and LUMO per unit cell. These molecules are located on two kinds of layers, where the layer 1 includes molecules 1 and 3, and the layer 2 includes molecules 2 and 4, respectively. The original unit cell is given by lattice vectors, $\boldsymbol{a}_o$, $\boldsymbol{b}_o$, and $\boldsymbol{c}_o$, while a new unit cell is introduced by a transformation, $\boldsymbol{a} = -(\boldsymbol{a}_o + \boldsymbol{c}_o)$, $\boldsymbol{b} = -\boldsymbol{b}_o$, and $\boldsymbol{c} = \boldsymbol{c}_o$ in the present band calculation. Thus, the a axis becomes parallel to layer 1 and layer 2.

A TB model corresponding to Figure 1 has been recently obtained using the crystal structure observed under pressure [36]. There are several kinds of transfer energies between two molecular orbitals, which are listed in Table 1. The inter-layer energies in the z direction are given by a (1 and 2 molecules, and 3 and 4 molecules), and c (1 and 4 molecules, and 2 and 3 molecules). The intra-layer energies in the a-b plane are given by p (1 and 3 molecules) and q (2 and 4 molecules) and b (perpendicular to the a-c plane). Further, these energies are classified by three kinds of transfer energies given by HOMO-HOMO (H), LUMO-LUMO (L), and HOMO-LUMO (HL).

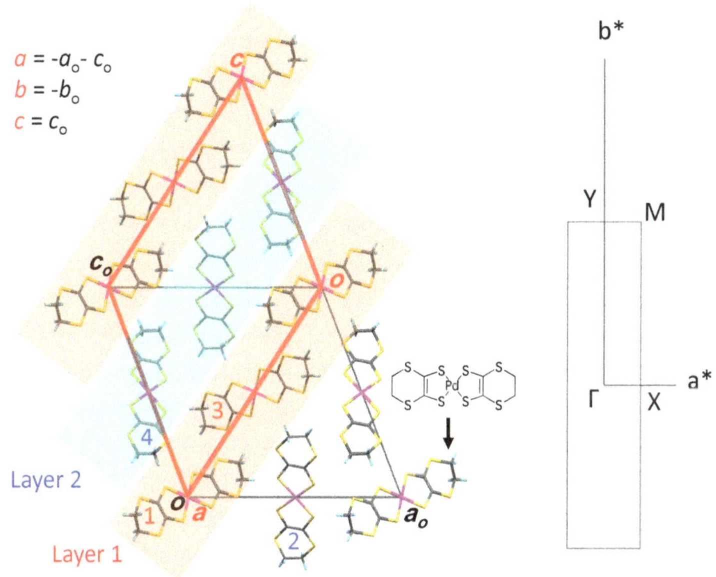

Figure 1. (Color online) Crystal structure of [Pd(dddt)$_2$] shown in the ac plane, where the molecule is stacked along the b direction perpendicular to the plane. Layer 1 (molecules 1 and 3) and Layer 2 (molecules 2 and 4) are parallel to the ab plane and alternated along the c direction. Lattice vectors $\boldsymbol{a}$, $\boldsymbol{b}$, and $\boldsymbol{c}$ form a new unit cell, where $\boldsymbol{a}$ is introduced for the simple representation of the energy band (see the main body). The location of the Pd atom in the unit cell is given by (x,y,z) = (0,0,0), (1/2,1/2,1/2),(1/2,1/2,0) and (1/2,0,1/2) for molecules 1, 2, 3, and 4, respectively. The right figure denotes the Fourier space, where $\boldsymbol{a}^*$ and $\boldsymbol{b}^*$ are reciprocal lattice vectors and Γ, X, Y, and M are TRIM (time reversal invariant momentum).

Table 1. Transfer energies for P = 5.9GPa [36] which are multiplied by 10^{-3} eV. The energy difference between the HOMO and LUMO is taken as ΔE = 0.696 eV.

	H − H	L − L	H − L	
$b1$	209.3	−1.9	−51.2	(stacking)
$p1(p)$	28.1	−12.4	19.9	Layer 1
$p2$	—	—	17.1	
$b2$	49.9	−80.4	−67.2	(stacking)
$q1(q)$	10.8	8.1	9.3	Layer 2
$q2$	—	—	9.2	
$a1$	−28.2	14.6	−20.1	
$a2$	2.2	1.3	−1.7	Interlayer
$c1$	15.4	12.7	14.1	
$c2$	−3.9	15.8	−11.8	

The TB model Hamiltonian is expressed as

$$H_{\rm TB} = \sum_{i,j=1}^{N} \sum_{\alpha,\beta} t_{i,j;\alpha,\beta} \left| i,\alpha \right\rangle \left\langle j,\beta \right| \ , \quad (1)$$

where $t_{i,j;\alpha,\beta}$ are transfer energies between nearest-neighbor sites and $|i,\alpha\rangle$ is a state vector. i and j are the lattice sites of the unit cell with N being the total number of the unit cells, α and β denote the 8 molecular orbitals given by HOMO (H1, H2, H3, H4) and LUMO (L1, L2, L3, L4). These energies in the unit of eV are listed in Table 1 where the gap between the energy of HOMO and that of LUMO is taken as $\Delta E = 0.696$ eV to reproduce the energy band of the first principle calculation [28].

Using a Fourier transform $|\alpha(\boldsymbol{k})\rangle = N^{-1/2}\sum_j \exp[-i\boldsymbol{k}\boldsymbol{r}_j]\ |j,\alpha\rangle$ with a wave vector $\boldsymbol{k} = (k_x, k_y, k_z)$, Equation (1) is rewritten as

$$H_{\rm TB} = \sum_{\boldsymbol{k}} |\Phi(\boldsymbol{k})\rangle\, \hat{H}(\boldsymbol{k})\, \langle\Phi(\boldsymbol{k})| \ , \quad (2)$$

where $\langle\Phi(\boldsymbol{k})| = (\langle {\rm H1}|, \langle {\rm H2}|, \langle {\rm H3}|, \langle {\rm H4}|, \langle {\rm L1}|, \langle {\rm L2}|, \langle {\rm L3}|, \langle {\rm L4}|)$. We take the lattice constant as unity and then $0 < |k_x|, |k_y|, |k_z| < \pi$ in the first Brillouin zone.

The matrix Hamiltonian $\hat{H}(\boldsymbol{k})$ is given in Appendix A. The nodal line has been found using $H(\boldsymbol{k})$ [28] where the Dirac point is supported by the existence of the inversion center [2]. The energy band $E_j(\boldsymbol{k})$ and the wave function $|\Psi_j(\boldsymbol{k})\rangle$, $(j = 1, 2, \cdots, 8)$ are calculated from

$$\hat{H}(\boldsymbol{k})\, |\Psi_j(\boldsymbol{k})\rangle = E_j(\boldsymbol{k})\, |\Psi_j(\boldsymbol{k})\rangle \ , \quad (3)$$

where $E_1 > E_2 > \cdots > E_8$ and

$$|\Psi_j(\boldsymbol{k})\rangle = \sum_{\alpha} d_{j,\alpha}(\boldsymbol{k})\, |\alpha\rangle \ , \quad (4)$$

with α = H1, H2, H3, H4, L1, L2, L3, and L4.

Since the electron close to the chemical potential is relevant for the electron-hole excitation, we consider only $E_4(\boldsymbol{k})$ and $E_5(\boldsymbol{k})$, i.e., the valence and conduction bands for the calculation of the conductivity. Thus, $E_4(\boldsymbol{k})$ and $E_5(\boldsymbol{k})$ are replaced by $E_+(\boldsymbol{k})$ and $E_-(\boldsymbol{k})$ for the calculation of the conductivity, while $E_\pm(\boldsymbol{k})$ represents not only the Dirac cone but also full dispersion of $E_4(\boldsymbol{k})$ and $E_5(\boldsymbol{k})$ in the first Brillouin zone. The present energy bands $E_\pm(\boldsymbol{k})$ provide a nodal line, i.e., a loop of the Dirac point $\boldsymbol{k}_{\rm D}$, which is obtained from

$$E_+(\boldsymbol{k}_{\rm D}) = E_-(\boldsymbol{k}_{\rm D}) \ . \quad (5)$$

The chemical potential $\mu = \mu(T)$ is determined self-consistently in the clean limit from

$$\frac{1}{N}\sum_{\boldsymbol{k}}\sum_{j=1}^{8} f(E_j(\boldsymbol{k}) - \mu(T))$$
$$= \int_{-\infty}^{\infty} d\omega\ D(\omega) f(\omega - \mu) = 4\,, \tag{6}$$

where $f(\omega) = 1/(\exp[\omega/T] + 1)$ with T being temperature in the unit of eV and $k_B = 1$. Equation (6) is the condition of the half-filled band due to the HOMO and LUMO bands.

$D(\omega)$ denotes a density of states (DOS) per spin and per unit cell, which is given by

$$D(\omega) = \frac{1}{N}\sum_{\boldsymbol{k}}\sum_{\gamma=\pm} \delta(\omega - E_\gamma(\boldsymbol{k}))\,, \tag{7}$$

where $\int d\omega D(\omega) = 8$.

Figure 2a shows a nodal line, a loop of Dirac point $\boldsymbol{k}_D$, which is obtained from Equation (5). The Dirac point $\boldsymbol{k}_D$ is symmetric with respect to the Γ point, $\boldsymbol{k} = (0,0,0)$. There is a mirror symmetry with respect to $k_y = 0$, which gives two typical Dirac points given by (0, 0.096, 0)π (A), and (−0.42, 0, 0.64)π (B). The energy $E_\pm(\boldsymbol{k}_D)$ increases from 0.5052 to 0.5056 as the Dirac point moves from (A) to (B) and the chemical potential μ = 0.5053 is found on the line between (A) and (B) suggesting the nodal line semi-metal close to a zero-gap state.

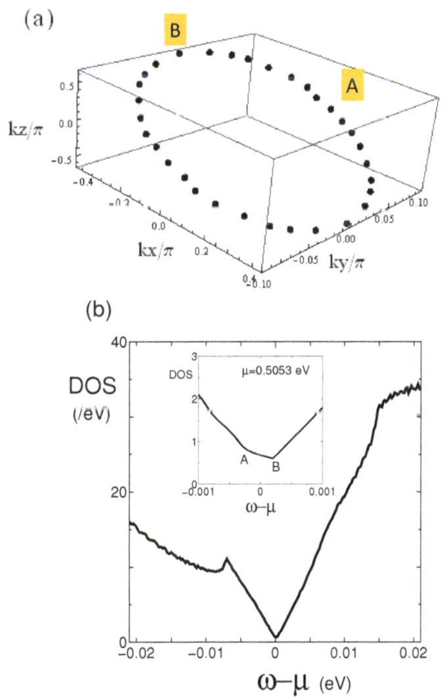

Figure 2. (Color online) (**a**) Nodal line in the three-dimensional momentum space, which connects Dirac points (symbols). (**b**) Density of states (DOS), as a function of $\tilde{\omega} = \omega - \mu$ with $\mu = \mu(0)$ = 0.5053 eV. The inset denotes the behavior close to $\tilde{\omega} = 0$, where A (B) corresponds to k_z/π = 0 (the maximum of $|k_z/\pi|$ in the loop.

Figure 2b shows the DOS, $D(\tilde{\omega})$, for $-0.02 < \tilde{\omega} < 0.02$, where the origin of ω is taken at the chemical potential $\mu = \mu(0)$, i.e., $\tilde{\omega} = \omega - \mu$. The peak for $\tilde{\omega} > 0$ corresponds to the maximum of HOMO while that of $\tilde{\omega} < 0$ also exists due to the minimum of LUMO, where both peaks are located at $\sim \pm 0.0025$ (not shown in the Figure). We note that there is a sufficient region of $\tilde{\omega}$, where $D(\tilde{\omega}) \propto |\tilde{\omega}|$. Such a behavior comes from two-dimensional character of the Dirac cone, which gives $D(\tilde{\omega}) = |\tilde{\omega}|/2\pi v^2$ with v being the averaged velocity of the cone along the nodal line. From the comparison, we obtain $v \sim 0.025$, which is about a half of α-(BEDT-TTF)$_2$I$_3$ [16]. The inset denotes $D(\omega)$ close to the chemical potential, which is not symmetric with respect to $\tilde{\omega} = 0$ and the minimum is located in the region of $\tilde{\omega} > 0$. Compared with the previous model [28] these behaviors are similar but $D(\mu)$ of the present case is much smaller suggesting a robust Dirac cone with the almost zero-gap state.

2.2. Total Hamiltonian

We consider a total Hamiltonian consisting of the electron and phonon systems:

$$H = H_0 + H_p + H_{e--p} + H_{imp} \,, \tag{8}$$

where the first three terms denote the Fröhlich Hamiltonian [37] applied to the Dirac electron system. The first term H_0 represents the energy band $E_\gamma(\boldsymbol{k})$ of single-component molecular conductor [Pd(dddt)$_2$] [10] and is given by Equation (1),

$$H_0 = \sum_{\boldsymbol{k}} \sum_{\gamma=\pm} E_\gamma(\boldsymbol{k}) a^\dagger_{\gamma,\boldsymbol{k}} a_{\gamma,\boldsymbol{k}} \,, \tag{9}$$

where $E_+(\boldsymbol{k})$ and $E_-(\boldsymbol{k})$ correspond to $E_4(\boldsymbol{k})$ and $E_5(\boldsymbol{k})$, respectively in Equation (3). The spin is ignored for simplicity. $a^\dagger_{\gamma,\boldsymbol{k}}$ is the creation operator of the electron with a wave vector $\boldsymbol{k}$ and the γ band. The second term H_p describes the acoustic phonon with a spectrum of $\omega_{\boldsymbol{q}} = v_s q$ and a creation of operator $b^\dagger_{\boldsymbol{q}}$, where $v \gg v_s$ with v being the velocity of the Dirac cone. The wave vector is defined by the Fourier transform on the square lattice with the lattice constant taken as unity. ve vector $\boldsymbol{k}$ of the γ band and acoustic phonon with a vector $\boldsymbol{q}$. γ = + (-) denotes a conduction (valence) band. The third term H_{e--p} represents an electron-phonon (e–p) interaction expressed as

$$H_{e--p} = \sum_{\boldsymbol{k},\gamma} \sum_{\boldsymbol{q}} \alpha_{\boldsymbol{q}} a^\dagger_{\gamma,\boldsymbol{k}+\boldsymbol{q}} a_{\gamma,\boldsymbol{k}} \phi_{\boldsymbol{q}} \,, \tag{10}$$

where $\phi_{\boldsymbol{q}} = b_{\boldsymbol{q}} + b^\dagger_{-\boldsymbol{q}}$. Later, we introduce a coupling constant $\lambda = |\alpha_{\boldsymbol{q}}|^2/\omega_{\boldsymbol{q}}$, which becomes independent of $|\boldsymbol{q}|$ for small $|\boldsymbol{q}|$. The scattering by the phonon is considered within the same band (i.e., intraband) due to the energy conservation with $v \gg v_s$. The fourth term H_{imp} denotes a normal impurity scattering, which is introduced to avoid the large conductivity in the presence of only the e–p interaction [38]. We take $k_B = \hbar = 1$.

2.3. Conductivity

The damping of the electron of the γ band, Γ_γ which is obtained from the Green function expressed as,

$$G_\gamma(\boldsymbol{k}, i\omega_n)^{-1} = i\omega_n - E_{\gamma,\boldsymbol{k}} + \mu + i\Gamma_\gamma \,, \tag{11a}$$

$$\Gamma_\gamma = \Gamma_0 + \Gamma^\gamma_{\text{ph}} \,, \tag{11b}$$

where $\Gamma^\gamma_{\text{ph}} = -\text{Im}\Sigma_\gamma(\boldsymbol{k}, E_{\gamma,\boldsymbol{k}} - \mu)$, and the real part can be neglected for small doping [33]. The quantity Γ_0, which is the damping by the impurity scattering, is taken as a parameter to scale the energy. Note that Γ_{ph} does not depend on Γ_0, and that the ratio $\Gamma^\gamma_{\text{ph}}/\Gamma_0$ is crucial to determine the T dependence of the conductivity. The quantity $\Sigma_\gamma(\boldsymbol{k},\omega) = \Sigma_\gamma(\boldsymbol{k}, i\omega_n)$ with $i\omega_n \to \omega + 0$ denotes a self-energy of the electron Green function estimated as [39]

$$\Sigma_\gamma(\boldsymbol{k}, i\omega_n) = T\sum_m \sum_q |\alpha_q|^2 \times \frac{1}{i\omega_{n+m} - \xi_{\gamma,\boldsymbol{k}+q}} \times \frac{2\omega_q}{\omega_m^2 + \omega_q^2}, \tag{12}$$

which is a product of electron and phonon Green functions. $\omega_n = (2n+1)\pi T$, $\omega_m = 2\pi m T$ with n and m being integers. $\xi_{\gamma,\boldsymbol{k}} = E_{\gamma,\boldsymbol{k}} - \mu$ where μ is a chemical potential.

Here we note that there is a following Dirac point $\boldsymbol{k}_{\rm D}$ on the nodal line. When $E_\pm(\boldsymbol{k})$ contributes to a linear dispersion of DOS in Figure 2b, the energy dispersion $E_+(\boldsymbol{k})$ and $E_-(\boldsymbol{k})$ are given by the Dirac cone as a function of $\boldsymbol{k} - \boldsymbol{k}_{\rm D}$, which gives a plane perpendicular to a tangent of the nodal line at $\boldsymbol{k}_{\rm D}$ [30]. In this case, the damping by the acoustic phonon scattering can be estimated based on the previous calculation of two-dimensional Dirac cone [33] which is given by $\Gamma_{\rm ph} \propto Tvk$ with $k = |\boldsymbol{k} - \boldsymbol{k}_{\rm D}|$. Such a result is extended to the present case by taking account of the variation of μ due to finite temperatures and the three-dimensional spectrum, $E_\pm(\boldsymbol{k})$. In fact, we obtain:

$$\Gamma^\gamma_{\rm ph} = CRT|\xi_{\gamma,\boldsymbol{k}}|, \tag{13a}$$

$$R = \frac{\lambda}{\lambda_0}, \tag{13b}$$

where $C = 12.5\ (\text{eV})^{-1}$ and $\xi_{\gamma,\boldsymbol{k}} = E_{\gamma,\boldsymbol{k}} - \mu$. The quantity λ denotes the coupling constant of the e–p interaction, where $\lambda = |\alpha_q|^2/\omega_q$ and is treated as a parameter. The quantity R denotes a normalization of λ by λ_0, where a typical value of $\lambda_0 = 0.03$ corresponding to a weak coupling is taken and is compared with the energy of the Dirac cone ~ 0.3.

Using $d_{\alpha\gamma}$ in Equation (4), the electric conductivity per spin and per unit cell is calculated. The conductivity is given by [31,34]

$$\sigma_\nu(T) = \int_{-\infty}^{\infty} d\omega \left(-\frac{\partial f(\omega)}{\partial \omega}\right) F_\nu(\omega), \tag{14}$$

$$F_\nu(\omega) = \frac{e^2}{\pi\hbar N}\sum_{\boldsymbol{k}}\sum_{\gamma=\pm}\sum_{\gamma'=\pm} \overline{v^\nu_{\gamma\gamma'}(\boldsymbol{k})} v^\nu_{\gamma'\gamma}(\boldsymbol{k})\Pi_{\gamma',\gamma},$$

$$\Pi_{\gamma',\gamma} = \frac{\Gamma_{\gamma'}}{(\omega - \xi_{\gamma',\boldsymbol{k}})^2 + \Gamma_{\gamma'}^2} \times \frac{\Gamma_\gamma}{(\omega - \xi_{\gamma,\boldsymbol{k}})^2 + \Gamma_\gamma^2}, \tag{15}$$

$$v^\nu_{\gamma\gamma'}(\boldsymbol{k}) = \sum_{\alpha\beta} \overline{d_{\alpha\gamma}(\boldsymbol{k})}\frac{\partial \tilde{H}_{\alpha\beta}}{\partial k_\nu} d_{\beta\gamma'}(\boldsymbol{k}), \tag{16}$$

where $\nu = x, y$, and z, and $h = 2\pi\hbar$. h and e denote a Plank's constant and electric charge, respectively. $\xi_{\boldsymbol{k}\gamma} = E_\gamma(\boldsymbol{k}) - \mu$ and μ denotes a chemical potential.

The previous calculation of the conductivity was performed by taking account of the impurity scattering of the damping only Γ_0 [35], while the damping by phonon scattering $\Gamma_{\rm ph}$ [33] is added to the new TB model [36] in the present calculation. The total number of the lattice site is given by $N = N_x N_y N_z$, where $N_x N_y$ is the number of the intralayer sites and N_z is the number of the layer. Note that the calculation of Equation (14) with the summation of k_z in the end, i.e., a two-dimensional conductivity for the fixed k_z is utilized to calculate the nodal line semimetal as shown previously [34]. It is noted that Equation (14) can be understood using DOS when the intraband contribution ($\gamma = \gamma'$) is dominant and $\boldsymbol{k}$ dependence of $v^\nu_{\gamma'\gamma}$ is small.

3. Electric Transport

The conductor of [Pd(dddt)$_2$] under pressure exhibits a Dirac electron system in three-dimension. The purpose of this section is to explain the almost temperature independent conductivity of such a system by taking account of the acoustic phonon scattering. The conductivity Equation (14) per spin is calculated, which is normalized by $e^2/\hbar$, i.e., $e^2/\hbar \to 1$. The unit of energy is taken as eV,

We examine the temperature T dependence of the conductivity σ_ν (ν = x, y, and z), which is determined by the combined effect of impurity and phonon scatterings. Both scattering decrease σ_ν but the combined effect is complicated at finite T. With increasing temperature, σ_ν increases due to the DOS of the Dirac cone but σ_ν is suppressed by the phonon scattering, which increases with increasing T. We first show the T dependence of σ_ν in the presence of only impurity scattering, i.e., R = 0. Next, T dependence of σ_ν in the presence of both impurity and phonon scatterings are shown in detail by varying parameters Γ_0 and R to compare with the T dependence found in the experiment. The numerical calculation of the conductivity is performed by choosing Γ_0 = 0.0005 and 0.0003, which are smaller than the previous one [35]. Hereafter, we use $\Gamma_0' = 10^4\Gamma_0$.

3.1. Conductivity by Impurity Scattering

In Figure 3, the anisotropic conductivity with only impurity scattering is shown by the solid line. Compared with the result obtained by the previous TB model [35], a noticeable result is an existence of T-linear dependence of σ_y, which is seen in a wide region of temperature due to the robust Dirac cone. As seen from Figure 1 and Table 1, the large magnitude of σ_y is obtained since the molecules are stacked along the y-axis. Further $\sigma_y(0)$ is smaller than the previous one due to smaller $D(0)$. Such a linear increase is also found for both $\sigma_x(T)$ and $\sigma_z(T)$, in which the linear increase begins to deviate at lower temperatures. There is a large anisotropy, where σ_y is much larger than σ_x and σ_z. We note that the anisotropy of σ_ν comes mainly from that of the velocity. In fact, v_x = 0.0094, v_y = 0.058, and v_z = 0.0072, where $v_\nu = \sqrt{< v_\nu^2 >}$ and $< \cdots >$ denotes an average by summation of γ, γ' and $\boldsymbol{k}$ of Equation (16). Note that the velocity obtained from DOS in Figure 2b is an average of these velocities. However the ratio of σ_ν is different from that of velocities and is more complicated, since Equation (14) is calculated as a combined effect of the velocity and a quantity associated with the electron Green function. The inset denotes the T dependence of the chemical potential $\mu(T)$, which decreases monotonously, suggesting the increase of hole close to the nodal line. A fitting formula of μ is given by $\mu - \mu(0) \simeq -0.45T^2/(T + 0.003)$. The dot-dashed line in the main figure denotes $\sigma_y(T)$, which is calculated with the fixed $\mu = \mu(0)$. The difference between the solid and dot-dashed lines is negligibly small even at finite temperature. However such a difference becomes noticeable for a moderate strength of the e–p interaction as shown in the next subsection.

3.2. Effect of Phonon Scattering on the Conductivity

Since σ_y is the largest one and the direction is the same as that of measured one [26] we examine $\sigma_y(T)$ in detail with some choices of R. Before the numerical result, we examine σ_y semi-analytically.

We show $\sigma_y(T)$ can be written as,

$$\sigma_y(T) \simeq a + \frac{X}{1 + bX^2/(1 + cX)}, \tag{17}$$

where $X = T/T_0$. Parameters T_0, a, b, and c in Equation (17) are determined by fitting σ_y in Figures 4 and 5. Equation (17), which describes well for both low and high temperatures, is obtained as follows. Noting that σ_y with only impurity scattering is written as $\sigma_y = T/\Gamma$ with $\Gamma \propto \Gamma_0$ [31].

In the presence of $\Gamma^{\gamma}_{\rm ph} = \Gamma_{\rm ph}$, σ_y may be written in the same form, where Γ_0 is replaced by $\Gamma = \Gamma_\gamma = \Gamma_0 + \Gamma_{\rm ph}$, i.e.,

$$\begin{aligned} \sigma_y &= a + a' \frac{T}{<\Gamma>} \\ &= a + a' \frac{T}{\Gamma_0 + <\Gamma_{\rm ph}>} \\ &= a + \frac{T/T_0}{1 + <\Gamma_{\rm ph}>/\Gamma_0}, \end{aligned} \tag{18a}$$

where $T_0 = \Gamma_0/a'$. Using Equation (13a) and a relation,

$$<|\xi|> \;=\; <|\xi_{\gamma,\boldsymbol{k}}|> = \frac{b'T}{1 + c(T/T_0)}, \tag{18b}$$

which is shown numerically in the inset of Figure 4, $<\Gamma_{\rm ph}>$ may be written as

$$\begin{aligned} \frac{<\Gamma_{\rm ph}>}{\Gamma_0} &= \frac{CRT <|\xi|>}{\Gamma_0} \\ &= \frac{b(T/T_0)^2}{1 + c(T/T_0)}, \end{aligned} \tag{18c}$$

where $b = CRb'T_0^2/\Gamma_0$. Thus Equation (17) is derived from Equations (18a)–(18c). As shown later, Equation (18b) can be verified by calculating a quantity

$$<\Gamma_{\rm ph}> = CRT <|\xi_{\gamma,\boldsymbol{k}}|> = \frac{F}{\sigma_y}, \tag{19}$$

where F is calculated from Equation (14) with $\Pi_{\gamma',\gamma}$ replaced by $\Gamma_{\rm ph}\Pi_{\gamma',\gamma}$ in Equation (15). Since $<|\xi_{\gamma,\boldsymbol{k}}|>$ denotes an average of energy close to the nodal line , we may write $<|\xi_{\gamma,\boldsymbol{k}}|> \sim T$, and $<\Gamma_{\rm ph}> \sim CRT <|\xi_{\gamma,\boldsymbol{k}}|> \sim CRT^2$. This gives a simpler expression compared with Equation (17), i.e., $\sigma_y \simeq a + (T/T_0)/(1 + b''(T/T_0)^2)$, with $b'' \sim CRT_0^2/\Gamma_0$.

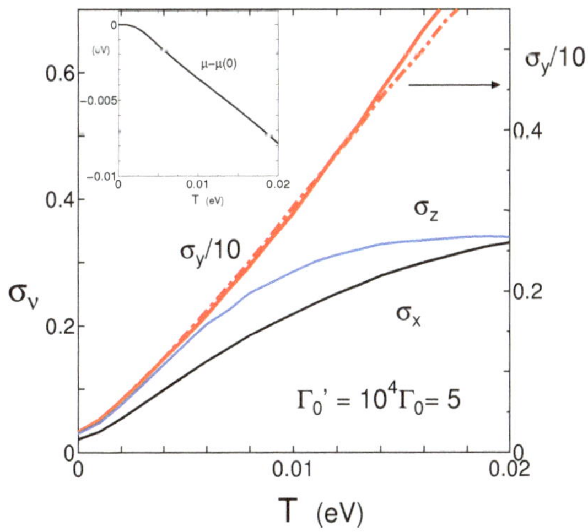

Figure 3. (Color online) Temperature dependence of conductivity of σ_ν ($\nu = x, y$, and z) with $\Gamma_0' = 10^4\Gamma_0 = 5$, where R = 0. The dot-dashed line denotes σ_y, which is calculated with the fixed $\mu = \mu(0)$ = 0.5053 eV. The inset denotes temperature dependence of $\mu(T)$.

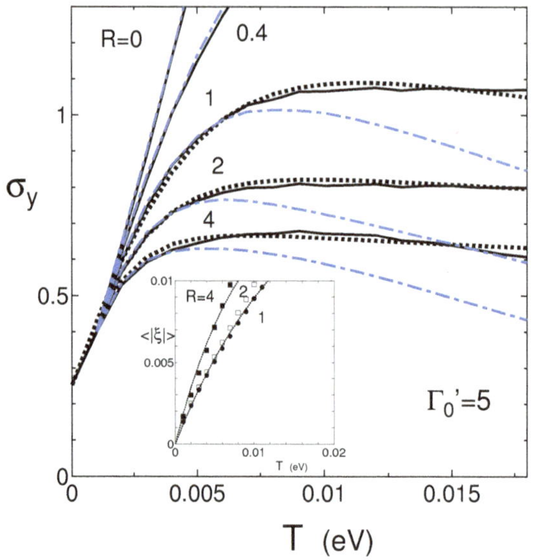

Figure 4. (Color online) T dependence of σ_y with $\Gamma_0' = 5$ for R = 0, 0.4, 1, 2, and 4 (solid line). The dot line denotes a fitting one obtained from Equation (17). The dot-dashed line denotes σ_y obtained for the fixed $\mu = \mu(0)$, where the deviation from the solid line starts at lower T for the larger R. The inset denotes T dependence of $< |\xi| >$ obtained from Equation (19) with a fitted line for R = 0.5, 1, and 4.

Now we examine σ_y using Equation (17) as a fitting line. Figure 4 shows T dependence of σ_y (solid line) for $\Gamma_0' = \Gamma_0 \times 10^4$ = 5 with the fixed R= 0, 0.4, 1, 2, and 4 (solid line). It is found that $\sigma_y(T)$ is reduced and takes a maximum in the presence of R, while $\sigma_y(T)$ with R = 0 increases monotonously. It is the main purpose of the present work to demonstrate that $\sigma_y(T)$ for R = 1 and 2 exhibit almost T independent behavior in a wide temperature region of $0.01 < T < 0.02$. This exotic T dependence exists for some choices of Γ_0' and R, as shown later. From the dot line corresponding to Equation (17), we obtain T_0 = 0.0055, a = 0.25, and (b, c) = (0.5,0.22), (1.39,0.6), and (2.6,0.81) for R = 1, 2, and 4 respectively. It is verified that b increases and σ_y decreases with increasing R. Note that the almost temperature indepednent σ_y comes from the existence of the c term, which weakens the effect of the b term. This corresponds to the suppression of $< |\xi_{\gamma,\boldsymbol{k}}| > /T$ as shown in the inset. In Figure 4, σ_y with the fixed $\mu(0)$ is also shown by the dot-dashed line. At high temperatures, σ_y is reduced noticeably due to the deviation of μ from the energy of the Dirac point. For the solid line, the T dependence of $< \Gamma_{\mathrm{ph}}^{\gamma} >$ in Equation (13a) is small due to $< E_\gamma(\boldsymbol{k}) > \sim \mu$, which comes from the electron-hole excitation in the conductivity. However, for the dot-dashed line, $< |E_\gamma(\boldsymbol{k}) - \mu(T)| >$ becomes large due to the fixed $\mu(0)$, resulting in the large suppression of σ_y, where the maximum is broadened. With increasing R, $T_{\max}$ corresponding to a maximum of σ_y decreases, while the reduction at high temperatures is almost parallel. The inset denotes $< |\xi| >$ obtained from Equation (17), for R = 0.5, 1, and 2, where the symbols are fitted by the dot lines given by $< |\xi| > \propto T/(1 + c'(T/T_0))$ with $c' = 0.4(0.2)$ for R = 4 (1), which corresponds well to that of the fitting of σ_y.

Figure 5 shows T dependence of σ_y for smaller $\Gamma_0' = 3$ with the fixed R = 0, 0.12, 0.24, 0.6, 1.2 and 2.4 (solid line). The dot line denotes a fitting one obtained from Equation (17), where parameters are estimated as T_0 = 0.0053, a = 0.27, and (b, c) = (0.15,0.06) (0.58,0.50), and (1.9,1.4) for R = 0.6, 1.2, and 2.4, respectively. Compared with Figure 4, we find as follows. A ratio of T_0 to Γ_0 increases with decreasing Γ_0, e.g., $T_0/\Gamma_0 \simeq 10$ and 16 for Γ_0' = 5 and 3, respectively. Although σ_y still increases, it is expected that the almost constant behavior occurs at higher temperature due to the increase of the phonon scattering. There is a quantitative difference between R = 0.12 and R = 0.36. With decreasing

Γ_0, σ_y increases followed by the increase of the crossover temperature from T linear behavior to that being almost constant. In the inset, $<|\xi|>$ obtained from Equation (18b) is shown by the symbols, where the dot line denotes the fitting line given by $<|\xi|>\propto T/(1+c'(T/T_0))$ with $c' = 0.17(0.18)$ for R = 2.4 (0.6).

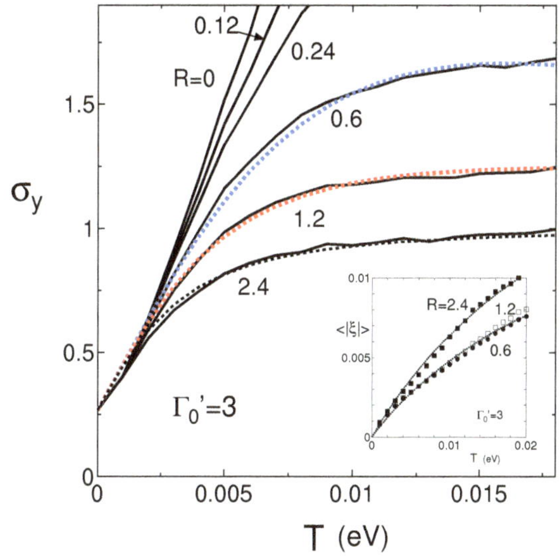

Figure 5. (Color online) T dependence of σ_y with $\Gamma_0' = 3$ for R = 0, 0.12, 0.24, 0.6, 1.2, and 2.4 (solid line). The dot line denotes a fitting one obtained from Equation (17). The symbols in the inset denotes $<|\xi|>$ obtained from Equation (18b), which are fitted by the dot line.

Figure 6 shows σ_y for some choices of Γ_0' with the fixed R = 2, where the lines with Γ_0' = 3 and 5 are the same as those in Figures 4 and 5. With increasing R, the T dependence changes from the monotonous increase to the constant behavior and a maximum is followed for $0 < T < 0.02$. This suggests the constant behavior is obtained for the moderate strength of Γ_0' in contrast to Figures 4 and 5, which are examined for the fixed Γ_0'. The fitting by Equation (17) is also valid, since Γ_0' = 3 and 5 are already shown in Figures 4 and 5. The line with Γ_0' = 10 shows a noticeable decrease of σ_y suggesting the dominant effect by the large e–p scattering at high temperatures. From Equation (17), it turns out that b suppresses σ_y but c enhances σ_y. The role of c is crucial to obtain the almost constant conductivity, since a relation $bX^2/(1+cX) \propto T$ at high T gives almost T independent σ_y in Equation (17). We found the almost T independent σ_y. for (Γ_0', R) = (4, 0.8), (4, 1.2), (4, 2). (5, 0.5), (5, 1), (5, 2), (6, 1), (6, 1.2), and (6, 2). It turns out that the almost T independent σ_y in the interval region of $0 < T < 0.02$ is realized for $4 < \Gamma_0' < 6$ and $1 < R < 2$.

Here we discuss the result of almost temperature independent conductivity in Dirac electrons, which is obtained by taking account of both impurity scattering (Γ_0) and the acoustic phonon scattering (Γ_{ph}). In the presence of only Γ_0, the conductivity increases linearly with increasing T due to the increase of the relevant DOS being linear in ω (Figure 2b). Such behavior is found in both the 2D Dirac cone at zero doping and the half-filled [Pd(dddt)$_2$] with a nodal line, where the chemical potential is located close to the nodal line and the energy variation along the nodal line is very small as seen from the inset of Figure 2b. However such an increase of the conductivity is suppressed by a moderate strength of the acoustic phonon scattering. The conductivity of the 2D Dirac cone shows a broad maximum as a function of T, while that of the nodal line further exhibits the wide temperature region followed by the almost temperature independent conductivity. Such a result is obtained by treating the damping of phonon scattering as $\Gamma_{ph} \propto T|E_{\gamma,\boldsymbol{k}} - \mu|$ (Equation (13a)), with $\mu \simeq -0.45T$ (the inset of

Figure 3) due to asymmetry of DOS, where a peak around $\omega - \mu \simeq 0.03$ is much larger than that of $\omega - \mu \simeq -0.03$. The suppression of the conductivity is understood by $< |E_{\gamma,k} - \mu| > \simeq T$ in the inset of Figure 4, which results in $\Gamma_{\rm ph}/\Gamma_0 \propto T^2$ due to the e–p excitation in the conductivity Equation (14). However Equation (18c) suggests a constant conductivity by the c-term in the denominator. The wide temperature region for the relevant c-term could be attributable to a nodal line with a half-filled band.

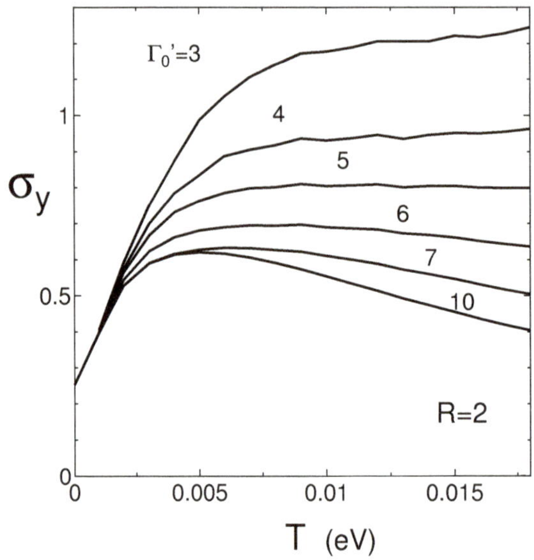

Figure 6. T dependence of σ_y, for Γ_0' = 3, 4, 5, 6, 7, and 10 with fixed R = 2.

3.3. Comparison with Experiment

Finally, in Figure 7, we compare the present theoretical results with those of the experiment [10,40]. The conductivity of a single crystal under quasi hydrostatic pressure was measured by the four-probe method using the DAC (Diamond Anvil Cell) technique. For the convenience of comparison, we show the resistivity $\rho_y (= 1/\sigma_y)$. Since the comparison of the absolute value is complicated both theoretically and experimentally, we use a simple method that the quantity is normalized at a temperature of T = 0.015. In Figure 7, the experimental result is shown by $\rho_y^{\rm I}$ and $\rho_y^{\rm II}$ corresponding to P = 12.6 GPa, and P = 13 GPa, respectively, which increase about 1.8 and 3.4 $\times \rho$ (0.015), respectively, with decreasing temperature. The results of the present calculation are shown for (Γ_0', R) = (4,1.6), (3,0.6), and (3,0.12). The line of (3,0.6) reproduces $\rho_y^{\rm I}$ well for $0.0008 < T$, and the line of (4,1.6) shows almost constant resistivity for $0.0007 < T < 0.015$. The rapid increase of $\rho_y^{\rm II}$ is compared with that of (3,0.12) for $0.01 < T < 0.015$, where the weaker e–p coupling is taken. Such behavior could be expected for higher pressures due to the increase of the velocity. The larger damping $\Gamma_0'(> 7)$ (not shown in the Figure) gives a minimum for a moderate choice of R with increasing T. Such a minimum, which could be relevant to that of α-(BEDT-TTF)$_2$I$_3$ [32] is obtained for a slightly large Γ_0'.

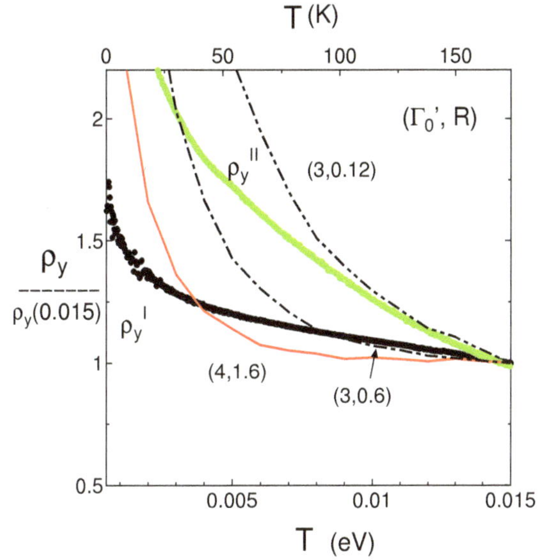

Figure 7. (Color online) T dependence of $\rho_y (= 1/\sigma_y)$ normalized by the value at T = 0.015 eV (174 K) in the interval range of 0.0005 eV $< T <$ 0.015 eV. The quantities ρ^{I} and ρ^{II} denote the normalized resistivity at P = 12.6 GPa [10] and 13 GPa [40] respectively, which are compared with those of $(\Gamma_0', R) = (4, 1.6), (3, 0.6)$, and $(3, 0.12)$. The line (4, 1.6) stays almost constant at high temperature.

4. Summary and Discussion

Using a TB model, which has been recently obtained, we examined the temperature dependence of the conductivity σ_y of [Pd(dddt)$_2$] with a nodal line semimetal. The main finding is the almost constant resistivity obtained by the acoustic phonon scattering, which plays an exotic role in the presence of the Dirac cone and nearly the half-filled band. The DOS shows linear behavior in the sufficient energy region, which suggests a robust Dirac cone even for the nodal line.

The conductivity σ_y is the largest one since the transfer energy is large along the stacking direction of y axis. The phonon scattering, which occurs in the half-filled Dirac electron system, has an effect of reducing the linear increase of the conductivity and shows almost temperature independent behavior at high temperatures. With increasing temperature, the conductivity increases linearly in the presence of only the impurity scattering, but the presence of acoustic phonon gives the strong suppression of the conductivity. The mechanism of damping by the phonon scattering is noticeable. The damping by phonon, i.e., the imaginary part of the self-energy is proportional to $T\,|\xi_{\gamma,\mathbf{k}}|$ with $\xi_{\gamma,\mathbf{k}}$ is the energy of the electron measured from the chemical potential.

This is in contrast to the conventional case with a Fermi surface corresponding to a large doping. For $\mu(0) \gg T, \sigma(0)$ becomes large but $\sigma(T)$ as a function of T decreases rapidly [33] due to the large effect of the phonon scattering with a damping being proportional to T [41].

Thus it is concluded that the present conductivity being almost temperature independent is an evidence of the Dirac electron system with a half-filled band.

Finally, we comment on the validity of the present TB model. In the previous work [10,28], transfer energies of the TB model have been estimated for both 0 GPa and 8 GPa (corresponding to 12.6 GPa of the experiment), where the former (latter) provides the insulating gap (Dirac points) in accordance with the experiment. Thus, the present TB model may be justified at lower pressures, but the state at higher pressure given by P = 13 GPa in Figure 7 shows a rapid increse of the normalized ρ_y^{II} being opposite to the conventional pressure-resistance relation suggesting a possible modification of the model. Actually, it is known that higher pressure induces disorder in the molecular structure [40].

Author Contributions: Investigation, Y.S., R.K. and M.O. All authors have read and agreed to the published version of the manuscript.

Funding: This work was supported by JSPS KAKENHI Grant Number JP16H06346.

Acknowledgments: One of the authors (Y.S.) thanks T. Tsumuraya for useful discussions on the nodal line semimetal.

Conflicts of Interest: The authors declare no conflict of interest.

Appendix A. TB Model

As shown in Table A1, the Hamiltonian is divided into the 4 × 4 matrix, $h_{\rm H,H}$, $h_{\rm H,L}$ (and $h_{\rm L,H}$), $h_{\rm L,L}$ corresponding to H-H, H-L and L-L components. In terms of $X = {\rm e}^{ik_x}$, $\bar{X} = {\rm e}^{-ik_x}$, $Y = {\rm e}^{ik_y}$, $\bar{Y} = {\rm e}^{-ik_y}$, and $Z = {\rm e}^{ik_z}$, $\bar{Z} = {\rm e}^{-ik_z}$, matrix elements for HOMO-HOMO (H-H) are evaluated [36]. Here we show the real matrix Hamiltonian $\tilde{H}$ from $\tilde{H} = U\hat{H}U^{-1}$, where a 8 × 8 matrix U has only the diagonal element given by $(U)_{\rm H1,H1}, \cdots, (U)_{\rm L4,L4} = (-i, -i(XYZ)^{1/2}, -i(XY)^{1/2}, -i\bar{Z}^{1/2}$ $1, (XYZ)^{1/2}, (XY)^{1/2}, \bar{Z}^{1/2})$ Using $\tilde{x} = k_x/2, \tilde{y} = k_y/2$, and $\tilde{z} = k_z/2$, Matrix elements $(\tilde{H})_{\alpha,\beta}/2 = \tilde{t}_{\alpha,\beta}$ are calculated as

$$\begin{aligned}
\tilde{t}_{\rm H1,H1} &= \tilde{t}_{\rm H3,H3} = b1_{\rm H} \sin 2\tilde{y}\,, \\
\tilde{t}_{\rm H1,H2} &= a1_{\rm H} \cos(\tilde{x} - \tilde{y} + \tilde{z}) + a2_{\rm H} \cos(\tilde{x} + \tilde{y} + \tilde{z})\,, \\
\tilde{t}_{\rm H1,H3} &= 2p_{\rm H} \cos \tilde{x} \cos \tilde{y}\,, \\
\tilde{t}_{\rm H1,H4} &= c1_{\rm H} \cos \tilde{z} + c2_{\rm H} \cos(2\tilde{y} + \tilde{z})\,, \\
\tilde{t}_{\rm H2,H2} &= \tilde{t}_{\rm H4,H4} = b2_{\rm H} \cos 2\tilde{y}\,, \\
\tilde{t}_{\rm H2,H3} &= c1_{\rm H} \cos \tilde{z} + c2_{\rm H} \cos(2\tilde{y} - \tilde{z})\,, \\
\tilde{t}_{\rm H2,H4} &= 2q_{\rm H} \cos \tilde{x} \cos \tilde{y}\,, \\
\tilde{t}_{\rm H3,H4} &= a1_{\rm H} \cos(\tilde{x} + \tilde{y} + \tilde{z}) + a2_{\rm H} \cos(\tilde{x} - \tilde{y} + \tilde{z})\,,
\end{aligned}$$

for HOMO-HOMO elements,

$$\begin{aligned}
\tilde{t}_{\rm H1,L1} &= h_{\rm H3,L3} = -b1_{\rm HL} \sin \tilde{y}\,, \\
\tilde{t}_{\rm H1,L2} &= -a1_{\rm HL} \sin(\tilde{x} - \tilde{y} + \tilde{z}) - a2_{\rm HL} \sin(\tilde{x} + \tilde{y} + z))\,, \\
\tilde{t}_{\rm H1,L3} &= -p1_{\rm HL} \sin(\tilde{x} + \tilde{y}) + p2_{\rm HL} \sin(-\tilde{x} + \tilde{y})\,, \\
\tilde{t}_{\rm H1,L4} &= c1_{\rm HL} \sin \tilde{z} + c2_{\rm HL} \sin(2\tilde{y} + \tilde{z})\,, \\
\tilde{t}_{\rm H2,L1} &= a1_{\rm HL} \sin(-\tilde{x} + \tilde{y} - \tilde{z}) + a2_{\rm HL} \sin(\tilde{x} + \tilde{y} + \tilde{z})\,, \\
\tilde{t}_{\rm H2,L2} &= h_{\rm H4,L4} = -b2_{\rm HL} \sin(2\tilde{y})\,, \\
\tilde{t}_{\rm H2,L3} &= -c1_{\rm HL} \sin(\tilde{z}) + c2_{\rm HL} \sin(-2\tilde{y} + \tilde{z})\,, \\
\tilde{t}_{\rm H2,L4} &= -q1_{\rm HL} \sin(\tilde{x} + \tilde{y}) - q2_{\rm HL} \sin(\tilde{x} - \tilde{y})\,, \\
\tilde{t}_{\rm H3,L1} &= -p1_{\rm HL} \sin(\tilde{x} - \tilde{y}) - p2_{\rm HL} \sin(\tilde{x} + \tilde{y})\,, \\
\tilde{t}_{\rm H3,L2} &= -c1_{\rm HL} \sin(\tilde{z}) + c2_{\rm HL} \sin(2\tilde{y} - \tilde{z})\,, \\
t_{\rm H3,L4} &= a1_{\rm HL} \sin(\tilde{x} + \tilde{y} + \tilde{z}) + a2_{\rm HL} \sin(\tilde{x} - \tilde{y} + \tilde{z})\,, \\
\tilde{t}_{\rm H4,L1} &= -c1_{\rm HL} \sin(\tilde{z})) + c2_{\rm HL} \sin(2\tilde{y} + \tilde{z})\,, \\
\tilde{t}_{\rm H4,L2} &= q1_{\rm HL} \sin(\tilde{x} - \tilde{y}) + q2_{\rm HL} \sin(\tilde{x} + \tilde{y})\,, \\
\tilde{t}_{\rm H4,L3} &= -a1_{\rm HL} \sin(\tilde{x} + \tilde{y} + \tilde{z}) + a2_{\rm HL} \sin(\tilde{x} - \tilde{y} + \tilde{z}))\,,
\end{aligned}$$

for HOMO-LUMO elements,

$$\begin{aligned}
\tilde{t}_{\mathrm{L1,L1}} &= h_{\mathrm{L3,L3}} = \Delta E/2 + b1_{\mathrm{L}}\cos(2\tilde{y})\,, \\
\tilde{t}_{\mathrm{L1,L2}} &= a1_{\mathrm{L}}\cos(\tilde{x}-\tilde{y}+\tilde{z}) + a2_{\mathrm{L}}\cos(\tilde{x}+\tilde{y}+\tilde{z})\,, \\
\tilde{t}_{\mathrm{L1,L3}} &= 2p_{\mathrm{L}}\cos\tilde{x}\cos\tilde{y}\,, \\
\tilde{t}_{\mathrm{L1,L4}} &= c1_{\mathrm{L}}\cos(\tilde{z}) + c2_{\mathrm{L}}\cos(2\tilde{y}+\tilde{z})\,, \\
\tilde{t}_{\mathrm{L2,L2}} &= h_{\mathrm{L4,L4}} = \Delta E/2 + b2_{\mathrm{L}}\cos(2\tilde{y})\,, \\
\tilde{t}_{\mathrm{L2,L3}} &= -c1_{\mathrm{L}}\cos(\tilde{z}) - c2_{\mathrm{L}}\cos(2\tilde{y}-\tilde{z})\,, \\
\tilde{t}_{\mathrm{L2,L4}} &= 2q_{\mathrm{L}}\cos\tilde{x}\cos\tilde{y}\,, \\
\tilde{t}_{\mathrm{L3,L4}} &= -a1_{\mathrm{L}}\cos(\tilde{x}+\tilde{y}+\tilde{z}) - a2_{\mathrm{L}}\cos(\tilde{x}-\tilde{y}+z)\,,
\end{aligned}$$

for LUMO-LUMO elements, and $h_{\alpha,\beta} = h_{\beta,\alpha}$. Note that matrix elements of HOMO-HOMO and LUMO-LUMO are the even function of $\boldsymbol{k}$ while those of of HOMO-LUMO is the odd function due to the difference in the symmetry of HOMO and LUMO around the Pd atom. These parities are used for constructing 2×2 effective Hamiltonian, where the diagonal element show the even function and the off-diagonal element shows the odd function with respect to $\boldsymbol{k} \rightarrow -\boldsymbol{k}$. These relations provide ingredients for obtaining a nodal line and Dirac points.

Table A1. Matrix elements of Hamiltonian: $h_{\beta,\alpha} = \overline{h_{\alpha,\beta}}$.

$h_{\mathrm{H,H}}$	H1	H2	H3	H4
H1	$h_{\mathrm{H1,H1}}$	$h_{\mathrm{H1,H2}}$	$h_{\mathrm{H1,H3}}$	$h_{\mathrm{H1,H4}}$
H2	$h_{\mathrm{H2,H1}}$	$h_{\mathrm{H2,H2}}$	$h_{\mathrm{H2,H3}}$	$h_{\mathrm{H2,H4}}$
H3	$h_{\mathrm{H3,H1}}$	$h_{\mathrm{H3,H2}}$	$h_{\mathrm{H3,H3}}$	$h_{\mathrm{H3,H4}}$
H4	$h_{\mathrm{H4,H1}}$	$h_{\mathrm{H4,H2}}$	$h_{\mathrm{H4,H3}}$	$h_{\mathrm{H4,H4}}$
$h_{\mathrm{H,L}}$	**L1**	**L2**	**L3**	**L4**
H1	$h_{\mathrm{H1,L1}}$	$h_{\mathrm{H1,L2}}$	$h_{\mathrm{H1,L3}}$	$h_{\mathrm{H1,L4}}$
H2	$h_{\mathrm{H2,L1}}$	$h_{\mathrm{H2,L2}}$	$h_{\mathrm{H2,L3}}$	$h_{\mathrm{H2,L4}}$
H3	$h_{\mathrm{H3,L1}}$	$h_{\mathrm{H3,L2}}$	$h_{\mathrm{H3,L3}}$	$h_{\mathrm{H3,L4}}$
H4	$h_{\mathrm{H4,L1}}$	$h_{\mathrm{H4,L2}}$	$h_{\mathrm{H4,L3}}$	$h_{\mathrm{H4,L4}}$
$h_{\mathrm{L,L}}$	**L1**	**L2**	**L3**	**L4**
L1	$h_{\mathrm{L1,L1}}$	$h_{\mathrm{L1,L2}}$	$h_{\mathrm{L1,L3}}$	$h_{\mathrm{L1,L4}}$
L2	$h_{\mathrm{L2,L1}}$	$h_{\mathrm{L2,L2}}$	$h_{\mathrm{L2,L3}}$	$h_{\mathrm{L2,L4}}$
L3	$h_{\mathrm{L3,L1}}$	$h_{\mathrm{L3,L2}}$	$h_{\mathrm{L3,L3}}$	$h_{\mathrm{L3,L4}}$
L4	$h_{\mathrm{L4,L1}}$	$h_{\mathrm{L4,L2}}$	$h_{\mathrm{L4,L3}}$	$h_{\mathrm{L4,L4}}$

References

1. Seo, H.; Hotta, C.; Fukuyama, H. Toward Systematic Understanding of Diversity of Electronic Properties in Low-Dimensional Molecular Solids. *Chem. Rev.* **2004**, *104*, 5005–5035. [CrossRef] [PubMed]
2. Herring, C. Accidental Degeneracy in the Energy Bands of Crystal. *Phys. Rev.* **1937**, *52*, 365–373. [CrossRef]
3. Fu, L.; Kane, C.L. Topological insulators with inversion symmetry. *Phys. Rev. B* **2007**, *76*, 045302. [CrossRef]
4. Novoselov, K.S.; Geim, A.K.; Morozov, S.V.; Jiang, D.; Katsnelson, M.I.; Grigorieva, I.V.; Dubonos, S.V.; Firsov, A.A. Two-dimensional gas of massless Dirac fermions in graphene. *Nature* **2005**, *438*, 197–200. [CrossRef] [PubMed]
5. Katayama, S; Kobayashi, A.; Suzumura, Y. Pressure-Induced Zero-Gap Semiconducting State in Organic Conductor α-(BEDT-TTF)$_2$I$_3$. *J. Phys. Soc. Jpn.* **2006**, *75*, 054705. [CrossRef]
6. Kajita, K.; Nishio, Y.; Tajima, N.; Suzumura, Y.; Kobayashi, A. Molecular Dirac Fermion Systems—Theoretical and Experimental Approaches. *J. Phys. Soc. Jpn.* **2014**, *83*, 072002. [CrossRef]
7. Murakami, S. Phase transition between the quantum spin Hall and insulator phases in 3D: Emergence of a topological gapless phase. *N. J. Phys.* **2007**, *9*, 356. [CrossRef]

8. Hirayama, M.; Okugawa, R.; Murakami, S. Topological Semimetals Studied by Ab Initio Calculations. *J. Phys. Soc. Jpn.* **2018**, *87*, 041002. [CrossRef]
9. Bernevig, A.; Weng, H.; Fang, Z.; Dai, X. Recent Progress in the Study of Topological Semimetals. *J. Phys. Soc. Jpn.* **2018**, *87*, 041001. [CrossRef]
10. Kato, R.; Cui, H.B.; Tsumuraya, T.; Miyazaki, T.; Suzumura, Y. Emergence of the Dirac Electron System in a Single-Component Molecular Conductor under High Pressure. *J. Am. Chem. Soc.* **2017**, *139*, 1770–1773. [CrossRef]
11. Kondo, R.; Kagoshima, S.; Harada, J. Crystal structure analysis under uniaxial strain at low temperature using a unique design of four-axis X-ray diffractometer with a fixed sample. *Rev. Sci. Instrum.* **2005**, *76*, 093902. [CrossRef]
12. Mori, T.; Kobayashi, A.; Sasaki, Y.; Koybayashi, H.; Saito, G.; Inokuchi, H. The Intermolecular Interaction of Tetrathiafulvalene and Bis(ethylenedithio)-tetrathiafulvalene in Organic Metals. Calculation of Orbital Overlaps and Model of Energy-band Structures. *Bull. Chem. Soc. Jpn.* **1984**, *57*, 627. [CrossRef]
13. Kino, H.; Fukuyama, H. Phase Diagram of Two-Dimensional Organic Conductors: (BEDT-TTF)2X. *J. Phys. Soc. Jpn.* **1996**, *65*, 2158–2169. [CrossRef]
14. Kobayashi, A.; Suzumura, Y.; Fukuyama, H. Hall Effect and Orbital Diamagnetism in Zerogap State of Molecular Conductor α-(BEDT-TTF)$_2$I$_3$. *J. Phys. Soc. Jpn.* **2008**, *77*, 064718. [CrossRef]
15. Tajima, N.; Kato, R.; Sugawara, S.; Nishio, Y.; Kajita, K. Interband effects of magnetic field on Hall conductivity in the multilayered massless Dirac fermion system α-(BEDT-TTF)$_2$I$_3$. *Phys. Rev. B* **2012**, *85*, 033401. [CrossRef]
16. Katayama, S.; Kobayashi, A.; Suzumura, Y. Electronic properties close to Dirac cone in two-dimensional organic conductor α-(BEDT-TTF)$_2$I$_3$. *Eur. Phys. J. B* **2009**, *67*, 139–148. [CrossRef]
17. Takano, Y.; Hiraki, K.; Takada, Y.; Yamamoto, H.M.; Takahashi, T. Local Spin Susceptibility Characteristic of Zero-Gap State of α-(BEDT-TTF)$_2$I$_3$ under Pressure. *J. Phys. Soc. Jpn.* **2010**, *79*, 104704. [CrossRef]
18. Hirata, M.; Ishikawa1, K.; Miyagawa1, K.; Tamura, M.; Berthier, C.; Basko, D.; Kobayashi, A.; Matsuno, G.; Kanoda, K. Observation of an anisotropic Dirac cone reshaping and ferrimagnetic spin polarization in an organic conductor. *Nat. Commun.* **2016**, *7*, 12666. [CrossRef]
19. Suzumura, Y.; Proskurin, I.; Ogata, M. Effect of Tilting on the In-Plane Conductivity of Dirac Electrons in Organic Conductor. *J. Phys. Soc. Jpn.* **2014**, *83*, 023701. [CrossRef]
20. Proskurin, I.; Ogata, M. Thermoelectric Transport Coefficients for Massless Dirac Electrons in Quantum Limit. *J. Phys. Soc. Jpn.* **2013**, *82*, 063712. [CrossRef]
21. Cui, H.; Brooks, J.S.; Kobayashi, A.; Kobayashi, H. Metallization of the single component molecular semiconductor [Ni(ptdt)$_2$] under very high pressure. *J. Am. Chem. Soc.* **2009**, *131*, 6358–6359.
22. Souto, M.; Cui, H.B.; Pena-Alvarez, M.; Baonza, V.G.; Jeschke, H.O.; Tomic, M.; Valenti, R.; Blasi, D.; Ratera, I.; Rovira, C.; Veciana, J. Pressure-Induced Conductivity in a Neutral Nonplanar Spin-Localized Radical. *J. Am. Chem. Soc.* **2016**, *138*, 11517–11525. [PubMed]
23. Souto, M.; Gullo, M.C.; Cui, H.B.; Casati, N.; Montisci, F.; Jeschke, H.O.; Valenti, R.; Ratera, I.; Rovira, C.; Veciana, J. Role of the Open-Shell Character on the Pressure-Induced Conductivity of an Organic Donor-Acceptor Radical Dyad. *Chem. Eur. J.* **2018**, *24*, 5500–5505. [PubMed]
24. Mailman, A.; Wong, J.W.L.; Winter, S.M.; Claridge, R.C.M.; Robertson, C.M.; Assoud, A.; Yong, W.; Steven, E.; Dube, P.A.; Tse, J.S.; et al. Fine Tuning the Performance of Multiorbital Radical Conductors by Substituent Effects. *J. Am. Chem. Soc.* **2017**, *139*, 1625–1635.
25. Tian, D.;Winter, S.M.; Mailman, A.; Wong, J.W.L.; Yong, W.; Yamaguchi, H.; Jia, Y.; Tse, J.S.; Desgreniers, S.; Secco, R.A.; et al. The Metallic State in Neutral Radical Conductors: Dimensionality, Pressure and Multiple Orbital Effects. *J. Am. Chem. Soc.* **2015**, *137*, 14136–14148. [PubMed]
26. Cui, H.B.; Tsumuraya, T.; Kawasugi, Y.; Kato R. Pressure Induced Superconductivity and Dirac Cone Formation in Single-Component Molecular Conductors. In Proceedings of the 17th International Conference on High Pressure in Semiconductor Physics (HPSP-17), Tokyo, Japan, 7–11 August 2016.
27. Tsumuraya, T.; Cui, H.B.; Miyazaki, T.; Okano. Y.; Kato, R. Pressure-Induced Metallic Conductivity in the Single-Component Molecular Crystal [Ni(dmit)$_2$]. *Eur. J. Inorg. Chem.* **2014**, *24*, 3837–3840.
28. Kato, R.; Suzumura, Y. Novel Dirac Electron in Single-Component Molecular Conductor [Pd(dddt)$_2$] (dddt = 5,6-dihydro-1,4-dithiin-2,3-dithiolate). *J. Phys. Soc. Jpn.* **2017**, *86*, 064705.

29. Liu, Z.; Wang, H.; Wang, Z.F.; Yang, J.; Liu, F. Pressure-induced organic topological nodal-line semimetal in the three-dimensional molecular crystal Pd(dddt)$_2$. *Phys. Rev. B* **2018**, *97*, 155138.
30. Tsumuraya, T.; Kato, R.; Suzumura, Y. Effective Hamiltonian of Topological Nodal Line Semimetal in Single-Component Molecular Conductor [Pd(dddt)$_2$] from First-Principles. *J. Phys. Soc. Jpn.* **2018**, *87*, 113701.
31. Katayama, S.; Kobayashi, A.; Suzumura, Y. Electric Conductivity of the Zero-Gap Semiconducting State in α-(BEDT-TTF)$_2$I$_3$ Salt. *J. Phys. Soc. Jpn.* **2006**, *75*, 023708.
32. Tajima, N.; Sugawara, S.; Tamura, M.; Kato, R.; Nishio, Y.; Kajita, K. Transport properties of massless Dirac fermions in an organic conductor α-(BEDT-TTF)$_2$I$_3$ under pressure. *EPL* **2007**, *80*, 47002. [CrossRef]
33. Suzumura, Y.; Oagata, M. Role of acoustic phonons in exotic conductivity of two-dimensional Dirac electrons. *Phys. Rev. B* **2018**, *98*, 161205. [CrossRef]
34. Suzumura, Y. Anisotropic Conductivity of Nodal Line Semimetal in Single-Component Molecular Conductor [Pd(dddt)$_2$]. *J. Phys. Soc. Jpn.* **2017**, *86*, 124710. [CrossRef]
35. Suzumura, Y.; Cui, H.B.; Kato, R. Conductivity and Resistivity of Dirac Electrons in Single-Component Molecular Conductor [Pd(dddt)$_2$]. *J. Phys. Soc. Jpn.* **2018**, *87*, 084702. [CrossRef]
36. Kato, R.; Yeung, H.H.-M.; Cui, H.B.; Minamidate, T.; Suzumura, Y. Electronic Structure of a Single-Component Molecular Conductor [Pd(dddt)$_2$] (dddt = 5,6-dihydro-1,4-dithiin-2,3-dithiolate) under High Pressure. *arXiv* **2020**, arXiv:2008.09277.
37. Fröhlich, H. On the theory of superconductivity: The one-dimensional case. *Proc. Phys. Soc. A* **1954**, *223*, 296–305.
38. Holstein, T. Theory of Transport Phenomena in an Electron-Phonon Gas. *Ann. Phys.* **1964**, *29*, 410–535. [CrossRef]
39. Abrikosov, A.A.; Gorkov, L.P.; Dzyaloshinskii, I.E. *Methods of Quantum Field Theory in Statistical Physics*; Prentice-Hall: Englewood Cliffs, NJ, USA, 1963.
40. Cui, H.B.; Yeung, H.H.-M.; Kawasugi, Y.; Minamidate, T.; Warren, M.R.; Kato, R. High Pressure Crystal Structure and Unusual Magnetoresistance of a Single-component Molecular conductor [Pd(dddt)$_2$] (dddt = 5,6-dihydro-1,4-dithiin-2,3-dithiolate). **2020**, submitted.
41. Sarma, S.D.; Adam, S.; Hwang, E.H.; Rossi, E. Electronic transport in two-dimensional graphene. *Rev. Mod. Phys.* **2011**, *83*, 407–470. [CrossRef]

crystals

MDPI

Article

Gas-Dependent Reversible Structural and Magnetic Transformation between Two Ladder Compounds

Jun Manabe [1], Kazuki Nishida [1], Xiao Zhang [1], Yuki Nakano [1], Masaru Fujibayashi [1,2], Goulven Cosquer [1,2], Katsuya Inoue [1,2,3], Seiya Shimono [4], Hiroki Ishibashi [5], Yoshiki Kubota [5], Misaki Shiga [6], Ryo Tsunashima [6], Yoko Tatewaki [7] and Sadafumi Nishihara [1,2,3,8,*]

1 Department of Chemistry, Graduate School of Science, Hiroshima University, 1-3-1, Kagamiyama, Higashi-hiroshima 739-8526, Japan; m195832@hiroshima-u.ac.jp (J.M.); bmgsrnp.chemistry@gmail.com (K.N.); zhang_x163@163.com (X.Z.); yuki.n.mg.jk@gmail.com (Y.N.); fujiba@hiroshima-u.ac.jp (M.F.); cosquer.goulven@gmail.com (G.C.); kxi@hiroshima-u.ac.jp (K.I.)
2 Graduate School of Advanced Science and Engineering, Hiroshima University, 1-3-1, Kagamiyama, Higashi-hiroshima 739-8526, Japan
3 Chirality Research Center & Institute for Advanced Materials Research, Hiroshima University, 1-3-1, Kagamiyama, Higashi-hiroshima 739-8526, Japan
4 Department of Materials Science and Engineering, National Defense Academy, 1-10-20 Hashirimizu,Yokosuka, Kanagawa 239-8686, Japan; sshimono@nda.ac.jp
5 Department of Physical Science, Graduate School of Science, Osaka Prefecture University, Sakai, Osaka 599-8531, Japan; hiroki@p.s.osakafu-u.ac.jp (H.I.); kubotay@p.s.osakafu-u.ac.jp (Y.K.)
6 Graduate School of Sciences and Technology for Innovation, Yamaguchi University, Yoshida, 1677-1, Yamaguchi 753-8512, Japan; b002wb@yamaguchi-u.ac.jp (M.S.); ryotsuna@yamaguchi-u.ac.jp (R.T.)
7 Department of Applied Chemistry, Graduate School of Engineering, Tokyo University of Agriculture and Technology 2-24-16, Nakacho, Koganei 184-8588, Japan; ytatewa@cc.tuat.ac.jp
8 JST, PRESTO, 4-1-8, Honcho, Kawaguchi, Saitama 332-0012, Japan
* Correspondence: snishi@hiroshima-u.ac.jp

Received: 26 August 2020; Accepted: 17 September 2020; Published: 19 September 2020

Abstract: We report reversible structural transformation that occurs in two ladder compounds: $Cu_2CO_3(ClO_4)_2(NH_3)_6$ (**1**) and $Cu_2CO_3(ClO_4)_2(NH_3)_5(H_2O)$ (**2**), when they are exposed to gaseous vapors. The ladder structures of both **1** and **2** consist of two Cu^{2+} ions and one $CO_3{}^{2-}$ ion. In **1**, the Cu^{2+} ions are coordinated by three NH_3 molecules on each side, while those in **2** are coordinated by three NH_3 molecules on one side, and two NH_3 molecules and one H_2O molecule on the other side. We demonstrated reversible transformation of **1** and **2** via the exposure of **1** to H_2O vapor and the exposure of **2** to NH_3 vapor using a simple bench-scale method. The minor structural change observed led to a significant difference in physical properties, which we observed using several methods.

Keywords: magnetic property; reversible transformation; spin ladder

1. Introduction

Materials with the capacity to undergo reversible structural and physical changes, usually thermal phase transitions, are required in devices such as sensors and memories [1,2]. Such materials exhibit different structures above and below the phase transition temperature. However, the physical properties of each structure cannot be examined across the entire temperature range. In contrast, light irradiation-induced reversible changes allow for the detailed examination of physical properties throughout the entire temperature range, both before and after the transformation. However, transformation induced by irradiation poses certain problems, like the influence of heat generated by irradiation and the limited surface area reached by the irradiating light. Therefore, a material that undergoes reversible change in response to an external stimulus other than temperature or light is practically applicable as sensors and memories.

In recent years, spin ladders have attracted significant attention in the field of superconductivity [3–15] due to the theoretical possibility, for an even-leg spin ladder system, to undergo a superconducting transition with doped carriers. In the past, we successfully synthesized two molecular even-leg ladder compounds: $Cu_2CO_3(ClO_4)_2(NH_3)_6$ (**1**) and $Cu_2CO_3(ClO_4)_2(NH_3)_5(H_2O)$ (**2**) [16]. The ladder structure of **1**, $[(H_3N)_3Cu\text{-}CO_3\text{-}Cu(NH_3)_3]_n$ is constructed by alternately stacking two Cu^{2+} ions and one CO_3^{2-} ion, with each Cu^{2+} ion being coordinated by three NH_3 molecules (Figure 1a,b). The ladder structure of **1** is magnetically isolated due to the presence of perchlorate ions between ladders. The temperature-dependent molar magnetic susceptibility of this compound can be reproduced using a magnetically isolated spin ladder model [17], whereby the magnetic exchange interactions of the ladder rung and leg were estimated to be $J_{rung}/k_B = -364$ K and $J_{leg}/k_B = -27.4$ K, respectively. Meanwhile, the ladder structure of **2**, $[(H_3N)_3Cu\text{-}CO_3\text{-}Cu(NH_3)_2(H_2O)]_n$, contains two Cu^{2+} ions and one CO_3^{2-} ion and its structure is similar to that of **1**. One of the two Cu^{2+} ions is, however, coordinated using two NH_3 molecules and one H_2O molecule, thus distinguishing it from **1** (Figure 1c,d). In addition, a small structural phase transition involving the reorientation of the perchlorate ions was observed in **2** at approximately 205 K, thus inducing adjustment in the ladder structure of **2**, without the destruction of the exchange interaction model. The CIF file of **2** at room temperature (HT) is attached to the supplementary materials. The temperature dependence of the molar magnetic susceptibility of **2** was reproduced using an alternating chain model [18], rather than the spin ladder model, with the following magnetic exchange interactions: $J_1/k_B = -7.26$ K, $J_2/k_B = -4.42$ K, and $J_3/k_B = 0$ K. Furthermore, **2** exhibited an antiferromagnetic transition at 3.4 K [19].

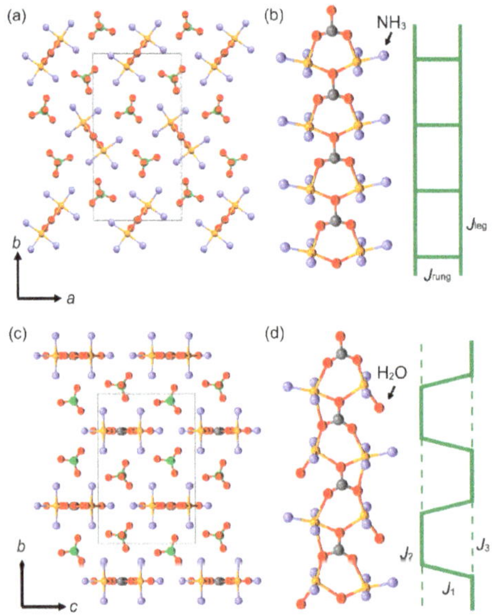

Figure 1. (**a**) The crystal structure of $Cu_2CO_3(ClO_4)_2(NH_3)_6$ (**1**), where the Cu^{2+} ions are coordinated on both sides by three NH_3 molecules, observed along the *c*-axis. (**b**) The ladder structure of **1** and schematic of the ladder configuration with ladder-rung (J_{rung}) and ladder leg (J_{leg}) magnetic interactions. (**c**) The crystal structure of $Cu_2CO_3(ClO_4)_2(NH_3)_5(H_2O)$ (**2**), where the Cu^{2+} ions are coordinated by three NH_3 molecules on one side and two NH_3 molecules and one H_2O molecule on the other side, observed along the *a*-axis. (**d**) The distorted ladder structure of **2** and schematic representation of the distorted ladder configuration with the ladder-rung (J_1) and two different ladder-leg (J_2 and J_3) magnetic interactions. (C—grey; N—light blue; O—red; Cl—yellow-green; Cu—yellow).

In the present study, we investigated the reversible structural and magnetic switching between **1** and **2** by exposing the compounds to H_2O and NH_3 vapors, respectively (Scheme 1). Such reversible magnetic switching may be applicable in sensors where they would be used to make distinctions between gases such as NH_3 and H_2O.

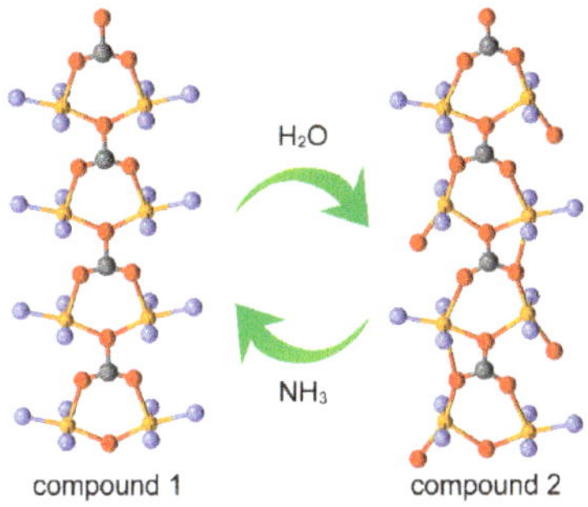

Scheme 1. Reversible structural transformation between **1** and **2** via the exchange of H_2O and NH_3.

2. Materials and Methods

Sample preparation: Single crystals of $Cu_2CO_3(ClO_4)_2(NH_3)_6$ (**1**) and $Cu_2CO_3(ClO_4)_2(NH_3)_5(H_2O)$ (**2**) were synthesized based on past studies [16,20]. Elemental analysis (calcd.) for **1**: C: 2.38 (2.46), H: 3.60 (3.72), N: 16.88 (17.21), and for **2**: C: 2.43 (2.46), H: 3.52 (3.50), N: 13.95 (14.32). To identify structural transformation, powder samples of **1** and **2** were exposed to H_2O or NH_3 vapors, respectively. In this experiment, the powder sample was spread over a glass plate, which was then placed on top of a glass stand inside a beaker containing either H_2O or 14.8 mol dm^{-3} NH_3 aqueous solution (20 mL) in a confined space at 303 K (Figure S1).

Powder X-ray diffraction measurement: Powder X-ray diffraction (PXRD) patterns were measured using a Rigaku RINT2100 diffractometer at r.t. for all samples. The measurements were performed via Cu Kα radiation (λ = 1.5418 Å) at a scanning rate of 2.0° min^{-1} under an applied electric voltage of 40 kV and a current of 40 mA.

Magnetic measurement: The temperature dependence of the magnetic susceptibility and field-dependent magnetization were measured using a Quantum Design MPMS-5S (Figure 2 and Figure S2), MPMS-XL (Figure S4), and MPMS-3 (Figure S4) superconducting quantum interference device (SQUID) magnetometer using single crystals or powdered samples contained in a gelatin capsule.

Thermal analysis: Thermogravimetric (TG) analyses were carried out on powdered samples using a SII TG/DTA 6200N instrument under N_2 flow. Measurements were conducted at a scanning rate of 5 K min^{-1}. Gas chromatography–mass spectrometry (GC–MS) analyses were performed using a SHIMADZU GCMS-QP2010 Ultra instrument at a scanning rate of 5 K min^{-1}.

Infrared spectroscopy: Infrared spectroscopy (IR) spectra were measured using KBr pellets and a JASCO FT/IR-660 Plus spectrometer within the range of 400–4000 cm^{-1}.

Heat capacities: The heat capacities were measured via the thermal relaxation method using a Quantum Design Physical Property Measurement System (Figure S3).

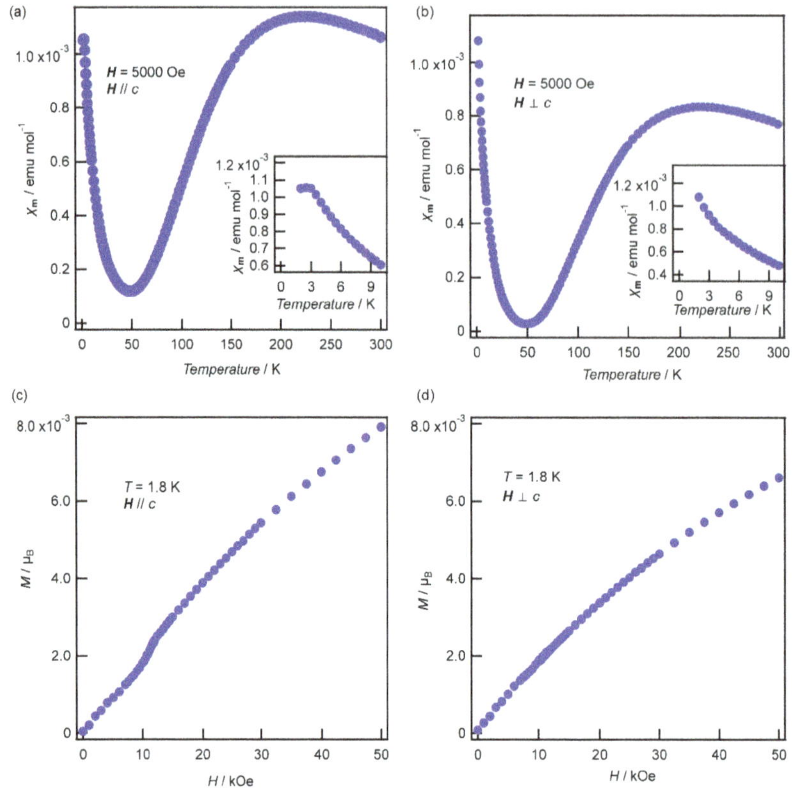

Figure 2. The temperature-dependent molar magnetic susceptibility of **1**, where fields H were applied (**a**) parallel to and (**b**) perpendicular to the c-axis. Insets show the plot in the low-temperature range below 10 K. M–H curves for **1** up to fields of 50 kOe, where fields H were applied (**c**) parallel to and (**d**) perpendicular to the c-axis at 1.8 K.

3. Results and Discussion

3.1. Magnetic Anomaly of Compound 1

Magnetic measurements were performed using a single crystal of **1** in magnetic fields H parallel (Figure 2a,c) or perpendicular (Figure 2b,d) to the c-axis. For H // c, the temperature dependence of the molar magnetic susceptibility (χ_m-T) of **1** in a field of 5 kOe exhibited unexpected magnetic behavior, hereafter called "anomaly", at approximately 3 K (Figure 2a). Additionally, a magnetic jump considered to be a spin-flop transition was observed at approximately 11 kOe in the field-dependent magnetization (M-H) curve for **1** at 1.8 K (Figure 2c). When magnetic fields of 3, 11, and 15 kOe were applied at the temperature dependence of magnetization measurements, the magnetic anomaly of the magnetic field of 15 kOe, which was larger than the spin-flop field, vanished (Figure S2). In contrast, such an anomaly was not observed in the χ_m-T and M-H measurements for H // c (Figure 2b,d). Heat capacity measurements were performed in magnetic fields below 30 kOe, parallel to the c-axis, to determine whether the anomaly was attributable to a magnetic phase transition (Figure S3). A peak (partially out of the measured temperature range) was observed below 3 K in the zero-field, and the peak broadened as the magnetic field increased. The results of the magnetic and heat capacity measurements showed that the anomaly of **1** at low temperature was attributable to an antiferromagnetic transition. A similar anomaly in an inorganic spin ladder, whereby an antiferromagnetic transition was detected in a spin ladder compound doped with magnetic impurities, $Sr(Cu_{1-x}Zn_x)_2O_3$, was reported by Azuma et al. [21–23]. In the present study, however, **1** was not doped, and the anomaly cannot be explained using this mechanism. To further analyze this phenomenon, we performed magnetic measurements of $H // c$ using a single crystal of **1** exposed to air for 10 days (Figure S4). Under this condition, the peak at 3 K became more prominent. To examine this change, PXRD measurement was performed on **1** after exposing the powder sample to air for 3 days (**1_air**). The PXRD pattern of **1_air** corresponded to a superposition of the PXRD patterns of **1** and **2** (Figure 3). Because the compositions of **1** and **2** only differ by a molecule (NH_3 or H_2O), these results suggest that NH_3 can be substituted by H_2O from the air in the crystal of **1**. Therefore, the anomaly may not be intrinsic in **1** but in the partial transformation of **1** into **2**.

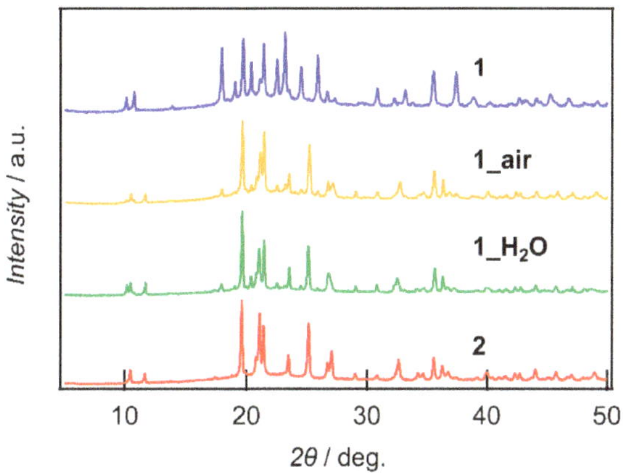

Figure 3. PXRD patterns of **1** (blue), **1_air** (**1** exposed to air for 3 days; yellow), **1_H₂O** (**1** exposed to H_2O vapor for 4.5 h; green), and **2** (red).

3.2. *Transformation from Compound* **1** *to* **2**

To further establish the aforementioned observations and implications, we evaluated the PXRD pattern of **1** exposed to H_2O vapor for 4.5 h (**1_H₂O**). The PXRD pattern of **1_H₂O** was a superposition of the patterns of **1** and **2** (Figure 3). This implies that in the basic formula $[(H_3N)_3Cu\text{-}CO_3\text{-}Cu(NH_3)_3]$ of **1**, one NH_3 molecule was partially replaced by one H_2O molecule. We confirmed the replacement of the coordinating molecules through TG analysis and GC–MS. Figure S5 shows weight loss in the samples, as well as MS intensity corresponding to the presence of H_2O (m/z = 18) across a temperature range of 340 to 450 K. Up to 400 K, the weight loss and MS intensity derived from H_2O was not observed in **1**. In contrast, weight loss was observed in **1_H₂O**, and MS intensity derived from H_2O was observed below 400 K. The analysis of **2** yielded results that were similar to those of **1_H₂O**. In addition, the elemental analysis of **1_H₂O** yielded intermediate values (elemental analysis: C: 2.49, H: 3.68, N: 15.46) between those of **1** and **2** (see the Materials and Methods section). IR spectrum in the 3000–3700 cm^{-1} region, mainly derived from NH_3 and H_2O ligands, for **1_H₂O** is closer to that of **2** than that of **1** (Figure S6). The magnetic behavior of **1_H₂O**, as shown in Figure 4b, was drastically different compared to that of **1** (Figure 4a) and tended toward that of **2** (Figure 4c). From these results, we conclude that the NH_3 in **1** was substituted by H_2O after exposure to H_2O vapor, and the anomaly observed in **1** at low temperature was caused by the transformation of **1** to **2** by H_2O, corresponding to the prediction mentioned earlier.

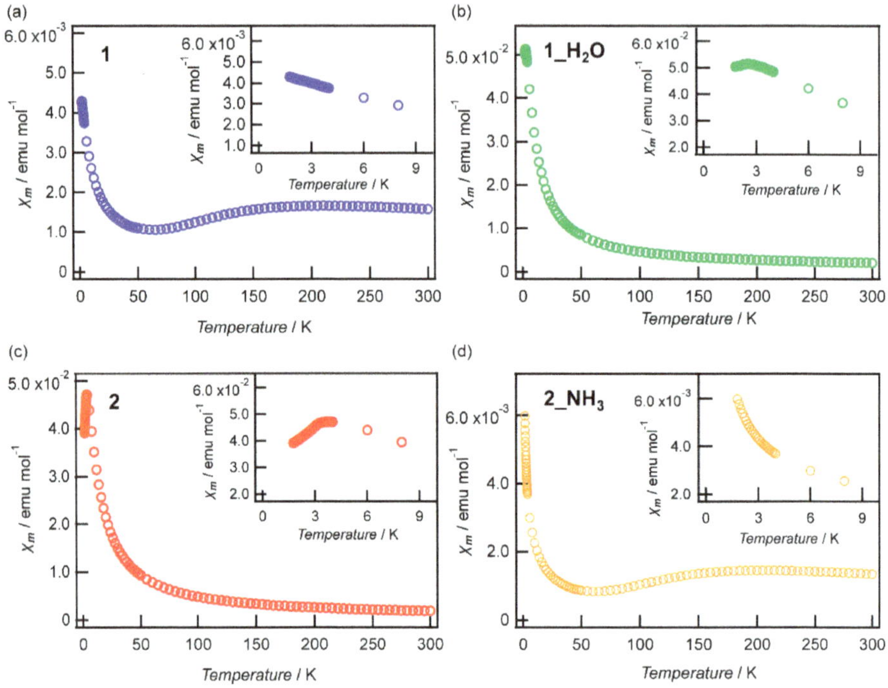

Figure 4. Temperature-dependent molar magnetic susceptibility in an applied field of 5 kOe for (**a**) **1**, (**b**) **1_H$_2$O** (**1** exposed to H_2O vapor for 4.5 h), (**c**) **2**, and (**d**) **2_NH$_3$** (**2** exposed to NH_3 vapor for 2 h). The insets show the data in the low-temperature range (below 10 K).

3.3. *Transformation from Compound* **2** *to* **1**

In order to investigate the reverse reaction from **2** to **1**, **2** was exposed to NH_3 vapor from an NH_3 aqueous solution for 2 h (**2_NH$_3$**). The PXRD pattern of **2_NH$_3$** exhibited characteristic peaks of both **1** and **2**, implying that the expected transformation had occurred (Figure 5). Therefore, TG analysis and GC–MS were performed for **2_NH$_3$** (Figure S5). Similar to **1**, the weight loss and MS intensity derived from H_2O was not observed until 400 K. In addition, the results of the elemental analysis of **2_NH$_3$** also corresponded to intermediate values (elemental analysis: C: 2.68, H: 3.61, N: 16.55) between those of **1** and **2** (see the Materials and Methods section). The IR spectrum of **2_NH$_3$** in the 3000–3700 cm^{-1} region also changed and was similar to the spectrum for **1** (Figure S6). The results of the magnetic measurements show that the magnetic behavior of **2_NH$_3$** (Figure 4d) differed from that of **2** (Figure 4c) and approached that of **1** (see Figure 4a). These results indicate that the exposure of **2** to NH_3 vapor transforms it into **1**.

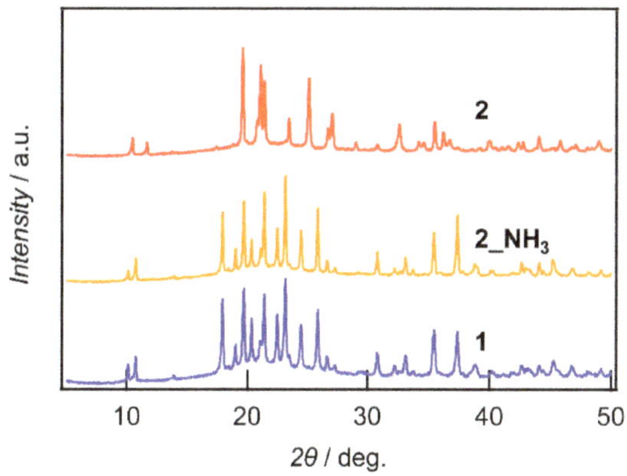

Figure 5. PXRD patterns of **2** (red), **2_NH₃** (**2** exposed to NH_3 vapor for 2 h; yellow) and **1** (blue).

3.4. *Reversible Transformation between Compounds* **1** *and* **2**

Reversible transformation between **1** and **2** was investigated. **1_H₂O**, which was obtained by exposing **1** to vapor for 4.5 h, was exposed to NH_3 vapor from an NH_3 aqueous solution for 2 h (**1_H₂O_NH₃**). The PXRD measurement of **1_H₂O_NH₃** was carried out to determine whether the change between **1** and **2** occurred reversibly depending on the gaseous vapors. Figure 6 shows the PXRD pattern of **1_H₂O_NH₃**, including the patterns of **1**, **1_H₂O**, and **2** for comparison. Given that the PXRD pattern of **1_H₂O_NH₃** was similar to that of **1**, we can conclude that **1** and **2** were reversibly changed by gaseous vapors.

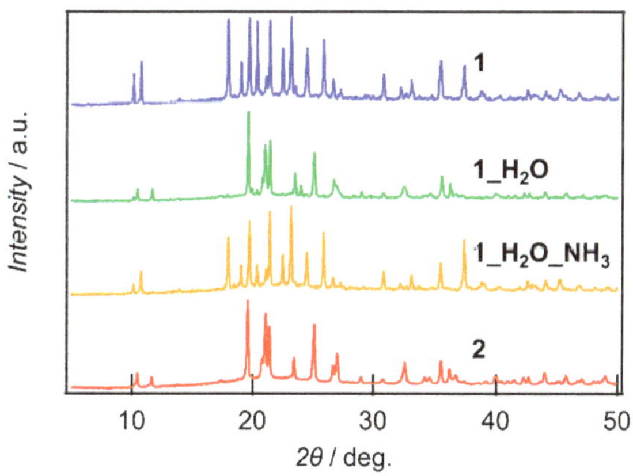

Figure 6. PXRD patterns of **1** (blue), **1_H₂O** (**1** exposed to H_2O vapor for 4.5 h; green), **1_H₂O_NH₃** (**1_H₂O** exposed to NH_3 vapor for 2 h; yellow), and **2** (red).

3.5. *Stability of Compounds* **1** *and* **2**

Based on these experiments, the transformation from compound **1** to **2** occurred when **1** was exposed to H_2O vapor, and the opposite transformation from compound **2** to **1** occurred when **2** was exposed to NH_3 vapor, despite the simultaneous presence of H_2O vapor due to the aqueous solution. The results indicated that **1** and **2** were stable in the presence of NH_3 and H_2O vapors, respectively. To verify this, we exposed **1** to NH_3 vapor for 2 h and **2** to H_2O vapor for 4.5 h. The PXRD measurements showed that **1** exposed to NH_3 vapor (**1′**) and **2** exposed to H_2O vapor (**2′**) underwent almost no structural changes (Figure S7). This confirmed that **1** and **2** maintained their original structure when exposed to NH_3 and H_2O vapor present in the air, respectively. Based on these results, the anomaly observed in the magnetic measurements of **1** at low temperature is assumed to be attributable to a contribution of the antiferromagnetic transition of **2**, produced by the partial transformation of **1** to the more stable structure of **2**, in the presence of H_2O vapor. The transformation did not completely progress to the point where all physical properties were consistent with those of the opposite compound after the transformation, owing to the decomposition of the crystal surface.

4. Conclusions

This study of gas-dependent reversible structural and magnetic transformation between two ladder compounds $Cu_2CO_3(ClO_4)_2(NH_3)_6$ (**1**) and $Cu_2CO_3(ClO_4)_2(NH_3)_5(H_2O)$ (**2**) was inspired by the observation of an anomaly in a single crystal of **1** at low temperature based on magnetic and heat capacity measurements. Because this anomaly became more prominent after the exposure of **1** to air, we realized that the exposure of **1** to H_2O vapor transformed it into **2**. This transformation was identified using PXRD, TG analyses, GC–MS, elemental analyses, IR spectroscopy, and magnetic measurements. Subsequently, we demonstrated the transformation of **2** to **1** by exposing **2** to NH_3 vapor. We also achieved a reversible transformation between **1** and **2**. In the particular cases where the moisture sensibility of **1** could be avoided or used as advantage, the reversible switching properties demonstrated in our study could allow the development of sensors and memories.

Supplementary Materials: The following are available online at http://www.mdpi.com/2073-4352/10/9/841/s1, Figure S1: Experimental setup and schematic diagram, Figure S2: Temperature-dependent magnetization of **1**, Figure S3: Temperature-dependent heat capacity of **1** under variable DC magnetic fields, Figure S4: Temperature-dependent magnetic susceptibility of **1** and **1** exposed to air, Figure S5: TG and GC–MS analysis, Figure S6: IR spectra, Figure S7: PXRD patterns of **1′** and **2′**. Compound2_HT is the CIF file of **2** at room temperature (HT).

Author Contributions: Conceptualization, S.N.; methodology, J.M., K.N., X.Z., Y.N.; validation, J.M., K.N., X.Z., Y.N.; formal analysis, J.M.; investigation, J.M., K.N., X.Z., Y.N., M.S., S.S.; resources, K.I., S.S., H.I., Y.K., R.T., S.N.; data curation, J.M., S.N.; writing—original draft preparation, J.M.; writing—review and editing, M.F., G.C., Y.T., S.N.; visualization, J.M.; supervision, K.I., S.N.; project administration, S.N.; funding acquisition, K.I., S.N. All authors have read and agreed to the published version of the manuscript.

Funding: This research was funded by the Japan Society for the Promotion of Science (JSPS) KAKENHI Grant-in-Aid for Scientific Research, grant number JP19H02799, and the JSPS Core-to-Core Program, Advanced Research Networks. This work was also supported by JST, PRESTO Grant Number JPMJPR19M8, Japan.

Acknowledgments: We thank M. Kubota of the Natural Science Center for Basic Research and Development (N-BARD), Hiroshima University, for performing the elemental analyses. We also thank N. Manada of the Organization for Research Initiatives, Yamaguchi University, for performing the gas chromatography–mass spectrometry analyses.

Conflicts of Interest: The authors declare no conflict of interest.

References

1. Potyrailo, R.A. Multivariable Sensors for Ubiquitous Monitoring of Gases in the Era of Internet of Things and Industrial Internet. *Chem. Rev.* **2016**, *116*, 11877–11923. [CrossRef]
2. Wuttig, M.; Yamada, N. Phase change materials for rewriteable data storage. *Nat. Mater.* **2007**, *6*, 824–832. [CrossRef] [PubMed]
3. Dagotto, E.; Riera, J.; Scalapino, D. Superconductivity in ladders and coupled planes. *Phys. Rev. B* **1992**, *45*, 5744–5747. [CrossRef] [PubMed]
4. Rice, T.M.; Gopalan, S.; Sigrist, M. Superconductivity, Spin Gaps and Luttinger Liquids in a Class of Cuprates. *Europhys. Lett.* **1993**, *23*, 445–449. [CrossRef]
5. Dagotto, E.; Rice, T.M. Surprises on the Way from One- to Two-Dimensional Quantum Magnets: The Ladder Materials. *Science* **1996**, *271*, 618–623. [CrossRef]
6. Azuma, M.; Hiroi, Z.; Takano, M.; Ishida, K.; Kitaoka, Y. Observation of a Spin Gap in $SrCu_2O_3$ Comprising Spin-$\frac{1}{2}$ Quasi-1D Two-Leg Ladders. *Phys. Rev. Lett.* **1994**, *73*, 3463–3466. [CrossRef]

7. Ishida, K.; Kitaoka, Y.; Asayama, K.; Azuma, M.; Hiroi, Z.; Takano, M. Spin Gap Behavior in Ladder-Type of Quasi-One-Dimensional Spin (S = 1/2) System $SrCu_2O_3$. *J. Phys. Soc. Jpn.* **1994**, *63*, 3222–3225. [CrossRef]
8. Azuma, M.; Yoshida, H.; Saito, T.; Yamada, T.; Takano, M. Pressure-Induced Buckling of Spin Ladder in $SrCu_2O_3$. *J. Am. Chem. Soc.* **2004**, *126*, 8244–8246. [CrossRef]
9. Kojima, K.; Keren, A.; Luke, G.M.; Nachumi, B.; Wu, W.D.; Uemura, Y.J.; Azuma, M.; Takano, M. Magnetic Behavior of the 2-Leg and 3-Leg Spin Ladder Cuprates $Sr_{n-1}Cu_{n+1}O_{2n}$. *Phys. Rev. Lett.* **1995**, *74*, 2812–2815. [CrossRef]
10. Rovira, C. Molecular Spin Ladders. *Chem. Eur. J.* **2000**, *6*, 1723–1729. [CrossRef]
11. Nishihara, S.; Akutagawa, T.; Hasegawa, T.; Nakamura, T. Formation of a molecular spin ladder induced by a supramolecular cation structure. *Chem. Commun.* **2002**, 408–409. [CrossRef] [PubMed]
12. Nishihara, S.; Akutagawa, T.; Hasegawa, T.; Fujiyama, S.; Nakamura, T.; Nakamura, T. Two Polymorphs of (Anilinium)(18-Crown-6)[Ni(dmit)$_2$]: Structure and Magnetic Properties. *J. Solid State Chem.* **2002**, *168*, 661–667. [CrossRef]
13. Nishihara, S.; Akutagawa, T.; Hasegawa, T.; Nakamura, T.; Fujiyama, S.; Nakamura, T. Magnetic and 1H-NMR spectroscopic studies of [Ph(NH$_3$)](18-crown-6) [Ni(dmit)$_2$] having molecular spin ladder structure. *Synth. Met.* **2003**, *137*, 1279–1280. [CrossRef]
14. Ichihashi, K.; Konno, D.; Date, T.; Nishimura, T.; Maryunina, K.Y.; Inoue, K.; Nakaya, T.; Toyoda, K.; Tatewaki, Y.; Akutagawa, T.; et al. Optimizing Lithium Ion Conduction through Crown Ether-Based Cylindrical Channels in [Ni(dmit)$_2$]$^-$ Salts. *Chem. Mater.* **2018**, *30*, 7130–7137. [CrossRef]
15. Ichihashi, K.; Konno, D.; Maryunina, K.Y.; Inoue, K.; Toyoda, K.; Kawaguchi, S.; Kubota, Y.; Tatewaki, Y.; Akutagawa, T.; Nakamura, T.; et al. Selective Ion Exchange in Supramolecular Channels in the Crystalline State. *Angew. Chem. Int. Ed.* **2019**, *131*, 4213–4216. [CrossRef]
16. Zhang, X.; Nishihara, S.; Nakano, Y.; Yoshida, E.; Kato, C.; Ren, X.-M.; Maryunina, K.Y.; Inoue, K. A magnetically isolated cuprate spin-ladder system: Synthesis, structures, and magnetic properties. *Dalton Trans.* **2014**, *43*, 12974–12981. [CrossRef]
17. Johnston, D.C.; Troyer, M.; Miyahara, S.; Lidsky, D.; Ueda, K.; Azuma, M.; Hiroi, Z.; Takano, M.; Isobe, M.; Ueda, Y.; et al. Magnetic Susceptibilities of Spin-1/2 Antiferromagnetic Heisenberg Ladders and Applications to Ladder Oxide Compounds. *arXiv* **2000**, arXiv:con-mat/0001147.
18. Kahn, O. *Molecular Magnetism*; VCH: New York, NY, USA, 1993.
19. Kobori, S.; Matsui, K.; Kuwahara, H.; Goto, T.; Zhang, X.; Nakano, Y.; Nishihara, S.; Inoue, K.; Sasaki, T. NMR study on the quasi one-dimensional quantum spin magnet with ladder structure. *Hyperfine Interact.* **2016**, *237*, 116. [CrossRef]
20. Zhang, X.; Nishihara, S.; Nakano, Y.; Maryunina, K.Y.; Inoue, K. A Cuprate Spin Ladder Linked by a Pyridyl Ligand. *Chem. Lett.* **2014**, *43*, 1713–1715. [CrossRef]
21. Azuma, M.; Fujishiro, Y.; Takano, M.; Nohara, M.; Takagi, H. Switching of the gapped singlet spin-liquid state to an antiferromagnetically ordered state in $Sr(Cu_{1-x}Zn_x)_2O_3$. *Phys. Rev. B* **1997**, *55*, R8658–R8661. [CrossRef]
22. Fujiwara, N.; Yasuoka, H.; Fujishiro, Y.; Azuma, M.; Takano, M. NMR Study of Zn Doping Effect in Spin Ladder System $SrCu_2O_3$. *Phys. Rev. Lett.* **1998**, *80*, 604–607. [CrossRef]
23. Ohsugi, S.; Tokunaga, Y.; Ishida, K.; Kitaoka, Y.; Azuma, M.; Fujishiro, Y.; Takano, M. Impurity-induced staggered polarization and antiferromagnetic order in spin-1/2 Heisenberg two-leg ladder compound $SrCu_2O_3$: Extensive Cu NMR and NQR studies. *Phys. Rev. B* **1999**, *60*, 4181–4190. [CrossRef]

crystals

MDPI

Article

Structures and Properties of New Organic Conductors: BEDT-TTF, BEST and BETS Salts of the $HOC_2H_4SO_3^-$ Anion

Hiroki Akutsu [1,*], Yuta Koyama [1], Scott S. Turner [2], Keigo Furuta [3] and Yasuhiro Nakazawa [1]

[1] Department of Chemistry, Graduate School of Science, Osaka University, 1-1 Machikaneyama, Toyonaka, Osaka 560-0043, Japan; icesnowkori@yahoo.co.jp (Y.K.); nakazawa@chem.sci.osaka-u.ac.jp (Y.N.)
[2] Department of Chemistry, University of Surrey, Guildford, Surrey GU2 7XH, UK; s.s.turner@surrey.ac.uk
[3] Department of Environment and System Sciences, Yokohama National University, 79-7 Tokiwadai, Hodogaya-ku, Yokohama 240-8501, Japan; keigofuruta_z@ybb.ne.jp
* Correspondence: akutsu@chem.sci.osaka-u.ac.jp; Tel.: +81-6-6850-5399

Received: 11 August 2020; Accepted: 28 August 2020; Published: 1 September 2020

Abstract: New bis(ethylenedithio)tetrathiafulvalene (BEDT-TTF)-, bis(ethylenediseleno) tetrathiafulvalene (BEST)- and bis(ethylenedithio) tetraselenafulvalene (BETS)-based organic charge-transfer (CT) salts—α-(BEDT-TTF)$_3$($HOC_2H_4SO_3$)$_2$ (**1**), β-(BEST)$_3$($HOC_2H_4SO_3$)$_2$·H_2O (**2**) and α-(BETS)$_2$($HOC_2H_4SO_3$)·H_2O (**3**)—have been prepared. Salts **1** and **2** show semiconducting behaviour. Salt **3**, which is almost isostructural to α-(BETS)$_2$I$_3$, shows metallic behaviour down to 70 K and then shows a broader metal–insulator transition than that of α-(BETS)$_2$I$_3$. The reason for the difference in behaviour is estimated by the comparison of the Madelung energies of the full set of patterns of possible donor's charge-ordered and anion's disordered states.

Keywords: organic conductors; bis(ethylenedithio) tetrathiafulvalene (BEDT-TTF); bis (ethylenediseleno) tetrathiafulvalene (BEST); bis(ethylenedithio) tetraselenafulvalene (BETS); electrical resistivity; magnetic susceptibility; X-ray analysis; charge-ordered state; quantum chemical calculations; Madelung energy

1. Introduction

Organic conductors have been synthesized by chemists [1,2] and their physical properties have been measured by physicists [3,4] for some time now. Chemists still now prepare many new types of organic conductors with novel structures and molecular features [5]. However, most physicists tend to continue to measure "old-fashioned" organic (super)conductors [3,4]. For example, popular salts are κ-(BEDT-TTF)$_2$Cu(NCS)$_2$ [6], where BEDT-TTF is bis(ethylenedithio)tetrathiafulvalene; κ-(BEDT-TTF)$_2$Cu[N(CN)$_2$]X (X = Cl, Br) [7,8]; θ-(BEDT-TTF)$_2$[MM′(SCN)$_4$] (M = Cs, Rb, Tl; M′ = Zn, Co) [9]; α-(BEDT-TTF)$_2$[AHg(SCN)$_4$] (A = K, NH_4) [10,11]; α-(D)$_2$I$_3$ (D = BEDT-TTF [12]; BETS [13], where BETS = bis(ethylenedithio)tetraselenafulvalene); and (TMTSF)$_2$X (X = PF_6, AsF_6, ClO_4, etc., where TMTSF = tetramethyltetrathiafulvalene) [14]. Most of these examples were first made for more than a quarter of a century ago. This may be because organic conductors prepared recently have had more complicated molecular and/or crystal structures, which are definitely interesting for chemists. However, more complication often implies bulkier molecules, which usually leads to poor conductivity. This might prevent physicists from making the measurements because of the difficulty of finding physical properties in which many physicists are interested. In this circumstance, chemists preparing simple organic conductors might be of interest, especially for physicists. Here, we report the BEDT-TTF, BEST (=bis(ethylenediseleno)tetrathiafulvalene) and BETS salts (Scheme 1) of a simple organic anion, isethionate ($HOC_2H_4SO_3^-$), the structures and properties of which are reported.

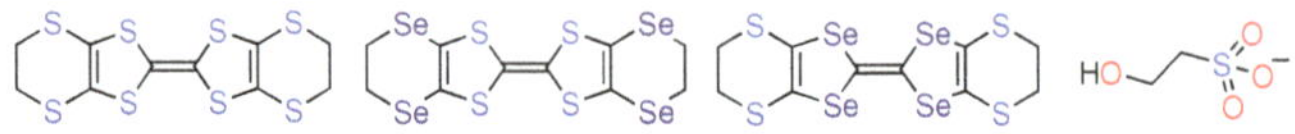

BEDT-TTF (ET) BEST BETS isethionate

Scheme 1. Structural formula of BEDT-TTF, BEST, BETS salts of a simple organic anion, isethionate ($HOC_2H_4SO_3^-$).

2. Materials and Methods

Electrical resistivity measurements were performed by the conventional four-probe method using a HUSO HECS 994C1 four channel resistivity meter with cooling and heating rates of ≈ 0.5 K/min. The magnetic susceptibilities of polycrystalline samples of the BETS salt (**3**) were measured from 2 to 300 K using a Quantum Design MPMS-2S SQUID magnetometer. The data were corrected for a contribution from the sample holder (aluminum foil), and the sample's diamagnetic contribution was estimated from Pascal's constants. X-ray diffraction data were collected with a Rigaku Rapid II imaging plate system with a MicroMax-007 HF/VariMax rotating-anode X-ray generator with confocally monochromatic Mo*K*α radiation.

BEDT-TTF and isethionic acid sodium salt ($HOC_2H_4SO_3Na$) were purchased from Tokyo Chemical Industry and Kanto Chemical, respectively, and used without any further purification. BEST and BETS were prepared according to the literature methods [15–17]. The electrocrystallisation of BEDT-TTF, BEST and BETS with $HOC_2H_4SO_3Na$ and 18-crown-6 ether in PhCl + 10% of EtOH produced black crystals.

The electronic states of the cationic and neutral donors and anions of **3** were calculated using the GAMESS software package [18] at the BY3LP 6-31G* level (SBKJC core potentials for iodine) using geometries determined by the X-ray analyses. All GAMESS input options used in these calculations are shown in Supporting Information. The input files were prepared using Facio [19,20]. Population analyses were performed using the Multiwfn software package (version 3.6, Beijing Kein Research Center for Natural Science, Beijing, China) [21], which calculated Hirshfeld populations from the output files of GAMESS. The resultant point charges of all the atoms of each molecule were used for Madelung energy calculations, which were performed with the Gulp program package [22] version 5.1, with a "molecule" option to restrict the calculations to only intermolecular Coulombic repulsions; namely, the effect of intramolecular repulsions was excluded.

3. Results

Table 1 shows the crystallographic data of α-$(BEDT\text{-}TTF)_3(HOC_2H_4SO_3)_2$ (1), β-$(BEST)_3(HOC_2H_4SO_3)_2{\cdot}H_2O$ (2) and α-$(BETS)_2(HOC_2H_4SO_3){\cdot}H_2O$ (3).

Table 1. Crystallographic data of **1**, **2** and **3**.

Compound	1	2	3
Composition	α-(BEDT-TTF)$_3$·$(HOC_2H_4SO_3)_2$	β-$(BEST)_3$ $(HOC_2H_4SO_3)_2{\cdot}H_2O$	α-$(BETS)_2$ $(HOC_2H_4SO_3){\cdot}H_2O$
Formula	$C_{34}H_{34}O_8S_{26}$	$C_{34}H_{36}O_9S_{14}Se_{12}$	$C_{22}H_{23}O_5S_9Se_8$
Fw	1404.20	1213.17	1287.64
Space Group	$P2_1/c$	$P2_1$	$P\bar{1}$
a (Å)	18.6954(7)	12.8225(14)	9.2422(5)
b (Å)	11.2106(4)	12.3690(13)	10.7806(5)
c (Å)	12.8995(5)	18.129(2)	18.6601(9)
α (°)	90.000	90.000	103.579(7)
β (°)	109.300(8)	106.653(8)	97.219(7)
γ (°)	90.000	90.000	90.455(6)
V (Å^3)	2551.6(2)	2754.7(5)	1791.57(17)

Table 1. *Cont.*

Compound	1	2	3
Z	2	2	2
T (K)	290	290	290
d_{calc} (g·cm^{-1})	1.828	2.393	2.387
μ (cm^{-1})	11.364	85.303	87.198
F(000)	1436	1888	1222
2θ range (°)	4–55	4–55	4–55
Total ref.	23,703	26,052	17,212
Unique ref.	5827	12,546	8101
R_{int}	0.0251	0.1753	0.0607
Parameters	307	505	478
R_1 ($I > 2\sigma(I)$)	0.044	0.114	0.042
wR_2 (all data)	0.103	0.320	0.109
S	1.083	0.970	1.064
$\Delta\rho_{max}$ (e Å^{-3})	0.85	1.68	1.48
$\Delta\rho_{min}$ (e Å^{-3})	−0.67	−1.43	−1.05
Flack		0.37(2)	
CCDC reference	2,022,621	2,022,623	2,022,624

3.1. *α-(BEDT-TTF)$_3$(HOC$_2$H$_4$SO$_3$)$_2$ (**1**)*

α-(BEDT-TTF)$_3$(HOC$_2$H$_4$SO$_3$)$_2$ (**1**) crystallizes in the space group $P2_1/c$. One and a half of BEDT-TTF and one isethionate are crystallographically independent. Figure 1a shows the crystal structure of **1**. The structure has alternating donor and anion layers propagating along the a axis. The structure of the two-dimensional conducting BEDT-TTF layer possessing an α-type packing motif is shown in Figure 1b. There are many interstack short S···S contacts, suggesting that the salt is a 2D conductor. However, the donor-to-anion ratio of 3:2 indicates that there is a BEDT-TTF trimer, in which two holes are located that form a spin dimer. Therefore, the structure suggests that the salt is a band insulator. Actually, the electrical resistivity measurements (Figure S1) indicate that the salt is a semiconductor with $\rho_{RT} = 1.15 \times 10^2$ Ω·cm and $E_a = 190$ meV. The molecular arrangement of isethionate in an anionic layer is shown in Figure 1c. Intermolecular hydrogen bonding between the hydroxyl group and an oxygen atom of a nearest neighbor's sulfo group (O···O = 2.861(2) Å) forms a 1D hydrogen bond chain along the b axis. In addition, there is an inversion centre in the anionic layer, indicating that the anion layer is not polar.

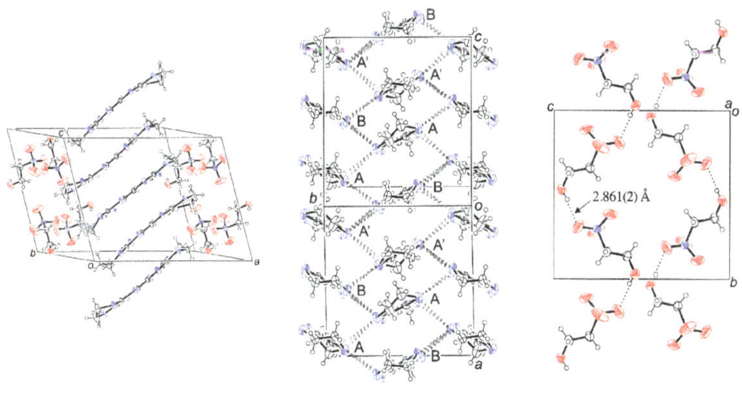

(a) (b) (c)

Figure 1. Crystal structure (**a**); arrangement of the donor layer (**b**), where dashed lines indicate short S···S contacts (<3.7 Å); and structure of the anionic layer (**c**) of **1**.

3.2. β-$(BEST)_3(HOC_2H_4SO_3)_2{\cdot}H_2O$ (**2**)

Compound **2** has similar cell parameters to **1**, as shown in Table 1, but has a different space group ($P2_1$) and is not isomorphous with **1**. There are three BEST molecules, two isethionate anions and one water molecule, in the asymmetric unit. The crystal structure is shown in Figure 2a. The unit cell consists of alternating donor/anion and water layers along the *c* axis. The donor layer (Figure 2b) has a β-type packing motif with both intra-and interstack short S···S contacts, suggesting that the salt is a 2D conductor. The ratio of donor to anion is 3:2, indicating that the BEST molecules form into trimers with two holes. In turn, the holes are arranged into spin dimers, which suggests that the salt is a band insulator. The resistivity measurements (Figure S2) indicate that the salt is, in fact, a semiconductor with higher conductivity than **1**. Salt **2** has a room temperature resistivity that is approximately two orders of magnitude smaller, ρ_{RT} of 4.47 Ω·cm, and approximately a half of the activation energy, E_a of 108 meV, when compared to **1**.

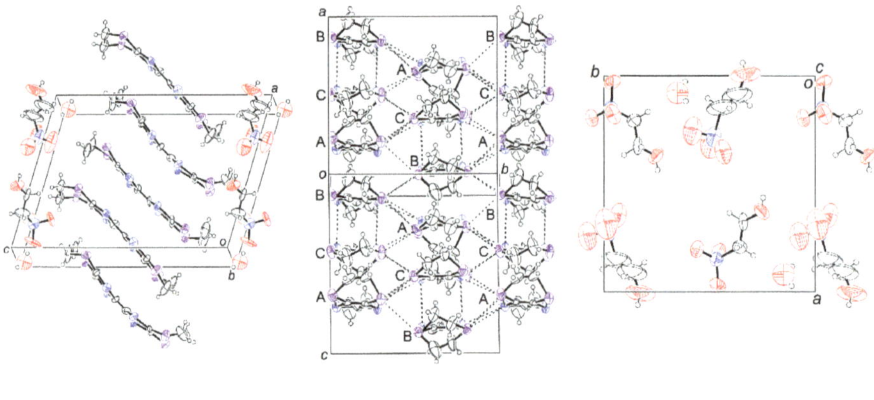

(**a**) (**b**) (**c**)

Figure 2. Crystal structure (**a**); arrangement of the donor layer (**b**), where dashed lines indicate short S···S, Se···S and Se···Se contacts (<3.70, 3.75 and 3.80 Å, respectively); and structure of the anionic layer (**c**) of **2**.

Figure 2c shows the structure of the anionic layer. The present R value of 11.6% is relatively high, so we cannot discuss some details of the structure. Several covalent bonds are distorted; therefore, bond length and angle restraints were applied. The space group, $P2_1$, is polar, so there are no inversion centres in the anionic layer. However, the anions are arranged to cancel almost all the dipole moments. The only small dipole moment seems to exist along the *b* axis (see Figure 2c). In addition, the X-ray diffraction measurement at 150 K gave a much more distorted structure, so we have not shown these data.

3.3. α-$(BETS)_2(HOC_2H_4SO_3){\cdot}H_2O$ (**3**)

The asymmetric unit of **3** consists of one (A in Figure 3c) and two halves (B and C) of BETS, one isethionate and one water molecule. The anion and the water molecule are disordered as schematically shown in Figure 3a, in which there are two anions having different conformations, which are superimposed at the same position. The crystal structure is shown in Figure 3b. The space group is $P\overline{1}$, so four BETS molecules, two anions and two water molecules exist in the unit cell. Figure 3c shows the packing arrangement of a BETS layer, where A′ is a centrosymmetrically inverted A. The layer has an α-type packing motif with short side-by-side chalcogen–chalcogen contacts, indicating that the salt has 2D conducting layers. The ratio of donor to anion is 2:1 which suggests that the salt would show low resistivity (*vide infra*). Figure 3d shows a packing arrangement of the anion/water layer. Because there are two molecular configurations (ρ and σ) in the same position and two anion positions (I and II), there are four possible patterns ((1)–(4) in Figure 3d), where ρ′ and σ′ are centrosymmetrically inverted ρ and σ, respectively. In addition, patterns (1) and (2) are centrosymmetric but (3) and (4) are not exactly centrosymmetric.

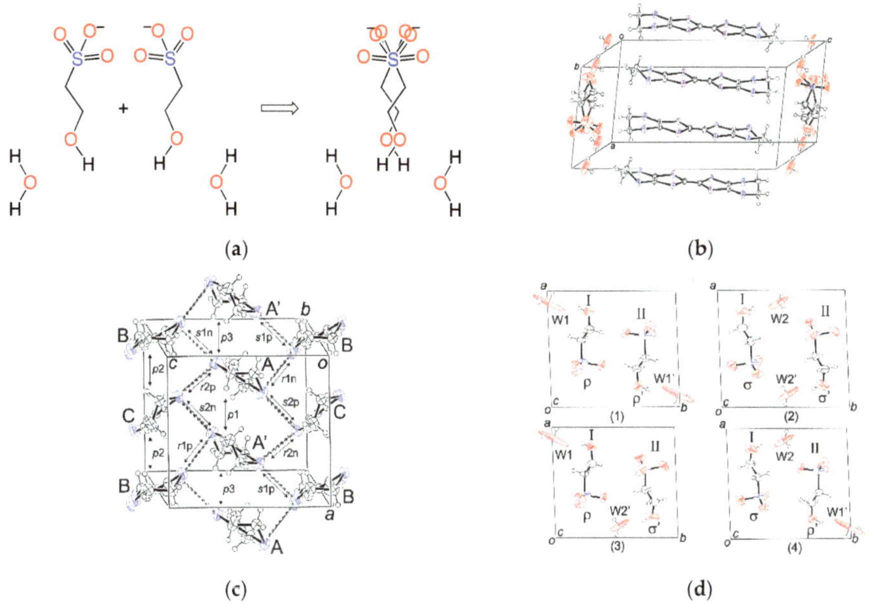

Figure 3. (**a**) A schematic diagram of the disorder of the isethionate anion. (**b**) Crystal structure. (**c**) Arrangement of the donor layer, in which donors A, B and C are crystallographically independent and A′ is the centrosymmetrically inverted A. Dashed lines indicate short S···S, Se···S and Se···Se contacts (<3.70, 3.75 and 3.80 Å, respectively). Transfer integrals are labelled as *p*1, *p*2, etc. (**d**) Four possible patterns of the donor layer of **3**. The unit cell has two disordered water molecules (W1 and W2) and two isethionate anions (I and II), the latter of which are disordered over two conformations (ρ and σ) as schematically shown in Figure 3a. ρ′ and σ′ are the symmetrically inverted ρ and σ.

The donor arrangement (α-type), the numbers of independent BETS molecules (one (A) and two halves (B and C)), and the cell parameters of **3** are quite similar to those of α-$(BETS)_2I_3$ [12], suggesting that the transport property of **3** would also be similar. α-$(BETS)_2I_3$ shows a metal–insulator (MI) transition at 50 K, and there is a shoulder at 40 K, suggesting another phase transition, below which the resistivity increases more rapidly [13]. The temperature-dependent electrical resistivity of **3** is shown in Figure 4a, where four crystals (Samples A–D) were measured at the same time; therefore, the cooling and heating rates are the same. However, the four samples show a wide variety of temperature dependences, and therefore, the samples were measured again. The resistivity curves of the second run are shown in Figure 4b. The behaviours of samples A, B and C are clearly different from each other. Moreover, each sample shows a different behaviour from the first measurement apart from sample C. The resistivity of almost all experiments shows MI transitions at around 70 K, and shoulders can be seen in some curves at around 30 K. The general behaviours are similar to those of α-$(BETS)_2I_3$. However, the ratio of the largest and smallest resistivity values (ρ_{max}/ρ_{min}) of sample C of **3**, which shows the sharpest MI transition of all the measured samples, and of α-$(BETS)_2I_3$ are 4.6 and 570, respectively. This observation indicates that the MI transition of **3** is much less sharp than that of α-$(BETS)_2I_3$.

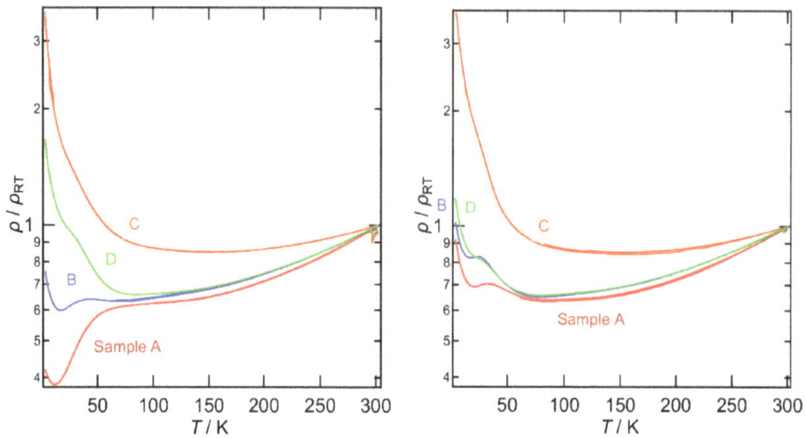

(**a**) (**b**)

Figure 4. Temperature dependences of electrical resistivity of four crystals of **3**, where (**a**) is the first and (**b**) is the second run.

The temperature-dependent magnetic susceptibility of α-$(BETS)_2I_3$ was fitted by $\chi^{\mathrm{Spin}} = C/T \exp(-\Delta/k_BT)$ (equation 1) with the gap Δ/k_B = 290 K and a Curie constant C = 0.29 emu mol^{-1} [23]. The magnetic behaviour of **3** (Figure 5) is similar to that of α-$(BETS)_2I_3$. The susceptibility decreases with decreasing temperature, where the differences in the susceptibilities between room and low temperatures are approximately 2.7×10^{-4} emu mol^{-1}, which is 1.0×10^{-4} emu mol^{-1} smaller than that of α-$(BETS)_2I_3$. The data of **3** cannot be fitted by equation 1 and are closer to a 1D Heisenberg model (green line in Figure 5) with J = −213 K. However, the large deviations do not suggest that the spins on the BETS layer forms a 1D chain. This may indicate that more spins or more itinerant electrons exist at low temperatures in salt **3** when compared to α-$(BETS)_2I_3$.

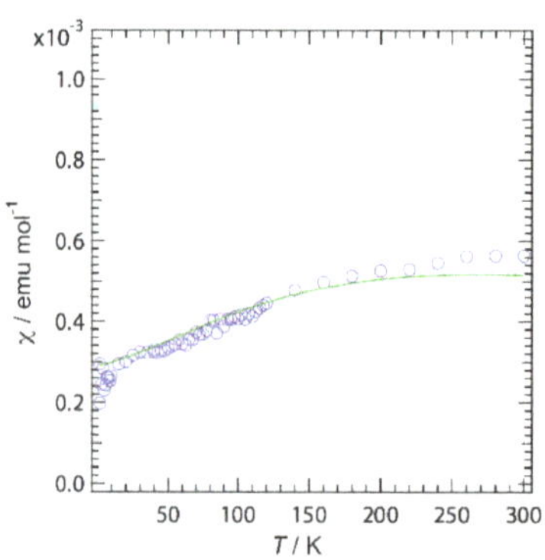

Figure 5. Temperature dependence of magnetic susceptibility of **3**. The green solid line is calculated on the basis of a 1D Heisenberg model.

The electronic structure was characterized using extended Hückel tight-binding band structure calculations of the donor layer [24] as shown in Figure 6. The Fermi surfaces are almost the same as for α-$(BETS)_2I_3$ [25]. Therefore, **3** and α-$(BETS)_2I_3$ have the same electronic structures.

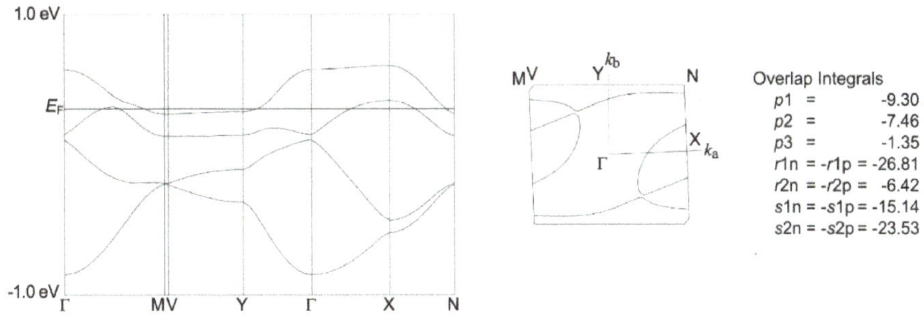

Figure 6. Band dispersions (**left**), Fermi surfaces (**centre**) and overlap integrals (**right**) of **3**. The donor···donor interactions ($p1$, $p2$, etc.) are labelled in Figure 3c.

4. Discussion

Although the molecular structures of the triiodide and isethionate BETS salts are very different, the cell parameters of **3** are quite similar, as are the donor arrangements and band structures. The α-(BETS)$_2$I$_3$ salt is isomorphous to α-(BEDT-TTF)$_2$I$_3$, which also shows a MI transition at 135 K (T_{MI}) [12]. Therefore, it is believed that both MI transition mechanisms are similar. The mechanism of the MI transition of α-(BEDT-TTF)$_2$I$_3$ has been widely studied ([12,13,25–28] and references therein). At room temperature, the positive charges are equally distributed in the BEDT-TTF donor layers. However, below T_{MI}, two BEDT-TTF molecules (A and B) become monocations and two become neutral (A′ and C), which is charge ordering (CO). In addition, the space group changes into the polar space group $P1$, and the previously crystallographically equivalent A and A′ molecules become independent. The crystal also becomes polar below the transition temperature [29]. Recently, J.-P. Pouget and co-workers reported a more consistent mechanism of the MI transition [27,28], where modulating the strength of the donor–anion hydrogen bonding modified the charges on the HOMOs of the donors; namely, I···H-C hydrogen bonds played a key role in the MI transition. Due to the similar structural and electronic properties, we expect that the salt **3** also shows a similar CO transition.

Figure 7a,b shows the structural relationship between -SO$_3^-$ of the HOC$_2$H$_4$SO$_3^-$ anions and the donors. There are no short contacts between donor C (Figure 3) and the anions, suggesting that the negatively charged sulfonate groups are not close to C but are close to A, A′ and B. Therefore, donor C is likely to be neutral or have a small positive charge in the CO state. In our previously reported sulfonate salts with CO ground states [30–32], the donor having the shortest S(BEDT-TTF)···O(-SO$_3^-$) contacts (especially less than 3.1 Å), which are more direct interactions between positive and negative charges than those via the hydrogen bonds [27,28], has the largest positive charge because the state has the largest Coulomb energy. Moreover, as shown in Table 2, of the two anions in the unit cell with ρ and ρ′ conformations, only donor B has two contacts shorter than 3.1 Å. Therefore, in this situation, for example, only donor B becomes charge-rich and A, A′ and C donors become charge-poor molecules. This suggests that the configuration of the anion is relevant in determining the electronic structure of the CO state. Figure 8 shows a schematic diagram of the CO state, which includes six possible CO patterns, (b)–(g), in which C is neutral in patterns (b), (d) and (g). Therefore, one of these three states is likely to be the ground state of **3**. In addition, the ground state of α-(BEDT-TTF)$_2$I$_3$ is represented by Figure 8b [26], suggesting that those of α-(BETS)$_2$I$_3$ and **3** would also be depicted by pattern (b). To strengthen this hypothesis, we calculated the Madelung energies of the CO patterns of (b)–(g) (also Table S1). In the calculations, the formula charges of each donor molecule (Q) were fixed to 0 or +1 for (b)–(g), which may be far from the real CO state, in which, normally, $0 < Q < 1$. However, we believe that the calculations point to the stability of each CO pattern. The charge distributions in the component molecules were calculated using GAMESS at the BY3LP 6-31G* level for A, B and C donors and all crystallographically independent anions of α-(BEDT-TTF)$_2$I$_3$, α-(BETS)$_2$I$_3$ and **3**, respectively. Then, each molecule's Hirshfeld population analysis was performed using Multiwfn [21].

The resultant point charges (where we did not use the point charges of all H_2O molecules in **3** because of the uncertainty of hydrogen atom positions in x-ray crystal structures. We also performed the calculations in which the charges of tentatively oriented all water molecules were included as shown in Table S1. The tendency of the results is similar to that shown in Table 3) were input into Gulp [22] to calculate each Madelung energy using an Ewald method, in which we calculated only the intermolecular Coulomb interactions. The resultant Madelung energies are summarized in Table 3. The anion in **3** has two disordered configurations on a general position, so the Madelung energies were calculated of all possible combinations of each CO pattern (a)–(g), as shown in Figure 8, and each anion arrangement pattern (1)–(4), as shown in Figure 3d.

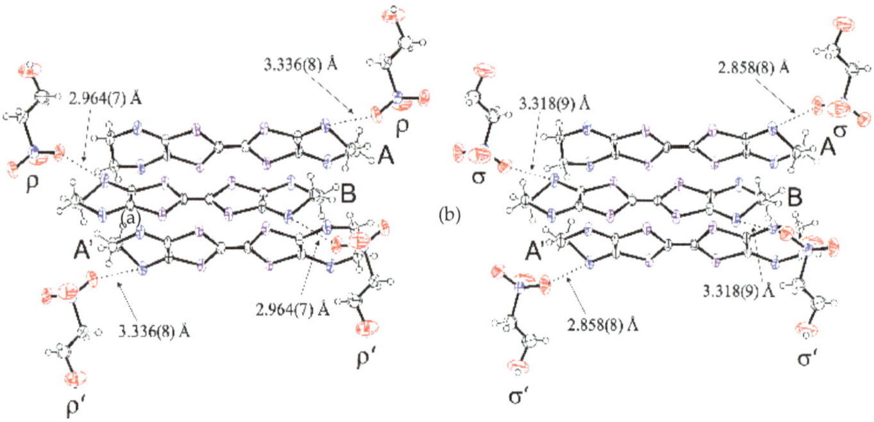

(**a**) (**b**)

Figure 7. Interactions between bis(ethylenedithio)tetraselenafulvalene (BETS) and each disordered conformer ((**a**) ρ and (**b**) σ) of isethionate in **3**, where dotted lines indicate S···O contacts shorter than the van der Waals distance (3.37 Å).

Table 2. Numbers (*N*) of effective short S···O contacts (<3.1 Å) of **3** for four possible anion arrangements.

Configurations of Anions		*N* with Donor A or A′	*N* with Donor B	*N* with Donor C
ρ	ρ′	0	2	0
σ	σ′	2	0	0
ρ	σ′	1	1	0
σ	ρ′	1	1	0

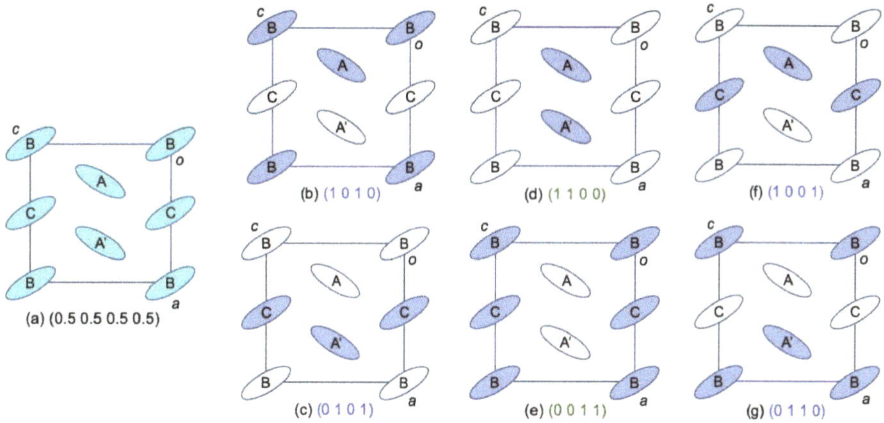

Figure 8. Schematic diagram of α-type packing arrangement of donor layer. (**b**–**g**) Schematics of the full set of six possible arrangements of neutral (white) and cationic (dark blue) donor molecules in the conduction plane, where ($w\ x\ y\ z$) means the molecular charges of A, A′, B and C molecules (see Figure 3c) of w, x, y and z, respectively. Blue ($w\ x\ y\ z$) and green ($w\ x\ y\ z$) indicate that the charge-ordered patterns are horizontal and vertical stripes, respectively. The same colours are used in Table 3. (**a**) A schematic diagram of the non-charge-ordered state, where each molecule has a +1/2 charge (light blue).

Table 3. Madelung energies of α-(BEDT-TTF)$_2$I$_3$, α-(BETS)$_2$I$_3$ and **3**.

α-(BEDT-TTF)$_2$I$_3$									
CO pattern *1	BEDT-TTF charge							Madelung E *4	ΔE *5
	A	A′	B	C					
(b) Horizontal *2	+1	0	+1	0				−442.8454	0.2540
(c) Horizontal	0	+1	0	+1				−443.0994	0.0000
(d) Vertical *3	+1	+1	0	0				−426.1212	16.9782
(e) Vertical	0	0	+1	+1				−424.6368	18.4626
(f) Horizontal	+1	0	0	+1				−443.0912	0.0082
(g) Horizontal	0	+1	+1	0				−442.8319	0.2675
α-(BETS)$_2$I$_3$									ΔE *5
CO pattern	BETS charge							Madelung E	
	A	A′	B	C					
(b) Horizontal	+1	0	+1	0				−414.7377	0.1976
(c) Horizontal	0	+1	0	+1				−414.9353	0.0000
(d) Vertical	+1	+1	0	0				−401.2690	13.6663
(e) Vertical	0	0	+1	+1				−399.0569	15.8784
(f) Horizontal	+1	0	0	+1				−414.9291	0.0062
(g) Horizontal	0	+1	+1	0				−414.7411	0.1942
	3, α-(BETS)$_2$(HOC$_2$H$_4$SO$_3$)·H$_2$O								ΔE *5
CO pattern	BETS charge				Configuration *6			Madelung E	
	A	A′	B	C		I	II		
(b) Horizontal	+1	0	+1	0	(1)	ρ	ρ′	−444.3595	21.9183
〃 *7	〃	〃	〃	〃	(2)	σ	σ′	−441.4648	24.8130
〃	〃	〃	〃	〃	(3)	ρ	σ′	−461.4680	4.8098
〃	〃	〃	〃	〃	(4)	σ	ρ′	−459.1554	7.1224
〃	〃	〃	〃	〃		Average		−451.6119	14.6659

Table 3. *Cont.*

(c) Horizontal	0	+1	0	+1	(1)	ρ	ρ′	−446.8165	19.4613
"	"	"	"	"	(2)	σ	σ′	−448.6298	17.648
"	"	"	"	"	(3)	ρ	σ′	−463.9675	2.3103
"	"	"	"	"	(4)	σ	ρ′	−466.2778	0.0000
"	"	"	"	"		Average		−456.4229	9.8549
(d) Vertical	+1	+1	0	0	(1)	ρ	ρ′	−443.9287	22.3491
"	"	"	"	"	(2)	σ	σ′	−416.8843	49.3935
"	"	"	"	"	(3)	ρ	σ′	−447.8066	18.4712
"	"	"	"	"	(4)	σ	ρ′	−447.8055	18.4723
"	"	"	"	"		Average		−439.1063	27.1715
(e) Vertical	0	0	+1	+1	(1)	ρ	ρ′	−416.4921	49.7857
"	"	"	"	"	(2)	σ	σ′	−442.4551	23.8227
"	"	"	"	"	(3)	ρ	σ′	−446.8737	19.4041
"	"	"	"	"	(4)	σ	ρ′	−446.8726	19.4052
"	"	"	"	"		Average		−438.1734	28.1044
(f) Horizontal	+1	0	0	+1	(1)	ρ	ρ′	−446.8152	19.4626
"	"	"	"	"	(2)	σ	σ′	−448.6285	17.6493
"	"	"	"	"	(3)	ρ	σ′	−466.2777	0.0001
"	"	"	"	"	(4)	σ	ρ′	−463.9651	2.3127
"	"	"	"	"		Average		−456.4216	9.8562
(g) Horizontal	0	+1	+1	0	(1)	ρ	ρ′	−444.3614	21.9164
"	"	"	"	"	(2)	σ	σ′	−441.4667	24.8111
"	"	"	"	"	(3)	ρ	σ′	−459.1583	7.1195
"	"	"	"	"	(4)	σ	ρ′	−461.4688	4.8090
"	"	"	"	"		Average		−451.6138	14.664

*1 CO patterns = charge-ordering patterns shown in Figure 8; *2 Horizontal = horizontal stripe; *3 Vertical = vertical stripe; *4 kJ mol^{-1} (unit cell); *5 $\Delta E = E - E$ (the lowest in the salt); *6 Configuration = Anion Configuration at I and II positions in the unit cell as shown in Figure 3d; *7 A ditto mark.

As shown in Table 3, the Madelung energies of α-(BEDT-TTF)$_2$I$_3$ of four horizontal stripe-type CO patterns (Horizontal) of (b), (c), (f) and (g) are similar and smaller than those of the vertical stripe-type CO patterns (Vertical) of (d) and (e), indicating that Horizontals are more stable than Verticals. However, the difference between the largest and smallest values of the four Horizontals are small (less than 0.25 kJ mol^{-1}) such that we cannot say which is the most stable state. Similarly, the energies of α-(BETS)$_2$I$_3$ of four Horizontals are similar and smaller than those of the Verticals, also indicating that Horizontals are more stable than Verticals, but it cannot be suggested which state is the most stable due to the small difference between the largest and smallest values (<0.20 kJ mol^{-1}). A similar tendency was also observed in **3**. The averages of the Madelung energies of (1)–(4) configurations of each CO pattern suggest that Horizontals are more stable than Verticals. However, the difference between the largest and smallest values of the averages of the Horizontals of (b), (c), (f) and (g) are larger than those of the two I$_3$ salts. Those of (c) and (f) are approximately 5 kJ mol^{-1} smaller than those of (b) and (g), suggesting that (c) or (f) are possible ground states of **3**. Indeed, the the difference between the largest and smallest values of the Madelung energies of the (1), (2), (3) and (4) of each CO pattern is 17.7–30.9 kJ mol^{-1}. This indicates that the conformational effect of the anion is energetically more effective than that of the CO patterns of the donor layers. Table 3 indicates that the (c)-(4) ($E = -466.2778$) and (f)-(3) states ($E = -466.2777$ kJ mol^{-1}) have the lowest energies, where (c)-(4) is (A′ C σ ρ′) and (f)-(3) is (A C ρ σ′), so centrosymmetircally inverted (c)-(4) is equivalent to (f)-(3) and vice versa. This indicates that the ground state of **3** is (c)-(4) = (f)-(3). However, if the CO pattern and the anion configuration are fixed to (c)-(4) = (f)-(3), at least above T_{MI}, the MI transition of **3** is likely to be as sharp as in α-(BETS)$_2$I$_3$, which is not the case.

The disorder of the anion as shown in Figure 3a is related to the puckering motion of the ethylene group of isethionate. Similar puckering motions are observed in several BEDT-TTF salts [33], in which a puckering motion is seen in the terminal ethylene groups of BEDT-TTF. The motion in ET salts is frozen at around 100 K, where a glass transition occurs, indicating that perfect ordering of the ethylene groups is not realized at the time scale. If the anionic puckering motion in **3** is also frozen to form a glassy state, this is a disordered state where the full set of the configurations of (1)–(4) exist below a certain temperature. Then, the CO pattern is frozen. This type of disorder would provide a relatively large random potential because the Madelung energies (1)–(4) show a larger energy variation than those of (b)–(g). This random potential would depend on the measurement and/or sample conditions. Therefore, it is speculated that this is the reason why the salt **3** shows a broad MI transition and has large variations in the resistivity measurements of repeated samples. In addition, the random potential caused by the disorder of the anion might provide paramagnetic and/or metallic islands below T_{MI}, which have a non-zero magnetic susceptibility. This agrees with the fact that the smallest magnetic susceptibility value at low temperature of **3** is larger than that of α-(BETS)$_2$I$_3$. We anticipate that some samples of **3** show either metallic or insulating behaviours when changing the cooling rates of the resistivity measurement. An emergent, passively even superconducting, phase may arise from the phase boundaries between the metallic and insulating phase, the investigation of which is now in progress.

5. Conclusions

Three purely organic charge-transfer salts—α-(BEDT-TTF)$_3$(HOC$_2$H$_4$SO$_3$)$_2$ (**1**), β-(BEST)$_3$(HOC$_2$H$_4$SO$_3$)$_2$·H$_2$O (2) and α-(BETS)$_2$(HOC$_2$H$_4$SO$_3$)·H$_2$O (**3**)—have been prepared Salts **1** and **2** are semiconducting with ρ_{RT} of 1.15 × 10^2 and 4.47 Ω·cm, and E_a of 190 and 108 meV, respectively. Salt **3** is metallic, with a broad MI transition at around 70 K. The HOC$_2$H$_4$SO$_3{}^-$ anion in **3** is disordered, the randomness of which appears to make the resistivity behaviour vary as a function of the measured sample, which is supported by Madelung energy calculations.

Supplementary Materials: The following are available online at http://www.mdpi.com/2073-4352/10/9/775/s1, GAMESS input options; Figure S1: Temperature dependence of electrical resistivity of **1**; Figure S2: Temperature dependence of electrical resistivity of **2**; Table S1: Madelung energies of **3** calculated excluding and including the point charges of water molecules.

Author Contributions: H.A. conceived and designed the experiments; K.F. synthesized the BETS molecule; Y.K. synthesized the BEST molecule and all the charge-transfer salts; H.A. and Y.K. participated in all the physical measurements; H.A., S.S.T. and Y.N. wrote and/or reviewed the paper with contributions from all authors. All authors have read and agreed to the published version of the manuscript.

Funding: This research was partially funded by JSPS KAKENHI, grant numbers 17K05751, 17H01144.

Acknowledgments: This article is dedicated to Professor Peter Day. The authors thank Shusuke Yamanaka of Osaka University for helpful discussions on Madelung energy calculations.

Conflicts of Interest: The authors declare no conflict of interest.

References

1. Williams, J.M.; Ferraro, J.R.; Thorn, R.J.; Carlson, K.D.; Geiser, U.; Wang, H.H.; Kini, A.M.; Whangbo, M.-H. *Organic Superconductor (Including Fullerenes) Synthesis, Structure, Properties and Theory*; Prentice-Hall: New York, NY, USA, 1992.
2. Mori, T. *Electronic Properties of Organic Conductors*; Springer: Tokyo, Japan, 2016.
3. Ishiguro, T.; Yamaji, K.; Saito, G. *Organic Superconductors*, 2nd ed.; Springer: Berlin/Heidelberg, Germany, 1998.
4. Lebed, A. *The Physics of Organic Superconductors and Conductors*; Springer: Berlin, Germany, 2008.
5. Yamada, J.; Nishikawa, H. Design of Conducting and Superconducting Organic Molecules. In *Encyclopedia of Physical Organic Chemistry*; Wang, Z., Wille, U., Juaristi, E., Eds.; John Wily & Sons Inc.: Hoboken, NJ, USA, 2017; Volume 3, pp. 2133–2188.
6. Urayama, H.; Yamochi, H.; Saito, G.; Nozawa, K.; Sugano, T.; Kinoshita, M.; Sato, S.; Oshima, K.; Kawamoto, A.; Tanaka, A. A new ambient pressure organic superconductor based on BEDT-TTF with T_C higher than 10 K (T_C = 10.4 K). *Chem. Lett.* **1988**, *17*, 55–58. [CrossRef]
7. Wang, H.H.; Geiser, U.; Williams, J.M.; Mason, J.M.; Perr, J.T.; Heindl, J.E.; Lathrop, M.W.; Love, B.J.; Watkins, D.M.; Yaconi, G.A. Phase selectivity in the simultaneous synthesis of the T_c = 12.8 K (0.3 kbar)

organic superconductor κ-$(BEDT\text{-}TTF)_2Cu[N(CN)_2]Cl$ or the semiconductor $(BEDT\text{-}TTF)Cu[N(CN)_2]_2$. *Chem. Mater.* **1992**, *4*, 247–249. [CrossRef]

8. Kini, A.M.; Geiser, U.; Wang, H.H.; Carlson, K.D.; Williams, J.M.; Kwok, W.K.; Vandervoort, K.G.; Thompson, J.E.; Stupka, D.L. A new ambient-pressure organic superconductor, κ-$(ET)_2Cu[N(CN)_2]Br$, with the highest transition temperature yet observed (Inductive Onset T_c = 11.6 K, Resistive Onset = 12.5 K). *Inorg. Chem.* **1990**, *29*, 2555–2557. [CrossRef]
9. Mori, H.; Tanaka, S.; Mori, T. Systematic study of the electronic state in theta-type BEDT-TTF organic conductors by changing the electronic correlation. *Phys. Rev. B* **1998**, *57*, 12023–12029. [CrossRef]
10. Oshima, M.; Mori, H.; Saito, G.; Oshima, K. Crystal Structures and Electrical Properties of BEDT-TTF Salts of Mercury(II) Thiocyanate with and without K Ion. *Chem. Lett.* **1989**, *18*, 1159–1162. [CrossRef]
11. Wang, H.H.; Carlson, K.D.; Geiser, U.; Kwok, W.K.; Vashon, M.D.; Thompson, J.E.; Larsen, N.F.; McCabe, G.D.; Hulscher, R.S.; Williams, J.M. A new ambient-pressure organic superconductor: $(BEDT\text{-}TTF)_2(NH_4)Hg(SCN)_4$. *Phys. C* **1990**, *166*, 57–61. [CrossRef]
12. Bender, K.; Hennig, I.; Schweitzer, D.; Dietz, K.; Endres, H.; Keller, H.J. Synthesis, Structure and Physical Properties of a Two-Dimensional Organic Metal, Di[bis(ethylenedithiolo)tetrathiofulvalene] triiodide, $(BEDT\text{-}TTF)^+{}_2I^-{}_3$. *Mol. Cryst. Liq. Cryst.* **1984**, *108*, 359–371. [CrossRef]
13. Inokuchi, M.; Tajima, H.; Kobayashi, A.; Ohta, T.; Kuroda, H.; Kato, R.; Naito, T.; Kobayashi, H. Electrical and Optical Properties of α-$(BETS)_2I_3$ and α-$(BEDT\text{-}STF)_2I_3$. *Bull. Chem. Soc. Jpn.* **1995**, *68*, 547–553. [CrossRef]
14. Jérome, D.; Schulz, H.J. Organic conductors and superconductors. *Adv. Phys.* **1982**, *31*, 299–490. [CrossRef]
15. Kato, R.; Kobayashi, H.; Kobayashi, A. Synthesis and Properties of Bis(ethylenedithio)tetraselenafulvalene (BEDT-TSeF) Compounds. *Synth. Met.* **1991**, *42*, 2093–2096. [CrossRef]
16. Courcet, T.; Malfant, I.; Pokhodnia, K.; Cassoux, P. Bis(ethylenedithio)tetraselenafulvalene: Short-cut synthesis, X-raycrystal structure andp-electron density distribution. *New J. Chem.* **1998**, *22*, 585–589. [CrossRef]
17. Kini, A.N.; Gates, B.D.; Beno, M.A.; Williams, J.M. Bis(ethylenediseleno)tetrathiafulvalene: Convenient One-pot Synthesis and X-Ray Crystal Structure. *J. Chem. Soc. Chem. Commun.* **1989**, 169–171. [CrossRef]
18. Schmidt, M.W.; Baldridge, K.K.; Boatz, J.A.; Elbert, S.T.; Gordon, M.S.; Jensen, J.H.; Koseki, S.; Matsunaga, N.; Nguyen, K.A.; Su, S.; et al. General Atomic and Molecular Electronic Structure System. *J. Comput. Chem.* **1993**, *14*, 1347–1363. [CrossRef]
19. Suenaga, M. Facio: New Computational Chemistry Environmrnt for PC GAMESS. *J. Comput. Chem. Jpn.* **2005**, *4*, 25–32. [CrossRef]
20. Suenaga, M. Development of GUI for GAMESS/FMO Calculation. *J. Comput. Chem. Jpn.* **2008**, *7*, 33–54. [CrossRef]
21. Lu, T.; Chen, F. Multiwfn: A Multifunctional Wavefunction Analyzer. *J. Comput. Chem.* **2012**, *33*, 580–592. [CrossRef]
22. Gale, J.D. GULP—A computer program for the symmetry adapted simulation of solids. *J. Chem. Soc. Faraday Trans.* **1997**, *93*, 629–637. [CrossRef]
23. Hiraki, K.; Harada, S.; Arai, K.; Takano, Y.; Takahashi, T.; Tajima, N.; Kato, R.; Naito, T. Local Spin Susceptibility of α-D_2I_3 (D = bis(ethylendithio)tetraselenafulvalene (BETS) and bis(ethylendithio)dithiadiselenafulvalene (BEDT-STF)) Studied by ^{77}Se NMR. *J. Phys. Soc. Jpn.* **2011**, *80*, 14715. [CrossRef]
24. Mori, T.; Kobayashi, A.; Sasaki, Y.; Kobayashi, H.; Saito, G.; Inokuchi, H. The Intermolecular Interaction of Tetrathiafulvalene and Bis(ethylenedithio)tetrathiafulvalene in Organic Metals. Calculation of Orbital Overlaps and Models of Energy-band Structures. *Bull. Chem. Soc. Jpn.* **1984**, *57*, 627–633. [CrossRef]
25. Kondo, R.; Kagoshima, S.; Tajima, N.; Kato, R. Crystal and Electronic Structures of the Quasi-Two-Dimensional Organic Conductor α-$(BEDT\text{-}TTF)_2I_3$ and Its Selenium Analogue α-$(BEDT\text{-}TSeF)_2I_3$ under Hydrostatic Pressure at Room Temperature. *J. Phys. Soc. Jpn.* **2009**, *78*, 114714. [CrossRef]
26. Kakiuchi, T.; Wakabayashi, Y.; Sawa, H.; Takahashi, T.; Nakamura, T. Charge Ordering in α-$(BEDT\text{-}TTF)_2I_3$ by Synchrotron X-ray Diffraction. *J. Phys. Soc. Jpn.* **2007**, *76*, 113702. [CrossRef]
27. Pouget, J.-P.; Alemany, P.; Canadell, E. Donor-anion interactions in quarter-filled low-dimensional organic conductors. *Mater. Horiz.* **2018**, *5*, 590–640. [CrossRef]
28. Alemany, P.; Pouget, J.-P.; Canadell, E. Essential role of anions in the charge ordering transition of α-$(BEDT\text{-}TTF)_2I_3$. *Phys. Rev. B* **2012**, *85*, 195118. [CrossRef]

29. Yamamoto, K.; Iwai, S.; Boyko, S.; Kashiwazaki, A.; Hiramatsu, F.; Okabe, C.; Nishi, N.; Yakushi, K. Strong Optical Nonlinearity and its Ultrafast Response Associated with Electron Ferroelectricity in an Organic Conductor. *J. Phys. Soc. Jpn.* **2008**, *77*, 074709. [CrossRef]
30. Akutsu, H.; Yamada, J.; Nakatsuji, S.; Turner, S.S. A novel BEDT-TTF-based purely organic magnetic conductor, α-(BEDT-TTF)$_2$(TEMPO-N(CH$_3$)COCH$_2$SO$_3$)·3H$_2$O. *Solid State Commun.* **2006**, *140*, 256–260. [CrossRef]
31. Akutsu, H.; Yamada, J.; Nakatsuji, S.; Turner, S.S. A New BEDT-TTF-Based Organic Charge Transfer Salt with a New Anionic Strong Acceptor, *N,N′*-Disulfo-1,4-benzoquinonediimine. *Crystals* **2012**, *2*, 182–192. [CrossRef]
32. Akutsu, H.; Ishihara, K.; Yamada, J.; Nakatsuji, S.; Turner, S.S.; Nakazawa, Y. A strongly polarized organic conductor. *CrystEngComm* **2016**, *18*, 8151–8154. [CrossRef]
33. Akutsu, H.; Saito, K.; Sorai, M. Phase behavior of the organic superconductors κ-(BEDT-TTF)$_2$Cu[N(CN)$_2$] *X* (*X* = Br and Cl) studied by ac calorimetry. *Phys. Rev. B* **2000**, *61*, 4346–4352. [CrossRef]

crystals

MDPI

Article

Formation of Three-Dimensional Electronic Networks Using Axially Ligated Metal Phthalocyanines as Stable Neutral Radicals

Ryoya Sato and Masaki Matsuda *

Department of Chemistry, Kumamoto University, 2-39-1 Kurokami, Kumamoto 860-8555, Japan; 207d1308@st.kumamoto-u.ac.jp
* Correspondence: masaki@kumamoto-u.ac.jp

Received: 3 August 2020; Accepted: 21 August 2020; Published: 24 August 2020

Abstract: Organic π-radical crystals are potential single-component molecular conductors, as they involve charge carriers. We fabricated new organic π-radical crystals using axially ligated metal phthalocyanine anions ($[M^{III}(Pc)L_2]^-$) as starting materials. Electrochemical oxidation of $[M^{III}(Pc)L_2]^-$ afforded single crystals of organic π-radicals of the type $M^{III}(Pc)Cl_2{\cdot}THF$ (M = Co or Fe, THF = tetrahydrofuran), where the π-conjugated macrocyclic phthalocyanine ligand is one-electron oxidized. The X-ray crystal structure analysis revealed that $M^{III}(Pc)Cl_2$ formed three-dimensional networks with π-π overlaps. The electrical resistivities of $Co^{III}(Pc)Cl_2{\cdot}THF$ and $Fe^{III}(Pc)Cl_2{\cdot}THF$ at room temperature along the a-axis were 6×10^2 and 6×10^3 Ω cm, respectively, and were almost isotropic, meaning that $M^{III}(Pc)Cl_2{\cdot}THF$ had three-dimensional electronic systems.

Keywords: organic π-radical; molecular conductor; phthalocyanine; three-dimensional network; three-dimensional electronic system

1. Introduction

Ordinary organic radicals are unstable due to their high chemical reactivity, which is caused by the presence of an unpaired electron. However, organic radicals can be considered to have charge carriers, because their highest occupied molecular orbital (HOMO) is singly occupied, leading to a half-filled band in the solid state. Recently, a nonconjugated organic radical crystal has been reported to be conductive, although the conductivity is low in the solid state [1]. This might be due to the low degree of intermolecular overlap for the molecular orbital in which the radical lies. On the other hand, organic π-radicals tend to stack with significant overlap between adjacent molecules, giving rise to conducting pathways. For this reason, organic π-radicals have attracted significant interest from those studying molecular conductors, and it has been reported that despite the expected half-filled band, most organic π-radical crystals behave as Mott insulators because of the larger on-site Coulomb repulsion energy (U) compared to the transfer energy (t) [2,3]. However, external changes such as pressure can control the competition between U and t, leading to drastic phase transitions from the insulating state to metallic or superconducting states being reported for various molecular Mott insulators [3,4]. This means that if a stable neutral π-radical crystal is fabricated, it could be possible to obtain a single-component molecular conductor, which is a hot topic in the study of molecular crystals [5–9].

Axially ligated metal phthalocyanine anions ($[M^{III}(Pc)L_2]^-$) are attractive components for the construction of neutral π-radical crystals as single-component molecular conductors, because the wide π-conjugated system of phthalocyanine (Pc) might suppress U. Inabe expected that the electrochemical oxidation of the π-conjugated macrocyclic ligand in $[M^{III}(Pc)L_2]^-$ would give rise to conducting crystals, while the $M^{III}(Pc)L_2$ unit could impart many kinds of dimensionality to the molecular arrangement as the axial ligand induces slipped stacks (Figure 1) [10]. Indeed, it was found that there were two different types of π-π overlaps between the $M^{III}(Pc)L_2$ units (Figure 2). Although the π-π overlaps are significantly reduced compared to those seen in M(Pc)X-type conductors having face-to-face stacking (the overlap integral of type A is about 40% of that of M(Pc)X-type conductors) [11], it is enough for electrical conduction. Consequently, a large number of molecular conducting crystals composed of $M^{III}(Pc)L_2$ units were fabricated, including π-neutral radical crystals as well as crystals of highly conducting partially oxidized salts, with molecular arrangements showing the formation of one-dimensional, two-dimensional, and also three-dimensional networks [12–14]. Although all π-neutral radical crystals composed of $M^{III}(Pc)L_2$ are Mott insulators, their resistivities are relatively small.

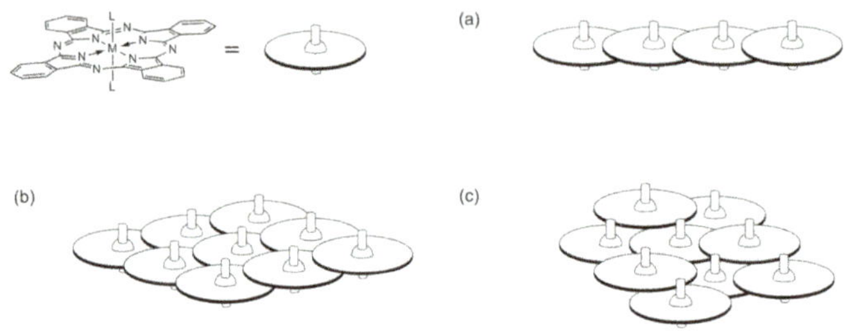

Figure 1. Possible dimensionalities of the molecular arrangements induced by slipped stacking of $M^{III}(Pc)L_2$: (**a**) one-dimensional arrangement, (**b**) two-dimensional arrangement, and (**c**) three-dimensional arrangement.

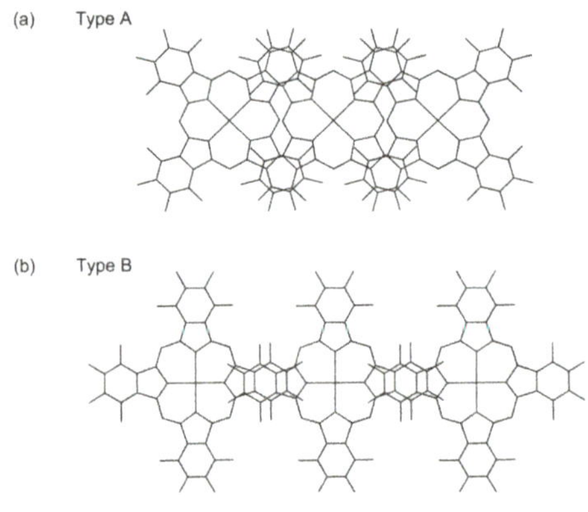

Figure 2. Two types of π-π overlaps between the $M^{III}(Pc)L_2$ units: (**a**) overlap involving two peripheral benzene rings per unit (type A); (**b**) overlap involving one peripheral benzene ring per unit (type B).

Herein, we focused on the dimensionality of $M^{III}(Pc)L_2$ crystals. It has been reported that the conductivity of $M^{III}(Pc)L_2$ radical crystals tends to increase as the dimensionality of the network is increased [14]; the first example of a three-dimensional neutral radical crystal of $Co^{III}(Pc)(CN)_2{\cdot}2H_2O$ exhibited a resistivity at room temperature of no more than 10^0 Ω cm with an activation energy of less than 0.1 eV [10]. However, only a few examples of three-dimensional systems exist at present [14]. We have fabricated a new three-dimensional system consisting of neutral radical crystals of $M^{III}(Pc)Cl_2{\cdot}THF$ (M = Co or Fe, THF = tetrahydrofuran). In this paper, the crystal structures and electrical properties of $M^{III}(Pc)Cl_2{\cdot}THF$ are reported.

2. Materials and Methods

2.1. Materials

Co^{II}(Pc) and Fe^{II}(Pc) were prepared by refluxing $CoCl_2 \cdot 6H_2O$ or $FeCl_2 \cdot 4H_2O$ with four molar equivalents of 1,2-dicyanobenzene in quinoline, and Co^{II}(Pc) or Fe^{II}(Pc) (3.0 g, 5.2 mmol) was treated with $SOCl_2$ (5 mL, 69 mmol) in nitrobenzene (70 mL) at 70 °C for 6 h to produce $Co^{III}(Pc)Cl_2$ (yield 3.2 g, 5.0 mmol, 95%) or $Fe^{III}(Pc)Cl_2$ (yield 2.1 g, 3.3 mmol, 62%) [15,16]. Then, $K[M^{III}(Pc)(SCN)_2]$ was prepared according to the literature [16,17], by reacting $M^{III}(Pc)Cl_2$ (2.0 g, 3.1 mmol) with KSCN (6.0 g, 62 mmol) in acetone (200 mL) at 25 °C for 48 h (yield 2.1 g, 2.9 mmol, 94%).

Electrochemical oxidation of $K[M^{III}(Pc)(SCN)_2]$ (30 mg) with two equivalents of tetraethylammonium chloride or tetrabutylammonium chloride was performed under a constant current of 1 μA in a THF:acetonitrile (30 mL, 1:1 *v*/*v*) solution at 25 °C using an electrocrystallization cell equipped with a glass frit between the two compartments, and yielded black square-pillar crystals of $M^{III}(Pc)Cl_2 \cdot$THF on the anode. The typical size of the crystals obtained was between 0.5 × 0.3 × 0.3 mm^3 and 1.0 × 0.5 × 0.5 mm^3. The obtained radical crystals were stable, and no decomposition was detected in the X-ray crystal structures or electrical resistivity measurements over several months. The Schematic route for the synthesis is shown in Figure 3.

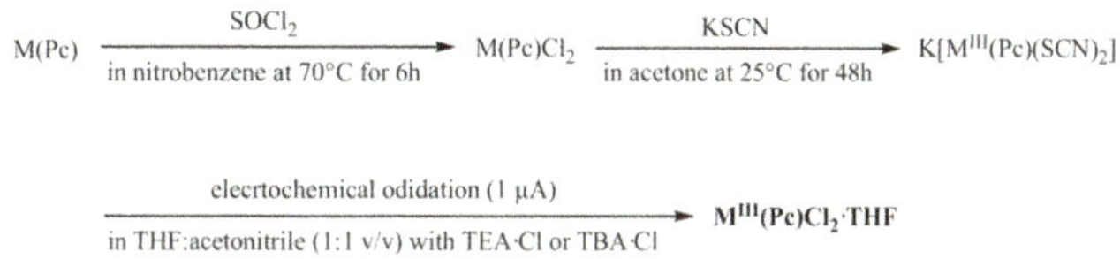

Figure 3. Schematic route for the synthesis of single crystals of $M^{III}(Pc)Cl_2 \cdot$THF.

2.2. X-ray Crystal Structure Determination

Crystal data for $M(Pc)Cl_2 \cdot$THF were collected at 298 K using an automated Rigaku SuperNova system (Rigaku, Tokyo, Japan) with monochromated Mo-Kα radiation (λ = 0.71073 Å). The structure was solved by using SIR2019 [18,19], and was refined by a full-matrix least-squares technique with SHELXL-2018/1 [20] using anisotropic and isotropic thermal parameters for non-hydrogen and hydrogen atoms, respectively. Orientational disorder was observed for the solvent site (THF molecule). We assigned C and O atoms according to the bond lengths.

Crystal data for $Co(Pc)Cl_2 \cdot$THF at 298 K: Formula, $C_{36}H_{24}Cl_2CoN_8O$, Formula weight 714.46, monoclinic, $P2_1/n$ (#14), a = 8.2526(2), b = 15.4184(6) Å, c = 11.8762(4) Å, β = 92.996(3)°, V = 1509.08(9) Å^3, Z = 2, d_{cal} = 1.572 g cm^{-3}, μ (MoKα) = 0.793 mm^{-1} (λ = 0.71073 Å), R_1 = 0.0545, wR_2 = 0.1852, GoF = 1.280. CCDC 2013803.

Crystal data for $Fe(Pc)Cl_2 \cdot$THF at 298 K: Formula, $C_{36}H_{24}Cl_2FeN_8O$, Formula weight 711.38, monoclinic, $P2_1/n$ (#14), a = 8.2062(1), b = 15.4115(1) Å, c = 11.9043(1) Å, β = 93.2238(9)°, V = 1503.15(2) Å^3, Z = 2, d_{cal} = 1.572 g cm^{-3}, μ (MoKα) = 0.727 mm^{-1} (λ = 0.71073 Å), R_1 = 0.0435, wR_2 = 0.1797, GoF = 1.503. CCDC 2013804.

CCDC 2013803 and 2013804 contain the supplementary crystallographic data for this paper. These data can be obtained free of charge via http://www.ccdc.cam.ac.uk/conts/retrieving.html (or from the CCDC, 12 Union Road, Cambridge CB2 1EZ, UK; Fax: +44-1223-336033; E-mail: deposit@ccdc.cam.ac.uk).

Extended Hückel calculations were performed using a CAESAR 2 software package (Prime Color Software, Inc., Charlotte, NC, USA) with the atomic parameters determined by the X-ray structure analysis. Default parameters were used.

2.3. Measurements

Electrical resistivity measurements along the *a*-axis of $M^{III}(Pc)Cl_2$·THF single crystals were performed by the standard four-probe method. The two-probe method was also adopted for the measurement with the current parallel and perpendicular to the *a*-axis. The electrical leads, which were gold wires with diameters of 25 μm, were attached to the crystal face using gold paste (Figure 4). The data are typical, as the resistivity and activation energies varied slightly depending on the individual crystal examined. However, we confirmed the reproducibility of the data by examining several crystals of both materials.

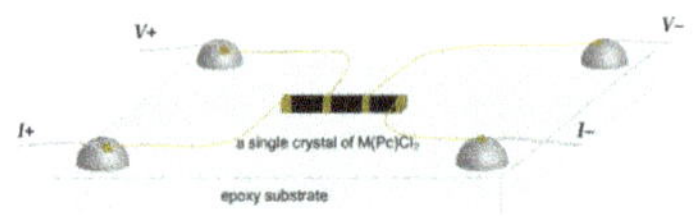

Figure 4. Schematic of the electrical resistivity measurements. Four or two gold wires were attached to a single crystal using gold paste.

3. Results

$Co^{III}(Pc)Cl_2$·THF and $Fe^{III}(Pc)Cl_2$·THF were found to have a monoclinic unit cell with the $P2_1/n$ space group, and were isostructural to each other (Table 1). Similar isostructural natures have been reported for $M(Pc)L_2$-based conductors [21–23], because the rigid geometry of the Pc framework is insensitive to the central ion. Figure 5 shows the crystal structure of $Co^{III}(Pc)Cl_2$·THF. Because the oxidation potentials of Co^{III} and Fe^{III} are higher than that of Pc^{2-} [10,24], the Pc ligand is oxidized to $Pc^{\bullet -}$, meaning $M^{III}(Pc)Cl_2$·THF is a neutral π-radical. The $M^{III}(Pc)Cl_2$ units form one-dimensional stacks along the *a*-axis via interactions between two peripheral benzene rings per unit, with interplanar distances of 3.49 and 3.54 Å (type A), and also form additional stacks along the $[11\bar{1}]$ and $[1\bar{1}\bar{1}]$ directions with one peripheral benzene ring per unit (type B) with interplanar distances of 3.43 and 3.44 Å, respectively, resulting in the formation of three-dimensional networks. To evaluate the effectiveness of the stacking, the overlap integrals between the HOMOs of adjacent Pcs along each stacking direction were calculated by the extended Hückel method based on the structural data (Table 2). Because the intermolecular overlap integral is usually regarded as being proportional to transfer energy, the estimation of the overlap integrals can be a useful index for investigating the anisotropies of molecular stacks [25], and it was found that the overlap integrals along the *a*-axis of type A stacks were almost twice those seen in other stacks.

Table 1. Crystal data of $Co^{III}(Pc)Cl_2$·THF and $Fe^{III}(Pc)Cl_2$·THF.

	$Co^{III}(Pc)Cl_2$·THF	$Fe^{III}(Pc)Cl_2$·THF
Space Group	$P2_1/n$	$P2_1/n$
a (Å)	8.2526(2)	8.2062(1)
b (Å)	15.4184(6)	15.4115(1)
c (Å)	11.8762(4)	11.9043(1)
β(°)	92.996(3)	92.2238(9)
Volume (Å^3)	1509.08(9)	1503.15(2)
Temperature (K)	298	298
CCDC	2013803	2013804

Figure 6 shows the temperature dependences of the electrical resistivities along the *a*-axes of $Co^{III}(Pc)Cl_2$·THF and $Fe^{III}(Pc)Cl_2$·THF. The electrical resistivities of $Co^{III}(Pc)Cl_2$·THF and $Fe^{III}(Pc)Cl_2$·THF at room temperature were 6.3×10^2 and 6.1×10^3 Ω cm, respectively, and both systems showed semi-conducting behavior, with activation energies of 0.20 eV for $Co^{III}(Pc)Cl_2$·THF and 0.24 eV for $Fe^{III}(Pc)Cl_2$·THF.

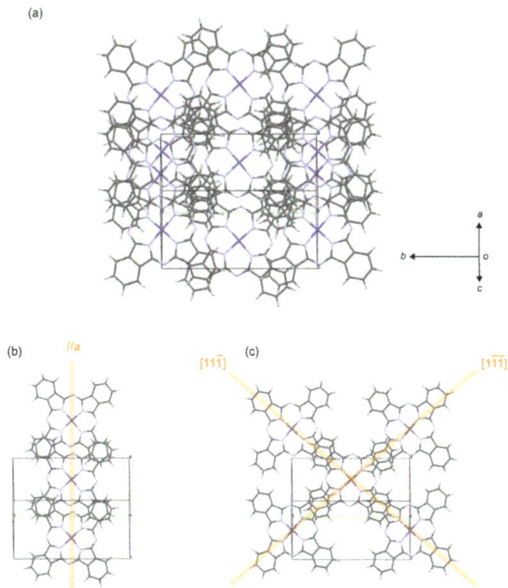

Figure 5. Crystal structure of $Co^{III}(Pc)Cl_2 \cdot THF$: (**a**) view perpendicular to Pc plane, (**b**) molecular arrangement along the *a*-axis, and (**c**) molecular arrangement along [11$\bar{1}$] and [1$\bar{1}\bar{1}$] directions. Tetrahydrofuran (THF) molecules have been omitted for clarity.

Table 2. Overlap integrals between HOMOs of adjacent Pcs in $M^{III}(Pc)Cl_2 \cdot THF$.

		[100] with Type A	[11$\bar{1}$] with Type B	[1$\bar{1}\bar{1}$] with Type B
$Co^{III}(Pc)Cl_2 \cdot THF$	Overlap Integral	2.0×10^{-3}	0.9×10^{-3}	0.9×10^{-3}
	Interplanar Distance (Å)	3.49, 3.54	3.43	3.44
$Fe^{III}(Pc)Cl_2 \cdot THF$	Overlap Integral	2.3×10^{-3}	0.9×10^{-3}	0.9×10^{-3}
	Interplanar Distance (Å)	3.49, 3.54	3.42	3.43

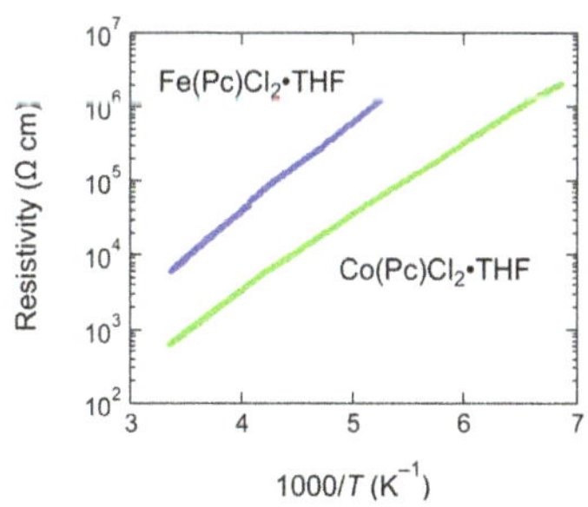

Figure 6. Temperature dependence of the electrical resistivities of $Co^{III}(Pc)Cl_2 \cdot THF$ (green) and $Fe^{III}(Pc)Cl_2 \cdot THF$ (blue) along the *a*-axis, measured by the standard four-probe method.

Figure 7 shows the temperature dependences of the electrical resistivities of $Co^{III}(Pc)Cl_2 \cdot THF$ and $Fe^{III}(Pc)Cl_2 \cdot THF$ along the *a*-axis (//*a*) and perpendicular to the *a*-axis ($\perp a$), as measured by the two-probe method. In both systems, the resistivity along the *a*-axis was smaller than that perpendicular to the *a*-axis, reflecting the difference in the overlap integrals; however, the current-direction dependence of the electrical resistivity was less than one order of magnitude. The activation energies of $Co^{III}(Pc)Cl_2$·THF and $Fe^{III}(Pc)Cl_2$·THF were almost independent of the direction of conduction.

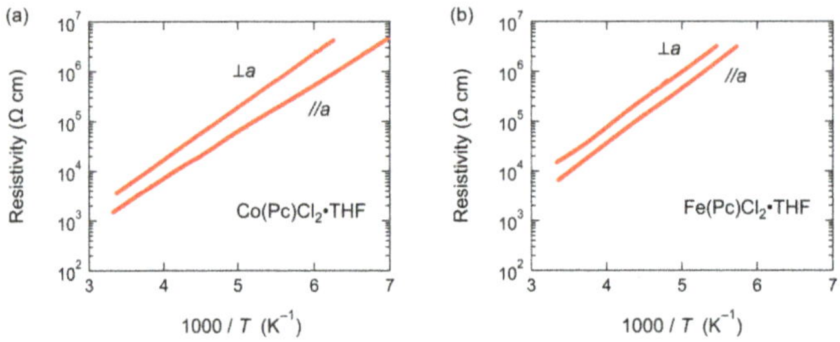

Figure 7. Temperature dependences of the electrical resistivities of (**a**) $Co^{III}(Pc)Cl_2$·THF and (**b**) $Fe^{III}(Pc)Cl_2$·THF along the *a*-axis (//*a*) and perpendicular to the *a*-axis (⊥*a*), as measured by the two-probe method.

4. Discussions

Despite the isostructural nature of the two compounds, the electrical resistivity of $Fe^{III}(Pc)Cl_2$·THF was almost one order of magnitude higher than that of $Co^{III}(Pc)Cl_2$·THF, and the activation energy was also higher. Similar enhancements of the electrical resistivity and activation energy have been reported for molecular conductors based on $M^{III}(Pc)L_2$ [21–23], and could be attributable to the different magnetic moments of the central metal ions. Low-spin Fe^{III} has a magnetic moment of $S = 1/2$, while low-spin Co^{III} has no magnetic moment ($S = 0$), and it has been reported that the magnetic moment enhances the electrical resistivity [26].

As shown in Figure 7, the electrical resistivity of $M^{III}(Pc)Cl_2$·THF depended on the current direction; however, the difference was quite small. In one- or two-dimensional systems, the resistivity along the stacking direction(s) was two orders of magnitude smaller than that perpendicular to the stacking direction(s) [27]. Therefore, the almost isotropic nature observed in the resistivity measurements in this study demonstrates that $M^{III}(Pc)Cl_2$·THF had a three-dimensional electronic system. However, the electrical resistivity of $Co^{III}(Pc)Cl_2$·THF at room temperature (6.3×10^2 Ω cm) was two orders of magnitude higher than that of $Co^{III}(Pc)(CN)_2{\cdot}2H_2O$, which was the first example of a three-dimensional neutral π-radical crystal of the type $M^{III}(Pc)L_2$ [10]. The activation energy of $M^{III}(Pc)Cl_2$·THF was also higher than that of $Co^{III}(Pc)(CN)_2{\cdot}2H_2O$. Therefore, we calculated the overlap integrals of $Co^{III}(Pc)(CN)_2{\cdot}2H_2O$ based on the reported crystal data, and compared the results with the overlap integrals of $Co^{III}(Pc)Cl_2$·THF (Table 3). Although the molecular arrangement of $M^{III}(Pc)Cl_2$ in $M^{III}(Pc)Cl_2$·THF appeared to be similar to that of $Co^{III}(Pc)(CN)_2$ in $Co^{III}(Pc)(CN)_2{\cdot}2H_2O$, the overlap integrals of $Co^{III}(Pc)(CN)_2{\cdot}2H_2O$ were quite different from those of $M^{III}(Pc)Cl_2$·THF. The overlap integral along the *c*-axis, where $Co^{III}(Pc)(CN)_2$ formed a type A one-dimensional stack, was 5.7×10^{-3}, while the overlap integrals along the [112] and [1$\bar{1}$2] directions, where $Co^{III}(Pc)(CN)_2$ formed type B stacks, were 1.7×10^{-3}. These values were more than twice those obtained for $Co^{III}(Pc)Cl_2$·THF. These discrepancies could be the cause of the difference in the electrical properties, meaning that slight changes in molecular arrangements could lead to drastic changes in the electrical properties in the neutral π-radical crystals composed of $M^{III}(Pc)L_2$.

Table 3. Comparison of the overlap integrals between similar three-dimensional systems of $Co^{III}(Pc)Cl_2$·THF and $Co^{III}(Pc)(CN)_2{\cdot}2H_2O$.

	with Type A	with Type B
$Co^{III}(Pc)Cl_2$·THF	2.0×10^{-3}	0.9×10^{-3}
$Co^{III}(Pc)(CN)_2{\cdot}H_2O$	5.7×10^{-3}	1.7×10^{-3}

5. Conclusions

We succeeded in fabricating single crystals of stable organic π-radicals of the type $M^{III}(Pc)Cl_2·THF$ (M = Co or Fe, THF = tetrahydrofuran). The overlap integrals between the HOMOs of adjacent Pcs were calculated based on the atomic parameters determined by the X-ray crystal structural analyses, and it was revealed that $M^{III}(Pc)Cl_2$ formed three-dimensional networks. Furthermore, the electrical resistivities measured were almost independent of the applied current direction, meaning that $M^{III}(Pc)Cl_2·THF$ had three-dimensional electronic systems, which are rarely observed in the study of molecular conductors. The resistivities of $Co^{III}(Pc)Cl_2·THF$ and $Fe^{III}(Pc)Cl_2·THF$ at room temperature were 6×10^2 and 6×10^3 Ω cm, respectively. These are similar to or significantly smaller than those reported for neutral radical conductors showing pressure-induced metallization [6,8], indicating that further study of three-dimensional Mott insulators composed of $M^{III}(Pc)L_2$ under the application of pressure could allow for phase transitions from insulating to metallic or superconducting states. Furthermore, it was revealed that a slight change in molecular arrangement led to drastic changes in the electrical properties. As $M^{III}(Pc)L_2$ shows various stacking types depending on the fabrication conditions, there is a strong possibility that new single-component molecular conductors composed of $M^{III}(Pc)L_2$ could be produced.

Author Contributions: Conceptualization, R.S. and M.M.; methodology, R.S.; software, R.S.; formal analysis, M.M.; writing—original draft preparation, M.M.; writing—review and editing, R.S. and M.M.; visualization, M.M.; supervision, M.M.; project administration, M.M.; funding acquisition, M.M. All authors have read and agreed to the published version of the manuscript.

Funding: This research was funded by a Grant-in-Aid for Scientific Research (B), grant number 19H02691, from the Japan Society for the Promotion of Science.

Acknowledgments: We dedicate this study to Tamotsu Inabe. We thank Akira Yoshiasa at Kumamoto University for his help with the X-ray diffraction measurements.

Conflicts of Interest: The authors declare no conflict of interest.

References

1. Yu, I.; Jo, Y.; Ko, J.; Kim, D.-Y.; Sohn, D.; Joo, Y. Making nonconjugated small-molecule organic radicals conduct. *Nano Lett.* **2020**, *20*, 5376–5382. [CrossRef] [PubMed]
2. Jérome, D. Organic conductors: From charge density wave TTF-TCNQ to superconducting $(TMTSF)_2PF_6$. *Chem. Rev.* **2004**, *104*, 5565–5591. [CrossRef] [PubMed]
3. Saito, G.; Yoshida, Y. Development of conductive organic molecular assemblies: Organic metals, superconductors, and exotic functional materials. *Bull. Chem. Soc. Jpn.* **2007**, *80*, 1–137. [CrossRef]
4. Ishiguro, T.; Yamaji, K.; Saito, G. *Organic Superconductors*, 2nd ed.; Springer: Berlin, Germany, 1998.
5. Wong, J.W.L.; Mailman, A.; Lekin, K.; Winter, S.M.; Yong, W.; Zhao, J.; Garimella, S.V.; Tse, J.S.; Secco, R.A.; Desgreniers, S.; et al. Pressure induced phase transitions and metallization of a neutral radical conductor. *J. Am. Chem. Soc.* **2014**, *136*, 1070–1081. [CrossRef]
6. Tian, D.; Winter, S.M.; Mailman, A.; Wong, J.W.L.; Yong, W.; Yamaguchi, H.; Jia, Y.; Tse, J.S.; Desgreniers, S.; Secco, R.A.; et al. The metallic state in neutral radical conductors: Dimensionality, pressure and multiple orbital effects. *J. Am. Chem. Soc.* **2015**, *137*, 14136–14148. [CrossRef]
7. Souto, M.; Cui, H.B.; Peña-Álvarez, M.; Baonza, V.G.; Jeschke, H.O.; Tomic, M.; Valenti, R.; Blasi, D.; Ratera, I.; Rovira, C.; et al. Pressure-induced conductivity in a neutral nonplanar spin-localized radical. *J. Am. Chem. Soc.* **2016**, *138*, 11517–11525. [CrossRef]
8. Souto, M.; Gullo, M.C.; Cui, H.B.; Casati, N.; Montisci, F.; Jeschke, H.O.; Valenti, R.; Ratera, I.; Rovira, C.; Veciana, J. Role of the open-shell character on the pressure-induced conductivity of an organic donor-acceptor radical dyad. *Chem. Eur. J.* **2018**, *24*, 5500–5505. [CrossRef]
9. Isono, T.; Kamo, H.; Ueda, A.; Takahashi, K.; Nakao, A.; Kumai, R.; Nakao, H.; Kobayashi, K.; Murakami, Y.; Mori, H. Hydrogen bond-promoted metallic state in a purely organic single-component conductor under pressure. *Nat. Commun.* **2013**, *4*, 1344. [CrossRef]

10. Inabe, T.; Maruyama, Y. Multi-dimensional stacking structure in phthalocyanine-based electrical conductors, K[Co(phthalocyaninato)$(CN)_2]_2 \cdot 5CH_3CN$ and Co(phthalocyaninato)$(CN)_2 \cdot 2H_2O$. *Bull. Chem. Soc. Jpn.* **1990**, *63*, 2273–2280. [CrossRef]
11. Hasegawa, H.; Naito, T.; Inabe, T.; Akutagawa, T.; Nakamura, T. A highly conducting partially oxidized salt of axially substituted phthalocyanine. Structure and physical properties of TPP[Co(Pc)$(CN)_2]_2$ {TPP = tetraphenylphosphonium, [Co(Pc)$(CN)_2$] = dicyano(phthalocyaninato)cobalt(III)}. *J. Mater. Chem.* **1998**, *8*, 1567–1570. [CrossRef]
12. Inabe, T.; Tajima, H. Phthalocyanines–Versatile components of molecular conductors. *Chem. Rev.* **2004**, *104*, 5503–5533. [CrossRef]
13. Inabe, T. Design of functional molecular crystal by controlling intermolecular interactions. *Bull. Chem. Soc. Jpn.* **2005**, *78*, 1373–1383. [CrossRef]
14. Inabe, T.; Ishikawa, M.; Asari, T.; Hasegawa, H.; Fujita, A.; Matsumura, N.; Naito, T.; Matsuda, M.; Tajima, H. Phthalocyanine conductors: New trend for crystal and functionality design. *Mol. Cryst. Liq. Cryst.* **2006**, *455*, 87–92. [CrossRef]
15. Myers, J.F.; Canham, G.W.R.; Lever, A.B.P. Higher oxidation level phthalocyanine complexes of chromium, iron, cobalt, and zinc. Phthalocyanine radical species. *Inorg. Chem.* **1975**, *14*, 461–468. [CrossRef]
16. Yu, D.E.C. Synthesis of Axially-Ligated Metallophthalocaynine Molecular Conductors and the Chemical and Physical Factors Affecting Their Solid-State Properties. Ph.D. Thesis, Hokkaido University, Sapporo, Japan, 2008.
17. Hedtmann-Rein, C.; Hanack, M.; Peters, K.; Peters, E.-M.; Schnering, H.G. Synthesis and properties of (phthlaocyaninato) and (tetrabenzoporphyrinato)cobalt(III) thiocyanate and isothiocyanate compounds. Crystal and molecular structures of (phthalocyaninato)(pyridine)(thiocyanato)cobalt(III). *Inorg. Chem.* **1987**, *26*, 2647–2651. [CrossRef]
18. Sir 2019. Available online: http://www.ba.ic.cnr.it/softwareic/sir/ (accessed on 25 May 2020).
19. Burla, M.C.; Caliandro, R.; Carrozzini, B.; Cuocci, C.; Mallamo, M.; Mazzone, A.; Polidori, G. Crystal structure determination and refinement via SIR 2014. *J. Appl. Cryst.* **2015**, *48*, 306–309. [CrossRef]
20. Sheldrick, G.M. Crystal structure refinement with SHELXL. *Acta Crystallogr. Sect. C Struct. Chem.* **2015**, *71*, 3–8. [CrossRef]
21. Matsuda, M.; Naito, T.; Inabe, T.; Hanasaki, N.; Tajima, H.; Otsuka, T.; Awaga, K.; Narymbetov, B.; Kobayashi, H. A one-dimensional macrocyclic π-ligand conductor carrying a magnetic center. Structure and electrical, optical and magnetic properties of TPP[Fe(Pc)$(CN)_2]_2$ {TPP = tetraphenylphosphonium and [Fe(Pc)$(CN)_2$] = dicyano(phthalocyaninato)iron(III)}. *J. Mater. Chem.* **2000**, *10*, 631–636. [CrossRef]
22. Matsuda, M.; Naito, T.; Inabe, T.; Hanasaki, N.; Tajima, H. Structure and electrical and magnetic properties of $(PTMA)_x$[M(Pc)$(CN)_2$]$\cdot y$(solvent) (PTMA = phenyltrimethylammonium and [M(Pc)$(CN)_2$] = dicyano(phthalocyaninato)M^{III} with M = Co and Fe). Partial oxidation by partial solvent occupation of the cation site. *J. Mater. Chem.* **2001**, *11*, 2493–2497. [CrossRef]
23. Matsuda, M.; Asari, T.; Naito, T.; Inabe, T.; Hanasaki, N.; Tajima, H. Structure and physical properties of low dimensional molecular conductors, [PXX][Fe^{III}(Pc)$(CN)_2$] and [PXX][Co^{III}(Pc)$(CN)_2$] (PXX = *peri*-xanthenoxanthene, Pc = phthalocyaninato). *Bull. Chem. Soc. Jpn.* **2003**, *76*, 1935–1940. [CrossRef]
24. Behnisch, R.; Hanack, M. Cyclic voltammetric and electrocrystallization studies of axially substituted biscyanophthalocyaninato metal complexes and related compounds. *Synth. Met.* **1990**, *36*, 387–397. [CrossRef]
25. Mori, T. Principles that govern electronic transport in organic conductors and transistors. *Bull. Chem. Soc. Jpn.* **2016**, *89*, 973–986. [CrossRef]
26. Ikeda, M.; Kanda, A.; Murakawa, H.; Matsuda, M.; Inabe, T.; Tajima, H.; Hanasaki, N. Effect of localized spin concentration on giant magnetoresistance in molecular conductors TPP[Fe_xCo_{1-x}(Pc)$(CN)_2]_2$. *J. Phys. Soc. Jpn.* **2016**, *85*, 024713. [CrossRef]
27. Morimoto, K.; Inabe, T. Molecular conductors based on axially substituted phthalocyanine neutral radicals. *Mol. Cryst. Liq. Cryst.* **1996**, *284*, 291–300. [CrossRef]

crystals

MDPI

Review

Modern History of Organic Conductors: An Overview

Toshio Naito [1,2,3]

[1] Graduate School of Science and Engineering, Ehime University, Bunkyo-cho 2-5, Matsuyama, Ehime 790-8577, Japan; tnaito@ehime-u.ac.jp
[2] Research Unit for Development of Organic Superconductors, Ehime University, Matsuyama 790-8577, Japan
[3] Geodynamics Research Center (GRC), Ehime University, Matsuyama 790-8577, Japan

Abstract: This short review article provides the reader with a summary of the history of organic conductors. To retain a neutral and objective point of view regarding the history, background, novelty, and details of each research subject within this field, a thousand references have been cited with full titles and arranged in chronological order. Among the research conducted over ~70 years, topics from the last two decades are discussed in more detail than the rest. Unlike other papers in this issue, this review will help readers to understand the origin of each topic within the field of organic conductors and how they have evolved. Due to the advancements achieved over these 70 years, the field is nearing new horizons. As history is often a reflection of the future, this review is expected to show the future directions of this research field.

Keywords: π–*d* system; Mott insulator; strongly correlated electron system; multiferroic; dielectric; photoconductor; Dirac electron system; single-component molecular conductor

Citation: Naito, T. Modern History of Organic Conductors: An Overview. *Crystals* **2021**, *11*, 838. https://doi.org/10.3390/cryst11070838

Academic Editor: Andrei Vladimirovich Shevelkov

Received: 22 June 2021
Accepted: 8 July 2021
Published: 20 July 2021

1. Introduction

At the beginning of this Special Issue of "Organic Conductors", we will briefly review the history of the organic-conductor research field (~70 years), during which tens of thousands of related papers have been published for readers to better understand the background and significance of the work collected in this issue. Note that the selected topics and references in this review are not exhaustive and that our purpose does not lie in comprehensively discussing each topic. Instead, we will limit ourselves to an overview of the course of the history, namely, how the studies concerning a specific topic were conducted and how they have been advanced. Although there still remains a number of important contributions and topics that are referred to in this review, many original papers, reviews, and books are cited, including those of closely related fields. Unlike the conventional manner of citing scientific literature, the references in this review were collected in a broad context and are listed in chronological order, rather than classifying them into particular topics. This was carried out to provide objective descriptions of the evolution and background of the various research topics in this field and the contributions of individual groups. Despite numerous references being cited, the reader can easily find the desired reference owing to their titles being included and their listing in chronological order. It is also expected that such a manner of citation would provide a comprehensive (underlying) connection between different studies, papers, and topics, i.e., how they have evolved by interacting with each other and different research fields. The reader can delve into the continuous, worldwide endeavors toward advancing the field of organic-conductor research by reading through the titles of the cited references, consequently developing an understanding of the field. After a brief summary of the older history of this field, we will provide an overview of the more recent progress that has been made in this field and the new trends that have developed in the last two decades. There are a number of other comprehensive reviews and books on the earlier stages of this field [1–21].

Constructing semiconducting materials comprised of organic compounds was first attempted by Akamatu, Inokuchi, and Matsunaga in 1954 [22], which has been recognized as the emergence of a new research field, i.e., "organic (semi)conductors." Although the first "organic conductor" was unstable in air and possessed an unknown structure, it impacted the scientific community so greatly that it was succeeded by a series of significant findings, including the first organic metallic conductor TTF-TCNQ (TTF = tetrathiafulvalene, TCNQ = tetracyanoquinodimethane) in 1973 [23–29], doped polyacetylene in 1977 [30] (awarded the Nobel Prize in Chemistry in 2000), superconducting TMTSF (TMTSF = tetramethyltetraselenafulvalene) in the 1980s [31–66], and BEDT-TTF (BEDT-TTF = bis(ethylenedithio)tetrathiafulvalene) salts throughout the 1980s and 1990s [67–163]. This was followed by the advent of doped-fullerene (fulleride) superconductors in 1991, which demonstrated transition temperatures (T_Cs) of ~18–30 K [164–177].

1.1. Renaissance of Organic Superconductors

This historical overview should begin with a brief summary of the "organic superconductor age", during which the field rapidly progressed and expanded. The period, which occurred from the 1980s to the 1990s, is characterized by the worldwide efforts to discover the first organic superconductor, as well as the discovery of organic superconductors that succeeded the first and exhibited higher T_Cs, which was due to the dimensions of their intermolecular interactions being enhanced (Figure 1) [178–255]. These intermolecular interactions, which are indispensable for electrical conduction, are based on the overlap integrals of adjacent molecular orbitals. These overlap integrals are expected to be enhanced by introducing highly polarizable atoms, such as chalcogen atoms, onto the periphery of π-conjugated molecules. Such a synthetic strategy also leads to the stabilization of radical ionic molecular species in the solid state of charge transfer (CT) complexes, and a great number of new π-conjugated molecules, including metal-complex derivatives, were synthesized in this period of the 1980s to the 1990s [178–298]. There were four major families of organic superconductors established in this period, except for the fulleride superconductors. In the order of their appearance, they are based on the TMTSF [31–66], BEDT-TTF [67–163], DMET (DMET = dimethyl(ethylenedithio)diselenadithiafulvalene) [179–184], and $M(dmit)_2$ (M = Ni, Pd, Pt, etc; dmit = dimercaptoisotrithione or 1,3-dithiol-2-thione-4,5-dithiolate) [256–298] molecules. To emphasize their metallic and superconducting properties, organic conductors were often called "organic metals and superconductors" as well as "synthetic metals." In addition to their metallic and superconducting properties, the enhancement in the dimensions of their intermolecular interactions completely changed the crystal and electronic structures of organic conductors; specifically, they wiped out the prevailed image that organic conductors possess one-dimensional (1D) columnar (stacking) structures comprising planar π-conjugated radical ionic molecular species. The diversity of molecular arrangements in BEDT-TTF salts, which usually take on two-dimensional (2D) arrangements of BEDT-TTF cation radical species, far exceeds that of the molecular arrangements in 1D-type salts, such as TTF and TMTSF salts. The crystal structure variety of a compound is illustrated by the Greek letters placed at the beginning of its chemical formula, such as α-BEDT-TTF_2X (X = anions). The structural and electronic features of BEDT-TTF salts provided rich and unique findings for areas of chemistry and physics, which established the research field of organic conductors. However, during this period (~1980–2000), we also learned that similar molecules did not always produce CT salts with similar structural and physical properties. As metal d-orbitals can mix with ligand π-orbitals, species such as $M(dmit)_2$ distinguish themselves with some of their chemical properties. First, they are easily oxidized to provide stable radical anions and sometimes neutral radical species. Second, they have much narrower band gaps (~1 eV) between their HOMOs and LUMOs (HOMO = the highest occupied molecular orbital, LUMO = the lowest unoccupied molecular orbital) than other components of molecular conductors. In other words, the physical properties of $M(dmit)_2$ species may be governed by their LUMO and/or HOMO, depending on their oxidation state, crystal structure, and the existing thermodynamic conditions. Owing to these features, $M(dmit)_2$ and related salts have led to many new and unique topics such as HOMO–LUMO inversion [273,274,284,285], single-component conductors, and Dirac electronic systems; the latter two are discussed below.

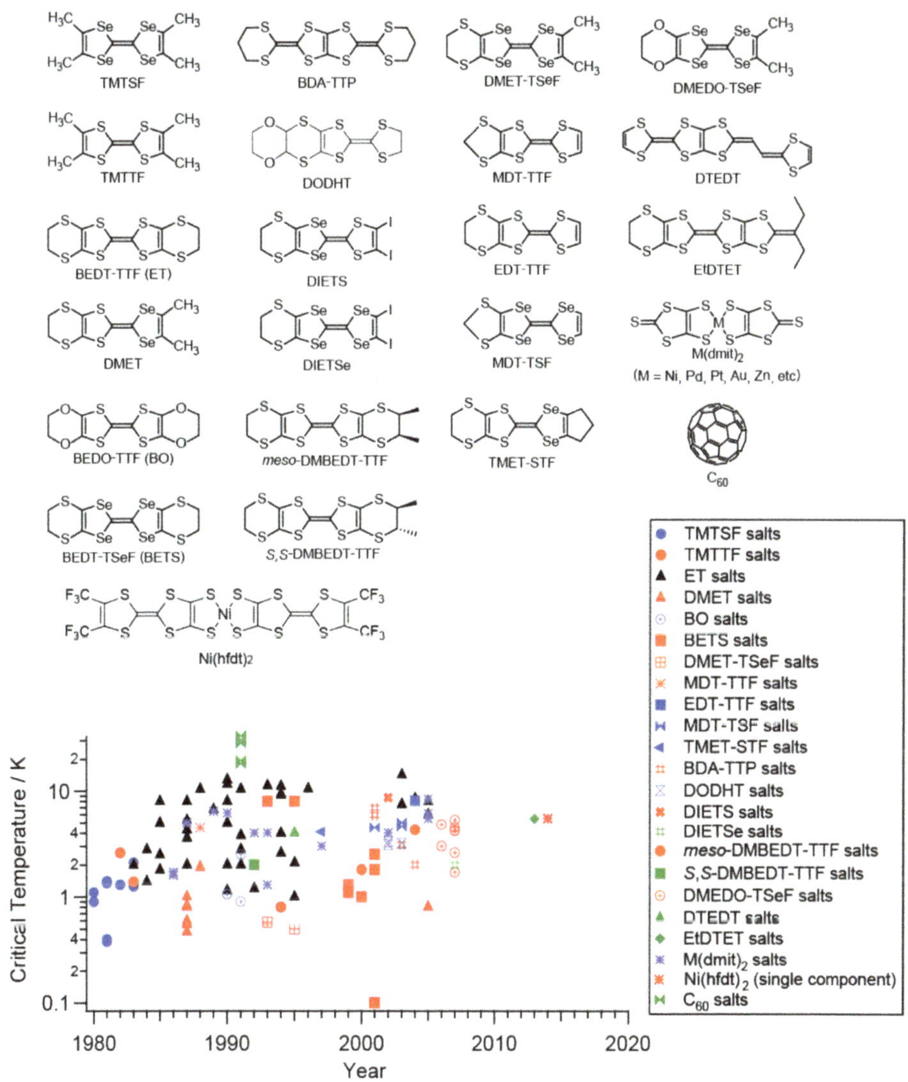

Figure 1. The history of the superconducting critical temperatures (T_Cs) of organic and related conductors (not an exhaustive list). Note that all the component molecules, except for M(dmit)$_2$ and C_{60}, belong to donor systems. Reproduced and modified from Ref. [21] with permission.

1.2. *Beyond π-Systems: d-Electrons*

1.2.1. The Rise of an Acceptor: DCNQI Salts

Toward the end of the "organic superconductor age", the merging of two extreme types of unpaired electrons, i.e., localized and delocalized electronic systems, was observed. The molecular conductors developed thus far in history (up until ~1985) were dominated by electron-donating molecular systems (donor systems), which are predominantly represented by the cation radical salts of TTF derivatives. This is natural when considering the high polarizability and large van der Waals radii shared by chalcogen atoms, both of which being indispensable for the stability of the resultant radical species and the intermolecular interactions providing electrical conduction. Beginning with the first example of a purely organic metallic conductor, TTF-TCNQ [23–29], a strategy for developing more thermodynamically stable organic materials with enhanced conduction involved the prosperity of π-conjugated donor molecules possessing many chalcogen atoms at their peripheries. Meanwhile, electron-acceptor (acceptor) molecules, such as TCNQ, inherently contain highly electronegative (EN) atoms and electron-withdrawing functional groups. Such a molecular structure tends to attract and localize the mobile electrons on particular parts of the acceptor molecules, which appears to be disadvantageous for high conductivity. However, highly EN atoms and electron-withdrawing functional groups generally share another important chemical feature that is advantageous for intermolecular interactions, namely, strong coordination ability, which chalcogen atoms generally do not have.

In 1986, a series of highly conducting copper salts of organic acceptors attracted enormous attention because their electrical conduction was based on both the *d*-levels of the copper ions and the π-bands of the acceptors, i.e., π–*d* mixed-band conduction. The acceptors were quinone derivatives called DCNQIs (2,5-disubstituted *N*,*N*′-dicyanoquionediimines) (Figure 2) [299–366]. They are distinguishable from the existing components in organic conductors because they can transform into stable radical anions and form covalent bonds with appropriate transition-metal cations, such as copper ions, to give CT complexes; this results in π–*d* mixed bands. Extensive studies on DCNQI salts revealed a number of unprecedented phenomena in the organic conductors they were a part of, including three-dimensional (3D) Fermi surfaces [302,331,334,341,346], metallic conductors with magnetic ordering [307,326,332,340,350], pressure-induced metal instability and reentrant metal–insulator transitions [301,303–305,309,310,326–328,337,360], dense Kondo effects [310,321,325,326,334], and charge/spin fluctuations [324,344]. Regarding such hybrid band structures, DCNQI salts contain characteristics of 1D and higher-dimensional conductors, where they exhibit one or both types of characteristics depending on their chemical compositions and the implemented thermodynamic conditions. This puzzling feature of DCNQI salts provided researchers with a deeper understanding of organic conductors, but more uncertainty also emerged. The studies conducted on organic conductors thus far have provided some current, important notions in related research fields, such as charge separation. This term originally referred to the deviation in the electron density of the conduction pathways, from evenly distributed, i.e., delocalized, to periodically localized patterns of metal–insulator (M–I) transitions. The term "charge separation" is more frequently used in the field of semiconductor devices such as field-effect transistors and solar cells [367–428] than in the field of organic crystalline conductors. During ~1985–2000, ideal samples and elaborate techniques for studies on the complicated behavior of DCNQI salts were timely provided one after another; for example, studies regarding alloys [309,313,325,326,333,355,357] and selectively deuterated [328–333,340,343,345,354] DCNQI derivatives and fine-tuning of delicate pressures [303–305]. In addition to these researchers' endeavors, other successful studies have provided advanced techniques for crystalline molecular conductors to exhibit novel and unique conducting properties; namely, they were soft and thus highly sensitive to perturbations, such as applied pressure or magnetic fields. These properties make these crystalline molecular conductors superior to conductors composed of harder materials, as the latter do not exhibit responses to such perturbations as clearly as the former do. The key to success was also based on a close and efficient collaboration between chemists and physicists, which has always been present at one time or another in this field. In this way, the "π–*d* interaction" became a popular research topic and a desirable research target in the field of organic conductors. This research trend was followed by π–*d* systems implemented in donor-based conductors [429–518], as shown in the next subsection.

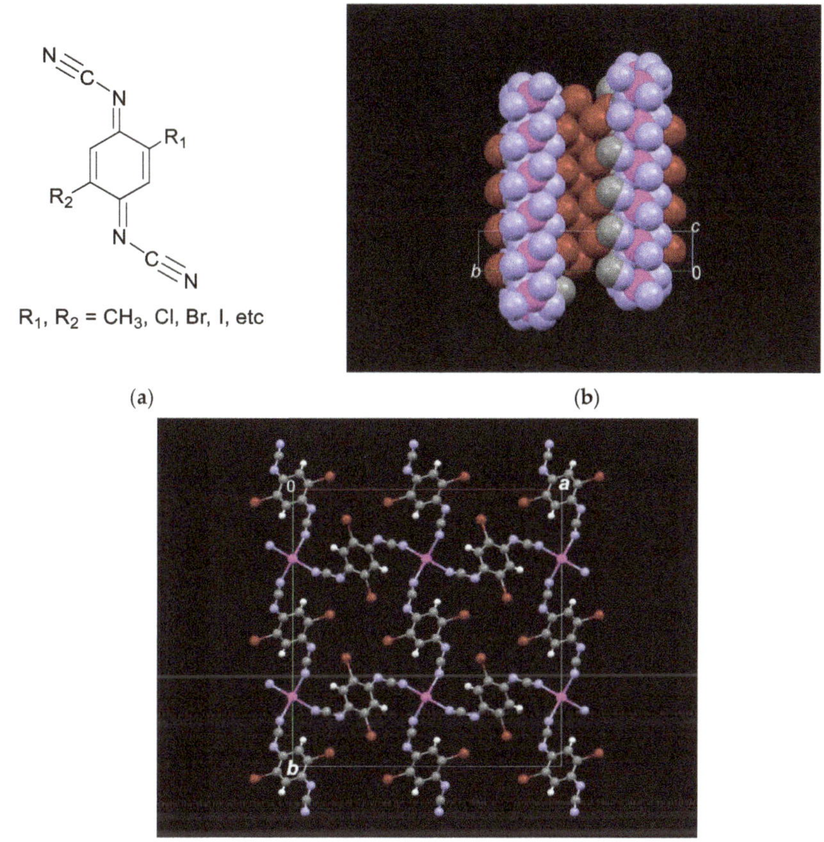

(c)

Figure 2. (**a**) The structure of the DCQNI (DCQNI = 2,5-disubstituted *N,N′*-dicyanoquionediimine) molecule; (**b**,**c**) A typical crystal structure of a Cu salt: Cu(Br_2-DCNQI)$_2$. The white, grey, purple, pink, and dark-red spheres designate H, C, N, Cu, and Br atoms, respectively. Note that every –NCN group at the end of DCNQI makes a coordination bond with a Cu ion in a tetrahedral geometry. Reproduced and modified from Refs. [519,520] with permission.

1.2.2. Comeback of the Donor Dynasty: BETS Salts with Magnetic Anions

In parallel with the extensive examination of DCNQI salts, many researchers have attempted to synthesize a similar system based on donor molecules since ~1990. The prominent examples of unique conductors based on *π–d* interactions are a series of CT salts of a BEDT-TTF derivative called BETS (bis(ethylenedithio)-tetraselenafulvalene) [451–518]. Various derivatives of BEDT-TTF have been extensively explored from an early stage of the field [451–462]. BETS was first discovered in 1983 [454]; however, it was not until 1993 that BETS was suddenly paid significant attention when λ-(BETS)$_2$GaCl$_4$ was found to be a new superconductor with nearly the highest T_C (~10 K) among those of the organic compounds known at the time [463]. Almost at the same time, λ-(BETS)$_2$FeCl$_4$ (Figure 3), an isostructural salt of λ-(BETS)$_2$GaCl$_4$, was reported [464] and garnered more attention than the preceding non-magnetic salt λ-(BETS)$_2$GaCl$_4$. This was because λ-(BETS)$_2$FeCl$_4$ and succeeding, related salts united the two main streams of organic superconductors and organic *π–d* systems. For example, some BETS salts, κ-BETS$_2$FeX$_4$ and λ-BETS$_2$FeX$_4$ (X = halogen atoms), have been able to achieve what DCNQI salts could not; examples include (high magnetic) field-induced superconductivity [465,481–483,485,486,489,490,494,497,500,501,504] and superconductors with magnetic ordering in their ground states [471–475,477,479,480,497,498,504]. The 2D conduction sheets in λ-$BETS_2FeX_4$ serve as ideal samples for studies on the Fulde–Ferrell–Larkin–Ovchinnikov (FLLO) state, a theoretically predicted superconducting state where the order parameter oscillates in real space [490,501,513,514,521–524]. The common key feature in these π–*d* systems is weak or moderate π–*d* interactions, without which the systems could not exhibit unusual behavior based on both localized (*d*-electrons as spins) and delocalized (π-electrons as carriers) electronic characteristics. However, these interactions are far more difficult to achieve than originally assumed. A great number of TTF-type donor salts with paramagnetic metal complex anions were synthesized and their electrical and magnetic properties were measured [429–450,464–518]. Similarly, a large number of TTF-type molecules and their cation radical salts bearing stable radical moieties, such as nitroxide derivatives, were synthesized and were examined. However, most of them turned out to be antiferromagnetic semiconductors or even diamagnetic insulators in the ground state, except for a limited number of successful examples [525–528]. It is substantially difficult to finely tune the interactions between localized and delocalized unpaired electrons for them to coexist. Since then, researchers have begun paying more attention to "crystal designing" than "molecular designing", the former being a trend that also emerged in related research fields such as supramolecular chemistry [529–546] and those involving metal–organic frameworks (MOFs) [547,548].

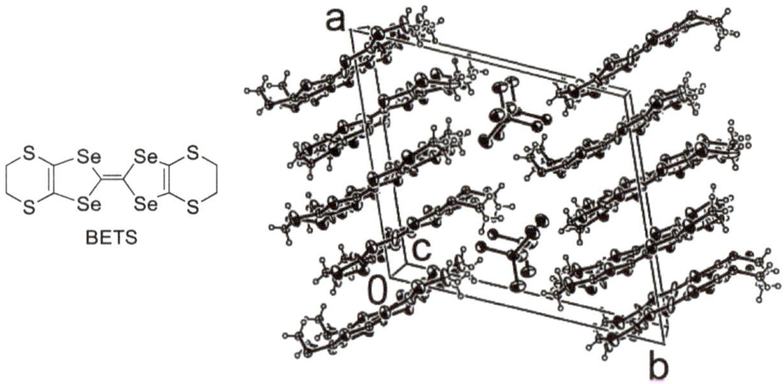

(A) (B)

Figure 3. (**A**) The chemical structure of the BETS (BETS = (bis(ethylenedithio)-tetraselenafulvalene)) molecule; (**B**) A typical crystal structure of the magnetic superconductors: λ-$(BETS)_2FeCl_4$. Note that one of the ethylene groups at the end of BETS is bent toward $FeCl_4^-$ to favor the π–*d* interaction. Reproduced from Ref. [511] with permission.

1.2.3. The Return of 1D Systems: Axially Ligated Iron(III) Phthalocyanines

There is another unique π–*d* system, where both delocalized π-electrons and localized *d*-electrons are located on the same molecules, i.e., transition-metal phthalocyanines (MPcs). Owing to the superior stability and ready availability of MPc derivatives, there have been many pioneering works, including those in the 1980s that investigated (pseudo)halogen- or oxygen-ligated Pc polymers [549–552], single-component conductors [553], π–*d* interactions in Fe(III)Pc complexes [554], electrochemically synthesized $(NiPc)_2AsF_6$ complexes [555], and iodine-doped CuPc complexes [556] toward 1D molecular conductors. Then, in 1990, hybrid derivatives were discovered, i.e., electrochemically synthesized radical cation salts of axially ligated MPcs ($M(Pc)X_2$; M = metal cation, Pc = phthalocyanines and their derivatives, X = (pseudo)halogen monoanions as axial ligands; Figure 4) [557–578]. The studies on $M(Pc)X_2$ complexes also evaluated their crystal designs, as mentioned above, but most of them exhibited a similar columnar stacking structure with a 1D conduction band. The most intensively studied examples include $TPP[M(Pc)(CN)_2]_2$ (TPP = tetraphenyl phosphonium,

M = Co(III), Fe(III)) complexes. Because of the steric hindrance of the axial ligands, the Pc ligands stack with each other via limited π–π overlap (~single benzene rings). This produced narrow bands and semiconducting behavior in a wide variety of the $M(Pc)X_2$ salts, whether the metal centers were paramagnetic or not. Their slipped stacking pattern distinguishes $M(Pc)X_2$ salts from other organic conductors composed of planar molecules, which results in a number of unique and puzzling electrical properties in the former. Despite limited π–π interactions existing exclusively along the 1D columns, these salts exhibit high conductivity and no obvious transitions at ~2–300 K, which apparently contradicts the idea of 1D-metal instability (Peierls instability) that was established in the early stages of this field [579]. The commonly observed semiconducting behaviors of $M(Pc)X_2$ salts are not of a thermally activated type, and their reflectance spectra show a Drude-type dispersion [558,559]. This semiconducting behavior suggests the presence of small energy gaps originating from strong electron correlations and/or other reasons. The ground state of a $Co(III)(Pc)X_2$ salt has been clarified as a charge-ordered (CO) state [568], which is the typical ground state of thermodynamically unstable 1D conductors. The CO state has been acknowledged since the early stages of this field and has been gaining increasing attention due to its connection with superconductivity [580–625]. The electrical behavior of complexes bearing paramagnetic metal ions, such as Fe(III), is more unusual than that of complexes bearing diamagnetic metals. Furthermore, π–*d* interactions manifest under giant negative magnetoresistance (GNMR), which is clearly observed generally at $T \leq 50$ K and $H \geq 10$ T and depend on the magnetic-field direction relative to the stacking axes [557–578]. Unlike BETS and DCNQI salts, $M(Pc)X_2$ salts apparently belong to 1D systems, which are known to be inherently unfavorable for metallic conduction at low temperatures [579]. However, GNMR is generally observed at low temperatures; thus, metal-like high conduction is required at low temperatures. Therefore, the observation of GNMR in such a series of 1D systems is surprising. Accordingly, these π–*d* systems have demonstrated that organic conductors are a sufficient choice for studying solid-state physics on a much wider scale than was originally possible regarding cooperative phenomena.

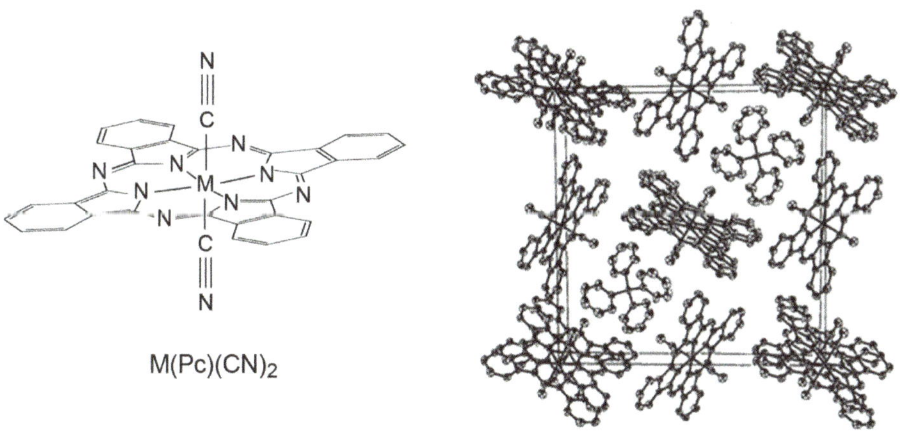

(a) (b)

Figure 4. *Cont.*

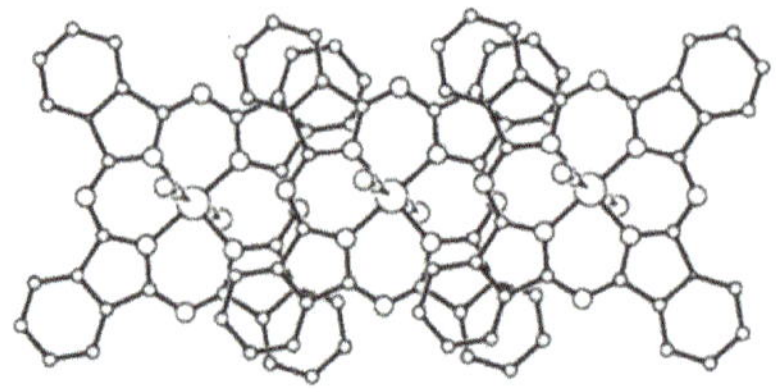

(c)

Figure 4. (**a**) The molecular structure of a metal phthalocyanine (MPc) $[M(Pc)(CN)_2]^n$ unit ($0 \leq n \leq 1$); (**b**) A typical crystal structure of $Cat[M(Pc)X_2]_2$ salts (Cat = onium monocations): $(C_6H_5)_4P[Fe(Pc)(CN)_2]_2$; (**c**) A close-up view of a 1D column of $[Fe(Pc)(CN)_2]^{0.5-}$. Reproduced from Ref. [559] with permission from The Royal Society of Chemistry.

1.3. To Be or Not to Be a Conductor—That Is the Problem: Fluctuation

Researchers that were active during the "organic superconductor age" had gradually realized that the enhancement of superconducting (SC) T_Cs could also provide an understanding of the insulating and/or magnetic phases of neighboring SC phases. In fact, it has often been pointed out by both theoretical and experimental studies that the fluctuation in charge and/or spin degrees of freedom could play an indispensable role in a potential, the universal mechanism behind SC transitions in organic and inorganic compounds [626–646]. As a result, the strategy for developing organic superconductors with higher T_Cs drastically changed from involving the stabilization of metallic states to employing the destabilization of metallic states [647–655]. In pursuing this strategy, synthetic efforts towards new conductors, even those where insulators had resulted instead, led to successive findings of unprecedented physical properties and various phase transitions [228,241]. All of these features originate from various types of degrees of freedom that are characteristic of molecular crystals, i.e., "molecular degrees of freedom." The variety of these degrees of freedom is important, in addition to their comparable thermodynamic stabilities and their states being incompatible with each other. These properties have been observed in various forms, such as fluctuation [626–646], hidden states [42,647], and field-induced cascade transitions [40,492].

As studies on organic conductors have progressed, so have the experimental tools/techniques used to observe electronic behavior; namely, rapid progress has been made in both the variety and specification of these tools, which has provided us with high magnetic fields [656–682]; high time- [683–710] (Figure 5), space-, and/or energy-resolutions [365, 711–721]; and high (hydrostatic or uniaxial) pressures [722–731]. As a result, the term "organic conductor" now includes a wide variety of compounds and components, such as organic polymers, inorganic ions, metal complexes, metal clusters, and organic molecules possessing electrical properties ranging from those of insulators to those of superconductors (depending on the circumstances). Thus, organic conductors are now better described as "molecular materials", which can be defined as a unique group of compounds with well-defined chemical formulae and crystal structures that are used in studies on cooperative phenomena.

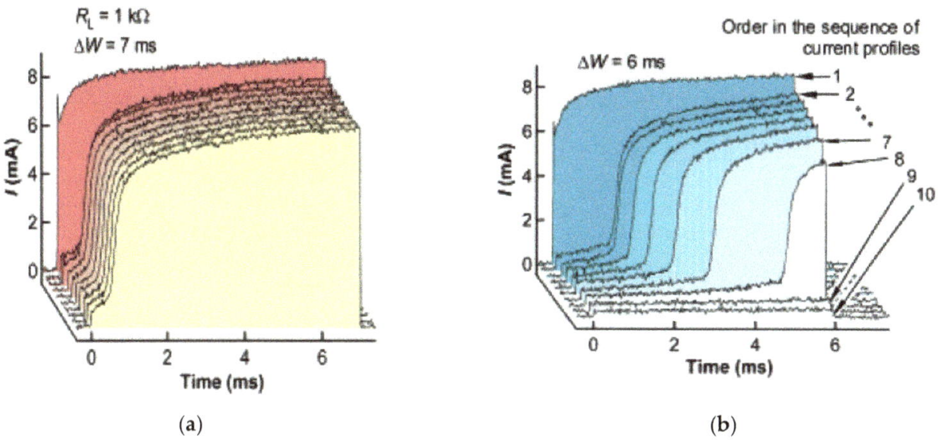

Figure 5. Sequential transient photocurrent measurements under synchronized voltage and photoirradiation pulses using the single crystal of α-(BEDT-TTF)$_2$I$_3$ and a four-probe method at 115 K, where the material is in an insulating phase. Photoirradiation was synchronized only at the first voltage pulse (11 V) and was turned off for the second and following pulses: the voltage pulse widths (ΔW) are (**a**) 7 ms and (**b**) 6 ms. Note that ΔW drastically affects the current, i.e., the relaxation time of the highly conducting state (HS) after the cessation of photoirradiation, while ΔW does not affect the resistivity in the HS. Adapted from Ref. [693] with permission.

Besides the basic studies on crystalline organic conductors, organic CT salts, and related compounds, semiconducting thin films in field-effect transistors (FET), batteries, luminescence devices, and non-linear optics have also been extensively studied since the 1970s [367–428]. Studies on such "organic devices" have become well established in various research fields of both academia and industry. Since there are a great number of reviews and books detailing such studies [382–385,397,398,401,402,417–428], we will not go into detail here.

Overall, the history of "organic conductors" involved the discovery of new, surprising materials and investigations into their unique properties. Selected landmark studies from the last two decades will be described in the following sections.

2. Recent Progress and New Trends

The field of organic conductor research has made steady progress over the years, beginning with organic semiconductors based on aromatic hydrocarbons and evolving to include organic metals, organic superconductors, and organic magnetic (super)conductors. Alongside such progress, the field has also evolved to include a broader range of topics. Advancements in the field have been achieved by incorporating heteroatoms, mainly chalcogen atoms and metal ions, in the aromatic hydrocarbons responsible for electrical conduction. Consequently, the component molecules of conductors become so diverse that they should be referred to as molecular conductors/materials rather than organic conductors. At the same time, experimental and theoretical tools used to study these types of materials have also made remarkable progress in recent decades. These new materials and new methods have accelerated the advancement of one another, which led the field in new directions. For example, since approximately the "organic superconductor age", particular types of organic insulators have been intensively studied for various reasons.

2.1. Mott Insulators: Mysterious Clues to Superconductors

Mott insulators [732–749], which are often closely related to high-T_C superconductors of both organic (Figure 6) and inorganic nature, have been paid particular attention because they are considered key compounds for clarifying a possibly universal high-T_C mechanism. Mott insulators are characterized by a paramagnetic half-filled band. Although such an electronic feature appears to be that of metals, the unpaired electrons cannot travel through these solids because of the strong electron correlation. As a result, they have energy gaps called Mott (Mott–Hubbard) gaps at the Fermi level, which makes them different from common insulators, i.e., band insulators. Band insulators are diamagnetic, so they do not have unpaired electrons. By applying the band theory, Mott and band insulators could correspond to high and low spin states of a given band structure and band filling, whose differences originate from the on-site Coulomb energy. However, Mott insulators are intrinsically beyond the band theory, as their electron correlation is due to many-body problems instead of Coulomb repulsion between a pair of electrons. Currently, many organic and inorganic compounds are known as Mott insulators. Owing to their unique insulating mechanisms and potential to serve as high-T_C organic superconductors, many researchers have attempted to control the conducting properties of such compounds using chemical ("chemical pressure") and/or physical (applied high pressure) methods [649–655,722–731]. This is generally referred to as bandwidth control. Band-filling control, which is typically performed by synthesizing mixed crystals (doping), is frequently attempted with inorganic compounds, while bandwidth control is rather typical of organic conductors owing to their soft structures. Meanwhile, except for limited kinds of salts [196,309,313,325,326,333,355,357], the doping of crystalline CT salts is rather difficult [750]; the mixing of isostructural compounds with different electron counts often resulted in unexpected, pure compounds with different crystal structures or, more frequently, failed to yield well-defined crystalline solids. To solve this problem, a new method using photochemical reactions has been established [21,519,520,751–759].

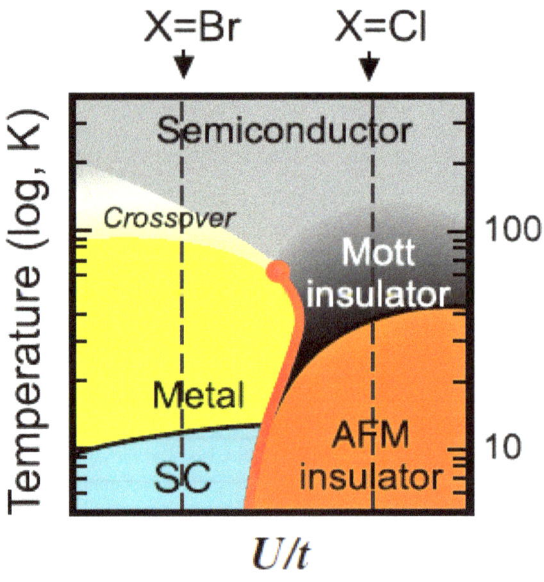

Figure 6. A schematic phase diagram of κ-(BEDT-TTF)$_2$Cu[N(CN)$_2$]X (X = Br, Cl) complexes, which are considered typical, organic Mott insulators and superconductors depending on the anion X. An increase in the horizontal axis (U/t) corresponds to a decrease in the actual pressure. Adapted from Ref. [704] with permission.

Similar to SC transitions, various kinds of M–I transitions have been extensively studied. Naturally occurring metals, i.e., the elemental metals, do not exhibit M–I transitions, except for tin. The insulating phases are generally induced by order–disorder transitions in a part of the crystal structures, charge and/or magnetic orderings, as well as changes in the temperature/pressure. Because there can be different (meta)stable states in organic conductors that exhibit nearly equal stabilities, and because they frequently have low-dimensional electronic structures, a fluctuation between different states [626–646] or even the disappearance of phase transitions [41,647,648] are often observed. Both are characteristic of organic conductors, in addition to the M–I transitions.

2.2. Between Electron and Lattice Systems: Proton Dynamics

As protons are most frequently included in organic compounds, they have been paid significant attention in attempts to control the electron dynamics and resultant physical properties of organic conductors [535,760–772]. A new series of dielectrics [773–783] and multiferroics [784] based on organic CT salts have been synthesized/discovered. Multiferroics are materials that combine multiple order parameters, such as ferroelectrics and ferromagnets, where the spontaneous ordering in the magnetic and dipole moments simultaneously occurs within the same temperature range. Generally, magnetic orderings are observed at low temperatures, while the orderings of electric dipoles are observed at high temperatures. Thus, their coexistence in a single organic or inorganic material at the same temperature seldom occurs. Although their performance and temperature ranges have yet to be improved for practical applications, they have been employed in representative and successful studies derived from organic conductors, where various phenomena often couple together. In relation to proton dynamics, organic proton conductors [782] and organic conductors bearing components that participate in hydrogen bonding have also been developed [770,771]. Regarding the latter, it is extremely difficult to make the conduction π-electrons and the protons of hydrogen bonds interact with each other, even if they are located on the same molecules. This is likely due to the differences in their energy scales. Such subtle material design requires the extremely fine-tuning of energy levels and crystal structures, similar to the π–d systems. In fact, an analysis of non-deuterated and deuterated samples revealed that deuteration qualitatively affects conduction behavior [771] (Figure 7), which was also observed for DCNQI salts. The mechanism of this is currently under study. Regarding dielectrics, an electronic mechanism is often noted for some molecular CT salts instead of proton/ion displacement [785–795]. The dielectrics with an electronic mechanism comprise the insulating, i.e., CO, phases of organic conductors. Such dielectrics are characteristic of organic conductors, exemplifying the rich variety of their electronic states.

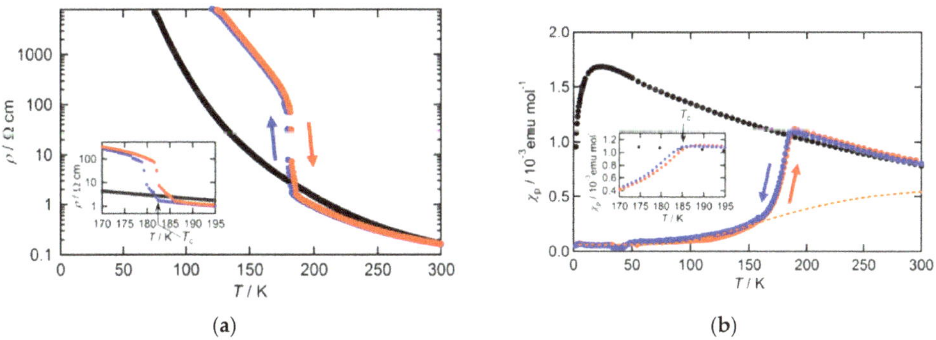

Figure 7. Temperature dependence of the physical properties of κ-H_3(Cat-EDT-TTF)$_2$ (κ-H) and κ-D_3(Cat-EDT-TTF)$_2$ (κ-D): (**a**) electrical resistivity measured using a single crystal; (**b**) magnetic susceptibility measured using a polycrystalline sample. In both figures, the blue and red circles denote the data observed in the cooling and heating processes, respectively, for κ-D, while the black circles denote the data for κ-H observed in the cooling processes. The orange broken curve in (**b**) represents the best fitting curve for the κ-D data using a singlet–triplet dimer model with an antiferromagnetic coupling of $2J/k_B \sim -600$ K. Adapted from Ref. [771] with permission.

2.3. Light Control: Unique Properties Otherwise Impossible

In addition, the extensive development and recent advancement of organic photoconductors and related materials are notable [796–826]. In contrast to the ground states of these materials, which most scientists in the field of the organic conductors have been interested in, their photoexcited states correspond to extremely high energy states that thermal excitation could not reach. Thus, photoirradiation may provide us with a new method for discovering new physical properties and structures. Therefore, instead of simply synthesizing photoconductors based on organic compounds, the development of these materials involved inducing novel physical properties and unique electronic phases by photoexcitation. This methodology was significant because many of these new properties had never been observed by simply controlling the temperature and/or pressure conditions. The development of these kinds of materials requires knowledge of organic conductors and the techniques used for their development, even though most of these materials are insulators/intrinsic semiconductors under dark conditions or without photoirradiation of appropriate wavelengths. Similar to photovoltaics and solar cells, the charge separation under photoirradiation and the lifetimes (relaxation times) of photocarriers produced thus far are crucial for ensuring the efficiency of photoconductors. Whether organic or inorganic, ordinary insulators typically exhibit an increase in conductivity of $\leq$20–30% at room temperature under ultraviolet-visible (UV-Vis) photoirradiation. Most of the increase in conductivity can be explained by heating effects involved with the irradiation, i.e., thermally activated carriers, and the net photoconductivity is generally ~2–3% of the original dark conductivity [21,520,759]. The relaxation of photoexcited carriers is generally quick, as long as the optical absorption occurs as resonance with UV-Vis (~10^{15} Hz), which is a typical energy range for HOMO–LUMO transitions in organic molecules used as conductors. Thus, a strategy for stabilizing photoexcited carriers is required, in addition to the design of conductors. In this sense, organic photoconductors can be regarded as an advanced design of organic conductors. Combining the typical component molecules of organic conductors, such as TTF derivatives, with well-known photosensitive molecules, such as bipyridyl derivatives, has resulted in various kinds of donor–acceptor type CT salts being reported for new organic photoconductors and related materials [796–826]. However, difficulties were encountered regarding the control of the donor–acceptor interactions toward producing a sufficient number of carriers, i.e., photoexcited electrons and holes, with sufficiently long relaxation times. Strong CT interactions lead to quick recombination between photoexcited electrons and holes, while weak CT interactions lead to an insufficient number of photoexcited electrons and holes. Additionally, if the photosensitive moieties are bulky, the formation of the conduction pathway is hindered. This is due to the close proximity of the π–π interactions between the main parts of the molecules that are responsible for conduction. Based on previous encounters with this kind of problem during the molecular and crystal design of π–d systems, several solutions have been proposed. One of the strategies involved insulating common organic CT salts with donor–acceptor mixed stacking structures, denoted as D_nA_m (n, m = 1, 2, 3, ...) [520]. For example, if n = 1 and m = 2, the unit cell is rich in acceptor A and an infinite ... A–A–A ... network may form, based on both stacking and side-by-side overlaps serving as conduction pathways (Figure 8). This situation is more favorable if the molecular size of A is larger than that of D. The requirement for actual conduction is to produce carriers in the conduction pathways, which is often called "doping" and can be performed by the photoexcitation of the CT bands between acceptors and donors. If the excited states are stabilized by the CT interactions, similar to exciplexes [424,425,428,799,817], the relaxation times of the resultant carriers can be sufficiently long. Based on this idea, (para)magnetic photoconductors have been developed based on stable diamagnetic insulators, where strong interactions between photoexcited localized spins and photocarriers manifest in a Kondo effect (T_K ~ 100–120 K) [21,520,759,808]. Using the unusual stability of the photoexcited states of some molecular CT salts, a novel type of material for photon energy storage is under investigation [823,825].

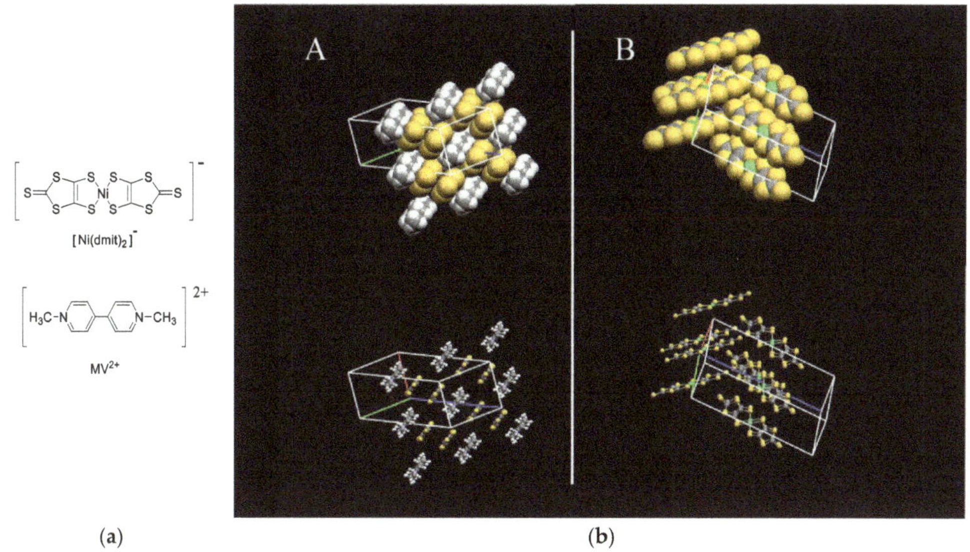

Figure 8. The crystal structure of a photomagnetic conductor, $MV[Ni(dmit)_2]_2$: (**a**) MV^{2+} and $[Ni(dmit)_2]^-$; (**b**) Molecular packing motif in van der Waals (upper) and ball-and-stick (lower) models; (**A**) Chemical structures of MV^{2+} and $[Ni(dmit)_2]^-$ and (**B**) 3D conduction pathways composed of $[Ni(dmit)_2]^-$. $MV[Ni(dmit)_2]_2$ is a diamagnetic insulator under dark conditions, but it turns into a metallic substance with localized spins on the MV^{2+} species under UV irradiation, thus exhibiting the Kondo effect at low temperature. The UV (~375 nm) irradiation triggers a CT transition between MV^{2+} and $[Ni(dmit)_2]^-$.

2.4. Single-Component Molecular Conductors: The Simplest and Most Difficult Molecular Conductors

Throughout the development of organic conductors, a molecular version of elemental metals, i.e., single-component molecular conductors (SCMC), has been a desirable target (Figure 9a) [729,827–882]. Unlike CT salts, SCMCs should have simpler crystal structures because they consist of a single type of molecular species. However, it turned out to be extremely difficult to produce such materials. Since the late 1980s, pioneering work towards these materials has been carried out by some independent groups. They obtained diamagnetic single crystals of neutral species [827–829] or paramagnetic and highly conducting/metallic polycrystalline samples of neutral species with unknown structures [830]. In 2001, i.e., nearly half a century since the beginning of this research field, the first SCMC with a well-defined crystal structure was developed based on a Ni–dithiolene complex molecule, $Ni(tmdt)_2$ (tmdt = trimethylenetetrathiafulvalenedithiolate) (Figure 9b) [831,832,838]. The SCMC exclusively contains neutral molecules of the same kind and produces carriers via the overlap between its HOMO and LUMO bands, similar to how semimetals produce carriers [837,873]. Because all the planar molecules are densely packed to ensure their equal contribution to conduction, the resultant conduction properties are 3D; however, there remains some anisotropy that is reflective of that of the component molecules. In principle, the crystal and electronic structures of SCMCs are uniform, and thus there should be usually no charge ordering or dimerization to make the material insulating. Consequently, most SCMCs are thermodynamically stable metals. However, SCMCs exhibiting an antiferromagnetic transition ($[Au(tmdt)_2]$) [843] and a coupled electric and magnetic transition ($[Cu(dmdt)_2]$) (dmdt = dimethyltetrathiafulvalenedithiolate) [857] are also known. Among the SCMCs, $[Ni(hfdt)_2]$ (hfdt = bis(trifluoromethyl)tetrathiafulvalenedithiolate) exhibits a superconducting transition (onset $T_C \leq 5.5$ K under 7.5–8.7 GPa) [729]. Recently, an increasing number of new SCMCs, most of which containing Au(III)–dithiolene complexes, have been reported [729,827–882]. Furthermore, new SCMC materials have been recently found to possess unique electronic band structures called Dirac cones [869–872,875,876,878,880–882], which will be discussed in the next section.

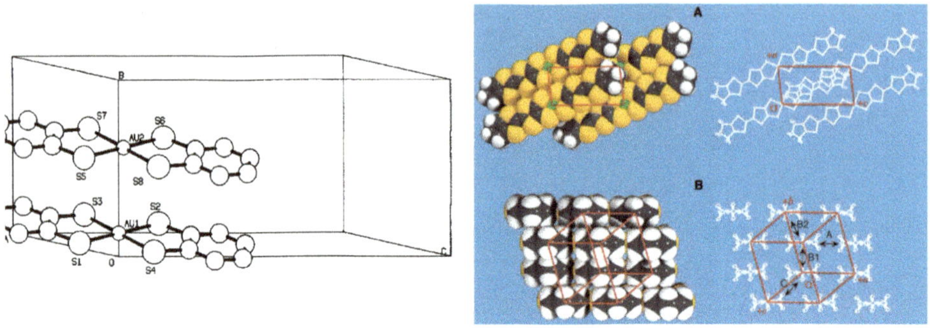

(a) (b)

Figure 9. Crystal structures of single component molecular materials: (**a**) A 1D soft Mott insulator, $Au(bdt)_2$ (bdt = benzene-1,2-dithiolate); (**b**) A 3D metallic conductor, $Ni(tmdt)_2$ (tmdt = trimethylenetetrathiafulvalenedithiolate), viewed along (**A**) the *b* axis and (**B**) the long molecular axis. A, B1, B2, and C denote intermolecular interactions responsible for conductivity. Reproduced from (**a**) Ref. [829] and (**b**) Ref. [831] with permission.

2.5. Dirac Electrons: Beyond Fermions

Interestingly, different fields with no obvious correlation with organic conductors have been known to catalyze the research field of organic conductors, thus broadening the scope of the field. Examples include Dirac electron systems [869–872,875,876,878,880–925] (Figure 10), which were originally a subject of particle physics but now are indicative of massless fermions in organic conductors. They occur when the materials contain a cone structure in the electronic bands, i.e., two cone-shaped bands touch at the apex of each cone, which are called Dirac points. When the Fermi levels are exactly located at the Dirac points, the materials are called zero-gap semiconductors. Because the physical properties are governed by the electrons/holes at the Fermi levels, unique behavior originating from the Dirac electrons is expected to be discovered in the zero-gap semiconductors. In fact, they were revealed in one of the oldest organic conductors, α-BEDT-TTF_2I_3, almost simultaneously by experimental and theoretical studies [883,884,886–889,891,892,895–900, 902–910,916–918]. The Dirac electrons are characterized by their negligibly small masses with Fermi velocities comparable to that of light. Coupled with the research interest in topological materials [885,890,893,894,901,911–915,919], organic Dirac electron systems have become an emergent topic not only in the field of organic conductors but also in the broad field of solid-state physics. Although Dirac electron systems have also been observed in inorganic compounds, such as the famous example of graphene (awarded the Nobel Prize in Physics in 2010) [885], the advantage of crystalline organic CT salts over other kinds of materials lies in their well-defined Fermi levels based on the well-defined crystal structures, chemical compositions (stoichiometries), and the electron count of their components. Accordingly, organic Dirac systems provide a platform for intriguing particles to be produced using small glassware instead of huge accelerators. Both organic and inorganic Dirac electron systems exhibit almost temperature-independent electrical resistivity, which has been extensively studied. The reflectance spectra of such systems have also been relatively well studied [893,894,901,911,921]. In the meantime, other physical properties common to Dirac electrons, such as magnetic behavior, appear to be controversial and require further study [872,878,888,900,921]. One of the difficulties in studying organic Dirac electron systems originates from the fact that the occurrence of Dirac cones (zero-gap semiconductors) requires high pressures in most of the organic compounds, which restricts the available experiments. In this sense, recent findings regarding organic Dirac electron systems existing at ambient pressure are important [878,920–925]. Additionally, the extremely small energy scales of organic and inorganic Dirac electron systems cause further difficulty in their study, as this makes the characteristic electronic states of zero-gap semiconductors qualitatively unclear. An understanding of the ground states of zero-gap semiconductors requires precise calculations that consider every possible interaction within the solid states as well as physical property data at the lowest temperature possible, both of which are currently being explored.

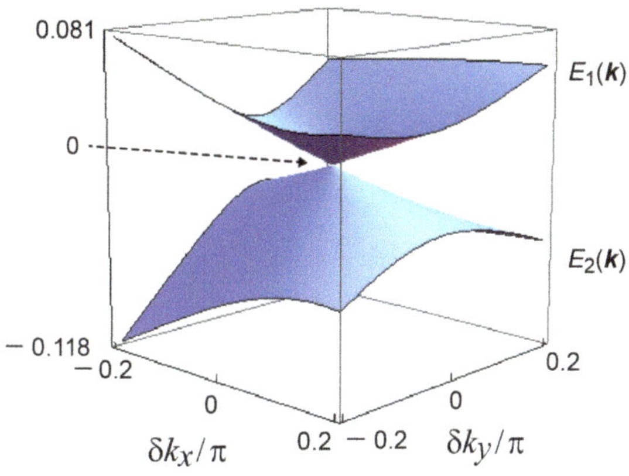

Figure 10. A characteristic curvature of the band structure, which is referred to as a Dirac cone. A close-up view around one of the two Dirac points, shown by 0 in the figure. The band structure was calculated for α STF_2I_3. Reproduced from Ref. [920] with permission.

2.6. *Chiral Conductors: Electrons in an Asymmetric Wonderland*

Chirality is one of the most widely known structural features in chemistry and physics [926–958]. The control of the chirality of molecular structures and the resultant effects on their physical properties have been extensively studied for a long time. However, it has been difficult to reveal some unique conducting properties directly associated with chirality, even for single crystals belonging to non-centrosymmetric space groups [951]. Recently, an increasing number of papers on such attempts can be found [926–947,949, 950,953–959]. Chirality control is an advanced stage of "crystal designing/engineering." During the initial stages of "crystal designing" (investigated since ~1990), researchers considered and controlled the arrangement of neighboring molecules based on interatomic interactions, such as hydrogen bonding, and coordination/supramolecular chemistry [952]. Such crystal designing is based largely on molecular designing. Upon investigating the chirality control of crystal structures (investigated since ~2000), one should consider the orientation and arrangement of all the molecules in the crystals, even though there is no direct interaction between distant molecules. Thus, the synthesis of single crystals with non-centrosymmetric space groups is difficult (Figure 11), making the discovery of their unique conducting properties associated with the chirality even more so. Still, in the last two decades, there have been frequent reports on various new chiral donor molecules and their CT salts that exhibit interesting resistivity behavior with chiral crystal structures [953–959]. This demonstrates that chiral molecular (super)conductors are ever-advancing toward a new direction of organic-conductors development.

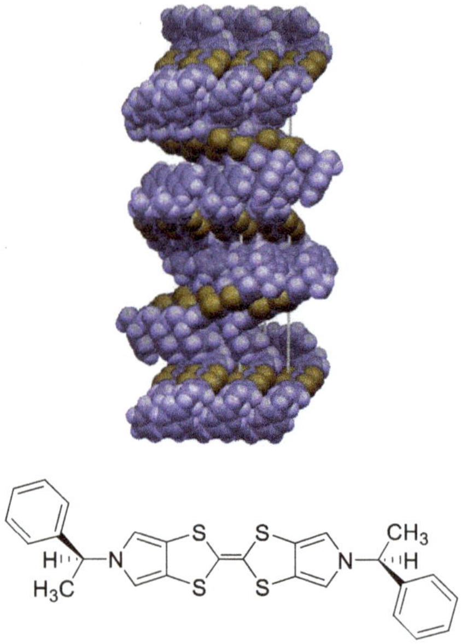

Figure 11. Organic crystal with a chiral molecular arrangement (upper; Space Group *P*1) containing a bis(pyrrolo)tetrathiafulvalene derivative (lower). The donor packing motif represents a 4_3 axis along the *c* axis (vertical). Reproduced and modified from Ref. [934] with permission from The Royal Society of Chemistry.

3. Concluding Remarks: Towards the New Age

In this review, we could refer to only a limited portion of the field of organic conductors. There are still a number of important topics that were not discussed herein. However, even from this brief review of the history of organic conductors, we can see that the field covers a wide spectrum of topics and possesses an ever-increasing potential to develop into a new field. The focus of such a field is unknown, but by maintaining close interactions with other research fields, the field of organic conductors will continue to grow and evolve, very likely until we can no longer define what "research on organic conductors" entails. Pursuing (super)conducting and magnetic properties similar to those in inorganic compounds, the early stage of organic conductors research can be characterized by the pursuit of what organic compounds can also do. In the meantime, the last two decades of the field are characterized by a new trend, i.e., the pursuit of what only organic compounds can do. Similar to some patterns of chirality, there are phenomena unique to molecules, which might produce unknown effects on conducting properties exclusively possible in molecular crystals. The new age should be a stage where the unique significance of molecular conductors is demonstrated. Without a doubt, all of the authors in this Special Issue will overjoy if this issue can help the understanding of the interesting and profound world of organic conductors, which is always ready to welcome new young generation to join us and add their new idea from different points of view.

Funding: A part of the work in this review was funded by Grant-in-Aid for Challenging Exploratory Research (18K19061) of JSPS, the Tokyo Chemical Industry Foundation, the Tokyo Ohka Foundation for the Promotion of Science and Technology, the Kato Foundation for Promotion of Science (KJ-2627), Iketani Science and Technology Foundation (ISTF; 0331005-A), and an Ehime University Grant for Project for the Promotion of Industry/University Cooperation, and the APC was funded by ISTF.

Informed Consent Statement: Not applicable.

Conflicts of Interest: The author declares no conflict of interest.

References

1. Pouget, J.-P.; Moret, R.; Comes, R.; Bechgaard, K.; Fabre, J.M.; Giral, L. X-ray diffuse scattering study of some $(TMTSF)_2X$ and $(TMTTF)_2X$ salts. *Mol. Cryst. Liq. Cryst.* **1981**, *79*, 129–143. [CrossRef]
2. Bryce, M.R.; Murphy, L.C. Organic Metals. *Nature* **1984**, *309*, 119–126. [CrossRef]
3. Williams, J.M.; Beno, M.A.; Wang, H.H.; Leung, P.C.W.; Emge, T.J.; Geiser, U.; Carlson, K.D. Organic superconductors: Structural aspects and design of new material. *Acc. Chem. Res.* **1985**, *18*, 261–267. [CrossRef]
4. Saito, G. Tetrachalcogenafulvalenes with outer chalcogeno substituents. Precursors of organic metals, superconductors, LB fims, etc. *Pure Appl. Chem.* **1987**, *59*, 999–1004. [CrossRef]
5. Schweitzer, D.; Gogu, E.; Henning, I.; Klutz, T.; Keller, H.J. Electrochemically prepared radical salts of BEDT-TTF: Molecular metals and superconductors. *Ber. Bunsenges. Phys. Chem.* **1987**, *91*, 890–896. [CrossRef]
6. Williams, J.M.; Wang, H.H.; Emje, T.J.; Geiser, U.; Beno, M.A.; Leung, P.C.W.; Carlson, K.D.; Thorn, R.J.; Schultz, A.J. Rational design of synthetic metal superconductors. *Prog. Inorg. Chem.* **1987**, *35*, 51–218. [CrossRef]
7. Inokuchi, H. New organic supercondcutors. *Angew. Chem. Int. Ed. Engl.* **1988**, *27*, 1747–1751. [CrossRef]
8. Yoshida, Z.-I.; Sugimoto, T. New donors for molecular organic (super)conductors and ferromagnets. *Angew. Chem. Int. Ed. Engl.* **1988**, *27*, 1573–1577. [CrossRef]
9. Ishiguro, T.; Yamaji, K. *Organic Superconductors*; Springer: New York, NY, USA, 1990.
10. Williams, J.M.; Schultz, A.J.; Geiser, U.; Carlson, K.D.; Kini, A.M.; Wang, H.H.; Kwok, W.-K.; Whangbo, M.-H.; Schirber, J.E. Organic superconductors-new benchmarks. *Science* **1991**, *252*, 1501–1508. [CrossRef]
11. Jérôme, D. The physics of organic superconductors. *Science* **1991**, *252*, 1509–1514. [CrossRef]
12. Bryce, M.R. Recent progress on conducting organic charge-transfer salts. *Chem. Soc. Rev.* **1991**, *20*, 355–390. [CrossRef]
13. Williams, J.M.; Ferraro, J.R.; Thorn, R.J.; Carlson, K.D.; Geiser, U.; Wang, H.H.; Kini, A.M.; Whangbo, M.-H. *Organic Superconductors (Including Fullerenes)*; Prentice Hall: Englewood Cliffs, NJ, USA, 1992.
14. Adam, M.; Müllen, K. Oligomeric Tetrathiafulvalenes: Extended donors for increasing the dimensionality of electrical conduction. *Adv. Mater.* **1994**, *6*, 439–459. [CrossRef]
15. Bryce, M.R. Current trends in tetrathiafulvalene chemistry: Towards increased dimensionality. *J. Mater. Chem.* **1995**, *5*, 1481–1496. [CrossRef]
16. Wzietek, P.; Mayaffre, H.; Jérôme, D.; Brazovskii, S. NMR in the 2D organic superconductors. *J. Phys. I France* **1996**, *6*, 2011–2041. [CrossRef]
17. Ishiguro, T.; Yamaji, K.; Saito, G. *Organic Superconductors*, 2nd ed.; Springer: Berlin, Germany, 1998.
18. Batail, P.; Boubekeur, K.; Fourmigué, M.; Gabriel, J.-C.P. Electrocrystallization, an Invaluable tool for the construction of ordered, electroactive molecular solids. *Chem. Mater.* **1998**, *10*, 3005–3015. [CrossRef]
19. Bryce, M.R. Tetrathiafulvalenes as π-electron donors for intramolecular charge-transfer materials. *Adv. Mater.* **1999**, *11*, 11–23. [CrossRef]
20. Molecular Conductors. *Chem. Rev.* **2004**, *104*, 4887–5782. [CrossRef]
21. Naito, T. Control of magnetism and conduction in organic materials by light. In *Functional Materials: Advances and Applications in Energy Storage and Conversion*; Naito, T., Ed.; Pan Stanford Publishing: Singapore, 2019; Chapter 1; pp. 1–82.
22. Akamatu, H.; Inokuchi, H.; Matsunaga, Y. Electrical conductivity of the perylene-bromine complex. *Nature* **1954**, *173*, 168–169. [CrossRef]
23. Cohen, M.J.; Coleman, L.B.; Garito, A.F.; Heeger, A.J. Electronic properties of tetrathiafulvalenium-tetracyanoquinodimethanide (TTF-TCNQ). *Phys. Rev. B* **1976**, *13*, 5111–5116. [CrossRef]
24. Thomas, G.A.; Schafer, D.E.; Wudl, F.; Horn, P.M.; Rimai, D.; Cook, J.W.; Glocker, D.A.; Skove, M.J.; Chu, C.W.; Groff, R.P.; et al. Electrical conductivity of terathiafulvalenium-tetracyanoquinodimethanide (TTF-TCNQ). *Phys. Rev. B* **1976**, *13*, 5105–5110. [CrossRef]
25. Pouget, J.-P.; Khanna, S.K.; Denoyer, F.; Comès, R.; Garito, A.F.; Heeger, A.J. X ray observation of $2k_F$ and $4k_F$ scatterings in tetrathiafulvalene- tetracyanoquinodimethane (TTF-TCNQ). *Phys. Rev. Lett.* **1976**, *37*, 437–440. [CrossRef]
26. Gutfreund, H.; Weger, M. Temperature dependence of the metallic conductivity of tetrathiafulvalene-tetracyanoquinodimethane (TTF-TCNQ). *Phys. Rev. B* **1977**, *16*, 1753–1755. [CrossRef]
27. Khanna, S.K.; Pouget, J.-P.; Comes, R.; Garito, A.F.; Heeger, A.J. X-ray studies of $2k_F$ and $4k_F$ anomalies in tetrathiafulvalene-tetracyanoquinodimethane (TTF-TCNQ). *Phys. Rev. B* **1977**, *16*, 1468–1479. [CrossRef]
28. Claessen, R.; Sing, M.; Schwingenschlögl, U.; Blaha, P.; Dressel, M.; Jacobsen, C.S. Spectroscopic signatures of spin-charge separation in the quasi-one-dimensional organic conductor TTF-TCNQ. *Phys. Rev. Lett.* **2002**, *88*, 096402. [CrossRef]
29. Sing, M.; Schwingenschlögl, U.; Claessen, R.; Blaha, P.; Carmelo, P.; Martelo, M.; Sacramento, D.; Dressel, M.; Jacobsen, S. Electronic structure of the quasi-one-dimensional organic conductor TTF-TCNQ. *Phys. Rev. B* **2003**, *68*, 125111. [CrossRef]
30. Shirakawa, H.; Louis, E.J.; MacDiarmid, A.G.; Chiang, C.K.; Heeger, A.J. Synthesis of electrically conducting organic polymers: Halogen derivatives of polyacetylene, $(CH)_x$. *J. Chem. Soc. Chem. Commun.* **1977**, 578–580. [CrossRef]
31. Jérome, D.; Mazaud, A.; Ribault, M.; Bechggard, K. Superconductivity in a synthetic organic conductor $(TMTSF)_2PF_6$. *J. Phys. Lett.* **1980**, *41*, 95–98. [CrossRef]
32. Ribault, M.; Pouget, J.-P.; Jérome, D.; Bechggard, K. Superconductivity and absence of a Kohn anomaly in the quasi-one-dimensional organic conductor: $(TMTSF)_2AsF_6$. *J. Phys. Lett.* **1980**, *41*, 607–610. [CrossRef]

33. Parkin, S.S.P.; Ribault, M.; Jerome, D.; Bechgaard, K. Three new superconducting members of the family of tetramethyltetraselenafulvalene (TMTSF) salts: $TMTSF_2ClO_4$, $TMTSF_2SbF_6$, $TMTSF_2TaF_6$. *J. Phys. C* **1981**, *14*, L445–L450. [CrossRef]
34. Parkin, S.S.P.; Ribault, M.; Jérome, D.; Bechggard, K. Superconductivity in a family of organic salts based on the teramethyltetraselenafulvalene (TMTSF) molecule: $(TMTSF)_2X$ ($X = ClO_4$, PF_6, AsF_6, SbF_6, TaF_6). *J. Phys. C* **1981**, *14*, 5305–5326. [CrossRef]
35. Bechgaard, K.; Carneiro, K.; Olsen, M.; Rasmussen, F.B.; Jacobsen, C.S. Zero-pressure organic superconductor: Di-(Tetramethyltetraselenafulvalenium)-perchlorate [$(TMTSF)_2ClO_4$]. *Phys. Rev. Lett.* **1981**, *46*, 852–855. [CrossRef]
36. Parkin, S.S.P.; Jérome, D.; Bechgaard, K. Pressure dependence of the metal-insulator and superconducting phase transitions in $(TMTSF)_2ReO_4$. *Mol. Cryst. Liq. Cryst.* **1981**, *79*, 213–224.
37. Pouget, J.-P.; Shirane, G.; Bechgaard, K.; Fabre, J.M. X-ray evidence of a structural phase transition in di-tetramethyltetraselenafulvalenium perchlorate [$(TMTSF)_2ClO_4$], pristine and slightly doped. *Phys. Rev. B* **1983**, *27*, 5203–5206. [CrossRef]
38. Schulz, H.J.; Bourbonnais, C. Quantum fluctuations in quasi-one-dimensional superconductors. *Phys. Rev. B* **1983**, *27*, 5856–5859. [CrossRef]
39. Lacoe, R.C.; Wolf, S.A.; Chaikin, P.M.; Wudl, F.; Aharon-Shalom, E. Metal-insulator transitions and superconductivity in ditetramethyltetraselenafulvalenium fluorosulfonate [$(TMTSF)_2FSO_3$]. *Phys. Rev. B* **1983**, *27*, 1947–1950. [CrossRef]
40. Pesty, F.; Garoche, P.; Bechgaard, K. Cascade of field-induced phase transitions in the organic metal tetramethyltetraselenafulvalenium perchlorate [$(TMTSF)_2ClO_4$]. *Phys. Rev. Lett.* **1985**, *55*, 2495–2498. [CrossRef]
41. Bourbonnais, C.; Caron, L.G. New mechanisms for phase transitions in quasi-one-dimensional conductors. *Europhys. Lett.* **1988**, *5*, 209–215. [CrossRef]
42. Whangbo, M.-H.; Canadell, E.; Foury, P.; Pouget, J.-P. Hidden Fermi surface nesting and charge density wave instability in low-dimensional metals. *Science* **1991**, *252*, 96–98. [CrossRef]
43. Kang, W.; Hannahs, S.T.; Chaikin, P.M. Toward a unified phase diagram in $(TMTSF)_2X$. *Phys. Rev. Lett.* **1993**, *70*, 3091–3094. [CrossRef] [PubMed]
44. Klemme, B.J.; Brown, S.E.; Wzietek, P.; Kriza, G.; Batail, P.; Jérome, D.; Fabre, J.M. Commensurate and incommensurate spin-density waves and a modified phase diagram of the bechgaard salts. *Phys. Rev. Lett.* **1995**, *75*, 2408–2411. [CrossRef] [PubMed]
45. Behnia, K.; Balicas, L.; Kang, W.; Jérome, D.; Carretta, P.; Fagot-Revurat, Y.; Berthier, C.; Horvatić, M.; Ségransan, P.; Hubert, L.; et al. Confinement in Bechgaard salts: Anomalous magnetoresistance and nuclear relaxation. *Phys. Rev. Lett.* **1995**, *74*, 5272–5275. [CrossRef] [PubMed]
46. Pouget, J.P.; Ravy, S. Structural aspects of the Bechgaard salts and related compounds. *J. Phys. I* **1996**, *6*, 1501–1525. [CrossRef]
47. Degiorgi, L.; Dressel, M.; Schwartz, A.; Alavi, B.; Grüner, G. Direct observation of the spin-density-wave gap in $(TMTSF)_2PF_6$. *Phys. Rev. Lett.* **1996**, *76*, 3838–3841. [CrossRef]
48. Dressel, M.; Schwartz, A.; Grüner, G. Deviations from drude response in low-dimensional metals: Electrodynamics of the metallic state of $(TMTSF)_2PF_6$. *Phys. Rev. Lett.* **1996**, *77*, 398–401. [CrossRef]
49. Pouget, J.P.; Ravy, S. X-Ray evidence of charge density wave modulations in the magnetic phases of $(TMTSF)_2PF_6$ and $(TMTTF)_2Br$. *Synth. Met.* **1997**, *85*, 1523–1528. [CrossRef]
50. Zwick, F.; Brown, S.; Margaritondo, G.; Merlic, C.; Onellion, M.; Voit, J.; Grioni, M. Absence of quasiparticles in the photoemission spectra of quasi-one-dimensional bechgaard salts. *Phys. Rev. Lett.* **1997**, *79*, 3982–3985. [CrossRef]
51. Moser, J.; Gabay, M.; Auban-Senzier, P.; Jérome, D.; Bechgaard, K.; Fabre, J.M. Transverse transport in $(TM)_2X$ organic conductors: Possible evidence for a Luttinger liquid. *Eur. Phys. J. B* **1998**, *1*, 39–46. [CrossRef]
52. Schwartz, A.; Dressel, M.; Grüner, G.; Vescoli, V.; Degiorgi, L. On-chain electrodynamics of metallic $(TMTSF)_2X$ salts: Observation of Tomonaga-Luttinger liquid response. *Phys. Rev. B* **1998**, *58*, 1261–1271. [CrossRef]
53. Bourbonnais, C.; Jérome, D. Electronic confinement in organic metals. *Science* **1998**, *281*, 1155–1156. [CrossRef]
54. Dumm, M.; Loidl, A.; Fravel, B.; Starkey, K.; Montgomery, L.; Dressel, M. Electron spin resonance studies on the organic linear-chain compounds $(TMTCF)_2X$ ($C = S$, Se; $X = PF_6$, AsF_6, ClO_4, Br). *Phys. Rev. B* **2000**, *61*, 511–521. [CrossRef]
55. Wilhelm, H.; Jaccard, D.; Duprat, R.; Bourbonnais, C.; Jérome, D.; Moser, J.; Carcel, C.; Fabre, J.M. The case for universality of the phase diagram of the Fabre and Bechgaard salts. *Eur. Phys. J. B* **2001**, *21*, 175–183. [CrossRef]
56. Lorenz, T.; Hofmann, M.; Grüninger, M.; Freimuth, A.; Uhrig, G.S.; Dumm, M.; Dressel, M. Evidence for spin-charge separation in quasi-one-dimensional organic conductors. *Nature* **2002**, *418*, 614–617. [CrossRef]
57. Jérome, D.; Schulz, H.J. Organic conductors and superconductors. *Adv. Phys.* **2002**, *51*, 293–479. [CrossRef]
58. Vuletić, T.; Auban-Senzier, P.; Pasquier, C.; Tomić, S.; Jérome, D.; Héritier, M.; Bechgaard, K. Coexistence of superconductivity and spin density wave orderings in the organic superconductor $(TMTSF)_2PF_6$. *Eur. Phys. J. B* **2002**, *25*, 319–331. [CrossRef]
59. Jérome, D. Organic conductors: From charge density wave TTF-TCNQ to superconducting $(TMTSF)_2PF_6$. *Chem. Rev.* **2004**, *104*, 5565–5591. [CrossRef] [PubMed]
60. Joo, N.; Auban-Senzier, P.; Pasquier, C.R.; Jérome, D.; Bechgaard, K. Impurity-controlled superconductivity/spin density wave interplay in the organic superconductor: $(TMTSF)_2ClO_4$. *Europhys. Lett.* **2005**, *72*, 645–651. [CrossRef]
61. Sakata, M.; Yoshida, Y.; Maesato, M.; Saito, G.; Matsumoto, K.; Hagiwara, R. Preparation of superconducting $(TMTSF)_2NbF_6$ by electrooxidation of TMTSF using ionic liquid as electrolyte. *Mol. Cryst. Liq. Cryst.* **2006**, *452*, 103–112. [CrossRef]
62. Shinagawa, J.; Kurosaki, Y.; Zhang, F.; Parker, C.; Brown, S.E.; Jérome, D.; Christensen, J.B.; Bechgaard, K. Superconducting state of the organic conductor $(TMTSF)_2ClO_4$. *Phys. Rev. Lett.* **2007**, *98*, 147002. [CrossRef] [PubMed]

63. Yonezawa, S.; Kusaba, S.; Maeno, Y.; Auban-Senzier, P.; Pasquier, C.; Bechgaard, K.; Jérome, D. Anomalous in-plane anisotropy of the onset of superconductivity in $(TMTSF)_2ClO_4$. *Phys. Rev. Lett.* **2008**, *100*, 117002. [CrossRef] [PubMed]
64. Powell, B.J. A phenomenological model of the superconducting state of the Bechgaard salts. *J. Phys. Cond. Mat.* **2008**, *20*, 345234. [CrossRef]
65. Doiron-Leyraud, N.; Auban-Senzier, P.; de Cotret, R.S.; Bourbonnais, C.; Jérome, D.; Bechgaard, K.; Taillefer, L. Correlation between linear resistivity and T_c in the Bechgaard salts and the pnictide superconductor Ba $(Fe_{1-x}\ Co_x)_2As_2$. *Phys. Rev. B* **2009**, *80*, 214531. [CrossRef]
66. Pouget, J.-P. The Peierls instability and charge density wave in one-dimensional electronic conductors. *Comp. Rend. Phys.* **2016**, *17*, 332–356. [CrossRef]
67. Mizuno, M.; Garito, A.F.; Cava, M.P. 'Organic metals': Alkylthio substitution effects in tetrathiafulvalene- tetracyanoquinodimethane charge-transfer complexes. *J. Chem. Soc. Chem. Commun.* **1978**, 18–19. [CrossRef]
68. Parkin, S.S.P.; Engler, E.M.; Schumaker, R.R.; Lagier, R.; Lee, V.Y.; Scott, J.C.; Greene, R.L. Superconductivity in a new family of organic conductors. *Phys. Rev. Lett.* **1983**, *50*, 270–273. [CrossRef]
69. Engler, E.M.; Lee, Y.V.; Schumaker, R.R.; Parkin, S.S.P.; Greene, R.L.; Scott, J.C. Synthesis of biethylenedithiolylene-tetrathiafulvalene donors (BEDT-TTF) and electrochemical preparation of their charge transfer complexes. *Mol. Cryst. Liq. Cryst.* **1984**, *107*, 19–31. [CrossRef]
70. Yagubskii, É.B.; Shchegolev, I.F.; Laukhin, V.N.; Kononovich, P.A.; Karatsovnik, M.V.; Zvarykina, A.V.; Buravov, L.I. Normal-pressure superconductivity in an organic metal $(BEDT\text{-}TTF)_2I_3$ [bis (ethylene dithio) tetrathiofulvalene triiodide. *JETP Lett.* **1984**, *39*, 12–16.
71. Yagubskii, É.B.; Shchegolev, I.F.; Laukhin, V.N.; Shibaeva, R.P.; Kostyuchenko, E.É.; Khomenko, A.G.; Sushko, Y.V.; Zvarykina, A.V. Superconducting transition in the dielectric α phase of iodine-doped $(BEDT\text{-}TTF)_2I_3$ compound. *JETP Lett.* **1984**, *40*, 1201–1204.
72. Williams, J.M.; Wang, H.H.; Beno, M.A.; Emge, T.J.; Sowa, L.M.; Copps, P.T.; Behroozi, F.; Hall, L.N.; Douglas Carlson, K.; Crabtree, G.W. Ambient-pressure superconductivity at 2.7 K and higher temperatures in derivatives of $(BEDT\text{-}TTF)_2IBr_2$: Synthesis, structure, and detection of superconductivity. *Inorg. Chem.* **1984**, *23*, 3839–3841. [CrossRef]
73. Laukhin, V.N.; Kostyuchenko, E.É.; Sushko, Y.V.; Shchegolev, I.F.; Yagubskii, É.B. Effect of pressure on the superconductivity of β-$(BEDT\text{-}TTF)_2I_3$ compound. *JETP Lett.* **1985**, *41*, 81–84.
74. Murata, K.; Tokumoto, M.; Anzai, H.; Bando, H.; Saito, G.; Kajimura, K.; Ishiguro, T. Superconductivity with the onset at 8 K in the organic conductor β-$(BEDT\text{-}TTF)_2I_3$ under pressure. *J. Phys. Soc. Jpn.* **1985**, *54*, 1236–1239. [CrossRef]
75. Murata, K.; Tokumoto, M.; Anzai, H.; Bando, H.; Kajimxjra, K.; Ishiguro, T. Pressure phase diagram of the organic superconductor β-$(BEDT\text{-}TTF)_2I_3$. *J. Phys. Soc. Jpn.* **1985**, *54*, 2084–2087. [CrossRef]
76. Wang, H.H.; Beno, M.A.; Geiser, U.; Firestone, M.A.; Webb, K.S.; Nuiiez, L.; Crabtree, G.W.; Carlson, K.D.; Williams, J.M.; Azevedo, L.J.; et al. Ambient-pressure superconductivity at the highest temperature (5 K) observed in an organic system: β-$(BEDT\text{-}TTF)_2AuI_2$. *Inorg. Chem.* **1985**, *24*, 2465–2466. [CrossRef]
77. Lyubovskaya, R.N.; Lyubovskii, R.B.; Shibaeva, R.P.; Aldoshina, M.Z.; Gol'denberg, L.M.; Rozenberg, L.P.; Khidekel, M.L.; Shul'pyakov, Y.F. Superconduictivity in a BEDT-TTF organic conductor with a chloromercurate anion. *JETP Lett.* **1985**, *42*, 468–472.
78. Shibaeva, R.P.; Kaminskii, V.P.; Yagubskii, E.B. Crystal structures of organic metals and superconductors of (BEDT-TTP)-I system. *Mol. Cryst. Liq. Cryst.* **1985**, *119*, 361–373. [CrossRef]
79. Baram, G.O.; Buravov, L.I.; Degtyarev, L.S.; Kozlov, M.E.; Laukhin, V.N.; Laukhina, E.E.; Onishchenko, V.G.; Pokhodnya, K.I.; Sheinkman, M.K.; Shibaeva, R.P.; et al. Transformation of the *α-phase* $(BEDT\text{-}TTF)_2I_3$ to the superconducting β phase with T_c = 6–7 K. *JETP Lett.* **1986**, *44*, 376–378.
80. Kobayashi, H.; Kato, R.; Kobayashi, A.; Nishio, Y.; Kajita, K.; Sasaki, W. A new molecular superconductor, $(BEDT\text{-}TTF)_2(I_3)_{1-x}(AuI_2)_x(x < 0.02)$. *Chem. Lett.* **1986**, *15*, 789–792. [CrossRef]
81. Varma, K.S.; Bury, A.; Harris, N.J.; Underhill, A.E. Improved synthesis of Bis(ethylenedithio)tetrathiafulvalene (BEDT-TTF): π-Donor for Synthetic Metals. *Synthesis* **1987**, *1987*, 837–838. [CrossRef]
82. Schweitzer, D.; Bele, P.; Brunner, H.; Gogu, E.; Haeberlen, U.; Hennig, I.; Klutz, I.; Świetlik, R.; Keller, H.J. A stable superconducting state at 8 K and ambient pressure in α_t-$(BEDT\text{-}TTF)_2I_3$. *Z. Phys. B Condens. Matter.* **1987**, *67*, 489–495. [CrossRef]
83. Kobayashi, A.; Kato, R.; Kobayashi, H.; Moriyama, S.; Nishio, Y.; Kajita, K.; Sasaki, W. Crystal and electronic structures of a new molecular superconductor, κ-$(BEDT\text{-}TTF)_2I_3$. *Chem. Lett.* **1987**, *16*, 459–462. [CrossRef]
84. Kato, R.; Kobayashi, H.; Kobayashi, A.; Moriyama, S.; Nishio, Y.; Kajita, K.; Sasaki, W. A new ambient-pressure superconductor, κ-$(BEDT\text{-}TTF)_2I_3$. *Chem. Lett.* **1987**, *16*, 507–510. [CrossRef]
85. Mori, T.; Inokuchi, H. Superconductivity in $(BEDT\text{-}TTF)_3Cl_22H_2O$. *Solid State Commun.* **1987**, *64*, 335–337. [CrossRef]
86. Lyubovskaya, R.N.; Zhilyaeva, E.A.; Zvarykina, A.V.; Laukhin, V.N.; Lyubovskii, R.B.; Pesotskii, S.I. Is the organic metal $(ET)_4Hg_3Br_8$ a quasi-2D superconductor? *JETP Lett.* **1987**, *45*, 530–533.
87. Lyubovskaya, R.N.; Zhilyaeva, E.I.; Pesotskii, S.I.; Lyubovskii, R.B.; Atovmyan, L.O.; D'yachenko, O.A.; Takhirov, T.G. Superconductivity of $(ET)_4Hg_{2.89}Br_8$ at atmospheric pressure and T_c = 4.3 K and the critical-field anisotropy. *JETP Lett.* **1987**, *46*, 188–191.

88. Urayama, H.; Yamochi, H.; Saito, G.; Nozawa, K.; Sugano, T.; Kinoshita, M.; Sato, S.; Oshima, K.; Kawamoto, A.; Tanaka, J. A new ambient pressure organic superconductor based on BEDT-TTF with T_C higher than 10 K (T_C = 10.4 K). *Chem. Lett.* **1988**, *17*, 55–58. [CrossRef]
89. Schirber, J.E.; Overmyer, D.L.; Venturini, E.L.; Wang, H.H.; Carlson, K.D.; Kwok, W.K.; Kleinjan, S.; Williams, J.M. Anomalous pressure dependence of the superconducting transition temperature of $(ET)_4Hg_{2.89}Br_8$. *Phys. C* **1989**, *161*, 412–414. [CrossRef]
90. June, D.; Evein, M.; Novoa, J.J.; Whangbc, M.-H.; Beno, M.A.; Kini, A.M.; Schultz, A.J.; Williams, J.M.; Nigrey, P.J. Similarities and differences in the structural and electronic properties of κ-Phase Organic Conducting and Superconducting Salts. *Inorg. Chem.* **1989**, *28*, 4516–4522. [CrossRef]
91. Larsen, J.; Lenoir, C. Synthesis of Bis(ethylenedithio)tetrathiafulvalene (BEDT-TTF). *Synthesis* **1989**, *1989*, 134. [CrossRef]
92. Reed, P.E.; Braam, J.M.; Sowa, L.M.; Barkhau, R.A.; Blackman, G.S.; Cox, D.D.; Ball, G.A.; Wang, H.H.; Williams, J.M. Synthesis of 5,5′,6,6′-tetrahydro-2,2′-Bi-1,3-dithiolo[4,5-*b*][1,4]dithiinylidene (BEDT-TTF). In *Inorganic Syntheses*; Wiley: New York, NY, USA, 1989; pp. 386–390.
93. Wang, H.H.; Carlson, K.D.; Geiser, U.; Kwok, W.K.; Vashon, M.D.; Thompson, J.E.; Larsen, N.F.; McCabe, G.D.; Hulscher, R.S.; Williams, J.M. A new ambient-pressure organic superconductor: $(BEDT\text{-}TTF)_2(NH_4)Hg(SCN)_4$. *Physica C* **1990**, *166*, 57–61. [CrossRef]
94. Mori, H.; Tanaka, S.; Oshima, M.; Saito, G.; Mori, T.; Maruyama, Y.; Inokuchi, H. Crystal and electronic structures of $(BEDT\text{-}TTF)_2[MHg(SCN)_4]$ (M = K and NH_4). *Bull. Chem. Soc. Jpn.* **1990**, *63*, 2183–2190. [CrossRef]
95. Kini, A.M.; Geiser, U.; Wang, H.H.; Carlson, K.D.; Williams, J.M.; Kwok, W.K.; Vandervoort, K.G.; Thompson, J.E.; Stupka, D.L.; Jung, D.; et al. A new ambient-pressure organic superconductor, $\kappa\text{-}(ET)_2Cu[N(CN)_2]Br$, with the highest transition temperature yet observed (inductive onset T_c = 11.6 K, resistive onset = 12.5 K). *Inorg. Chem.* **1990**, *29*, 2555–2557. [CrossRef]
96. Williams, J.M.; Kini, A.M.; Wang, H.H.; Carlson, K.D.; Geiser, U.; Montgomery, L.K.; Pyrka, G.J.; Watkins, D.M.; Kommers, J.M.; Boryschuk, S.J.; et al. From semiconductor-semiconductor transition (42 K) to the highest-T_c organic superconductor, $\kappa\text{-}(ET)_2Cu[N(CN)_2]Cl$ (Tc = 12.5 K). *Inorg. Chem.* **1990**, *29*, 3272–3274. [CrossRef]
97. Mori, H.; Hirabayashi, I.; Tanaka, S.; Mori, T.; Inokuchi, H. A new ambient-pressure organic superonductor, $\kappa\text{-}(BEDT\text{-}TTF)_2Ag(CN)_2H_2O$ (T_C = 5.0 K). *Solid State Commun.* **1990**, *76*, 35–37. [CrossRef]
98. Geiser, U.; Wang, H.H.; Carlson, K.D.; Williams, J.M.; Charlier, H.A., Jr.; Heindl, J.E.; Yaconi, G.A.; Love, B.J.; Lathrop, M.W.; Schirber, J.E.; et al. Superconductivity at 2.8 K and 1.5 kbar in $\alpha\text{-}(BEDT\text{-}TTF)_2Cu_2(CN)_3$: The first organic superconductor containing a polymeric copper cyanide anion. *Inorg. Chem.* **1991**, *30*, 2586–2588. [CrossRef]
99. Komatsu, T.; Nakamura, T.; Matsukawa, N.; Yamochi, H.; Saito, G.; Ito, H.; Ishiguro, T.; Kusunoki, M.; Sakaguchi, K.-I. New ambient-pressure organic superconductors based on BEDT-TTF, Cu, $N(CN)_2$ and CN with T_c = 10.7 K and 3.8 K. *Solid State Commun.* **1991**, *80*, 843–847. [CrossRef]
100. Mori, H.; Hirabayashi, I.; Tanaka, S.; Mori, T.; Maruyama, Y.; Inokuchi, H. Superconductivity in $(BEDT\text{-}TTF)_4Pt(CN)_4H_2O$. *Solid State Commun.* **1991**, *80*, 411–415. [CrossRef]
101. Singleton, J.; Pratt, F.L.; Doporto, M.; Janssen, T.J.B.M.; Kurmoo, M.; Perenboom, J.A.A.J.; Hayes, W.; Day, P. Far-infrared cyclotron resonance study of electron dynamics in $(BEDT\text{-}TTF)_2KHg(SCN)_4$. *Phys. Rev. Lett.* **1992**, *68*, 2500–2503. [CrossRef]
102. Pratt, F.L.; Singleton, J.; Doporto, M.; Fisher, A.J.; Janssen, T.J.B.M.; Perenboom, J.A.A.J.; Kurmoo, M.; Hayes, W.; Day, P. Magnetotransport and Fermi-surface topology of $[bis(ethylenedithio)tetrathiafulvalene]_2KHg(SCN)_4$. *Phys. Rev. B* **1992**, *45*, 13904–13912. [CrossRef] [PubMed]
103. Brooks, J.S.; Agosta, C.C.; Klepper, S.J.; Tokumoto, M.; Kinoshita, N.; Anzai, H.; Uji, S.; Aoki, H.; Perel, A.S.; Athas, G.J.; et al. Novel interplay of Fermi-surface behavior and magnetism in a low-dimensional organic conductor. *Phys. Rev. Lett.* **1992**, *69*, 156–159. [CrossRef] [PubMed]
104. Mori, T.; Kato, K.; Maruyama, Y.; Inokuchi, H.; Mori, H.; Hirabayashi, I.; Tanaka, S. Structural and physical properties of a new organic superconductor, $(BEDT\text{-}TTF)_4Pd(CN)_4H_2O$. *Solid State Commun.* **1992**, *82*, 177–181. [CrossRef]
105. Yamochi, H.; Nakamura, T.; Komastu, T.; Matsukawa, N.; Inoue, T.; Saito, G.; Mori, T. Crystal and electronic structures of the organic superconductors, $\kappa\text{-}(BEDT\text{-}TTF)_2Cu(CN)[N(CN)_2]$ and $\kappa'\text{-}(BEDT\text{-}TTF)_2Cu_2(CN)_3$. *Solid State Commun.* **1992**, *82*, 101–105. [CrossRef]
106. Yamochi, H.; Komatsu, T.; Matsukawa, N.; Saito, G.; Mori, T.; Kusunoki, M.; Sakaguchi, K.-I. Structural aspects of the ambient-pressure BEDT-TTF superconductors. *J. Am. Chem. Soc.* **1993**, *115*, 11319–11327. [CrossRef]
107. Kushch, N.D.; Buravov, L.I.; Khomenko, A.G.; Yagubskii, E.B.; Rosenberg, L.P.; Shibaeva, R.P. Novel organic superconductor $\kappa\text{-}(ET)_2Cu[N(CN)_2]Cl_{0.5}Br_{0.5}$ with T_C ~ 11.3 K. *Synth. Met.* **1993**, *53*, 155–160. [CrossRef]
108. Achkir, D.; Poirier, M.; Bourbonnais, C.; Quirion, G.; Lenoir, C.; Batail, P.; Jérome, D. Microwave surface impedance of $\kappa\text{-}(BEDT\text{-}TTF)_2Cu(NCS)_2$, where BEDT-TTF is bis(ethylenedithio)tetrathiafulvalene: Evidence for unconventional superconductivity. *Phys. Rev. B* **1993**, *47*, 11595–11598. [CrossRef]
109. Ito, H.; Kaneko, H.; Ishiguro, T.; Ishimoto, H.; Kono, K.; Horiuchi, S.; Komatsu, T.; Saito, G. On superconductivity of the organic conductor $\alpha\text{-}(BEDT\text{-}TTF)_2KHg(SCN)_4$. *Solid State Commun.* **1993**, *85*, 1005–1009. [CrossRef]
110. Klepper, S.J.; Brooks, J.S.; Chen, X.; Bradaric, I.; Tokumoto, M.; Kinoshita, N.; Tanaka, Y.; Agosta, C.C. Pressure-induced nesting in the low-dimensional organic superconductor $\alpha\text{-}(BEDT\text{-}TTF)_2NH_4Hg(SCN)_4$. *Phys. Rev. B* **1993**, *48*, 9913–9916. [CrossRef]

111. Kahlich, S.; Schweitzer, D.; Rovira, C.; Paradis, J.A.; Whangbo, M.-H.; Heinen, I.; Keller, H.J.; Nuber, B.; Bele, P.; Brunner, H.; et al. Characterisation of the Fermi surface and phase transitions of (BEDO-TTF)$_2$ReO$_4$·(H$_2$O) by physical property measurements and electronic band structure calculations. *Z. Phys. B* **1994**, *94*, 39–47. [CrossRef]
112. Schlueter, J.A.; Geiser, U.; Williams, J.M.; Wang, H.H.; Kwok, W.-K.; Fendrich, J.A.; Carlson, K.D.; Achenbach, C.A.; Dudek, J.D.; Naumann, D.; et al. The first organic cation-radical salt superconductor (T_c = 4 K) with an organometallic anion: Superconductivity, synthesis and structure of κ_L-(BEDT-TTF)$_2$Cu(CF$_3$)$_4$·TCE. *J. Chem. Soc. Chem. Commun.* **1994**, 1599–1600. [CrossRef]
113. Schlueter, J.A.; Carlson, K.D.; Williams, J.M.; Geiser, U.; Wang, H.H.; Welp, U.; Kwok, W.-K.; Fendrich, J.A.; Dudek, J.D.; Achenbach, C.A.; et al. A new 9 K superconducting organic salt composed of the bis (ethylenedithio) tetrathiafulvalene (ET) electron-donor molecule and the tetrakis (trifluoromethyl) cuprate (III) anion, [Cu(CF$_3$)$_4$]$^-$. *Phys. C* **1994**, *230*, 378–384. [CrossRef]
114. Schlueter, J.A.; Carlson, K.D.; Geiser, U.; Wang, H.H.; Williams, J.M.; Kwok, W.-K.; Fendrich, J.A.; Welp, U.; Keane, P.M.; Dudek, J.D.; et al. Superconductivity up to 11.1 K in three solvated salts composed of [Ag(CF$_3$)$_4$]$^-$ and the organic electron-donor molecule bis (ethylenedithio) tetrathiafulvalene (ET). *Phys. C* **1994**, *233*, 379–386. [CrossRef]
115. Caulfield, J.; Lubczynski, W.; Pratt, F.L.; Singleton, J.; Ko, D.Y.K.; Hayes, W.; Kurmoo, M.; Day, P. Magnetotransport studies of the organic superconductor kappa -(BEDT-TTF)$_2$Cu(NCS)$_2$ under pressure: The relationship between carrier effective mass and critical temperature. *J. Phys. Cond. Mat.* **1994**, *6*, 2911–2924. [CrossRef]
116. Dressel, M.; Klein, O.; Grüner, G.; Carlson, K.D.; Wang, H.H.; Williams, J.M. Electrodynamics of the organic superconductor κ-(BEDT-TTF)$_2$Cu(NCS)$_2$ and κ-(BEDT-TTF)$_2$Cu[N(CN)$_2$]Br. *Phys. Rev. B* **1994**, *50*, 13603–13615. [CrossRef]
117. Pratt, F.L.; Sasaki, T.; Toyota, N.; Nagamine, K. Zero field muon spin relaxation study of the low temperature state in α-(BEDT-TTF)$_2$KHg(SCN)$_4$. *Phys. Rev. Lett.* **1995**, *74*, 3892–3895. [CrossRef]
118. Brooks, J.S.; Chen, X.; Klepper, S.J.; Valfells, S.; Athas, G.J.; Tanaka, Y.; Kinoshita, T.; Kinoshita, N.; Tokumoto, M.; Anzai, H.; et al. Pressure effects on the electronic structure and low-temperature states in the α-(BEDT-TTF)$_2$*M*Hg(SCN)$_4$ organic-conductor family (*M* = K, Rb, Tl, NH$_4$). *Phys. Rev. B* **1995**, *52*, 14457. [CrossRef] [PubMed]
119. Schlueter, J.A.; Williams, J.M.; Geiser, U.; Dudek, J.D.; Sirchio, S.A.; Kelly, M.E.; Gregar, J.S.; Kwok, W.H.; Fendrich, J.A.; Schirber, J.E.; et al. Synthesis and characterization of two new organic superconductors, κ_L- and κ_H-[bis(ethylenedisulfanyl) tetrathiafulvalene]$_2$Au(CF$_3$)$_4$·(1,1,2- trichloroethane) via microelectrocrystallization. *J. Chem. Soc. Chem. Commun.* **1995**, 1311–1312. [CrossRef]
120. Kurmoo, M.; Graham, A.W.; Day, P.; Coles, S.J.; Hursthouse, M.B.; Caulfield, J.L.; Singleton, J.; Pratt, F.L.; Hayes, W.; Ducasse, L.; et al. Superconducting and semiconducting magnetic charge transfer salts: (BEDT-TTF)$_4$AFe(C$_2$O$_4$)$_3$·C$_6$H$_5$CN (A = H$_2$O, K, NH$_4$). *J. Am. Chem. Soc.* **1995**, *117*, 12209–12217. [CrossRef]
121. Campos, C.E.; Brooks, J.S.; van Bentum, P.J.M.; Perenboom, J.A.A.J.; Klepper, S.J.; Sandhu, P.S.; Valfells, S.; Tanaka, Y.; Kinoshita, T.; Kinoshita, N.; et al. Uniaxial-stress-induced superconductivity in organic conductors. *Phys. Rev. B* **1995**, *52*, R7014–R7017. [CrossRef]
122. Schlueter, J.A.; Williams, J.M.; Geiser, U.; Dtulek, J.D.; Kelly, M.E.; Sirchio, S.A.; Carlson, K.D.; Naumann, D.D.; Roy, T.; Campana, C.F. Seven new organic superconductors in the system (ET)$_2$M(CF$_3$)$_4$(solvent) (M = Cu, Ag): Effect of solvent replacement. *Adv. Mater.* **1995**, *7*, 634–639. [CrossRef]
123. Schlueter, J.A.; Geiser, U.; Wang, H.H.; Kelly, M.E.; Dudek, J.D.; Williams, J.M.; Naumann, D.; Roy, T. Synthesis and physical properties of a novel, highly tunable family of organic superconductors: (ET)$_2$M(CF$_3$)$_4$(1,1,2-trihaloethane) (M = Cu, Ag, Au). *Mol. Cryst. Liq. Cryst.* **1996**, *284*, 195–202. [CrossRef]
124. Schlueter, J.A.; Williams, J.M.; Geiser, U.; Wang, H.H.; Kini, A.M.; Kelly, M.E.; Dudek, J.D.; Naumann, D.; Roy, T. New organic superconductors in the system (ET)$_2$*M*(CF$_3$)$_4$(solvent) (*M* = Cu, Ag, Au): Dramatic effects of organometallic anion and solvent replacement. *Mol. Cryst. Liq. Cryst.* **1996**, *285*, 43–50. [CrossRef]
125. Geiser, U.; Schlueter, J.A.; Hau Wang, H.; Kini, A.M.; Williams, J.M.; Sche, P.P.; Zakowicz, H.I.; VanZile, M.L.; Dudek, J.D.; Nixon, P.G.; et al. Superconductivity at 5.2 K in an electron donor radical salt of bis(ethylenedithio)tetrathiafulvalene (BEDT-TTF) with the novel polyfluorinated organic anion SF$_5$CH$_2$CF$_2$SO$_3$$^-$. *J. Am. Chem. Soc.* **1996**, *118*, 9996–9997. [CrossRef]
126. Meline, R.L.; Elsenbaumer, R.L. A high yield conversion of tetrathiafulvalene into bis(ethylenedithio)tetrathiafulvalene and derivatives. *J. Chem. Soc. Perkin Trans.* **1997**, *1*, 3575–3576. [CrossRef]
127. Guionneau, P.; Kepert, C.J.; Bravic, G.; Chasseau, D.; Truter, M.R.; Kurmoo, M.; Day, P. Determining the charge distribution in BEDT-TTF salts. *Synth. Met.* **1997**, *86*, 1973–1974. [CrossRef]
128. Martin, L.; Turner, S.S.; Day, P.; Mabbs, F.E.; McInnes, E.J.L. New molecular superconductor containing paramagnetic chromium(III) ions. *Chem. Commun.* **1997**, 1367–1368. [CrossRef]
129. Lee, S.L.; Pratt, F.L.; Blundell, S.J.; Aegerter, C.M.; Pattenden, P.A.; Chow, K.H.; Forgan, E.M.; Sasaki, T.; Hayes, W.; Keller, H. Investigation of vortex behavior in the organic superconductor κ-(BEDT-TTF)$_2$Cu(SCN)$_2$ using muon spin rotation. *Phys. Rev. Lett.* **1997**, *79*, 1563–1566. [CrossRef]
130. Mori, T. Structural genealogy of BEDT-TTF-based organic conductors I. Parallel molecules: β and β'' phases. *Bull. Chem. Soc. Jpn.* **1998**, *71*, 2509–2526. [CrossRef]
131. Mori, T.; Mori, H.; Tanaka, S. Structural genealogy of BEDT-TTF-based organic conductors II. Inclined molecules: θ, α, and κ phases. *Bull. Chem. Soc. Jpn.* **1999**, *72*, 179–197. [CrossRef]
132. Mori, T. Structural genealogy of BEDT-TTF-based organic conductors III. Twisted molecules: δ and α' phases. *Bull. Chem. Soc. Jpn.* **1999**, *72*, 2011–2027. [CrossRef]

133. Carrington, A.; Bonalde, I.J.; Prozorov, R.; Giannetta, R.W.; Kini, A.M.; Schlueter, J.; Wang, H.H.; Geiser, U.; Williams, J.M. Low-temperature penetration depth of κ-$(ET)_2Cu[N(CN)_2]Br$ and κ-$(ET)_2Cu(NCS)_2$. *Phys. Rev. Lett.* **1999**, *83*, 4172–4175. [CrossRef]
134. Lefebvre, S.; Wzietek, P.; Brown, S.; Bourbonnais, C.; Jérome, D.; Mézière, C.; Fourmigué, M.; Batail, P. Mott transition, antiferromagnetism, and unconventional superconductivity in layered organic superconductors. *Phys. Rev. Lett.* **2000**, *85*, 5420–5423. [CrossRef]
135. Kushch, N.D.; Tanatar, M.A.; Yagubskii, E.B.; Ishiguro, T. Supercondcutivity of κ-$(ET)_2Cu[N(CN)_2]I$ under pressure. *JETP Lett.* **2001**, *73*, 429–431. [CrossRef]
136. Rashid, S.; Turner, S.S.; Day, P.; Howard, J.A.K.; Guionneau, P.; McInnes, E.J.L.; Mabbs, F.E.; Clark, R.J.H.; Firth, S.; Biggs, T. New superconducting charge-transfer salts $(BEDT\text{-}TTF)_4[A{\cdot}M(C_2O_4)_3]{\cdot}C_6H_5NO_2$ ($A = H_3O$ or NH_4, M = Cr or Fe, BEDT-TTF = bis(ethylenedithio)tetrathiafulvalene). *J. Mater. Chem.* **2001**, *11*, 2096–2102. [CrossRef]
137. Singleton, J.; Mielke, C. Quasi-two-dimensional organic superconductors: A review. *Contemp. Phys.* **2002**, *43*, 63–96. [CrossRef]
138. Müller, J.; Lang, M.; Steglich, F.; Schlueter, J.A.; Kini, A.M.; Sasaki, T. Evidence for structural and electronic instabilities at intermediate temperatures in κ-$(BEDT\text{-}TTF)_2X$ for $X = Cu[N(CN)_2]Cl$, $Cu[N(CN)_2]Br$ and $Cu(NCS)_2$: Implications for the phase diagram of these quasi-two-dimensional organic superconductors. *Phys. Rev. B* **2002**, *65*, 144521. [CrossRef]
139. Singleton, J.; Goddard, P.A.; Ardavan, A.; Harrison, N.; Blundell, S.J.; Schlueter, J.A.; Kini, A.M. Test for interlayer coherence in a quasi-two-dimensional superconductor. *Phys. Rev. Lett.* **2002**, *88*, 037001. [CrossRef]
140. Akutsu, H.; Akutsu-Sato, A.; Turner, S.S.; Le Pevelen, D.; Day, P.; Laukhin, V.; Klehe, A.-K.; Singleton, J.; Tocher, D.A.; Probert, M.R.; et al. Effect of included guest molecules on the normal state conductivity and superconductivity of β''-$(ET)_4$ $[(H_3O)Ga(C_2O_4)_3]{\cdot}G$ (G = pyridine, nitrobenzene). *J. Am. Chem. Soc.* **2002**, *124*, 12430–12431. [CrossRef]
141. Tajima, N.; Ebina-Tajima, A.; Tamura, M.; Nishio, Y.; Kajita, K. Effects of uniaxial strain on transport properties of organic conductor α-$(BEDT\text{-}TTF)_2I_3$ and discovery of superconductivity. *J. Phys. Soc. Jpn.* **2002**, *71*, 1832–1835. [CrossRef]
142. Dressel, M.; Drichko, N.; Schlueter, J.; Merino, J. Proximity of the layered organic conductors α-$(BEDT\text{-}TTF)_2MHg(SCN)_4$ (M = K, NH_4), to a charge-ordering transition. *Phys. Rev. Lett.* **2003**, *90*, 167002. [CrossRef] [PubMed]
143. Taniguchi, H.; Miyashita, M.; Uchiyama, K.; Satoh, K.; Môri, N.; Okamoto, H.; Miyagawa, K.; Kanoda, K.; Hedo, M.; Uwatoko, Y. Superconductivity at 14.2 K in layered organics under extreme pressure. *J. Phys. Soc. Jpn.* **2003**, *72*, 468–471. [CrossRef]
144. Shibaeva, R.P.; Yagubskii, E.B. Molecular conductors and superconductors based on trihalides of BEDT-TTF and some of its analogues. *Chem. Rev.* **2004**, *104*, 5347–5378. [CrossRef]
145. Yamada, J., Sugimoto, T., (Eds.) *TTF Chemistry: Fundamentals and Applications of Tetrathiafulvalene*; Springer: Berlin, Germany, 2004.
146. Uchiyama, K.; Miyashita, M.; Taniguchi, H.; Satoh, K.; Môri, N.; Miyagawa, K.; Kanoda, K.; Hedo, M.; Uwatoko, Y. Characterization of transport and magnetic properties of a Mott insulator, β'-$(BEDT\text{-}TTF)_2IBrCl$. *J. Phys. IV* **2004**, *114*, 387–389. [CrossRef]
147. Lang, M.; Müller, J. Organic superconductors. In *The Physics of Superconductors*; Bennemann, K.H., Ketterson, J.B., Eds.; Springer: Berlin/Heidelberg, Germany, 2004. [CrossRef]
148. Geiser, U.; Schlueter, J.A. Conducting organic radical cation salts with organic and organometallic anions. *Chem. Rev.* **2004**, *104*, 5203–5241. [CrossRef]
149. Coronado, E.; Curreli, S.; Giménez-Saiz, C.; Gómez-García, C.J. A novel paramagnetic molecular superconductor formed by bis(ethylenedithio)tetrathiafulvalene, tris(oxalato)ferrate(III) anions and bromobenzene as guest molecule: $ET_4[(H_3O)Fe(C_2O_4)_3]{\cdot}C_6H_5Br$. *J. Mater. Chem.* **2005**, *15*, 1429–1436. [CrossRef]
150. Coronado, E.; Curreli, S.; Giménez-Saiz, C.; Gómez-García, C.J. New magnetic conductors and superconductors based on BEDT-TTF and BEDS-TTF. *Synth. Met.* **2005**, *154*, 245–248. [CrossRef]
151. Mori, H. Materials viewpoint of organic superconductors. *J. Phys. Soc. Jpn.* **2006**, *75*, 051003. [CrossRef]
152. Saito, G.; Yoshida, Y. Development of conductive organic molecular assemblies: Organic metals, superconductors, and exotic functional materials. *Bull. Chem. Soc. Jpn.* **2007**, *80*, 1–137. [CrossRef]
153. Mori, T.; Kawamoto, T. Organic conductors—From fundamentals to nonlinear conductivity. *Annu. Rep. Prog. Chem. Sect. C Phys. Chem.* **2007**, *103*, 134–172. [CrossRef]
154. Wosnitza, J. Quasi-two-dimensional organic supercondcutors. *J. Low Temp. Phys.* **2007**, *146*, 641–667. [CrossRef]
155. Monthoux, P.; Pines, D.; Lonzarich, G.G. Superconductivity without phonons. *Nature* **2007**, *450*, 1177–1183. [CrossRef]
156. Wolter, A.U.B.; Feyerherm, R.; Dudzik, E.; Süllow, S.; Strack, C.; Lang, M.; Schweitzer, D. Determining ethylene group disorder levels in κ-$(BEDT\text{-}TTF)_2Cu[N(CN)_2]Br$. *Phys. Rev. B* **2007**, *75*, 104512. [CrossRef]
157. Lang, M.; Müller, J. Organic superconductors. In *Superconductivity*; Bennemann, K.H., Ketterson, J.B., Eds.; Springer: Berlin/Heidelberg, Germany, 2008; pp. 1155–1223. [CrossRef]
158. Naito, T.; Yamada, Y.; Inabe, T.; Toda, Y. Carrier dynamics in κ-type organic superconductors: Time-resolved observation. *J. Phys. Soc. Jpn.* **2008**, *77*, 064709. [CrossRef]
159. Naito, T.; Inabe, T.; Toda, Y. Carrier dynamics in organic superconductors. In *Molecular Electronic and Related Materials—Control and Probe with Light*; Naito, T., Ed.; Transworld Research Network: Kerala, India, 2010; pp. 1–36.
160. Kawamoto, T.; Mori, T.; Yamaguchi, T.; Uji, S.; Graf, D.; Brooks, J.S.; Shirahata, T.; Kibune, M.; Yoshino, H.; Imakubo, T. Fermi surface and in-plane anisotropy of the layered organic superconductor κ_L-$(DMEDO\text{-}TSeF)_2[Au(CN)_4](THF)$ with domain structures. *Phys. Rev. B* **2011**, *83*, 012505. [CrossRef]

161. Ardavan, A.; Brown, S.; Kagoshima, S.; Kanoda, K.; Kuroki, K.; Mori, H.; Ogata, M.; Uji, S.; Wosnitza, J. Recent topics of organic superconductors. *J. Phys. Soc. Jpn.* **2012**, *81*, 011004. [CrossRef]
162. Saito, G.; Yoshida, Y. Frontiers of organic conductors and superconductors. *Top. Curr. Chem.* **2012**, *312*, 67–126. [CrossRef] [PubMed]
163. Mori, T. *Electronic Properties of Organic Conductors*; Springer: Tokyo, Japan, 2016. [CrossRef]
164. Hebard, A.F.; Rosseinsky, M.J.; Haddon, R.C.; Murphy, D.W.; Glarum, S.H.; Palstra, T.T.M.; Ramirez, A.P.; Kortan, A.R. Superconductivity at 18 K in potassium-doped C_{60}. *Nature* **1991**, *350*, 600–601. [CrossRef]
165. Zhou, O.; Fischer, J.E.; Coustel, N.; Kycia, S.; Zhu, Q.; McGhie, A.R.; Romanow, W.J.; McCauley, J.P., Jr.; Smith, A.B., III; Cox, D.E. Structure and bonding in alkali-metal-doped C60. *Nature* **1991**, *351*, 462–464. [CrossRef]
166. Tanigaki, K.; Ebbesen, T.W.; Saito, S.; Mizuki, J.; Tsai, J.S.; Kubo, Y.; Kuroshima, S. Superconductivity at 33 K in $Cs_xRb_yC_{60}$. *Nature* **1991**, *352*, 222–223. [CrossRef]
167. Kelty, S.P.; Chen, C.-C.; Lieber, C.M. Superconductivity at 30 K in caesium-doped C_{60}. *Nature* **1991**, *352*, 223–225. [CrossRef]
168. Uemura, Y.J.; Keren, A.; Le, L.P.; Luke, G.M.; Sternlieb, B.J.; Wu, W.D.; Brewer, J.H.; Whetten, R.L.; Huang, S.M.; Lin, S.; et al. Magnetic-field penetration depth in K_3C_{60} measured by muon spin relaxation. *Nature* **1991**, *352*, 605–607. [CrossRef]
169. Stephens, P.W.; Mihaly, L.; Lee, P.L.; Whetten, R.L.; Huang, S.-M.; Kaner, R.; Deiderich, F.; Holczer, K. Structure of single-phase superconducting K_3C_{60}. *Nature* **1991**, *352*, 632–634. [CrossRef]
170. Fleming, R.M.; Ramirez, A.P.; Rosseinsky, M.J.; Murphy, D.W.; Haddon, R.C.; Zahurak, S.M.; Makhija, A.V. Relation of structure and superconducting transition temperatures in A_3C_{60}. *Nature* **1991**, *352*, 787–788. [CrossRef]
171. Wang, H.H.; Kini, A.M.; Savall, B.M.; Carlson, K.D.; Williams, J.M.; Lykke, K.R.; Wurz, P.; Parker, D.H.; Pellin, M.J. First easily reproduced solution-phase synthesis and confirmation of superconductivity in the fullerene K_xC_{60} (T_c = 18.0 ± 0.1 K). *Inorg. Chem.* **1991**, *30*, 2838–2839. [CrossRef]
172. Schirber, J.E.; Overmyer, D.L.; Wang, H.H.; Williams, J.M.; Carlson, K.D.; Kini, A.M.; Welp, U.; Kwok, W.-K. Pressure dependence of the superconducting transition temperature of potassium fullerene, K_xC_{60}. *Phys. C* **1991**, *178*, 137–139. [CrossRef]
173. Wang, H.H.; Kini, A.M.; Carlson, K.D.; Williams, J.M.; Pellin, M.J.; Schirber, J.E.; Savall, B.M.; Lathrop, M.W.; Lykke, K.R.; Parker, D.H.; et al. Superconductivity at 28.6 K in a rubidium-C60 fullerene compound, Rb_xC_{60}, synthesized by a solution-phase technique. *Inorg. Chem.* **1991**, *30*, 2962–2963. [CrossRef]
174. Holczer, K.; Klein, O.; Huang, S.-M.; Kaner, R.B.; Fu, K.-J.; Whetten, R.L.; Diederich, F. Alkali-fulleride superconductors: Synthesis, composition, and diamagnetic shielding. *Science* **1991**, *252*, 1154–1157. [CrossRef]
175. Sparn, G.; Thompson, J.D.; Huang, S.-M.; Kaner, R.B.; Diederich, F.; Whetten, R.L.; Grüner, G.; Holczer, K. Pressure dependence of superconductivity in single-phase K_3C_{60}. *Science* **1991**, *252*, 1829–1831. [CrossRef]
176. Rosseinsky, M.J.; Ramirez, A.P.; Glarum, S.H.; Murphy, D.W.; Haddon, R.C.; Hebard, A.F.; Palstra, T.T.M.; Kortan, A.R.; Zahurak, S.M.; Makhija, A.V. Superconductivity at 28 K in Rb_xC_{60}. *Phys. Rev. Lett.* **1991**, *66*, 2830–2832. [CrossRef]
177. McCauley, J.P., Jr.; Zhu, Q.; Coustel, N.; Zhou, O.; Vaughan, G.; Idziak, S.H.J.; Fischer, J.E.; Tozer, S.W.; Groski, D.M.; Bykovetz, N.; et al. Synthesis, structure, and superconducting properties of single-phase Rb_3C_{60}. A new, convenient method for the preparation of M_3C_{60} superconductors. *J. Am. Chem. Soc.* **1991**, *113*, 8537–8538. [CrossRef]
178. Sugimoto, T.; Awaji, H.; Misaki, Y.; Yoshida, Z.-I.; Kai, Y.; Nakagawa, H.; Kasai, N. Tetrakis(1,3-dithiol-2-ylidene)cyclobutane: A novel and promising electron donor for organic metals. *J. Am. Chem. Soc.* **1985**, *107*, 5792–5793. [CrossRef]
179. Kikuchi, K.; Kikuchi, M.; Namiki, T.; Saito, K.; Ikemoto, I.; Murata, K.; Ishiguro, T.; Kobayashi, K. New organic superconductor, $(DMET)_2Au(CN)_2$. *Chem. Lett.* **1987**, *16*, 931–932. [CrossRef]
180. Kikuchi, K.; Honda, Y.; Namiki, T.; Saito, K.; Ikemoto, I.; Murata, K.; Anzai, H.; Ishiguro, T.; Kobayashi, K. Superconductivity in $(DMET)_2AuCl_2$ and $(DMET)_2AuI_2$. *J. Phys. Soc. Jpn.* **1987**, *56*, 4241–4244. [CrossRef]
181. Kikuchi, K.; Honda, Y.; Namiki, T.; Saito, K.; Ikemoto, I.; Ishiguro, T.; Murata, K.; Kobayashi, K. Superconductivity and the possibility of semiconductor-metal transition in $(DMET)_2AuBr_2$. *J. Phys. Soc. Jpn.* **1987**, *56*, 2627–2628. [CrossRef]
182. Kikuchi, K.; Honda, Y.; Namiki, T.; Saito, K.; Ikemoto, I.; Murata, K.; Ishiguro, T.; Kobayashi, K. On ambient-pressure superconductivity in organic conductors: Electrical properties of $(DMET)_2I_3$, $(DMET)_2I_2Br$ and $(DMET)_2IBr_2$. *J. Phys. Soc. Jpn.* **1987**, *56*, 3436–3439. [CrossRef]
183. Kikuchi, K.; Murata, K.; Klkuchi, M.; Honda, Y.; Takahashi, T.; Oyama, T.; Ikemoto, I.; Ishiguro, T.; Kobayashi, K. Superconductivity and Surrounding Phase of Organic Conductor, $(DMET)_2Au(CN)_2$. *Jap. J. Appl. Phys.* **1987**, *26*, 1369–1370. [CrossRef]
184. Kikuchi, K.; Honda, Y.; Ishikawa, Y.; Saito, K.; Ikemoto, I.; Murata, K.; Anzai, H.; Ishiguro, T. Polymorphism and electrical conductivity of the organic superconductor $(DMET)_2AuBr_2$. *Solid State Commun.* **1988**, *66*, 405–408. [CrossRef]
185. Papavassiliou, G.C.; Mousdis, G.A.; Zambounis, J.S.; Terzis, A.; Hountas, A.; Hilti, B.; Mayer, C.W.; Pfeiffer, J. Low temperature measurements of the electrical conductivities of some charge transfer salts with the asymmetric donors MDT-TTF, EDT-TTF and EDT-DSDTF. $(MDT\text{-}TTF)_2AuI_2$, a new superconductor (T_c = 3.5 K at ambient pressure). *Synth. Met.* **1988**, *27*, 379–383. [CrossRef]
186. Sugimoto, T.; Awaji, H.; Sugimoto, I.; Misaki, Y.; Kawase, T.; Yoneda, S.; Yoshida, Z.-I.; Anzai, H. Ethylene analogs of tetrathiafulvalene and tetraselenafulvalene: New donors for organic metals. *Chem. Mater.* **1989**, 535–537. [CrossRef]
187. Suzuki, T.; Yamochi, H.; Srdanov, G.; Hinkelmann, K.; Wudl, F. Bis(ethylenedioxy)tetrathiafulvalene: The first oxygen substituted tetrathiafulvalene. *J. Am. Chem. Soc.* **1989**, *111*, 3108–3109. [CrossRef]
188. Wudl, F.; Yamochi, H.; Suzuki, T.; lsotalo, H.; Fite, C.; Kasmai, H.; Liou, K.; Srdanov, G.; Coppens, P.; Maly, K.; et al. $(BEDO)_{2.4}I_3$: The First Robust Organic Metal of BEDO-TTF. *J. Am. Chem. Soc.* **1990**, *112*, 2461–2462. [CrossRef]

189. Beno, M.A.; Wang, H.H.; Kini, A.M.; Carlson, K.D.; Geiser, U.; Kwok, W.K.; Thompson, J.E.; Williams, J.M.; Ren, J.; Whangbo, M.-H. The first ambient pressure organic superconductor containing oxygen in the donor molecule, β_m-(BEDO-TTF)$_3$Cu$_2$(NCS)$_3$, T_C = 1.06 K. *Inorg. Chem.* **1990**, *29*, 1599–1601. [CrossRef]
190. Kahlich, S.; Schweitzer, D.; Heinen, I.; En Lan, S.; Nuber, B.; Keller, H.J.; Winzer, K.; Helberg, H.W. (BEDO-TTF)$_2$ReO$_4$·(H$_2$O): A new organic superconductor. *Soliod State Comun.* **1991**, *80*, 191–195. [CrossRef]
191. Naito, T.; Miyamoto, A.; Kobayashi, H.; Kato, R.; Kobayashi, A. Structure and electrical properties of θ- and κ-type BEDT-TSeF salts with bromomercurate anions. *Chem. Lett.* **1991**, *20*, 1945–1948. [CrossRef]
192. Kobayashi, H.; Bun, K.; Miyamoto, A.; Naito, T.; Kato, R.; Kobayashi, A.; Williams, J.A. Superconducting transition of a grease-coated crystal of κ-(BEDT-TFT)$_2$Cu[N(CN)$_2$]Cl. *Chem. Lett.* **1991**, *20*, 1997–2000. [CrossRef]
193. Zambounis, J.S.; Mayer, C.W.; Hauenstein, K.; Hilti, B.; Hofherr, W.; Pfeiffer, J.; Bürkle, M.; Rihs, G. Crystal structure and electrical properties of κ-((*S*,*S*)-DMBEDT-TTF)$_2$ClO$_4$. *Adv. Mater.* **1992**, *4*, 33–35. [CrossRef]
194. Misaki, Y.; Nishikawa, H.; Kawakami, K.; Uehara, T.; Yamabe, T. Bis(2-methylidene-1,3-dithiolo[4,5-d])tetrathiafulvalene (BDT-TTF): A tetrathiafulvalene condensed with 1,3-dithiol-2-ylidene moieties. *Tetrahedron Lett.* **1992**, *33*, 4321–4324. [CrossRef]
195. Misaki, Y.; Nishikawa, H.; Fujiwara, H.; Kawakami, K.; Yamabe, T.; Yamochi, H.; Saito, G. (2-Methylidene-1,3-dithiolo[4,5-d])tetrathiafulvalene (DT-TTF): New unsymmetrical TTFs condensed with 1,3-dithiol-2-ylidene moieties. *J. Chem. Soc. Chem. Commun.* **1992**, 1408–1409. [CrossRef]
196. Naito, T.; Miyamoto, A.; Kobayashi, H.; Kato, R.; Kobayashi, A. Superconducting transition temperature of the organic alloy system: κ-[(BEDT-TTF)$_{1-x}$(BEDT-STF)$_x$]$_2$Cu[N(CN)$_2$]Br. *Chem. Lett.* **1992**, *21*, 119–122. [CrossRef]
197. Kato, R.; Aonuma, S.; Okano, Y.; Sawa, H.; Tamura, M.; Kinoshita, M.; Oshima, K.; Kobayashi, A.; Bun, K.; Kobayashi, H. Metallic and superconducting salts based on an unsymmetrical π-donor dimethyl(ethylenedithio)tetraselenafulvalene (DMET-TSeF). *Synth. Met.* **1993**, *61*, 199–206. [CrossRef]
198. Sallé, M.; Jubault, M.; Gorgues, A.; Boubekeur, K.; Fourmigué, M.; Batail, P.; Canadell, E. Bis- and Tetrakis(1,4-dithiafulven-6-yl)-Substituted Tetrathiafulvalenes and Dihydrotetrathiafulvalenes: A novel class of planar donor molecules with multiple redox functionalities and the demonstration of a novel type of two-dimensional association in the solid state. *Chem. Mater.* **1993**, *5*, 1196–1198. [CrossRef]
199. Fourmigué, M.; Johannsen, I.; Boubekeur, K.; Nelson, C.; Batail, P. Tetrathiafulvalene- and dithiafulvene-substituted Mesitylenes, new π-Donor molecules with 3-fold symmetry and the formation of an unprecedented new class of electroactive polymers. *J. Am. Chem. Soc.* **1993**, *115*, 3752–3759. [CrossRef]
200. Balicas, L.; Behnia, K.; Kang, W.; Canadell, E.; Auban-Senzier, P.; Jérome, D.; Ribault, M.; Fabre, J.M. Supercondcutivity and magnetic field induced spin density waves in the (TMTTF)$_2$X family. *J. Phys. I* **1994**, *4*, 1539–1550. [CrossRef]
201. Naito, T.; Tateno, A.; Udagawa, T.; Kobayashi, H.; Kato, R.; Kobayashi, A.; Nogami, T. Synthesis, structures and electrical properties of the charge-transfer salts of 4,5-ethylenedithio-4′,5′-(2-oxatrimethylenedithio) diselenadithiafulvalene (EOST) with linear anions [I$_3^-$, IBr$_2^-$, ICl$_3^-$, I$_2$Br$^-$, AuBr$_2^-$, Au(CN)$_2^-$]. *J. Chem. Soc. Faraday Trans.* **1994**, *90*, 763–771. [CrossRef]
202. Tateno, A.; Udagawa, T.; Naito, T.; Kobayashi, H.; Kobayashi, A.; Nogami, T. Crystal structures and electrical properties of the radical salts of the unsymmetrical donor EOTT [4,5-ethylenedithio-4′,5′-(2- oxatrimethylenedithio)tetrathiafulvalene]. *J. Mater. Chem.* **1994**, *4*, 1559–1569. [CrossRef]
203. Kobayashi, H.; Tomita, H.; Naito, T.; Tanaka, H.; Kobayashi, A.; Saito, T. A new organic superconductor, λ-BETS$_2$GaBrCl$_3$ [BETS = bis(ethylenedithio)tetraselenafulvalene]. *J. Chem. Soc. Chem. Commun.* **1995**, 1225–1226. [CrossRef]
204. Kobayashi, H.; Kawano, K.; Naito, T.; Kobayashi, A. Electronic band structure and superconducting transition of κ-(BEDT-TTF)$_2$I$_3$. *J. Mater. Chem.* **1995**, *5*, 1681–1687. [CrossRef]
205. Inokuchi, M.; Tajima, H.; Kobayashi, A.; Ohta, T.; Kuroda, H.; Kato, R.; Naito, T.; Kobayashi, H. Electrical and optical properties of α-(BETS)$_2$I$_3$ and α-(BEDT-STF)$_2$I$_3$. *Bull. Chem. Soc. Jpn.* **1995**, *68*, 547–553. [CrossRef]
206. Misaki, Y.; Ohta, T.; Higuchi, N.; Fujiwara, H.; Yamabe, T.; Mori, T.; Mori, H.; Tanaka, S. A vinylogue of bis-fused tetrathiafulvalene: Novel π-electron framework for two-dimensional organic metals. *J. Mater. Chem.* **1995**, *5*, 1571–1579. [CrossRef]
207. Misaki, Y.; Higuchi, N.; Fujiwara, H.; Yamabe, T.; Mori, T.; Mori, H.; Tanaka, S. (DTEDT)[Au(CN)$_2$]$_{0.4}$: An organic superconductor based on the novel π-electron framework of vinylogous bis-fused tetrathiafulvalene. *Angew. Chem. Int. Ed. Engl.* **1995**, *34*, 1222–1225. [CrossRef]
208. Oshima, K.; Okuno, H.; Kato, K.; Maruyama, R.; Kato, R.; Kobayashi, A.; Kobayashi, H. Superconductivity and field induced states in DMET-TSeF family. *Synth. Met.* **1995**, *70*, 861–862. [CrossRef]
209. Naito, T.; Kobayashi, H.; Kobayashi, A.; Underhill, A.E. New synthetic metals based on a thiadiazole network. *Chem. Commun.* **1996**, 521–522. [CrossRef]
210. Horiuchi, S.; Yamochi, H.; Saito, G.; Sakaguchi, K.-I.; Kusunoki, M. Nature and origin of stable metallic state in organic charge-transfer complexes of bis(ethylenedioxy)tetrathiafulvalene. *J. Am. Chem. Soc.* **1996**, *118*, 8604–8622. [CrossRef]
211. Naito, T.; Kobayashi, H.; Kobayashi, A. The Electrical Behavior of Charge-Transfer Salts Based on an Unsymmetrical Donor Bis(ethylenedithio)diselenadithiafulvalene (STF): Disorder Effect on the Transport Properties. *Bull. Chem. Soc. Jpn.* **1997**, *70*, 107–114. [CrossRef]
212. Mori, T.; Kawamoto, T.; Yamaura, J.; Enoki, T.; Misaki, Y.; Yamabe, T.; Mori, H.; Tanaka, S. Metal-insulator transition in the organic metal (TTM-TTP)I$_3$ with a one-dimensional half-filled band. *Phys. Rev. Lett.* **1997**, *79*, 1702–1705. [CrossRef]

213. Kato, R.; Yamamoio, K.; Okano, Y.; Tajima, H.; Sawa, H. A new ambient-pressure organic superconductor (TMET-STF)$_2$BF$_4$ [TMET-STF = trimethylene(ethylenedithio)diselenadithiafulvalene]. *Chem. Commun.* **1997**, 947–948. [CrossRef]
214. Sakata, J.-I.; Sato, H.; Miyazaki, A.; Enoki, T.; Okano, Y.; Kato, R. Superconductivity in new organic conductor κ-(BEDSe-TTF)$_2$CuN(CN)$_2$Br. *Solid State Commun.* **1998**, *108*, 377–381. [CrossRef]
215. Kawamoto, T.; Aragaki, M.; Mori, T.; Misaki, Y.; Yamabe, T. Crystal structure and physical properties of (TTM-TTP)AuI$_2$. *J. Mater. Chem.* **1998**, *8*, 285–288. [CrossRef]
216. Heuzé, K.; Fourmigué, M. The crystal chemistry of amide-functionalized ethylenedithiotetrathiafulvalenes: EDT-TTF-CONRR′ (R,R′ = H, Me). *J. Mater. Chem.* **1999**, *9*, 2373–2379. [CrossRef]
217. Kondo, R.; Hasegawa, T.; Mochida, T.; Kagoshima, S.; Iwasa, Y. Donor-acceptor type superconductor, (BETS)$_2$(Cl$_2$TCNQ). *Chem. Lett.* **1999**, *28*, 333–334. [CrossRef]
218. Okano, Y.; Iso, M.; Kashimura, Y.; Yamaura, J.; Kato, R. New synthesis of Se-containing TTF derivatives. *Synth. Met.* **1999**, *102*, 1703–1704. [CrossRef]
219. Adachi, T.; Ojima, E.; Kato, K.; Kobayashi, H.; Miyazaki, T.; Tokumoto, M.; Kobayashi, A. Superconducting transition of (TMTTF)$_2$PF$_6$ above 50 kbar [TMTTF = Tetramethyltetrathiafulvalene]. *J. Am. Chem. Soc.* **2000**, *122*, 3238–3239. [CrossRef]
220. Drozdova, O.; Yamochi, H.; Yakushi, K.; Uruichi, M.; Horiuchi, S.; Saito, G. Determination of the charge on BEDO-TTF in its complexes by Raman spectroscopy. *J. Am. Chem. Soc.* **2000**, *122*, 4436–4442. [CrossRef]
221. Dehnet, A.; Batail, P.; Misaki, Y.; Auban-Senzier, P.; Canadell, E. Donor slab robustness and band filling variations in BDT-TTP-based molecular conductors: β-(BDT-TTP)$_6$[Re$_6$S$_6$Cl$_8$]·(CH$_2$Cl-CH$_2$Cl$_2$)$_2$ and β-(BDT-TTP)$_6$[Mo$_6$Cl$_{14}$]·(CH$_2$Cl-CHCl$_2$)$_2$. *Adv. Mater.* **2000**, *12*, 436–439. [CrossRef]
222. Tanaka, H.; Ojima, E.; Fujiwara, H.; Nakazawa, Y.; Kobayashi, H.; Kobayashi, A. A new κ-type organic superconductor based on BETS molecules, κ-(BETS)$_2$GaBr$_4$ [BETS = bis(ethylenedithio)tetraselenafulvalene]. *J. Mater. Chem.* **2000**, *10*, 245–247. [CrossRef]
223. Jaccard, D.; Wilhelm, H.; Jérome, D.; Moser, J.; Carcel, C.; Fabre, J.M. From spin-Peierls to superconductivity: (TMTTF)$_2$PF$_6$ under high pressure. *J. Phys. Condens. Matter.* **2001**, *13*, L89–L95. [CrossRef]
224. Yamada, J.-I.; Watanabe, M.; Akutsu, H.; Nakasuji, S.; Nishikawa, H.; Ikemoto, I.; Kikuchi, K. New organic superconductors β-(BDA-TTP)$_2$X [BDA-TTP = 2,5-bis(1,3-dithian-2-ylidene)-1,3,4,6-tetrathiapentalene; X^- = SbF_6^-, AsF_6^-, and PF_6^-]. *J. Am. Chem. Soc.* **2001**, *123*, 4174–4180. [CrossRef]
225. Mielke, C.; Singleton, J.; Nam, M.-S.; Harrison, N.; Agosta, C.C.; Fravel, B.; Montgomery, L.K. Superconducting properties and Fermi-surface topology of the quasi-two-dimensional organic superconductor λ-(BETS)$_2$GaCl$_4$ (BETS ≡ bis(ethylene-dithio)tetraselenafulvalene). *J. Phys. Cond. Mat.* **2001**, *13*, 8325–8345. [CrossRef]
226. Gritsenko, V.; Tanaka, H.; Kobayashi, H.; Kobayashi, A. A new molecular superconductor, κ-(BETS)$_2$TlCl$_4$ [BETS = bis(ethylenedithio)tetraselenafulvalene]. *J. Mater. Chem.* **2001**, *11*, 2410–2411. [CrossRef]
227. Takimiya, K.; Kataoka, Y.; Aso, Y.; Otsubo, T.; Fukuoka, H.; Yamanaka, S. Quasi one-dimensional organic superconductor MDT-TSF·AuI$_2$ with T_c = 4.5 K at ambient pressure. *Angew. Chem. Int. Ed. Engl.* **2001**, *40*, 1122–1125. [CrossRef]
228. Ota, A.; Yamochi, H.; Saito, G. A novel metal-insulator phase transition observed in (EDO-TTF)$_2$PF$_6$. *J. Mater. Chem.* **2002**, *12*, 2600–2602. [CrossRef]
229. Kawamoto, T.; Mori, T.; Takimiya, K.; Kataoka, Y.; Aso, Y.; Otsubo, T. Organic superconductor with an incommensurate anion structure: (MDT-TSF)(AuI$_2$)$_{0.44}$. *Phys. Rev. B* **2002**, *65*, 140508. [CrossRef]
230. Imakubo, T.; Tajima, N.; Tamura, M.; Kato, R.; Nishio, Y.; Kajita, K. A supramolecular superconductor θ-(DIETS)$_2$[Au(CN)$_4$]. *J. Mater. Chem.* **2002**, *12*, 159–161. [CrossRef]
231. Nishikawa, H.; Morimoto, T.; Kodama, T.; Ikemoto, I.; Kikuchi, K.; Yamada, J.-I.; Yoshino, H.; Murata, K. New organic superconductors consisting of an unprecedented π-electron donor. *J. Am. Chem Soc.* **2002**, *124*, 730–731. [CrossRef]
232. Shimojo, Y.; Ishiguro, T.; Toita, T.; Yamada, J.-I. Superconductivity of layered organic compound β-(BDA-TTP)$_2$SbF$_6$, where BDA-TTP is 2,5-bis(1,3-dithian-2-ylidene)-1,3,4,6-tetrathiapentalene. *J. Phys. Soc. Jpn.* **2002**, *71*, 717–720. [CrossRef]
233. Kodani, M.; Takamori, A.; Takimiya, K.; Aso, Y.; Otsubo, T. Novel conductive radical cation salts based on methylenediselenotetraselenafulvalene (MDSe-TSF): A sign of superconductivity in κ-(MDSe-TSF)$_2$Br below 4 K. *J. Solid State Chem.* **2002**, *168*, 582–589. [CrossRef]
234. Takimiya, K.; Takamori, A.; Aso, Y.; Otsubo, T.; Kawamoto, T.; Mori, T. Organic superconductors based on a new electron donor, methylenedithio-diselenadithafulvalene (MDT-ST). *Chem. Mater.* **2003**, *15*, 1225–1227. [CrossRef]
235. Takimiya, K.; Kodani, M.; Kataoka, Y.; Aso, Y.; Otsubo, T.; Kawamoto, T.; Mori, T. New organic superconductors with an incommensurate anion. Lattice consisting of polyhalide chains (MDT-TSF)X$_y$ (MDT-TSF = methylenedithiotetraselenafulvalene; X = halogen; y = 1.27–1.29). *Chem. Mater.* **2003**, *15*, 3250–3255. [CrossRef]
236. Auban-Senzier, P.; Pasquier, C.; Jérome, D.; Carcel, C.; Fabre, J.M. From Mott insulator to superconductivity in (TMTTF)$_2$BF$_4$: High pressure transport measurements. *Synth. Met.* **2003**, *133–134*, 11–14. [CrossRef]
237. Nishikawa, H.; Machida, A.; Morimoto, T.; Kikuchi, K.; Kodama, T.; Ikemoto, I.; Yamada, J.-I.; Yoshino, H.; Murata, K. A new organic superconductor, (DODHT)$_2$BF$_4$·H$_2$O. *Chem. Commun.* **2003**, *3*, 494–495. [CrossRef] [PubMed]
238. Yamada, J.-I.; Toita, T.; Akutsu, H.; Nakasuji, S.; Nishikawa, H.; Ikemoto, I.; Kikuchi, K.; Choi, E.S.; Graf, D.; Brooks, J.S. A new organic superconductor, β-(BDA-TTP)$_2$GaCl$_4$ [BDA-TTP = 2,5-bis(1,3-dithian-2-ylidene)-1,3,4,6-tetrathiapentalene]. *Chem. Commun.* **2003**, *3*, 2230–2231. [CrossRef]

239. Choi, E.S.; Graf, D.; Brooks, J.S.; Yamada, J.; Akutsu, H.; Kikuchi, K.; Tokumoto, M. Pressure-dependent ground states and fermiology in β-(BDA-TTP)$_2$MCl$_4$ (M = Fe,Ga). *Phys. Rev. B* **2004**, *70*, 024517. [CrossRef]
240. Zhilyaeva, E.I.; Torunova, S.A.; Lyubovskaya, R.N.; Mousdis, G.A.; Papavassiliou, G.C.; Perenboom, J.A.A.J.; Pesotskii, S.I.; Lyubovskii, R.B. New ambient pressure organic superconductor with T_c=8.1 K based on unsymmetrical donor molecule, ethylenedithiotetrathiafulvalene: (EDT-TTF)$_4$Hg$_{3-\delta}$I$_8$, δ ~ 0.1–0.2. *Synth. Met.* **2004**, *140*, 151–154. [CrossRef]
241. Drozdova, O.; Yakushi, K.; Yamamoto, K.; Ota, A.; Yamochi, H.; Saito, G.; Tashiro, H.; Tanner, D.B. Optical characterization of $2k_F$ bond-charge-density wave in quasi-one-dimeesional 3/4-filled (EDO-TTF)$_2X$ (X = PF$_6$ and AsF$_6$). *Phys. Rev. B* **2004**, *70*, 075107. [CrossRef]
242. Mori, T. Organic conductors with unusual band fillings. *Chem. Rev.* **2004**, *104*, 4947–4969. [CrossRef]
243. Kimura, S.; Maejima, T.; Suzuki, H.; Chiba, R.; Mori, H.; Kawamoto, T.; Mori, T.; Moriyama, H.; Nishio, Y.; Kajita, K. A new organic superconductor β-(*meso*-DMBEDT-TTF)$_2$PF$_6$. *Chem. Commun.* **2004**, 2454–2455. [CrossRef]
244. Takimiya, K.; Kodani, M.; Niihara, N.; Aso, Y.; Otsubo, T.; Bando, Y.; Kawamoto, T.; Mori, T. Pressure-induced superconductivity in (MDT-TS)(AuI$_2$)$_{0.441}$ [MDT-TS = 5H-2-(1,3-diselenol-2-ylidene)-1,3,4,6-tetrathiapentalene]: A new organic superconductor possessing an incommensurate anion lattice. *Chem. Mater.* **2004**, *16*, 5120–5123. [CrossRef]
245. Nishikawa, H.; Sato, Y.; Kikuchi, K.; Kodama, T.; Ikemoto, I.; Yamada, J.-I.; Oshio, H.; Kondo, R.; Kagoshima, S. Charge ordering and pressure-induced superconductivity in β"-(DODHT)$_2$PF$_6$. *Phys. Rev. B* **2005**, *72*, 052510. [CrossRef]
246. Ito, H.; Suzuki, D.; Yokochi, Y.; Kuroda, S.; Umemiya, M.; Miyasaka, H.; Sugiura, K.I.; Yamashita, M.; Tajima, H. Quasi-one-dimensional electronic structure of (DMET)$_2$CuCl$_2$. *Phys. Rev. B* **2005**, *71*, 212503. [CrossRef]
247. Shirahata, T.; Kibune, M.; Maesato, M.; Kawashima, T.; Saito, G.; Imakubo, T. New organic conductors based on dibromo- and diiodo-TSeFs with magnetic and non-magnetic MX$_4$ counter anions (M = Fe, Ga; X = Cl, Br). *J. Mater. Chem.* **2006**, *16*, 3381–3390. [CrossRef]
248. Yamada, J.-I.; Fujimoto, K.; Akutsu, H.; Nakatsuji, S.; Miyazaki, A.; Aimatsu, M.; Kudo, S.; Enoki, T.; Kikuchi, K. Pressure effect on the electrical conductivity and superconductivity of β-(BDA-TTP)$_2$I$_3$. *Chem. Commun.* **2006**, 1331–1333. [CrossRef] [PubMed]
249. Shirahata, T.; Kibune, M.; Imakubo, T. New ambient pressure organic superconductors κ_H- and κ_L-(DMEDO-TSeF)$_2$[Au(CN)$_4$](THF). *Chem. Commun.* **2006**, 1592–1594. [CrossRef]
250. Itoi, M.; Kano, M.; Kurita, N.; Hedo, M.; Uwatoko, Y.; Nakamura, T. Pressure-induced superconductivity in the quasi-one-dimensional organic conductor (TMTTF)$_2$AsF$_6$. *J. Phys. Soc. Jpn.* **2007**, *76*, 053703. [CrossRef]
251. Araki, C.; Itoi, M.; Hedo, M.; Uwatoko, Y.; Mori, H. Electrical resistivity of (TMTTF)$_2$PF$_6$ under high pressure. *J. Phys. Soc. Jpn.* **2007**, *76* (Suppl. A), 198–199. [CrossRef]
252. Itoi, M.; Araki, C.; Hedo, M.; Uwatoko, Y.; Nakamura, T. Anomalously wide superconducting phase of one-dimensional organic conductor (TMTTF)$_2$SbF$_6$. *J. Phys. Soc. Jpn.* **2008**, *77*, 023701. [CrossRef]
253. Misaki, Y. Tetrathiapentalene-based organic conductors. *Sci. Tech. Adv. Mat.* **2009**, *10*, 024301. [CrossRef] [PubMed]
254. Lorcy, D.; Bellec, N.; Fourmigué, M.; Avarvari, N. Tetrathiafulvalene-based group XV ligands: Synthesis, coordination chemistry and radical cation salts. *Coord. Chem. Rev.* **2009**, *253*, 1398–1438. [CrossRef]
255. Shikama, T.; Shimokawa, T.; Lee, S.; Isono, T.; Ueda, A.; Takahashi, K.; Nakao, A.; Kumai, R.; Nakao, H.; Kobayashi, K.; et al. Magnetism and pressure-induced superconductivity of checkerboard-type charge-ordered molecular conductor β-(*meso*-DMBEDT-TTF)$_2$X (X = PF$_6$ and AsF$_6$). *Crystals* **2012**, *2*, 1502–1513. [CrossRef]
256. Steimecke, G.; Sieler, H.-J.; Kirmse, R.; Hoyer, E. 1,3-Dithiol-2-thion-4,5-dithiolat aus Schwefelkohlenstoff und Alkalimetall. *Phosphorus Sulfur* **1979**, *7*, 49–55. [CrossRef]
257. Kirmse, R.; Stach, J.; Dietzsch, W.; Steimecke, G.; Hoyer, E. Single-Crystal EPR Studies on Nickel(III), Palladium(III), and Platinum(III) Dithiolene Chelates Containing the Ligands Isotrithionedithiolate, *o*-Xylenedithiolate, and Maleonitriledithiolate. *Inorg. Chem.* **1980**, *19*, 2679–2685. [CrossRef]
258. Alvarez, S.; Vicente, R.; Hoffmann, R. Dimerization and Stacking in Transition-Metal Bisdithiolenes and Tetrathiolates. *J. Am. Chem. Soc.* **1985**, *107*, 6253–6277. [CrossRef]
259. Valade, L.; Legros, J.; Bousseau, M.; Cassoux, P.; Garbauskas, M.; Interrante, L.V. Molecular structure and solid-state properties of the two-dimensional conducting mixed-valence complex [NBu$_4$]$_{0.29}$[Ni(dmit)$_2$] and the neutral [Ni(dmit)$_2$] (H$_2$dmit = 4,5-dimercapto-1,3-dithiole-2-thione); members of an electron-transfer series. *J. Chem. Soc. Dalton Trans.* **1985**, 783–794. [CrossRef]
260. Brossard, L.; Ribault, M.; Bousseau, M.; Valade, L.; Cassoux, P. A new type of molecular superconductor: TTF[Ni(dmit)$_2$]$_2$. *C. R. Acad. Sci. Paris Ser. II* **1986**, *302*, 205–210.
261. Bousseau, M.; Valade, L.; Legros, J.-P.; Cassoux, P.; Garbauskas, M.; Interrante, L.V. Highly Conducting Charge-Transfer Compounds of Tetrathiafulvalene and Transition Metal–"dmit" Complexes. *J. Am. Chem. Soc.* **1986**, *108*, 1908–1916. [CrossRef]
262. Brossard, L.; Ribault, M.; Valade, L.; Cassoux, P. The first 3D molecular superconductor under pressure?: TTF [Ni(dmit)$_2$]$_2$. *Phys. B+C* **1986**, *143*, 378–380. [CrossRef]
263. Kobayashi, A.; Kim, H.; Sasaki, Y.; Kato, R.; Kobayashi, H.; Moriyama, S.; Nishio, Y.; Kajita, K.; Sasaki, W. The first molecular conductors based on π-acceptor molecules and closed-shell cations, [(CH$_3$)$_4$N][Ni(dmit)$_2$]$_2$, low-temperature X-ray studies and superconducting transition. *Chem. Lett.* **1987**, *16*, 1819–1822. [CrossRef]
264. Kajita, K.; Nishio, Y.; Moriyama, S.; Kato, R.; Kobayashi, H.; Sasaki, W.; Kobayashi, A.; Kim, H.; Sasaki, Y. Transport properties of ((CH$_3$)$_4$N) (Ni(dmit)$_2$)$_2$: A new organic superconductor. *Solid State Commun.* **1988**, *65*, 361–363. [CrossRef]

265. Brossard, L.; Hurdequint, H.; Ribault, M.; Valade, L.; Legros, J.P.; Cassoux, P. Pressure-temperature phase diagram of α'-TTF $[Pd(Dmit)_2]_2$. *Synth. Met.* **1988**, *27*, 157–162. [CrossRef]
266. Brossard, L.; Ribault, M.; Valade, L.; Cassoux, P. Pressure induced superconductivity in molecular $TTF(Pd(dmit)_2)_2$. *J. Phys. Fr.* **1989**, *50*, 1521–1534. [CrossRef]
267. Kato, R.; Kobayashi, H.; Kobayashi, A.; Naito, T.; Tamura, M.; Tajima, H.; Kuroda, H. New molecular conductors, α- and β-(EDT-TTF)$[Ni(dmit)_2]$ metal with anomalous resistivity maximum vs. semiconductor with mixed stacks. *Chem. Lett.* **1989**, *18*, 1839–1842. [CrossRef]
268. Brossard, L.; Ribault, M.; Valade, L.; Cassoux, P. Simultaneous competition and coexistence between charge-density waves and reentrant superconductivity in the pressure-temperature phase diagram of the molecular conductor $TTF[Ni(dmit)_2]_2$ (TTF is tetrathiafulvalene and dmit is the 1,3-dithia-2-thione-4,5-dithiolato group). *Phys. Rev. B* **1990**, *42*, 3935–3943. [CrossRef]
269. Canadell, E.; Ravy, S.; Pouget, J.P.; Brossard, L. Concerning the band structure of $D(M(dmit)_2)_2$ (D = TTF,Cs,NMe_4); M = Ni,Pd) molecular conductors and superconductors: Role of the $M(dmit)_2$ Homo and Lumo. *Solid State Commun.* **1990**, *75*, 633–638. [CrossRef]
270. Kobayashi, A.; Kim, H.; Sasaki, Y.; Murata, K.; Kato, R.; Kobayashi, H. Crystal and electronic structures of new molecular conductors tetramethylammonium and tetramethylarsonium complexes of $Pd(dmit)_2$. *J. Chem. Soc. Faraday Trans.* **1990**, *86*, 361–369. [CrossRef]
271. Cassoux, P.; Valade, L.; Kobayashi, H.; Kobayashi, A.; Clark, R.A.; Underhill, A.E. Molecular metals and superconductors derived from metal complexes of 1,3-dithiol-2-thione-4,5-dithiolate (dmit). *Coord. Chem. Rev.* **1991**, *110*, 115–160. [CrossRef]
272. Kobayashi, A.; Kobayashi, H.; Miyamoto, A.; Kato, R.; Clark, R.A.; Unerhill, A.E. New molecular superconductor, β-$[(CH_3)_4N][Pd(dmit)_2]_2$. *Chem. Lett.* **1991**, *20*, 2163–2166. [CrossRef]
273. Underhill, A.E.; Clark, R.A.; Marsden, I.; Allan, M.; Friend, R.H.; Tajima, H.; Naito, T.; Tamura, M.; Kuroda, H.; Kobayashi, A.; et al. Structural and electronic properties of $Cs(Pd(dmit)_2)_2$. *J. Phys. Cond. Mat.* **1991**, *3*, 933–954. [CrossRef]
274. Tajima, H.; Naito, T.; Tamura, M.; Kobayashi, A.; Kuroda, H.; Kato, R.; Kobayashi, H.; Clark, R.A.; Underhill, A.E. Energy level inversion in strongly dimerized $[Pd(dmit)_2]$ salts. *Solid State Commun.* **1991**, *79*, 337–341. [CrossRef]
275. Kobayashi, H.; Bun, K.; Naito, T.; Kato, R.; Kobayashi, A. New molecular superconductor, $[Me_2Et_2N][Pd(dmit)_2]_2$. *Chem. Lett.* **1992**, *21*, 1909–1912. [CrossRef]
276. Olk, R.-M.; Olk, B.; Dietzsch, W.; Kirmse, R.; Hoyer, E. The chemistry of 1,3-dithiole-2-thione-4,5-dithiolate (dmit). *Coord. Chem. Rev.* **1992**, *117*, 99–131. [CrossRef]
277. Tajima, H.; Inokuchi, M.; Kobayashi, A.; Ohta, T.; Kato, R.; Kobayashi, H.; Kuroda, H. First ambient-pressure superconductor based on $Ni(dmit)_2$, α-EDT-TTF$[Ni(dmit)_2]$. *Chem. Lett.* **1993**, *22*, 1235–1238. [CrossRef]
278. Kobayashi, A.; Kato, R.; Clark, R.A.; Underhill, A.E.; Miyamoto, A.; Bun, K.; Naito, T.; Kobayashi, H. New molecular superconductors, β-$[(CH_3)_4N][PD(dmit)_2]_2$ and $[(CH_3)_2(C_2H_5)_2N][Pd(dmit)_2]_2$. *Synth. Met.* **1993**, *56*, 2927–2932. [CrossRef]
279. Kobayashi, A.; Naito, T.; Kobayashi, H. Crystal and electronic structures of the two-dimensional transition-metal-complex molecule -$[(CH_3)_2(C_2H_5)_2N]$ $[Ni(dmit)_2]_2$ (dmit = 1,3-dithiol-2-thione-4,5-dithiolate). *Phys. Rev. B* **1995**, *51*, 3198–3201. [CrossRef]
280. Naito, T.; Sato, A.; Kawano, K.; Tateno, A.; Kobayashi, H.; Kobayashi, A. The new synthetic metals of $M(dmise)_2$:$[Me_3HN]$ $[Ni(dmise)_2]_2$ and (EDT-TTF)$[Ni(dmise)_2]$. *J. Chem. Soc. Chem. Commun.* **1995**, 351–352. [CrossRef]
281. Kobayashi, A.; Sato, A.; Kawano, K.; Naito, T.; Kobayashi, H.; Watanabe, T. Origin of the resistivity anomalies of (EDT-TTF)$[M(dmit)_2]$ (M = Ni, Pd). *J. Mater. Chem.* **1995**, *5*, 1671–1679. [CrossRef]
282. Svenstrup, N.; Becher, J. The organic chemistry of 1,3-dithiole-2-thione-4,5-dithiolate (DMIT). *Synthesis* **1995**, 215–235. [CrossRef]
283. Inokuchi, M.; Tajima, H.; Ohta, T.; Kuroda, H.; Kobayashi, A.; Sato, A.; Naito, T.; Kobayashi, H. Electrical Resistivity under High Pressure and Upper Critical Magnetic Field of the Molecular Superconductor α-(EDT-TTF)$[Ni(dmit)_2]$. *J. Phys. Soc. Jpn.* **1996**, *65*, 538–544. [CrossRef]
284. Canadell, E. Electronic structure of two-band molecular conductors. *New J. Chem.* **1997**, *21*, 1147–1159.
285. Kato, R.; Liu, Y.-L.; Hosokoshi, Y.; Aonuma, S.; Sawa, H. Se-substitution and cation effects on the high-pressure molecular superconductor, β-$Me_4N[Pd(dmit)_2]_2$–A unique two-band system. *Mol. Cryst. Liq. Cryst.* **1997**, *296*, 217–244. [CrossRef]
286. Sato, A.; Kobayashi, H.; Naito, T.; Sakai, F.; Kobayashi, A. Enhancement of the Dimensionality of Molecular π Conductors by the selone substitution of $M(dmit)_2$ (M = Ni, Pd) systems: Newly synthesized dmise compounds $[Me_xH_{4-x}N][Ni(dmise)_2]_2$ (x = 1–3) and $Cs[Pd(dmise)_2]_2$ (dmise = 4,5-Dimercapto-1,3-dithiole-2-selone). *Inorg. Chem.* **1997**, *36*, 5262–5269. [CrossRef]
287. Kato, R.; Kashimura, Y.; Aonuma, S.; Hanasaki, N.; Tajima, H. A new molecular superconductor β'-$Et_2Me_2P[Pd(dmit)_2]_2$ (dmit = 2-thioxo-1,3-dithiole-4,5-dithiolate). *Solid State Commun.* **1998**, *105*, 561–565. [CrossRef]
288. Pullen, A.E.; Olk, R.-M. The coordination chemistry of 1,3-dithiole-2-thione-4,5-dithiolate (dmit) and isologs. *Coord. Chem. Rev.* **1999**, *188*, 211–262. [CrossRef]
289. Akutagawa, T.; Nakamura, T. $[Ni(dmit)_2]$ salts with supramolecular cation structure. *Coord. Chem. Rev.* **2000**, *198*, 297–311. [CrossRef]
290. Naito, T.; Inabe, T.; Kobayashi, H.; Kobayashi, A. A new molecular metal based on $Pd(dmit)_2$: Synthesis, structure and electrical properties of $(C_7H_{13}NH)[Pd(dmit)_2]_2$($dmit^{2-}$ = 2-thioxo-1,3-dithiole-4,5-dithiolate). *J. Mater. Chem.* **2001**, *11*, 2200–2205. [CrossRef]
291. Kato, R.; Tajima, N.; Tamura, M.; Yamaura, J.-I. Uniaxial strain effect in a strongly correlated two-dimensional system β'-$(CH_3)_4As[Pd\ (dmit)_2]_2$. *Phys. Rev. B* **2002**, *66*, 020508. [CrossRef]

292. Robertson, N.; Cronin, L. Metal bis-1,2-dithiolene complexes in conducting or magnetic crystalline assemblies. *Coord. Chem. Rev.* **2002**, *227*, 93–127. [CrossRef]
293. Ribas, X.; Dias, J.C.; Morgado, J.; Wurst, K.; Molins, E.; Ruiz, E.; Almeida, M.; Veciana, J.; Rovira, C. Novel Cu^{III} Bis-1,2-dichalcogenene complexes with tunable 3D framework through alkaline cation coordination: A structural and theoretical study. *Chem. Eur. J.* **2004**, *10*, 1691–1704. [CrossRef]
294. Kato, R. Conducting metal dithiolene complexes: Structural and electronic properties. *Chem. Rev.* **2004**, *104*, 5319–5346. [CrossRef] [PubMed]
295. Tajima, A.; Nakao, A.; Kato, R. Uniaxial strain effects in the conducting $Pd(dmit)_2$ system (dmit = 1,3-dithiol-2-thione-4,5-dithiolate). *J. Phys. Soc. Jpn.* **2005**, *74*, 412–416. [CrossRef]
296. Sarangi, R.; George, S.D.; Rudd, D.J.; Szilagyi, R.K.; Ribas, X.; Rovira, C.; Almeida, M.; Hodgson, K.O.; Hedman, B.; Solomon, E.I. Sulfur K-edge X-ray absorption spectroscopy as a probe of ligand-metal bond covalency: Metal vs. ligand oxidation in copper and nickel dithiolene complexes. *J. Am. Chem. Soc.* **2007**, *129*, 2316–2326. [CrossRef] [PubMed]
297. Deplano, P.; Pilia, L.; Espa, D.; Mercuri, M.L.; Serpe, A. Square-planar d^8 metal mixed-ligand dithiolene complexes as second order nonlinear optical chromophores: Structure/property relationship. *Coord. Chem. Rev.* **2010**, *254*, 1434–1447. [CrossRef]
298. Kato, R. Development of π-electron systems based on $[M(dmit)_2]$ (M = Ni and Pd; dmit: 1,3-dithiole-2-thione-4,5-dithiolate) anion radicals. *Bull. Chem. Soc. Jpn.* **2014**, *87*, 355–374. [CrossRef]
299. Aumüller, A.; Hünig, S. Multistep reversible redox systems, XLVI[1)] *N,N′*-Dicyanoquinonediimines—A new class of compounds, I: Synthesis and general properties. *Liebigs Ann. Chem.* **1986**, 142–164. [CrossRef]
300. Aumüller, A.; Erk, P.; Klebe, G.; Hünig, S.; von Schütz, J.U.; Werner, H.-P. A radical anion salt of 2,5-Dimethyl-*N,N′*-dicyanoquinonediimine with extremely high electrical conductivity. *Angew. Chem. Int. Ed. Engl.* **1986**, *25*, 740–741. [CrossRef]
301. Mori, T.; Imaeda, K.; Kato, R.; Kobayashi, A.; Kobayashi, H.; Inokuchi, H. Pressure-induced one-dimensional instability in $(DMDCNQI)_2Cu$. *J. Phys. Soc. Jpn.* **1987**, *56*, 3429–3431. [CrossRef]
302. Kobayashi, A.; Kato, R.; Kobayashi, H.; Mori, T.; Inokuchi, H. The organic π-electron metal system with interaction through mixed-valence metal cation: Electronic and structural properties of radical salts of dicyano-quinodiimine, $(DMe\text{-}DCNQI)_2Cu$ and $(MeCl\text{-}DCNQI)_2Cu$. *Solid State Commun.* **1987**, *64*, 45–51. [CrossRef]
303. Tomić, S.; Jérome, D.; Aumüller, A.; Erk, P.; Hünig, S.; von Schütz, J.U. Pressure-temperature phase diagram of the organic conductor $(DM\text{-}DCNQI)_2Cu$. *Synth. Met.* **1988**, *27*, B281–B288. [CrossRef]
304. Tomić, S.; Jérome, D.; Aumüller, A.; Erk, P.; Hünig, S.; von Schutz, J.U. The pressure-temperature phase diagram of the organic conductor $(2{,}5\ DM\text{-}DCNQI)_2Cu$. *J. Phys. C Solid State Phys.* **1988**, *21*, L203–L207. [CrossRef]
305. Tomić, S.; Jérome, D.; Aumüller, A.; Erk, P.; Hünig, S.; von Schutz, J.U. Pressure-induced metal-to-insulator phase transitions in the organic conductor $(2{,}5\ DM\text{-}DCNQI)_2Cu$. *EPL* **1988**, *5*, 553–558. [CrossRef]
306. Mori, T.; Inokuchi, H.; Kobayashi, A.; Kato, R.; Kobayashi, H. Electrical conductivity, thermoelectric power, and ESR of a new family of molecular conductors, dicyanoquinonediimine-metal $[(DCNQI)_2M]$ compounds. *Phys. Rev. B* **1988**, *38*, 5913–5923. [CrossRef]
307. Kobayashi, H.; Kato, R.; Kobayashi, A.; Mori, T.; Inokuchi, H. The first molecular metals with ordered spin structures, R_1,R_2-$DCNQI_2Cu$ (R_1, R_2 = CH_3, CH_3O, Cl, Br)—Jahn-Teller distortion, CDW instability and antiferromagnetic spin ordering. *Solid State Commun.* **1988**, *65*, 1351–1354. [CrossRef]
308. Werner, H.P.; von Schütz, J.U.; Wolf, H.C.; Kremer, R.; Gehrke, M.; Aumüller, A.; Erk, P.; Hünig, S. Radical anion salts of *N,N′*-dicyanoquinonediimine (DCNQI): Conductivity and magnetic properties. *Solid State Commun.* **1988**, *65*, 809–813. [CrossRef]
309. Kobayashi, A.; Kato, R.; Kobayashi, H. Reentrant behavior of the temperature dependence of resistivity of DCNQI-Cu alloy system, $[(DMe)_{1-x}(MeBr)_x\text{-}DCNQI]_2Cu$. *Chem. Lett.* **1989**, *18*, 1843–1846. [CrossRef]
310. Kobayashi, H.; Miyamoto, A.; Kato, R.; Kobayashi, A.; Nishio, Y.; Kajita, K.; Sasaki, W. Reentrant behavior in the pressure-temperature dependence of the resistivity of $(DMeO\text{-}DCNQI)_2Cu$. *Solid State Commun.* **1989**, *72*, 1–5. [CrossRef]
311. Karutz, F.O.; von Schutz, J.U.; Wachtel, H.; Wolf, H.C. Optically reversed Peierls trasition in crystals of $Cu(dicyanoquinonediimine)_2$. *Phys. Rev. Lett.* **1998**, *81*, 140–143. [CrossRef]
312. Kato, R.; Kobayashi, H.; Kobayashi, A. Crystal and electronic structures of conductive anion-radical salts, $(2{,}5\text{-}R_1{,}R_2\text{-}DCNQI)_2Cu$ (DCNQI = *N,N′*-Dicyanoquinonediimine; R_1, R_2 = CH_3, CH_3O, Cl, Br). *J. Am. Chem. Soc.* **1989**, *111*, 5224–5232. [CrossRef]
313. Erk, P.; Hünig, S.; Meixner, H.; Gross, H.-J.; Langohr, U.; Werner, H.-P.; von Schütz, J.U.; Wolf, H.C. Binary alloys of 2,5-disubstituted DCNQI radical anion salts of copper and their electrical conductivity. *Angew. Chem. Int. Ed. Engl.* **1989**, *28*, 1245–1246. [CrossRef]
314. Koch, W. Extended-huckel energy band structures of organometallic compounds with one-dimensional crystal geometries. Computational results for bis(2,5-dimethyl-*N,N′*-dicyanoquinonediimine)copper, -silver, and -lithium. *Z. Naturforsch.* **1990**, *45a*, 148–156. [CrossRef]
315. Yakushi, K.; Ugawa, A.; Ojima, G.; Ida, T.; Tajima, H.; Kuroda, H.; Kobayashi, A.; Kato, R.; Kobayashi, H. Polarized reflectance spectra of DCNQI salts. *Mol. Cryst. Liq. Cryst.* **1990**, *181*, 217–231. [CrossRef]
316. Ermer, O. Sevenfold diamond structure and conductivity of copper dicyanoquinonediimines $Cu(DCNQI)_2$. *Adv. Mater.* **1991**, *3*, 608–611. [CrossRef]

317. Erk, P.; Meixner, H.; Metzenthin, T.; Hünig, S.; Langohr, U.; von Schütz, J.U.; Werner, H.-P.; Wolf, H.C.; Burkert, R.; Helberg, H.W.; et al. A guidance for stable metallic conductivity in copper salts of *N,N′*-dicyanobenzoquinonediimines (DCNQIs). *Adv. Mater.* **1991**, *3*, 311–315. [CrossRef]
318. Hünig, S.; Erk, P. DCNQIs—new electron acceptors for charge-transfer complexes and highly conducting radical anion salts. *Adv. Mater.* **1991**, *3*, 225–236. [CrossRef]
319. Lunardi, G.; Pecile, C. *N,N′*-dicyanoquinonediimines as a molecular constituent of organic conductors: Vibrational behavior and electron-molecular vibration coupling. *J. Chem. Phys.* **1991**, *95*, 6911–6923. [CrossRef]
320. Kagoshima, S.; Sugimoto, N.; Osada, T.; Kobayashi, A.; Kato, R.; Kobayashi, H. Magnetic and structural properties of mixed-valence molecular conductors $(DMeDCNQI)_2Cu$ and $(DMeODCNQI)_2Cu$. *J. Phys. Soc. Jpn.* **1991**, *60*, 4222–4229. [CrossRef]
321. Miyamoto, A.; Kobayashi, H.; Kato, R.; Kobayashi, A.; Nishio, Y.; Kajita, K.; Sasaki, W. Metal instability of $(DMe\text{-}DCNQI)_2Cu$ induced by uniaxial stress and enhancement of electron mass. *Chem. Lett.* **1992**, *21*, 115–118. [CrossRef]
322. Fukuyama, H. $(DCNQI)_2Cu$: A Luttinger-Peierls system. *J. Phys. Soc. Jpn.* **1992**, *61*, 3452–3456. [CrossRef]
323. Suzumura, Y.; Fukuyama, H. Mean-field theory of mixed-valence conductors $(R_1,R_2\text{-}DCNQI)_2Cu$. *J. Phys. Soc. Jpn.* **1992**, *61*, 3322–3330. [CrossRef]
324. Inoue, I.H.; Kakizaki, A.; Namatame, H.; Fujimori, A.; Kobayashi, A.; Kato, R.; Kobayashi, H. Copper valence fluctuation in the organic conductor (dimethyl-*N,N′*-dicyanoquinonediimine)$_2$Cu studied by x-ray photoemission spectroscopy. *Phys. Rev. B* **1992**, *45*, 5828–5833. [CrossRef] [PubMed]
325. Nishio, Y.; Kajita, K.; Sasaki, W.; Kato, R.; Kobayashi, A.; Kobayashi, H. Thermal and magnetic properties in organic metals $(DMe\text{-}DCNQI)_2Cu$, $(DMeO\text{-}DCNQI)_2Cu$ and $(DMe_{1-x}\text{-}MeBr_x\text{-}DCNQI)_2Cu$: Enhancement of density of states. *Solid State Commun.* **1992**, *81*, 473–476. [CrossRef]
326. Kobayashi, H.; Miyamoto, A.; Kato, R.; Sakai, F.; Kobayashi, A.; Yamakita, Y.; Furukawa, Y.; Tasumi, M.; Watanabe, T. Mixed valency of Cu, electron-mass enhancement, and three-dimensional arrangement of magnetic sites in the organic conductors (R_1,R_2-*N,N′*-dicyanoquinonediimine)$_2$Cu (where R_1,R_2 = CH_3,CH_3O,Cl,Br). *Phys. Rev. B* **1993**, *47*, 3500–3510. [CrossRef] [PubMed]
327. Kobayashi, H.; Sawa, H.; Aonuma, S.; Kato, R. Evidence for reentrant structural-phase transition in DCNQI-copper system. *J. Am. Chem. Soc.* **1993**, *115*, 7870–7871. [CrossRef]
328. Bauer, D.; von Schütz, J.U.; Wolf, H.C.; Hünig, S.; Sinzger, K.; Kremer, R.K. Alloyed deuterated copper-DCNQI salts: Phase transitions and reentry of conductivity, giant hysteresis effects, and coexistence of metallic and semiconducting modes. *Adv. Mater.* **1993**, *5*, 829–834. [CrossRef]
329. Sinzger, K.; Hünig, S.; Jopp, M.; Bauer, D.; Bietsch, W.; von Schütz, J.U.; Wolf, H.C.; Kremer, R.K.; Metzenthin, T.; Bau, R.; et al. The organic metal $(Me_2\text{-}DCNQI)_2Cu$: Dramatic changes in solid-state properties and crystal structure due to secondary deuterium effects. *J. Am. Chem. Soc.* **1993**, *115*, 7696–7705. [CrossRef]
330. Aonuma, S.; Sawa, H.; Kato, R.; Kobayashi, H. Giant metal-insulator-metal transition induced by selective deuteration of the molecular conductor, $(DMe\text{-}DCNQI)_2Cu$ (DMe-DCNQI = 2,5-dimethyl-*N,N′*-dicyanoquinonediimine). *Chem. Lett.* **1993**, *22*, 513–516. [CrossRef]
331. Sawa, H.; Tamura, M.; Aonuma, S.; Kato, R.; Kinoshita, M.; Kobayashi, H. Novel electronic states of partially deuterated $(DMe\text{-}DCNQI)_2Cu$. *J. Phys. Soc. Jpn.* **1993**, *62*, 2224–2228. [CrossRef]
332. Tamura, M.; Sawa, H.; Aonuma, S.; Kato, R.; Kinoshita, M.; Kobayashi, H. Weak ferromagnetism and magnetic anisotropy in copper salt of fully deuterated DMe-DCNQI, (DMe-DCNQI-d_8)$_2$Cu. *J. Phys. Soc. Jpn.* **1993**, *62*, 1470–1473. [CrossRef]
333. Kato, R.; Sawa, H.; Aonuma, S.; Tamura, M.; Kinoshita, M.; Kobayashi, H. Preparation and physical properties of an alloyed $(DMe\text{-}DCNQI)_2Cu$ with fully deuterated DMe-DCNQI (DMe-DCNQI = 2,5-dimethyl-*N,N′*-dicyanoquinonediimine). *Solid State Commun.* **1993**, *85*, 831–835. [CrossRef]
334. Uji, S.; Terashima, T.; Aoki, H.; Brooks, J.S.; Kato, R.; Sawa, H.; Aonuma, S.; Tamura, M.; Kinoshita, M. Coexistence of one- and three-dimensional Fermi surfaces and heavy cyclotron mass in the molecular conductor $(DMe\text{-}DCNQI)_2Cu$. *Phys. Rev. B* **1994**, *50*, 15597–15601. [CrossRef] [PubMed]
335. Sawa, H.; Tamura, M.; Aonuma, S.; Kinoshita, M.; Kato, R. Charge-transfer-controlled phase transition in a molecular conductor, $(DMe\text{-}DCNQI)_2Cu$—Doping effect. *J. Phys. Soc. Jpn.* **1994**, *63*, 4302–4305. [CrossRef]
336. Yamakita, Y.; Furukawa, Y.; Kobayashi, A.; Tasumi, M.; Kato, R.; Kobayashi, H. Vibrational studies on electronic structures in metallic and insulating phases of the Cu complexes of substituted dicyanoquinonediimines (DCNQI). A comparison with the cases of the Li and Ba complexes. *J. Chem. Phys.* **1994**, *100*, 2449–2457. [CrossRef]
337. Kashimura, Y.; Sawa, H.; Aonuma, S.; Kato, R.; Takahashi, H.; Mori, N. Anomalous pressure-temperature phase diagram of the molecular conductor, $(DI\text{-}DCNQI)_2Cu$ (DI-DCNQI = 2,5-diiodo-*N,N′*-dicyanoquinonediimine). *Solid State Commun.* **1995**, *93*, 675–679. [CrossRef]
338. Uji, S.; Terashima, T.; Aoki, H.; Kato, R.; Sawa, H.; Aonuma, S.; Tamura, M.; Kinoshita, M. Fermi surface and absence of additional mass enhancement near the insulating phase in $(DMe\text{-}DCNQI)_2Cu$. *Solid State Commun.* **1995**, *93*, 203–207. [CrossRef]
339. Hünig, S. *N,N′*-dicyanoquinonediimines (DCNQIs): Unique acceptors for conducting materials. *J. Mater. Chem.* **1995**, *5*, 1469–1479. [CrossRef]

340. Hiraki, K.; Kobayashi, Y.; Nakamura, T.; Takahashi, T.; Aonuma, S.; Sawa, H.; Kato, R.; Kobayashi, H. Magnetic structure in the antiferromagnetic state of the organic conductor, (DMe-DCNQI[3,3:1]d_7)$_2$Cu: ^{1}H-NMR analysis. *J. Phys. Soc. Jpn.* **1995**, *64*, 2203–2211. [CrossRef]
341. Miyazaki, Y.; Terakura, K.; Morikawa, Y.; Yamasaki, T. First-principles theoretical study of metallic states of DCNQI-(Cu,Ag) systems: Simplicity and variety in complex systems. *Phys. Rev. Lett.* **1995**, *74*, 5104–5107. [CrossRef]
342. Tamura, M.; Kashimura, Y.; Sawa, H.; Aonuma, S.; Kato, R.; Kinoshita, M. Enhanced magnetic susceptibility of (DI-DCNQI)$_2$Cu. *Solid State Commun.* **1995**, *93*, 585–588. [CrossRef]
343. Aonuma, S.; Sawa, H.; Kato, R. Chemical pressure effect by selective deuteration in the molecular-based conductor, 2,5-dimethyl-*N,N′*-dicyano-*p*-benzoquinone imine-copper salt, (DMe-DCNQI)$_2$Cu. *J. Chem. Soc. Perkin Trans.* **1995**, *2*, 1541–1549. [CrossRef]
344. Sekiyama, A.; Fujimori, A.; Aonuma, S.; Sawa, H.; Kato, R. Fermi-liquid versus Luttinger-liquid behavior and metal-insulator transition *N,N′*-dicyanoquinonediimine-Cu salt studied by photoemission. *Phys. Rev. B* **1995**, *51*, 13899–13902. [CrossRef]
345. Takahashi, T.; Yokoya, T.; Chainani, A.; Kumigashira, H.; Akaki, O. Cooperative effects of electron correlation and charge ordering on the metal-insulator transition in quasi-one-dimensional deuterated (DMe-DCNQI)$_2$Cu. *Phys. Rev. B* **1996**, *53*, 1790–1794. [CrossRef] [PubMed]
346. Miyazaki, Y.; Terakura, K. First-principles theoretical study of metallic states of DCNQI-(Cu,Ag,Li) systems. *Phys. Rev. B* **1996**, *54*, 10452–10464. [CrossRef]
347. Gómez, D.; von Schütz, J.U.; Wolf, C.H.; Hünig, S. Tunable phase transitions in conductive Cu(2,5-dimethyl-dicyanoquinonediimine)$_2$ radical ion salts. *J. Phys. I Fr.* **1996**, *6*, 1655–1671. [CrossRef]
348. Ogawa, T.; Suzumura, Y. Electronic properties of strongly correlated states in dicyanoquinonediimine-Cu organic conductors. *Phys. Rev. B* **1996**, *53*, 7085–7093. [CrossRef] [PubMed]
349. Ogawa, T.; Suzumura, Y. Effect of strong correlation on metal-insulator transition of DCNQI-Cu salts—Rigorous treatment of the local constraint–. *J. Phys. Soc. Jpn.* **1997**, *66*, 690–702. [CrossRef]
350. Seo, H.; Fukuyama, H. Antiferromagnetic phases of one-dimensional quarter-filled organic conductors. *J. Phys. Soc. Jpn.* **1997**, *66*, 1249–1252. [CrossRef]
351. Yonemitsu, K. Renormalization-group approach to the metal-insulator transitions in (DCNQI)$_2M$ (DCNQI is *N,N′*-dicyanoquinonediimine and M = Ag, Cu). *Phys. Rev. B* **1997**, *56*, 7262–7276. [CrossRef]
352. Nogami, Y.; Hayashi, S.; Date, T.; Oshima, K.; Hiraki, K.; Kanoda, K. High pressure structures of organic low dimensional conductor DCNQI compounds. *Rev. High Pressure Sci. Technol.* **1998**, *7*, 404–406. [CrossRef]
353. Uwatoko, Y.; Hotta, T.; Matsuoka, E.; Mori, H.; Ohki, T.; Sarraot, J.L.; Thompson, J.D.; Möri, N.; Oomi, G. High pressure apparatus for magnetization measurements. *Rev. High Pressure Sci. Technol.* **1998**, *7*, 1508–1510. [CrossRef]
354. Kawamoto, A.; Miyagawa, K.; Kanoda, K. ^{13}C NMR study of the metal-insulator transition in (DMe-DCNQI)$_2$Cu systems with partial deuteration. *Phys. Rev. B* **1998**, *58*, 1243–1251. [CrossRef]
355. Hünig, S.; Kemmer, M.; Meixner, H.; Sinzger, K.; Wenner, H.; Bauer, T.; Tillmanns, E.; Lux, F.R.; Hollstein, M.; Groß, H.-G.; et al. Multistep reversible redox systems, LXVII 2,5-Disubstituted *N,N′*-dicyanobenzoquinonediimines (DCNQIs): Charge-transfer complexes and radical-anion salts and copper salts with ligand alloys: Syntheses, structures and conductivities. *Eur. J. Inorg. Chem.* **1999**, 899–916. [CrossRef]
356. Miyagawa, K.; Kawamoto, A.; Kanoda, K. *π-d* orbital hybridization in the metallic state of organic-inorganic complexes seen by ^{13}C and ^{15}N NMR at selective sites. *Phys. Rev. B* **1999**, *60*, 14847–14851. [CrossRef]
357. Yamamoto, T.; Tajima, H.; Yamaura, J.-I.; Aonuma, S.; Kato, R. Reflectance spectra and electrical resistivity of (Me$_2$-DCNQI)$_2$Li$_{1-x}$Cu$_x$. *J. Phys. Soc. Jpn.* **1999**, *68*, 1384–1391. [CrossRef]
358. Kato, R. Conductive Copper Salts of 2,5-Disubstituted-*N,N′*-dicyanoquinonediimines (DCNQIs): Structural and physical properties. *Bull. Chem. Soc. Jpn.* **2000**, *73*, 515–534. [CrossRef]
359. Yonemitsu, K.; Kishine, J. Charge gap and dimensional crossovers in quasi-one-dimensional organic conductors. *J. Phys. Chem. Solids* **2000**, *62*, 99–104. [CrossRef]
360. Nishio, Y.; Tamura, M.; Kajita, K.; Aonuma, S.; Sawa, H.; Kato, R.; Kobayashi, H. Thermodynamic study of (DMe-DCNQI)$_2$Cu system—Mechanism of reentrant metal-insulator transition–. *J. Phys. Soc. Jpn.* **2000**, *69*, 1414–1422. [CrossRef]
361. Pinterić, M.; Vuletić, T.; Tomić, S.; von Schütz, J.U. Complex low-frequency dielectric relaxation of the charge-density wave state in the (2,5(OCH$_3$)$_2$DCNQI)$_2$Li. *Eur. Phys. J. B* **2001**, *22*, 335–341. [CrossRef]
362. Hünig, S.; Herberth, E. *N,N′*-Dicyanoquinone Diimines (DCNQIs): Versatile acceptors for organic conductors. *Chem. Rev.* **2004**, *104*, 5535–5563. [CrossRef]
363. Tanaka, Y.; Ogata, M. Effects of charge ordering on the spin degrees of freedom in one-dimensional extended Hubbard model. *J. Phys. Soc. Jpn.* **2005**, 3283–3287. [CrossRef]
364. Kanoda, K. Metal-insulator transition in κ-(ET)$_2$X and (DCNQI)$_2$M: Two contrasting manifestation of electron correlation. *J. Phys. Soc. Jpn.* **2006**, *75*, 051007. [CrossRef]
365. Takahashi, T.; Nogami, Y.; Yakushi, K. Charge ordering in organic conductors. *J. Phys. Soc. Jpn.* **2006**, *75*, 051008. [CrossRef]
366. Shinohara, Y.; Kazama, S.; Mizoguchi, K.; Hiraoka, M.; Sakamoto, H.; Masubuchi, S.; Kato, R.; Hiraki, K.; Takahashi, T. Spin density distribution and electronic states in (DMe-DCNQI)$_2$M (M = Li,Ag,Cu) from high-resolution solid state NMR. *Phys. Rev. B* **2007**, *76*, 35128. [CrossRef]

367. Miyasaka, T.; Watanabe, T.; Fujishima, A.; Honda, K. Light energy conversion with chlorophyll monolayer electrodes. In vitro electrochemical simulation of photosynthetic primary processes. *J. Am. Chem. Soc.* **1978**, *100*, 6657–6665. [CrossRef]
368. Miyasaka, T.; Watanabe, T.; Fujishima, A.; Honda, K. Highly efficient quantum conversion at chlorophyll a-lecithin mixed monolayer coated electrodes. *Nature* **1979**, *277*, 638–640. [CrossRef]
369. Liu, Y.Q.; Wu, X.L.; Wang, X.H.; Yang, D.L.; Zhu, D.B. Conducting Langmuir-Blodgett films based on unsymmetrical alkylthiotetrathiafulvalene and alkylammonium-metal $(dmit)_2$ complexes. *Synth. Met.* **1991**, *42*, 1529–1533. [CrossRef]
370. Zhu, D.; Yang, C.; Liu, Y.; Xu, Y. Syntheses and Langmuir-Blodgett film formation of donor-acceptor molecules. *Thin Solid Film* **1992**, *210–211*, 205–207. [CrossRef]
371. Miyasaka, T.; Koyama, K.; Itoh, I. Quantum conversion and image detection by a bacteriorhodopsin-based artificial photoreceptor. *Science* **1992**, *255*, 342–344. [CrossRef]
372. Mitzi, D.B.; Feild, C.A.; Harrison, W.T.A.; Guloy, A.M. Conducting tin halides with a layered organic-based perovskite structure. *Nature* **1994**, *369*, 467–469. [CrossRef]
373. Koyama, K.; Yamaguchi, N.; Miyasaka, T. Antibody-mediated bacteriorhodopsin orientation for molecular device architectures. *Science* **1994**, *265*, 762–765. [CrossRef]
374. Liu, Y.; Xu, Y.; Zhu, D.; Wada, T.; Sasabe, H.; Liu, L.; Wang, W. Langmuir-Blodgett films of an asymmetrically substituted metal-free phthalocyanine and the second-order non-linear optical properties. *Thin Solid Films* **1994**, *244*, 943–946. [CrossRef]
375. Mitzi, D.B.; Wang, S.; Feild, C.A.; Chess, C.A.; Guloy, A.M. Conducting layered organic-inorganic halides containing <110>-oriented perovskite sheets. *Science* **1995**, *267*, 1473–1476. [CrossRef]
376. Mitzi, D.B. Synthesis, crystal structure, and optical and thermal properties of $(C_4H_9NH_3)_2MI_4$ (M = Ge, Sn, Pb). *Chem. Mater.* **1996**, *8*, 791–800. [CrossRef]
377. Idota, Y.; Kubota, T.; Matsufuji, A.; Maekawa, Y.; Miyasaka, T. Tin-based amorphous oxide: A high-capacity lithium-ion-storage material. *Science* **1997**, *276*, 1395–1397. [CrossRef]
378. Kagan, C.R.; Mitzi, D.B.; Dimitrakopoulos, C.D. Organic-inorganic hybrid materials as semiconducting channels in thin-film field-effect transistors. *Science* **1999**, *286*, 945–947. [CrossRef]
379. Otsubo, T.; Aso, Y.; Takimiya, K. Functional oligothiophenes as advanced molecular electronic materials. *J. Mater. Chem.* **2002**, *12*, 2565–2575. [CrossRef]
380. Chen, J.; Law, C.C.W.; Lam, J.W.Y.; Dong, Y.; Lo, S.M.F.; Williams, I.D.; Zhu, D.; Tang, B.Z. Synthesis, light emission, nanoaggregation, and restricted intramolecular rotation of 1,1-subsituted 2,3,4,5-tetraphenylsiloles. *Chem. Mater.* **2003**, *15*, 1535–1546. [CrossRef]
381. Yu, G.; Yin, S.; Liu, Y.; Shuai, Z.; Zhu, D. Structures, electronic states, and electroluminescent properties of a Zinc(II) 2-(2-hydroxyphenyl)benzothiazolate complex. *J. Am. Chem. Soc.* **2003**, *125*, 14816–14824. [CrossRef] [PubMed]
382. Rovira, C. Bis(ethylenedithio)tetrathiafulvalene (BET-TTF) and related dissymmetrical electron donors: From the molecule to functional molecular materials and devices (OFETs). *Chem. Rev.* **2004**, *104*, 5289–5317. [CrossRef]
383. Mas-Torrent, M.; Hadley, P.; Bromly, S.T.; Ribas, X.; Tarrés, J.; Mas, M.; Molins, E.; Veciana, J.; Rovira, C. Correlation between crystal structure and mobility in organic filed-effect transistors based on single crystals of tetrathiafulvalene derivatives. *J. Am. Chem. Soc.* **2004**, *126*, 8546–8553. [CrossRef] [PubMed]
384. Takahashi, A.; Adachi, C. Development of highly efficient thermally activated delayed fluorescent porphyrins and its application to the polymer OLEDs. In Proceedings of the Frontiers in Optics 2005, Tuscon, AZ, USA, 16–21 October 2005; OSA Publishing: Washington, DC, USA, 2005.
385. Sun, Y.; Liu, Y.; Zhu, D. Advances in organic filed-effect transistors. *J. Mater. Chem.* **2005**, *15*, 53–65. [CrossRef]
386. Takahashi, Y.; Hasegawa, J.; Abe, Y.; Tokura, Y.; Nishimura, K.; Saito, G. Tuning of electron injections for *n*-type organic transistor based on charge-transfer compounds. *Appl. Phys. Lett.* **2005**, *86*, 063504. [CrossRef]
387. Tang, Q.; Li, H.; He, M.; Hu, W.; Liu, C.; Chen, K.; Wang, C.; Liu, Y.; Zhu, D. Low threshold voltage transistors based on individual single-crystalline submicrometer-sized ribbons of copper phthalocyanine. *Adv. Mater.* **2006**, *18*, 65–68. [CrossRef]
388. Takahashi, Y.; Hasegawa, T.; Abe, Y.; Tokura, Y.; Saito, G. Organic metal electrodes for controlled *p*- and *n*-type carrier injections in organic field-effect transistors. *Appl. Phys. Lett.* **2006**, *88*, 073504. [CrossRef]
389. Takahashi, Y.; Hasegawa, T.; Horiuchi, S.; Kumai, R.; Tokura, Y.; Saito, G. High mobility organic field-effect transistor based on hexamethylenetetrathiafulvalene with organic metal electrodes. *Chem. Mater.* **2007**, *19*, 6382–6384. [CrossRef]
390. Takimiya, K.; Kunugi, Y.; Otsubo, T. Development of new semiconducting materials for durable high-performance air-stable organic field-effect transistors. *Chem. Lett.* **2007**, *36*, 578–583. [CrossRef]
391. Torrent, M.-M.; Rovira, C. Novel small molecules for organic field-effect transistors: Towards processability and high performances. *Chem. Soc. Rev.* **2008**, *37*, 827–838. [CrossRef] [PubMed]
392. Kojima, A.; Teshima, K.; Shirai, Y.; Miyasaka, T. Organometal halide perovskites as visible-light sensitizers for photovoltaic cells. *J. Am. Chem. Soc.* **2009**, *131*, 6050–6051. [CrossRef]
393. Haas, S.; Takahashi, Y.; Takimiya, K.; Hasegawa, T. High-performance dinaphtho-thieno-thiophene single crystal field-effect transistors. *Appl. Phys. Lett.* **2009**, *95*, 022111. [CrossRef]
394. Endo, A.; Ogasawara, M.; Takahashi, A.; Yokoyama, D.; Kato, Y.; Adachi, C. Thermally activated delayed fluorescence from Sn^{4+}-porphyrin complexes and their application to organic light-emitting diodes—A novel mechanism for electroluminescence. *Adv. Mater.* **2009**, *21*, 4802–4806. [CrossRef] [PubMed]

395. Zhan, X.; Zhu, D. Conjugated polymers for high-efficiency organic photovoltaics. *Polym. Chem.* **2010**, *1*, 409–419. [CrossRef]
396. Takahashi, Y.; Obara, R.; Lin, Z.-Z.; Takahashi, Y.; Naito, T.; Inabe, T.; Ishibashi, S.; Terakura, K. Charge-transport in tin-iodide perovskite $CH_3NH_3SnI_3$: Origin of high conductivity. *Dalton Trans.* **2011**, *40*, 5563–5568. [CrossRef] [PubMed]
397. Mas-Torrent, M.; Rovira, C. Role of molecular order and solid-state structure in organic filed-effect transistors. *Chem. Rev.* **2011**, *111*, 4833–4856. [CrossRef]
398. Takimiya, K.; Shinamura, S.; Osaka, I.; Miyazaki, E. Thienoacene-based organic semiconductors. *Adv. Mater.* **2011**, *23*, 4347–4370. [CrossRef] [PubMed]
399. Endo, A.; Sato, K.; Yoshimura, K.; Kai, T.; Kawada, A.; Miyazaki, H.; Adachi, C. Efficient up-conversion of triplet excitons into a singlet state and its application for organic light emitting diodes. *Appl. Phys. Lett.* **2011**, *98*, 083302. [CrossRef]
400. Lee, M.M.; Teuscher, J.; Miyasaka, T.; Murakami, T.N.; Snaith, H.J. Efficient hybrid solar cells based on meso-superstructured organometal halide perovskites. *Science* **2012**, *338*, 643–647. [CrossRef]
401. Mas-Torrent, M.; Crivillers, N.; Rovira, C.; Veciana, J. Attaching persistent organic free radicals to surfaces: How and why. *Chem. Rev.* **2012**, *112*, 2506–2527. [CrossRef]
402. Wang, C.; Dong, H.; Hu, W.; Liu, Y.; Zhu, D. Semiconducting π-conjugated systems in field-effect transistors: A material odyssey of organic electronics. *Chem. Rev.* **2012**, *112*, 2208–2267. [CrossRef] [PubMed]
403. Inatomi, Y.; Hojo, N.; Yamamoto, T.; Watanabe, S.-I.; Misaki, Y. Construction of rechargeable batteries using multifused tetrathiafulvalene systems as cathode materials. *ChemPlusChem* **2012**, *77*, 973–976. [CrossRef]
404. Youn Lee, S.; Yasuda, T.; Nomura, H.; Adachi, C. High-efficiency organic light-emitting diodes utilizing thermally activated delayed fluorescence from triazine-based donor-acceptor hybrid molecules. *Appl. Phys. Lett.* **2012**, *101*, 093306. [CrossRef]
405. Nakagawa, T.; Ku, S.-Y.; Wong, K.-T.; Adachi, C. Electroluminescence based on thermally activated delayed fluorescence generated by a spirobifluorene donor-acceptor structure. *Chem. Commun.* **2012**, *48*, 9580–9582. [CrossRef] [PubMed]
406. Zhang, Q.; Li, J.; Shizu, K.; Huang, S.; Hirata, S.; Miyazaki, H.; Adachi, C. Design of efficient thermally activated delayed fluorescence materials for pure blue organic light emitting diodes. *J. Am. Chem. Soc.* **2012**, *134*, 14706–14709. [CrossRef] [PubMed]
407. Tanaka, H.; Shizu, K.; Miyazaki, H.; Adachi, C. Efficient green thermally activated delayed fluorescence (TADF) from a phenoxazine–triphenyltriazine (PXZ–TRZ) derivative. *Chem. Commun.* **2012**, *48*, 11392–11394. [CrossRef] [PubMed]
408. Takahashi, Y.; Hasegawa, H.; Takahashi, Y.; Inabe, T. Hall mobility in tin iodide perovskite $CH_3NH_3SnI_3$: Evidence for a doped semiconductor. *J. Solid State Chem.* **2013**, *205*, 39–43. [CrossRef]
409. Zhang, F.; Hu, Y.; Schuettfort, T.; Di, C.; Gao, X.; McNeil, C.R.; Thomsen, L.; Mannsfeld, S.C.B.; Yuan, W.; Sirringhaus, H.; et al. Critical role of alkyl chain branching of organic semiconductors in enabling solution-processed N-channel organic thin-film transistors with mobility of up to 3.50 $cm^2\ V^{-1}\ s^{-1}$. *J. Am. Chem. Soc.* **2013**, *135*, 2338–2349. [CrossRef] [PubMed]
410. Takimiya, K.; Nakano, M.; Kang, M.J.; Miyazaki, E.; Osaka, I. Thienannulation: Efficient synthesis of π-extended thienoacenes applicable to organic semiconductors. *Eur. J. Org. Chem.* **2013**, 217–227. [CrossRef]
411. Zhang, Q.; Li, B.; Huang, S.; Nomura, H.; Tanaka, H.; Adachi, C. Efficient blue organic light-emitting diodes employing thermally activated delayed fluorescence. *Nat. Photonics* **2014**, *8*, 326–332. [CrossRef]
412. Kato, M.; Senoo, K.-I.; Yao, M.; Misaki, Y. A pentakis-fused tetrathiafulvalene system extended by cyclohexene-1,4- diylidenes: A new positive electrode material for rechargeable batteries utilizing ten electron redox. *J. Mater. Chem. A* **2014**, *2*, 6747–6754. [CrossRef]
413. Takimiya, K.; Osaka, I.; Nakano, M. π-building blocks for organic electronics: Revaluation of "inductive" and "resonance" effects of π-electron deficient units. *Chem. Mater.* **2014**, *26*, 587–593. [CrossRef]
414. Saiki, T.; Mori, S.; Ohara, K.; Naito, T. Capacitor-like behavior of molecular crystal β-DiCC[Ni(dmit)$_2$]. *Chem. Lett.* **2014**, *43*, 1119–1121. [CrossRef]
415. Miyasaka, T. Perovskite photovoltaics: Rare functions of organo lead halide in solar cells and optoelectronic devices. *Chem. Lett.* **2015**, *44*, 720–729. [CrossRef]
416. Huang, X.; Sheng, P.; Tu, Z.; Zhang, F.; Wang, J.; Geng, H.; Zou, Y.; Di, C.-A.; Yi, Y.; Sun, Y.; et al. A two-dimensional π-d conjugated coordination polymer with extremely high electrical conductivity and ambipolar transport behavior. *Nat. Commun.* **2015**, *6*, 7408. [CrossRef]
417. Osaka, I.; Takimiya, K. Backbone orientation in semiconducting polymers. *Polymer* **2015**, *59*, A1–A15. [CrossRef]
418. Takimiya, K.; Nakano, M.; Sugino, H.; Osaka, I. Design and elaboration of organic molecules for high filed-effect-mobility semiconductors. *Synth. Met.* **2016**, *217*, 68–78. [CrossRef]
419. Mori, T. Principles that govern electronic transport in organic conductors and transistors. *Bull. Chem. Soc. Jpn.* **2016**, *89*, 973–986. [CrossRef]
420. Wong, M.Y.; Zysman-Colman, E. Purely organic thermally activated delayed fluorescence materials for organic light-emitting diodes. *Adv. Mater.* **2017**, *29*, 1605444. [CrossRef]
421. Correa-Baena, J.-P.; Saliba, M.; Buonassisi, T.; Grätzel, M.; Abate, A.; Tress, W.; Hagfeldt, A. Promises and challenges of perovskite solar cells. *Science* **2017**, *358*, 739–744. [CrossRef]
422. Nakano, M.; Takimiya, K. Sodium-sulfide promoted thiophene-annulations: Powerful tools for elaborating organic semiconducting materials. *Chem. Mater.* **2017**, *29*, 256–264. [CrossRef]
423. Osaka, I.; Takimiya, K. Naphthobischalcogenadiazole conjugated polymers: Emerging materials for organic electronics. *Adv. Mat.* **2017**, *29*, 1605218. [CrossRef] [PubMed]

424. Wang, Q.; Tian, Q.-S.; Zhang, Y.-L.; Tang, X.; Liao, L.-S. High-efficiency organic light-emitting diodes with exciplex hosts. *J. Mater. Chem. C* **2019**, *7*, 11329–11360. [CrossRef]
425. Wang, Z.; Wang, C.; Zhang, H.; Liu, Z.; Zhao, B.; Li, W. The application of charge transfer host based exciplex and thermally activated delayed fluorescence materials in organic light-emitting diodes. *Org. Elec.* **2019**, *66*, 227–241. [CrossRef]
426. Wang, S.; Zhang, H.; Zhang, B.; Xie, Z.; Wong, W.-Y. Towards high-power-efficiency solution-processed OLEDs: Material and device perspectives. *Mater. Sci. Eng. R* **2020**, *140*, 100547. [CrossRef]
427. Kim, M.; Ryu, S.U.; Park, S.A.; Choi, K.; Kim, T.; Chung, D.; Park, T. Donor-acceptor-conjugated polymer for high-performance organic field-effect transistors: A progress report. *Adv. Func. Mater.* **2020**, *30*, 1904545. [CrossRef]
428. Schmidbaur, H.; Raubenheimer, H.G. Excimer and exciplex formation in gold(I) complexes preconditioned by aurophilic interactions. *Angew. Chem. Int. Ed. Engl.* **2020**, *59*, 14748–14771. [CrossRef]
429. Mallah, T.; Hollis, C.; Bott, S.; Kurmoo, M.; Day, P.; Allan, M.; Friend, R.H. Crystal structures and physical properties of bis(ethylenedithio)-tetrathiafulvalene charge-transfer salts with FeX_4^- (X = Cl or Br) anions. *J. Chem. Soc. Dalton Trans.* **1990**, 859–865. [CrossRef]
430. Day, P.; Kurmoo, M.; Mallah, T.; Marsden, I.R.; Friend, R.H.; Pratt, F.L.; Hayes, W.; Chasseau, D.; Gaultier, J.; Bravic, G.; et al. Structure and properties of tris[bis(ethylenedithio)tetrathiafulvalenium]tetrachlorocopper(II) hydrate, (BEDT-TTF)$_3$CuCl$_4$·H$_2$O: First evidence for coexistence of localized and conduction electrons in a metallic charge-transfer salt. *J. Am. Chem. Soc.* **1992**, *114*, 10722–10729. [CrossRef]
431. Gama, V.; Henriques, R.T.; Bonfait, G.; Almeida, M.; Meetsma, A.; van Smaalen, S.; de Boer, J.L. (Perylene)Co(mnt)$_2$(CH$_2$Cl$_2$)$_{0.5}$: A mixed molecular and polymeric conductor. *J. Am. Chem. Soc.* **1992**, *114*, 1986–1989. [CrossRef]
432. Gama, V.; Henriques, R.; Bonfait, G.; Pereira, L.; Waerenborgh, J.C.; Santos, I.; Teresa Duarte, M.; Cabral, J.; Almeida, M. Low-dimensional molecular metals (Per)$_2$M(mnt)$_2$ (M = Fe and Co). *Inorg. Chem.* **1992**, *31*, 2598–2604. [CrossRef]
433. Gómez-García, C.J.; Ouahab, L.; Giménez-Saiz, C.; Triki, S.; Coronado, E.; Delhaés, P. Coexistence of mobile and localized electrons in bis(ethylene)dithiotetrathiafulvalene (BEDT-TTF) radical salts with paramagnetic polyoxometalates: Synthesis and physical properties of (BEDT-TTF)$_8$[CoW$_{12}$O$_{40}$]·5.5H$_2$O. *Angew. Chem. Int. Ed. Engl.* **1994**, *33*, 223–226. [CrossRef]
434. Graham, A.W.; Kurmoo, M.; Day, P. β″-(bedt-ttf)$_4$[(H$_2$O)Fe(C$_2$O$_4$)$_3$]·PhCN: The first molecular superconductor containing paramagnetic metal ions. *J. Chem. Soc. Chem. Commun.* **1995**, 2061–2062. [CrossRef]
435. Galán-Mascarós, J.R.; Giménez-Saiz, C.; Triki, S.; Gómez-García, C.J.; Coronado, E.; Ouahab, L. A novel chainlike heteropolyanion formed by Keggin units: Synthesis and structure of (ET)$_{8n}$[PMnW$_{11}$O$_{39}$]$_n$·2nH$_2$O. *Angew. Chem. Int. Ed. Engl.* **1995**, *34*, 1460–1462. [CrossRef]
436. Ouahab, L. Organic/inorganic supramolecular assemblies and synergy between physical properties. *Chem. Mater.* **1997**, *9*, 1909–1926. [CrossRef]
437. Ouahab, L. Coordination complexes in conducting and magnetic molecular materials. *Coord. Chem. Rev.* **1998**, *178–180*, 1501–1531. [CrossRef]
438. Ribera, E.; Rovira, C.; Veciana, J.; Tarrés, J.; Canadell, E.; Rousseau, R.; Molins, E.; Mas, M.; Schoeffel, J.-P.; Pouget, J.-P.; et al. The [(DT-TTF)$_2$M(mnt)$_2$] family of radical ion salts: From a spin ladder to delocalised conduction electrons that interact with localised magnetic moments. *Chem. Eur. J.* **1999**, *5*, 2025–2039. [CrossRef]
439. Coronado, E.; Gómez-García, C.J. Polyoxometalate-based molecular materials. *Chem. Rev.* **1998**, *98*, 273–296. [CrossRef]
440. Coronado, E.; Galán-Mascarós, J.R.; Gómez-García, C.J.; Laukhin, V. Coexistence of ferromagnetism and metallic conductivity in a molecule-based layered compound. *Nature* **2000**, *408*, 447–449. [CrossRef] [PubMed]
441. Naito, T.; Inabe, T.; Takeda, K.; Awaga, K.; Akutagawa, T.; Hasegawa, T.; Nakamura, T.; Kakiuchi, T.; Sawa, H.; Yamamoto, T.; et al. β″-(ET)$_3$(MnCl$_4$)(1,1,2 C$_2$H$_3$Cl$_3$) (ET = bis(ethylenedithio)tetrathiafulvalene); a pressure-sensitive new molecular conductor with localized spins. *J. Mater. Chem.* **2001**, *11*, 2221–2227. [CrossRef]
442. Prokhorova, T.G.; Khasanov, S.S.; Zorina, L.V.; Burabov, L.I.; Tkacheva, V.A.; Baskakov, A.A.; Morgunov, R.B.; Gener, M.; Canadell, E.; Shibaeva, R.P.; et al. Molecular metals based on BEDT-TTF radical cation salts with magnetic metal oxalates as counterions: β″-(BEDT-TTF)$_4$A[M(C$_2$O$_4$)$_3$]·DMF (A = NH$_4^+$, K$^+$; M = Cr^{III}, Fe^{III}). *Adv. Func. Mater.* **2003**, *13*, 403–411. [CrossRef]
443. Naito, T.; Inabe, T. Molecular hexagonal perovskite: A new type of organic-inorganic hybrid conductor. *J. Solid State Chem.* **2003**, *176*, 243–249. [CrossRef]
444. Coronado, E.; Galán-Mascarós, J.R.; Giménez-Saiz, C.; Gómez-García, C.J.; Martínez-Ferrero, E.; Almeida, M.; Lopes, E.B. Metallic conductivity in a polyoxovanadate radical salt of bis(ethylenedithio)tetrathiafulvalene (BEDT-TTF): Synthesis, structure, and physical characterization of β″-(BEDT-TTF)$_5$[H$_3$V$_{10}$O$_{28}$]·4H$_2$O. *Adv. Mat.* **2004**, *16*, 324–327. [CrossRef]
445. Coronado, E.; Day, P. Magnetic molecular conductors. *Chem. Rev.* **2004**, *104*, 5419–5448. [CrossRef]
446. Ouahab, L.; Enoki, T. Multiproperty molecular materials: TTF-based conducting and magnetic molecular materials. *Eur. J. Inorg. Chem.* **2004**, 933–941. [CrossRef]
447. Naito, T.; Inabe, T. Structural, electrical, and magnetic properties of α-(ET)$_7$[MnCl$_4$]$_2$·(1,1,2-C$_2$H$_3$Cl$_3$)$_2$ (ET = bis(ethylenedithio) tetrathiafulvalene). *Bull. Chem. Soc. Jpn.* **2004**, *77*, 1987–1995. [CrossRef]
448. Coronado, E.; Galán-Mascarós, J.R. Hybrid molecular conductors. *J. Mater. Chem.* **2005**, *15*, 66–74. [CrossRef]
449. Fujiwara, H.; Wada, K.; Hiraoka, T.; Hayashi, T.; Sugimoto, T.; Nakazumi, H.; Yokogawa, K.; Teramura, M.; Yasuzuka, S.; Murata, K.; et al. Stable metallic behavior and antiferromagnetic ordering of Fe(III) d spins in (EDO-TTFVO)$_2$·FeCl$_4$. *J. Am. Chem. Soc.* **2005**, *127*, 14166–14167. [CrossRef]

450. Kushch, N.D.; Kazakova, A.V.; Dubrovskii, A.D.; Shilov, G.V.; Buravov, L.I.; Morgunov, R.B.; Kurganova, E.V.; Tanimoto, Y.; Yagubskii, E.B. Molecular magnetism semiconductors formed by cationic and anionic networks: $(ET)_2Mn[N(CN)_2]_3$ and $(ET)_2CuMn[N(CN)_2]_4$. *J. Mater. Chem.* **2007**, *17*, 4407–4413. [CrossRef]
451. Engler, E.M.; Patel, V.V. Anomalous reaction of selenium and carbon disulfide with sodium acetylide. Synthesis of selenium analogs of 1,3-Dithiole-2-thione. *J. Org. Chem.* **1975**, *40*, 387–389. [CrossRef]
452. Engler, E.M.; Patel, V.V. Synthesis of *cis*- and *trans*-diselenadithiafulvalene and its highly conducting charge-transfer salt with tetracyano-*p*-quinodimethane. *J. Chem. Soc. Chem. Commun.* **1975**, 671–672. [CrossRef]
453. Engler, E.M.; Patel, V.V.; Schumaker, R.R. Triselenathiafulvalenes: A novel sulphur-selenium interchange on trimethyl phosphite coupling of substituted 1,3-diselenole-2-thiones. *J. Chem. Soc. Chem. Commun.* **1977**, 835–836. [CrossRef]
454. Schumaker, R.R.; Lee, Y.V.; Engler, E.M. New synthetic approaches to tetrathiafulvalene derivatives: Systematic modifications of BEDT-TTF and TMTTF donors. *J. Phys. Colloques* **1983**, *44*. [CrossRef]
455. Lee, V.Y.; Engler, E.M.; Schumaker, R.R.; Parkin, S.S.P. Bis(ethylenediseleno)tetraselenafulvalene (BEDSe-TSeF). *J. Chem. Soc. Chem. Commun.* **1983**, 235–236. [CrossRef]
456. Bryce, M.R.; Moore, A.J.; Lorcy, D.; Dhindsa, A.S.; Robert, A. Unsymmetrical and highly-conjugated tetrathiafulvalene and selenatrithiafulvalene derivatives: Synthesis and reactions of novel heterocyclic Wittig-Horner reagents. *J. Chem. Soc. Chem. Commun.* **1983**, 470–472. [CrossRef]
457. Schumaker, R.R.; Lee, V.Y.; Engler, E.M. Noncoupling synthesis of tetrathiafulvalenes. *J. Org. Chem.* **1984**, *49*, 564–566. [CrossRef]
458. Kato, R.; Kobayashi, H.; Kobayashi, A. Synthesis and properties of bis(ethylenedithio)tetraselenafulvalene (BEDT-TSeF) compounds. *Synth. Met.* **1991**, *42*, 2093–2096. [CrossRef]
459. Moore, A.J.; Bryce, M.R. Highly conjugated π-electron donors for organic metals: Synthesis and redox chemistry of new 1,3-dithiole and 1,3-selenathiole derivatives. *J. Chem. Soc. Perkin Trans.* **1991**, 157–168. [CrossRef]
460. Moore, A.J.; Bryce, M.R.; Ando, D.J.; Hursthouse, M.B. New bis(ethylenedithio)tetrathiafulvalene derivatives with low oxidation potentials. *J. Chem. Soc. Chem. Commun.* **1991**, 320–321. [CrossRef]
461. Bryce, M.R.; Coffin, M.A.; Hursthouse, M.B.; Karaulov, A.I.; Müllen, K.; Scheich, H. Synthesis, x-ray crystal structure and multistage redox properties of a severely-distorted tetrathiafulvalene donor. *Tetrahedron Lett.* **1991**, *32*, 6029–6032. [CrossRef]
462. Montgomery, L.K.; Burgin, T.; Husting, C.; Tilley, L.; Huffman, J.C.; Carlson, K.D.; Dudek, J.D.; Yaconi, G.A.; Geiser, U.; Williams, J.M. Synthesis and characterization of radical cation salts derived from tetraselenafulvalene and bis(ethylenedithio)tetraselenafulvaler *Mol. Cryst. Liq. Cryst.* **1992**, *211*, 283–288. [CrossRef]
463. Kobayashi, A.; Udagawa, T.; Tomita, H.; Naito, T.; Kobayashi, H. A New Organic Superconductor, λ-$(BEDT\text{-}TSF)_2GaCl_4$. *Chem. Lett.* **1993**, *22*, 1559–1562. [CrossRef]
464. Kobayashi, H.; Udagawa, T.; Tomita, H.; Bun, K.; Naito, T.; Kobayashi, A. New organic metals based on BETS compounds with MX_4^- Anions (BETS = bis(ethylenedithio)tetraselenafulvalene; M = Ga, Fe, In; X = Cl, Br). *Chem. Lett.* **1993**, *22*, 2179–2182. [CrossRef]
465. Goze, F.; Laukhin, V.N.; Brossard, L.; Audouard, A.; Ulmet, J.P.; Askenazy, S.; Naito, T.; Kobayashi, H.; Kobayashi, A.; Tokumoto, M.; et al. Magnetotransport measurements on the λ-phase of the organic conductors $(BETS)_2MCl_4$ (M = Ga, Fe). Magnetic-field-restored highly conducting state in λ-$(BETS)_2FeCl_4$. *EPL* **1994**, *28*, 427–431. [CrossRef]
466. Kobayashi, H.; Tomita, H.; Naito, T.; Kobayashi, A.; Sakai, F.; Watanabe, T.; Cassoux, P. New BETS superconductors with magnetic anions (BETS = bis(ethylenedithio)tetraselenafulvalene). *J. Am. Chem. Soc.* **1996**, *118*, 368–377. [CrossRef]
467. Kobayashi, H.; Akutsu, H.; Arai, E.; Tanaka, H.; Kobayashi, A. Electric and magnetic properties and phase diagram of a series of organic superconductors λ-$BETS_2GaX_zY_{4-z}$ [BETS = bis(ethylenedithio)tetraselenafulvalene, X, Y = F, Cl, Br; $0 < z < 2$]. *Phys. Rev. B* **1997**, *56*, R8526–R8529. [CrossRef]
468. Courcet, T.; Malfant, I.; Pokhodnia, K.; Cassoux, P. Bis(ethylenedithio)tetraselenafulvalene: Short-cut synthesis, X-ray crystal structure and π-electron density distribution. *New J. Chem.* **1998**, *22*, 585–589. [CrossRef]
469. Brossard, L.; Clerac, L.; Coulon, C.; Tokumoto, M.; Ziman, T.; Petrov, D.K.; Laukhin, D.N.; Naughton, M.J.; Audouard, A.; Goze, F.; et al. Interplay between chains of $S = 5/2$ localised spins and two-dimensional sheets of organic donors in the synthetically built magnetic multilayer λ-$BETS_2FeCl_4$. *Eur. Phys. J. B* **1998**, *1*, 439–452. [CrossRef]
470. Sato, A.; Ojima, E.; Akutsu, H.; Kobayashi, H.; Kobayashi, A.; Cassoux, P. Temperature-composition phase diagram of the organic alloys, λ-$BETS_2(Fe_xGa_{1-x})Cl_4$ with mixed magnetic and non-magnetic anions. *Chem. Lett.* **1998**, 673–674. [CrossRef]
471. Akutsu, H.; Kato, K.; Arai, E.; Kobayashi, H.; Kobayashi, A.; Tokumoto, M.; Brossard, L.; Cassoux, P. A coupled metal-insulator and antiferromagnetic transition of λ-$BETS_2FeCl_4$ under high-pressure and magnetic field [BETS = bis(ethylenedithio)tetraselenafulvalene]. *Solid State Commun.* **1998**, *105*, 485–489. [CrossRef]
472. Ojima, E.; Fujiwara, H.; Kato, K.; Kobayashi, H.; Tanaka, H.; Kobayashi, A.; Tokumoto, M.; Cassoux, P. Antiferromagnetic organic metal exhibiting superconducting transition, κ-$BETS_2FeBr_4$ [BETS = bis(ethylenedithio)tetraselenafulvalene]. *J. Am. Chem. Soc.* **1999**, *121*, 5581–5582. [CrossRef]
473. Tanaka, H.; Adachi, T.; Ojima, E.; Fujiwara, H.; Kato, K.; Kobayashi, H.; Kobayashi, A.; Cassoux, P. Pressure-induced superconducting transition of λ-$BETS_2FeCl_4$ with π-d coupled antiferromagnetic insulating ground state at ambient pressure [BETS = bis(ethylenedithio)tetraselenafulvalene]. *J. Am. Chem. Soc.* **1999**, *121*, 11243–11244. [CrossRef]
474. Otsuka, T.; Kobayashi, A.; Miyamoto, Y.; Kiuchi, J.; Wada, N.; Ojima, E.; Fujiwara, H.; Kobayashi, H. Successive antiferromagnetic and superconducting transitions in an organic metal, κ-$BETS_2FeCl_4$. *Chem. Lett.* **2000**, 732–733. [CrossRef]

475. Balicas, L.; Brooks, J.S.; Storr, K.; Graf, D.; Uji, S.; Shinagawa, H.; Ojima, E.; Fujiwara, H.; Kobayashi, H.; Kobayashi, A.; et al. Schubnikov-de-Haas effect and Yamaji oscillations in the antiferromagnetically ordered organic superconductor κ-$(BETS)_2FeBr_4$: A fermiology study. *Solid State Commun.* **2000**, *116*, 557–562. [CrossRef]
476. Mazumdar, S.; Clay, R.T.; Campbell, D.K. Bond-order and charge-density waves in the isotropic interacting two-dimensional quarter-filled and the insulating state proximate to organic superconductivity. *Phys. Rev. B Cond. Matter Mater. Phys.* **2000**, *62*, 13400–13425. [CrossRef]
477. Kobayashi, H.; Kobayashi, A.; Cassoux, P. BETS as a source of molecular magnetic superconductors (BETS = bis(ethylenedithio) tetraselenafulvalene). *Chem. Soc. Rev.* **2000**, *29*, 325–333. [CrossRef]
478. Hotta, C.; Fukuyama, H. Effects of localized spins in quasi-two dimensional organic conductors. *J. Phys. Soc. Jpn.* **2000**, *69*, 2577–2596. [CrossRef]
479. Otsuka, T.; Kobayashi, A.; Miyamoto, Y.; Kiuchi, J.; Nakamura, S.; Wada, N.; Fujiwara, E.; Fujiwara, H.; Kobayashi, H. Organic antiferromagnetic metals exhibiting superconducting transitions κ-$(BETS)_2FeX_4$ (X = Cl, Br): Drastic effect of halogen substitution on the successive phase transitions. *J. Solid State Chem.* **2001**, *159*, 407–412. [CrossRef]
480. Fujiwara, H.; Fujiwara, E.; Nakazawa, Y.; Narymbetov, B.Z.; Kato, K.; Kobayashi, H.; Kobayashi, A.; Tokumoto, M.; Cassoux, P. A novel antiferromagnetic organic superconductor κ-$BETS_2FeBr_4$ [where BETS = bis(ethylenedithio)tetraselenafulvalene]. *J. Am. Chem. Soc.* **2001**, *123*, 306–314. [CrossRef]
481. Uji, S.; Shinagawa, H.; Terashima, T.; Yakabae, T.; Terai, Y.; Tokumoto, M.; Kobayashi, A.; Tanaka, H.; Kobayashi, H. Magnetic-field-induced superconductivity in a two-dimensional organic conductor. *Nature* **2001**, *410*, 908–910. [CrossRef]
482. Balicas, L.; Brooks, J.S.; Storr, K.; Uji, S.; Tokumoto, M.; Tanaka, H.; Kobayashi, H.; Kobayashi, A.; Barzykin, V.; Gor'kov, L.P. Superconductivity in an organic insulator at very high magnetic fields. *Phys. Rev. Lett.* **2001**, *87*, 670021–670024. [CrossRef] [PubMed]
483. Uji, S.; Shinagawa, H.; Terakura, C.; Terashima, T.; Yakabe, T.; Terai, Y.; Tokumoto, M.; Kobayashi, A.; Tanaka, H.; Kobayashi, H. Fermi surface studies in the magnetic-field-induced superconductor λ-$BETS_2FeCl_4$. *Phys. Rev. B* **2001**, *64*, 024531. [CrossRef]
484. Takimiya, K.; Jigami, T.; Kawashima, M.; Kodani, M.; Aso, Y.; Otsubo, T. Synthetic procedure for various selenium-containing electron donors of the bis(ethylenedithio)tetrathiafulvalene (BEDT-TTF) type. *J. Org. Chem.* **2002**, *67*, 4218–4227. [CrossRef] [PubMed]
485. Uji, S.; Kobayashi, H.; Balicas, L.; Brooks, J.S. Superconductivity in an organic conductor stabilized by a high magnetic field. *Adv. Mater.* **2002**, *14*, 243–245. [CrossRef]
486. Cépas, O.; McKenzie, R.H.; Merino, J. Magnetic-field induced superconductivity in layered organic molecular crystals with localized magnetic moments. *Phys. Rev. B* **2002**, *65*, 100502(R). [CrossRef]
487. Mori, T.; Katsuhara, M. Estimation of the πd-interactions in organic conductors including magnetic anions. *J. Phys. Soc. Jpn.* **2002**, *71*, 826–844. [CrossRef]
488. Uji, S.; Terakura, C.; Terashima, T.; Yakabe, T.; Terai, Y.; Tokumoto, M.; Kobayashi, A.; Sakai, F.; Tanaka, H.; Kobayashi, H. Fermi surface and internal magnetic field of the organic conductors λ-$BETS_2Fe_xGa_{1-x}Cl_4$. *Phys. Rev. B Cond. Matter Mater. Phys.* **2002**, *65*, 113101. [CrossRef]
489. Fujiwara, H.; Kobayashi, H.; Fujiwara, E.; Kobayashi, A. An indication of magnetic-field-induced superconductivity in a bifunctional layered organic conductor κ-$BETS_2FeBr_4$. *J. Am. Chem. Soc.* **2002**, *124*, 6816–6817. [CrossRef]
490. Shimahara, H. Fulde-Ferrell-Larkin-Ovchinnikov state and field-induced superconductivity in an organic superconductor. *J. Phys. Soc. Jpn.* **2002**, *71*, 1644–1647. [CrossRef]
491. Zhang, B.; Tanaka, H.; Fujiwara, H.; Kobayashi, H.; Fujiwara, E.; Kobayashi, A. Dual-action molecular superconductors with magnetic anions. *J. Am. Chem. Soc.* **2002**, *124*, 9982–9983. [CrossRef]
492. Houzet, M.; Buzdin, A.; Bulaevskii, L.; Maley, M. New superconducting phases in field-induced organic superconductor λ-$BETS_2FeCl_4$. *Phys. Rev. Lett.* **2002**, *88*, 227001. [CrossRef]
493. Alberola, A.; Coronado, E.; Galán-Mascarós, J.R.; Giménez-Saiz, C.; Gómez-García, C.J. A molecular ferromagnet from the organic donor bis(ethylenedithio)tetraselenafulvalene and bimetallic oxalate complexes. *J. Am. Chem. Soc.* **2003**, *125*, 10774–10775. [CrossRef]
494. Uji, S.; Terashima, T.; Terakura, C.; Yakabe, T.; Terai, Y.; Yasuzuka, S.; Imanaka, Y.; Tokumoto, M.; Kobayashi, A.; Sakai, F.; et al. Global phase diagram of the magnetic-field-induced superconductors λ-$BETS_2Fe_xGa_{1-x}Cl_4$. *J. Phys. Soc. Jpn.* **2003**, *72*, 369–373. [CrossRef]
495. Kobayashi, H.; Cui, H.; Kobayashi, A. Organic metals and superconductors based on BETS (BETS = bis(ethylenedithio) tetraselenafulvalene). *Chem. Rev.* **2004**, *104*, 5265–5288. [CrossRef] [PubMed]
496. Pilia, L.; Malfant, I.; de Caro, D.; Senocq, F.; Zwick, A.; Valade, L. Conductive thin films of θ-$BETS_4[Fe(CN)_5NO]$ on silicon electrodes—New perspectives on charge transfer salts. *New J. Chem.* **2004**, *28*, 52–55. [CrossRef]
497. Konoike, T.; Uji, S.; Terashima, T.; Nishimura, M.; Yasuzuka, S.; Enomoto, K.; Fujiwara, H.; Zhang, B.; Kobayashi, H. Magnetic-field-induced superconductivity in the antiferromagnetic organic superconductor κ-$(BETS)_2FeBr_4$. *Phys. Rev. B* **2004**, *70*, 094514. [CrossRef]
498. Fujiwara, H.; Kobayashi, H. Development of an antiferromagnetic organic superconductor κ-$(BETS)_2FeBr_4$. *Bull. Chem. Soc. Jpn.* **2005**, *78*, 1181–1196. [CrossRef]

499. Powell, B.J.; McKenzie, R.H. Half-filled layers organic superconductors and the resonating-valence-bond theory of the Hubbard-Heisenberg model. *Phys. Rev. Lett.* **2005**, *94*, 047004. [CrossRef]
500. Uji, S.; Brooks, J.S. Magnetic-field-induced superconductivity in organic conductors. *J. Phys. Soc. Jpn.* **2006**, *75*, 051014. [CrossRef]
501. Uji, S.; Terashima, T.; Nishimura, M.; Takahide, Y.; Konoike, T.; Enomoto, K.; Cui, H.; Kobayashi, H.; Kobayashi, A.; Tanaka, H.; et al. Vortex dynamics and the Fulde-Ferrell-Larkin-Ovchinnikov state in a magnetic-field-induced organic superconductor. *Phys. Rev. Lett.* **2006**, *97*, 157001. [CrossRef]
502. Powell, B.J.; McKenzie, R.H. Strong electronic correlations in superconducting organic charge transfer salts. *J. Phys. Condens. Matter* **2006**, *18*, R827–R866. [CrossRef]
503. Zhang, B.; Wang, Z.; Zhang, Y.; Takahashi, K.; Okano, Y.; Cui, H.; Kobayashi, H.; Inoue, K.; Kurmoo, M.; Pratt, F.L.; et al. Hybrid organic-inorganic conductor with a magnetic chain anion: κ-$BETS_2[Fe^{III}(C_2O_4)Cl_2]$ [BETS = bis(ethylenedithio)tetraselenafulvalene]. *Inorg. Chem.* **2006**, *45*, 3275–3280. [CrossRef] [PubMed]
504. Hiraki, K.-I.; Mayaffre, H.; Horvatić, M.; Berthier, C.; Uji, S.; Yamaguchi, T.; Tanaka, H.; Kobayashi, A.; Kobayashi, H.; Takahashi, T. ^{77}Se NMR evidence for the Jaccarino-Peter mechanism in the field induced superconductor, λ-$BETS_2FeCl_4$. *J. Phys. Soc. Jpn.* **2007**, *76*, 124708. [CrossRef]
505. Kushch, N.D.; Yagubskii, E.B.; Kartsovnik, M.V.; Buravov, N.I.; Dubrovskii, A.D.; Chekhlov, A.N.; Biberacher, W. π-donor BETS based bifunctional superconductorwith polymeric dicyanamidomanganate(II) anion layer:κ-$BETS_2Mn[N(CN)_2]_3$. *J. Am. Chem. Soc.* **2008**, *130*, 7238–7240. [CrossRef] [PubMed]
506. Coronado, E.; Curreli, S.; Gimenez-Saiz, C.; Gómez-García, C.J.; Alberola, A.; Canadell, E. Molecular conductors based on the mixed-valence polyoxometallates $[SMo_{12}O_4O]^{n-}$ (n = 3 and 4) and the organic donors bis(ethylenedithio)tetrathiafulvalene and bis(ethylenedithio)tetraselenafulvalene. *Inorg. Chem.* **2009**, *48*, 11314–11324. [CrossRef] [PubMed]
507. Waerenborgh, J.C.; Rabaça, S.; Almeida, M.; Lopes, E.B.; Kobayashi, A.; Zhou, B.; Brooks, J.S. Mössbauer spectroscopy and magnetic transition of λ-$BETS_2FeCl_4$. *Phys. Rev. B* **2010**, *81*, 060413(R). [CrossRef]
508. Zverev, V.N.; Kartsovnik, M.V.; Biberacher, W.; Khasanov, S.S.; Shibaeva, R.P.; Ouahab, L.; Toupet, L.; Kushch, N.D.; Yagubskii, E.B.; Canadell, E. Temperature-pressure phase diagram and electronic properties of the organic metal κ-$BETS_2Mn[N(CN)_2]_3$. *Phys. Rev. B* **2010**, *82*, 155123. [CrossRef]
509. Vyaselev, O.M.; Kartsovnik, M.V.; Biberacher, W.; Zorina, L.V.; Kushch, N.D.; Yagubskii, E.B. Magnetic transformations in the organic conductor κ-$(BETS)_2Mn[N(CN)_2]_3$ at the metal-insulator transition. *Phys. Rev. B* **2011**, *83*, 094425. [CrossRef]
510. Kobayashi, H.; Kobayashi, A.; Tajima, H. Studies on molecular conductors: From organic semiconductors to molecular metals and superconductors. *Chem. Asian J.* **2011**, *6*, 1688–1704. [CrossRef]
511. Naito, T.; Matsuo, S.; Inabe, T.; Toda, Y. Carrier dynamics in a series of organic magnetic superconductors. *J. Phys. Chem. C* **2012**, *116*, 2588–2593. [CrossRef]
512. Akiba, H.; Shimada, K.; Tajima, N.; Kajita, K.; Nishio, Y. Paramagnetic metal-antiferromagnetic insulator transition in π-d system λ-$BETS_2FeCl_4$, BETS = bis(ethylenedithio)tetraselenafulvalene. *Crystals* **2012**, *2*, 984–995. [CrossRef]
513. Uji, S.; Kodama, K.; Sugii, K.; Terashima, T.; Takahide, Y.; Kurita, N.; Tsuchiya, S.; Kimata, M.; Kobayashi, A.; Zhou, B.; et al. Magnetic torque studies on FFLO phase in magnetic-field-induced superconductor λ-$BETS_2FeCl_4$. *J. Phys. Soc. Jpn.* **2012**, *85*, 174530. [CrossRef]
514. Uji, S.; Kodama, K.; Sugii, K.; Terashima, T.; Yamaguchi, T.; Kurita, N.; Tsuchiya, S.; Kimata, M.; Konoike, T.; Kobayashi, A.; et al. Orbital effect on FFLO phase and energy dissipation due to vortex dynamics in magnetic-field-induced superconductor λ-$BETS_2FeCl_4$. *J. Phys. Soc. Jpn.* **2013**, *82*, 034715. [CrossRef]
515. Lyubovskaya, R.; Zhilyaeva, E.; Shilov, G.; Audouard, A.; Vignolles, D.; Canadell, E.; Pesotskii, S.; Lyubovskii, R. Dual-layered quasi-two-dimensional organic conductors with presumable incoherent electron transport. *Eur. J. Inorg. Chem.* **2014**, 3820–3836. [CrossRef]
516. Kushch, N.D.; Buravov, L.I.; Kushch, P.P.; Shilov, G.V.; Yamochi, H.; Ishikawa, M.; Otsuka, A.; Shakin, A.A.; Maximova, O.V.; Volkova, O.S.; et al. Multifunctional compound combining conductivity and single-molecule magnetism in the same temperature range. *Inorg. Chem.* **2018**, *57*, 2386–2389. [CrossRef] [PubMed]
517. Fukuoka, S.; Fukuchi, S.; Akutsu, H.; Kawamoto, A.; Nakazawa, Y. Magnetic and electronic properties of *π-d* interacting molecular magnetic superconductor κ-$(BETS)_2FeX_4$ (X = Cl, Br) studied by angle-resolved heat capacity measurements. *Crystals* **2019**, *9*, 66. [CrossRef]
518. Ramazashvili, R.; Grigoriev, P.D.; Helm, T.; Kollmannsberger, F.; Kunz, M.; Biberacher, W.; Kampert, E.; Fujiwara, H.; Erb, A.; Wosnitza, J.; et al. Experimental evidence for Zeeman spin-orbit coupling in layered antiferromagnetic conductors. *Npj. Quantum Mater.* **2021**, *6*, 11. [CrossRef]
519. Naito, T.; Kakizaki, A.; Wakeshima, M.; Hinatsu, Y.; Inabe, T. Photochemical modification of magnetic properties in organic low-dimensional conductors. *J. Solid State Chem.* **2009**, *182*, 2733–2742. [CrossRef]
520. Naito, T. Development of control method of conduction and magnetism in molecular crystals. *Bull. Chem. Soc. Jpn.* **2017**, *90*, 89–136. [CrossRef]
521. Coniglio, W.A.; Winter, L.E.; Cho, K.; Agosta, C.C.; Fravel, B.; Montgomery, L.K. Superconducting phase diagram and FFLO signature in λ-$BETS_2GaCl_4$ from rf penetration depth measurements. *Phys. Rev. B* **2011**, *83*, 224507. [CrossRef]

522. Agosta, C.C.; Jin, J.; Coniglio, W.A.; Smith, B.E.; Cho, K.; Stroe, I.; Martin, C.; Tozer, S.W.; Murphy, T.P.; Palm, E.C.; et al. Experimental and semiempirical method to determine the Pauli-limiting field in quasi-two-dimensional superconductors as applied to κ-(BEDT-TTF)$_2$Cu(NCS)$_2$: Strong evidence of a FFLO state. *Phys. Rev. B* **2012**, *85*, 214514. [CrossRef]
523. Koutroulakis, G.; Kühne, H.; Schlueter, J.A.; Wosnitza, J.; Brown, S.E. Microscopic study of the Fulde-Ferrell-Larkin-Ovchinnikov state in an all-organic superconductor. *Phys. Rev. Lett.* **2016**, *116*, 067003. [CrossRef] [PubMed]
524. Uji, S.; Iida, Y.; Sugiura, S.; Isono, T.; Sugii, K.; Kikugawa, N.; Terashima, T.; Yasuzuka, S.; Akutsu, H.; Nakazawa, Y.; et al. Ferrell-Larkin-Ovchinnikov superconductivity in the layered organic superconductor β″-(BEDT-TTF)$_4$[(H$_3$O)Ga(C$_2$O$_4$)$_3$]C$_6$H$_5$NO$_2$. *Phys. Rev. B* **2018**, *97*, 144505. [CrossRef]
525. Akutsu, H.; Yamada, J.-I.; Nakasuji, S.; Turner, S.S. A novel BEDT-TTF-based purely organic magnetic conductor, α-(BEDT-TTF)$_2$(TEMPO-N(CH$_3$)COCH$_2$SO$_3$)·3H$_2$O. *Solid State Commun.* **2006**, *140*, 256–260. [CrossRef]
526. Akutsu, H.; Ohnishi, R.; Yamada, J.-I.; Nakasuji, S.; Turner, S.S. Novel bis(ethylenedithio)tetrathiafulvalene-based organic conductor with 1,1′-ferrocenedisulfonate. *Inorg. Chem.* **2007**, *46*, 8472–8474. [CrossRef]
527. Akutsu, H.; Yamashita, S.; Yamada, J.-I.; Nakasuji, S.; Hosokoshi, Y.; Turner, S.S. A purely organic paramagnetic metal, κ-β″-(BEDT-TTF)$_2$(PO-CONHC$_2$H$_4$SO$_3$), where PO = 2,2,5,5-tetramethyl-3-pyrrolin-1-oxyl free radical. *Chem. Mater.* **2011**, *23*, 762–764. [CrossRef]
528. Akutsu, H.; Yamada, J.-I.; Nakasuji, S. New BEDT-TTF-based organic conductor including an organic anion derived from the TEMPO radical, α-(BEDT-TTF)$_3$(TEMPO-NHCOCH$_2$SO$_3$)$_2$·6H$_2$O. *Chem. Lett.* **2003**, *32*, 1118–1119. [CrossRef]
529. Blanchard, P.; Boubekeur, K.; Sallé, M.; Duguay, G.; Jubault, M.; Gorgues, A.; Martin, J.D.; Canadell, E.; Auban-Senzier, P.; Jérome, D.; et al. A construction principle of the κ-phase based on the efficient (O–H)$_{donor}$···O$_{anion}$ structural functionality: The examples of κ-(EDT-TTF(CH$_2$OH))$_2$X (X = ClO$_4^-$ and ReO$_4^-$). *Adv. Mater.* **1992**, *4*, 579–581. [CrossRef]
530. Pénicaud, A.; Boubekeur, K.; Batail, P.; Canadell, E.; Auban-Senzier, P.; Jérome, D. Hydrogen-bond tuning of macroscopic transport properties from the neutral molecular component site along the series of metallic organic-inorganic solvates (BEDT-TTF)$_4$Re$_6$Se$_5$Cl$_9$[guest], [guest = DMF, THF, dioxane]. *J. Am. Chem. Soc.* **1993**, *115*, 4101–4112. [CrossRef]
531. Imakubo, T.; Sawa, H.; Kato, R. Synthesis and crystal structure of the molecular metal based on iodine-bonded π-donor, (IEDT)[Pd(dmit)$_2$]. *J. Chem. Soc. Chem. Commun.* **1995**, 1097–1098. [CrossRef]
532. Imakubo, T.; Sawa, H.; Kato, R. Novel molecular conductors, (DIETS)$_4$M(CN)$_4$ (M = Ni, Pd, Pt): Highly reticulated donor··· anion contacts by –I··· NC–interaction. *J. Chem. Soc. Chem. Commun.* **1995**, 1667–1668. [CrossRef]
533. Naito, T.; Kobayashi, N.; Inabe, T. Synthesis of new Ni-complexes with a chalcogen donor ligand and cyano groups. *Chem. Lett.* **1998**, *27*, 723–724. [CrossRef]
534. Yamamoto, H.M.; Yamaura, J.-I.; Kato, R. Multicomponent molecular conductors with supramolecular assembly: Iodine-containing neutral molecules as building blocks. *J. Am. Chem. Soc.* **1998**, *120*, 5905–5913. [CrossRef]
535. Heuzé, K.; Fourmigué, M.; Batail, P.; Canadell, E.; Auban-Senzier, P. Directing the structures and collective electronic properties of organic conductors: The interplay of π-overlap interactions and hydrogen bonds. *Chem. Eur. J.* **1999**, *5*, 2971–2976. [CrossRef]
536. Dautel, O.J.; Fourmigué, M. Fluorinated tetrathiafulvalenes with preserved electron-donor properties and segregated fluorous bilayer structures based on F···F nonbonded interactions. *J. Org. Chem.* **2000**, *65*, 6479–6486. [CrossRef] [PubMed]
537. Domercq, B.; Devic, T.; Fourmigué, M.; Auban-Senzier, P.; Canadell, E. Hal···Hal interactions in a series of three isostructural salts of halogenated tetrathiafulvalenes. Contribution of the halogen atoms to the HOMO-HOMO overlap interactions. *J. Mater. Chem.* **2001**, *11*, 1570–1575. [CrossRef]
538. Fourmigué, M.; Batail, P. Activation of hydrogen- and halogen-bonding interactions in tetrathiafulvalene-based crystalline molecular conductors. *Chem. Rev.* **2004**, *104*, 5379–5418. [CrossRef]
539. Imakubo, T.; Shirahata, T.; Hervé, K.; Ouahab, L. Supramolecular organic conductors based on diiodo-TTFs and spherical halide ion X$^-$ (X = Cl, Br). *J. Mater. Chem.* **2006**, *16*, 162–173. [CrossRef]
540. Imakubo, T.; Shirahata, T.; Kibune, M.; Yoshino, H. Hybrid organic/inorganic supramolecular conductors D$_2$[Au(CN)$_4$] [D = diiodo(ethylenedichalcogeno)tetrachalcogenofulvalene], including a new ambient pressure superconductor. *Eur. J. Inorg. Chem.* **2007**, 4727–4735. [CrossRef]
541. Réthoré, C.; Madalan, A.; Fourmigué, M.; Canadell, E.; Lopes, E.B.; Almeida, M.; Clérac, R.; Avarvari, N. O··· S vs. N··· S intramolecular nonbonded interactions in neutral and radical cation salts of TTF-oxazoline derivatives: Synthesis, theoretical investigations, crystalline structures, and physical properties. *New J. Chem.* **2007**, *31*, 1468–1483. [CrossRef]
542. Fourmigué, M. Halogen bonding in conducting or magnetic molecular materials. In *Structure and Bonding*; Metrangolo, P., Resnati, G., Eds.; Springer: Berlin/Heidelberg, Germany, 2008; Volume 126, pp. 181–207. [CrossRef]
543. Fourmigué, M. Halogen bonding: Recent advances. *Curr. Opin. Solid State Mater. Sci.* **2009**, *13*, 36–45. [CrossRef]
544. Lieffrig, J.; Jeannin, O.; Frąckowiak, A.; Olejniczak, I.; Świetlik, R.; Dahaoui, S.; Aubert, E.; Espinosa, E.; Auban-Senzier, P.; Fourmigué, M. Charge-assisted halogen bonding: Donor-acceptor complexes with variable ionicity. *Chem. Eur. J.* **2013**, *19*, 14804–14813. [CrossRef] [PubMed]
545. Brezgunova, M.E.; Lieffrig, J.; Aubert, E.; Dahaoui, S.; Fertey, P.; Lebègue, S.; Ángyán, J.G.; Fourmigué, M.; Espinosa, E. Chalcogen bonding: Experimental and theoretical determinations from electron density analysis. Geometrical preferences driven by electrophilic-nucleophilic interactions. *Cryst. Growth Des.* **2013**, *13*, 3283–3289. [CrossRef]
546. Huynh, H.-T.; Jeannin, O.; Fourmigué, M. Organic selenocyanates as strong and directional chalcogen bond donors for crystal engineering. *Chem. Commun.* **2017**, *53*, 8467–8469. [CrossRef]

547. Coronado, E.; Galán Mascarós, J.R.; Giménez-López, M.C.; Almeida, M.; Waerenborgh, J.C. Spin crossover Fe^{II} complexes as templates for bimetallic oxalate-based 3D magnets. *Polyhedron* **2007**, *26*, 1838–1844. [CrossRef]
548. Zhang, G.; Jin, L.; Zhang, R.; Bai, Y.; Zhu, R.; Pang, H. Recent advances in the development of electronically and ionically conductive metal-organic frameworks. *Coord. Chem. Rev.* **2021**, *439*, 213915. [CrossRef]
549. Nohr, R.S.; Kuznesof, P.M.; Kenney, M.E.; Siebenmanu, P.G.; Wynne, K.J. Highly conducting linear stacked polymers: Iodine-doped fluoroaluminum and fluorogallium phthalocyanines. *J. Am. Chem. Soc.* **1981**, *103*, 4371–4377. [CrossRef]
550. Metz, J.; Hanack, M. Synthesis, characterization, and conductivity of (μ-Cyano)(phthalocyaninato)cobalt(III). *J. Am. Chem. Soc.* **1983**, *105*, 828–830. [CrossRef]
551. Dirk, C.W.; Inabe, T.; Schoch, K.F.; Marks, T.J. Cofacial assembly of partially oxidized metallomacrocycles as an approach to controlling lattice architecture in low-dimensional molecular solids. chemical and architectural properties of the "face-to-face" polymers [M(phthalocyaninato)0], where M = Si, Ge, and Sn. *J. Am. Chem. Soc.* **1983**, *105*, 1539–1550. [CrossRef]
552. Hanack, M.; Deger, S.; Lange, A. Bisaxially coordinated macrocyclic transition metal complexes. *Coord. Chem. Rev.* **1988**, *83*, 115–136. [CrossRef]
553. Andre, J.-J.; Holczer, K.; Petit, P.; Riou, M.-T.; Clarisse, C.; Even, R.; Fourmigue, M.; Simon, J. Electrical and magnetic properties of thin films and single crystals of bis(phthalocyaninato)lutetium. *Chem. Phys. Lett.* **1985**, *115*, 463–466. [CrossRef]
554. Kennedy, B.J.; Murray, K.S.; Zwack, P.R.; Homborg, H.; Kalz, W. Spin states in iron(III) phthalocyanines studied by Mössbauer, magnetic susceptibility, and ESR measurements. *Inorg. Chem.* **1986**, *25*, 2539–2545. [CrossRef]
555. Yakushi, K.; Sakuda, M.; Hamada, I.; Kuroda, H.; Kawamoto, A.; Tanaka, J.; Sugano, T.; Kinoshita, M. Preparation, structure and properties of metallic (phthalocyanato)nickel salts. *Synth. Met.* **1987**, *19*, 769–774. [CrossRef]
556. Ogawa, M.Y.; Martinsen, J.; Palmer, S.M.; Stanton, J.L.; Tanaka, J.; Greene, R.L.; Hoffman, B.M.; Ibers, J.A. Cu(pc)I: A molecular metal with a one-dimensional array of local moments embedded in a "Fermi Sea" of charge carriers. *J. Am. Chem. Soc.* **1987**, *109*, 1115–1121. [CrossRef]
557. Inabe, T.; Maruyama, Y. Multi-dimensional stacking structures in phthalocyanine-based electrical conductors, K[Co(phthalocyaninato)$(CN)_2]_2{\cdot}5CH_3CN$ and Co(phthalocyaninato)$(CN)_2{\cdot}2H_2O$. *Bull. Chem. Soc. Jpn.* **1990**, *63*, 2273–2280. [CrossRef]
558. Hasegawa, H.; Naito, T.; Inabe, T.; Akutagawa, T.; Nakamura, T. A highly conducting partially oxidized salt of axially substituted phthalocyanine. Structure and physical properties of TPP[Co(Pc)$(CN)_2]_2$ {TPP = tetraphenylphosphonium, [Co(Pc)$(CN)_2$] = dicyano(phthalocyaninate)cobalt(III)}. *J. Mater. Chem.* **1998**, *8*, 1567–1570. [CrossRef]
559. Matsuda, M.; Naito, T.; Inabe, T.; Hanasaki, N.; Tajima, H.; Otsuka, T.; Awaga, K.; Narymbetov, B.; Kobayashi, H. A one-dimensional macrocyclic π-ligand conductor carrying a magnetic center. Structure and electrical, optical and magnetic properties of TPP[Fe(Pc)$(CN)_2]_2$ {TPP = tetraphenylphosphonium and [Fe(Pc)$(CN)_2$] = dicyano(phthalocyaninato)iron(III)}. *J. Mater. Chem.* **2000**, *10*, 631–636. [CrossRef]
560. Hanasaki, N.; Tajima, H.; Matsuda, M.; Naito, T.; Inabe, T. Giant negative magnetoresistance in quasi-one-dimensional conductor TPP[Fe(Pc)$(CN)_2]_2$: Interplay between local moments and one-dimensional conduction electrons. *Phys. Rev. B* **2000**, *62*, 5839–5842. [CrossRef]
561. Matsuda, M.; Naito, T.; Inabe, T.; Hanasaki, N.; Tajima, H. Structure and electrical and magnetic properties of $(PTMA)_x$[M(Pc)$(CN)_2$] y(solvent) (PTMA = phenyltrimethylammonium and [M(Pc)$(CN)_2$] = dicyano(phthalocyaninato)M^{III} with M = Co and Fe). Partial oxidation by partial solvent occupation of the cationic site. *J. Mater. Chem.* **2001**, *11*, 2493–2497. [CrossRef]
562. Tajima, H.; Hanasaki, N.; Matsuda, M.; Sakai, F.; Naito, T.; Inabe, T. Magnetoresistance study on TPP[*M*(Pc)$(CN)_2]_2$ (*M* = Fe, Co, $Fe_{0.30}Co_{0.70}$) salts. *J. Solid State Chem.* **2002**, *168*, 509–513. [CrossRef]
563. Hanasaki, N.; Matsuda, M.; Tajima, H.; Naito, T.; Inabe, T. Contribution of degenerate molecular orbitals to molecular orbital angular momentum in molecular magnet Fe(Pc)$(CN)_2$. *J. Phys. Soc. Jpn.* **2003**, *72*, 3226–3230. [CrossRef]
564. Inabe, T.; Tajima, H. Phthalocyanines—Versatile components of molecular conductors. *Chem. Rev.* **2004**, *104*, 5503–5533. [CrossRef] [PubMed]
565. Matsuda, M.; Hanasaki, N.; Tajima, H.; Naito, T.; Inabe, T. Anisotropic giant magnetoresistance originating from the π-d interaction in a molecule. *J. Phys. Chem. Solids* **2004**, *65*, 749–752. [CrossRef]
566. Hotta, C.; Ogata, M.; Fukuyama, H. Interaction of the ground state of quarter-filled one-dimensional strongly correlated electronic system with localized spins. *Phys. Rev. Lett.* **2005**, *95*, 216402. [CrossRef]
567. Hanasaki, N.; Matsuda, M.; Tajima, H.; Ohmichi, E.; Osada, T.; Naito, T.; Inabe, T. Giant negative magnetoresistance reflecting molecular symmetry in dicyano(phthalocyaninato)iron compounds. *J. Phys. Soc. Jpn.* **2006**, *75*, 033703. [CrossRef]
568. Hanasaki, N.; Masuda, K.; Kodama, K.; Matsuda, M.; Tajima, H.; Yamazaki, J.; Takigawa, M.; Yamaura, J.; Ohmichi, E.; Osada, T.; et al. Charge disproportionation in highly one-dimensional molecular conductor TPP[Co(Pc)$(CN)_2]_2$]. *J. Phys. Soc. Jpn.* **2006**, *75*, 104713. [CrossRef]
569. Tajima, H.; Yoshida, G.; Matsuda, M.; Nara, K.; Kajita, K.; Nihsio, Y.; Hanasaki, N.; Naito, T.; Inabe, T. Magnetic torque and heat capacity measurements on TPP[Fe(Pc)$(CN)_2]_2$. *Phys. Rev. B Cond. Matter Mater. Phys.* **2008**, *78*, 064424. [CrossRef]
570. Yu, D.E.C.; Matsuda, M.; Tajima, H.; Kikuchi, A.; Taketsugu, T.; Hanasaki, N.; Naito, T.; Inabe, T. Variable magnetotransport properties in the TPP[Fe(Pc)$L_2]_2$ system (TPP = tetraphenylphosphonium, Pc = phthalocyaninato, L = CN, Cl, and Br). *J. Mater. Chem.* **2009**, *19*, 718–723. [CrossRef]

571. Tajima, H.; Yoshida, G.; Matsuda, M.; Yamaura, J.-I.Y.; Hanasaki, N.; Naito, T.; Inabe, T. Magnetic torque and ac and dc magnetic susceptibility measurements on $PTMA_{0.5}[Fe(Pc)(CN)_2]CH_3CN$: Origin of spontaneous magnetization in $[Fe(Pc)(CN)_2]$ molecular conductors. *Phys. Rev. B Cond. Matter Mater. Phys.* **2009**, *80*, 024424. [CrossRef]
572. Kimata, M.; Takahide, Y.; Harada, A.; Satsukawa, H.; Hazama, K.; Terashima, T.; Uji, S.; Naito, T.; Inabe, T. Interplay between magnetism and conductivity in the one-dimensional organic conductor $TPP[Fe(Pc)(CN)_2]_2$. *Phys. Rev. B* **2009**, *80*, 085110. [CrossRef]
573. Ishikawa, M.; Yamashita, S.; Naito, T.; Matsuda, M.; Tajima, H.; Hanasaki, N.; Akutagawa, T.; Nakamura, T.; Inabe, T. Non-linear transport phenomena in highly one-dimensional $M^{III}(Pc)(CN)_2$ chains with π-d interaction (M = Co and Fe and Pc = phthalocyaninato). *J. Phys. Soc. Jpn.* **2009**, *78*, 104709. [CrossRef]
574. Hotta, C. Interplay of strongly correlated electrons and localized Ising moments in one dimension. *Phys. Rev. B* **2010**, *81*, 245104. [CrossRef]
575. Otsuka, Y.; Seo, H.; Motome, Y. Charge ordering due to π-d coupling in one-dimensional system. *Phys. B Cond. Mat.* **2010**, *405*, S317–S320. [CrossRef]
576. Ishikawa, M.; Asari, T.; Matsuda, M.; Tajima, H.; Hanasaki, N.; Naito, T.; Inabe, T. Giant magnetoresistance response by the π–d interaction in an axially ligated phthalocyanine conductor with two-dimensional π–π stacking structure. *J. Mater. Chem.* **2010**, *20*, 4432–4438. [CrossRef]
577. Hanasaki, N.; Tateishi, T.; Tajima, H.; Kimata, M.; Tokunaga, M.; Matsuda, M.; Kanda, A.; Muralawa, H.; Naito, T.; Inabe, T. Metamagnetic transition and its related magnetocapacitance effect in phthalocyanine-molecular conductor exhibiting giant magnetoresistance. *J. Phys. Soc. Jpn.* **2013**, *82*, 094713. [CrossRef]
578. Torizuka, K.; Tajima, H.; Inoue, M.; Hanasaki, N.; Matsuda, M.; Yu, D.E.C.; Naito, T.; Inabe, T. Magnetic torque experiments on $TPP[Fe(Pc)L_2]_2$ (L = Br and Cl): Antiferromagnetic short range ordering of *d* electrons, antiferromagnetic ordering of π electrons, and the anisotropy energy. *J. Phys. Soc. Jpn.* **2013**, *82*, 034719. [CrossRef]
579. Peierls, R.E. *Quantum Theory of Solids*; Clarendon Press: Oxford, UK, 1955; pp. 101–114.
580. Soos, Z.G.; Mazumdar, S. Neutral-ionic interface in organic charge-transfer salts. *Phys. Rev. B* **1978**, *18*, 1991–2003. [CrossRef]
581. Soos, Z.G.; Bondeson, S.R.; Mazumdar, S. Magnetic analog of mott transition. *Chem. Phys. Lett.* **1979**, *65*, 331–334. [CrossRef]
582. Torrance, J.B.; Vazquez, J.E.; Mayerle, J.J.; Lee, V.Y. Discovery of a neutral-to-ionic phase transition in organic materials. *Phys. Rev. Lett.* **1981**, *46*, 253–257. [CrossRef]
583. Torrance, J.B.; Girlando, A.; Mayerle, J.J.; Crowley, J.I.; Lee, V.Y.; Batail, P.; LaPlaca, S.J. Anomalous nature of neutral-to-ionic phase transition in tetrathiafulvalene chloranil. *Phys. Rev. Lett.* **1981**, *47*, 1747–1750. [CrossRef]
584. Mazumdar, S.; Soos, Z.G. Valence-bond analysis of extended Hubbard models: Charge-transfer excitations of molecular conductors. *Phys. Rev. B* **1981**, *23*, 2810–2823. [CrossRef]
585. Girlando, A.; Marzola, F.; Pecile, C.; Torrance, J.B. Vibrational spectroscopy of mixed stack organic semiconductors: Neutral and ionic phases of tetrathiafulvalene-chloranil (TTF-CA) charge transfer complex. *J. Chem. Phys.* **1983**, *79*, 1075–1085. [CrossRef]
586. Mazumdar, S.; Dixit, S.N. Coulomb effects on one-dimensional Peierls instability: The Peierls-Hubbard model. *Phys. Rev. Lett.* **1983**, *51*, 292–295. [CrossRef]
587. Mazumdar, S.; Bloch, A.N. Systematic trends in short-range coulomb effects among nearly one-dimensional organic conductors. *Phys. Rev. Lett.* **1983**, *50*, 207–211. [CrossRef]
588. Mazumdar, S.; Dixit, S.N.; Bloch, A.N. Correlation effects on charge-density waves in narrow-band one-dimensional conductors. *Phys. Rev. B* **1984**, *30*, 4842–4845. [CrossRef]
589. Dixit, S.N.; Mazumdar, S. Electron-electron interaction effects on Peierls dimerization in a half-filled band. *Phys. Rev. B* **1984**, *29*, 1824–1839. [CrossRef]
590. Painelli, A.; Girlando, A. Electron-molecular vibration (e-mv) coupling in charge-transfer compounds and its consequences on the optical spectra: A theoretical framework. *J. Chem. Phys.* **1985**, *84*, 5655–5671. [CrossRef]
591. Mazumdar, S.; Dixit, S.N. Unified theory of segregated-stack organic charge-transfer solids: Magnetic properties. *Phys. Rev. B* **1986**, *34*, 3683–3699. [CrossRef]
592. Ung, K.C.; Mazumdar, S.; Toussaint, D. Metal-insulator and insulator-insulator transitions in the quarter-filled band organic conductors. *Phys. Rev. Lett.* **1994**, *73*, 2603–2606. [CrossRef]
593. Mazumdar, S.; Ramasesha, S.; Clay, R.T.; Campbell, D.K. Theory of coexisting charge- and spin-density waves in $(TMTTF)_2Br$, $(TMTSF)_2PF_6$ and α-$(BEDT\text{-}TTF)_2MHg(SCN)_4$. *Phys. Rev. Lett.* **1999**, *82*, 1522–1525. [CrossRef]
594. Girlando, A.; Masino, M.; Visentini, G.; della Valle, R.G.; Brillante, A.; Venuti, E. Lattice dynamics and electron-phonon coupling in the β-$(BEDT\text{-}TTF)_2I_3$ organic superconductor. *Phys. Rev. B* **2000**, *62*, 14476–14486. [CrossRef]
595. Moser, J.; Cooper, J.R.; Jérome, D.; Alavi, B.; Brown, S.E.; Bechgaard, K. Hall effect in the normal phase of the organic superconductor $(TMTSF)_2PF_6$. *Phys. Rev. Lett.* **2000**, *84*, 2674–2677. [CrossRef]
596. Chow, D.S.; Zamborszky, F.; Alavi, B.; Tantillo, D.J.; Baur, A.; Merlic, C.A.; Brown, S.E. Charge ordering in the TMTTF family of molecular conductors. *Phys. Rev. Lett.* **2000**, *85*, 1698–1701. [CrossRef]
597. Girlando, A.; Masino, M.; Brillante, A.; Della Valle, R.G.; Venuti, E. BEDT-TTF organic superconductors: The role of phonons. *Phys. Rev. B* **2002**, *66*, 100507. [CrossRef]
598. Zamborszky, F.; Yu, W.; Raas, W.; Brown, S.E.; Alavi, B.; Merlic, C.A.; Baur, A. Competition and coexistence of bond and charge orders in $(TMTTF)_2AsF_6$. *Phys. Rev. B* **2002**, *66*, 081103. [CrossRef]

599. Lee, I.J.; Brown, S.E.; Clark, W.G.; Strouse, M.J.; Naughton, M.J.; Kang, W.; Chaikin, P.M. Triplet superconductivity in an organic superconductor probed by NMR Knight shift. *Phys. Rev. Lett.* **2002**, *88*, 017004. [CrossRef] [PubMed]
600. Clay, R.T.; Mazumdar, S.; Campbell, D.K. Pattern of charge ordering in quasi-one-dimensional organic charge-transfer solids. *Phys. Rev. B* **2003**, *67*, 115121. [CrossRef]
601. Dressel, M.; Drichko, N. Optical properties of two-dimensional organic conductors: Signatures of charge ordering and correlation effects. *Cherm. Rev.* **2004**, *104*, 5689–5715. [CrossRef]
602. Yu, W.; Zhang, F.; Zamborszky, F.; Alavi, B.; Baur, A.; Merlic, C.A.; Brown, S.E. Electron-lattice coupling and broken symmetries of the molecular salt $(TMTTF)_2SbF_6$. *Phys. Rev. B* **2004**, *70*, 121101. [CrossRef]
603. Yamamoto, T.; Yakushi, K.; Shimizu, Y.; Saito, G. Infrared and Raman study of the charge-ordered state of θ-$(ET)_2Cu_2CN[N(CN)_2]_2$. *J. Phys. Soc. Jpn.* **2004**, *73*, 2326–2332. [CrossRef]
604. Yamamoto, T.; Uruichi, M.; Yamamoto, K.; Yakushi, K.; Kawamoto, A.; Taniguchi, H. Examination of the charge-sensitive vibrational modes in bis(ethylenedithio)tetrathiafulvalene. *J. Phys. Chem. B* **2005**, *109*, 15226–15235. [CrossRef] [PubMed]
605. Bangura, A.F.; Coldea, A.I.; Singleton, J.; Ardavan, A.; Akutsu-Sato, A.; Akutsu, H.; Turner, S.S.; Day, P.; Yamamoto, T.; Yakushi, K. Robust superconducting state in the low-quasiparticle-density organic metals β''-$(BEDT\text{-}TTF)_4[(H_3O)M(C_2O_4)_3]Y$: Superconductivity due to proximity to a charge-ordered state. *Phys. Rev. B* **2005**, *72*, 014543. [CrossRef]
606. Seo, H.; Merino, J.; Yoshioka, H.; Ogata, M. Theoretical aspects of charge ordering in molecular conductors. *J. Phys. Soc. Jpn.* **2006**, *75*, 051009. [CrossRef]
607. Drichko, N.; Dressel, M.; Kuntscher, C.A.; Pashkin, A.; Greco, A.; Merino, J.; Schlueter, J. Electronic properties of correlated metals in the vicinity of a charge-order transition: Optical spectroscopy of α-$(BEDT\text{-}TTF)_2MHg(SCN)_4$ ($M = NH_4$, Rb, Tl). *Phys. Rev. B* **2006**, *74*, 235121. [CrossRef]
608. Merino, J.; Greco, A.; Drichko, N.; Dressel, M. Non-Fermi liquid behavior in nearly charge ordered layered metals. *Phys. Rev. Lett.* **2006**, *96*, 216402. [CrossRef]
609. Dressel, M. Ordering phenomena in quasi-one-dimensional organic conductors. *Naturwissenschaften* **2007**, *94*, 527–541. [CrossRef] [PubMed]
610. De Souza, M.; Foury-Leylekian, P.; Moradpour, A.; Pouget, J.-P.; Lang, M. Evidence for lattice effects at the charge-ordering transition in $(TMTTF)_2X$. *Phys. Rev. Lett.* **2008**, *101*, 216403. [CrossRef]
611. Yamamoto, T.; Yamamoto, H.M.; Kato, R.; Uruichi, M.; Yakushi, K.; Akutsu, H.; Sato-Akutsu, A.; Kawamoto, A.; Turner, S.S.; Day, P. Inhomogeneous site charges at the boundary between the insulating, superconducting, and metallic phases of β''-type bis-ethylenedithio-tetrathiafulvalene molecular charge-transfer salts. *Phys. Rev. B* **2008**, *77*, 205120. [CrossRef]
612. Drichko, N.; Kaiser, S.; Sun, Y.; Clauss, C.; Dressel, M.; Mori, H.; Schlueter, J.; Zhyliaeva, E.I.; Torunova, S.A.; Lyubovskaya, R.N. Evidence for charge order in organic superconductors obtained by vibrational spectroscopy. *Phys. B Cond. Mat.* **2009**, *404*, 490–493. [CrossRef]
613. Sawa, H.; Kakiuchi, T. Study of the novel charge ordering state in molecular conducting using synchrotron radiation X-ray diffraction. In *Molecular Electronic and Related Materials—Control and Probe with Light*; Naito, T., Ed.; Transworld Research Network: Kerala, India, 2010; pp. 117–134.
614. Girlando, A. Charge sensitive vibrations and electron-molecular vibration coupling in bis(ethylenedithio)-tetrathiafulvalene (BEDT-TTF). *J. Phys. Chem. C* **2011**, *115*, 19371–19378. [CrossRef]
615. Dayal, S.; Clay, R.T.; Li, H.; Mazumdar, S. Paired electron crystal: Order from frustration in the quarter-filled band. *Phys. Rev. B* **2011**, *83*, 245106. [CrossRef]
616. Yoshimi, K.; Seo, H.; Ishibashi, S.; Brown, S.E. Tuning the magnetic dimensionality by charge ordering in the molecular TMTTF salts. *Phys. Rev. Lett.* **2012**, *108*, 096402. [CrossRef]
617. Sedlmeier, K.; Elsässer, S.; Neubauer, D.; Beyer, R.; Wu, D.; Ivek, T.; Tomić, S.; Schlueter, J.A.; Dressel, M. Absence of charge order in the dimerized κ-phase BEDT-TTF salts. *Phys. Rev. B* **2012**, *86*, 245103. [CrossRef]
618. Dressel, M.; Dumm, M.; Knoblauch, T.; Masino, M. Comprehensive optical investigations of charge order in organic chain compounds $(TMTTF)_2X$. *Crystals* **2012**, *2*, 528–578. [CrossRef]
619. Girlando, A.; Masino, M.; Schlueter, J.A.; Drichko, N.; Kaiser, S.; Dressel, M. Charge-order fluctuations and superconductivity in two-dimensional organic metals. *Phys. Rev. B* **2014**, *89*, 174503. [CrossRef]
620. Drichko, N.; Beyer, R.; Rose, E.; Dressel, M.; Schlueter, J.A.; Turunova, S.A.; Zhilyaeva, E.I.; Lyubovskaya, R.N. Metallic state and charge-order metal-insulator transition in the quasi-two-dimensional conductor κ-$(BEDT\text{-}TTF)_2Hg(SCN)_2Cl$. *Phys. Rev. B* **2014**, *89*, 075133. [CrossRef]
621. Mazumdar, S.; Clay, R.T. The chemical physics of unconventional superconductivity. *Int. J. Quant. Chem.* **2014**, *114*, 1053–1059. [CrossRef]
622. Pustogow, A.; Peterseim, T.; Kolatschek, S.; Engel, L.; Dressel, M. Electronic correlations versus lattice interactions: Interplay of charge and anion orders in $(TMTTF)_2X$. *Phys. Rev. B* **2016**, *94*, 195125. [CrossRef]
623. Masino, M.; Castagnetti, N.; Girlando, A. Phenomenology of the neutral-ionic valence instability in mixed stack charge-transfer crystals. *Crystals* **2017**, *7*, 108. [CrossRef]
624. Mazumdar, S. Valence transition model of the pseudogap, charge order, and superconductivity in electron-doped and hole-doped copper oxides. *Phys. Rev. B* **2018**, *98*, 205153. [CrossRef]

625. Clay, R.T.; Mazumdar, S. From charge- and spin-ordering to superconductivity in the organic charge-transfer solids. *Phys. Rep.* **2019**, *788*, 1–89. [CrossRef]
626. Kurmoo, M.; Rosseinsky, M.J.; Day, P.; Auban, P.; Kang, W.; Jérôme, D.; Batail, P. Competition between localisation and superconductivity in (BEDT-TTF)$_3$Cl$_2$.2H$_2$O. *Synth. Met.* **1988**, *27*, A425–A431. [CrossRef]
627. Lang, M.; Toyota, N.; Sasaki, T.; Sato, H. Magnetic penetration depth of κ-(BEDT-TTF)$_2$Cu(NCS)$_2$ strong evidence for conventional cooper pairing. *Phys. Rev. Lett.* **1992**, *69*, 1443–1446. [CrossRef]
628. Lang, M.; Steglich, F.; Toyota, N.; Sasaki, T. Fluctuation effects and mixed-state properties of the layered organic superconductors κ-(BEDT-TTF)$_2$Cu(NCS)$_2$ and κ-(BEDT-TTF)$_2$Cu[N(CN)$_2$]Br. *Phys. Rev. B* **1994**, *49*, 15227–15234. [CrossRef] [PubMed]
629. Pintschovius, L.; Rietschel, H.; Sasaki, T.; Mori, H.; Tanaka, S.; Toyota, N.; Lang, M.; Steglich, F. Observation of superconductivity-induced phonon frequency changes in the organic superconductor κ-(BEDT-TTF)$_2$Cu(NCS)$_2$. *Eur. Phys. Lett.* **1997**, *37*, 627–632. [CrossRef]
630. Chow, D.S.; Wzietek, P.; Fogliatti, D.; Alavi, B.; Tantillo, D.J.; Merlic, C.A.; Brown, S.E. Singular behavior in the pressure-tuned competition between Spin-Peierls and antiferromagnetic ground states of (TMTTF)$_2$PF$_6$. *Phys. Rev. Lett.* **1998**, *81*, 3984–3987. [CrossRef]
631. Schmalian, J. Pairing due to spin fluctuations in layered organic superconductors. *Phys. Rev. Lett.* **1998**, *81*, 4232–4235. [CrossRef]
632. Kondo, H.; Moriya, T. Spin fluctuation-induced superconductivity in organic compounds. *J. Phys. Soc. Jpn.* **1998**, *67*, 3695–3698. [CrossRef]
633. Kino, H.; Kontani, H. Phase diagram of superconductivity on the anisotropic triangular lattice hubbard model: An effective model of κ-(BEDT-TTF) salts. *J. Phys. Soc. Jpn.* **1998**, *67*, 3691–3694. [CrossRef]
634. Müller, J.; Lang, M.; Steglich, F.; Schlueter, J.; Kini, A.; Geiser, U. Comparative thermal-expansion study of β''-(ET)$_2$SF$_5$CH$_2$CF$_2$SO$_3$ and κ-(ET)$_2$Cu(NCS)$_2$: Uniaxial pressure coefficients of T_C and upper critical fields. *Phys. Rev. B* **2000**, *61*, 11739–11744. [CrossRef]
635. Merino, J.; McKenzie, R.H. Superconductivity mediated by charge fluctuations in layered molecular crystals. *Phys. Rev. Lett.* **2001**, *87*, 237002. [CrossRef] [PubMed]
636. Yanase, Y.; Jujo, T.; Nomura, T.; Ikeda, H.; Hotta, T.; Yamada, K. Theory of superconductivity in strongly correlated electron systems. *Phys. Rep.* **2003**, *387*, 1–149. [CrossRef]
637. Miyagawa, K.; Kanoda, K.; Kawamoto, A. NMR studies on two-dimensional molecular conductors and superconductors: Mott transition in κ-(BEDT-TTF)$_2$X. *Chem. Rev.* **2004**, *104*, 5635–5653. [CrossRef] [PubMed]
638. Lee, I.J.; Brown, S.E.; Yu, W.; Naughton, M.J.; Chaikin, P.M. Coexistence of superconductivity and antiferromagnetism probed by simultaneous nuclear magnetic resonance and electrical transport in (TMTSF)$_2$PF$_6$ system. *Phys. Rev. Lett.* **2005**, *94*, 197001. [CrossRef]
639. Kuroki, K. Pairing symmetry competition in organic superconductors. *J. Phys. Soc. Jpn.* **2006**, *75*, 051013. [CrossRef]
640. Nam, M.-S.; Ardavan, A.; Blundell, S.J.; Schlueter, J.A. Fluctuating superconductivity in organic molecular metals close to the Mott transition. *Nature* **2007**, *449*, 584–587. [CrossRef]
641. Gantmakher, V.F.; Dolgopolov, V.T. Superconductor-insulator quantum phase transition. *Phys. Uspekhi* **2010**, *53*, 1–49. [CrossRef]
642. Dressel, M. Quantum criticality in organic conductors? Fermi liquid versus non-Fermi-liquid behavior. *J. Phys. Cond. Mat.* **2011**, *23*, 293201. [CrossRef]
643. Müller, J. Fluctuation spectroscopy: A new approach for studying low-dimensional molecular metals. *ChemPhysChem* **2011**, *12*, 1222–1245. [CrossRef] [PubMed]
644. Wright, J.A.; Green, E.; Kuhns, P.; Reyes, A.; Brooks, J.; Schlueter, J.; Kato, R.; Yamamoto, H.; Kobayashi, M.; Brown, S.E. Zeeman-driven phase transition within the superconducting state of κ-(BEDT-TTF)$_2$Cu(NCS)$_2$. *Phys. Rev. Lett.* **2011**, *107*, 087002. [CrossRef]
645. Agosta, C.C. Inhomogeneous superconductivity in organic and related superconductors. *Crystals* **2018**, *8*, 285. [CrossRef]
646. Pustogow, A.; Saito, Y.; Rohwer, A.; Schlueter, J.A.; Dressel, M. Coexistence of charge order and superconductivity in β″-(BEDT-TTF)$_2$SF$_5$CH$_2$CF$_2$SO$_3$. *Phys. Rev. B* **2019**, *99*, 140509. [CrossRef]
647. Tomić, S.; Jérome, D. A hidden low-temperature phase in the organic conductor (TMTSF)$_2$ReO$_4$. *J. Phys. Condens. Matter.* **1989**, *1*, 4451–4456. [CrossRef]
648. Kagawa, F.; Oike, H. Quenching of charge and spin degrees of freedom in condensed matter. *Adv. Mater.* **2017**, *29*, 1601979. [CrossRef]
649. Mori, H.; Tanaka, S.; Mori, T. Systematic study of the electronic state in θ-type BEDT-TTF organic conductors by changing the electronic correlation. *Phys. Rev. B* **1998**, *57*, 12023–12029. [CrossRef]
650. Yamada, J.-I.; Akutsu, H.; Nishikawa, H.; Kikuchi, K. New trends in the synthesis of π-electron donors for molecular conductors and superconductors. *Chem. Rev.* **2004**, *104*, 5057–5083. [CrossRef] [PubMed]
651. Yamada, J.-I. New approach to the achievement of organic superconductivity. *J. Mater. Chem.* **2004**, *14*, 2951–2953. [CrossRef]
652. Faltermeier, D.; Barz, J.; Dumm, M.; Dressel, M.; Drichko, N.; Petrov, B.; Semkin, V.; Vlasova, R.; Meźière, C.; Batail, P. Bandwidth-controlled Mott transition in κ-(BEDT-TTF)$_2$Cu[N(CN)$_2$]Br$_x$Cl$_{1-x}$: Optical studies of localized charge excitations. *Phys. Rev. B* **2007**, *76*, 165113. [CrossRef]
653. Dumm, M.; Faltermeier, D.; Drichko, N.; Dressel, M.; Mézière, C.; Batail, P. Bandwidth-controlled Mott transition in κ-(BEDT-TTF)$_2$Cu[N(CN)$_2$]Br$_x$Cl$_{1-x}$: Optical studies of correlated carriers. *Phys. Rev. B* **2009**, *79*, 195106. [CrossRef]

654. Kaiser, S.; Dressel, M.; Sun, Y.; Greco, A.; Schlueter, J.A.; Gard, G.L.; Drichko, N. Bandwidth tuning triggers interplay of charge order and superconductivity in two-dimensional organic materials. *Phys. Rev. Lett.* **2010**, *105*, 206402. [CrossRef]
655. Yamada, J.-I.; Akutsu, H. Chemical modifications of BDH-TTP [2,5-bis(1,3-dithiolan-2-ylidene)-1,3,4,6-tetrathiapentalene]: Control of electron correlation. *Crystals* **2012**, *2*, 812–844. [CrossRef]
656. Chamberlin, R.V.; Naughton, M.J.; Yan, X.; Chiang, L.Y.; Hsu, S.-Y.; Chaikin, P.M. Extreme quantum limit in a quasi-two-dimensional organic conductor. *Phys. Rev. Lett.* **1988**, *60*, 1189–1192. [CrossRef]
657. Cooper, J.R.; Kang, W.; Auban, P.; Montambaux, G.; Jérome, D.; Bechgaard, K. Quantized Hall effect and a new field-induced phase transition in the organic superconductor $(TMTSF)_2PF_6$. *Phys. Rev. Lett.* **1989**, *63*, 1984–1987. [CrossRef] [PubMed]
658. Kang, W.; Montambaux, G.; Cooper, J.R.; Jérome, D.; Batail, P.; Lenoir, C. Observation of giant magnetoresistance oscillations in the high-T_C phase of the two-dimensional organic conductor β-$(BEDT\text{-}TTF)_2I_3$. *Phys. Rev. Lett.* **1989**, *62*, 2559–2562. [CrossRef] [PubMed]
659. Kushch, N.D.; Buravov, L.I.; Kartsovnik, M.V.; Laukhin, V.N.; Pesotskii, S.I.; Shibaeva, R.P.; Rozenberg, L.P.; Yagubskii, E.B.; Zvarikina, A.V. Resistance and magnetoresistance anomaly in a new stable organic metal $(ET)_2TlHg(SCN)_4$. *Synth. Met.* **1992**, *46*, 271–276. [CrossRef]
660. Dupuis, N.; Montambaux, G.; Sá de Melo, C.A.R. Quasi-one-dimensional superconductors in strong magnetic field. *Phys. Rev. Lett.* **1993**, *70*, 2613–2616. [CrossRef]
661. Kovalev, A.E.; Kartsovnik, M.V.; Shibaeva, R.P.; Rozenberg, L.P.; Schegolev, I.F.; Kushch, N.D. Angular magnetoresistance oscillations in the organic conductor α-$(ET)_2KHg(SCN)_4$ above and below the phase transition. *Solid State Commun.* **1994**, *89*, 575–578. [CrossRef]
662. Dupuis, N.; Montambaux, G. Superconductivity of quasi-one-dimensional conductors in a high magnetic field. *Phys. Rev. B* **1994**, *49*, 8993–9008. [CrossRef] [PubMed]
663. Kartsovnik, M.V.; Logvenov, G.Y.; Ishiguro, T.; Biberacher, W.; Anzai, H.; Kushch, N.D. Direct Observation of the magnetic-breakdown induced quantum interference in the quasi-two-dimensional organic metal κ-$(BEDT\text{-}TTF)_2CU(NCS)_2$. *Phys. Rev. Lett.* **1996**, *77*, 2530–2533. [CrossRef]
664. Kartsovnik, M.V.; Laukhin, V.N. Magnetotransport in quasi-two-dimensional organic conductors based on BEDT-TTF and its derivatives. *J. Phys. I France* **1996**, *6*, 1753–1786. [CrossRef]
665. Herlach, F.; Agosta, C.C.; Bogaerts, R.; Boon, W.; Deckers, I.; de Keyser, A.; Harrison, N.; Lagutin, A.; Li, L.; Trappeniers, L.; et al. Experimental techniques for pulsed magnetic fields. *Phys. B Cond. Mat.* **1996**, *216*, 161–165. [CrossRef]
666. Hill, S. Semiclassical description of cyclotron resonance in quasi-two-dimensional organic conductors: Theory and experiment. *Phys. Rev. B* **1997**, *55*, 4931–4940. [CrossRef]
667. Kartsovnik, M.V.; Biberacher, W.; Steep, E.; Christ, P.; Andres, K.; Jansen, A.G.M.; Müller, H. High-field studies of the *H-T* phase diagram of α-$(BEDT\text{-}TTF)_2KHg(SCN)_4$. *Synth. Met.* **1997**, *86*, 1933–1936. [CrossRef]
668. Chaikin, P.M.; Chashechkina, E.I.; Lee, I.J.; Naughton, M.J. Field-induced electronic phase transitions in high magnetic fields. *J. Phys. Condens. Matter* **1998**, *10*, 11301–11314. [CrossRef]
669. Weiss, H.; Kartsovnik, M.V.; Biberacher, W.; Steep, E.; Balthes, E.; Jansen, A.G.M.; Andres, K.; Kushch, N.D. Magnetotransport studies of the Fermi surface in the organic superconductor κ-$(BEDT\text{–}TTF)_2Cu[N(CN)_2]Br$. *Phys. Rev. B* **1999**, *59*, 12370–12378. [CrossRef]
670. Singleton, J. Studies of quasi-two-dimensional organic conductors based on BEDT-TTF using high magnetic fields. *Rep. Prog. Phys.* **2000**, *63*, 1111–1207. [CrossRef]
671. Christ, P.; Biberacher, W.; Kartsovnik, M.V.; Steep, E.; Balthes, E.; Weiss, H.; Müller, H. Magnetic field-temperature phase diagram of the organic conductor α-$(BEDT\text{-}TTF)_2KHg(SCN)_4$. *JETP Lett.* **2000**, *71*, 303–306. [CrossRef]
672. Mola, M.; Hill, S.; Goy, P.; Gross, M. Instrumentation for millimeter-wave magnetoelectrodynamic investigations of low-dimensional conductors and superconductors. *Rev. Sci. Instrum.* **2000**, *71*, 186–200. [CrossRef]
673. Andres, D.; Kartsovnik, M.V.; Biberacher, W.; Weiss, H.; Balthes, E.; Müller, H.; Kushch, N. Orbital effect of a magnetic field on the low-temperature state in the organic metal α-$(BEDT\text{-}TTF)_2KHg(SCN)_4$. *Phys. Rev. B* **2001**, *64*, 161104. [CrossRef]
674. Kartsovnik, M.V.; Grigoriev, P.D.; Biberacher, W.; Kushch, N.D.; Wyder, P. Slow Oscillations of magnetoresistance in quasi-two-dimensional metals. *Phys. Rev. Lett.* **2002**, *89*, 126802. [CrossRef]
675. Andres, D.; Kartsovnik, M.V.; Grigoriev, P.D.; Biberacher, W.; Müller, H. Orbital quantization in the high-magnetic-field state of a charge-density-wave system. *Phys. Rev. B* **2003**, *68*, 201101(R). [CrossRef]
676. Maki, K.; Dóra, B.; Kartsovnik, M.; Virosztek, A.; Korin-Hamzić, B.; Basletić, M. Unconventional charge-density wave in the organic conductor α-$(BEDT\text{-}TTF)_2KHg(SCN)_4$. *Phys. Rev. Lett.* **2003**, *90*, 256402. [CrossRef]
677. Graf, D.; Choi, E.S.; Brooks, J.S.; Matos, M.; Henriques, R.T.; Almeida, M. High magnetic field induced charge density wave state in a quasi-one-dimensional organic conductor. *Phys. Rev. Lett.* **2004**, *93*, 076406. [CrossRef]
678. Graf, D.; Brooks, J.S.; Choi, E.S.; Uji, S.; Dias, J.C.; Almeida, M.; Matos, M. Suppression of a charge-density-wave ground state in high magnetic fields: Spin and orbital mechanisms. *Phys. Rev. B* **2004**, *69*, 125113. [CrossRef]
679. Kartsovnik, M.V. High magnetic fields: A tool for studying electronic properties of layered organic metals. *Chem. Rev.* **2004**, *104*, 5737–5781. [CrossRef]
680. Takahashi, S.; Hill, S. Rotating cavity for high-field angle-dependent microwave spectroscopy of low-dimensional conductors and magnets. *Rev. Sci. Instrum.* **2005**, *76*, 023114. [CrossRef]

681. Cho, K.; Smith, B.E.; Coniglio, W.A.; Winter, L.E.; Agosta, C.C.; Schlueter, J.A. Upper critical field in the organic superconductor β''-$(ET)_2SF_5CH_2CF_2SO_3$: Possibility of Fulde-Ferrell-Larkin-Ovchinnikov state. *Phys. Rev. B* **2009**, *79*, 220507(R). [CrossRef]
682. Agosta, C.C.; Fortune, N.A.; Hannahs, S.T.; Gu, S.; Liang, L.; Park, J.-H.; Schleuter, J.A. Calorimetric Measurements of magnetic-field-induced inhomogeneous superconductivity above the paramagnetic limit. *Phys. Rev. Lett.* **2017**, *118*, 267001. [CrossRef]
683. Koshihara, S.; Tokura, Y.; Mitani, T.; Saito, G.; Koda, T. Photoinduced valence instability in the organic molecular compound tetrathiafulvalene-*p*-chloranil (TTF-CA). *Phys. Rev. B* **1990**, *42*, 6853–6856. [CrossRef] [PubMed]
684. Koshihara, S.-Y.; Takahashi, Y.; Sakai, H.; Tokura, Y.; Luty, T. Photoinduced cooperative charge transfer in low-dimensional organic crystals. *J. Phys. Chem. B* **1999**, *103*, 2592–2600. [CrossRef]
685. Chollet, M.; Guerin, L.; Uchida, N.; Fukaya, S.; Shimoda, H.; Ishikawa, T.; Matsuda, K.; Hasegawa, T.; Ota, A.; Yamochi, H.; et al. Gigantic photoresponse in 1/4-filled-band organic salt $(EDO\text{-}TTF)_2PF_6$. *Science* **2005**, *307*, 86–89. [CrossRef]
686. Tajima, N.; Fujisawa, J.-I.; Nakay, N.; Ishihara, T.; Kato, R.; Nishio, Y.; Kajita, K. Photo-induced insulator-metal transition in an organic conductor α-$(BEDT\text{-}TTF)_2I_3$. *J. Phys. Soc. Jpn.* **2005**, *74*, 511–514. [CrossRef]
687. Tajima, N.; Sugawara, S.; Tamura, M.; Nishio, Y.; Kajita, K. Electronic phases in an organic conductor α-$(BEDT\text{-}TTF)_2I_3$: Ultra narrow gap semiconductor, superconductor, metal, and charge-ordered insulator. *J. Phys. Soc. Jpn.* **2006**, *75*, 051010. [CrossRef]
688. Iwai, S.; Okamoto, H. Ultrafast phase control in one-dimensional correlated electron systems. *J. Phys. Soc. Jpn.* **2006**, *75*, 011007. [CrossRef]
689. Okamoto, H.; Matsuzaki, H.; Wakabayashi, T.; Takahashi, Y.; Hasegawa, T. Photoinduced metallic state mediated by spin-charge separation in a one-dimensional organic mott insulator. *Phys. Rev. Lett.* **2007**, *98*, 037401. [CrossRef]
690. Iwai, S.; Yamamoto, K.; Kashiwazaki, A.; Hiramatsu, F.; Nakaya, H.; Kawakami, Y.; Yakushi, K.; Okamoto, H.; Mori, H.; Nishio, Y. Photoinduced melting of a stripe-type charge-order and metallic domain formation in a layered BEDT-TTF-based organic salt. *Phys. Rev. Lett.* **2007**, *98*, 097402. [CrossRef] [PubMed]
691. Iimori, T.; Ohta, N.; Naito, T. Molecular-based light-activated thyristor. *Appl. Phys. Lett.* **2007**, *90*, 262103. [CrossRef]
692. Iimori, T.; Naito, T.; Ohta, N. Photoinduced phase transition in the organic conductor α-$(BEDT\text{-}TTF)_2I_3$ at temperatures near the metal-insulator phase transition. *Chem. Lett.* **2007**, *36*, 536–537. [CrossRef]
693. Iimori, T.; Naito, T.; Ohta, N. A memory effect controlled by a pulsed voltage in photoinduced conductivity switching in an organic charge-transfer salt. *J. Am. Chem. Soc.* **2007**, *129*, 3486–3487. [CrossRef]
694. Onda, K.; Ogihara, S.; Yonemitsu, K.; Maeshima, N.; Ishikawa, T.; Okimoto, Y.; Shao, X.; Nakano, Y.; Yamochi, H.; Saito, G.; et al. Photoinduced change in the charge order pattern in the quarter-filled organic conductor $(EDO\text{-}TTF)_2PF_6$ with a strong electron phonon interaction. *Phys. Rev. Lett.* **2008**, *101*, 067403. [CrossRef]
695. Yonemitsu, K.; Nasu, K. Theory of photoinduced phase transitions in itinerant electron systems. *Phys. Rep.* **2008**, *465*, 1–60. [CrossRef]
696. Iwai, S.; Nakaya, H.; Kawakami, Y. Ultrafast photo-induced insulator to metal transition in layered BEDT-TTF based salts. In *Molecular Electronic and Related Materials—Control and Probe with Light*; Naito, T., Ed.; Transworld Research Network: Kerala, India, 2010; pp. 37–58.
697. Okamoto, H. Ultrafast photoinduced phase transitions in one-dimensional organic correlated electron systems. In *Molecular Electronic and Related Materials—Control and Probe with Light*; Naito, T., Ed.; Transworld Research Network: Kerala, India, 2010; pp. 37–58.
698. Tajima, N.; Fujisawa, J.-I.; Kato, R. Photoswitching between charge-ordered insulator and metal phases in an organic conductor α-$(BEDT\text{-}TTF)_2I_3$. In *Molecular Electronic and Related Materials—Control and Probe with Light*; Naito, T., Ed.; Transworld Research Network: Kerala, India, 2010; pp. 155–165.
699. Iimori, T.; Naito, T.; Ohta, N. Synergy effects of photoirradiation and applied voltage on electrical conductivity of α-$(BEDT\text{-}TTF)_2I_3$. In *Molecular Electronic and Related Materials—Control and Probe with Light*; Naito, T., Ed.; Transworld Research Network: Kerala, India, 2010; pp. 167–184.
700. Yonemitsu, K. Theory of photoinduced phase transitions in quasi-one-dimensional organic conductors. In *Molecular Electronic and Related Materials—Control and Probe with Light*; Naito, T., Ed.; Transworld Research Network: Kerala, India, 2010; pp. 305–320.
701. Iimori, T.; Naito, T.; Ohta, N. Unprecedented optoelectronic function in organic conductor: Memory effect of photoswitching controlled by voltage pulse width. *J. Phys. Chem. C* **2009**, *113*, 4654–4661. [CrossRef]
702. Iimori, T.; Naito, T.; Ohta, N. Time-resolved measurement of the photoinduced change in the electrical conductivity of the organic superconductor κ-$(BEDT\text{-}TTF)_2Cu[N(CN)_2]Br$. *J. Phys. Chem. C* **2010**, *114*, 9070–9075. [CrossRef]
703. Iimori, T.; Sabeth, F.; Naito, T.; Ohta, N. Time-resolved photoresponse measurements of the electrical conductivity of the quasi-two-dimensional organic superconductor β-$(BEDT\text{-}TTF)_2I_3$ using a nanosecond laser pulse. *J. Phys. Chem. C* **2011**, *115*, 23998–24003. [CrossRef]
704. Toda, Y.; Mertelj, T.; Naito, T.; Mihailovic, D. Femtosecond carrier relaxation dynamics and photoinduced phase separation in κ-$(BEDT\text{-}TTF)_2Cu[N(CN)_2]X$ (X = Br, Cl). *Phys. Rev. Lett.* **2011**, *107*, 227002. [CrossRef] [PubMed]
705. Yonemitsu, K. Theory of photoinduced phase transitions in molecular conductors: Interplay between correlated electrons, lattice phonons and molecular vibrations. *Crystals* **2012**, *2*, 56–77. [CrossRef]
706. Iwai, S. Photoinduced phase transitions in α-, θ-, and κ-type ET salts: Ultrafast melting of the electronic ordering. *Crystals* **2012**, *2*, 590–617. [CrossRef]

707. Kakinuma, T.; Kojima, H.; Kawamoto, T.; Mori, T. Giant phototransistor response in dithienyltetrathiafulvalene derivatives. *J. Mater. Chem. C* **2013**, *1*, 2900–2905. [CrossRef]
708. Ishikawa, T.; Hayes, S.A.; Keskin, S.; Corthey, G.; Hada, M.; Pichugin, K.; Marx, A.; Hirscht, J.; Shionuma, K.; Onda, K.; et al. Direct observation of collective modes coupled to molecular orbital-driven charge transfer. *Science* **2015**, *350*, 1501–1505. [CrossRef]
709. Morimoto, T.; Miyamoto, T.; Okamoto, H. Ultrafast electron and molecular dynamics in photoinduced and electric-field-induced neutral-ionic transitions. *Crystals* **2017**, *7*, 132. [CrossRef]
710. Smit, B.; Hüwe, F.; Payne, N.; Olaoye, O.; Bauer, I.; Pflaum, J.; Schwoerer, M.; Schwoerer, H. Ultrafast pathways of the photoinduced insulator-metal transition in a low-dimensional organic conductor. *Adv. Mater.* **2019**, *31*, 1900652. [CrossRef]
711. Bai, C.; Dai, C.; Zhu, C.; Chen, Z.; Huang, G.; Wu, X.; Zhu, D. Scanning tunneling microscopy of silver containing salt of bis(ethylenedithio)tetrathiafulvalene. *J. Vac. Sci. Tech.* **1990**, *8*, 484–487. [CrossRef]
712. Wang, H.H.; Ferraro, J.R.; Williams, J.M.; Geiser, U.; Schlueter, J.A. Rapid Raman spectroscopic determination of the stoichiometry of microscopic quantities of BEDT-TTF-based organic conductors and superconductors. *J. Chem. Soc. Chem. Commun.* **1994**, 1893–1894. [CrossRef]
713. Shigekawa, H.; Miyake, K.; Miyauchi, A.; Ishida, M.; Oigawa, H.; Nannichi, Y.; Yoshizaki, R.; Mori, T. Surface superstructures of quasi-one-dimensional organic conductor β-(BEDT-TTF)$_2$PF$_6$ crystal studied by scanning tunneling microscopy. *Phys. Rev. B* **1995**, *52*, 16361–16364. [CrossRef]
714. Shigekawa, H.; Miyake, K.; Oigawa, H.; Nannichi, Y.; Mori, T.; Saito, Y. Molecular structure of a crystal phase coexisting with κ-(BEDT-TTF)$_2$Cu(NCS)$_2$ studied by scanning tunneling microscopy. *Phys. Rev. B* **1995**, *50*, 15427–15430. [CrossRef] [PubMed]
715. Arai, T.; Ichimura, K.; Nomura, K.; Takasaki, S.; Yamada, J.; Nakatsuji, S.; Anzai, H. Tunneling spectroscopy on the organic superconductor κ-(BEDT-TTF)$_2$Cu(NCS)$_2$ using STM. *Phys. Rev. B* **2001**, *63*, 104518. [CrossRef]
716. Taylor, O.J.; Carrington, A.; Schlueter, J.A. Specific-heat measurements of the gap structure of the organic superconductors κ-(ET)$_2$Cu[N(CN)$_2$]Br and κ-(ET)$_2$Cu(NCS)$_2$. *Phys. Rev. Lett.* **2007**, *99*, 057001. [CrossRef] [PubMed]
717. Claessen, R.; Schäfer, J.; Sing, M. Photoemission on quasi-one-dimensional solids: Peierls, Luttinger & Co. In *Very High Resolution Photoelectron Spectroscopy. Lecture Notes in Physics*; Hüfner, S., Ed.; Springer: Berlin/Heidelberg, Germany, 2007; Volume 715, pp. 115–146. [CrossRef]
718. Sasaki, T. Infrared imaging in the strongly correlated molecular conductors. In *Molecular Electronic and Related Materials—Control and Probe with Light*; Naito, T., Ed.; Transworld Research Network: Kerala, India, 2010; pp. 99–116.
719. Mori, E.; Usui, H.; Sakamoto, H.; Mizoguchi, K.; Naito, T. Charge distribution in the surface BEDT-TTF layer of α-(BEDT-TTF)$_2$I$_3$ at room temperature with scanning tunneling microscopy. *J. Phys. Soc. Jpn.* **2012**, *81*, 014707. [CrossRef]
720. Sakamoto, H.; Mori, E.; Arimoto, H.; Namai, K.; Tahara, H.; Naito, T.; Hiramatsu, T.; Yamochi, H.; Mizoguchi, K. Wavefunction Analysis of STM Image: Surface Reconstruction of Organic Charge Transfer Salts. In *Microscopy and Analysis*; Stanciu, S.G., Ed.; IntechOpen: London, UK, 2016; Chapter 14. [CrossRef]
721. Pustogow, A.; McLeod, A.S.; Saito, Y.; Basov, D.N.; Dressel, M. Internal strain tunes electronic correlations on the nanoscale. *Sci. Adv.* **2018**, *4*, eaau9123. [CrossRef]
722. Maesato, M.; Kaga, Y.; Kondo, R.; Kagoshima, S. Uniaxial strain method for soft crystals: Application to the control of the electronic properties of organic conductors. *Rev. Sci. Instrum.* **2000**, *71*, 176–181. [CrossRef]
723. Adachi, T.; Tanaka, H.; Kobayashi, H.; Miyazaki, T. Electrical resistivity measurements on fragile organic single crystals in the diamond anvil cell. *Rev. Sci. Instrum.* **2001**, *72*, 2358–2360. [CrossRef]
724. Müller, J.; Lang, M.; Helfrich, R.; Steglich, F.; Sasaki, T. High-resolution ac-calorimetry studies of the quasi-two-dimensional organic superconductor κ-(BEDT-TTF)$_2$Cu(NCS)$_2$. *Phys. Rev. B* **2002**, *65*, 14509. [CrossRef]
725. Kagoshima, S.; Kondo, R. Control of electronic properties of molecular conductors by uniaxial strain. *Chem. Rev.* **2004**, *104*, 5593–5608. [CrossRef]
726. Murata, K.; Kagoshima, S.; Yasuzuka, S.; Yoshino, H.; Kondo, R. High-pressure research in organic conductors. *J. Phys. Soc. Jpn.* **2006**, *75*, 051015. [CrossRef]
727. Boldyreva, E.V. High-pressure diffraction studies of molecular organic solids. A personal view. *Acta Cryst.* **2008**, *64*, 218–231. [CrossRef]
728. Iwase, F.; Miyagawa, K.; Kanoda, K. High-frequency nuclear quadrupole resonance apparatus for use in pressure cell. *Rev. Sci. Instrum.* **2012**, *83*, 064704. [CrossRef]
729. Cui, H.; Kobayashi, H.; Ishibashi, S.; Sasa, M.; Iwase, F.; Kato, R.; Kobayashi, A. A single-component molecular superconductor. *J. Am. Chem. Soc.* **2014**, *136*, 7619–7622. [CrossRef]
730. Shen, G.; Mao, H.K. High-pressure studies with x-rays using diamond anvil cells. *Rep. Prog. Phys.* **2017**, *80*, 016101. [CrossRef]
731. Nakazawa, Y.; Imajo, S.; Matsumura, Y.; Yamashita, S.; Akutsu, H. Thermodynamic picture of dimer-Mott organic superconductors revealed by heat capacity measurements with external and chemical pressure control. *Crystals* **2018**, *8*, 143. [CrossRef]
732. Mott, N.F. Metal-insulator transition. *Rev. Mod. Phys.* **1968**, *40*, 677–683. [CrossRef]
733. Kanoda, K. Recent progress in NMR studies on organic conductors. *Hyperfine Interact.* **1997**, *104*, 235–249. [CrossRef]
734. Sasaki, T.; Yoneyama, N.; Matsuyama, A.; Kobayashi, N. Magnetic and electronic phase diagram and superconductivity in the organic superconductors κ-(ET)$_2$X. *Phys. Rev. B* **2002**, *65*, 060505. [CrossRef]
735. Limelette, P.; Georges, A.; Jérome, D.; Wzietek, P.; Metcalf, P.; Honig, J.M. Universality and critical behavior at the Mott transition. *Science* **2003**, *302*, 88–92. [CrossRef] [PubMed]

736. Limelette, P.; Wzietek, P.; Florens, S.; Georges, A.; Costi, T.A.; Pasquier, C.; Jérome, D.; Mézière, C.; Batail, P. Mott transition and transport crossovers in the organic compound κ-(BEDT-TTF)$_2$Cu[N(CN)$_2$]Cl. *Phys. Rev. Lett.* **2003**, *91*, 016401. [CrossRef]
737. Fournier, D.; Poirier, M.; Castonguay, M.; Truong, K.D. Mott transition, compressibility divergence, and the *P-T* phase diagram of layered organic superconductors: An ultrasonic investigation. *Phys. Rev. Lett.* **2003**, *90*, 127002. [CrossRef] [PubMed]
738. Heuzé, K.; Fourmigué, M.; Batail, P.; Couion, C.; Clérac, R.; Canadell, E.; Auban-Senzier, P.; Ravy, S.; Jérome, D. A genuine quarter-filled band mott insulator, (EDT-TTF-CONMe$_2$)$_2$AsF$_6$: Where the chemistry and physics of weak intermolecular interactions act in unison. *Adv. Mat.* **2003**, *15*, 1251–1254. [CrossRef]
739. Kagawa, F.; Miyagawa, K.; Kanoda, K. Unconventional critical behaviour in a quasi-two-dimensional organic conductor. *Nature* **2005**, *436*, 534–537. [CrossRef]
740. Scheffler, M.; Dressel, M.; Jourdan, M.; Adrian, H. Extremely slow Drude relaxation of correlated electrons. *Nature* **2005**, *438*, 1135–1137. [CrossRef]
741. Sasaki, T.; Yoneyama, N.; Suzuki, A.; Kobayashi, N.; Ikemoto, Y.; Kimura, H. Real space imaging of the metal-insulator phase separation in the band width controlled organic Mott system κ-(BEDT-TTF)$_2$Cu[N(CN)$_2$]Br. *J. Phys. Soc. Jpn.* **2005**, *74*, 2351–2360. [CrossRef]
742. Sasaki, T.; Yoneyama, N.; Suzuki, A.; Ito, I.; Kobayashi, N.; Ikemoto, Y.; Kimura, H.; Hanasaki, N.; Tajima, H. Electrical inhomogeneity at the Mott transition in the band width controlled κ-(BEDT-TTF)$_2$Cu[N(CN)$_2$Br. *J. Low Temp. Phys.* **2006**, *142*, 377–382. [CrossRef]
743. De Souza, M.; Brühl, A.; Strack, C.; Wolf, B.; Schweitzer, D.; Lang, M. Anomalous lattice response at the Mott transition in a quasi-2D organic conductor. *Phys. Rev. Lett.* **2007**, *99*, 037003. [CrossRef]
744. Merino, J.; Dumm, M.; Drichko, N.; Dressel, M.; McKenzie, R.H. Quasiparticles at the verge of localization near the mott metal-insulator transition in a two-dimensional material. *Phys. Rev. Lett.* **2008**, *100*, 086404. [CrossRef]
745. Zorina, L.; Simonov, S.; Mézière, C.; Canadell, E.; Suh, S.; Brown, S.E.; Foury-Leylekian, P.; Fertey, P.; Pouget, J.-P.; Batail, P. Charge ordering, symmetry and electronic structure issues and Wigner crystal structure of the quarter-filled band Mott insulators and high pressure metals δ-(EDT-TTF-CONMe$_2$)$_2$X, X = Br and AsF$_6$. *J. Mater. Chem.* **2009**, *19*, 6980–6994. [CrossRef]
746. Basov, D.N.; Averitt, R.D.; van Der Marel, D.; Dressel, M.; Haule, K. Electrodynamics of correlated electron materials. *Rev. Mod. Phys.* **2011**, *83*, 471–541. [CrossRef]
747. Wall, S.; Brida, D.; Clark, S.R.; Ehrke, H.P.; Jaksch, D.; Ardavan, A.; Bonora, S.; Uemura, H.; Takahashi, Y.; Hasegawa, T.; et al. Quantum interference between charge excitation paths in a solid-state Mott insulator. *Nat. Phys.* **2011**, *7*, 114–118. [CrossRef]
748. Sasaki, T. Mott Anderson transition in molecular conductors: Influence of randomness on strongly correlated electrons in the κ-(BEDT-TTF)$_2$X system. *Crystals* **2012**, *2*, 374–392. [CrossRef]
749. Pinterić, M.; Lazić, P.; Pustogow, A.; Ivek, T.; Kuveždić, M.; Milat, O.; Gumhalter, B.; Basletić, M.; Čulo, M.; Korin-Hamzić, B.; et al. Anion effects on electronic structure and electrodynamic properties of the Mott insulator κ-(BEDT-TTF)$_2$Ag$_2$(CN)$_3$. *Phys. Rev. B* **2016**, *94*, 161105. [CrossRef]
750. Mori, H.; Kamiya, M.; Haemori, M.; Suzuki, H.; Tanaka, S.; Nishio, Y.; Kajita, K.; Moriyama, H. First systematic band-filling control in organic conductors. *J. Am. Chem. Soc.* **2002**, *124*, 1251–1260. [CrossRef]
751. Naito, T.; Inabe, T.; Niimi, H.; Asakura, K. Light-induced transformation of molecular materials into devices. *Adv. Mater.* **2004**, *16*, 1786–1790. [CrossRef]
752. Yamamoto, H.M.; Ito, H.; Shigeto, K.; Tsukagoshi, K.; Kato, R. Direct formation of micro/nanocrystalline 2,5-dimethyl-*N*,*N*′-dicyanoquinonediimine complexes on SiO$_2$/Si substrates and multiprobe measurement of conduction properties. *J. Am. Chem. Soc.* **2006**, *128*, 700–701. [CrossRef]
753. Naito, T.; Sugawara, H.; Inabe, T.; Kitajima, Y.; Miyamoto, T.; Niimi, H.; Asakura, K. UV-vis-induced vitrification of a molecular crystal. *Adv. Func. Mater.* **2007**, *17*, 1663–1670. [CrossRef]
754. Naito, T.; Sugawara, H.; Inabe, T. Mechanism of spatially resolved photochemical control of resistivity of a molecular crystalline solid. *Nanotechnology* **2007**, *18*, 424008. [CrossRef]
755. Miyamoto, T.; Niimi, H.; Chun, W.-J.; Kitajima, Y.; Sugawara, H.; Inabe, T.; Naito, T.; Asakura, K. Chemical states of Ag in Ag(DMe-DCNQI)$_2$ photoproducts and a proposal for its photoinduced conductivity change mechanism. *Chem. Lett.* **2007**, *36*, 1008–1009. [CrossRef]
756. Miyamoto, T.; Kitajima, Y.; Sugawara, H.; Naito, T.; Inabe, T.; Asakura, K. Origin of photochemical modification of the resistivity of Ag(DMe-DCNQI)$_2$ studied by X-ray absorption fine structure. *J. Phys. Chem. C* **2009**, *113*, 20476–20480. [CrossRef]
757. Naito, T. Spatially resolved control of electrical resistivity in organic materials—Development of a new fabrication method of junction structures. In *Nanotechnology: Nanofabrication, Patterning, and Self Assembly*; Dixon, C.J., Curtines, O.W., Eds.; Nova Science Publishers: Hauppauge, NY, USA, 2010; Chapter 7; pp. 275–292.
758. Naito, T.; Kakizaki, A.; Inabe, T.; Sakai, R.; Nishibori, E.; Sawa, H. Growth of nanocrystals in a single crystal of different materials: A way of giving function to molecular crystals. *Cryst. Growth Design* **2011**, *11*, 501–506. [CrossRef]
759. Naito, T. Optical control of electrical properties in molecular crystals; states of matter beyond thermodynamic restrictions. *Chem. Lett.* **2018**, *47*, 1441–1452. [CrossRef]

760. Heuzé, K.; Mézière, C.; Fourmigué, M.; Batail, P.; Coulon, C.; Canadell, E.; Auban-Senzier, P.; Jérome, D. An efficient, redox-enhanced pair of hydrogen-bond tweezers for chloride anion recognition, a key synthon in the construction of a novel type of organic metal based on the secondary amide-functionalized ethylenedithiotetrathiafulvalene, β″-(EDT-TTF-CONHMe)$_2$[Cl·H$_2$O]. *Chem. Mater.* **2000**, *12*, 1898–1904. [CrossRef]
761. Hirose, T.; Imai, H.; Naito, T.; Inabe, T. Charge carrier doping in the Ni(dmit)$_2$ simple salts by hydrogen-bonding pyridinium cations (dmit = 1,3-dithiol-2thione-4,5-dithiolate). *J. Solid State Chem.* **2002**, *168*, 535–546. [CrossRef]
762. Akutagawa, T.; Hasegawa, T.; Nakamura, T.; Saito, G. Hydrogen-bonded supramolecular (2,2′-bi-1*H*-benzimidazole)(2-(2-1*H*-benzimidazolyl)-1*H*-benzimidazolium$^+$)$_2$(Cl$^-$) as an electron donor in a TCNQ complex. *CrystEngComm* **2003**, *5*, 54–57. [CrossRef]
763. Baudron, S.A.; Avarvari, N.; Batail, P.; Coulon, C.; Clérac, R.; Canadell, E.; Auban-Senzier, P. Singular crystalline β′-layered topologies directed by ribbons of self-complementary amide···amide ring motifs in [EDT-TTF-(CONH$_2$)$_2$]$_2$X (X = HSO$_4^-$, ClO$_4^-$, ReO$_4^-$, AsF$_6^-$): Coupled activation of ribbon curvature, electron interactions, and magnetic susceptibility. *J. Am. Chem. Soc.* **2003**, *125*, 11583–11590. [CrossRef]
764. Devic, T.; Avarvari, N.; Batail, P. A series of redox active, tetrathiafulvalene-based amidopyridines and bipyridines ligands: Syntheses, crystal structures, a radical cation salt and group 10 transition-metal complexes. *Chem. Eur. J.* **2004**, *10*, 3697–3707. [CrossRef] [PubMed]
765. Baudron, S.A.; Avarvari, N.; Canadell, E.; Auban-Senzier, P.; Batail, P. Structural isomerism in crystals of redox-active secondary ortho-diamides: The role of competing intra- and intermolecular hydrogen bonds in directing crystalline topologies. *Chem. Eur. J.* **2004**, *10*, 4498–4511. [CrossRef]
766. Akutsu-Sato, A.; Akutsu, H.; Turner, S.S.; Day, P.; Probert, M.R.; Howard, J.A.K.; Akutagawa, T.; Takeda, S.; Nakamura, T.; Mori, T. The first proton-conducting metallic ion-radical salts. *Angew. Chem. Int. Ed. Engl.* **2004**, *44*, 292–295. [CrossRef]
767. Akutagawa, T.; Takeda, S.; Hasegawa, T.; Nakamura, T. Proton transfer and a dielectric phase transition in the molecular conductor (HDABCO$^+$)$_2$(TCNQ)$_3$. *J. Am. Chem. Soc.* **2004**, *126*, 291–294. [CrossRef]
768. Réthoré, C.; Fourmigué, M.; Avarvari, N. Tetrathiafulvalene-hydroxyamides and -oxazolines: Hydrogen bonding, chirality, and a radical cation salt. *Tetrahedron* **2005**, *61*, 10935–10942. [CrossRef]
769. Baudron, S.A.; Batail, P.; Coulon, C.; Clérac, R.; Canadell, E.; Laukhin, V.; Melzi, R.; Wzietek, P.; Jérome, D.; Auban-Senzier, P.; et al. (EDT-TTF-CONH$_2$)$_6$[Re$_6$Se$_8$(CN)$_6$], a metallic Kagome-type organic-inorganic hybrid compound: Electronic instability, molecular motion, and charge localization. *J. Am. Chem. Soc.* **2005**, *127*, 11785–11797. [CrossRef]
770. Isono, T.; Kamo, H.; Ueda, A.; Takahashi, K.; Nakao, A.; Kumai, R.; Nakao, H.; Kobayashi, K.; Murakami, Y.; Mori, H. Hydrogen bond-promoted metallic state in a purely organic single-component conductor under pressure. *Nat. Commun.* **2013**, *4*, 1344. [CrossRef]
771. Ueda, A.; Yamada, S.; Isono, T.; Kamo, H.; Nakao, A.; Kumai, R.; Nakao, H.; Murakami, Y.; Yamamoto, K.; Nishio, Y.; et al. Hydrogen-bond-dynamics-based switching of conductivity and magnetism: A phase transition caused by deuterium and electron transfer in a hydrogen-bonded purely organic conductor crystal. *J. Am. Chem. Soc.* **2014**, *136*, 12184–12192. [CrossRef] [PubMed]
772. Makhotkina, O.; Lieffrig, J.; Jeannin, O.; Fourmigué, M.; Aubert, E.; Espinosa, E. Cocrystal or salt: Solid state-controlled iodine shift in crystalline halogen-bonded systems. *Cryst. Growth Design* **2015**, *15*, 3464–3473. [CrossRef]
773. Horiuchi, S.; Tokura, Y. Organic ferroelectrics. *Nat. Mater.* **2008**, *7*, 357–366. [CrossRef]
774. Akutagawa, T.; Koshinaka, H.; Sato, D.; Takeda, S.; Noro, S.-I.; Takahashi, H.; Kumai, R.; Tokura, Y.; Nakamura, T. Ferroelectricity and polarity control in solid-state flip-flop supramolecular rotators. *Nat. Mater.* **2009**, *8*, 342–347. [CrossRef] [PubMed]
775. Kawamoto, T.; Mori, T.; Graf, D.; Brooks, J.S.; Takahide, Y.; Uji, S.; Shirahata, T.; Imakubo, T. Interlayer charge disproportionation in the layered organic superconductor κ_H-(DMEDO-TSeF)$_2$[Au(CN)$_4$](THF) with polar dielectric insulating layers. *Phys. Rev. Lett.* **2012**, *109*, 147005. [CrossRef]
776. Tomić, S.; Dressel, M. Ferroelectricity in molecular solids: A review of electrodynamic properties. *Rep. Prog. Phys.* **2015**, *78*, 096501. [CrossRef]
777. Harada, J.; Shimojo, T.; Oyamaguchi, H.; Hasegawa, H.; Takahashi, Y.; Satomi, K.; Suzuki, Y.; Kawamata, J.; Inabe, T. Directionally tunable and mechanically deformable ferroelectric crystals from rotating polar globular ionic molecules. *Nat. Chem.* **2016**, *8*, 946–952. [CrossRef]
778. Akutsu, H.; Ishihara, K.; Yamada, J.-I.; Nakatsuji, S.; Turner, S.S.; Nakazawa, Y. A strongly polarized organic conductor. *CrystEngComm* **2016**, *18*, 8151–8154. [CrossRef]
779. Akutsu, H.; Ishihara, K.; Ito, S.; Nishiyama, F.; Yamada, J.-I.; Nakasuji, S.; Turner, S.S.; Nakazawa, Y. Anion polarity-induced self-doping in a purely organic paramagnetic conductor, α′-α′-(BEDT-TTF)$_2$(PO-CONH-m-C$_6$H$_4$SO$_3$)·H$_2$O where BEDT-TTF is bis(ethylenedithio)tetrathiafulvalene and PO is the radical 2,2,5,5-Tetramethyl-3-pyrrolin-1-oxyl. *Polyhedron* **2017**, *136*, 23–29. [CrossRef]
780. Huang, Y.; Wang, Z.; Chen, Z.; Zhang, Q. Organic cocrystals: Beyond electrical conductivities and field-effect transistors (FETs). *Angew. Chem. Int. Ed. Engl.* **2019**, *58*, 9696–9711. [CrossRef] [PubMed]
781. Harada, J.; Kawamura, Y.; Takahashi, Y.; Uemura, Y.; Hasegawa, T.; Taniguchi, H.; Maruyama, K. Plastic/ferroelectric crystals with easily switchable polarization: Low-voltage operation, unprecedentedly high pyroelectric performance, and large piezoelectric effect in polycrystalline forms. *J. Am. Chem. Soc.* **2019**, *141*, 9349–9357. [CrossRef] [PubMed]
782. Lim, D.-W.; Kitagawa, H. Proton transport in metal-organic frameworks. *Chem. Rev.* **2020**, *120*, 8416–8467. [CrossRef]

783. Yoshimoto, R.; Yamashita, S.; Akutsu, H.; Nakazawa, Y.; Kusamoto, T.; Oshima, Y.; Nakano, T.; Yamamoto, H.M.; Kato, R. Electric dipole induced bulk ferromagnetism in dimer Mott molecular compounds. *Sci. Rep.* **2021**, *11*, 1332. [CrossRef]
784. Lunkenheimer, P.; Müller, J.; Krohns, S.; Schrettle, F.; Loidl, A.; Hartmann, B.; Rommel, R.; de Souza, M.; Hotta, C.; Schlueter, J.A.; et al. Multiferroicity in an organic charge-transfer salt that is suggestive of electric-dipole-driven magnetism. *Nat. Mater.* **2012**, *11*, 755–758. [CrossRef]
785. Rothaemel, B.; Forro, L.; Cooper, J.R.; Schilling, J.S.; Weger, M.; Bele, P.; Brunner, H.; Schweitzer, D.; Keller, H.J. Magnetic susceptibility of α and β phases of di[bis(ethylenediothiolo) tetrathiafulvalene] tri-iodide [(BEDT-TTF)$_2$I$_3$] under pressure. *Phys. Rev. B* **1986**, *34*, 704–712. [CrossRef] [PubMed]
786. Moldenhauer, J.; Horn, C.H.; Pokhodnia, K.I.; Schweitzer, D.; Heinen, I.; Keller, H.J. FT-IR absorption spectroscopy of BEDT-TTF radical salts: Charge transfer and donor-anion interaction. *Synth. Met.* **1993**, *60*, 31–38. [CrossRef]
787. Kino, H.; Fukuyama, H. On the phase transition of α-(ET)$_2$I$_3$. *J. Phys. Soc. Jpn.* **1995**, *64*, 1877–1880. [CrossRef]
788. Seo, H. Charge ordering in organic ET compounds. *J. Phys. Soc. Jpn.* **2000**, *69*, 805–820. [CrossRef]
789. Takano, Y.; Hiraki, K.; Yamamoto, H.M.; Nakamura, T.; Takahashi, T. Charge disproportionation in the organic conductor, α-(BEDT-TTF)$_2$I$_3$. *J. Phys. Chem. Solids* **2001**, *62*, 393–395. [CrossRef]
790. Wojciechowski, R.; Yamamoto, K.; Yakushi, K.; Inokuchi, M.; Kawamoto, A. High-pressure Raman study of the charge ordering in α-(BEDT-TTF)$_2$I$_3$. *Phys. Rev. B* **2003**, *67*, 224105. [CrossRef]
791. Kakiuchi1, T.; Wakabayashi, Y.; Sawa, H.; Takahashi, T.; Nakamura, T. Charge ordering in α-(BEDT-TTF)$_2$I$_3$ by synchrotron X-ray diffraction. *J. Phys. Soc. Jpn.* **2007**, *76*, 113702. [CrossRef]
792. Yamamoto, K.; Kowalska, A.A.; Yakushi, K. Direct observation of ferroelectric domains created by Wigner crystallization of electrons in α-[Bis (ethylenedithio) tetrathiafulvalene]$_2$I$_3$. *Appl. Phys. Lett.* **2010**, *96*, 122901. [CrossRef]
793. Yue, Y.; Nakano, C.; Yamamoto, K.; Uruichi, M.; Wojciechowski, R.; Inokuchi, M.; Yakushi, K.; Kawamoto, A. Charge order-disorder phase transition in α′-[bis(ethylenedithio) tetrathiafulvalene]$_2$IBr$_2$ [α′-(BEDT-TTF)$_2$IBr$_2$]. *J. Phys. Soc. Jpn.* **2009**, *78*, 044701. [CrossRef]
794. Yamamoto, K.; Yakushi, K. Second-harmonic generation study of ferroelectric organic conductors α-(BEDT-TTF)$_2$X (X = I$_3$ and I$_2$Br). In *Molecular Electronic and Related Materials—Control and Probe with Light*; Naito, T., Ed.; Transworld Research Network: Kerala, India, 2010; pp. 185–201.
795. Ivek, T.; Korin-Hamzić, B.; Milat, O.; Tomić, S.; Clauss, C.; Drichko, N.; Schweitzer, D.; Dressel, M. Electrodynamic response of the charge ordering phase: Dielectric and optical studies of α-(BEDT-TTF)$_2$I$_3$. *Phys. Rev. B* **2011**, *83*, 165128. [CrossRef]
796. Potember, R.S.; Poehler, T.O.; Cowan, D.O. Electrical switching and memory phenomena in Cu-TCNQ thin films. *Appl. Phys. Lett.* **1979**, *34*, 405–407. [CrossRef]
797. Bässler, H. Charge transport in disordered organic photoconductors a Monte Carlo simulation study. *Phys. Stat. Sol.* **1993**, *175*, 15–56. [CrossRef]
798. Nishikawa, H.; Kojima, S.; Kodama, T.; Ikemoto, I.; Suzuki, S.; Kikuchi, K.; Fujitsuka, M.; Luo, H.; Araki, Y.; Ito, O. Photophysical study of new methanofullerene-TTF diads: An obvious intramolecular charge transfer in the ground states. *J. Phys. Chem. A* **2004**, *108*, 1881–1890. [CrossRef]
799. Mataga, N.; Chosrowjan, H.; Taniguchi, S. Ultrafast charge transfer in excited electronic states and investigations into fundamental problems of exciplex chemistry: Our early studies and recent developments. *J. Photochem. Photobio. C: Photochem. Rev.* **2005**, *6*, 37–79. [CrossRef]
800. Loosli, C.; Jia, C.; Liu, S.-X.; Haas, M.; Dias, M.; Levillain, E.; Neels, A.; Labat, G.; Hauser, A.; Decurtins, S. Synthesis and electrochemical and photophysical studies of tetrathiafulvalene-annulated phthalocyanines. *J. Org. Chem.* **2005**, *70*, 4988–4992. [CrossRef]
801. Shigehiro, T.; Yagi, S.; Maeda, T.; Nakazumi, H.; Fujiwara, H.; Sakurai, Y. Novel 10,13-disubstitutedd dipyrido[3,2-a:2′,3′-c]phenazines and their platinum(II) complexes: Highly luminescent ICT-type fluorophores based on D–A–D structures. *Tetrahed. Lett.* **2005**, *55*, 5195–5198. [CrossRef]
802. Fujiwara, H.; Tsujimoto, K.; Sugishima, Y.; Takemoto, S.; Matsuzaka, H. New fluorene-substituted TTF derivatives as photofunctional materials. *Phys. B* **2010**, *405*, S12–S14. [CrossRef]
803. Fujiwara, H.; Yokota, S.; Hayashi, S.; Takemoto, S.; Matsuzaka, H. Development of photofunctional materials using TTF derivatives containing a 1,3-benzothiazole ring. *Phys. B* **2010**, *405*, S15–S18. [CrossRef]
804. Wenger, S.; Bouit, P.-A.; Chen, Q.; Teuscher, J.; di Censo, D.; Humphry-Baker, R.; Moser, J.-E.; Delgado, J.L.; Martín, N.; Zakeeruddin, S.M.; et al. Efficient electron transfer and sensitizer regeneration in stable π-extended tetrathiafulvalene-sensitized solar cells. *J. Am. Chem. Soc.* **2010**, *132*, 5164–5169. [CrossRef]
805. Lemmetyinen, H.; Tkachenko, N.V.; Efimov, A.; Niemi, M. Photoinduced intra- and intermolecular electron transfer in solutions and in solid organized molecular assemblies. *Phys. Chem. Chem. Phys.* **2011**, *13*, 397–412. [CrossRef]
806. Furukawa, K.; Sugishima, Y.; Fujiwara, H.; Nakamura, T. Photoinduced triplet states of photoconductive TTF derivatives including a fluorescent group. *Chem. Lett.* **2011**, *40*, 292–294. [CrossRef]
807. Naito, T.; Karasudani, T.; Mori, S.; Ohara, K.; Konishi, K.; Takano, T.; Takahashi, Y.; Inabe, T.; Nishihara, S.; Inoue, K. Molecular photoconductor with simultaneously photocontrollable localized spins. *J. Am. Chem. Soc.* **2012**, *134*, 18656–18666. [CrossRef]
808. Naito, T.; Karasudani, T.; Ohara, K.; Takano, T.; Takahashi, Y.; Inabe, T.; Furukawa, K.; Nakamura, T. Simultaneous control of carriers and localized spins with light in organic materials. *Adv. Mater.* **2012**, *24*, 6153–6157. [CrossRef]

809. Brunetti, F.G.; López, J.L.; Atienza, C.; Martín, N. π-extended TTF: A versatile molecule for organic electronics. *J. Mater. Chem.* **2012**, *22*, 4188–4205. [CrossRef]
810. Maeda, T.; Mineta, S.; Fujiwara, H.; Nakao, H.; Yagi, S.; Nakazumi, H. Conformationla effect of symmetrical squaraine dyes on the performance of dye-sensitized solar cells. *J. Mater. Chem. A* **2013**, *1*, 1303–1309. [CrossRef]
811. Tsujimoto, K.; Ogasawara, R.; Fujiwara, H. Photocurrent generation based on new tetrathiafulvalene-BODIPY dyads. *Tetrahedron Lett.* **2013**, *54*, 1251–1255. [CrossRef]
812. Takubo, N.; Tajima, N.; Yamamoto, H.M.; Cui, H.; Kato, R. Lattice distortion stabilizes the photoinduced metallic phase in the charge-ordered organic salts (BEDT-TTF)$_3$X$_2$ (X = ReO$_4$, ClO$_4$). *Phys. Rev. Lett.* **2013**, *110*, 227401. [CrossRef]
813. Naito, T.; Karasudani, T.; Nagayama, N.; Ohara, K.; Konishi, K.; Mori, S.; Takano, T.; Takahashi, Y.; Inabe, T.; Kinose, S.; et al. Giant photoconductivity in NMQ[Ni(dmit)$_2$]. *Eur. J. Inorg. Chem.* **2014**, 4000–4009. [CrossRef]
814. Noma, H.; Ohara, K.; Naito, T. [Cu(dmit)$_2$]$^{2-}$ Building block for molecular conductors and magnets with photocontrollable spin distribution. *Chem. Lett.* **2014**, *43*, 1230–1232. [CrossRef]
815. Tsujimoto, K.; Ogasawara, R.; Kishi, Y.; Fujiwara, H. TTF-fluorene dyads and their M(CN)$_2^-$ (M = Ag, Au) salts designed for photoresponsive conducting materials. *New J. Chem.* **2014**, *38*, 406–418. [CrossRef]
816. Tsujimoto, K.; Ogasawara, R.; Nakagawa, T.; Fujiwara, H. Photofunctional conductors based on TTF-BODIPY dyads bearing *p*-phenylene and *p*-phenylenevinylene spacers. *Eur. J. Inorg. Chem.* **2014**, *2014*, 3960–3972. [CrossRef]
817. Ng, T.-W.; Lo, M.-F.; Fung, M.-K.; Zhang, W.-J.; Lee, C.-S. Charge-transfer complexes and their role in exciplex emission and near-infrared photovoltaics. *Adv. Mater.* **2014**, *26*, 5569–5574. [CrossRef] [PubMed]
818. Nagayama, N.; Yamamoto, T.; Naito, T. Activation energy for photoconduction in molecular crystals. *Chem* **2015**, *2*, 74–80.
819. Mitrano, M.; Cantaluppi, A.; Nicoletti, D.; Kaiser, S.; Perucchi, A.; Lupi, S.; di Pietro, P.; Pontiroli, D.; Riccò, M.; Clark, S.R.; et al. Possible light-induced superconductivity in K$_3$C$_{60}$ at high temperature. *Nature* **2016**, *530*, 461–464. [CrossRef]
820. Noma, H.; Ohara, K.; Naito, T. Direct control of spin distribution and anisotropy in Cu-dithiolene complex anions by light. *Inorganics* **2016**, *4*, 7. [CrossRef]
821. Yamamoto, R.; Yamamoto, T.; Ohara, K.; Naito, T. Dye-sensitized molecular charge transfer complexes: Magnetic and conduction properties in the photoexcited states of Ni(dmit)$_2$ salts containing photosensitive dyes. *Magnetochemistry* **2017**, *3*, 20. [CrossRef]
822. Naito, T.; Yamamoto, T.; Yamamoto, R.; Zhang, M.Y.; Yamamoto, T. A possibly highly conducting state in an optically excited molecular crystal. *J. Mater. Chem. C* **2019**, *7*, 9175–9183. [CrossRef]
823. Naito, T.; Watanabe, N.; Sakamoto, Y.; Miyaji, Y.; Shirahata, T.; Misaki, Y.; Kitou, S.; Sawa, H. A molecular crystal with an unprecedentedly long-lived photoexcited state. *Dalton Trans.* **2019**, *48*, 12858–12866. [CrossRef]
824. Mogensen, J.; Michaels, H.; Roy, R.; Broløs, L.; Kilde, M.D.; Freitag, M.; Nielsen, M.B. Indenofluorene-Extended tetrathiafulvalene scaffolds for dye-sensitized solar cells. *Eur. J. Org. Chem.* **2020**, *2020*, 6127–6134. [CrossRef]
825. Naito, T. Prototype material for new strategy of photon energy storage. *Inorganics* **2020**, *8*, 53. [CrossRef]
826. Tsujimoto, K.; Yamamoto, S.; Fujiwara, H. Synthesis and physical properties of tetrathiafulvalene-8-quinolinato zinc(II) and nickel(II) complexes. *Inorganics* **2021**, *9*, 11. [CrossRef]
827. Schultz, A.J.; Wang, H.H.; Soderholm, L.C.; Sifter, T.L.; Williams, J.M.; Bechgaard, K.; Whangbo, M.-H. Crystal structures of [Au(DDDT)$_2$]^0 and [(*n*-Bu)$_4$N][Ni(DDDT)$_2$] and the ligandlike character of the isoelectronic radicals [Au(DDDT)$_2$]^0 and [Ni(DDDT)$_2$]$^-$. *Inorg. Chem.* **1987**, *26*, 3757–3761. [CrossRef]
828. Rindorf, G.; Thorup, N.; Bjørnholm, T.; Bechgaard, K. Structure of bis(benzene-1,2-dithiolato)gold(IV). *Acta Crystallogr. Sec. C* **1990**, *46*, 1437–1439. [CrossRef]
829. Schiødt, N.C.; Bjørnholm, T.; Bechgaard, K.; Neumeier, J.J.; Allgeier, C.; Jacobsen, C.S.; Thorup, N. Structural, electrical, magnetic, and optical properties of bis-benzene-1,2-dithiolate-Au(IV) crystals. *Phys. Rev. B* **1996**, *53*, 1773–1778. [CrossRef]
830. Belo, D.; Alves, H.; Lopes, E.B.; Duarte, M.T.; Gama, V.; Henriques, R.T.; Almeida, M.; Perez-Benitez, A.; Rovira, C.; Veciana, J. Gold complexes with dithiothiphene ligands: A metal based on a neutral molecule. *Chem. Eur. J.* **2001**, *7*, 511–519. [CrossRef]
831. Tanaka, H.; Okano, Y.; Kobayashi, H.; Suzuki, W.; Kobayashi, A. A three-dimensional synthetic metallic crystal composed of single-component molecules. *Science* **2001**, *291*, 285–287. [CrossRef] [PubMed]
832. Kobayashi, A.; Tanaka, H.; Kobayashi, H. Molecular design and development of single-component molecular metals. *J. Mater. Chem.* **2001**, *11*, 2078–2088. [CrossRef]
833. Dautel, O.J.; Fourmigué, M.; Canadell, E.; Auban-Senzier, P. Fluorine segregation controls the solid-state organization and Electronic properties of Ni and Au dithiolene complexes: Stabilization of a conducting single-component gold dithiolene complex. *Adv. Funct. Mater.* **2002**, *12*, 693–698. [CrossRef]
834. Tanaka, H.; Kobayashi, H.; Kobayashi, A. A conducting crystal based on a single-component paramagnetic molecule, [Cu(dmdt)$_2$] (dmdt = dimethyltetrathiafulvalenedithiolate). *J. Am. Chem. Soc.* **2002**, *124*, 10002–10003. [CrossRef]
835. Suzuki, W.; Fujiwara, E.; Kobayashi, A.; Fujishiro, Y.; Nishibori, E.; Takata, M.; Sakata, M.; Fujiwara, H.; Kobayashi, H. Highly conducting crystals based on single-component gold complexes with extended-TTF dithiolate ligands. *J. Am. Chem. Soc.* **2003**, *125*, 1486. [CrossRef]
836. Kobayashi, A.; Sasa, M.; Suzuki, W.; Fujiwara, E.; Tanaka, H.; Tokumoto, M.; Okano, Y.; Fujiwara, H.; Kobayashi, H. Infrared electronic absorption in a single-component molecular metal. *J. Am. Chem. Soc.* **2004**, *126*, 426–427. [CrossRef] [PubMed]
837. Kobayashi, A.; Fujiwara, E.; Kobayashi, H. Single-component molecular metals with extended-TTF dithiolate ligands. *Chem. Rev.* **2004**, *104*, 5243–5264. [CrossRef] [PubMed]

838. Tanaka, H.; Tokumoto, M.; Ishibashi, S.; Graf, D.; Choi, E.S.; Brooks, J.S.; Yasuzuka, S.; Okano, Y.; Kobayashi, H.; Kobayashi, A. Observation of three-dimensional Fermi surface in a single-component molecular metal, [Ni(tmdt)$_2$]. *J. Am. Chem. Soc.* **2004**, *126*, 10518–10519. [CrossRef] [PubMed]
839. Lusar, R.; Uriel, S.; Vicent, C.; Clemente-Juan, J.M.; Coronado, E.; Gómez-García, C.J.; Braïda, B.; Canadell, E. Single-component magnetic conductors based on Mo_3S_7 trinuclear clusters with outer dithiolate ligands. *J. Am. Chem. Soc.* **2004**, *126*, 12076–12083. [CrossRef]
840. Ishibashi, S.; Tanaka, H.; Kohyama, M.; Tokumoto, M.; Kobayashi, A.; Kobayashi, H.; Terakura, K. Ab initio electronic structure calculation for single-component molecular conductor Au(tmdt)$_2$ (tmdt = trimethylenetetrathiafulvalenedithiolate). *J. Phys. Soc. Jpn.* **2005**, *74*, 843–846. [CrossRef]
841. Sasa, M.; Fujiwara, E.; Kobayashi, A.; Ishibashi, S.; Terakura, K.; Okano, Y.; Fujiwara, H.; Kobayashi, H. Crystal structures and physical properties of single-component molecular conductors consisting of nickel and gold complexes with (trifluoromethyl)tetrathiafulvalenedithiolate ligands. *J. Mater. Chem.* **2005**, *15*, 155–163. [CrossRef]
842. Kobayashi, A.; Zhou, B.; Kobayashi, H. Development of metallic crystals composed of single-component molecules. *J. Mater. Chem.* **2005**, *15*, 3449–3451. [CrossRef]
843. Zhou, B.; Shimamura, M.; Fujiwara, E.; Kobayashi, A.; Higashi, T.; Nishibori, E.; Sakata, M.; Cui, H.B.; Takahashi, K.; Kobayashi, H. Magnetic transitions of single-component molecular metal [Au(tmdt)$_2$] and its alloy systems. *J. Am. Chem. Soc.* **2006**, *128*, 3872–3873. [CrossRef]
844. Kobayashi, A.; Okano, Y.; Kobayashi, H. Molecular design and physical properties of single-component molecular metals. *J. Phys. Soc. Jpn.* **2006**, *75*, 051002. [CrossRef]
845. Nunes, J.P.M.; Figueira, M.J.; Belo, D.; Santos, I.C.; Ribeiro, B.; Lopes, E.B.; Henriques, R.T.; Vidal-Gancedo, J.; Veciana, J.; Rovira, C.; et al. Transition metal bisdithiolene complexes based on extended ligands with fused terathiafulvalene and thiophene moieties: New single-component molecular metals. *Chem. Eur. J.* **2007**, *13*, 9841–9849. [CrossRef]
846. Llusar, R.; Triguero, S.; Polo, V.; Vicent, C.; Gómez-García, C.J.; Jeannin, O.; Fourmigué, M. Trinuclear Mo_3S_7 clusters coordinated to dithiolate or diselenolate ligands and their use in the preparation of magnetic single component molecular conductors. *Inorg. Chem.* **2008**, *47*, 9400–9409. [CrossRef]
847. Seo, H.; Ishibashi, S.; Okano, Y.; Kobayashi, H.; Kobayashi, A.; Fukuyama, H.; Terakura, K. Single-componenet molecular metals as multiband π-d systems. *J. Phys. Soc. Jpn.* **2008**, *77*, 023714. [CrossRef]
848. Hara, Y.; Miyagawa, K.; Kanoda, K.; Shimamura, M.; Zhou, B.; Kobayashi, A.; Kobayashi, H. NMR evidence for antiferromagnetic transition in the single component molecular conductor, [Au(tmdt)$_2$] at 110 K. *J. Phys. Soc. Jpn.* **2008**, *77*, 053706. [CrossRef]
849. Tenn, N.; Bellec, N.; Jeannin, O.; Piekara-Sady, L.; Auban-Senzier, P.; Iniguez, J.; Canadell, E.; Lorcy, D. A single-component molecular metal based on a thiazole dithiolate gold complex. *J. Am. Chem. Soc.* **2009**, *131*, 16961–16967. [CrossRef]
850. Zhou, B.; Kobayashi, A.; Okano, Y.; Nakashima, T.; Aoyagi, S.; Nishibori, E.; Sakata, M.; Tokumoto, M.; Kobayashi, H. Single-component molecular conductor [Pt(tmdt)$_2$] (tmdt = trimethylenetetrathiafulvalenedithiolate)—An advanced molecular metal exhibiting high metallicity. *Adv. Mat.* **2009**, *21*, 3596–3600. [CrossRef]
851. Mercuri, M.L.; Deplano, P.; Pilia, L.; Serpe, A.; Artizzu, F. Interactions modes and physical properties in transition metal chalcogenolene-based molecular materials. *Coord. Chem. Rev.* **2010**, *254*, 1419–1433. [CrossRef]
852. Garreau-de Bonneval, B.; Ching, K.I.M.-C.; Alary, F.; Bui, T.-T.; Valade, L. Neutral d^8 metals bis-dithiolene complexes: Synthesis, electronic properties and applications. *Coord. Chem. Rev.* **2010**, *254*, 1457–1467. [CrossRef]
853. Belo, D.; Almeida, M. Transition metal complexes based on thiophene-dithiolene ligands. *Coord. Chem. Rev.* **2010**, *254*, 1479–1492. [CrossRef]
854. Zhou, B.; Yajima, H.; Kobayashi, A.; Okano, Y.; Tanaka, H.; Kumashiro, T.; Nishibori, E.; Sawa, H.; Kobayashi, H. Single-component molecular conductor [Cu(tmdt)$_2$] containing an antiferromagnetic Heisenberg chain. *Inorg. Chem.* **2010**, *49*, 6740–6747. [CrossRef]
855. Perochon, R.; Davidson, P.; Rouzière, S.; Camerel, F.; Piekara-Sady, L.; Guizouarn, T.; Fourmigué, M. Probing magnetic interactions in columnar phases of a paramagnetic gold dithiolene complex. *J. Mater. Chem.* **2011**, *21*, 1416–1422. [CrossRef]
856. Takagi, R.; Miyagawa, K.; Kanoda, K.; Zhou, B.; Kobayashi, A.; Kobayashi, H. NMR evidence for antiferromagnetic transition in the single-component molecular system [Cu(tmdt)$_2$]. *Phys. Rev. B* **2012**, *85*, 184424. [CrossRef]
857. Zhou, B.; Idobata, Y.; Kobayashi, A.; Cui, H.; Kato, R.; Takagi, R.; Mitagawa, K.; Kanoda, K.; Kobayashi, H. Single-component molecular conductor [Cu(dmdt)$_2$] with three-dimensionally arranged magnetic moments exhibiting a coupled electric and magnetic transition. *J. Am. Chem. Soc.* **2012**, *134*, 12724–12731. [CrossRef]
858. Yzambart, G.; Bellec, N.; Nasser, G.; Jeannin, O.; Roisnel, T.; Fourmigué, M.; Auban-Senzier, P.; Íñiguez, J.; Canadell, E.; Lorcy, D. Anisotropic chemical pressure effects in single-component molecular metals based on radical dithiolene and diselenolene gold complexes. *J. Am. Chem. Soc.* **2012**, *134*, 17138–17148. [CrossRef]
859. Zhou, B.; Yajima, H.; Idobata, Y.; Kobayashi, A.; Kobayashi, T.; Nishibori, E.; Sawa, H.; Kobayashi, H. Single-component layered molecular conductor, [Au(ptdt)$_2$]. *Chem. Lett.* **2012**, *41*, 154–156. [CrossRef]
860. Papavassiliou, G.C.; Anyfantis, G.C.; Mousdis, G.A. Neutral metal 1,2-dithiolenes: Preparations, properties and possible applications of unsymmetrical in comparison to the symmetrical. *Crystals* **2012**, *2*, 762–811. [CrossRef]

861. Filatre-Furcate, A.; Bellec, N.; Jeannin, O.; Auban-Senzier, P.; Fourmigué, M.; Vacher, A.; Lorcy, D. Radical or not radical: Compared structures of metal (M = Ni, Au) bis-dithiolene complexes with a thiazole backbone. *Inorg. Chem.* **2014**, *53*, 8681–8690. [CrossRef] [PubMed]
862. Le Gal, Y.; Roisnel, T.; Auban-Senzier, P.; Guizouarn, T.; Lorcy, D. Hydrogen-bonding interactions in a single-component molecular conductor: A hydroxyethyl-substituted radical gold dithiolene. *Inorg. Chem.* **2014**, *53*, 8755–8761. [CrossRef]
863. Higashino, T.; Jeannin, O.; Kawamoto, T.; Lorcy, D.; Mori, T.; Fourmigué, M. A single-component conductor based on a radical gold dithiolene complex with alkyl-substituted thiophene-2,3-dithiolene ligand. *Inorg. Chem.* **2015**, *54*, 9908–9913. [CrossRef] [PubMed]
864. Mebrouk, K.; Kaddour, W.; Auban-Senzier, P.; Pasquier, C.; Jeannin, O.; Camerel, F.; Fourmigué, M. Molecular alloys of neutral nickel/gold dithiolene complexes in single-component semiconductors. *Inorg. Chem.* **2015**, *54*, 7454–7460. [CrossRef] [PubMed]
865. Filatre-Furcate, A.; Bellec, N.; Jeannin, O.; Auban-Senzier, P.; Fourmigué, M.; Íñiguez, J.; Canadell, E.; Brière, B.; Lorcy, D. Single-component conductors: A sturdy electronic structure generated by bulky substituents. *Inorg. Chem.* **2016**, *55*, 6036–6046. [CrossRef] [PubMed]
866. Branzea, D.G.; Pop, F.; Auban-Senzier, P.; Clérac, R.; Alemany, P.; Canadell, E.; Avarvari, N. Localization versus delocalization in chiral single component conductors of gold bis(dithiolene) complexes. *J. Am. Chem. Soc.* **2016**, *138*, 6838–6851. [CrossRef] [PubMed]
867. Pop, F.; Avarvari, N. Chiral metal-dithiolene complexes. *Coord. Chem. Rev.* **2017**, *346*, 20–31. [CrossRef]
868. Filatre-Furcate, A.; Roisnel, T.; Fourmigué, M.; Jeannin, O.; Bellec, N.; Auban-Senzier, P.; Lorcy, D. Subtle steric differences impact the structural and conducting properties of radical gold bis(dithiolene) complexes. *Chem. Eur. J.* **2017**, *23*, 16004–16013. [CrossRef]
869. Kato, R.; Suzumura, Y. Novel Dirac electron in single-component molecular conductor [Pd(dddt)$_2$] (dddt = 5,6-dihydro-1,4-dithiin-2,3-dithiolate). *J. Phys. Soc. Jpn.* **2017**, *86*, 064705. [CrossRef]
870. Kato, R.; Cui, H.-B.; Tsumuraya, T.; Miyazaki, T.; Suzumura, Y. Emergence of the Dirac electron system in a single-component molecular conductor under high pressure. *J. Am. Chem. Soc.* **2017**, *139*, 1770–1773. [CrossRef]
871. Suzumura, Y. Anisotropic conductivity of nodal line semimetal in single-component molecular conductor [Pd(dddt)$_2$]. *J. Phys. Soc. Jpn.* **2017**, *86*, 124710. [CrossRef]
872. Suzumura, Y.; Kato, R. Magnetic susceptibility of Dirac electrons in single-component molecular conductor [Pd(dddt)$_2$] under pressure. *Jpn. J. Appl. Phys.* **2017**, *56*, 05FB02. [CrossRef]
873. Kobayashi, Y.; Terauchi, T.; Sumi, S.; Matsushita, Y. Carrier generation and electronic properties of a single-component pure organic metal. *Nat. Mater.* **2017**, *16*, 109–114. [CrossRef] [PubMed]
874. Le Gal, Y.; Roisnel, T.; Auban-Senzier, P.; Bellec, N.; Íñiguez, J.; Canadell, E.; Lorcy, D. Stable Metallic state of a neutral-radical single-component conductor at ambient pressure. *J. Am. Chem. Soc.* **2018**, *140*, 6998–7004. [CrossRef] [PubMed]
875. Tsumuraya, T.; Kato, R.; Suzumura, Y. Effective hamiltonian of topological nodal line semimetal in single-component molecular conductor [Pd(dddt)$_2$] from first-principles. *J. Phys. Soc. Jpn.* **2018**, *87*, 113701. [CrossRef]
876. Suzumura, Y.; Cui, H.; Kato, R. Conductivity and resistivity of Dirac electrons in single-component molecular conductor [Pd(dddt)$_2$]. *J. Phys. Soc. Jpn.* **2018**, *87*, 084702. [CrossRef]
877. Silva, R.A.L.; Vieira, B.J.C.; Andrade, M.M.A.; Santos, I.C.; Rabaça, S.; Lopes, E.B.; Coutinho, J.T.; Pereira, L.C.J.; Almeida, M.; Belo, D. Gold and nickel extended thiophenic-TTF bisdithiolene complexes. *Molecules* **2018**, *23*, 424. [CrossRef] [PubMed]
878. Zhou, B.; Ishibashi, S.; Ishii, T.; Sekine, T.; Takehara, R.; Miyagawa, K.; Kanoda, K.; Nishibori, E.; Kobayashi, A. Single-component molecular conductor [Pt(dmdt)$_2$]—A three-dimensional ambient-pressure molecular Dirac electron system. *Chem. Commun.* **2019**, *55*, 3327–3330. [CrossRef] [PubMed]
879. Hachem, H.; Bellec, N.; Fourmigué, M.; Lorcy, D. Hydrogen bonding interactions in single component molecular conductors based on metal (Ni, Au) bis(dithiolene) complexes. *Dalton Trans.* **2020**, *49*, 6056–6064. [CrossRef]
880. Suzumura, Y.; Kato, R.; Ogata, M. Electric transport of nodal line semimetals in single-component molecular conductors. *Crystals* **2020**, *10*, 862. [CrossRef]
881. Kato, R.; Cui, H.; Minamidate, T.; Yeung, H.H.-M.; Suzumura, Y. Electronic structure of a single-component molecular conductor [Pd(dddt)$_2$] (dddt = 5,6-dihydro-1,4-dithiin-2,3-dithiolate) under high pressure. *J. Phys. Soc. Jpn.* **2020**, *89*, 124706. [CrossRef]
882. Cui, H.; Yeung, H.H.-M.; Kawasugi, Y.; Minamidate, T.; Saunders, L.K.; Kato, R. High-pressure crystal structure and unusual magnetoresistance of a single-component molecular conductor [Pd(dddt)$_2$] (dddt = 5,6-dihydro-1,4-dithiin-2,3-dithiolate). *Crystals* **2021**, *11*, 534. [CrossRef]
883. Tajima, N.; Sugawara, S.; Tamura, M.; Kato, R.; Nishio, Y.; Kajita, K. Transport properties of massless Dirac fermions in an organic conductor α-(BEDT-TTF)$_2$I$_3$ under pressure. *EPL* **2007**, *80*, 47002-p1-p5. [CrossRef]
884. Kobayashi, A.; Suzumura, Y.; Fukuyama, H. Hall effect and orbital diamagnetism in zerogap state of molecular conductor α-(BEDT-TTF)$_2$I$_3$. *J. Phys. Soc. Jpn.* **2008**, *77*, 064718. [CrossRef]
885. Castro Neto, A.H.; Guinea, F.; Peres, N.M.R.; Novoselov, K.S.; Geim, A.K. The electronic properties of graphene. *Rev. Mod. Phys.* **2009**, *81*, 109–162. [CrossRef]
886. Montambaux, G.; Piéchon, F.; Fuchs, J.-N.; Goerbig, M.O. Merging of Dirac points in a two-dimensional crystal. *Phys. Rev. B* **2009**, *80*, 153412. [CrossRef]
887. Hirata, M.; Ishikawa, K.; Miyagawa, K.; Kanoda, K.; Tamura, M. ^{13}C NMR study on the charge-disproportionated conducting state in the quasi-two-dimensional organic conductor α-(BEDT-TTF)$_2$I$_3$. *Phys. Rev. B* **2011**, *84*, 125133. [CrossRef]

888. Hiraki, K.-I.; Harada, S.; Arai, K.; Takano, Y.; Takahashi, T.; Tajima, N.; Kato, R.; Naito, T. Local spin susceptibility of α-D_2I_3 (D = bis(ethylendithio)tetraselenafulvalene (BETS) and bis(ethylendithio) dithiadiselenafulvalene (BEDT-STF)) studied by ^{77}Se NMR. *J. Phys. Soc. Jpn.* **2011**, *80*, 014715. [CrossRef]
889. Suzumura, Y.; Kobayashi, A. Berry curvature of the dirac particle in α-(BEDT-TTF)$_2$I$_3$. *J. Phys. Soc. Jpn.* **2011**, *80*, 104701. [CrossRef]
890. Hosur, P.; Parameswaran, S.A.; Vishwanash, A. Charge transport in Weyl semimetals. *Phys. Rev. Lett.* **2012**, *108*, 046602. [CrossRef] [PubMed]
891. Tajima, N.; Kato, R.; Sugawara, S.; Nishio, Y.; Kajita, K. Interband effects of magnetic field on Hall conductivity in the multilayered massless Dirac fermion system α-(BEDT-TTF)$_2$I$_3$. *Phys. Rev. B* **2012**, *85*, 033401. [CrossRef]
892. Suzumura, Y.; Kobayashi, A. Theory of Dirac electrons in organic conductors. *Crystals* **2012**, *2*, 266–283. [CrossRef]
893. Bácsi, Á.; Virosztek, A. Low-frequency optical conductivity in graphene and in other scale-invariant two-band systems. *Phys. Rev. B* **2013**, *87*, 125425. [CrossRef]
894. Timusk, T.; Carbotte, J.P.; Homes, C.C.; Bosov, D.N.; Sharapov, S.G. Three-dimensional Dirac fermions in quasicrystals seen via optical conductivity. *Phys. Rev. B* **2013**, *87*, 235121. [CrossRef]
895. Monteverde, M.; Goerbig, M.O.; Auban-Senzier, P.; Navarin, F.; Henck, H.; Pasquier, C.R.; Mézière, C.; Batail, P. Coexistence of Dirac and massive carriers in α-(BEDT-TTF)$_2$I$_3$ under hydrostatic pressure. *Phys. Rev. B* **2013**, *87*, 245110. [CrossRef]
896. Suzumura, Y.; Kobayashi, A. Effects of zero line and ferrimagnetic fluctuation on nuclear magnetic resonance for Dirac electrons in molecular conductor α-(BEDT-TTF)$_2$I$_3$. *J. Phys. Soc. Jpn.* **2013**, *82*, 054715. [CrossRef]
897. Kajita, K.; Nishio, Y.; Tajima, N.; Suzumura, Y.; Kobayashi, A. Molecular Dirac fermion systems—Theoretical and experimental approaches. *J. Phys. Soc. Jpn.* **2014**, *83*, 072002. [CrossRef]
898. Wehling, T.O.; Black-Schaffer, A.M.; Balatsky, A.V. Dirac materials. *Adv. Phys.* **2014**, *63*, 1–76. [CrossRef]
899. Wang, J.; Deng, S.; Liu, Z.; Liu, Z. The rare two-dimensional materials with Dirac cones. *Natl. Sci. Rev.* **2015**, *2*, 22–39. [CrossRef]
900. Fuseya, Y.; Ogata, M.; Fukuyama, H. Transport properties and diamagnetism of Dirac electrons in bismuth. *J. Phys. Soc. Jpn.* **2015**, *84*, 012001. [CrossRef]
901. Neubauer, D.; Carbotte, J.P.; Nateprov, A.A.; Löhle, A.; Dressel, M.; Pronin, A.V. Interband optical conductivity of the [001]-oriented Dirac semimetal Cd_3As_2. *Phys. Rev. B* **2016**, *93*, 121202(R). [CrossRef]
902. Miyagawa, K.; Sata, Y.; Taniguchi, T.; Hirata, M.; Liu, D.; Tamura, M.; Kanoda, K. Transition from a metal to a massless–Dirac-fermion phase in an organic conductor investigated by ^{13}C NMR. *J. Phys. Soc. Jpn.* **2016**, *85*, 073710. [CrossRef]
903. Matsuno, G.; Omori, Y.; Eguchi, T.; Kobayashi, A. Topological domain wall and valley Hall effect in charge ordered phase of molecular Dirac fermion system α-(BEDT-TTF)$_2$I$_3$. *J. Phys. Soc. Jpn.* **2016**, *85*, 094710. [CrossRef]
904. Hirata, M.; Ishikawa, K.; Miyagawa, K.; Tamura, M.; Berthier, C.; Basko, D.; Kobayashi, A.; Matsuno, G.; Kanoda, K. Observation of an anisotropic Dirac cone reshaping and ferrimagnetic spin polarization in an organic conductor. *Nat. Commun.* **2016**, *7*, 1–14. [CrossRef] [PubMed]
905. Liu, D.; Ishikawa, K.; Takehara, R.; Miyagawa, K.; Tamura, M.; Kanoda, K. Insulating nature of strongly correlated massless Dirac fermions in an organic crystal. *Phys. Rev. Lett.* **2016**, *116*, 226401. [CrossRef]
906. Beyer, R.; Dengl, A.; Peterseim, T.; Wackerow, S.; Ivek, T.; Pronin, A.V.; Schweitzer, D.; Dressel, M. Pressure-dependent optical investigations of α-(BEDT-TTF)$_2$I$_3$: Tuning charge order and narrow gap towards a Dirac semimetal. *Phys. Rev. B* **2016**, *93*, 195116. [CrossRef]
907. Zhang, C.; Jiao, Y.; Ma, F.; Bottle, S.; Zhao, M.; Chen, Z.; Du, A. Predicting a graphene-like WB_4 nanosheet with a double Dirac cone, an ultra-high Fermi velocity and significant gap opening by spin-orbit coupling. *Phys. Chem. Chem. Phys.* **2017**, *19*, 5449–5453. [CrossRef] [PubMed]
908. Hirata, M.; Ishikawa, K.; Matsuno, G.; Kobayashi, A.; Miyagawa, K.; Tamura, M.; Berthier, C.; Kanoda, K. Anomalous spin correlations and excitonic instability of interacting 2D Weyl fermions. *Science* **2017**, *358*, 1403–1406. [CrossRef] [PubMed]
909. Suzumura, Y. Effect of long-range coulomb interaction on NMR shift in massless Dirac electrons of organic conductor. *J. Phys. Soc. Jpn.* **2018**, *87*, 024705. [CrossRef]
910. Tajima, N. Effects of carrier doping on the transport in the Dirac electron system α-(BEDT-TTF)$_2$I$_3$ under high pressure. *Crystals* **2018**, *8*, 126. [CrossRef]
911. Neubauer, D.; Yaresko, A.; Li, W.; Löhle, A.; Hübner, R.; Schilling, M.B.; Shekhar, C.; Felser, C.; Dressel, M.; Pronin, A.V. Optical conductivity of the Wyle semimetal NbP. *Phys. Rev. B* **2018**, *98*, 195203. [CrossRef]
912. Huang, C.; Li, Y.; Wang, N.; Xue, Y.; Zuo, Z.; Liu, H.; Li, Y. Progress in research into 2D graphdiyne-based materials. *Chem. Rev.* **2018**, *118*, 7744–7803. [CrossRef] [PubMed]
913. Molle, A.; Grazianetti, C.; Tao, L.; Taneja, D.; Alam, M.H.; Akinwande, D. Silicene, silicene derivatives, and their device applications. *Chem. Soc. Rev.* **2018**, *47*, 6370–6387. [CrossRef] [PubMed]
914. Zheng, H.; Hasan, M.Z. Quasiparticle interference on type-I and type-II Weyl semimetal surfaces: A review. *Adv. Phys. X* **2018**, *3*, 569–591. [CrossRef]
915. Osada, T. Topological properties of τ-type organic conductors with a checkerboard lattice. *J. Phys. Soc. Jpn.* **2018**, *88*, 114707. [CrossRef]
916. Suzumura, Y.; Tsumuraya, T.; Kato, R.; Matsuura, H.; Ogata, M. Role of velocity field and principal axis of tilted Dirac cones in effective Hamiltonian of non-coplanar nodal loop. *J. Phys. Soc. Jpn.* **2019**, *88*, 124704. [CrossRef]

917. Tani, T.; Tajima, N.; Kobayashi, A. Field-angle dependence of interlayer magnetoresistance in organic Dirac electron system α-(BEDT-TTF)$_2$I$_3$. *Crystals* **2019**, *9*, 212. [CrossRef]
918. Li, W.; Uykur, E.; Kuntscher, C.A.; Dressel, M. Optical signatures of energy gap in correlated Dirac fermions. *NPJ Quantum Mater.* **2019**, *4*, 19.
919. Mandal, I.; Saha, K. Thermopower in an anisotropic two-dimensional Weyl semimetal. *Phys. Rev. B* **2020**, *101*, 045101. [CrossRef]
920. Naito, T.; Doi, R.; Suzumura, Y. Exotic Dirac cones on the band structure of α-STF$_2$I$_3$ at ambient temperature and pressure. *J. Phys. Soc. Jpn.* **2020**, *89*, 023701. [CrossRef]
921. Naito, T.; Doi, R. Band structure and physical properties of α-STF$_2$I$_3$: Dirac electrons in disordered conduction sheets. *Crystals* **2020**, *10*, 270. [CrossRef]
922. Kobara, R.; Igarashi, S.; Kawasugi, Y.; Doi, R.; Naito, T.; Tamura, M.; Kato, R.; Nishio, Y.; Kajita, K.; Tajima, N. Universal behavior of magnetoresistance in organic Dirac electron systems. *J. Phys. Soc. Jpn.* **2020**, *89*, 113703. [CrossRef]
923. Ohki, D.; Yoshimi, K.; Kobayashi, A. Transport properties of the organic Dirac electron system α-(BEDT-TSeF)$_2$I$_3$. *Phys. Rev. B* **2020**, *102*, 235116. [CrossRef]
924. Tsumuraya, T.; Suzumura, Y. First-principles study of the effective Hamiltonian for Dirac fermions with spin-orbit coupling in two-dimensional molecular conductor α-(BETS)$_2$I$_3$. *Eur. Phys. J. B* **2021**, *94*, 17. [CrossRef]
925. Kitou, S.; Tsumuraya, T.; Sawahata, H.; Ishii, F.; Hiraki, K.-I.; Nakamura, T.; Katayama, N.; Sawa, H. Ambient-pressure Dirac electron system in the quasi-two-dimensional molecular conductor α-(BETS)$_2$I$_3$. *Phys. Rev. B* **2021**, *103*, 035135. [CrossRef]
926. Martin, L.; Turner, S.S.; Day, P.; Guionneau, P.; Haward, J.A.K.; Hibbs, D.E.; Light, M.E.; Hoursthouse, M.B.; Uruichi, M.; Yakushi, K. Crystal chemistry and physical properties of superconducting and semiconducting charge transfer salts of the type (BEDT-TTF)$_4$[A^{I}M^{III}(C$_2$O$_4$)$_3$]·PhCN (A^I = H$_3$O, NH$_4$, K.; M^{III} = Cr, Fe, Co, Al; BEDT-TTF = bis(ethylenedithio)tetrathiafulvalene). *Inorg. Chem.* **2001**, *40*, 1363–1371. [CrossRef]
927. Minguet, M.; Luneau, D.; Lhotel, E.; Villar, V.; Paulsen, C.; Amabilino, D.B.; Veciana, J. An enantiopure molecular ferromagnet. *Angew. Chem. Int. Ed. Engl.* **2002**, *41*, 586–589. [CrossRef]
928. Réthoré, C.; Fourmigué, M.; Avarvari, N. Tetrathiafulvalene based phosphino-oxazolines: A new family of redox active chiral ligands. *Chem. Commun.* **2004**, *4*, 1384–1385. [CrossRef] [PubMed]
929. Coronado, E.; Galán-Mascarós, J.R.; Gómez-García, C.J.; Murcia-Martínez, A.; Canadell, E. A chiral molecular conductor: Synthesis, structure, and physical properties of [ET]$_3$[Sb$_2$(L-tart)$_2$]·CH$_3$CN (ET = bis(ethylendithio)tetrathiafulvalene; L-tart = (2*R*,3*R*)-(+)-tartrate). *Inorg. Chem.* **2004**, *43*, 8072–8077. [CrossRef] [PubMed]
930. Coronado, E.; Galán-Mascarós, J.R.; Gómez-García, C.J.; Martínez-Ferrero, E.; Almeida, M.; Waerenborgh, J.C. Oxalate-based 3D chiral magnets: The series [Z^{II}(bpy)$_3$][ClO$_4$][M^{II}Fe^{III} (ox)$_3$] (Z^{II} = Fe, Ru; M^{II} = Mn, Fe; bpy = 2,2′-bipyridine; ox = oxalate dianion). *Eur. J. Inorg. Chem.* **2005**, 2064–2070. [CrossRef]
931. Réthoré, C.; Avarvari, N.; Canadell, E.; Auban-Senzier, P.; Fourmigué, M. Chiral molecular metals: Syntheses, structures, and properties of the AsF$_6^-$ salts of racemic (±)-, (*R*)-, and (*S*)-tetrathiafulvalene-oxazoline derivatives. *J. Am. Chem. Soc.* **2005**, *127*, 5748–5749. [CrossRef]
932. Martin, L.; Day, P.; Akutsu, H.; Yamada, J.-I.; Nakatsuji, S.; Clegg, W.; Harrington, R.W.; Horton, P.N.; Hursthouse, M.B.; McMillan, P.; et al. Metallic molecular crystals containing chiral or racemic guest molecules. *CrystEngComm* **2007**, *9*, 865–867. [CrossRef]
933. Avarvari, N.; Wallis, J.D. Strategies towards chiral molecular conductors. *J. Mater. Chem.* **2009**, *19*, 4061–4076. [CrossRef]
934. Yang, S.; Brooks, A.C.; Martin, L.; Day, P.; Li, H.; Horton, P.; Male, L.; Wallis, J.D. Novel enantiopure bis(pyrrolo)tetrathiafulvalene donors exhibiting chiral crystal packing arrangements. *CrystEngComm* **2009**, *11*, 993–996. [CrossRef]
935. Martin, L.; Day, P.; Nakatsuji, S.; Yamada, J.; Akutsu, H.; Horton, P. A molecular charge transfer salt of BEDT-TTF containing a single enantiomer of tris(oxalate)chromate(III) crystallised from a chiral solvent. *CrystEngComm* **2010**, *12*, 1369–1372. [CrossRef]
936. Martin, L.; Day, P.; Horton, P.; Nakatsuji, S.; Yamada, J.; Akutsu, H. Chiral conducting salts of BEDT-TTF containing a single enantiomer tris(oxalate)chromate(III) crystallised from a chiral solvent. *J. Mater. Chem.* **2010**, *20*, 2738–2742. [CrossRef]
937. Madalan, A.M.; Réthoré, C.; Fourmigué, M.; Canadell, E.; Lopes, E.B.; Almeida, M.; Auban-Senzier, P.; Avarvari, N. Order versus disorder in chiral tetrathiafulvalene-oxazoline radical-cation salts: Structural and theoretical investigations and physical properties. *Chem. Eur. J.* **2010**, *16*, 528–537. [CrossRef]
938. Yang, S.; Brooks, A.C.; Martin, L.; Day, P.; Pilkington, M.; Clegg, W.; Harrington, R.W.; Russo, L.; Wallis, J.D. New chiral organosulfur donors related to bis(ethylenedithio)tetrathiafulvalene. *Tetrahedron* **2010**, *66*, 6977–6989. [CrossRef]
939. Awheda, I.; Krivickas, S.J.; Yang, S.; Martin, L.; Guziak, M.A.; Brooks, A.C.; Pelletier, F.; Le Kerneau, M.; Day, P.; Horton, P.N.; et al. Synthesis of new chiral organosulfur donors with hydrogen bonding functionality and their first charge transfer salts. *Tetrahedron* **2013**, *69*, 8738–8750. [CrossRef]
940. Biet, T.; Fihey, A.; Cauchy, T.; Vanthuyne, N.; Roussel, C.; Crassous, J.; Avarvari, N. Ethylenedithio-tetrathiafulvalene-helicenes: Electroactive helical precursors with switchable chiroptical properties. *Chem. Eur. J.* **2013**, *19*, 13160–13167. [CrossRef] [PubMed]
941. Pop, F.; Auban-Senzier, P.; Frąckowiak, A.; Ptaszyński, K.; Olejniczak, I.; Wallis, J.D.; Canadell, E.; Avarvari, N. Chirality driven metallic versus semiconducting behavior in a complete series of radical cation salts based on dimethyl-ethylenedithio-tetrathiafulvalene (DM-EDT-TTF). *J. Am. Chem. Soc.* **2013**, *135*, 17176–17186. [CrossRef]
942. Pop, F.; Laroussi, S.; Cauchy, T.; Gomez-Garcia, C.J.; Wallis, J.D.; Avarvari, N. Tetramethyl-bis(ethylenedithio)-tetrathiafulvalene (TM-BEDT-TTF) revisited: Crystal structures, chiroptical properties, theoretical calculations, and a complete series of conducting radical cation salts. *Chirality* **2013**, *25*, 466–474. [CrossRef] [PubMed]

943. Pop, F.; Auban-Senzier, P.; Canadell, E.; Rikken, G.L.J.A.; Avarvari, N. Electrical magnetochiral anisotropy in a bulk chiral molecular conductor. *Nat. Commun.* **2014**, *5*, 3757. [CrossRef] [PubMed]
944. Yang, S.; Pop, F.; Melan, C.; Brooks, A.C.; Martin, L.; Horton, P.; Auban-Senzier, P.; Rikken, G.L.J.A.; Avarvari, N.; Wallis, J.D. Charge transfer complexes and radical cation salts of chiral methylated organosulfur donors. *CrystEngComm* **2014**, *16*, 3906–3916. [CrossRef]
945. Martin, L.; Akutsu, H.; Horton, P.N.; Hursthouse, M.B. Chirality in charge-transfer salts of BEDT-TTF of tris(oxalato)chromate(III). *CrystEngComm* **2015**, *17*, 2783–2790. [CrossRef]
946. Martin, L.; Akutsu, H.; Horton, P.N.; Hursthouse, M.B.; Harrington, R.W.; Clegg, W. Chiral radical-cation salts of BEDT-TTF containing a single enantiomer of tris(oxalato)aluminate(III) and –chromate(III). *Eur. J. Inorg. Chem.* **2015**, *2015*, 1865–1870. [CrossRef]
947. Atzori, M.; Pop, F.; Auban-Senzier, P.; Clérac, R.; Canadell, E.; Mercuri, M.L.; Avarvari, N. Complete series of chiral paramagnetic molecular conductors based on tetramethyl-bis(ethylenedithio)-tetrathiafulvalene (TM-BEDT-TTF) and chloranilate-bridged heterobimetallic honeycomb layers. *Inorg. Chem.* **2015**, *54*, 3643–3653. [CrossRef]
948. Togawa, Y.; Kousaka, Y.; Inoue, K.; Kishine, J.-I. Symmetry, structure, and dynamics of monoaxial chiral magnets. *J. Phys. Soc. Jpn.* **2016**, *85*, 112001. [CrossRef]
949. Martin, L. Molecular conductors of BEDT-TTF with tris(oxalate)metallate anions. *Coord. Chem. Rev.* **2018**, *376*, 277–291. [CrossRef]
950. Pop, F.; Zigon, N.; Avarvari, N. Main-group-based electro- and photoactive chiral materials. *Chem. Rev.* **2019**, *119*, 8435–8478. [CrossRef] [PubMed]
951. Chen, T.; Tomita, T.; Minami, S.; Fu, M.; Koretsune, T.; Kitatani, M.; Muhammad, I.; Nishio-Hamane, D.; Ishii, R.; Ishii, F.; et al. Anomalous transport due to Wyle fermions in the chiral antiferromagnets Mn_3X, X = Sn, Ge. *Nat. Commun.* **2021**, *12*, 572. [CrossRef]
952. Pouget, J.-P.; Alemany, P.; Canadell, E. Donor-anion interactions in quarter-filled low-dimensional organic conductors. *Mater. Horiz.* **2018**, *5*, 590–640. [CrossRef]
953. Mroweh, N.; Auban-Senzier, P.; Vanthuyne, N.; Canadell, E.; Avarvari, N. Chiral EDT-TTF precursors with one stereogenic centre: Substituent size modulation of the conducting properties in the (R-EDT-TTF)$_2$PF$_6$ (R = Me or Et) series. *J. Mater. Chem. C* **2019**, *7*, 12664–12673. [CrossRef]
954. Mroweh, N.; Pop, F.; Mézière, C.; Allain, M.; Auban-Senzier, P.; Vanthuyne, N.; Alemany, P.; Canadell, E.; Avarvari, N. Combining chirality and hydrogen bonding in methylated ethylenedithio-tetrathiafulvalene primary diamide precursors and radical cation salts. *Cryst. Growth Des.* **2020**, *20*, 2516–2526. [CrossRef]
955. Mroweh, N.; Mézière, C.; Pop, F.; Auban-Senzier, P.; Alemany, P.; Canadell, E.; Avarvari, N. In search of chiral molecular superconductors: κ-[(*S*,*S*)-DM-BEDT-TTF]$_2$ClO$_4$ Revisited. *Adv. Mater.* **2020**, *32*, 2002811. [CrossRef] [PubMed]
956. Mroweh, N.; Mézière, C.; Allain, M.; Auban-Senzier, P.; Canadell, E.; Avarvari, N. Conservation of structural arrangements and 3:1 stoichiometry in a series of crystalline conductors of TMTTF, TMTSF, BEDT-TTF, and chiral DM-EDT-TTF with the oxo-bis[pentafluorotantalate(v)] dianion. *Chem. Sci.* **2020**, *11*, 10078–10091. [CrossRef] [PubMed]
957. Mroweh, N.; Auban-Senzier, P.; Vanthuyne, N.; Lopes, E.B.; Almeida, M.; Canadell, E.; Avarvari, N. Chiral conducting Me-EDT-TTF and Et-EDT-TTF-based radical cation salts with the perchlorate anion. *Crystals* **2020**, *10*, 1069. [CrossRef]
958. Short, J.I.; Blundell, T.J.; Krivickas, S.J.; Yang, S.; Wallis, J.D.; Akutsu, H.; Nakazawa, Y.; Martin, L. Chiral molecular conductor with an insulator-metal transition close to room temperature. *Chem. Commun.* **2020**, *56*, 9497–9500. [CrossRef] [PubMed]
959. Blundell, T.J.; Brannan, M.; Nishimoto, H.; Kadoya, T.; Yamada, J.-I.; Akutsu, H.; Nakazawa, Y.; Martin, L. Chiral metal down to 4.2 K—A BDH-TTP radical-cation salt with spiroboronate anion B(2-chloromandelate)$_2^-$. *Chem. Commun.* **2021**, *57*, 5406–5409. [CrossRef] [PubMed]

crystals

MDPI

Review

Interplay between Vortex Dynamics and Superconducting Gap Structure in Layered Organic Superconductors

Syuma Yasuzuka

Research Center for Condensed Matter Physics, Hiroshima Institute of Technology, Hiroshima 731-5193, Japan; yasuzuka@cc.it-hiroshima.ac.jp; Tel.: +81-82-921-6930

Abstract: Layered organic superconductors motivate intense investigations because they provide various unexpected issues associated with their low dimensionality and the strong electron correlation. Since layered organic superconductors possess simple Fermi surface geometry and they often share similarities to the high temperature oxide superconductors and heavy fermion compounds, research on layered organic superconductors is suitable for understanding the essence and nature of strongly correlated electron systems. In strongly correlated electron systems, one of the central problems concerning the superconducting (SC) state is the symmetry of the SC gap, which is closely related to the paring mechanism. Thus, experimental determination of the SC gap structure is of essential importance. In this review, we present the experimental results for the in-plane angular variation of the flux-flow resistance in layered organic superconductors κ-$(ET)_2Cu(NCS)_2$, β″-$(ET)_2SF_5CH_2CF_2SO_3$, and λ-$(BETS)_2GaCl_4$. The interplay between the vortex dynamics and nodal structures is discussed for these superconductors.

Keywords: layered organic conductor; unconventional superconductivity; vortex dynamics; *d*-wave pairing symmetry; superconducting gap structure; magnetic field; flux-flow resistivity

Citation: Yasuzuka, S. Interplay between Vortex Dynamics and Superconducting Gap Structure in Layered Organic Superconductors. *Crystals* **2021**, *11*, 600. https://doi.org/10.3390/cryst11060600

Academic Editor: Toshio Naito

Received: 24 March 2021
Accepted: 20 May 2021
Published: 26 May 2021

1. Introduction

Physics of organic conductors has provided the fascinating and rich issues associated with the low dimensionality and effect of electron correlation [1]. One of the most interesting systems are the layered organic conductors composed of the ET donor molecule, where ET denotes bis(ethylenedithio)tetrathiafulvalene. The ET donor molecules form two-dimensional (2D) conducting layers that are separated by insulating anion layers with a monovalent ion X^-. Figure 1a shows schematic structure of ET (or BETS) donor conducting layers and anion insulating layers in a layered organic conductor, where BETS denotes bis(ethylenedithio)tetraselenafulvalene. The molecular structures of ET and BETS are shown in the upper right side of Figure 1a. In layered organic superconductors, there are several types (labeled by Greek letters) in packed ET donor layers. The κ-, β″-, and λ-type arrangements in donor layers are shown in Figure 1b–d. In the κ type compounds, κ-$(ET)_2X$, the measurements of Shubnikov-de Haas (SdH) and de Haas-van Alphen (dHvA) oscillations have elucidated the presence of a well-defined Fermi surface (FS) with simple structures [2]. Moreover. the moderately heavy effective mass revealed by SdH and dHvA experiments suggests that electron correlation plays a significant role on determining the physical properties of the normal state as well as superconducting (SC) state. It was suggested that superconductivity appears in proximity to the antiferromagnetic insulating state in the electronic phase diagram [3,4]. Since some of these unusual physical properties suggest similarities with high-T_c cuprates, many researchers have pointed out that the spin fluctuations play a vital role for the appearance of SC state [5–8].

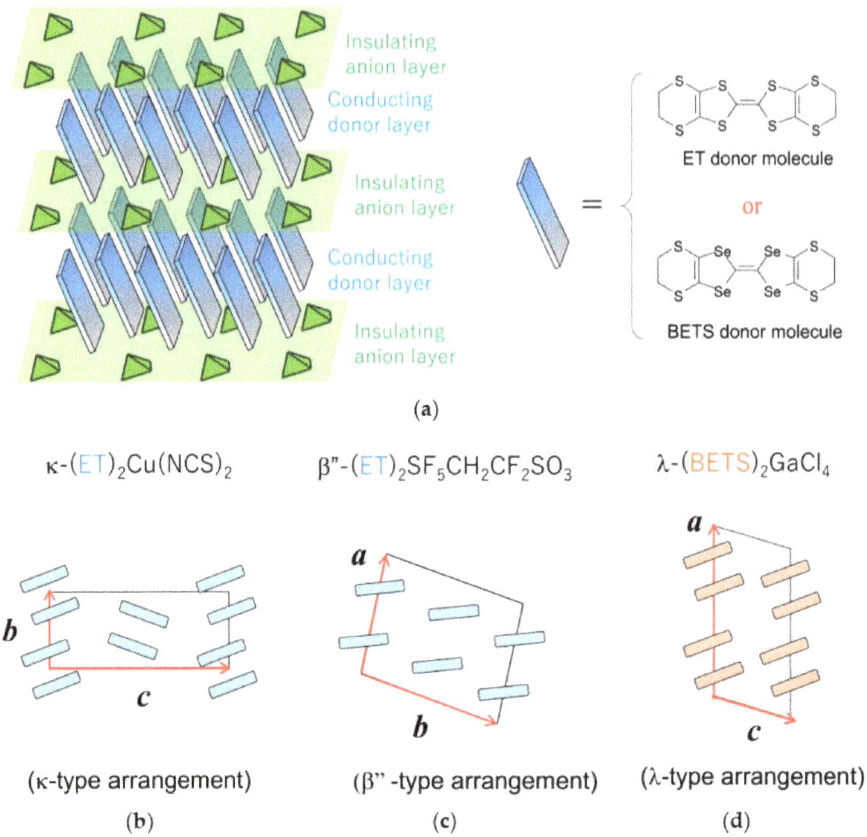

Figure 1. (**a**) (Upper left side) Schematic structure of donor conducting layers and anion insulating layers in a layered organic conductor. (Upper right side) Structures of ET and BETS donor molecules. (Lower side) Three patterns within packed donor conducting layers for (**b**) κ-, (**c**) β″-, and (**d**) λ-type arrangements.

For layered organic superconductors, many experimental studies have pointed out the existence of nodes in the SC gap [9–15]. The SC gap structure in unconventional superconductors mostly have nodes parallel to certain orientations in the momentum space. The confirmation of point or line nodes is often discussed from power-law temperature (T) dependences in physical quantities such as the heat capacity or the nuclear spin relaxation rate. However, determination of the SC gap structure is a harder task. For example, the SC gap symmetry of d_{xy} (with nodes along vertical and horizontal directions) differs from $d_{x^2-y^2}$ (diagonal direction) only in the location of the line nodes ($\pi/4$ rotation of the latter becomes identical with the former). Therefore, direction-sensitive experimental methods are highly needed to distinguish between these gap structures. Since the SC gap structure is intimately associated with the pairing mechanism [16], it is of fundamental importance to elucidate the SC gap structure.

In the vortex state of d-wave superconductors, gapless quasiparticles (QPs) extending outside the vortex core play an important role for determining their physical properties at magnetic fields [17,18]. As an effect of applied magnetic field on d-wave superconductors, Volovik [17] theoretically introduced a "Doppler shifted state" in the energy spectrum of the QP which is induced by superfluid flow around the vortices. The QPs with low energy are induced by the application of magnetic field, mainly around the nodes where the QP velocity is perpendicular to the magnetic field orientation. The QP density depends on the relative direction of the magnetic field with respect to the nodes. Derived from the Doppler sifted state that depend on the magnetic field orientation, the dependence of the heat capacity and the thermal conductivity on applied field orientation was demonstrated to be a useful probe for detecting the nodal directions in a SC gap structure [14,15,19].

Up to now, we have focused on the interplay between vortex dynamics and the SC gap structure in layered organic superconductors because the flux-flow resistivity is a measure of QP dissipation in the vortex dynamics [20–24]. To investigate the correlation between vortex dynamics and the SC gap structure, we measured the in-plane anisotropy of the flux-flow resistivity for the layered organic superconductors κ-$(ET)_2Cu(NCS)_2$, β″-$(ET)_2SF_5CH_2CF_2SO_3$, and λ-$(BETS)_2GaCl_4$ [25–27]. SC parameters for these superconductors are listed in Table 1. As shown below, we clearly observed the flux-flow resistivity with fourfold-symmetric anisotropy, owning to the *d*-wave SC gap symmetry in κ-$(ET)_2Cu(NCS)_2$ and β″-$(ET)_2SF_5CH_2CF_2SO_3$. On the other hand, twofold symmetric anisotropy was found in λ-$(BETS)_2GaCl_4$ although λ-$(BETS)_2GaCl_4$ possesses the similar FS and SC gap structures in the former two superconductors. Interplay between in-plane anisotropy of vortex dynamics and nodal SC gap structures for these superconductors are discussed below.

Table 1. Superconducting parameters for κ-$(ET)_2Cu(NCS)_2$ [2], β″-$(ET)_2SF_5CH_2CF_2SO_3$ [2,28–30] and λ-$(BETS)_2GaCl_4$ [2,31,32].

Properties	Unit	κ-$(ET)_2Cu(NCS)_2$	β″-$(ET)_2SF_5CH_2CF_2SO_3$	λ-$(BETS)_2GaCl_4$
T_c	K	8.7–10.4	5.2	5–8
$B_{c1\parallel}$	mT	0.2	0.006	5.2
$B_{c1\perp}$	mT	6.5	2	8.2
$B_{c2\parallel}$	T	30–35	10.4	12
$B_{c2\perp}$	T	6	1.4	3
$\lambda_{\parallel}$	Å	5100–20,000	10,000–20,000	1500
$\lambda_{\perp}$	μm	40–200	400–800	
$\xi_{\parallel}$	Å	74	144	105
$\xi_{\perp}$	Å	5–9	7.9	9–14
$\kappa_{\parallel}$	-	100–200	59	107

2. Experimental Methods

Single crystals of κ-$(ET)_2Cu(NCS)_2$, β″-$(ET)_2SF_5CH_2CF_2SO_3$, and λ-$(BETS)_2GaCl_4$ were synthesized electrochemically. The interlayer resistance was measured by a four-terminal ac method, where the electric current is parallel to the least conducting direction. To investigate the in-plane angular variation of the flux flow resistance for layered organic superconductors, it is crucial to rotate applied field parallel to the conducting layers with high accuracy because a slight field-angle misalignment gives rise to twofold anisotropy of the magnetoresistance related to the huge H_{c2} anisotropy [33]. To check the reproducibility, four samples were simultaneously mounted on a two-axis rotator in a ^{4}He cryostat with a 17-T SC magnet. By using the rotator, it is possible to sweep the θ angle with a resolution of $\Delta\theta = 0.1°$, where θ is the polar angle between the least conducting axis and magnetic field direction. In addition, we can discretely control the plane of rotation, which is represented by the azimuthal angle ϕ with intervals of $\Delta\phi = 5°$ or 10 within the conducting layers.

To investigate the in-plane anisotropy of the flux-flow resistance at a constant magnetic field, the resistivity as a function of the angle θ at various fixed ϕ was measured because we can easily find the position of the *H* parallel to the conducting plane at any ϕ. For example, the interlayer resistivity of κ-$(ET)_2Cu(NCS)_2$ is shown in Figure 2a [25]. For the magnetic field exactly applied parallel to the conducting plane, the ρ(θ) curve shows a sharp drop due to the SC transition at the lowest current density of 0.5 mA/cm^2. It is remarkable that the resistivity at $\theta = 90°$ depends strongly on the current density. A sharp peak is clearly observed at $\theta = 90°$ for highest current density of 100 mA/cm^2. The resistivity for $|\theta - 90°| > 5°$ is independent of current density of up to 100 mA/cm^2. Thus, the effect of Joule heating is negligibly small. Instead, the sharp peaks are due to the vortex dynamics [25,33–38]. Similar features are found for β″-$(ET)_2SF_5CH_2CF_2SO_3$ in Figure 2b and λ-$(BETS)_2GaCl_4$ in Figure 2c [26,27].

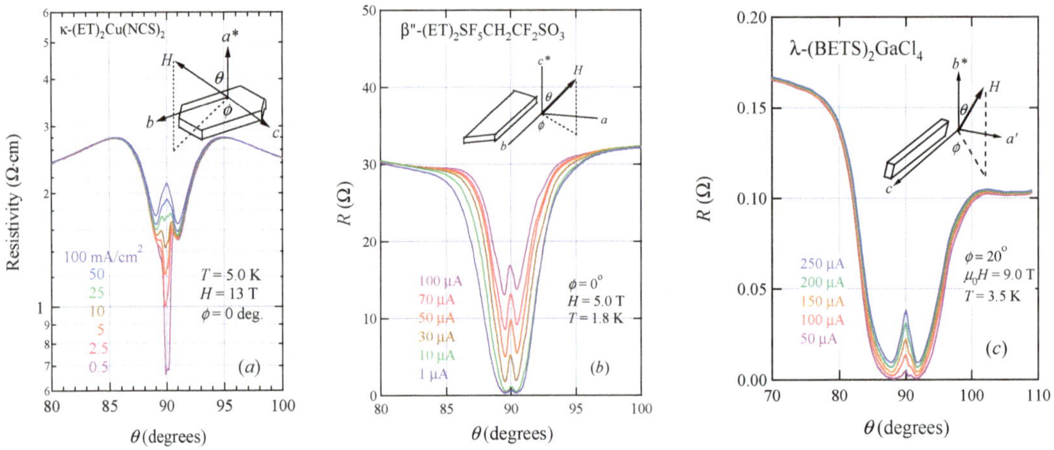

Figure 2. Interlayer resistance as a function of polar angle θ for κ-$(ET)_2Cu(NCS)_2$ (**a**), β″-$(ET)_2SF_5CH_2SF_2SO_3$ (**b**), and λ-$(BETS)_2GaCl_4$ (**c**). The insets show the definition of θ and ϕ. Adapted from refs. [25–27].

3. Results and Discussion

3.1. κ-$(ET)_2Cu(NCS)_2$

The nature of the κ-type organic superconductor κ-$(ET)_2X$ possesses similar features to that of high-T_c cuprates, such as quasi-2D electronic structure and the competition of its SC phase with the antiferromagnetic insulating phase [2–4]. To elucidate the pairing mechanism in the κ-$(ET)_2X$ system, its SC gap symmetry has been extensively investigated from both experimental and theoretical points of view [9–14]. Various experiments such as NMR [9], heat capacity [10,11], and mm-wave transmission measurements [12] suggested presence of *d*-wave SC gap in κ-$(ET)_2Cu(NCS)_2$. As for the nodal direction, scanning tunneling microscopy (STM) [13] and thermal conductivity experiments [14] suggested the line node gap rotated 45° relative to the *b* and *c* axes ($d_{x^2-y^2}$ symmetry). Malone et al. [15] argued in favor of d_{xy} symmetry (the line nodes along the *b* and *c* axes) from the heat capacity measurement. These results have demonstrated the *d*-wave superconductivity in the κ-$(ET)_2X$ system, but the location of gap nodes is still controversial. In this section, we present in-plane anisotropy of upper critical field and vortex flow resistivity for κ-$(ET)_2Cu(NCS)_2$ and discuss the nodal structure.

3.1.1. In-Plane Anisotropy of Upper Critical Field in κ-$(ET)_2Cu(NCS)_2$

To discuss the in-plane anisotropy of upper critical field, measurements of the interlayer resistance were performed for $\phi = 0°$ (H // c), $\phi = 45°$, $\phi = 90°$ (H // b). Figure 3a presents the temperature dependence of interlayer resistance for various values of magnetic field applied along the *b-axis* ($\phi = 90°$) up to 14.8 T. Due to a broad transition into the SC state, the following four criteria are used here to extract the SC transition temperature at each magnetic field. These are shown in Figure 3a, where the four criteria of the SC transition temperature: a "junction" T_J, a "midpoint" T_M, "a zero resistance extrapolation" T_X (ignoring the tail near $R = 0$), and "zero resistance point" T_Z are defined. A similar fashion was confirmed for $\phi = 45°$ and $\phi = 0°$ (H // **c**).

Figure 2b presents a *T–H* phase diagram for three magnetic field directions, using the four criteria. All criteria clearly show highly isotropic behavior within our experimental error, suggesting the lack of in-plane anisotropy of H_{c2} in spite of many suggestions of *d*-wave pairing symmetry in κ-(ET)$_2$Cu(NCS)$_2$. This is consistent with previous magnetoresistance study by Nam et al. [39]. Lower temperatures will be needed for confirmation of an expected *d*-wave nature in H_{c2}. In addition, we find change of slope in H_{c2} line near T_c, suggesting the dimensional crossover from anisotropic 3D SC to Josephson coupled 2D SC state [40–42]. It means that despite short coherence length along the interlayer direction, the SC state can be regarded as anisotropic 3D SC state only near H_{c2} line.

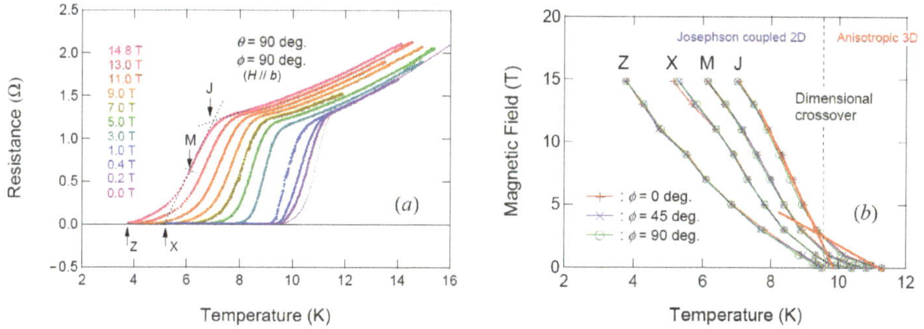

Figure 3. (**a**) Temperature dependence of interlayer resistance for various magnetic fields H // b (ϕ = 90°) in κ-(ET)$_2$Cu(NCS)$_2$. (**b**) H–T phase diagram derived from (**a**), where four criteria, J (junction), M (midpoint), X ($R \to 0$), and Z ($R = 0$) are plotted for $T_c(H)$.

3.1.2. In-Plane Anisotropy of Vortex Dynamics in κ-(ET)$_2$Cu(NCS)$_2$

Figure 4a presents the interlayer resistivity as a function of polar angle θ at various ϕ values with an interval of $\Delta\phi$ = 5°. As seen in Figure 4a, there is the peak structure at θ = 90° at all ϕ values, which confirms the vortex motion for all in-plane field directions. The peak structure gradually changes with ϕ. To discuss the in-plane anisotropic effect in flux-flow resistance, we display the ϕ dependence of the interlayer resistivity at θ = 90° at various temperatures in Figure 4b. In the entire temperature range measured, the $\rho(\phi,T)$ curves show minima at ϕ = 0° (H // c) and 90° (H // b). At temperatures beyond 7.8 K, a sinusoidal angular variation was observed. At elevated temperature, the amplitude of oscillatory behavior is weakened but survives even at 18 K, where the SC phase is vanished. At temperatures below 7.8 K, we find a non-sinusoidal fourfold angular-variation: cusp-like minima are found at ϕ = 0° and 90° that are independent of temperature.

As seen in Figure 4b, the ρ(ϕ) curves seem to be described by a fourfold symmetric origin. Takanaka and Kuboya theoretically showed that the angular variation of the upper critical field is described as $H_{c2}(\phi) \propto \cos 4\phi$ in the *d*-wave superconductivity [43–45]. The ρ(ϕ) curves are naively consistent with this theory; the maximum and minimum of H_{c2} give the minimum and maximum resistivities, respectively. When $d_{x^2-y^2}$ pairing is assumed, $H_{c2}(\phi)$ has maxima at ϕ = 0° (H // c) and ϕ = 90° (H // b), and a minimum at ϕ = 45°, whereas d_{xy} pairing leads to H_{c2} with maxima and minima reversed with respect to $d_{x^2-y^2}$ pairing. Thus, the fourfold symmetry in Figure 3b is not inconsistent with the H_{c2} anisotropy originating from the $d_{x^2-y^2}$ pairing state. However, detailed magnetoresistance studies at 1.5 and 4.2 K [39] as well as Figure 3 showed the lack of in-plane anisotropy of H_{c2}. In addition, the cusp-like minima that appear below 7.8 K cannot be represented by a simple cos4ϕ dependence. Thus, another mechanism to describe these oscillation patterns is necessary.

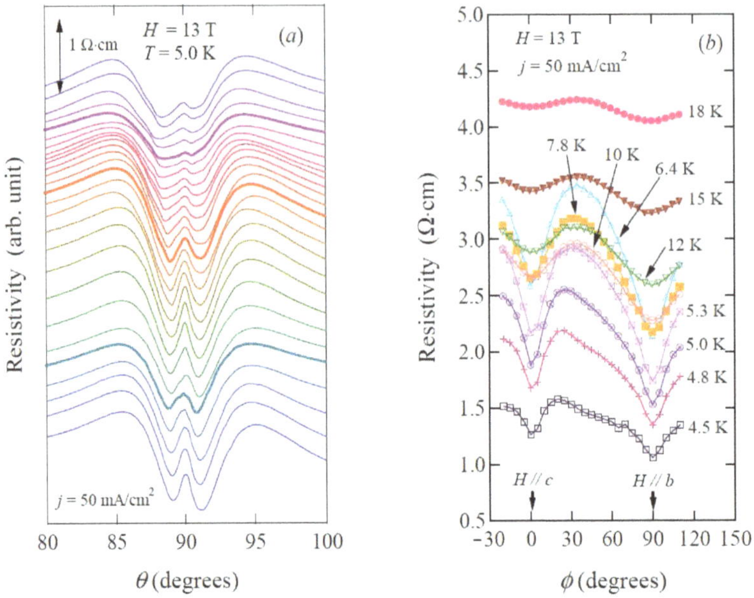

Figure 4. (**a**) Polar angle dependence of interlayer resistance at 5.0 K under a magnetic field of 13 T for various values of ϕ. The curves are measured in intervals of $\Delta\phi = 5°$ between 120° (top curve) and −20° (bottom). (**b**) Dependence of resistivity at $\theta = 90°$ at various temperatures. (Reprinted from [25]. Copyright 2013 The Physical Society of Japan.)

As seen in Figure 4b, we observe sharp peaks related to dissipative feature arising from the vortex dynamics at any azimuth direction within the conducting plane. Therefore, the origin of the cusp-like minima is discussed in terms of the anisotropic vortex dynamics. Since the Josephson vortex essentially does not have a normal core, the pinning effect arising from lattice defects and/or impurities is negligibly small for vortex motion parallel to the conducting layers. In this situation, the friction force is largely affected by the damping viscosity, which depends on the energy dissipation processes of QPs. In the simple description of flux-flow behavior, the Lorentz force on a vortex is balanced by a viscous drag force. Since the Lorentz force keeps constant during the field direction changed, the anisotropic flux-flow resistivity observed here suggests a remarkable variation of the QP excitations by the applied field rotating within the conducting layers because flux-flow resistance can be regarded as a measure of QP dissipation in the vortex dynamics [20–24].

Let us discuss the mechanism of the observed fourfold oscillation in terms of the Doppler effect originally proposed by Volovik [17,18]. In the *d*-wave paring state, a nonzero QP density-of-state (DOS) is generated on the Fermi surface by the effect of Doppler shifted state. Theoretical studies [18] clearly show that the DOS possesses a broad maximum for the field parallel to the antinodal direction and sharp minimum for the field along the node. When an applied field is parallel to the antinodal orientation, all four nodes affect the DOS. On the other hand, when the field is parallel to a nodal orientation, QPs at that node do not affect the DOS. Since a damping of the Josephson vortex motion by QPs is expected, the stronger damping will occur when the magnetic field is applied parallel to the antinodal orientations, leading to the suppression of the flux-flow resistivity for the antinodal orientation. When the $d_{x^2-y^2}$ pairing is assumed, $\rho(\phi)$ should become minima at $\phi = 0°$ (H // c) and $\phi = 90°$ (H // b) and maximum at $\phi = 45°$, which agrees with Figure 4b. Thus, we consider that our result concerned with flux-flow resistance is consistent with $d_{x^2-y^2}$ pairing state, as suggested by STM [13] and thermal conductivity experiments [14].

The above consideration is essentially based on the semiclassical theory [17,18]. Unfortunately, this semiclassical treatment is valid only at low temperatures and low fields and does not consider the scattering effect between the QPs and moving vortices. For wider temperature and field ranges, Vorontsov and Vekhter [46,47] developed a microscopic theory for the angular dependent properties of thermal conductivity and heat capacity in d-wave SC state. They found that application of the magnetic field parallel to the SC gap nodes may result in maxima or minima in their angular dependence, depending upon the location in the T–H plane. This variation is attributable to the scattering of the QPs on vortices. Such sign reversal oscillation was observed and discussed in the heavy fermion superconductor $CeCoIn_5$ [48]. A similar problem between the thermal conductivity and heat capacity measurements was discussed for κ-$(ET)_2Cu(NCS)_2$ [14,15]. According to the T–H diagrams in [46,47], field-angle dependences of heat capacity and thermal conductivity do not show sign reversal pattern at $H/H_{c2} \sim 0.43$ and $T/T_c > 0.48$, where our experimental study was done. Although our results are not inconsistent with no sign reversal oscillation for a wide temperature region, further experimental study such as confirmation of sign reversal pattern at lower fields will be needed to elucidate the effect of the Doppler shifted state on the vortex dynamics.

3.2. β″-$(ET)_2SF_5CH_2CF_2SO_3$

The crystal structure of β″-$(ET)_2SF_5CH_2CF_2SO_3$ (T_c = 5.2 K) has triclinic symmetry and the ET donor molecules form 2D conduction layers parallel to the a-b plane [49]. The insulating $SF_5CH_2CF_2SO_3^-$ layer is intercalated between the ET donor layers, which makes the c*-axis the least conducting direction. Both a 2D Fermi pocket and a pair of quasi-1D Fermi sheets is predicted from the band-structure calculation [50]. The SdH effect and AMRO (angular-dependent magnetoresistance oscillation) studies clearly show one small 2D FS with an area of 5% of the first Brillouin zone [50,51]. Reflecting the layered structure, its GL coherence length perpendicular to the conducting layers, $\xi_\perp(0)$ (~ 7.9 ± 1.5 Å), is shorter than the interlayer spacing d of 17.5 Å [28].

There are three intriguing properties in β″-$(ET)_2SF_5CH_2CF_2SO_3$. First, high field AMRO showed the nature of incoherent interlayer transport. This means that the FS of β″-$(ET)_2SF_5CH_2CF_2SO_3$ is regarded as the highly 2D confined electron system [52]. Second, the FS structure is very similar to that of κ-$(ET)_2Cu(NCS)_2$ [49]. Third and most importantly, the SC state of β″-$(ET)_2SF_5CH_2CF_2SO_3$ may be mediated by charge fluctuations because a pressure-induced charge ordering state is observed at around 1 GPa [53]. For β″ type organic conductors, a large intersite Coulomb repulsion has been theoretically predicted [6,8,54]. Moreover, d-wave superconductivity mediated by charge fluctuations has been proposed based on above theoretical study [8,55], but not confirmed experimentally. In this section, we discuss the relationship between in-plane angular variation of H_{c2} and vortex dynamics in β″-$(ET)_2SF_5CH_2CF_2SO_3$.

3.2.1. Anisotropy of Upper Critical Field in β″-$(ET)_2SF_5CH_2CF_2SO_3$

To study the upper critical field, $H_{c2}(\phi)$, within the conducting a-b plane, the interlayer resistance as a function of magnetic field at various fixed ϕ was investigated as shown by Figure 5a. The resistance curves were taken in intervals of $\Delta\phi = 10°$. At around the SC transition, the resistance gradually reaches zero with decreasing magnetic field. Figure 5b presents the in-plane angular variation of H_{c2}. We define H_{c2}, when R/R_n = 0.9, 0.7, and 0.5, where R_n is the normal state resistance given by extrapolating $R(H)$ in the higher field region of $13 \leqq \mu_0 H \leqq 14$ T. It is clear that the angular variation of H_{c2} exhibits maxima at around $\phi = \pm 90°$ and $\phi = 0°$. Although the values of H_{c2} are changed by the different criteria, the fourfold oscillation pattern of H_{c2} itself remains.

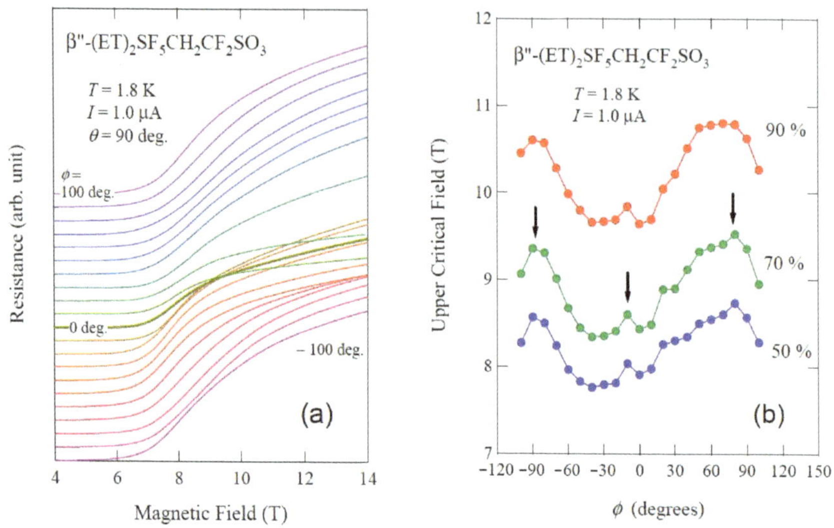

Figure 5. (**a**) Magnetic field dependence of interlayer resistance at various fixed ϕ in the ohmic regime, where applied field is parallel to the conducting plane. The curves are taken from $\phi = 100°$ (top curve) to $-100°$ (bottom) with intervals of $\Delta\phi = 10°$. (**b**) In-plane angular variation of H_{c2} determined from the resistive transition. The H_{c2} values are defined as the fields at which the resistance of the measured sample has reached 90%, 70%, and 50% of its normal-state value. (Reprinted from [27]. Copyright 2015 The Physical Society of Japan.)

For an unconventional superconductor, $H_{c2}(\phi)$ minima occur for applied field parallel to the nodal directions. Thus, it is considered that the SC gap possesses its node (or minimum) at approximately $\pi/4$ from the *b*-axis. This result is in favor of a $d_{x^2-y^2}$ gap symmetry [38].

3.2.2. In-Plane Anisotropy of Vortex Dynamics in β″-$(ET)_2SF_5CH_2CF_2SO_3$

To further discuss observed *d*-wave like anisotropy of H_{c2}, the in-plane anisotropy of the vortex dynamics is next shown. Figure 6 presents the polar angle dependence of the interlayer resistance at various ϕ values. The structures around $\theta = 90°$ (i.e., dip or peak) depend on ϕ, showing the anisotropic vortex dynamics within the conducting layers. To see the anisotropic field effect, the ϕ-dependence of the interlayer resistance at $\theta = 90°$ for various currents is shown in Figure 7a. At 100 μA, we observe a fourfold angular oscillation: cusp-like minima are observed at $\phi = 20°$ and $\pm 90°$. With increasing current, the amplitude increases, showing the remarkable non-ohmic transport phenomena. Figure 7b presents ϕ-dependence of the interlayer resistance at $\theta = 90°$ for several magnetic fields. As shown by Figure 7b, the effect of magnetic field on the flux-flow resistance is very similar to Figure 7a. With increasing magnetic field, a non-sinusoidal fourfold angular pattern is found. The cusp-like minima are observed at $\phi = 20°$ and $\pm 90°$ that are the same as Figure 7a. The current dependence of the flux-flow resistance at a fixed field strength (Figure 7a) is very similar to the field dependence of that at a fixed current (Figure 7b), suggesting that combination of high magnetic field and large current, that is, Lorentz force plays an important role for appearance of the fourfold pattern in the flux-flow resistance. Similar fourfold pattern has been found in the flux-flow resistance in κ-$(ET)_2Cu(NCS)_2$ [25].

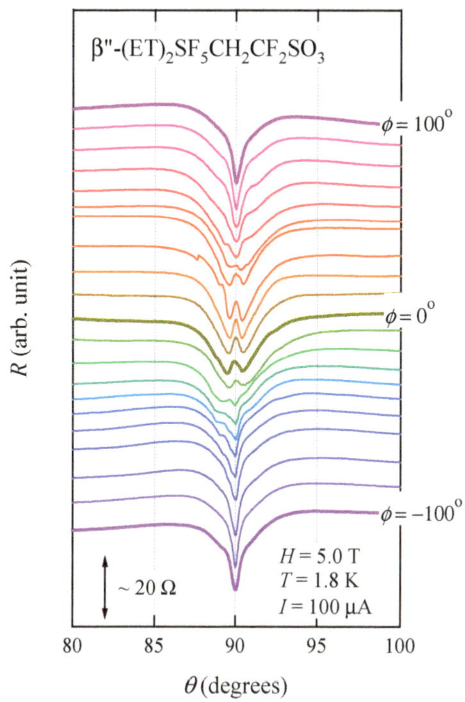

Figure 6. Polar angle dependence of interlayer resistance for various fixed ϕ values. The curves are taken in intervals of Δϕ = 10° between ϕ = 100° (top curve) and −100° (bottom). (Reprinted from [27]. Copyright 2015 The Physical Society of Japan.).

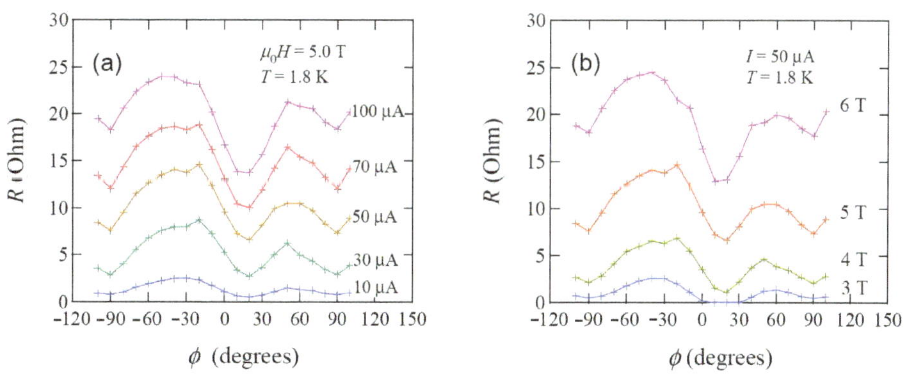

Figure 7. Azimuth angle dependence of the flux-flow resistance for several values of current (**a**) and of magnetic field (**b**). (Reprinted from [27]. Copyright 2015 The Physical Society of Japan.)

According to theoretical studies [18,46,47], the Doppler effect in the *d*-wave paring state gives rise to remarkable response of QP generation on the Fermi surface at low temperatures. It is known that as a field is parallel to the antinodal orientation, QPs are excited at four nodes that contribute to the DOS. On the other hand, when the field is directed along a nodal orientation, QPs at that node do not contribute to the DOS, and therefore the Doppler shift vanishes at these points. Since the QPs are expected to dampen the vortex motion, the flux-flow resistance associated with the vortex motion will be suppressed for the magnetic field applied parallel to the antinodal directions, giving rise to the weaker FFR for the antinodal orientation. For β″-$(ET)_2SF_5CH_2CF_2SO_3$, when $d_{x^2-y^2}$ pairing, as discussed in Figure 4b, is assumed, R(ϕ) should have minima at around ϕ = ±90° and ϕ = 0° and it should have maxima at around ϕ = ±45° orientations, and therefore this is consistent with Figure 7a,b.

Finally, a comment is given on the paring mechanism for the SC state in β″-$(ET)_2SF_5CH_2CF_2SO_3$. The nodal orientations discussed here are far from the orientations of most effective nesting vector on the Fermi surface predicted from the band-structure calculation, as well as determined from the AMRO experiments [26,50,51]. If a spin fluctuation scenario is assumed, it is natural to expect the nodes that are parallel to the antiferromagnetic nesting vector, to be along the b^* and a^* (or a) orientations. It is intriguing to point out that superconductivity with another d-wave symmetry is theoretically suggested based on the charge fluctuation scenario [55]. Further investigation is needed to elucidate which mechanism (i.e., spin fluctuations versus charge fluctuations) is more likely.

3.3. λ-$(BETS)_2GaCl_4$

The layered organic conductor λ-$(BETS)_2GaCl_4$ undergoes a SC transition at T_c of ~8 K [56]. The BETS donor molecules are stacked along the a- and c-axes. The insulating $GaCl_4^-$ anion layers are inserted between the BETS conducting layers. Thus, the 2D conducting layers are formed in λ-$(BETS)_2GaCl_4$. Reflecting the layered structure, its GL coherence length perpendicular to the layers is shorter than the interlayer spacing of 18.6 Å [57]. λ-$(BETS)_2GaCl_4$ is known as a good candidate for realizing the Fulde–Ferrell–Larkin–Ovchinnikov state [57,58]. Another intriguing point is that its isostructural compound λ-$(BETS)_2FeCl_4$ shows a field-induced SC transition [59]. Band-structure calculation [56] predicts the existence of one closed 2D Fermi pocket and two 1D Fermi sheets that are topologically the same as κ-$(ET)_2Cu(NCS)_2$ [50,51]. Measurements of the SdH and AMROs are qualitatively consistent with the band calculation [60].

As for the SC gap structure, a STM experiment suggested d_{xy}–wave symmetry with the line nodes along the a^*- and c^*-axes [61]. From systematic investigations by chemical substitution in the anions [62,63] or by selecting different donor molecules [64], the SC phase is suggested to exist next to the Mott insulating phase, which is similar to the κ-$(ET)_2X$ system [65]. An NMR study showed the development of spin fluctuations beyond the SC phase transition temperature [66]. A heat capacity study showed a d-wave pairing state [67], whereas a μSR study clamed a mixture of the extended s- and d-wave SC gap [68]. In this section, we discuss the interplay between in-plane anisotropy of vortex dynamics and the SC gap structure for λ-$(BETS)_2GaCl_4$ [26].

In-Plane Anisotropy of Vortex Dynamics in λ-$(BETS)_2GaCl_4$

Figure 8 shows the polar angle dependence of the interlayer resistance at various fixed ϕ values. The sharp peak is clearly observed at θ = 90° at all ϕ values, which shows the vortex dynamics for all ϕ-directions. Figure 9 shows the azimuth angle dependence of the interlayer resistance at θ = 90° for various currents. The in-plane angular dependence is mainly represented by the twofold symmetry with the sharp dip at ϕ = 0° (H // c). This feature differs from the cases of κ-$(ET)_2Cu(NCS)_2$ [25] and β″-$(ET)_2SF_5CH_2CF_2SO_3$ [27] even though the FS geometry of λ-$(BETS)_2GaCl_4$ [56] is similar to that of κ-$(ET)_2Cu(NCS)_2$ [60] and β″-$(ET)_2SF_5CH_2CF_2SO_3$ [50,51].

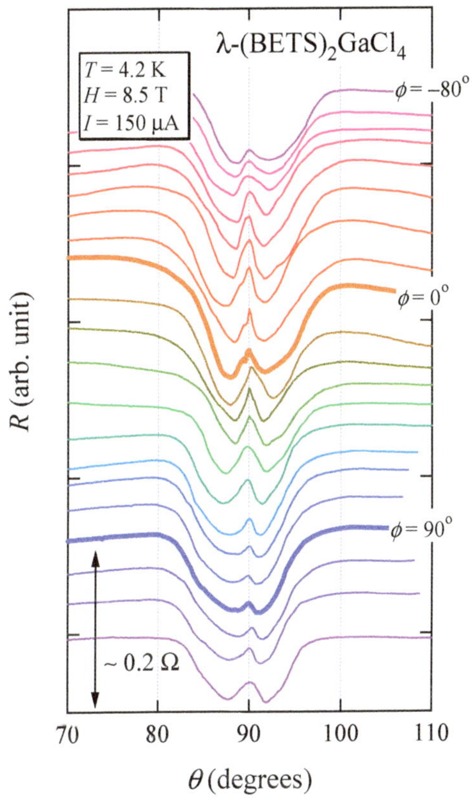

Figure 8. Polar angle dependence of interlayer resistance under rotating field of 8.5 T for various fixed ϕ. The curves are taken in intervals of $\Delta\phi = 10°$ between $-80°$ (top curve) and $120°$ (bottom). (Reprinted from [26]. Copyright 2014 The Physical Society of Japan.)

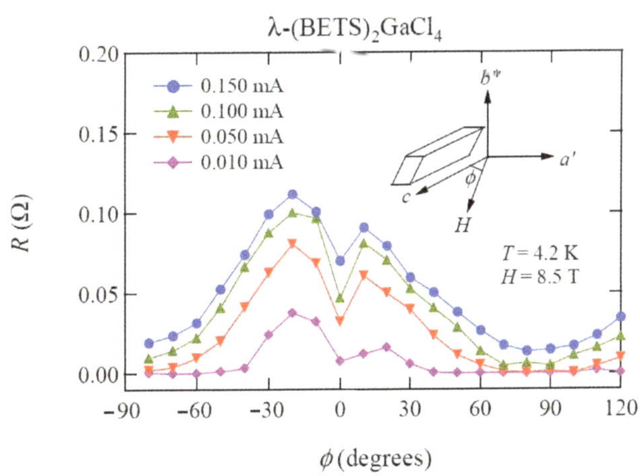

Figure 9. Azimuth angle dependence of flux-flow resistance in $\lambda-(BETS)_2GaCl_4$ at various currents. A sharp minimum is observed at $\phi = 0°$ (H // c). (Reprinted from [26]. Copyright 2014 The Physical Society of Japan.)

In spite of the similar FS geometry, we found an in-plane twofold flux-flow resistance anisotropy in λ-$(BETS)_2GaCl_4$ while fourfold-symmetric flux-flow resistance for β″-$(ET)_2SF_5CH_2CF_2SO_3$ and κ-$(ET)_2Cu(NCS)_2$. As a possible explanation, we consider that the different anisotropic feature may be related to the difference of the interlayer coupling strength [31]. The anisotropy parameter Γ, given by $(\xi_{//}/\xi_{\perp})^2$, for β″-$(ET)_2SF_5CH_2CF_2SO_3$ is Γ ~ 330 that is larger the those of κ-$(ET)_2Cu(NCS)_2$ (Γ ~ 100) and λ-$(BETS)_2GaCl_4$ (Γ ~ 60) [27]. Since λ-$(BETS)_2GaCl_4$ is more three dimensional than κ-$(ET)_2Cu(NCS)_2$ and β″-$(ET)_2SF_5CH_2CF_2SO_3$, an orbital pair breaking effect in λ-$(BETS)_2GaCl_4$ is stronger than in κ-$(ET)_2Cu(NCS)_2$ and β″-$(ET)_2SF_5CH_2CF_2SO_3$.

Due to the orbital pair breaking effect in λ-$(BETS)_2GaCl_4$, the superposition of the normal-state magnetoresistance cannot be avoided in the in-plane anisotropy of the flux-flow resistance. Figure 10 presents the in-plane angular dependence of normal-state magnetoresistance in the magnetic field of 14.8 T [26]. The normal-state magnetoresistance possesses the twofold symmetry with broad maximum at around $\phi = -20°$. The result is consistent with Tanatar et al. [57]. The twofold symmetric normal-state magnetoresistance can be understood in terms of Fermi-surface anisotropy if it mainly originates from the ellipsoidal 2D pocket elongated along the c-axis. In the small Γ system, the large twofold component in the normal-state magnetoresistance may mask the fourfold ones.

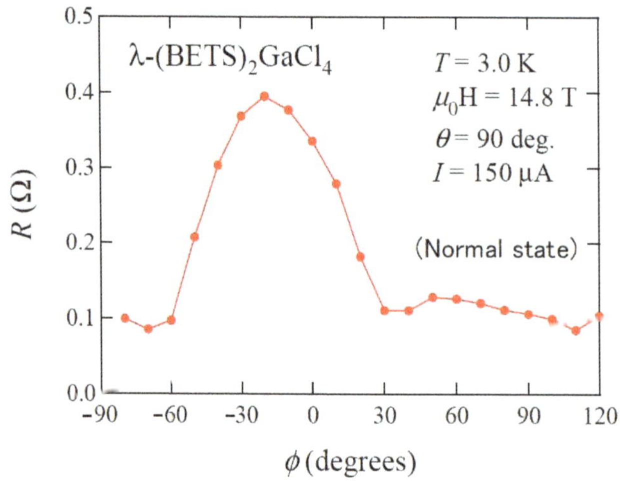

Figure 10. Azimuth angle dependence of normal-state magnetoresistance in λ−$(BETS)_2GaCl_4$ within the conducting plane. (Reprinted from [26]. Copyright 2014 The Physical Society of Japan.)

The origins of the sharp dip ($\phi = -20°$) and the broad minimum (around $\phi = 90°$) in Figure 9 are next discussed. Similar dip structures are found in κ-$(ET)_2Cu(NCS)_2$ and β″-$(ET)_2SF_5CH_2CF_2SO_3$. The origin of the dip in λ-$(BETS)_2GaCl_4$ should be the same as that in κ-$(ET)_2Cu(NCS)_2$ and β″-$(ET)_2SF_5CH_2CF_2SO_3$. To theoretically discuss the SC gap structure for λ-$(BETS)_2GaCl_4$, Aizawa et al. [69] performed first-principles band calculation. Considering the spin-fluctuation-mediated mechanism, they discussed the SC gap function by applying the random phase approximation. They showed that the obtained SC gap changes its sign four times along the Fermi surface, suggesting a d-wave SC gap in λ-$(BETS)_2GaCl_4$. Reflecting the low symmetry of the crystal structure in λ-$(BETS)_2GaCl_4$, however, the SC gap has only twofold symmetry. It means that the predicted SC gap has a large gap between narrow opening nodes with an acute angle (around the steep node structure) and a small gap between wide opening nodes with an obtuse angle. Recent magnetic-field-angle-resolved heat capacity study [70] is consistent with theoretically predicted SC gap function. The large gap is located along the c-axis [69] which agrees with the position of sharp dip [26]. The small gap exists at around a-axis [69], where we observed broad minimum in flux-flow resistance [26]. Thus, experimental results of flux-flow resistance in λ-$(BETS)_2GaCl_4$ [26] are consistent with the *d*-wave gap structure theoretically discussed by Aizawa [69].

4. Summary

In order to discuss the relationship between vortex dynamics and the SC gap structure with *d*-wave paring symmetry, we investigated in-plane angular variation of vortex dynamics for the layered organic superconductors κ-$(ET)_2Cu(NCS)_2$, β″-$(ET)_2SF_5CH_2CF_2SO_3$, and λ-$(BETS)_2GaCl_4$. We observed clear fourfold-symmetric anisotropy in the interlayer flux-flow resistance for κ-$(ET)_2Cu(NCS)_2$ and β″-$(ET)_2SF_5CH_2CF_2SO_3$, while only twofold symmetry in λ-$(BETS)_2GaCl_4$. For κ-$(ET)_2Cu(NCS)_2$ and β″-$(ET)_2SF_5CH_2CF_2SO_3$, flux-flow resistivity showing fourfold oscillation can be consistently explained by assuming the enhanced viscous motion of vortices by QPs arising from the Doppler effect. Absence of the fourfold anisotropy in λ-$(BETS)_2GaCl_4$ is discussed in the two regimes. The first regime is related to the stronger interlayer coupling in the λ-$(BETS)_2GaCl_4$ system as compared with κ-$(ET)_2Cu(NCS)_2$ and β″-$(ET)_2SF_5CH_2CF_2SO_3$. The second regime is discussed in terms of recent theoretical study by Aizawa et al. In this scenario, flux-flow resistivity with twofold anisotropy may be associated with the crystal structure with low symmetry, which is rather different from those of κ-$(ET)_2Cu(NCS)_2$ and β″-$(ET)_2SF_5CH_2CF_2SO_3$.

For these three organic superconductors, the origin of the in-plane anisotropy of flux-flow resistance with sharp minima is discussed in terms of the effect of Doppler shifted state. Based on these results, we claim that angular dependence of the vortex dynamics strongly depends on the SC gap structure. We hope that further understanding of the vortex dynamics presented here leads to clarifying the mechanism of the unconventional superconductors in various strongly correlated electron systems.

Funding: These works were partly supported by Grants-in-Aid for Scientific Research on Innovative Areas (Grant No. 20110004) from the Ministry of Education, Culture, Sports, Science, and Technology (MEXT) of Japan, and Grants-in-Aid for Scientific Research (C) (Grant No. 25400383) from the Japan Society for the Promotion of Science (JSPS).

Institutional Review Board Statement: Not applicable.

Informed Consent Statement: Not applicable.

Data Availability Statement: The data presented in this study are available within the article.

Acknowledgments: The author thanks Shinya Uji, Taichi Terashima, Takako Konoike, Satoshi Tsuchiya, Kaori Sugii, Takayuki Isono, Yoritsugu Iida, Kazuya Saito, Yasuhisa Yamamura, Motoi Kimata, Hidetaka Satsukawa, Jun-ichi Yamada, Biao Zhou, Akiko Kobayashi, Hayao Kobayashi, John A. Schlueter, Hirohito Aizawa, Kazuhiko Kuroki, Nobuhiko Hayashi, and Yoichi Higashi for useful discussions and suggestions.

Conflicts of Interest: The author declares no conflict of interest.

References

1. Ardavan, A.; Brown, S.; Kagoshima, S.; Kanoda, K.; Kuroki, K.; Mori, H.; Ogata, M.; Uji, S.; Wosnitza, J. Recent Topics of Organic Superconductors. *J. Phys. Soc. Jpn.* **2012**, *81*, 011012. [CrossRef]
2. Lang, M.; Muller, J. Organic Superconductors. In *Superconductivity*; Bennemann, K.H., Ketterson, J.B., Eds.; Springer: Berlin/Heidelberg, Germany, 2008; Volume II, pp. 1155–1223.
3. Kanoda, K. Electron correlation, metal-insulator transition and superconductivity in quasi-2D organic systems, $(ET)_2X$. *Physica C* **1997**, *282–287*, 299–302. [CrossRef]
4. Sasaki, T.; Yoneyama, N.; Matsuyama, A.; Kobayashi, N. Magnetic and electronic phase diagram and superconductivity in the organic superconductors κ-$(ET)_2X$. *Phys. Rev. B* **2002**, *65*, 060505. [CrossRef]
5. Clay, R.T.; Li, H.; Mazumdar, S. Absence of Superconductivity in the Half-Filled Band Hubbard Model on the Anisotropic Triangular Lattice. *Phys. Rev. Lett.* **2008**, *101*, 166403. [CrossRef] [PubMed]
6. Dayal, S.; Clay, R.T.; Li, H.; Mazumdar, S. Paired electron crystal: Order from frustration in the quarter-filled band. *Phys. Rev. B* **2011**, *83*, 245106. [CrossRef]

7. Dayal, S.; Clay, R.T.; Mazumdar, S. Absence of long-range superconducting correlations in the frustrated half-filled-band Hubbard model. *Phys. Rev. B* **2012**, *85*, 165141. [CrossRef]
8. Qin, M.; Chung, C.-M.; Shi, H.; Vitali, E.; Hubig, C.; Schollwöck, U.; White, S.R.; Zhang, S. Absence of Superconductivity in the Pure Two-Dimensional Hubbard Model. *Phys. Rev. X* **2020**, *10*, 031016. [CrossRef]
9. Mayaffre, H.; Wzietek, P.; Jérome, D.; Batail, P. Superconducting State of κ-$(ET)_2Cu[N(CN)_2]Br$ Studied by ^{13}C NMR: Evidence for Vortex-Core-Induced Nuclear Relaxation and Unconventional Pairing. *Phys. Rev. Lett.* **1995**, *75*, 4122–4125. [CrossRef] [PubMed]
10. Nakazawa, Y.; Kanoda, K. Low-temperature specific heat of κ-(BEDT-TTF)$_2$Cu[N(CN)$_2$]Br in the superconducting state. *Phys. Rev. B* **1986**, *55*, R8670–R8673. [CrossRef]
11. Taylor, O.J.; Carrington, A.; Schlueter, J.A. Specific-Heat Measurements of the Gap Structure of the Organic Superconductors κ-$(ET)_2Cu[N(CN)_2]Br$ and κ-$(ET)_2Cu(NCS)_2$. *Phys. Rev. Lett.* **2007**, *99*, 057001. [CrossRef] [PubMed]
12. Schrama, J.M.; Rzepniewski, E.; Edwards, R.S.; Singleton, J.; Ardvan, A.; Kurmoo, M.; Day, P. Millimeter-Wave Magneto-optical Determination of the Anisotropy of the Superconducting Order Parameter in the Molecular Superconductor κ-(BEDT-TTF)$_2$Cu(NCS)$_2$. *Phys. Rev. Lett.* **1999**, *83*, 3041–3044. [CrossRef]
13. Arai, T.; Ichimura, K.; Nomura, K.; Takasaki, S.; Yamada, J.; Nakatsuji, S.; Anzai, H. Tunneling spectroscopy on the organic superconductor κ-(BEDT-TTF)$_2$Cu(NCS)$_2$ using STM. *Phys. Rev. B* **2001**, *63*, 104518. [CrossRef]
14. Izawa, K.; Yamaguchi, H.; Sasaki, T.; Matsuda, Y. Superconducting Gap Structure of κ-(BEDT-TTF)$_2$Cu(NCS)$_2$ Probed by Thermal Conductivity Tensor. *Phys. Rev. Lett.* **2001**, *88*, 027002. [CrossRef]
15. Malone, L.; Taylor, O.J.; Schlueter, J.A.; Carrington, A. Location of gap nodes in the organic superconductors κ-$(ET)_2Cu(NCS)_2$ and κ-$(ET)_2Cu[N(CN)_2]Br$ determined by magnetocalorimetry. *Phys. Rev. B* **2010**, *82*, 014522. [CrossRef]
16. Sigrist, M.; Ueda, K. Phenomenological theory of unconventional superconductivity. *Rev. Mod. Phys.* **1991**, *63*, 239–311. [CrossRef]
17. Volovik, G.E. Superconductivity with lines of gap nodes: Density of states in the vortex. *JETP Lett.* **1993**, *58*, 469–473.
18. Vekhter, I.; Hirschfeld, P.J.; Carbotte, J.P.; Nicol, E.J. Anisotropic thermodynamics of *d*-wave superconductors in the vortex state. *Phys. Rev. B* **1999**, *59*, R9023–R9026. [CrossRef]
19. Deguchi, K.; Mao, Z.Q.; Yaguchi, H.; Maeno, Y. Gap Structure of the Spin-Triplet Superconductor Sr_2RuO_4 Determined from the Field-Orientation Dependence of the Specific Heat. *Phys. Rev. Lett.* **2004**, *92*, 047002. [CrossRef]
20. Higashi, Y.; Nagai, Y.; Machida, M.; Hayashi, N. Field-angle dependence of the quasiparticle scattering inside a vortex core in unconventional superconductors. *Physica C* **2011**, *471*, 828–830. [CrossRef]
21. Higashi, Y.; Nagai, Y.; Machida, M.; Hayashi, N. Effect of uniaxially anisotropic Fermi surface on the quasiparticle scattering inside a vortex core in unconventional superconductors. *Physica C* **2013**, *484*, 97–99. [CrossRef]
22. Higashi, Y.; Nagai, Y.; Machida, M.; Hayashi, N. Field-angle resolved flux-flow resistivity as a phase-sensitive probe of unconventional Cooper pairing. *Phys. Rev. B* **2013**, *88*, 224511. [CrossRef]
23. Clem, J.R.; Coffey, M.W. Viscous flux motion in a Josephson-coupled layer model of high-T_c superconductors. *Phys. Rev. B* **1990**, *42*, 6209–6216. [CrossRef]
24. Koshelev, A.E. Role of in-plane dissipation in dynamics of a Josephson vortex lattice in high-temperature superconductors. *Phys. Rev. B* **2000**, *62*, R3616–R3619. [CrossRef]
25. Yasuzuka, S.; Saito, K.; Uji, S.; Kimata, M.; Satsukawa, H.; Terashima, T.; Yamada, J.-I. Anisotropic Josephson-Vortex Dynamics in Layered Organic Superconductor with *d*-Wave Pairing Symmetry. *J. Phys. Soc. Jpn.* **2013**, *82*, 064716. [CrossRef]
26. Yasuzuka, S.; Uji, S.; Terashima, T.; Tsuchiya, S.; Sugii, K.; Zhou, B.; Kobayashi, A.; Kobayashi, H. In-Plane Anisotropy of Flux-Flow Resistivity in Layered Organic Superconductor λ-$(BETS)_2GaCl_4$. *J. Phys. Soc. Jpn.* **2014**, *83*, 013705. [CrossRef]
27. Yasuzuka, S.; Uji, S.; Terashima, T.; Sugii, K.; Isono, T.; Iida, Y.; Schlueter, J.A. In-Plane Anisotropy of Upper Critical Field and Flux-Flow Resistivity in Layered Organic Superconductor β″-$(ET)_2SF_5CH_2CF_2SO_3$. *J. Phys. Soc. Jpn.* **2015**, *84*, 094709. [CrossRef]
28. Müller, J.; Lang, M.; Steglich, F.; Schlueter, J.A.; Kini, A.M.; Geiser, U.; Mohtasham, J.; Winter, R.W.; Gard, G.L.; Sasaki, T.; et al. Comparative thermal-expansion study of β″-$(ET)_2SF_5CH_2CF_2SO_3$ and κ-$(ET)_2Cu(NCS)_2$: Uniaxial pressure coefficients of T_c and upper critical fields. *Phys. Rev. B* **2000**, *61*, 11739–11744. [CrossRef]
29. Wanka, S.; Hagel, J.; Beckmann, D.; Wosnitza, J.; Schlueter, J.A.; Williams, J.M.; Nixon, P.G.; Winter, R.W.; Garal, G.L. Specific heat and critical field of the organic superconductor β″-$(ET)_2SF_5CH_2CF_2SO_3$. *Phys. Rev. B* **1997**, *57*, 3084–3088. [CrossRef]
30. Prozorov, R.; Giannetta, R.W.; Schlueter, J.A.; Kini, A.M.; Mohtasham, J.; Winter, R.W.; Gard, G.L. Unusual temperature dependence of the London penetration depth in all-organic β″-$(ET)_2SF_5CH_2SF_2SO_3$ single crystals. *Phys. Rev. B* **2001**, *63*, 052506. [CrossRef]
31. Mielke, C.; Singleton, J.; Nam, M.-S.; Harrison, N.; Agosta, C.C.; Fravel, B.; Montgomery, L.K. Superconducting properties and Fermi-surface topology of the quasi-two-dimensional organic superconductor λ-$(BETS)_2GaCl_4$. *J. Phys. Condens. Matter* **2001**, *13*, 8325–8345. [CrossRef]
32. Sari, D.P.; Naito, R.; Hiraki, K.; Nakano, T.; Hagiwara, M.; Nozue, Y.; Kusakawa, T.; Hori, A.; Watanabe, I.; Ishii, Y. Anisotropy of Lower Critical Field in Organic Layered Superconductor λ-$(BETS)_2GaCl_4$. *Key Eng. Mater.* **2020**, *860*, 137–141. [CrossRef]
33. Saito, G.; Yamochi, H.; Nakamura, T.; Komatsu, T.; Ishiguro, T.; Nogami, Y.; Ito, Y.; Mori, H.; Oshima, K.; Nakashima, M.; et al. Overview of organic superconductor κ-(BEDT-TTF)$_2$[Cu(NCS)$_2$] and its related materials. *Synth. Met.* **1991**, *42*, 1993–1998. [CrossRef]
34. Chaparala, M.; Chung, O.H.; Ren, Z.F.; White, M.; Coppens, P.; Wang, J.H.; Hope, A.P.; Naughton, M.J. Vortex-state resistance near parallel orientation in layered superconductors. *Phys. Rev. B* **1996**, *53*, 5818. [CrossRef] [PubMed]

35. Uji, S.; Terashima, T.; Nishimura, M.; Konoike, T.; Enomoto, K.; Cui, H.; Kobayashi, H.; Kobayashi, A.; Tanaka, H.; Tokumoto, M.; et al. Vortex Dynamics and the Fulde-Ferrell-Larkin-Ovchinnikov State in a Magnetic-Field-Induced Organic Superconductor. *Phys. Rev. Lett.* **2006**, *97*, 157001. [CrossRef]
36. Yasuzuka, S.; Uji, S.; Satsukawa, H.; Kimata, M.; Terashima, T.; Koga, H.; Yamamura, Y.; Saito, K.; Akutsu, H.; Yamada, J. Anisotropic Josephson-vortex dynamics in layered organic superconductors. *Physica B* **2010**, *405*, S288–S290. [CrossRef]
37. Feinberg, D.; Villard, C. Intrinsic pinning and lock-in transition of flux lines in layered type-II superconductors. *Phys. Rev. Lett.* **1990**, *65*, 919–922. [CrossRef] [PubMed]
38. Mansky, P.A.; Chaikin, P.M.; Haddon, R.C. Vortex lock-in state in a layered superconductor. *Phys. Rev. Lett.* **1993**, *70*, 1323–1326. [CrossRef]
39. Nam, M.-S.; Symington, J.A.; Singleton, J.; Blundell, S.; Ardavan, A.; Perenboom, J.A.A.J.; Kurmoo, M.; Day, P. Angle dependence of the upper critical field in the layered organic superconductor κ-(BEDT-TTF)$_2$Cu(NCS)$_2$ (BEDT-TTF ≡ bis(ethylenedithio)tetrathiafulvalene). *J. Phys. Condens. Matter.* **1999**, *11*, L477–L484. [CrossRef]
40. Klemm, R.A.; Luther, A.; Beasley, M.R. Theory of the upper critical field in layered superconductors. *Phys. Rev. B* **1975**, *12*, 877–891. [CrossRef]
41. Schneider, T.; Schmidt, A. Dimensional crossover in the upper critical field of layered superconductors. *Phys. Rev. B* **1993**, *47*, 5915–5921. [CrossRef]
42. Oshima, K.; Urayama, H.; Yamochi, H.; Saito, G. Peculiar Critical Field Behaviour in the Recectly Discovered Ambient Pressure Organic Superconductor (BEDT-TTF)$_2$Cu (NCS)$_2$ (T_c = 10.4 K). *J. Phys. Soc. Jpn.* **1988**, *57*, 730–733. [CrossRef]
43. Takanaka, K.; Kuboya, K. Anisotropy of Upper Critical Field and Pairing Symmetry. *Phys. Rev. Lett.* **1995**, *75*, 323–325. [CrossRef]
44. Yasuzuka, S.; Koga, H.; Yamamura, Y.; Saito, K.; Uji, S.; Terashima, T.; Akutsu, H.; Yamada, J.-I. Dimensional Crossover and Its Interplay with In-Plane Anisotropy of Upper Critical Field in β-(BDA-TTP)$_2$SbF$_6$. *J. Phys. Soc. Jpn.* **2017**, *86*, 084704. [CrossRef]
45. Yasuzuka, S.; Uji, S.; Sugiura, S.; Terashima, T.; Nogami, Y.; Ichimura, K.; Tanda, S. Highly Isotropic In-plane Upper Critical Field in the Anisotropic *s*-Wave Superconductor 2*H*-NbSe$_2$. *J. Supercond. Nov. Magn.* **2020**, *33*, 953–958. [CrossRef]
46. Vorontsov, A.B.; Vekhter, I. Unconventional superconductors under a rotating magnetic field. I. Density of states and specific heat. *Phys. Rev. B* **2007**, *75*, 224501. [CrossRef]
47. Vorontsov, A.B.; Vekhter, I. Unconventional superconductors under a rotating magnetic field. II. Thermal transport. *Phys. Rev. B* **2007**, *75*, 224502. [CrossRef]
48. An, K.; Sakakibara, T.; Settai, R.; Onuki, Y.; Hiragi, M.; Ichioka, M.; Machida, K. Sign Reversal of Field-Angle Resolved Heat Capacity Oscillations in a Heavy Fermion Superconductor CeCoIn$_5$ and $d_{x^2-y^2}$ Pairing Symmetry. *Phys. Rev. Lett.* **2010**, *104*, 037002. [CrossRef]
49. Geiser, U.; Schlueter, J.A.; Wang, H.H.; Kini, A.M.; Williams, J.M.; Sche, P.P.; Zakowicz, H.I.; VanZile, M.L.; Dudek, J.D.; Nixon, P.G.; et al. Superconductivity at 5.2 K in an Electron Donor Radical Salt of Bis(ethylenedithio)tetrathiafulvalene (BEDT-TTF) with the Novel Polyfluorinated Organic Anion SF$_5$CH$_2$CF$_2$SO$_3^-$. *J. Am. Chem. Soc.* **1996**, *118*, 9996–9997. [CrossRef]
50. Beckmann, D.; Wanka, S.; Wosnitza, J.; Schlueter, J.A.; Williams, J.M.; Nixon, P.G.; Winter, R.W.; Gard, G.L.; Ren, J.; Whangbo, M.-H. Characterization of the Fermi surface of the organic superconductor β''-(ET)$_2$SF$_5$CH$_2$CF$_2$SO$_3$ by measurements of Shubnikov-de Haas and angle-dependent magnetoresistance oscillations and by electronic band-structure calculations. *Eur. Phys. J. B* **1998**, *1*, 295–300. [CrossRef]
51. Brooks, J.S.; Williams, V.; Choi, E.S.; Graf, D.; Tokumoto, M.; Uji, S.; Zou, F.; Wosnitza, J.; Schlueter, J.A.; Davis, H.; et al. Fermiology and superconductivity at high magnetic fields in a completely organic cation radical salt. *New J. Phys.* **2006**, *8*, 255. [CrossRef]
52. Wosnitza, J.; Hagel, J.; Qualls, J.S.; Brooks, J.S.; Balthes, E.; Schweitzer, D.; Schlueter, J.A.; Geiser, U.; Mohtasham, J.; Winter, R.W.; et al. Coherent versus incoherent interlayer transport in layered metals. *Phys. Rev. B* **2000**, *65*, R180506. [CrossRef]
53. Hagel, J.; Wosnitza, J.; Pfleiderer, C.; Schlueter, J.A.; Mohtasham, J.; Gard, G.L. Pressure-induced insulating state in an organic superconductor. *Phys. Rev. B* **2003**, *68*, 104504. [CrossRef]
54. Calandra, M.; Merino, J.; Mckenzie, R.H. Metal-insulator transition and charge ordering in the extended Hubbard model at one-quarter filling. *Phys. Rev. B* **2002**, *66*, 195102. [CrossRef]
55. Merino, J.; Mckenzie, R.H. Superconductivity Mediated by Charge Fluctuations in Layered Molecular Crystals. *Phys. Rev. Lett.* **2001**, *87*, 237002. [CrossRef]
56. Kobayashi, H.; Udagawa, T.; Tomita, H.; Bun, K.; Naito, T.; Kobayashi, A. A New Organic Superconductor, λ-(BEDT-TSF)$_2$GaCl$_4$. *Chem. Lett.* **1993**, *22*, 1559–1562. [CrossRef]
57. Tanatar, M.A.; Ishiguro, T.; Tanaka, H.; Kobayashi, H. Magnetic field–temperature phase diagram of the quasi-two-dimensional organic superconductor λ-(BETS)$_2$GaCl$_4$ studied via thermal conductivity. *Phys. Rev. B* **2002**, *66*, 134503. [CrossRef]
58. Uji, S.; Kodama, K.; Sugii, K.; Terashima, T.; Yamaguchi, T.; Kurita, N.; Tsuchiya, S.; Konoike, T.; Kimata, M.; Kobayashi, A.; et al. Vortex Dynamics and Diamagnetic Torque Signals in Two Dimensional Organic Superconductor λ-(BETS)$_2$GaCl$_4$. *J. Phys. Soc. Jpn.* **2015**, *84*, 104709. [CrossRef]
59. Uji, S.; Shinagawa, H.; Terashima, T.; Yakabe, T.; Terai, Y.; Tokumoto, M.; Kobayashi, A.; Tanaka, H.; Kobayashi, H. Magnetic-field-induced superconductivity in a two-dimensional organic conductor. *Nature* **2001**, *410*, 908–910. [CrossRef]

60. Goddard, P.A.; Blundell, S.J.; Singleton, J.; McDonald, R.D.; Ardavan, A.; Narduzzo, A.; Schlueter, J.A.; Kini, A.M.; Sasaki, T. Angle-dependent magnetoresistance of the layered organic superconductor κ-$ET_2Cu(NCS)_2$: Simulation and experiment. *Phys. Rev. B* **2004**, *69*, 174509. [CrossRef]
61. Clark, K.; Hassanien, A.; Khan, S.; Braun, K.-F.; Tanaka, H.; Hla, S.-W. Superconductivity in just four pairs of $(BETS)_2GaCl_4$ molecules. *Nat. Nanotechnol.* **2010**, *5*, 261–265. [CrossRef]
62. Kobayashi, H.; Akutsu, H.; Arai, E.; Tanaka, H.; Kobayashi, A. Electric and magnetic properties and phase diagram of a series of organic superconductors $\lambda-BETS_2GaX_zY_{4-z}$ [BETS = bis(ethylenedithiotetraselenafulvalene; X, Y = F, Cl, Br; $0 < z < 2$)]. *Phys. Rev. B* **1997**, *56*, R8526–R8529.
63. Tanaka, H.; Kobayashi, A.; Sato, A.; Akutsu, H.; Kobayashi, H. Chemical Control of Electrical Properties and Phase Diagram of a Series of λ-Type BETS Superconductors, λ-$(BETS)_2GaBr_xCl_{4-x}$. *J. Am. Chem. Soc.* **1999**, *121*, 760–768. [CrossRef]
64. Mori, H.; Okano, T.; Kamiya, M.; Haemori, M.; Suzuki, H.; Tanaka, S.; Nishio, Y.; Kajita, K.; Moriyama, H. Bandwidth and band filling control in organic conductors. *Physica C* **2001**, *357–360*, 103–107. [CrossRef]
65. Kanoda, K. Metal–Insulator Transition in κ-$(ET)_2X$ and $(DCNQI)_2M$: Two Contrasting Manifestation of Electron Correlation. *J. Phys. Soc. Jpn.* **2006**, *75*, 051007. [CrossRef]
66. Kobayashi, T.; Kawamoto, A. Evidence of antiferromagnetic fluctuation in the unconventional superconductor λ-$(BETS)_2GaCl_4$ by ^{13}C NMR. *Phys. Rev. B* **2017**, *96*, 125115. [CrossRef]
67. Imajo, S.; Kanda, N.; Yamashita, S.; Akutsu, H.; Nakazawa, Y.; Kumagai, H.; Kobayashi, T.; Kawamoto, A. Thermodynamic Evidence of *d*-Wave Superconductivity of the Organic Superconductor λ-$(BETS)_2GaCl_4$. *J. Phys. Soc. Jpn.* **2016**, *85*, 043705. [CrossRef]
68. Sari, D.P.; Asih, R.; Mohm-Tajudin, S.S.; Adam, N.; Hiraki, K.; Ishii, Y.; Takahashi, T.; Nakano, T.; Nozue, Y.; Sulaiman, S.; et al. μSR Study of Organic Superconductor λ-$(BETS)_2GaCl_4$. *IOP Conf. Ser. Mater. Sci. Eng.* **2017**, *196*, 012047. [CrossRef]
69. Aizawa, H.; Koretsune, T.; Kuroki, K.; Seo, H. Electronic Structure Calculation and Superconductivity in λ-$(BETS)_2GaCl_4$. *J. Phys. Soc. Jpn.* **2018**, *87*, 093701. [CrossRef]
70. Imajo, S.; Yamashita, S.; Akutsu, H.; Kumagai, H.; Kobayashi, T.; Kawamoto, A.; Nakazawa, Y. Gap Symmetry of the Organic Superconductor λ-$(BETS)_2GaCl_4$ Determined by Magnetic-Field-Angle-Resolved Heat Capacity. *J. Phys. Soc. Jpn.* **2019**, *88*, 023702. [CrossRef]

MDPI
St. Alban-Anlage 66
4052 Basel
Switzerland
Tel. +41 61 683 77 34
Fax +41 61 302 89 18
www.mdpi.com

Crystals Editorial Office
E-mail: crystals@mdpi.com
www.mdpi.com/journal/crystals

www.ingramcontent.com/pod-product-compliance
Ingram Content Group UK Ltd
Pitfield, Milton Keynes, MK11 3LW, UK
UKHW060121010826
14089UKWH00006B/103

9 783036 545592